W9-CPF-650

Table of Contents

List of Figures

List of Tables

Preface

This book presents a number of selected papers on multimode analysis. Many of these studies emanate from the work and influence of Ledyard R Tucker; his explication of three-mode factor analysis, his incorporation of individual differences in multidimensional scaling, and his insights into the important relationships of these two sometimes separate analytic traditions are landmarks in the development of multimode methodology. His career efforts clearly mark the second generation in the evolution of factor analysis and multidimensional scaling.

This book focuses on the new multimode methodologies that seek to extend and refine the structural accounts in life science research. It provides a coherent picture of the progress to date, thereby stimulating dialogue among the contributors and other methodologists as well as increasing relevant applications by substantive researchers. In chapter 1, the editors present an overview of the area, reviewing developments in terms of their interpretive features. Chapter 2, by Joseph Kruskal, offers a unifying framework for the consideration of these developments in terms of their statistical formulations. The multimode methods are divided into two traditions—factor analysis and multidimensional scaling. There remain considerable conceptual differences between these two camps, but many statistical problems and resolutions are shared by their methods. Indeed, some initiators, like Kruskal and Harshman, have crossed the boundaries of both traditions. Although the division has been retained in the organization of the chapters of this book, their combined appearance may further encourage cross-fertilization of statistical procedures and continued discussion of conceptual issues.

Chapter 3, by Pieter Kroonenberg, presents the most direct "descendant" of Ledyard Tucker's classic 1966 three-mode factor analysis method; improved statistical foundations and procedures render the Tucker approach an attractive general explorative method. In chapter 4, Bruce Bloxom recasts Tucker's model in terms of hierarchical decomposition principles, perhaps strengthening its conceptual basis and offering new possibilities for its refinement.

Chapters 5 and 6, by Richard Harshman and Margaret Lundy, document the extensive conceptual and empirical work that has gone into PARAFAC—parallel factors analysis—a trilinear analytic method that was born in the wake of Tucker's three-mode factor analysis. Although it suffered some hard times in the empirical arena, it now emerges—fortified with a new rationale for data scaling and its associated procedures for preprocessing—as an exciting and useful approach. The coverage given to PARAFAC is rationalized in terms of its powerful application capabilities and the interesting interrelationships it reveals. It lies at the crossroads of traditional two-mode factor analysis; its direct three-mode extension has roots in the classical debates on factorial invariance (specifically based on Cattell's principle of parallel proportional profiles) and extends its influence into the field of multidimensional scaling. In addition, these technical considerations are enhanced by the personage of Richard Harshman, who has clearly linked his career to the exploration of multimode methods. His experience with and understanding of multimode research and statistical issues is evident in these chapters.

Chapters 7 and 8, by Rod McDonald and Hariharan Swaminathan, respectively, reflect the sophistication and comprehensiveness of the analysis of covariance structures perspective. Tied to classical conceptualizations, but fitted with new insights and refinements, these general models offer an alternative to those with a singular-value-decomposition orientation, such as PARAFAC and Tucker. The COSAN program (written by Colin Fraser)—which operationalizes McDonald's invariant factors analysis and Swaminathan's longitudinal factor analysis models—allows very flexible parameter specifications and "confirmatory" evaluation by maximum likelihood methods. Chapter 9 presents the insightful contributions of Ray Cattell and his colleagues to this emerging area. His early recognition of the importance of multimode structural design and analysis and his central involvement in early explorations of these different perspectives were key contextual features in the eventual discovery of unified multimode models.

One of the most important developments in multimode analysis, marked by the work of Carroll and Chang in 1970, was the creative combination of the alternating least-squares algorithm with the trilinear model and its multilinear extension. Chapter 10, by Doug Carroll and Sandra Pruzansky, presents the CANDECOMP-CANDELINC family of models and methods that have evolved from the Carroll and Chang canonical decomposition approach. Doug Carroll must stand as one of the most significant pioneers in the incorporation of individual differences into multidimensional scaling and in pointing to the importance of the uniqueness property of the weighted Euclidean model. His approach still remains one of the most widely used methods for dealing with complex structural problems involving individual differences. Chapter 11, by Jim Lingoes and Ingwer Borg, presents an innovative exploratory approach for the same type of problem. In the search for "common" structure, the question always arises as to the similarity of individual structures to the common one; this chapter demonstrates a useful descriptive approach for the ascertainment of the commonality in a set of data.

In chapter 12, Forrest Young provides a unifying framework for the explication of individual differences models that can arise in multidimensional scaling and unfolding. This general model neatly portrays the range of possibilities and, when coupled with optimal scaling features, the method and associated computer program should find wide use. In chapter 13, by Yoshio Takane, multidimensional scaling notions are merged with a conjoint measurement approach to produce a weighted additive model for dealing with individual differences, which represents a very interesting special development. Chapter 14, by Doug Carroll and

Phipps Arabie, extends the INDSCAL concern for individual differences to the problems of cluster analysis, providing the possibility for a discrete rather than continuous representation of three-way data. This is another quite interesting special development.

The final two chapters reflect on these developments. In chapter 15, Peter Gould appraises the sophisticated statistical armament required for multimode analysis and the restrictive assumptions frequently accepted without question in this endeavor. He reminds us of the importance of performing careful conceptual analysis before undertaking empirical structural forays, and he challenges the necessity of *n*-way classifications. (This paper was included to sober the enthusiast!) Finally, chapter 16 ties up some of the loose ends and points to areas and issues for future work.

Three appendices are included for those who may use these methods. Appendix A, by Richard Harshman, provides a helpful guide to the interpretive problems that may arise in multimode analysis. It points to the necessity of validating analytic solutions by demonstrating reliability and it illustrates the use of diagnostics to guide the performance of the analysis itself. In appendix B, John Gower discusses the display possibilities for multimode results, and in appendix C, Richard Harshman and Wayne De Sarbo describe an application of PARAFAC in the small sample case. This paper demonstrates both the guiding and validating role of some of the diagnostics described in appendix A.

The bibliography was developed from the annotated bibliography of three-mode factor analysis published in the *British Journal of Mathematical and Statistical Psychology,* by Pieter Kroonenberg. He graciously supplied his notes, which were supplemented by further computer searches. This listing presents a complete view of multimode analysis. Perusal of the entries will provide some idea of the possibilities for application and the emerging developments in the field of multimode data analysis.

There are clearly many complicated equations and concepts contained in this book. Since many researchers are now familiar with factor analysis and metric multidimensional scaling, this book should be useful for those individuals who wish to extend their methods to additional data classification sets. The Snyder (Law and Hattie), Kruskal, Kroonenberg, Harshman (and Lundy), and Carroll (and Pruzansky) chapters are potentially useful studies for the beginner. Some of the more complicated approaches—including Bloxom, McDonald, Swaminathan, Young, and Takane—will probably be most useful to those already familiar with three-mode factor analysis and individual differences multidimensional scaling.

An edited book is obviously the work of many people. We were fortunate to have the assistance of Janice Fittell, Vince Murdoch, Peter Pamment, and Joclynn Ware, whose perseverance and concern were important ingredients in this endeavor. Their text-editing and general computer skills were absolutely essential in the difficult task of collecting these papers, and accuracy in the numerous equations was possible only because of their patience and attentiveness. We are very grateful for their enthusiasm and hard work. Furthermore, we are indebted to Colin Fraser (University of New England) for his general criticism and insightful comments about this project. He frequently clarified issues, provided mathematical and computer programming assistance, and offered many useful suggestions.

Preparation of this manuscript was performed by the typing staff of the Department of Psychology, University of Queensland. They entered the text into the computer from a variety of input formats, reflecting the different styles and preferences of these many authors. We are grateful for their cooperation under difficult and frequently rushed circumstances. In the last minute push to bring the whole project to fruition, Gail and Catrina Law also provided invaluable help in typing, photocopying, and general clerical work. We are thankful for their willing assistance at a critical time. Checking the page proofs was greatly facilitated by the expert assistance of Dale Caird.

Finally, we want to thank our departments. They supported our efforts, withstood our virtual dominance of the University of Queensland Psychology Department computer system, funded many international phone calls, and allowed us to arrange meetings at our different resident locations to discuss this book.

Henry G. Law
Conrad W. Snyder, Jr.
John A. Hattie
Roderick P. McDonald

List of Contributors

Phipps Arabie, Department of Psychology, University of Illinois, 603 East Daniel Street, Champaign, Illinois, 61820, United States

Daniel D. Blaine, College of Education, University of Hawaii at Manoa, 1776 University Avenue, Honolulu, Hawaii, 96822, United States

Bruce M. Bloxom, Department of Psychology, Vanderbilt University, Nashville, Tennessee, 37240, United States

Ingwer Borg, Department of Psychology, Justus-Liebig University, Glessen, West Germany

Jerry M. Brennan, College of Education, University of Hawaii at Manoa, 1776 University Avenue, Honolulu, Hawaii, 96822, United States

J. Douglas Carroll, AT & T Bell Laboratories, 600 Mountain Avenue, Murray Hill, New Jersey, 07974, United States

Raymond B. Cattell, Department of Psychology, University of Hawaii at Manoa, 2430 Campus Road, Honolulu, Hawaii, 96822, United States

Wayne S. De Sarbo, AT & T Bell Laboratories, 600 Mountain Avenue, Murray Hill, New Jersey, 07974, United States

Peter Gould, Department of Geography, The Pennsylvania State University, 302 Walker Building, University Park, Pennsylvania, 16802, United States

John C. Gower, Rothamsted Experimental Station, Harpenden, Herts, AL5 2JQ, England

Richard A. Harshman, Department of Psychology, University of Western Ontario, London, Ontario, N6A 5C2, Canada

John A. Hattie, Center for Behavioral Studies in Education, University of New England, Armidale, New South Wales, 2351, Australia

Pieter M. Kroonenberg, Rijksunlversiteit Lelden, Subfakulteit der Pedagogische, en Andragogische Wetenschappen, Schuttersveld 9, 2316 XG, Leiden, The Netherlands

Joseph B. Kruskal, AT & T Bell Laboratories, 600 Mountain Avenue, Murray Hill, New Jersey, 07974, United States

Henry G. Law, Psychology Department, University of Queensland, St. Lucia, Queensland, 4067, Australia

James C. Lingoes, Department of Psychology and Research Scientist, Computing Center Station, 1005 N. University Building, The University of Michigan, Ann Arbor, Michigan, 48104, United States

Margaret E. Lundy, Department of Psychology, University of Western Ontario, London, Ontario, NGA 5C2, Canada

Roderick P. McDonald, School of Education, Macquarie University, North Ryde, New South Wales, 2113, Australia

Sandra Pruzansky, AT & T Bell Laboratories, 600 Mountain Avenue, Murray Hill, New Jersey, 07974, United States

Conrad W. Snyder, Jr., Psychology Department, University of Queensland, St. Lucia, Queensland, 4067, Australia

Hariharan Swaminathan, School of Education, University of Massachusetts, Amherst, Massachusetts, 01003, United States

Yoshio Takane, Department of Psychology, McGill University, Stewart Biological Sciences Building, 1205 Docteur Penfield Avenue, Montreal, Quebec, H3A 1B1, Canada

Forrest W. Young, The L. L. Thurstone Psychometric Laboratory, The University of North Carolina at Chapel Hill, Davie Hall 013A, Chapel Hill, North Carolina, 27514, United States

PART I
Introduction

1

Overview of Multimode Analytic Methods

Conrad W. Snyder, Jr., Henry G. Law, and John A. Hattie

The researcher in the life sciences faces a myriad of information that is intricately interwoven and seemingly ephemeral. Order is imposed through creative generalizations, exchanging unrelated chronicles of events for simplifying normative principles. The emergent higher-level constructs, representing the commonality of events, enable some sense of control and anticipation (see Kelly 1955). Unfortunately, the reliance on single-occasion, single-measure explorative research designs in the life sciences has failed to grapple with the inherently complex nature of human and animal activity. The resulting simple hypothetical constructions have rarely yielded the hoped-for revelations. Although there is no replacement for ingenuity and creativity, insight frequently emerges from explorative methodology that offers new perspectives on empirical events. In fact, it could be argued that the role of empirical research is to reveal new possibilities for theory generation and development. Without some probing methodology to unravel the intricacies of complex life patterns, germinative perspectives may elude us.

Psychology—faced with the inherent complexity and reflexive quality of its constructs—has developed a specific methodology called *factor analysis* for the development and nurturance of these abstractions. Less obvious, hidden relations are *recovered* from empirically correlated variables. From a mathematical viewpoint correlation is merely an ordinary index for the assessment of covariation; however, Charles Spearman, one of the creators of factor analysis, endowed the term with greater meaning. He regarded correlated variables "as depending on a common factor" (1927, i), which was interpreted as some concrete entity or "organ." This extended idea is valid in modern references to

---

We wish to acknowledge the assistance of Douglas Carroll, Ross Gayler, Richard Harshman, Joseph Kruskal, Rod McDonald, and Vince Murdoch in reading and commenting upon earlier drafts of this chapter.

functional unities, latent influences, "true" sources, traits, or other terms that convey "the system of correlation . . . as expressing the *hidden underlying cause* of the variations investigated" (Spearman 1904a, 74–75 [his emphasis]). Although there is considerable debate over the semantic extension of the notion of a factor, it seems generally agreed that a factor represents a higher-level structure of relations—that is, in McDonald's terms, it is an *abstractive attribute* and may or may not be a cause. As pointed out by Burt, "The special value of a factor in psychology is that it enables us to hold together in thought a definite but complex pattern of characteristics" (1940, 237).

Given the wide application of factor analysis in the life sciences, researchers find the method to be useful in recovering higher-level relations within multivariate data designs. Traditional factor analysis and metric multidimensional scaling were applied to correlation or similarity (dissimilarity) coefficients obtained for individuals responding to some set of stimuli. When this two-way cross-classification of observations is extended to include additional identifying sets, a new methodology is required (unless one is willing to ignore the higher-level interactional patterns). Common examples of three-way analyses include: semantic differential studies (for example, Levin 1965; Miron and Osgood 1966; Snyder and Wiggins 1970; Tucker 1965; Tzeng 1975); multitrait-multimethod analyses (such as, Levin 1965; Tucker 1965); multivariate learning studies (for instance, Bridgman, Snyder, and Law 1981; Snyder 1976; Snyder, Bridgman, and Law 1981; Tucker 1967); and situation-response investigations (including Firth and Snyder 1979; Kjerulff and Wiggins 1976; Leah, Law, and Snyder 1979; Levin 1965; Thomas and Bain 1982; Tucker 1965). The methodology common to these references is *three-mode factor analysis,* as developed by Ledyard R. Tucker (1963, 1964, 1966a, 1972). Offering a new conceptual and analytic basis to factor analysis and scaling, Tucker's ingenious approach to the recovery of abstractive relations in complex data designs remains one of the more general and interesting proposals for explorative structural analysis. This chapter will examine some selected multimode analytic methods that have evolved since Tucker's seminal work. Emphasis will be given to their interpretive or conceptual foundations.

MULTIMODE METHODS

Social, behavioral, biological, and political scientists are primarily concerned with the articulation of theoretical propositions that relate to their particular substantive disciplines. These propositions are claimed to validly account for observed *relations* among specified properties of *entities,* where entities refer to the things or objects of experience. Science pertains to a special objective knowledge about these entities—called *fundaments* by Spearman (indicative of their fundamental nature)—and their relations. As pointed out by Cattell, "The recognition of these two basic classes of external 'fact' is woven into all languages, namely in nouns (for objects) and adjectives, or initially verbs (for attributes and relations)" (1966, 172).

The methodology of science provides an empirical, evidential basis for the validation of the theoretical propositions. At an empirical level, all scientific data sets employ selected entities in order to reflect the hypothesized properties of those entities. Simple relations among entities are implicitly assumed within the measurement or scaling decisions of the scientific investigations (see Coombs [1964] 1976). Sets of simple relations are encapsulated in *attributes* and quantified as variables. (The more complex relations are the subject of further analyses.) For all sciences, "Entities and relations (or the attributes distilled from relations) are thus the primary perceptions or proto-constructs out of which more refined constructs and scientific concepts are derived and elaborated" (Cattell 1966, 72).

Traditionally, in factor-analytic applications, two sets of entities were selected—often persons and tests—and the relation was indicated by a score for each person on each test. Therefore, each score was identifiable by a person and a test and provided information about the relation between them. These sets of entities were termed "modes" or "modes of classification" by Tucker (1966a). Thus, in this example, there are two modes of identifying classifications: persons and tests. The data classification is given by the Cartesian product of the sets of classification indices. As additional independent sets of entities are added—for example situations (conditions) or occasions or both—the number of modes is increased and each datum is more complexly and hopefully more precisely classified. Multimode analytic methods are designed to provide a systematic representation of the relations in these multiclassified data.

Factor-Analytic Developments

In traditional factor analysis, the parameters of a common factor space are estimated from the sample, bivariate linear relationships among a single attribute set (where the attributes are related over a set of entities—persons—from a two-mode data design). The factors are defined by those observable measures that align themselves with each of the reference vectors (dimensions) in the factor space. Once named in accordance with the semantic content of these salient variables, each factor is considered a *latent* construct, reflecting the covariation of the lower-level empirical variables. Individual differences on these abstractions are called *factor scores* and are fallibly estimated from the narrow perspective of the particular attribute and individual entity sets selected for the study. Accordingly, the factors "belong" to the observed measures (vectors) from which they are identified, and the factor scores reflect the positions of the individuals (points) in the particular factor space.

Extending the data set to include another mode poses some difficulty for the classical conceptualization of a factor. Tucker's insightful solution to this problem was to recast the two-mode factor model in terms of the Eckart-Young decomposition principle (1936). With this solid analytic basis, Tucker then generalized the approach to a multimode perspective utilizing the Kronecker product operator to handle the additional modes. The multimode generalization subtly alters the conceptual fabric of classical factor analysis (Harshman 1970; Levin 1965). Whereas classical factor analysis identifies a single set of latent influences for a two-mode data design, Tucker's method allows a different underlying structure to emerge from each mode, regardless of the number of classification sets. The interactional possibilities—in terms of process—of differing influences specific to particular modes are assessed in a higher-level *inner core*. The core is bounded by the sets of factors (derivational modes) obtained for each of the identifying classifications of the original observational modes. Within Tucker's framework, factors are conceptualized as *idealized entities* or *types,* and these derivational mode entities are related across modes in the inner core. The focus of the approach is not on the separate mode analyses but on the interactional relations revealed in the core. As expressed by Tucker:

> The values in the core box . . . tend to bring out newer statements of relation. One might take the view that the mode 1 and 2 factor matrices are dealing with relations of a more surface type, and that the core box is dealing with deeper and more subtle relations. These relations in the core box should have more general effects on the phenomena being observed. One might argue, then, that the relations indicated in the core boxes should be of greater general interest in understanding the phenomena. (1966b, 378)

Early presentations of this approach (Levin 1965; Tucker 1963, 1964, 1965, 1966a; and more recently extended to four modes by Lastovicka 1981) emphasized a principal-components-type analytic algorithm with supplementary rotational procedures on the reduced set of characteristic vectors for each mode. The core is obtained from the results of the separate mode component analyses and reflects a lower-dimensional representation of the observational modes. Conceptually, the core can also be interpreted in terms of component weights, where entries indicate the basic relations between the derivational combination-attribute-modes and the idealized individuals. This ingenious conceptual modification results in a very general and intuitively interesting multimode method. The approach exploits the information from each of the data classification sets and weights their unique underlying influences according to their mutual, higher-level relations.

Emerging from a "psychometric" orientation, the observational modes of Tucker's method were considered fixed and complete samples of their respective domains. In 1968, Bloxom presented a revised formulation for Tucker's approach that specified a randomly-selected individuals mode, thus opening the way for statistical estimation procedures. Several statistical treatments have evolved (for instance, Bentler and Lee 1978, 1979; Carroll and Chang 1970; Carroll, Pruzansky, and Kruskal 1980; Harshman 1970, 1981; Kroonenberg and de Leeuw 1980; McDonald 1978, 1980; Sands and Young 1980; Takane, Young, and de Leeuw 1980).

A diagrammatic representation of Tucker's three-mode component analysis method is given in Figure 1–1. This particular representation is based on Kroonenberg's TUCKALS3 computer program (1981a), since the program is readily available and has been frequently applied (see chapter 3). Using an alternating least-squares computational algorithm, the resulting TUCKALS3 core is directly interpretable in terms of the overall fit of the model, partitioning the total explainable variation into parts attributable to various cross-mode component combinations. Like Tucker, Kroonenberg focuses on the interpretation of the core relations, which are assumed to portray all the essential characteristics of the observational modes from which they are derived. This statistical formulation of the Tucker method (1966a) retains the generality of Tucker's model and yields a sophisticated and apparently very useful approach to multimode problems.

Figure 1–1. Three-Mode Principal Component Analysis

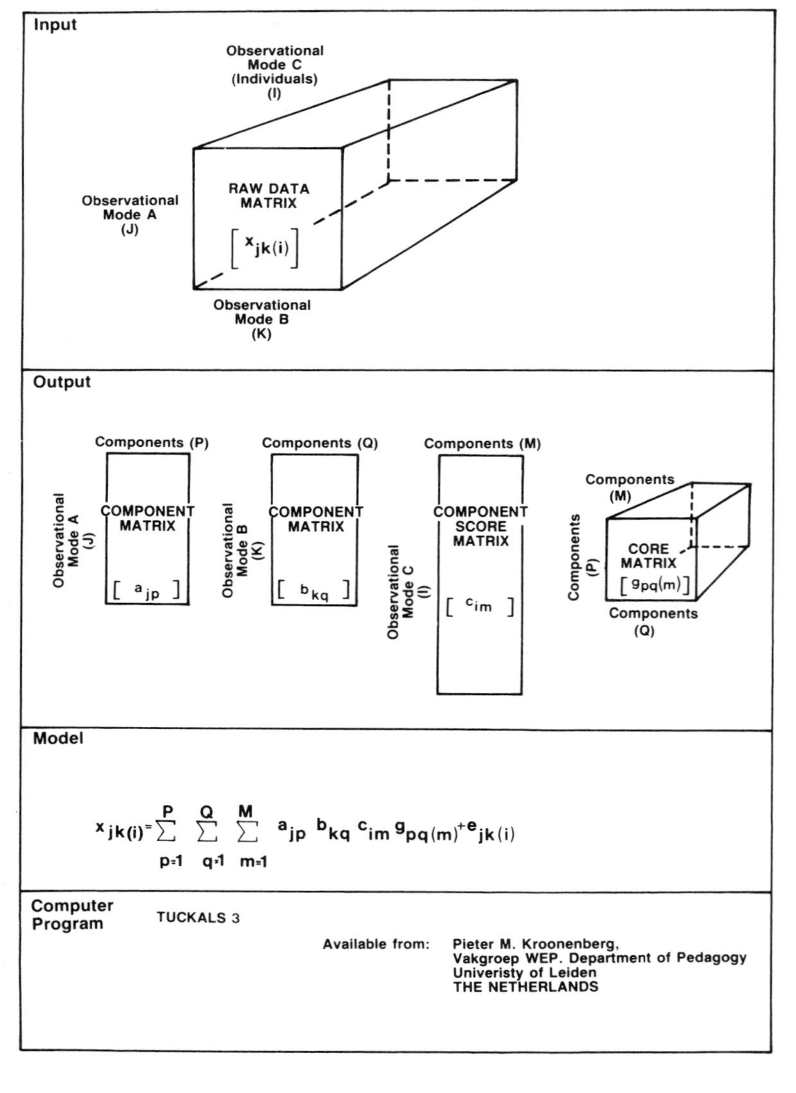

An interesting variant of the Tucker approach (1972) equates one of the derivational weights matrices with the identity, thus retaining the original classification units in the corresponding mode of the core box. This has been called the *extended core method* (Kroonenberg and de Leeuw 1980). When the identity is used for the individuals' mode, the formulation is algebraically equivalent to (in the case in which there is symmetry in the other two modes, in particular when data are scalar products derived from symmetric proximities data) Schonemann's (1972) *generalized subjective metrics model* for multidimensional scaling, which is in turn equivalent to Carroll and Chang's (1972) IDIOSCAL model (Sands and Young 1980). Figure 1–2 displays the features of Kroonenberg's TUCKALS2 computer program (1981b) for this method. This approach is particularly useful for those applications in which one of the observational modes cannot be meaningfully reduced (for example, time series, particular individuals, or specific conditions). Interpretational focus is directed to the interactional values in the frontal planes of the core matrix across the extended mode.

The extended core method can also be interpreted in terms of hierarchical (second-order) decomposition, where the first-mode analysis yields first-order factors (components) that are then related in the core (second-order) for the second mode at each level of the third classification set. Bloxom (chapter 4) has followed this expository approach in recasting Tucker's original model in terms of third-order composition rules. Using this translation, the factors of each successive mode are interpreted as higher-level latent influences, a view that is more consistent with traditional factorial conceptions (see also chapter 16). This modification of Tucker's formulation is diagrammed in Figure 1–3. At the present time, it is a reconceptualization that has not been explicitly applied but does offer a new framework for the development of multimode models.

The special feature of Tucker's multimode method is the inner core box. The generality of his approach is attributable to this ingenious conceptual device that combines the idealized entities associated with each of the modes. Despite the scope and conceptual sophistication of Tucker's method, applications are difficult because of the interpretational difficulties associated with the abstract core entities. Many of the new statistical estimation approaches apply constraints to the core to clarify its meaning. These researchers are interested in limited perspectives of the core but wish to retain the interactional notion (see Bentler and Lee 1978, 1979). Many other researchers have chosen to reconceptualize the multimode problem, applying simplifying assumptions that preclude the necessity for the complicated core. These models are more restrictive but perhaps more easily applied and interpreted.

One of the most interesting developments has been Harshman's parallel factors model, PARAFAC (1970, 1981; chapter 5), a model that formalizes Cattell's insightful solution to the factorial rotation problem. PARAFAC, while it includes many preprocessing options not part of that technique, is mathematically equivalent to the general three-way case of Carroll and Chang's (1970) CANDECOMP. The interpretive rationale of PARAFAC is derived from

Figure 1–2. Three-Mode Principal Component Analysis (Extended Core Method)

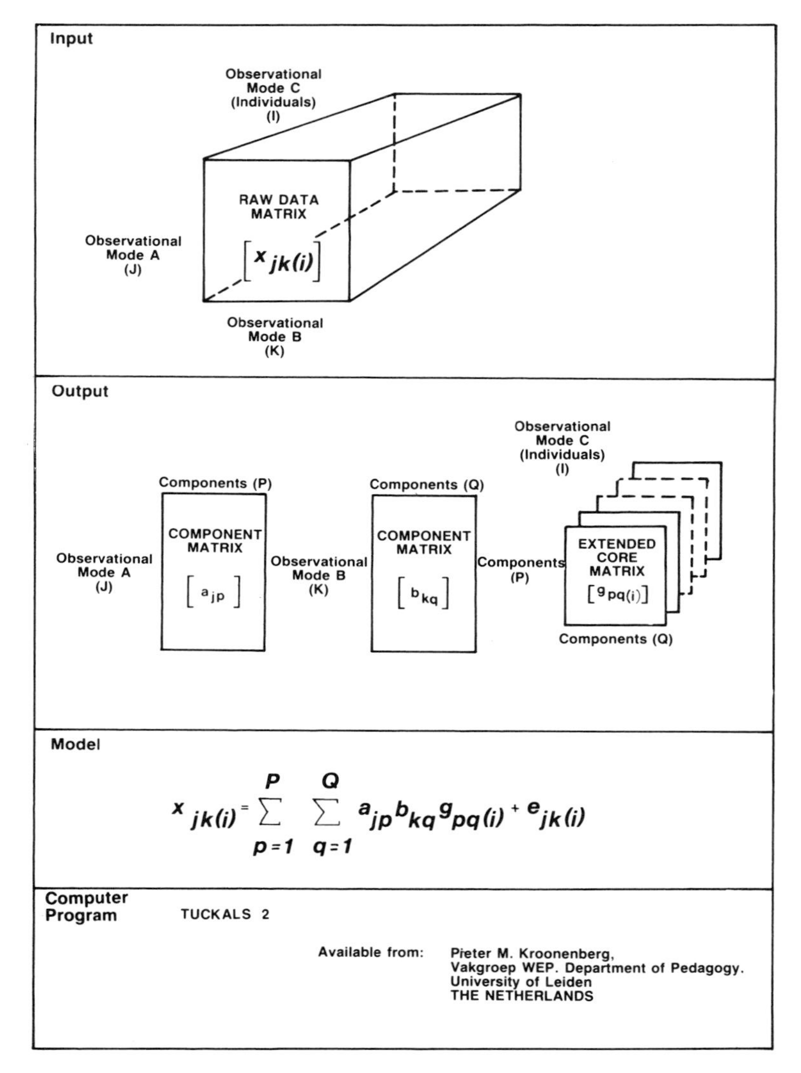

Cattell's *principle of proportional profiles* (1944), which Cattell called "the most fundamental principle" for the recovery of meaningful factors. Cattell contended that "real" factors and associated factor scores would retain their pattern but change proportionally from condition to condition (and study to study) according to the relative importance of the factors in explaining the covariation in each of the conditions. Accordingly, Cattell advocated systematic selection of conditions to ensure salient changes in factorial influence so that congruence could be indisputably established across studies. Harshman's notable contribution was the translation of Cattell's principle into a functional

Figure 1–3. Three-Mode Hierarchical Decomposition

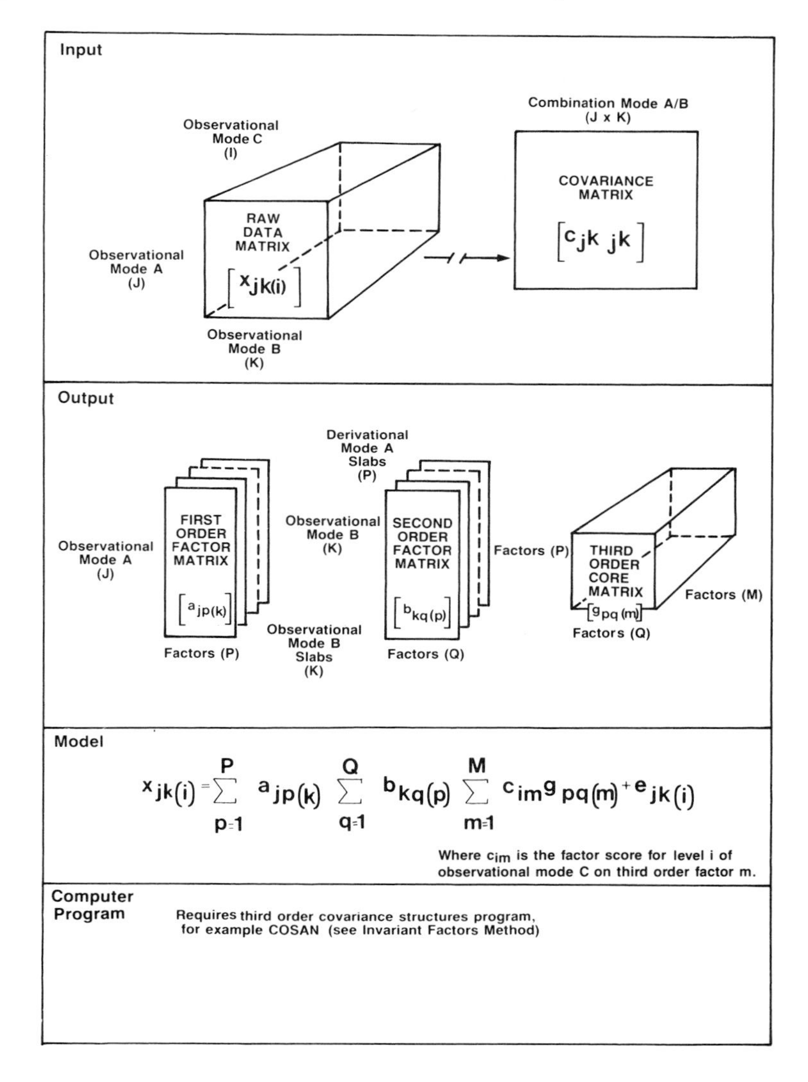

$$x_{jk(i)} = \sum_{p=1}^{P} a_{jp(k)} \sum_{q=1}^{Q} b_{kq(p)} \sum_{m=1}^{M} c_{im} g_{pq(m)} + e_{jk(i)}$$

three-mode model in which the conditions formed the third mode. Released from the narrowing foci of two-mode matrices and directly exploiting the systematic variation across the third mode, the parallel factors model yields a unique, meaningful solution.

In addition to this firm conceptual basis, PARAFAC has been used on a wide range of data, providing valuable insights about this model's strengths and weaknesses. Early versions proved restrictive and impractical. However, Harshman's obvious enthusiasm for data analysis has led to the development of three supplementary notions for PARAFAC application.

First, PARAFAC is based on a multiplicative model so the data level requirements are stringent. In order for the proportionality feature to be directly meaningful, the data should really be of ratio-type. Harshman and Lundy (chapter 6), joined by Kruskal (chapter 2), have advanced strategies for data preprocessing that make it easier to satisfy these data requirements. This alone is a major step forward because the measurement issues in structural analysis—although recognized—have eluded adequate rationalization, resting instead on precedent for their justification. (However, see Young, de Leeuw, and Takane 1980 for an excellent discussion on optimal scaling.)

Second, PARAFAC can be used on small samples, making split-sample analyses feasible. Harshman has revived the often praised (but rarely implemented) idea of cross-validation. With the use of a split-sample approach, the presence of extrinsic, nonsystemic variation can be detected to test the adequacy and appropriateness of the PARAFAC model. Given the small sample capability of PARAFAC, cross-validation is more likely to be utilized.

Third, when coupled with appropriate preprocessing, PARAFAC seems to be self-monitoring. When the model does not adequately represent the data, Harshman claims that it degenerates rather than misleads. Employment of intrinsic orthogonality constraints reduces the likelihood of degeneration and offers some insight into the limits of model fit. Therefore, stepwise application of PARAFAC with increasingly stringent, intrinsic rotational constraints provides an explorative evaluation of the presence of systemic variation. Unlike Tucker's conceptualization of factors as idealized entities underlying and belonging to the modes of classification, PARAFAC factors are undifferentiated with respect to mode of origin and are considered inherent influences of the *total situation*. The multimode cross-classification of data becomes a prerequisite condition for the unique identification of these latent variables. The additional mode delimits the situation for factor recovery rather than add new influences. This conceptualization is closer to the classical interpretation of factors, in terms of the identification of a single set of latent attributes.

Figure 1–4 illustrates the PARAFAC *direct fit* method. Starting with raw data—or, more likely, preprocessed data—PARAFAC evaluates the pattern of values across the three modes in search of the single set of latent influences that best accounts for the design entries. The simultaneous evaluation of the three-mode variation is sufficiently constrained by the third-mode information and the factor loading proportionality to obtain a unique solution for any selected dimensionality. Model parameters are estimated using an alternating least-squares algorithm (Jennrich in Harshman 1970; Jennrich 1972). (See also Carroll and Chang 1970; or Carroll and Pruzansky, chapter 10, for details on CANDECOMP, the canonical decomposition procedure equivalent to that used in PARAFAC.)

Since this method is directly fit to the three-mode data set (as advocated by Horst 1965; Kruskal 1978), the results may differ from an indirect method that seeks the least-squares fit on covariation patterns. An indirect PARAFAC is illustrated in Figure 1–5. In this case, the individuals' mode is lost in the computations of the covariances, thus limiting the output to the weights

Figure 1–4. Parallel Factors Analysis (Direct Method)

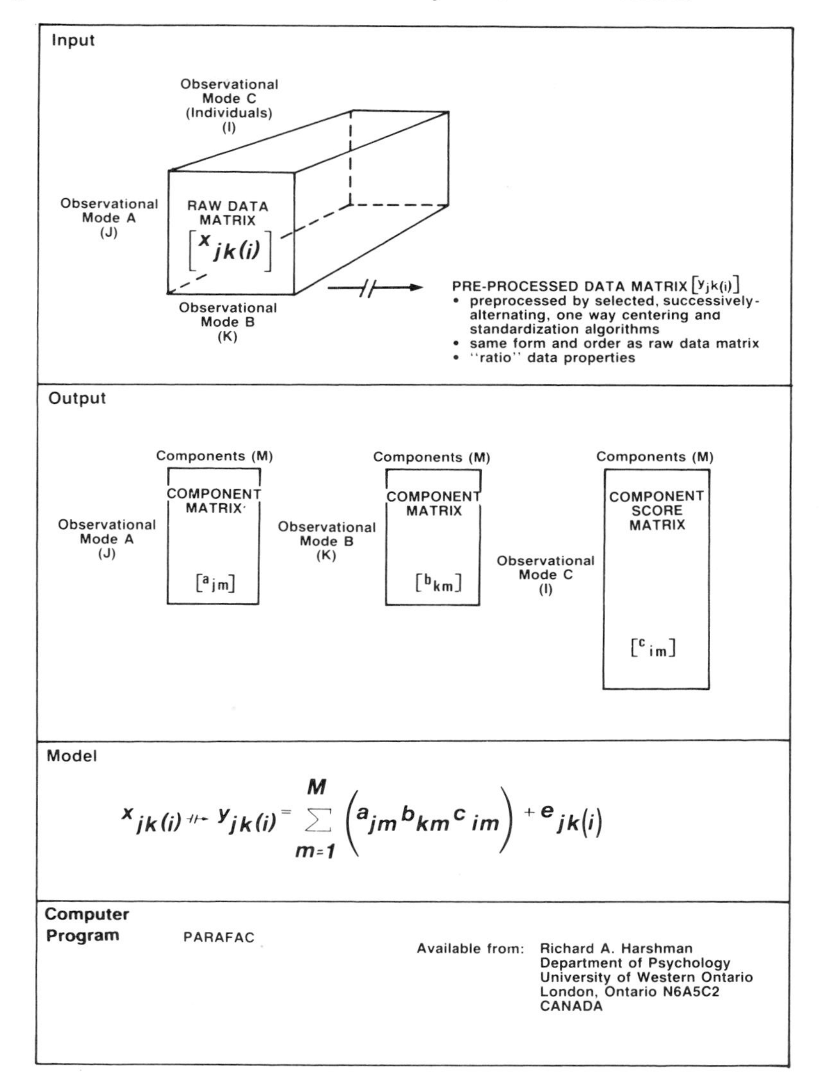

for the other two modes. Factors are conceived in much the same way for the direct and indirect methods, but those derived indirectly may also reflect influences *within particular individuals* as well as systemic variations. Factors identified within a subset of individuals are considered characteristic types of influences rather than single latent sources, presumably originating in the particular person rather than in some common system (Harshman 1970; Harshman and Berenbaum 1981). These differences in factor interpretation can cause the solutions from the two methods to be quite discrepant. Successive application of direct and indirect methods may be a useful strategy to sort out these

Figure 1–5. Parallel Factors Analysis (Indirect Method)

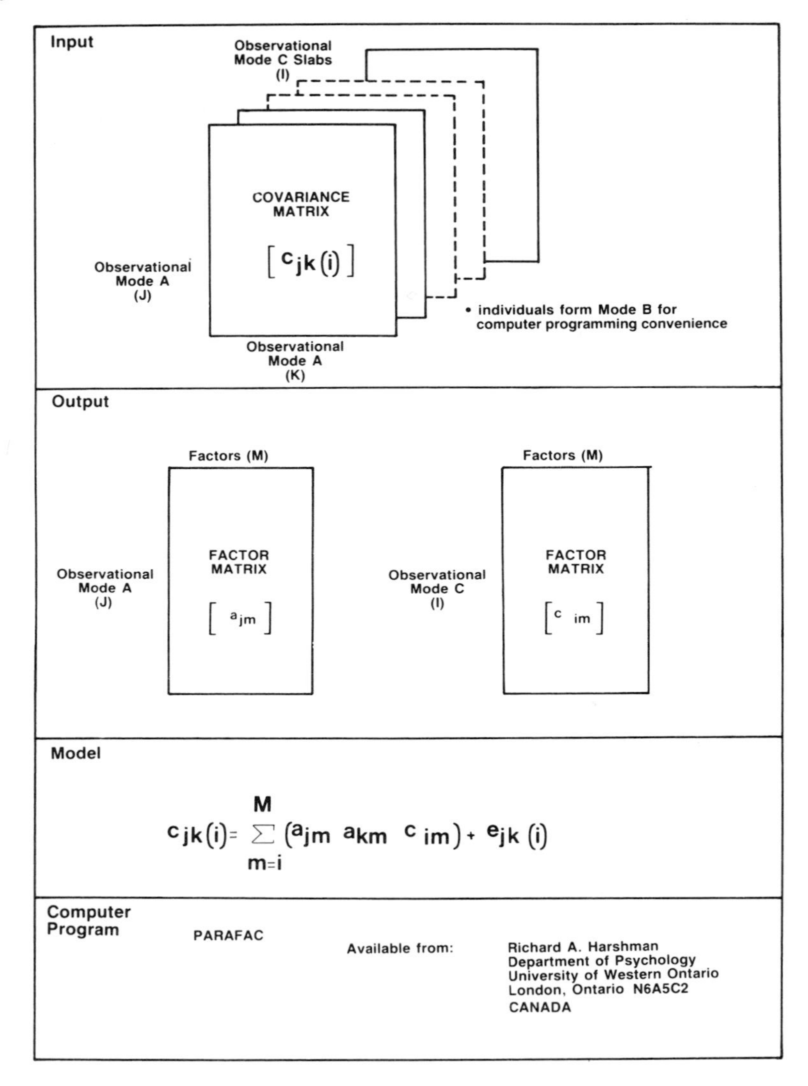

individual type and systemic influences (see Cattell 1978, 444–51, for an interesting discussion on within- and between-type factoring). Although care must be exercised in data preprocessing, PARAFAC promises to become a valuable and practical analytic approach.

Inherent in the proportionality requirement is the assumption of an invariant correlation matrix of the factor scores across the third mode, indicative of the systematic influence of the factors. Therefore, the factors "shrink" or "stretch" in proportion to their influence in each condition. This conceptual model underlies the power of both the direct and indirect PARAFAC methods.

Unfortunately, proportionality does not always hold, even under controlled conditions (as demonstrated by Cattell and Cattell 1955). It should be pointed out that in the indirect PARAFAC method, instead of requiring the profile of factor scores on each occasion to be proportional to that on every other occasion, it is only required that the correlations of these profiles be constant. The difference between the two approaches is that in the indirect case, one is permitted to drop one restrictive assumption (proportionality at the individual case level) and substitute a different one (orthogonality of factors).

McDonald (chapter 7) proposes a model that is closely related to PARAFAC but has a slightly more conventional basis. Traditionally, one aim of factor analysis has been the identification of an *invariant* factor structure; most researchers have relied on the simple structure principle to orient their factor solutions (see chapter 5 for Harshman and Lundy's comments). Factor scores, however, have not been restricted in terms of their relationships, while the influences of conditions are presumed to be reflected in changes in factor score means and distributions. McDonald's invariant factors model incorporates this traditional conceptualization, reducing the second derivational mode to the status of a transformation matrix for the changes in factor score covariances and relying on prescribed simple structure for the identification of the model. The output provides factor score means and, with some calculations, correlations that reflect the changes across the second mode. In this framework, factors are interpreted as abstractive attributes, belonging to the attribute mode and invariant over the conditions mode. Accordingly, the modes are not conceptually symmetric, as in the Tucker approach.

Because McDonald's invariant factors model is embedded within his general model for covariance structures analysis of the *n*th order, COSAN (1978, 1980; Fraser 1980), various forms of the model can be investigated; thus, the particulars of the various possibilities cannot be easily depicted diagrammatically. One important feature is the possibility of incorporating patterned residuals within the model. It is the structuring of "error" (in terms of uniquenesses) that gives factor analysis its special role among analytic methodologies. In the three-mode case, the likelihood of unique conditional factors is an important consideration that is explicitly recognized in patterned residuals. A second feature of the COSAN approach is the flexibility in model specification, even allowing for prescribed nonlinear functions. As indicated, PARAFAC restricts the factor score correlational structure from condition to condition, whereas classical factor analysis restricts the attribute factor structure, usually in accordance with the simple structure principle. Since the invariant factors method allows either type of specification, the indirect PARAFAC model can be explicitly tested in empirical investigations within the invariant factors framework. Furthermore, knowledge about the factor structure can be applied in the parameter specifications of the attribute mode, reflecting a more confirmatory approach to the analysis. Maximum likelihood estimates and an asymptotic chi-square test are available as COSAN output. Coupled with the patterned residuals, the COSAN invariant factors method is potentially a powerful analytic approach for three-mode data.

While Cattell's original idea of "proportional profiles" was formulated in terms of changes in the factor pattern, PARAFAC's even-handed treatment of all three modes allows the proportionality to be conceptualized as applying to any of the modes. Thus, the PARAFAC model can be interpreted in particular as the product of an invariant factor loading matrix and a varying factor score matrix. With this interpretation, PARAFAC does not differ from the McDonald model with respect to the traditional aim of identifying an invariant factor structure, although it does not impose the traditional simple structure requirement. The difference between the two models is in their assumptions about factor (or component) scores. While PARAFAC allows factor score variances to change across occasions, it makes specific assumptions about the way the scores vary, assuming either proportional variations (in the direct fitting case) or orthogonality (in the indirect case). These assumptions determine a basis which in general is incompatible with simple structure, while McDonald's model needs an additional principle such as simple structure to determine a basis. Another difference, already noted, is that McDonald's model includes a patterned residual matrix and is thus statistically falsifiable.

Both the PARAFAC and the invariant factors assumptions can be restrictive if the particular investigative area is uncharted and the nature of change is unknown. Corballis and Traub (1970, extended by Corballis 1973) have provided a model that allows both factor scores and patterns to vary across time; Swaminathan (chapter 8) has elaborated this approach. Maximum likelihood estimates and associated chi-squares are available for several general hypotheses: invariant factor pattern; invariant factor scores (as generated by a first order auto-regressive series); and invariant unique scores. Tested in sequence, information is accumulated on the presence of equivalent factor patterns, the possibility of proportional factor loading changes, and/or the completeness of the factor model. At the present time, Swaminathan's model offers an interesting conceptual and statistical development for future research application.

The McDonald invariant factor model was developed after intensive examination of the Tucker, Bloxom, and Harshman approaches. Although its roots lie in an earlier formulation (McDonald 1980), the character of the model was clearly forged as an alternative to these other conceptualizations. Swaminathan's maximum likelihood longitudinal factor analysis method enhances the hypothesis-testing capability of the Corballis-Traub approach (see also Harshman and Lundy, chapter 5) and is couched in the same general COSAN computer program as McDonald's analysis.

Despite the important differences among these approaches, they do share similar concerns and some common (or at least compatible) resolutions to multimode analytic issues. An apparently independent approach has been taken by Cattell (1966, 1978, 1980a, 1980b) in his *n*-way factoring. Emanating from his long-term attention to basic data relations and "representative" design considerations, Cattell developed two different approaches to the analysis of complex data designs. The *personality centered* factor analysis method yields *conjoint factors* of the transposed combination mode grids, which are strung-out data arrays across two modes (see Figure 1–6). The factor loadings are split into two parts: (a) an involvement index, indicating how much the particular personality trait is involved in a particular response; and (b) a modulator index, indicating how much the ambient situation modulates the particular trait in that setting. As it turns out, this model is algebraically equivalent to the PARAFAC-CANDECOMP model, as pointed out by Richard Harshman, but there is a conceptual difference. The conjoint factors are interpreted as resident in the person (personality)—a trait

Figure 1–6. Conjoint (Grid) *n*-Way Factor Analysis (Personality Centered)

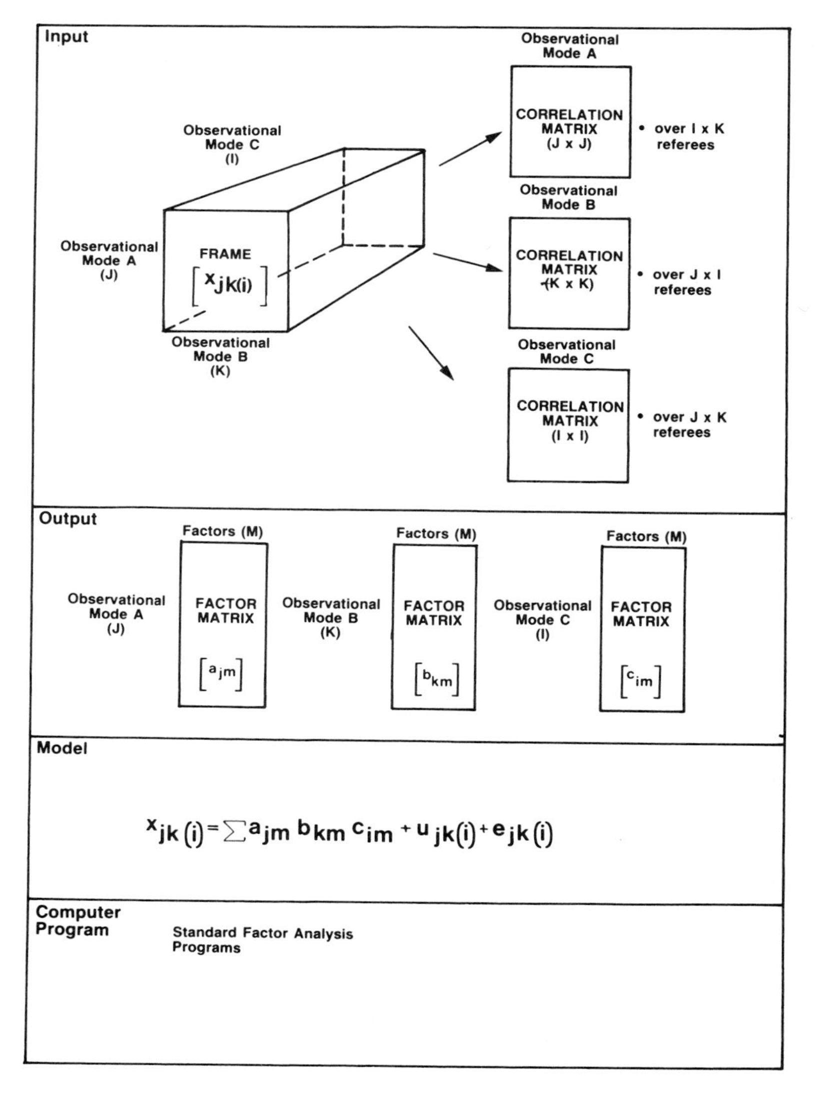

view—rather than the situation—a systemic view—as taken by the PARAFAC model. Cattell's second model, called the *additive id* or *additive attribute* model (see Figure 1–7), represents the data in terms of *disjunct factors,* expressed as the sum of two-way multiplicative terms. In this method, classical two-mode factor analysis is applied to each *face* of the data design, where a face is a two-way data array with mean entries calculated over the third classification set. Taken together, these disjunct factors tap the various common aspects associated with each mode combination. The factors extracted presume constant loadings across

Figure 1–7. Disjunct (Face) *n*-Way Factor Analysis (Additive Id/Attribute Method)

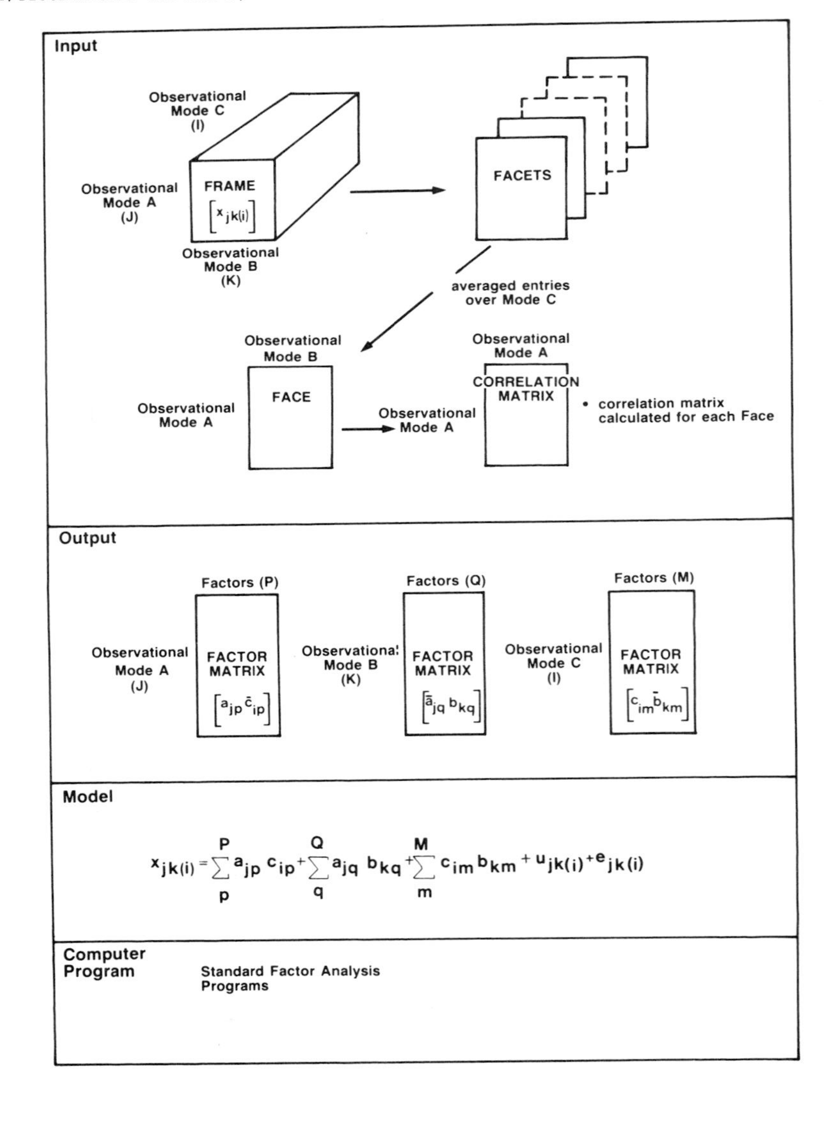

the conditions/occasions mode, similar to the invariant factors conceptualization.

Both of Cattell's *n*-way models are "psychometric" in orientation, as was Tucker's, serving as a reminder that valid perspectives ultimately depend on adequate representation of the relevant sets of entities. Comprehensive classification resulting from more precise dissections of the ambient setting for human activity will inevitably lead to larger data designs. As indicated by Cattell, "The road to an adequate and meaningful inclusion of a taxonomy of human environment will demand uncommon fortitude in experiment" (1980a, 246).

Scaling and Related Developments

Classical multidimensional scaling entails the analysis of the two-way matrix of stimulus similarities (and dissimilarities) from one individual. The aim is to recover the dimensions or configurations of stimuli in multidimensional space that are relevant to the perception of these stimuli. Configurational interpretations focus on clusters, regions, or geometric patterns of the stimuli in the multidimensional space. Unlike the classical two-mode factor-analytic analog, which explores patterns of individual differences, variations in multidimensional scaling due to these differences were not considered until somewhat recently. For many years, Tucker and Messick's "points of view" analysis (1963), which combined factor analysis and classical multidimensional scaling, was the primary method for the assessment of individual differences in multidimensional scaling. In this approach, the correlations between individuals—in terms of their similarity judgments—were factor analyzed to identify clusters of individuals about the centroids of this person space. The person closest to the centroid or the average similarity judgments within the cluster was designated as an "idealized" person, after which the similarity judgments matrix of each idealized person was analyzed by classical multidimensional scaling.

In reaction to the inherent difficulties of the "points of view" analysis and following some conceptual developments due to Horan (1969), Carroll and Chang (1970) proposed an individual differences model for MDS that included an *n*-way generalization of the Eckart-Young principle based on the method of canonical decomposition (later called CANDECOMP). It should be pointed out that Carroll and Chang's development of the INDSCAL model was independent of (though closely related to, conceptually) Horan's work. This generalization is similar to Tucker's three-mode principal components analysis (1966a, 1972) approach but leads to an exact least-squares fit. Algebraically, CANDECOMP is a special case of three-mode components analysis with two major constraints: equal dimensionality across the derivational modes, and a restricted core, "a kind of 3-way analogue of an identity matrix." When CANDECOMP is applied to scalar products data generated from an inherently three-mode design, the model is mathematically and conceptually equivalent to Harshman's indirect PARAFAC model. However, in practice, this application would not possess the useful properties of Harshman's preprocessing procedures. CANDECOMP—in its general three-mode form—is mathematically and conceptually equivalent to "direct" PARAFAC.

The most widely used application of CANDECOMP is INDSCAL (see Figure 1–8), a particular model for individual differences scaling with multiple sets of symmetric proximities (three-way) data. In the INDSCAL data box, each face represents relations within one mode and the slabs are usually the individuals; so these data are two-mode—individuals and stimuli—and three-way—individuals and the cross-classified relations between stimuli. Common "factors" for all individuals are assumed to underlie

Figure 1–8. Symmetric Trilinear Analysis (Weighted Individual Differences)

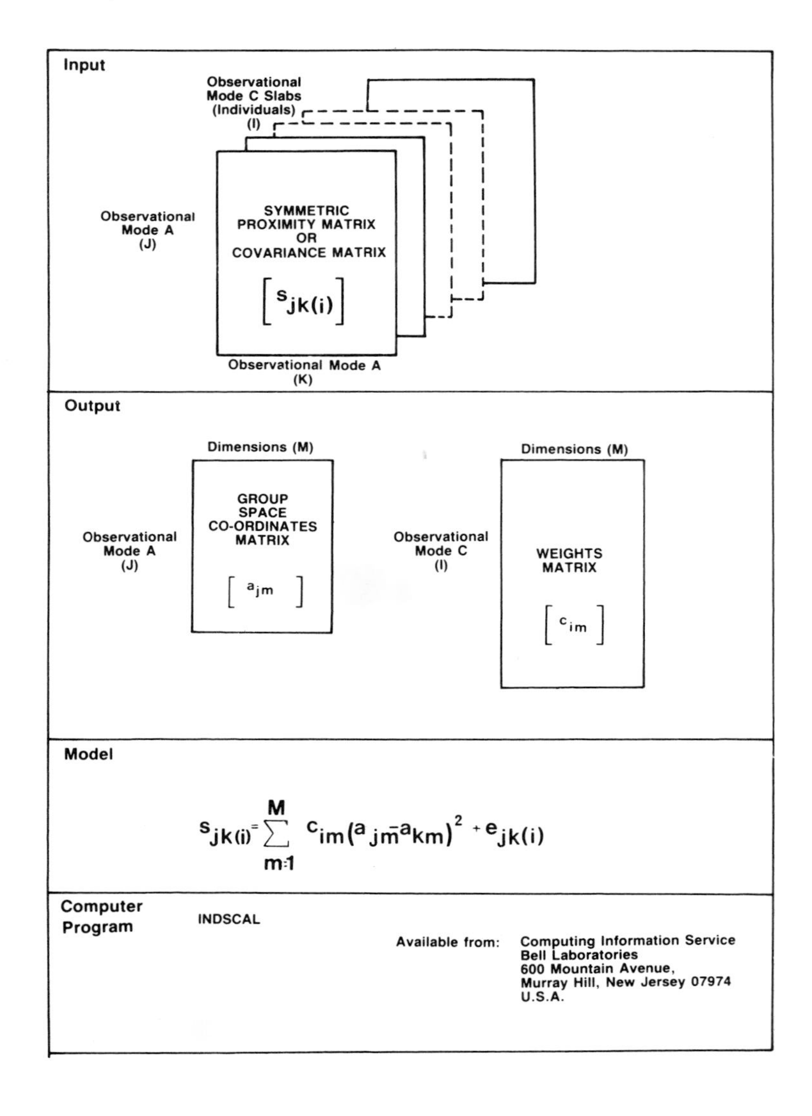

the stimuli in a *group stimulus space,* but each individual can have a different pattern of weights in a *private perceptual space,* where the common dimensions are stretched or shrunk accordingly. Arising from a mathematically unique coordinate system, these factors are assumed to be *fundamental* psychological dimensions and are treated like traditional common factors. The difference is that the error term is an unstructured residual of model fit in this case.

Foreshadowed in the Carroll and Chang CANDECOMP paper (1970) is the application of the canonical decomposition method to factorially designed experiments, in the analysis of variance sense. The ensuing CANDELINC (Canonical Decomposition with Linear Constraints) method (Carroll, Green, and Carmone 1976; Carroll, Pruzansky, and Kruskal 1980) makes it possible to linearly constrain each or some combination of the mode parameters. If, for example, the individuals (or stimuli or response scales) are selected according to some blocking variable—such as sex, age, or educational level, in the case of individuals—the parameters can be constrained to be perfectly described by an additive, main-effects-only ANOVA model. These constraints improve the identifiability of the model and enhance the external validity of the explorative investigation, which necessarily entails extrapolation to new elements of the classification sets. In its most general multilinear form, CANDECOMP can also fit Lazarsfeld's (1950) latent class model to *n*-way contingency tables or Gollob's (1968) factorial analysis of variance approach for the analysis of interaction effects.

The intent of INDSCAL and related methods is very similar to that of factor analysis—the identification of a relevant and reduced geometric space within which to examine empirical relations among entities. A major difference between factor analysis and scaling is that in the former, a researcher must have an a priori conception of the factors or dimensions appropriate to the domain. The nature of the common space onto which the variable vectors are projected is restricted interpretively by the semantic entailments of the measures selected. Multidimensional scaling relies less on these operationalized preconceptions, determining *dimensional salience* directly from stimulus pairs by way of a set of proximity indices (although there is an implicit selection of entity sets).

> The most fundamental idea in this context is to decompose each individual's perceptual space into a structural component that he shares with all or some other subjects and another component that is specific to him. . . . The unifying feature of these [MDS] approaches is that the common space is defined as some average configuration over all individuals who are related to it by various transformations. The commonality of an individual space is, therefore, a function of the properties of the average configuration and the respective Procrustean transformation. (Borg 1979, 609)

Horan (1969) had noted that the salience or importance of the dimensions determined for the *average* person may vary for any *particular* person. INDSCAL provides linear transformations, constrained to a diagonal matrix, that rescale each dimension to differentially stretch or shrink in accordance with the weights associated with different individuals. Further transformation, in the form of an orthogonal rotation of the group space, is possible in a later extension of INDSCAL, called IDIOSCAL (Individual Difference In Orientation SCALing) (Carroll and Chang 1972). However, neither of these methods provides information about potentially more complicated individual difference relationships beyond the dimensional salience model. An important cousin of the INDSCAL family—PINDIS (Procrustean INdividual Differences Scaling) (Lingoes and Borg 1976; also see Gower 1975; chapter 11)—does provide this information, as depicted in Figure 1–9. PINDIS applies transformations of the following type: similarity, dimensional weighting (like INDSCAL), dimensional weighting with idiosyncratic frames of reference (like IDIOSCAL), vector weighting with a fixed origin in the centroid configuration, and vector weighting with an idiosyncratic origin. The latter two transformations are indicative of more complex relations among individual configurations than is conveyed by a straightforward dimensional alteration. Lingoes and Borg have called them the *perspective* model, representing an individualistic orientation or perspective of the centroid structure and/or indicating few but substantial displacements among the stimulus representations. This model is particularly important in the evaluation of the dimensional salience model (which of course underlies PARAFAC factor analysis as well) to ascertain the representivity of some group space. As is well known in univariate analyses, averages can reflect the confounding of essentially different individual values, thus failing to represent any individual in the sample. In the same sense, a common space may mask importantly different individual structures. Since PINDIS operates on coordinates derived from either multidimensional scaling analysis or factor analysis, the method represents an important explorative approach for the evaluation of the more restrictive multimode analysis frameworks.

The INDSCAL and PINDIS class of methods seeks a set of continuous dimensions that are common to but differentially weighted by the sample individuals. An interesting alternative—which is particularly appropriate for "conceptual" stimulus domains from varying semantic fields (see Pruzansky, Tversky, and Carroll 1982)—is the INDCLUS (Individual Differences Clustering) model proposed by Carroll and Arabie (chapter 14). INDCLUS yields a discrete structure in which individuals differentially weight a common set of clusters. Thus, INDCLUS is the cluster analog of INDSCAL and is useful when a discrete network model is preferred to the continuous spatial (Euclidean) model (see Figure 1–10).

Young (chapter 12) has proposed a general framework for the specification of multidimensional models called GEM (the General Euclidean Model). Within this framework, another interesting alternative to INDSCAL emerges. In INDSCAL, each individual is represented by a diagonal weight matrix that indicates the differential importance the individual attaches to the group stimulus space dimensions. A different conceptual perspective is offered

Figure 1–9. Procrustean Individual Differences Scaling

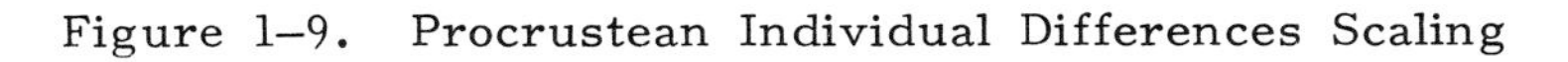

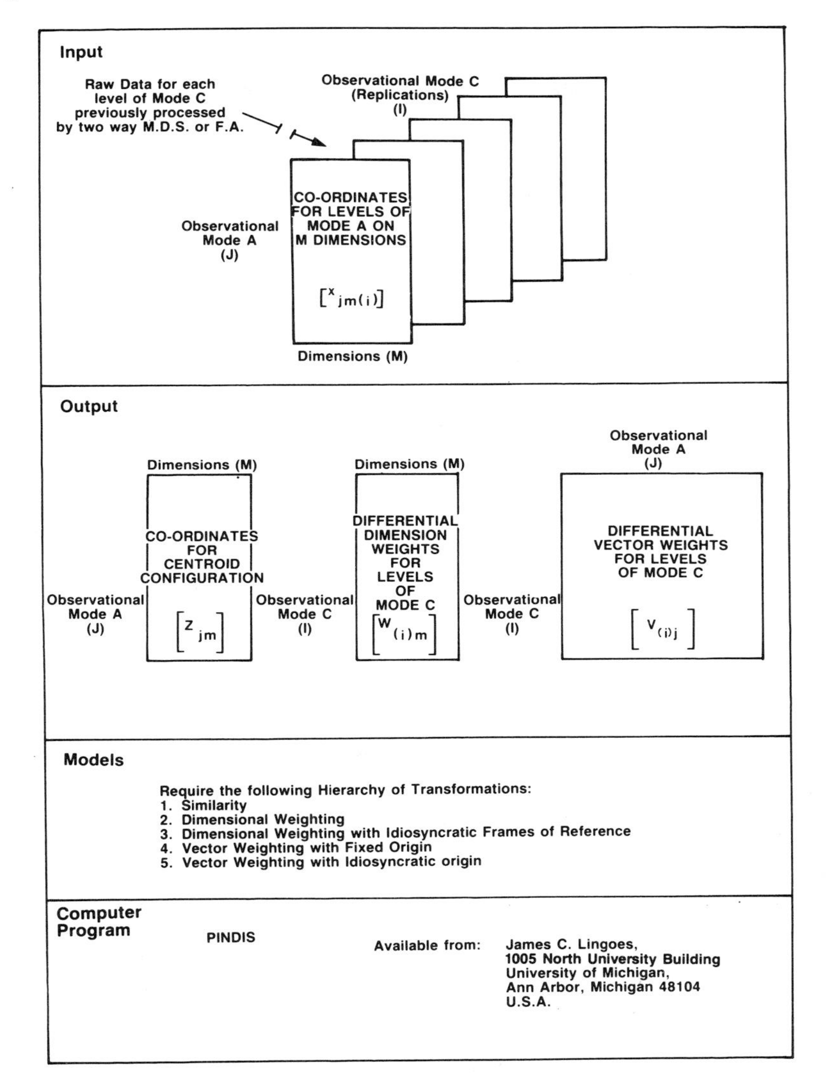

by GEMSCAL (see Figure 1–11), which assumes that an individual occupies *principal directions* in joint (individuals-stimuli) space rather than a separate person space. Individuals are represented by vectors; the orientation (from the origin) of the vector indicates the salient directions for the individual and the length of the vector indicates the relative importance of the principal directions. When the number of salient directions for all individuals equals the number of dimensions, GEMSCAL corresponds to IDIOSCAL, and when these directions coincide with the dimensions, GEMSCAL is INDSCAL. Like PINDIS, GEMSCAL offers an interesting alternative to those investigative situations in which

Figure 1–10. Individual Differences Clustering

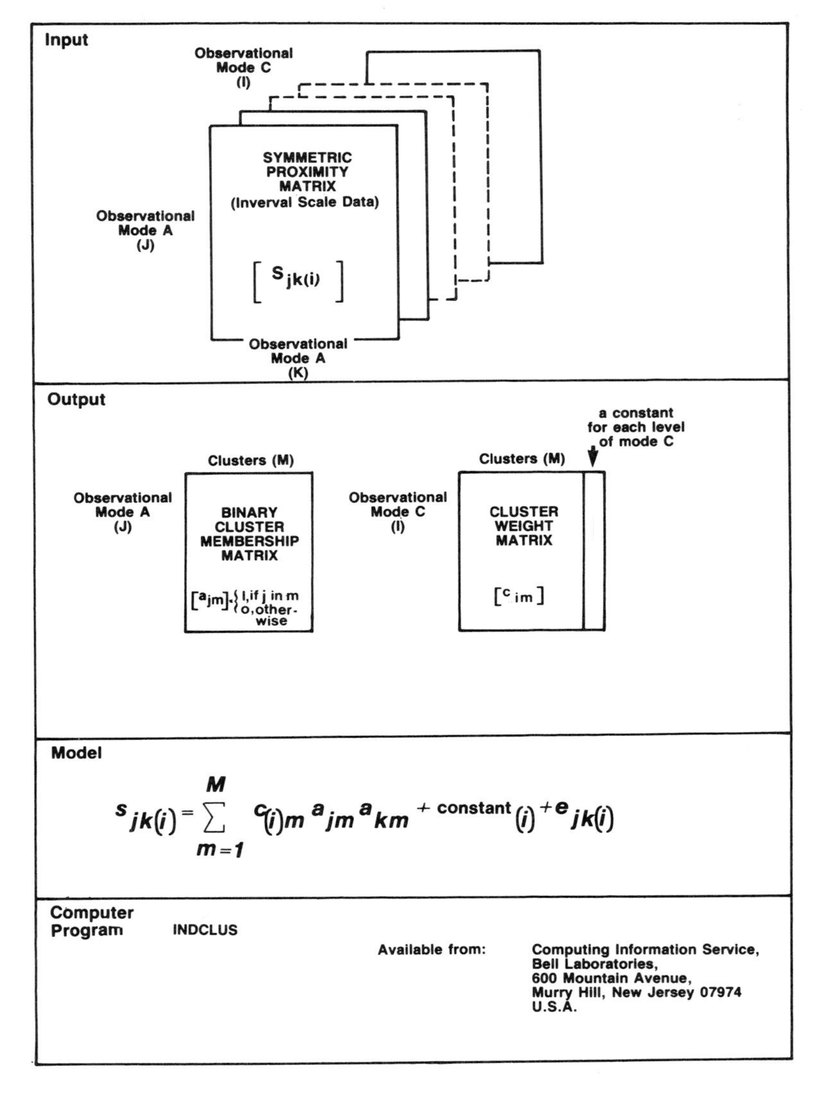

the dimensional salience model appears overly restrictive.

The INDSCAL, PINDIS, or GEMSCAL-type analysis is applied to sets of symmetric matrices. When the data are encased in nonsymmetric or rectangular three-mode arrays, the appropriate model is called a *replicated* or *weighted* individual differences unfolding model (see Carroll and Pruzansky, chapter 10). In unfolding, preferences are modeled in terms of *idealized points*. The smaller the Euclidean distance between a stimulus point in the multidimensional space and an individual's ideal point, the more preferred the stimulus is to that individual. Individual vectors are projected through the space according to these hypothesized

Figure 1–11. General Euclidean Model Scaling

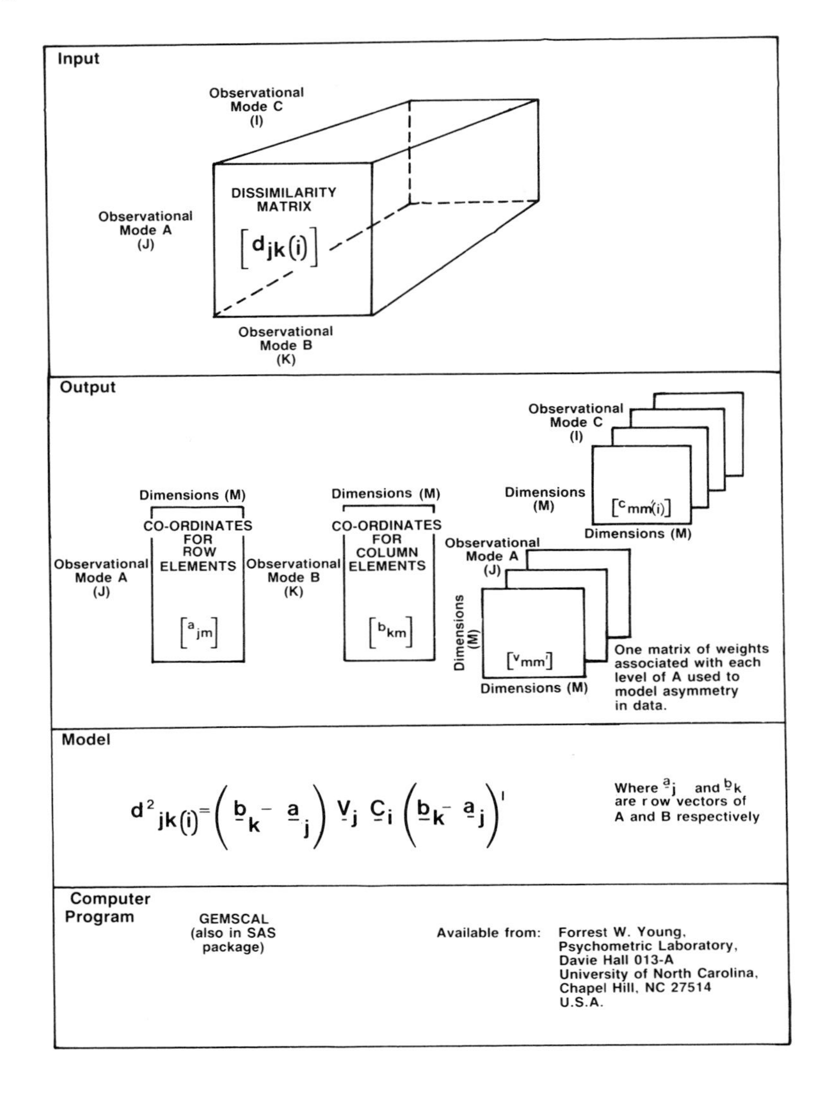

idealizations, and the stimuli are ordered along the vectors according to the particular preferences expressed by each individual. Extension to the three-mode case entails weighted distances between the stimuli and ideal points for each situation or condition (the third mode). Figure 1–12 outlines the approach developed by De Sarbo and Carroll (1981), which generalizes the Coombs ([1964] 1976) and Schonemann (1970) unfolding models. Young has categorized the models for the analysis of three-mode preference data into two families within GEM. One type assumes that the matrix of row weights usually associated with individuals is equal to the identity; individuals are treated as replications

Figure 1–12. Nonsymmetric Trilinear Analysis: Three-Way Unfolding

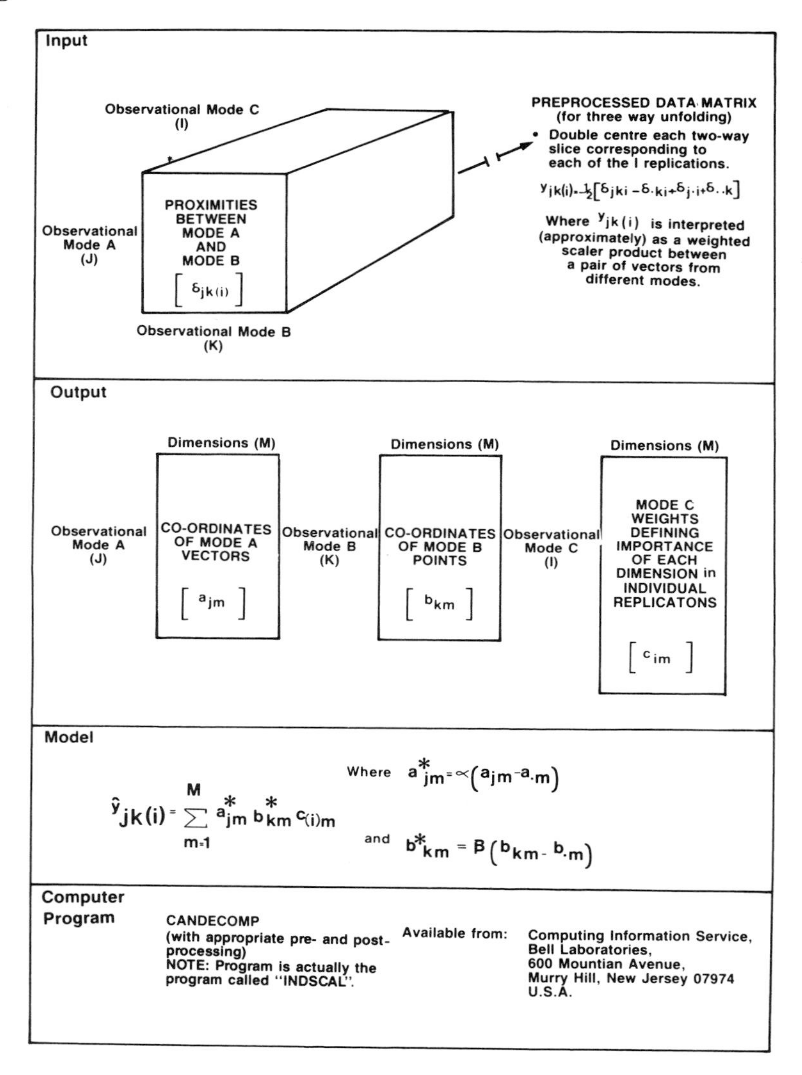

only. The second type refers to the generalized form of GEM, in which the weighting matrices associated with individuals and conditions (or occasions) are restricted to positive semidefinite matrices. Unfortunately, in practice, the general analysis usually degenerates because of insufficient constraints for identifiability.

All the methods discussed so far have dealt with metric data. Classical MDS, INDSCAL, and various unfolding models are now fit by a more general computer program called ALSCAL (Alternating Least-Squares Scaling) (Takane, Young, and de Leeuw 1977), which includes optimal scaling features to extend applications to nonmetric data. In its most general form, INDSCAL-CANDECOMP is a trilinear or multilinear or multiplicative model. More recently, Takane and his colleagues have developed methods for handling analysis-of-variance-type additivity models for nonmetric data. Within this framework, the focus is to explain a criterion variable in terms of the additive effects of two or more influencing factors. These factors are differentially weighted for each individual, or any other replicate, thus representing the common and unique aspects of the conjoint data in a fashion analogous to the INDSCAL approach. The difference between this method and the others discussed is that these factors are requantified attributes rather than latent constructs. Requantification is intended to improve the numerical assignment of values to the attribute in order to maximize the linear relationship with some criterion. Individual differences reflect either differences in the quantification of these new attribute variables—in which case a simple additive model is fit for each individual—or differences in the relative importance of these variables—where quantification is assumed constant across individuals. The features of the WAM (Weighted Additive Model) (Takane, Young, and de Leeuw 1980; Takane 1982) are illustrated in Figure 1–13.

GENERAL CONSIDERATIONS

Whenever a complex model is employed, the ultimate meaningfulness of the solution becomes vulnerable. Given the constraints introduced to identify and interpret such models, it is increasingly important for the research effort to grapple with the thematic measurement requirements. Although measurement issues are not the focus of this book, the usefulness of multimode models ultimately rests on their resolution within any particular research context. Entries in the multimode data box must be inextricably tied to their conceptual themes in order to avoid a multimode labyrinth.

The implications of measurement decisions are considerable. Not only are the central research themes delineated by the measurement procedures adopted, but the relational system that specifies the theoretical associations of those themes is also made explicit by the procedures. Regardless of the orientation of the research—manipulative or correlational—the operationalization of the research themes, in terms of an experimental design, focuses the research and thereby constrains the kinds of results that can be observed. As pointed out by Thomas Kuhn in his critical review of the function of measurement in modern physical science:

> New laws of nature are so very seldom discovered simply by inspecting the results of measurements made without advance knowledge of those laws. Because most scientific laws have so few quantitative points of contact with nature, because investigations of those contact points usually demand such laborious instrumentation and approximation, and because nature itself needs to be forced to yield the appropriate results, the route from theory or law to measurement can almost never be travelled backward. Numbers gathered without some knowledge of the regularity to be expected

almost never speak for themselves. Almost certainly they remain just numbers. (1977, 197–98)

. . . To discover quantitative regularity one must normally know what regularity one is seeking and one's instruments must be designed accordingly; even then nature may not yield consistent or generalizable results without a struggle. (1977, 219)

The life sciences function similarly, with the added difficulties of fewer contact points and less stable instrumentation. Investigators select and record the observations of interest from a

Figure 1–13. Weighted Additive Model

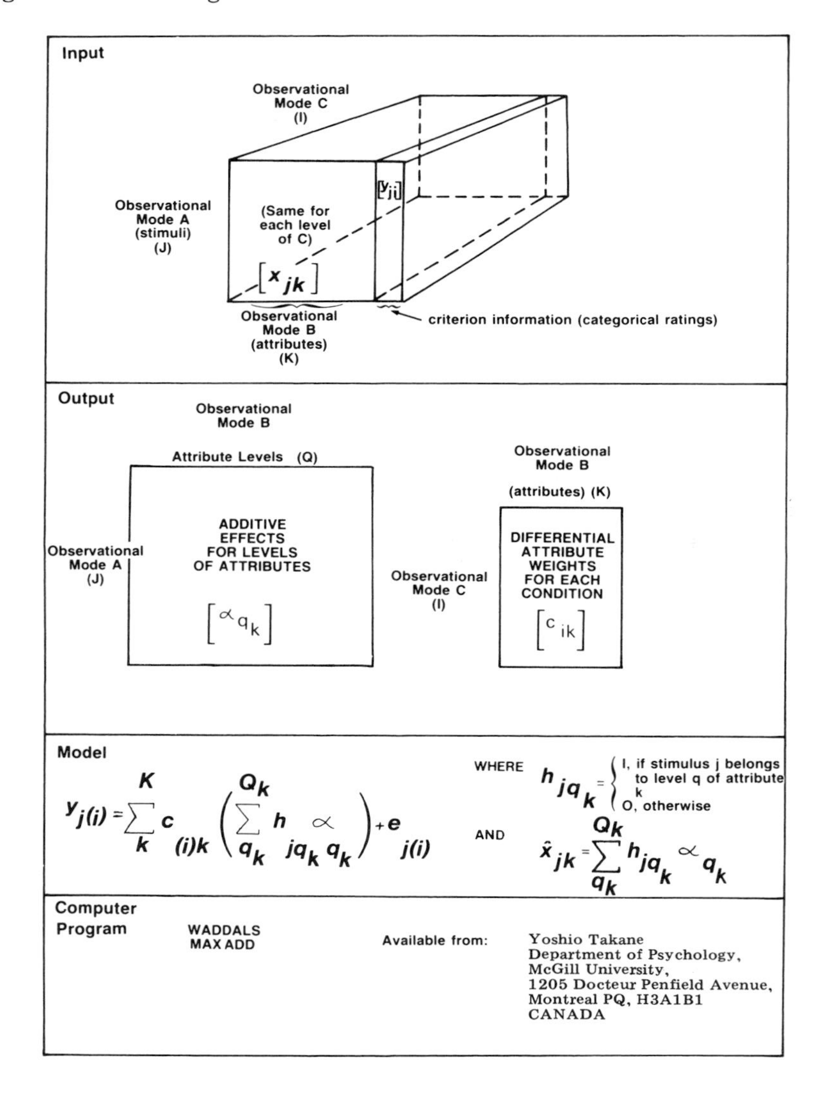

nearly infinite array of potential observations. This problem has bedeviled the factor analysis literature. (See Spearman's 1904 papers expressing his concern about the fallibility of observations, an issue that is still relevant today.) Although factor analysis has admittedly yielded considerable benefits when employed by seasoned data analysts, problems arise in its indiscriminate application:

> Most users of factor analysis employ the model in an exploratory fashion. In doing so, they do not begin with a set of relatively clear psychological concepts and then select variables for their study that on the basis of theory should contain or be determined by these concepts. Nor, on the other hand, do they begin with a clearly defined behavior domain (from which all investigators would recognize equally how to develop representative measures) and then select from it a reasonable number of variables to represent the domain in a balanced fashion. Instead, they begin with a number of variables in which they are interested, which might or might not correspond to a general preconception of a behavior domain of tests that "resemble" the tests chosen. Their hope is that when these variables are subjected to a factor analysis, the psychological attributes that determine the correlations among the variables will reveal themselves. (McDonald and Mulaik 1979, 304–5)

This is not to decry the usefulness of exploration of data by factor analysis and other multimode methods but to narrow the "blindness" connotation of these empirical search approaches. Exploration, even in its least planned form, involves the comparative validation (albeit informal) of a restricted range of hypothesized outcomes, the possibilities of which are at least tacitly recognized as inherent in the data collection and preprocessing operations. Some confusion may have arisen in (the vectors of) the minds of factor analysts due to the early emphasis on *exploratory* methodology. But exploration referred to the newness and unknown dimensionality of the conceptual domain and not to a prescribed directionless research approach that feigned some ultimate objectivity. As articulated by Thurstone:

> When there are no hypotheses to guide the study of a new domain, the experimenter tries to cover the domain with tests or measures that are diversified within the domain. The factorial result is likely to give an incomplete simple structure at best, but the exploratory study is justified if one or more functions or parameters appear in the interpretation that can be used as starting-points for more crucial subsequent studies. The second kind is the experiment in which some hypotheses are available and in which tests are specially constructed to represent the presence and absence of each postulated factor or parameter. Such studies are likely to be more productive of definitive results. (1947, 340)

The early notion of exploration, therefore, was the initial phase of confirmatory research. It has heuristic value in the generation of hypotheses amenable to more probing and precise analysis.

Factor analysis and multidimensional scaling are quantitative filters, sifting out noninformative "error" from systematic information. What constitutes error varies from method to method; the form of the information follows directly from the choice of filter. Qualified by the underlying quantitative model, the implicit interpretative decisions are intended to yield confirmatory evidence about suspected relationships. Even in those cases in which refutation of a fully specified model appears as an explicit goal, there are plausible alternative hypotheses that are implicitly confirmed by the argument. Without some hunch or speculation, the decisions are aimless and usually fruitless.

Far from suggesting only confirmatory approaches, the selected range of multimode methods offered herein attests to a preference for *reasoned* and *guided* exploration of their rich data bases. Since the variety of multimode methods does not reflect different paths to the same objective but different premises and objectives, the choice of analytic method should rest on the conceptual bases of the empirical investigation and the data assumptions invoked to quantify the focal relations.

Although advocating knowledge of the restricted hypotheses with which each model can deal and its appropriateness for any particular data set, there is the recognition that no research is strictly confirmatory of its theoretical underpinnings. The operable objective is "to improve the measure of 'reasonable agreement' characteristic of the theory in a given application and . . . to open up new areas of application and establish new measures of 'reasonable agreement' applicable to them" (Kuhn 1977, 192). Theory confirmation through research reduces to reasonable agreement, suggesting no higher achievement than to provide an empirical demonstration of a previously implicit but uncertain agreement between theory and the world. As pointed out by Rychlak: "Methods are vehicles for the exercise of evidence" (1981, 43).

PURPOSE

Nearly forty years ago, Cattell (1946) alerted psychologists to the analytic possibilities and conceptual advantages of multimode designs for structural considerations. Through the "covariation chart," graphically displayed as a three-dimensional parallelepiped design for data collection, the importance of extended data classifications for the adequate representation of psychological experience was illuminated. (See Buss 1979 for an interesting discussion of the data collection and psychometric issues involved with various slices of the covariation chart.) However, until Tucker's explication of three-mode factor analysis, the choice of statistical methodology was limited to two-mode analytic forays, which usually explicitly ignored the depths of the focal phenomenon. Since theory is embodied in stable sets of relations, multimode analysis widens the empirical horizons of nomothetic investigations by providing the mathematical structure for more truly representative designs.

The methods presented in this book have evolved from two somewhat separate traditions: factor analysis and multidimensional scaling. Despite their unilateral history, in their metric forms—from which the multimode developments have emanated—factor analysis and multidimensional scaling have somewhat common aims and ancestors. They both attempt to represent data in a geometric space in order to provide some reduced and presumably higher-level perspective of the phenomenon. Accordingly, they presuppose the applicability of geometric representation.

Subject to empirical validation of the definitional criteria (measurement procedures) for the numbers to be assigned to the relation, and thus the applicability of geometric models (see Drosler 1981), factor analysis or metric multidimensional scaling proceed by the principles of partial correlation and/or regression to decompose the observational data into fundamental elements. Representation based upon regression or multiple correlation aspects reflects maximum variance considerations (for example, metric MDS, image analysis, principal components analysis), whereas representation based upon partial correlation aspects reflects covariation or correlation (for example, common factor analysis, analysis of covariance structures). The complexity of these representations depends upon the semantic extension of the conceptual domain and the number of empirical measures linked to the concept. Broadly defined concepts are likely to require a wide range of measures to adequately reflect their full meaning. Simpler, more narrowly defined concepts may be adequately operationalized by few measures. As the complexity increases, so also may the "noise" increase in the representation, thus masking or distorting structures.

To the extent that the empirical relationships are known beforehand, the variance or covariation representations may be less useful. In the case of factor analysis, Thurstone warned:

> Factor analysis has its principal usefulness at the border line of science. It is naturally superseded by rational formulations in terms of the science involved. Factor analysis is useful, especially in those domains where basic and fruitful concepts are essentially lacking and where crucial experiments have been difficult to conceive. The new methods have a humble role. They enable us to make only the crudest first map of a new domain. (1947, 56)

There appear to be two situations in which these representations are most useful: (a) if there exists a promising conceptual definition about which there are empirical uncertainties; or (b) if there exists a set of representative measures for a domain about which there are conceptual uncertainties. In the first case, the researcher formulates a model for the measures in accordance with an hypothesized structure. Unspecified parameters are estimated to optimize goodness of fit, and the total model estimates are then compared to the observed data, or preprocessed variants. In the second case, the researcher explores the data with a set of competitive, disconfirmable models for some "best" choice, or the researcher seeks some approximate description of the data for further insight.

These quantitative methodologies owe their philosophical heritage, in large part, to British empiricism, as they were developed by men like Galton, Pearson, Yule, and Spearman. Although they have specific methodological objectives, they share the hope that the data can be adequately represented in some reduced, parsimonious form. Traditionally, parsimony has been synonymous with low dimensionality; but with the new generation of statistical modeling, parsimony is associated with the model involving the fewest parameters to be estimated. These are compatible conceptions but different emphases. In the exploratory approach, for example, the number of parameters will decrease the fewer dimensions involved. However, it may not be the case that low dimensionality "fits" the data (for instance, when there are simplex or circumplex relationships), so parsimony is sought among the statistically adequate models. With this modified notion of parsimony and the development of general models for the analysis of covariance structures, a wide range of interesting methods—other than factor analysis or scaling—are available for the decomposition of complex data designs. In this book, the focus is on explicitly *reductionistic* approaches, such as the multimode generalizations of factor analysis and multidimensional scaling. Parsimony still reigns as a critical criterion for model selection.

These quantitative methodologies derive their appeal from their explicit nomothetic character. They involve the study of entities as universals, with a view to formulating general propositions, specifically concerning grouping or ordering relations. Although entities (individuals) differ in many apparent ways, these variations are considered quantitative differences along common relations. Individual differences methodology explores these variations by searching and testing for their structure and lawfulness. The intent is to account for the multivariate commonality in the intra- and interindividual differences. Therefore, multimode analysis provides a stringent framework for the empirical generalization of nomothetic propositions. Generality entails the extension of the class of nouns and verbs—entities and relations—implicated by the proposition. If data collection and analysis methods take cognizance of multiclassifications, the evidential base of the research is extended to include these classifications and the challenge offered by the nomothetic search is engaged.

REFERENCES

Bentler, P. M., and S-Y. Lee. 1978. Statistical aspects of a three-mode factor analysis model. *Psychometrika* 43:343–52.

———. 1979. A statistical development of three-mode factor analysis. *British Journal of Mathematical and Statistical Psychology* 32:87–104.

Bloxom, B. 1968. A note on invariance in three-mode factor analysis. *Psychometrika* 33:347–50.

Borg, I. 1979. Geometric representation of individual differences. In *Geometric Representations of Relational Data: Readings in Multidimensional Scaling,* ed. J. C. Lingoes, R. E. Roskam, and I. Borg. Ann Arbor, Mich.: Mathesis Press.

Bridgman, R. P., C. W. Snyder, Jr., and H. G. Law. 1981.

Individual differences in conceptual behavioral following manipulated controllability. *Personality and Individual Differences* 2:197–205.

Burt, C. 1940. *The Factors of the Mind: An Introduction to Factor-Analysis in Psychology.* London, England: University of London Press.

Buss, A. R. 1979. Toward a unified framework for psychmetric concepts in the multivariate developmental situation: Intraindividual change and inter- and intraindividual differences. In *Longitudinal Research in the Study of Behavior and Development,* ed. J. R. Nesselroade and P. B. Baltes. New York: Academic Press.

Carroll, J. D., and J. J. Chang. 1970. Analysis of individual differences in multidimensional scaling via an N-way generalization of "Eckart-Young" decomposition. *Psychometrika* 35:283–319.

Carroll, J. D., and J. J. Chang. 1972. IDIOSCAL (Individual Differences In Orientation SCALing): A generalization of INDSCAL allowing IDIOsyncratic reference systems as well as an analytic approximation to INDSCAL. Manuscript, Bell Laboratories, Murray Hill, New Jersey. Presented at 1972 spring meeting of Psychometric Society, Princeton, New Jersey, March.

Carroll, J. D., P. E. Green, and F. J. Carmone. 1976. CANDELINC: A new method for multidimensional analysis with constrained solutions. Unpublished. Bell Laboratories: Murray Hill, New Jersey.

Carroll, J. D., S. Pruzansky, and J. B. Kruskal. 1980. CANDELINC: A general approach to multidimensional analysis of many-way arrays with linear constraints on parameters. *Psychometrika* 45:3–24.

Cattell, R. B. 1944. "Parallel Proportional Profiles" and other principles for determining the choice of factors by rotation. *Psychometrika* 9:267–83.

———. 1946. *The Description and Measurement of Personality.* New York: Harcourt, Brace and World.

———. 1966. The data box: Its ordering of total resources in terms of possible relational systems. In *Handbook of Multivariate Experimental Psychology,* ed. R. B. Cattell. Chicago: Rand McNally.

———. 1978. *The Scientific Use of Factor Analysis in Behavioral and Life Sciences.* New York: Plenum.

———. 1980a. Guest editorial: Two basic models for personality-environment interaction and the need for their substantive investigation. *Multivariate Behavioral Research* 15:243–47.

———. 1980b. The separation and evaluation of personal and environmental contributions to behavior by the person-centered model (PCER). *Multivariate Behavioral Research* 15:371–402.

Cattell, R. B., and A. K. S. Cattell. 1955. Factor rotation for proportional profiles: Analytical solution and an example. *British Journal of Statistical Psychology* 8:83–92.

Coombs, C. H. [1964] 1976. *A Theory of Data.* [New York: Wiley] Reprinted, Ann Arbor, Mich.: Mathesis Press.

Corballis, M. C. 1973. A factor model for analysing change. *British Journal of Mathematical and Statistical Psychology*

26:90–97.

Corballis, M. C., and R. E. Traub. 1970. Longitudinal factor analysis. *Psychometrika* 35:79–98.

De Sarbo, W. S., and J. D. Carroll. 1981. Three-way metric unfolding. *Proceedings of the Third ORSA/TIMS Special Interest Conference on Market Measurement and Analysis,* pp. 157–83.

Drosler, J. 1981. The empirical validity of multidimensional scaling. In *Multidimensional Data Representations: When and Why,* ed. I. Borg. Ann Arbor, Mich.: Mathesis Press.

Eckart, C., and G. Young. 1936. The approximation of one matrix by another of lower rank. *Psychometrika* 1:211–18.

Firth, P. M., and C. W. Snyder, Jr. 1979. Three-mode factor analysis of self-reported difficulty in assertiveness. *Australian Journal of Psychology* 31:125–35.

Fraser, C. 1980. *COSAN User's Guide.* Toronto, Ontario: The Ontario Institute for Studies in Education.

Gollob, H. F. 1968. A statistical model which combines features of factor analytic and analysis of variance techniques. *Psychometrika* 33:73–116.

Gower, J. C. 1975. Generalized procrustean analysis. *Psychometrika* 40:33–51.

Harshman, R. A. 1970. Foundations of the PARAFAC procedure: Models and conditions for an "explanatory" multi-modal factor analysis. *UCLA Working Papers in Phonetics* 16:1–84 (University Microfilms No. 10,085).

————. 1981. *Partial Reference Manual for the Preliminary Release of the PARAFAC Analysis Package.* London, Ontario: Scientific Software Associates.

Harshman, R. A., and S. A. Berenbaum. 1981. Basic concepts underlying the PARAFAC-CANDECOMP three-way factor analysis model and its application to longitudinal data. In *Present and Past in Middle Life,* ed. D. H. Eichorn, J. A. Clausen, N. Haan, M. P. Honzik, and P. H. Mussen. New York: Academic Press.

Horan, C. B. 1969. Multidimensional scaling: Combining observations when individuals have different perceptual structures. *Psychometrika* 34:139–65.

Horst, P. 1965. *Factor Analysis of Data Matrices.* New York: Holt, Rinehart and Winston.

Jennrich, R. A. 1972. A generalization of the multidimensional scaling model of Carroll and Chang. *UCLA Working Papers in Phonetics* 22:45–47.

Kelly, G. A. 1955. *The Psychology of Personal Constructs.* New York: Norton.

Kjerulff, K., and N. H. Wiggins. 1976. Graduate student styles for coping with stressful situations. *Journal of Educational Psychology* 68:247–54.

Kroonenberg, P. M. 1981a. *Users Guide to TUCKALS3. A Program for Three-Mode Principal Component Analysis (WEP-Reeks, WR 81-6-RP).* Leiden, The Netherlands: University of Leiden.

————. 1981b. *User's Guide to TUCKALS2. A Program for Three-Mode Principal Component Analysis with Extended Core Matrix (WEP-Reeks, WR-35-RP).* Leiden, The Netherlands:

University of Leiden.
Kroonenberg, P. M., and J. de Leeuw. 1980. Principal components analysis of three-mode data by means of alternating least-squares algorithms. *Psychometrika* 45:69–97.
Kruskal, J. B. 1978. Factor analysis and principal components: Bilinear methods. In *International Encyclopedia of Statistics,* ed. W. H. Kruskal and J. M. Tanur. New York: Free Press.
Kuhn, T. S. 1977. *The Essential Tension: Selected Studies in Scientific Tradition and Change.* Chicago: University of Chicago Press.
Lastovicka, J. 1981. The extension of component analysis to four-mode matrices. *Psychometrika* 46:47–57.
Lazarsfeld, P. F. 1950. The logical and mathematical foundation of latent structure analysis. In *Measurement and Prediction,* ed. S. A. Stouffer et al. Princeton, N.J.: Princeton University Press.
Leah, J. A., H. G. Law, and C. W. Snyder, Jr. 1979. The structure of self-reported difficulty in assertiveness: An application of three-mode common factor analysis. *Multivariate Behavioral Research* 14:443–62.
Levin, J. 1965. Three-mode factor analysis. *Psychological Bulletin* 64:442–52.
Lingoes, J. C., and I. Borg. 1976. Procrustean individual differences scaling: PINDIS. *Journal of Marketing Research* 13:406–7.
McDonald, R. P. 1978. A simple comprehensive model for the analysis of covariance structures. *British Journal of Mathematical and Statistical Psychology* 31:59–72.
————. 1980. A simple comprehensive model for the analysis of covariance structures: Some remarks on applications. *British Journal of Mathematical and Statistical Psychology* 33:161–83.
McDonald, R. P., and S. A. Mulaik. 1979. Determinacy of common factors: A nontechnical review. *Psychological Bulletin* 86:297–306.
Miron, M. S., and C. E. Osgood. 1966. Language behavior: The multivariate structure of qualification. In *Handbook of Multivariate Experimental Psychology,* ed. R. B. Cattell. Chicago: Rand McNally.
Pruzansky, S., A. Tversky, and J. D. Carroll. 1982. Spatial versus tree representations of proximity data. *Psychometrika* 47:3–24.
Rychlak, J. F. 1981. *A Philosophy of Science for Personality Theory.* 2nd ed. Malabar, Fla.: Krieger.
Sands, R., and F. W. Young. 1980. Component models for three-way data: An alternating least squares algorithm with optimal scaling features. *Psychometrika* 45:39–67.
Schonemann, P. H. 1970. On metric multidimensional unfolding. *Psychometrika* 35:349–66.
————. 1972. An algebraic solution for a class of subjective metrics models. *Psychometrika* 37:441–51.
Snyder, C. W., Jr. 1976. Multivariate analysis of intrinsic individual differences in disjunctive conceptual behavior. *Multivariate Behavioral Research* 11:195–216.
Snyder, C. W., Jr., R. P. Bridgman, and H. G. Law. 1981.

Three-mode factor analytic reference curves for concept identification. *Personality and Individual Differences* 2:265–72.
Snyder, F. W., and N. Wiggins. 1970. Affective meaning systems: A multivariate approach. *Multivariate Behavioral Research* 5:453–68.
Spearman, C. 1904a. The proof and measurement of association between two things. *American Journal of Psychology* 15: 72–101.
————. 1904b. General intelligence, objectively determined and measured. *American Journal of Psychology* 15:201–93.
————. 1927. *The Abilities of Man: Their Nature and Measurement.* London, England: Macmillan.
Takane, Y. 1982. Maximum likelihood additivity analysis. *Psychometrika* 47:225–41.
Takane, Y., F. W. Young, and J. de Leeuw. 1977. Nonmetric individual differences multidimensional scaling: An alternating least-squares method with optimal scaling features. *Psychometrika* 42:7–67.
————. 1980. An individual differences additive model: An alternating least-squares method with optimal scaling features. *Psychometrika* 45:183–209.
Thomas, P. R. and J. D. Bain. 1982. Consistency in learning strategies. *Higher Education* 11:249–59.
Thurstone, L. L. 1947. *Multiple-Factor Analysis: A Development and Expansion of the Vectors of Mind.* Chicago: University of Chicago Press.
Tucker, L. R. 1963. Implications of factor analysis of three-way matrices for measurement of change. In *Problems in Measuring Change,* ed. C. W. Harris. Madison, Wis.: University of Wisconsin Press.
————. 1964. The extension of factor analysis to three-dimensional matrices. In *Contributions to Mathematical Psychology,* ed. N. Frederiksen and H. Gulliksen. New York: Holt, Rinehart and Winston.
————. 1965. Experiments in multi-mode factor analysis. In *Testing Problems in Perspective, Washington D.C.: American Council on Education, 1966. Reprinted from: Proceedings of the 1964 Invitational Conference on Testing Problems,* ed. A. Anastasi. Princeton, N.J.: Educational Testing Service.
————. 1966a. Some mathematical notes on three-mode factor analysis. *Psychometrika* 31:279–311.
————. 1966b. Experiments in multimode factor analysis. In *Testing Problems in Perspective,* ed. A. Anastasi. Washington, D.C.: American Council on Education.
————. 1967. Three-mode factor analysis of Parker-Fleishman complex tracking behavior data. *Multivariate Behavioral Research* 2:139–51.
————. 1972. Relations between multidimensional scaling and three-mode factor analysis. *Psychometrika* 37:3–27.
Tucker, L. R., and S. Messick. 1963. An individual difference model for multidimensional scaling. *Psychometrika* 28:333–67.
Tzeng, O. C. S. 1975. Differentiation of affective and denotative meaning systems and their influence in personality ratings. *Journal of Personality and Social Psychology* 32:978–88.
Young, F. W., J. de Leeuw, and Y. Takane. 1980. Quantifying

qualitative data. In *Similarity and Choice,* ed. E. D. Lantermann and H. Feger. Bern: Hans Huber.

2

Multilinear Methods

Joseph B. Kruskal

This chapter is about *structural models* which are *bilinear, trilinear,* or *multilinear* of higher order, used to analyze *2-way arrays, 3-way arrays,* or *many-way arrays* of higher order. We will also discuss the geometrical meaning which many of these models have, using primarily the geometrical concepts of *inner product* and *distance*. Finally, we will present an application, to illustrate these ideas. All the terms in italics will be carefully defined and illustrated.

By a structural model, I mean the structural part of a statistical model. For example, if x_i is the height and y_i is the weight of person i, then a statistical model for the relationship might be

$$y_i = a + bx_i + \varepsilon_i ,$$

where ε_i is normal with mean 0 and variance σ^2.

The error term ε_i is the random part (or probabilistic part or stochastic part) of the model. The structural part of this model is

$$y_i \cong a + bx_i .$$

This chapter is printed with the kind permission of the Behaviormetric Society and of the American Mathematical Society, for which we are grateful. Most of the chapter first appeared in *Behaviormetrika,* 1981, No. 10, pp. 1–20. The form in which it appears here is virtually identical to its form in an American Mathematical Society Short Course, which was published by the A.M.S. as Gnanadesikan (1983). The changes from the *Behaviormetrika* version to the American Mathematical Society version occur primarily in sections from "The Rotation Problem" through "The Singular Value Decomposition," which are new or greatly changed.

A structural model means an equation like this one, where the error term or other error mechanism is not specified. Only one property is assumed about the error: it is small enough that it does not completely hide the relationship given by the structural model.

In this equation, the approximate equality sign $\cong$ is used to indicate that the equation is not expected to hold exactly, due to random error, but the nature of the random error is left unspecified. This casual attitude towards the random error may be appropriate when one central question is the form of the structural model and there is considerable uncertainty about it, e.g., we might be trying to choose among an additive model, multiplicative model, and other unspecified alternatives.

Roughly speaking, a model is linear or bilinear or trilinear, etc., if it is built on expressions of the corresponding type. Examples and more details are given below. To explain bilinear expressions, trilinear expressions, etc., suppose that a_i is one set of variables, b_j is a second set, and c_k is a third set. A 1-linear or linear expression is a sum of terms where each term has the form

$$(\text{some constant}) \times (\text{some } a).$$

A 2-linear or bilinear expression is a sum of terms where each term has the form

$$(\text{some constant}) \times (\text{some } a) \times (\text{some } b).$$

A 3-linear or trilinear expression is a sum of terms where each term has the form

$$(\text{some constant}) \times (\text{some } a) \times (\text{some } b) \times (\text{some } c).$$

Quadrilinear expressions and so forth may be defined in a similar manner. An expression is multilinear if it is n-linear for some n. (Usually, we use the word multilinear only if $n \geqq 2$, but strictly speaking the case $n = 1$ is included also.)

Ordinarily we understand that the sets of variables (the a_i, the b_j, etc.) are all disjoint. When the a_i are permitted to share some or all variables with the b_j, we refer to bilinear models in *the extended sense*, and similarly for trilinear models, etc.

To emphasize an important point, we illustrate how a model is built on expressions in the linear case, before going on to the bilinear and trilinear cases. Suppose that height x and weight y are observed variables, and suppose that b_0, b_1, and perhaps b_2 are variables to be determined. Two alternative linear models are given by

$$y = b_0 + b_1 x \qquad \text{and} \qquad y = b_0 + b_1 x + b_2 x^2 .$$

Of course, the last equation is not linear in x. However, both these equations are linear *in the variables to be determined.** We say that a model is linear if the expressions it uses are linear in the variables to be determined. Thus for us, both these models are linear. Other linear models include

$$y \cong b_0 + b_1x_1 + b_2x_2 + \ldots, \qquad \text{or}$$

$$y \cong b_0 + b_1x_1 + b_2x_2 + \ldots + b_{12}x_1x_2 + \ldots, \qquad \text{etc.}$$

Another linear model will help introduce the bilinear models. Suppose we are interested in how well people read (literacy) as a function of education and age. For each person in a survey, we determine literacy, formal education (years in school), and age. One linear model we might use is based on dividing people into (say) 6 categories according to education, and into (say) 5 categories according to age. Associate with these categories variables to be determined, namely, $a_1, \ldots, a_6$ and $b_1, \ldots, b_5$. One possible model is that a person in education category i and age category j has literacy

$$x_{ij} \cong a_i + b_j \; .$$

The variables to be determined are the a's and the b's, and the model is linear in these variables.

Until 10 or 15 years ago, almost all *widely-used general-purpose models* for data analysis were *linear*. This includes analysis of variance, regression, time series models, and models of many other kinds. The only exceptions were principal components analysis and factor analysis, which are bilinear. Today, there is a growing list of exceptions, almost all of which are multilinear, at least in some approximate sense.

A BASIC BILINEAR MODEL

To introduce bilinear models we start by describing a model which is extremely simple, and which will be helpful in introducing some others of great importance. As in the last linear model, the observed data form a 2-way array x_{ij}, where $i = 1$ to I and $j = 1$ to J, and a_i and b_j are the variables to be determined. The model involves not a sum but a product,

$$x_{ij} \cong a_i b_j \; .$$

This expression is not linear, because it involves the product of one variable to be determined, a_i, times another, b_j. Of course, we could take logarithms (if the a_i and b_j are positive) and obtain the transformed model

---

*Readers who are familiar with classical mathematical statistics may wonder about the distinction between parameters and variables. We do not make this distinction. The distinction is not, of course, a fact of nature but merely part of the formalism. Furthermore, it is difficult to find published discussions which give arguments in support of the distinction.

$$x_{ij}^* \cong a_i^* + b_j^* ,$$

where $x_{ij}^* = \log(x_{ij})$, $a_i^* = \log(a_i)$, $b_j^* = \log(b_j)$.

Transforming in this way is an important and useful technique, and models which can be made linear by using the logarithm transformation are an important class called log-linear models (Bishop, Fienberg, and Holland 1975). However the other bilinear models we will discuss cannot be transformed in this way.

Now we describe a bilinear model of great importance, which we shall call $\mathbf{B}_0(R)$ or $\mathbf{B}_0$, and illustrate in Figure 2–1. It is formed by adding R terms like those in the preceding model. Once again, the observed data are x_{ij}. The variables to be determined are a_{ir} and b_{jr}, where $r = 1$ to R. Each r corresponds to one "factor," as shown in Figure 2–1, where the factors are separated by dashed lines. We shall see later that the number R, which is usually small, corresponds to the number of factors in factor analysis, and to the number of components in principal components analysis. It also corresponds to the number of dimensions in multidimensional scaling. In case $R = 1$, this model (namely, $\mathbf{B}_0(1)$) is the model given in the preceding paragraph.

Figure 2–1. The Model $\mathbf{B}_0$

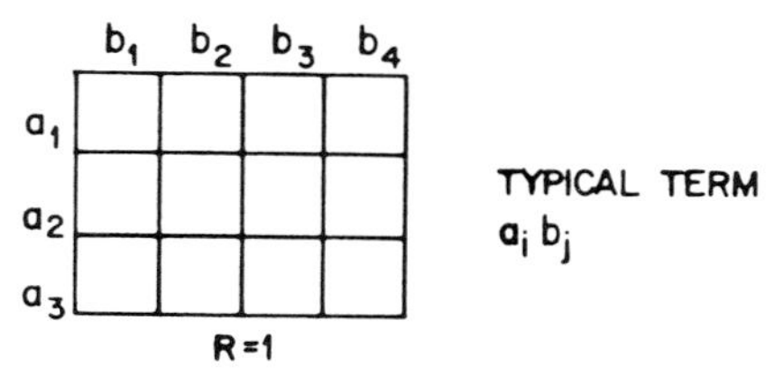

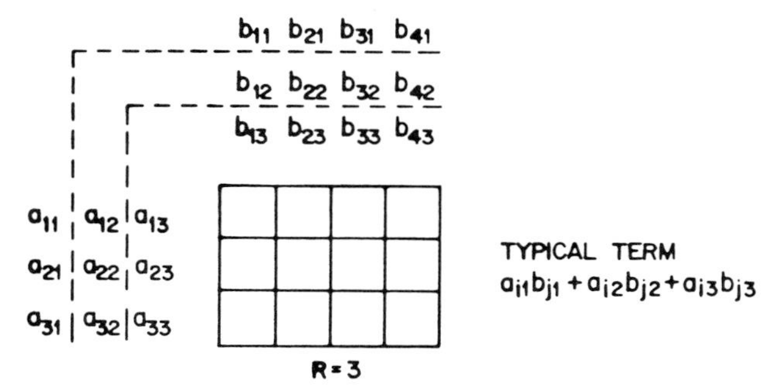

We present model $\mathbf{B}_0(R)$ in several different forms because of its great importance:

$$x_{ij} \cong a_{i1}b_{j1} + \ldots + a_{iR}b_{jR} ,$$

$$x_{ij} \cong \sum_{r=1}^{R} a_{ir}b_{jr} ,$$

$$x_{ij} \cong (a_{i1}, \ldots, a_{iR}) \cdot (b_{j1}, \ldots, b_{jR}) ,$$

$$\mathbf{X} \cong \mathbf{AB}' ,$$

where $\mathbf{X}$ is the I by J matrix x_{ij},
$\mathbf{A}$ is the I by R matrix a_{ir},
$\mathbf{B}$ is the J by R matrix b_{jr}.

The third equation shows this model using the concept of the inner product of two vectors, which is a geometrical concept. The fourth equation shows this model in matrix form.

We define a model to be bilinear if the value it permits for every observed variable is a bilinear expression in terms of the variables to be determined. Thus to verify that the model $\mathbf{B}_0(R)$ is bilinear, it is enough to verify that the expression $\Sigma\, a_{ir}b_{jr}$ is bilinear. This means first that the variables to be determined can be divided into two groups (in this case, the a_{ir} form one group and the b_{jr} the other group). It also means that the expression is a sum of terms (in this case, the terms are $a_{ir}b_{jr}$). Finally, it means that each term has the form

(some constant)
× (some variable in first group)
× (some variable in the second group)

(in this case, the terms obviously have this form where the constants are all 1).

Suppose that each a_i is constant and does not vary: then the expression above is linear in the b_j. On the other hand, suppose that each b_j is constant: then the expression is linear in the a_i. Thus the expression is linear in the a_i alone and it is linear in the b_j alone, so it is linear twice, which is the reason for the name 2-linear or bilinear.

Model $\mathbf{B}_0$ underlies factor analysis and principal components analysis. (For a fuller description see Kruskal 1978.) For example, suppose there are I people labelled by $i = 1$ to I, and J tests labelled by $j = 1$ to J, and x_{ij} is the score of person i on test j. Suppose we remove the mean of each test:

$$\bar{x}_{.j} = \frac{1}{I} \sum x_{ij} ,$$

$$x^*_{ij} = x_{ij} - \bar{x}_{.j} .$$

Suppose we fit model $\mathbf{B}_0$ to the x^*_{ij} ,

$$x^*_{ij} \cong \sum a_{ir}b_{jr} .$$

Depending on certain technical details, this analysis is one type of factor analysis or it is one type of principal components analysis, and

R = the number of factors or the number of components,

b_{jr} = the factor loadings or the coefficients with respect to the principal components (or the principal component loadings),

a_{ir} = the factor scores or the principal components (or the principal component scores).

THE ROTATION PROBLEM

There is a famous difficulty in using factor analysis and principal components analysis for some applications, called "the rotation problem." We are particularly interested in this fundamental difficulty because some of the very new trilinear models which we discuss later are able to avoid it. Suppose we write the model in matrix form $\mathbf{X} \cong \mathbf{AB}'$. The *fitted values* in this model are the entries of $\hat{\mathbf{X}} \equiv \mathbf{AB}'$, and the residuals are the entries of $\mathbf{X} - \hat{\mathbf{X}}$. ($\mathbf{X}$ and $\hat{\mathbf{X}}$ are I by J, $\mathbf{A}$ is I by R, and $\mathbf{B}$ is J by R.) Suppose $\mathbf{T}_1$ is any R by R matrix which is nonsingular. Then the model can be rewritten

$$\mathbf{X} \cong \hat{\mathbf{X}} = \mathbf{A}\mathbf{T}_1\mathbf{T}_1^{-1}\mathbf{B}'$$

$$\mathbf{X} \cong \hat{\mathbf{X}} = \mathbf{A}_1\mathbf{B}'_1\ , \qquad \text{where } \mathbf{A}_1 = \mathbf{A}\,\mathbf{T}_1,\ \mathbf{B}_1 = \mathbf{B}\,(\mathbf{T}_1^{-1})'.$$

Thus from one solution or factorization we obtain a second one which has the same fitted values and residuals, and therefore appears in many ways to be just as good as the first. Since there is a large class of nonsingular matrices $\mathbf{T}_1, \mathbf{T}_2, \ldots,$ there are many different factorizations:

$$\mathbf{X} \cong \hat{\mathbf{X}} = \left\{ \begin{array}{l} \mathbf{AB}' \\ \mathbf{A}_1\mathbf{B}'_1 \\ \mathbf{A}_2\mathbf{B}'_2 \\ \cdot \\ \cdot \\ \cdot \end{array} \right. .$$

Under mild conditions on $\mathbf{X}$, every factorization of $\hat{\mathbf{X}}$ can be obtained in this way. From many points of view, these solutions all fit the data equally well. (However, using the maximum likelihood approach to fitting, different factorizations may have different likelihoods, and hence implicitly may not be considered equally good.)

The existence of many different solutions which fit the data equally well is the *fundamental indeterminacy*. Any of the matrices $\mathbf{T}_1, \mathbf{T}_2, \ldots,$ may be referred to as the *linear transformation matrix* and denoted by $\mathbf{T}$. In factor analysis, the matrix $\mathbf{T}$ is often referred to as a *rotation matrix* (particularly if $\mathbf{T}$ has the additional property that $\mathbf{T}\mathbf{T}'$ is the identity matrix, or at least the weaker property that the diagonal part of $\mathbf{T}\mathbf{T}'$ is the identity matrix), and the process of converting from one solution to another with the aid of some $\mathbf{T}$ is referred to as *rotation*. Choosing a solution which is appropriate for some particular purpose is referred to as the *rotation problem*.

Restrictions are almost invariably imposed on the factorization to reduce the amount of indeterminacy. For example, the columns of **A** may be required to have length 1, or the columns may be required to be orthogonal (i.e., perpendicular), or both at once. Similar restrictions may be applied to **B**, or restrictions may be applied to both **A** and **B**. In addition, quite a number of other more elaborate restrictions may be used, a few of which are described below.

Restrictions are imposed for at least three different reasons. First, it generally seems better to resolve (or partially resolve) the indeterminacy by consciously chosen restrictions, even if they have no clear rationale, rather than leaving the indeterminacy to be resolved by fortuitous aspects of data, computer languages, compilers, and hardware. Second, for the purpose of comparing a solution with something else (either a solution from another data set, or some data of a different type), it is usually very important to select an appropriate rotation. Third, in some applications it is desired to discover the "true underlying factors." (The operational meaning of this phrase is discussed below.) Obviously there is no hope of finding such factors without a careful resolution of the fundamental indeterminacy.

When using restrictions, however, it is important to remember that they are not justified in the same sense that the basic fitting principle, such as least-squares or maximum-likelihood, is justified. Some of the factorizations which are excluded by any given restrictions may perhaps be just as useful or interesting as those retained.

RESTRICTIONS USED TO AID COMPARISON

Let us consider restrictions for two different comparison purposes. Suppose the rows of **X** correspond to people and its columns to tests of various different skills, so that each row of **A** describes a person and each row of **B** describes one skill (in terms of the R assumed factors). Suppose we wish to compare two data sets, taken from different populations (e.g., female students and male students), and wish to determine whether or not the two populations have the same distribution of skills. Even if the two populations have identical distributions, it may be difficult due to indeterminacy to discover this fact from the two solutions, since quite different factorizations may have been selected. (In principle, the same difficulty exists even if the two *samples* are identical, though the difficulty would not arise in this case if the same computer program, run on the same computer, were used for both analyses.)

To resolve the indeterminacy for this purpose, it would be appropriate to place restrictions on **B**, since it is reasonable to hope that approximately the same description will apply in both data sets, while the distribution of skills may differ between the populations. Typically, **B** would be restricted to a *suborthogonal* matrix: since **B** has at least as many rows as columns, this means that the columns of **B** are orthogonal and have length 1. This still leaves indeterminacy in the direction of each column vector, i.e., any column of **B** can be multiplied by -1 without alteration of the properties, and if the same change is made in **A** there is no change in $\mathbf{AB'}$. Usually this remaining amount of indeterminacy is accepted, since it causes little trouble, but an easy additional restriction can be used to resolve it, namely, that the sum of the cubes (or alternatively, the first element) in each column of **B** should be ≧ 0. (If the sum of cubes of some column is 0, this additional restriction does not resolve the indeterminacy. Furthermore, if **X** has two or more equal "singular values" as defined below, the restriction that **B** is suborthogonal does not reduce the indeterminacy as much as indicated above. Fortunately, both of these events occur with very small probability for real data sets, though the latter event is echoed by practical computational difficulties in finding a factorization if two singular values are sufficiently close to each other.)

On the other hand, suppose the two data sets are taken from a single population, but are based on two different sets of tests which may cover similar or at least overlapping domains of skill. Suppose we wish to compare how each skill covered in one data set relates to the skills in the other. Even if the skills in one data set are identical one by one with those in the other data set, it may be difficult due to the indeterminacy to discover this fact. To resolve indeterminacy for this purpose, it would be appropriate to place restrictions on the statistical structure of **A**, since it is reasonable to hope that approximately the same statistical distribution of underlying factors will apply to two samples from the same population, under the working hypothesis that the skill domains are sufficiently similar to have the same underlying factors. Typically, **A** would be required to be suborthogonal.

RESTRICTIONS TO FIND THE TRUE UNDERLYING FACTORS

In some applications it is desired to find the "true underlying factors." In practice, this phrase refers to a factorization, which may or may not exist, for which the factors are (1) intuitively interpretable and (2) have "convergent validity." By the latter phrase I mean that the factors correspond or agree with factors obtained from other sets of data, perhaps using other measurement methods, other populations of objects, and other methods of analysis. The better the numerical agreement, and the wider the domain of data, population and methods over which agreement exists, the greater the degree of convergent validity.

There are two different restrictions which are most often used to provide some chance that the factorization will correspond to true underlying factors. We discuss them each in turn. In addition, there are several methods for seeking a factorization which agrees with external variables or factors from other data, a step which is important in attempting to establish convergent validity, but they will not be discussed here.

The so-called *simple structure* restrictions were proposed specifically in order to help find true underlying factors, and have been widely used for this purpose. The rationale and a precise description of simple structure are outside the scope of this chapter, but can be found in any standard text on factor analysis. Very crudely, the idea is that most elements of **B** should either be relatively large or else different from 0 only due to sampling variation (and hence be near 0); thus **B** should have few elements of medium size. *Orthogonal simple structure* as generally used means that these restrictions are applied while simultaneously requiring that the columns of **A** should be orthogonal. I join a number of other authors in the belief that this form of simple structure is unrealistically severe and undesirable. *Oblique simple structure* omits the orthogonality requirements, and seems far more appealing to me. There are a great many versions of the simple structure concept, of which *Varimax* is perhaps the most widely used one. To me, the rationale for why simple structure should yield true underlying factors seems a little weak, but the method has had a moderate degree of success in practical applications.

The *principal components rotation* was not intended for the purpose of finding true underlying factors, but is often used for that purpose anyhow. There is an amazing variety of different restrictions which lead to the same rotation. An easily stated one is simply to require that **B** is suborthogonal, while simultaneously requiring that **A** is suborthogonal. The most frequently stated one is that the columns of **B** are chosen sequentially, and each one is selected to explain as much of the variance not yet accounted for as possible, subject to having length 1 and being orthogonal to the previously chosen columns of **B**. The rationale for why this factorization should yield true underlying factors seems very weak to me, but the method does seem to meet this goal now and then anyhow.

SINGULAR VALUE DECOMPOSITION

Since it is sometimes convenient to require **A** or **B** to be suborthogonal, it is natural to consider suborthogonality for *both* **A** and **B** at the same time. In general, however, this restriction is impossible, but a factorization which *almost* achieves it is possible and has great importance to our topic and to statistics, computer science, and mathematics generally. It is called the *singular value decomposition* or SVD for short.

If **X** is any I by J matrix with rank M, it is possible to write

$$\mathbf{X} = \mathbf{PDQ}'$$

where **P** is suborthogonal and I by M,
Q is suborthogonal and J and M,
D is diagonal, M by M, and has diagonal elements all positive and arranged in order of decreasing magnitude.

This is called the singular value decomposition or SVD of **X**. (Some authors use a similar decomposition in which $M = \min(I,J)$ instead, but the extra diagonal elements involved are all 0 and the difference in meaning is not very important.)

The elements on the diagonal of **D** are called the *singular values* of **X**, while the columns of **P** and **Q** are referred to respectively as the left singular vectors and the right singular vectors of **X**. If the singular values of **X** are all distinct, then the SVD is unique with one minor exception, namely, coupled changes of sign in the singular vectors. Specifically, the only possible change is to multiply any left singular vector and the corresponding right singular vector by -1. (If there are equal singular values, a weaker form of uniqueness can be stated.) The nonuniqueness of sign could be removed by requiring, for example, that the sum of the cubes of each left singular vector be ≥ 0, but this is not customary.

The SVD has some very close links to the concepts of eigenvalue and eigenvector. If **X** is symmetric (and hence square), then its SVD decomposition is the same as its eigenvalue decomposition. On the other hand, if it is square but not symmetric, there is no simple relationship between the SVD and the eigenvalue decomposition. For any **X**, simple manipulation shows that the nonzero eigenvalues of $\mathbf{X}\mathbf{X}'$ and of $\mathbf{X}'\mathbf{X}$ are the same as the *squares* of the singular values of X, and that the eigenvectors of $\mathbf{X}\mathbf{X}'$ and of $\mathbf{X}'\mathbf{X}$ are the same as the left singular vectors and right singular vectors of **X** respectively, except that eigenvectors corresponding to the eigenvalue 0 are omitted. In statistics, the use of eigenstructures seems to be much more important for symmetric matrices than for nonsymmetric matrices.

There is a natural additive decomposition of **X** based on the SVD. Let the diagonal elements of **D** be d_m, the columns of **P** (left singular vectors of **X**) be p_m, and the columns of **Q** (the right singular vectors of **X**), be q_m, for $m = 1$ to M. Then

$$\mathbf{X} = \mathbf{P}\,\mathbf{D}\,\mathbf{Q}' = \sum_{m=1}^{M} d_m p_m q_m' \ ,$$

where each term involves the outer product $p_m q_m'$, a rank 1 matrix which has the form of a "multiplication table."

If we multiply any one term by the transpose of any other term, in either order, the product is the 0 matrix, due to the orthogonality of columns. As a result, the sum of squares of the elements of **X**, which can also be described using the trace operator as $\mathrm{tr}(\mathbf{X}'\mathbf{X}) = \mathrm{tr}(\mathbf{X}\mathbf{X}')$, can be decomposed as follows:

$$\mathrm{tr}(\mathbf{X}'\mathbf{X}) = \sum d^2 \ \mathrm{tr}(q_m q_m') = \sum d_m^2 \ .$$

This is the basis for an analysis of the variance of the data values in **X**.

Note that **X** can be expressed as a product $\mathbf{A}\mathbf{B}'$ where **A** and **B** have R columns if and only if $\hat{\mathbf{X}}$ has rank $\leq R$. Thus finding the least-squares fit of **X** by a product $\mathbf{A}\mathbf{B}'$ is almost the same thing as finding the least-squares approximation to **X** by a matrix $\hat{\mathbf{X}}$ of rank $\leq R$. It can be proved that the solution to the latter problem is formed by using only the first R terms in the additive decomposition, i.e.,

$$\hat{\mathbf{X}} = \sum_{m=1}^{R} d_m p_m q_m' \ .$$

To say this in another way, let $\mathbf{D}_R$ be the M by M diagonal matrix which is the same as **D** for the first R diagonal entries, and which has 0 in the remaining diagonal positions. Then

$$\hat{\mathbf{X}} = \mathbf{P}\,\mathbf{D}_R\mathbf{Q}' \ .$$

Therefore the least-squares fit to **X** by a product $\mathbf{A}\mathbf{B}'$ involving R columns is given by factorizations such as

$\mathbf{A} = \mathbf{P}\,\mathbf{D}_R$, $\mathbf{B} = \mathbf{Q}$, or

$\mathbf{A} = \mathbf{P}$, $\mathbf{B} = \mathbf{Q}\,\mathbf{D}_R$, or

$\mathbf{A} = \mathbf{P}\,\mathbf{D}^*$, $\mathbf{B} = \mathbf{Q}\mathbf{D}^{**}$, where $\mathbf{D}^*$, $\mathbf{D}^{**}$ are diagonal and $\mathbf{D}^*\mathbf{D}^{**} = \mathbf{D}_R$.

To satisfy the restriction that $\mathbf{B}$ is suborthogonal, we can use the first factorization above, and to satisfy the restriction that $\mathbf{A}$ is suborthogonal we can use the second.

SOME MORE BILINEAR MODELS

Now we introduce some models closely related to $\mathbf{B}_0$. (For easy comparison, we include $\mathbf{B}_0$ with them.) Although these models are not bilinear in the sense described above, their most interesting part is bilinear, so we will introduce a broader concept to cover them. In particular, a *mixed bilinear expression* means an expression which is the sum of a linear expression and a bilinear expression. Ordinarily, the same variables are not permitted to occur in both expressions (though we might want to permit that in a "mixed bilinear expression in the extended sense"). We may use *pure bilinear expression* when it is necessary to emphasize that the more restricted definition given earlier is being used.

These models are:

$$\mathbf{B}_0: \quad x_{ij} \cong \sum a_{ir} b_{jr} \ ,$$

$$\mathbf{B}_1: \quad x_{ij} \cong b_{j0} + \sum a_{ir} b_{jr} \ ,$$

$$\mathbf{B}_1': \quad x_{ij} \cong a_{i0} + \sum a_{ir} b_{jr} \ ,$$

$$\mathbf{B}_2: \quad x_{ij} \cong m + a_{i0} + b_{j0} + \sum a_{ir} b_{jr} \ ,$$

$\mathbf{B}_3$: $\quad X \cong \mathbf{A}_0\mathbf{U}' + \mathbf{V}\mathbf{B}_0' + \mathbf{A}\mathbf{B}'$, where $\mathbf{U}$ and $\mathbf{V}$ are fixed and known in advance, and $\mathbf{A}_0$, $\mathbf{B}_0$, $\mathbf{A}$, and $\mathbf{B}$ are to be determined.

Suppose that the subscript i and the rows of the matrix $\mathbf{X}$ correspond to objects or other units, and suppose that subscript j and the columns of $\mathbf{X}$ correspond to variables which we measure on these objects. Then the model $\mathbf{B}_1$ is the model which underlies factor analysis and principal components analysis. The term b_{j0} corresponds conceptually to the sample mean of the jth variable. Furthermore, if we fit $\mathbf{B}_1$ by least-squares, or if we fit it by maximum likelihood using very familiar assumptions, then it will turn out that in a best-fitting solution,

$$b_{j0} = \text{sample mean of the } j\text{th variable.}$$

The model underlying factor analysis (see Harman 1960 or Thurstone 1947) is not usually described as $\mathbf{B}_1$. Several alternatives are used instead. One of them starts by removing sample means from the data (in the manner described earlier) for no stated reason, and then fitting model $\mathbf{B}_0$ to what is left. Most often, factor analysis is described as a method of analyzing a correlation matrix or a covariance matrix but the *motivation* given with the description is almost always based on removal of means and then fitting $\mathbf{B}_0$. (For example, see Harmon 1967, equation 2.9 on page 15; or Thurstone 1947, equation 3 on page 71.) While fitting $\mathbf{B}_1$ is mathematically equivalent to removal of means and fitting $\mathbf{B}_0$ in some very familiar situations, as just indicated, conceptually there is a difference. To me, model $\mathbf{B}_1$ seems like a much more sensible way to think about the methods.

The model $\mathbf{B}_1'$ is dual to $\mathbf{B}_1$, in that it has an additive term for each row instead of for each column. Model $\mathbf{B}_1'$ might be used if the meaning of the rows and the columns are reversed, so that rows correspond to variables and the columns to objects.

Model $\mathbf{B}_2$ can be thought of as underlying the method called FANOVA (Gollob 1968) in the same sense that model $\mathbf{B}_1$ can be thought of as underlying factor analysis, even though that paper describes FANOVA in a different manner. The name is derived from "factor analysis—analysis of variance," and the method is described as consisting of two steps. First, remove the grand mean and row and column effects as usual in analysis of variance (the terms removed correspond to m, a_{i0}, b_{j0} in model $\mathbf{B}_2$). Second, analyze what is left by factor analysis. Thus FANOVA accomplishes a decomposition of the observed data like that shown in $\mathbf{B}_2$, so it is natural to consider fitting the model $\mathbf{B}_2$ all at once instead of in stages, although the paper does not take notice of this possibility. It can be proved that the FANOVA procedure, which uses least-squares fitting or a procedure mathematically equivalent to it at each of the two stages, yields a least-squares solution to model $\mathbf{B}_2$.

Model $\mathbf{B}_3$ is a rather general model, which contains the others shown as special cases. (For details, see appendix 2–1). It is included here as the most general bilinear model for which a property like that just mentioned above is known. Specifically, it can be proved that if we fit model $\mathbf{B}_3$ by two-stage least-squares, the fitted values are also a true single-state least-squares solution to the model. (The first stage is to fit only the linear part of the model

$$\mathbf{X} \cong \mathbf{A}_0\mathbf{U}' + \mathbf{V}\mathbf{B}_0'$$

by least-squares. The second stage is to remove these terms, and fit $\mathbf{B}_0$ by least-squares to what is left.) The same property (i.e., that two-stage fitting yields true single-stage least-squares fitted values) does *not* hold for a wide variety of similar models. We give only one simple example, namely, the model

$$x_{ij} \cong m + \sum_r a_{ir} b_{jr} \ .$$

(Note that this model is *not* a special case of model $\mathbf{B}_2$, because omitting terms, or restricting variables-to-be-determined to 0, is a genuine change in a model, and does not yield a special case. For further information, see appendix 2–1.) If we fill a matrix x_{ij} with random numbers, even a 2 by 2 matrix, it is very unlikely that the true least-squares solution to this model will have the grand mean for the value of m.

Note that the bilinear part of all the models shown is not a general bilinear expression, but is a very special bilinear expression. A general pure bilinear expression involving the same variables would have the form

$$\sum_k \sum_r \sum_m \sum_s a_{kr} w_{ij;krms} b_{ms}$$

where $w_{ij;krms}$ is fixed and known in advance. The bilinear expression used above can be derived from the general expression by letting

$$w_{ij;krms} = \begin{cases} 1 \text{ if } i = k,\ j = m,\ r = s\ , \\ 0 \text{ otherwise.} \end{cases}$$

Perhaps in the future other bilinear expressions will be found useful for data analysis.

The bilinear part of these models has a very important and widely used geometrical interpretation. To describe this, suppose $R = 2$. This means that there are 2 factors in the case of factor analysis or 2 components in the case of principal components analysis. We may picture the ith row of the matrix a_{ir} and the jth row of the matrix b_{ij} as a point or vector in the plane. A very common convention is to use vectors or arrows for variables, and to use points for the objects or stimuli and so forth. The connection between this picture and the data values is based on the inner product (or dot product), specifically, $\vec{a}_i \cdot \vec{b}_j$ = the (i,j) data value.

Some other bilinear models include the following:

- Hayashi's e_{ijk} quantification method uses a bilinear model for a 3-way array (Hayashi 1972);
- Chino's ASYMSCAL is bilinear (Chino 1978);
- Carroll-Green-Pruzansky-Kruskal method CANDELINC includes bilinear models (Carroll, Pruzansky, and Kruskal 1980);
- multidimensional scaling of one simple type is bilinear in the extended sense (Carroll and Kruskal 1978; Kruskal and Wish 1980; Saito 1980);
- multidimensional scaling of most types is approximately bilinear (in the extended sense).

TRILINEAR MODELS

Not very many trilinear models have received serious consideration in the literature. Most of them, perhaps all, are listed here:

- Harshman's PARAFAC uses a trilinear model for 3-way arrays (Harshman 1970); } $\mathbf{T}_0$
- Carroll and Chang's CANDECOMP (in the 3-way case) uses a trilinear model for 3-way arrays (Carroll and Chang 1970); } $\mathbf{T}_0$
- Carroll and Chang's INDSCAL uses a trilinear model in the extended sense for 3-way arrays which are symmetric in the sense that $x_{ijk} = x_{jik}$ (Carroll and Chang 1970);
- Hayashi's e_{ijkl} quantification uses a trilinear model in the

extended sense for 4-way arrays.

All except one is for use with 3-way arrays.

PARAFAC and CANDECOMP both rest on a basic trilinear model $\mathbf{T}_0(R)$ or $\mathbf{T}_0$ (which is analogous to the bilinear model $\mathbf{B}_0$), though there are some differences as we shall see. I sometimes use the single name PARAFAC-CANDECOMP because of this close relationship. INDSCAL, though a distinctly different model, has a very close relationship to PARAFAC-CANDECOMP (as discussed in Carroll and Chang 1970; Carroll and Kruskal 1978; Kruskal 1978).

Figure 2–2. Relationship Among Several Models

	MODELS BASED ON INNER PRODUCTS	MODELS BASED ON DISTANCES	
2-WAY DATA	FACTOR ANALYSIS PRINCIPAL COMPONENTS ANALYSIS $B_0, B_1, B_2, B_3, \ldots$	MULTIDIMENSIONAL SCALING (MDS)	ROTATION PROBLEM
3-WAY DATA	PARAFAC-CANDECOMP $T_0, \ldots$	INDSCAL	NO ROTATION PROBLEM

As indicated in Figure 2–2, PARAFAC-CANDECOMP is the same kind of method for 3-way data which factor analysis and principal components analysis are for 2-way data. Similarly, INDSCAL is the same kind of method for 3-way data which multidimensional scaling is for 2-way data. However, PARAFAC-CANDECOMP and INDSCAL differ from the corresponding 2-way methods in a very important manner, namely, they do not have the fundamental indeterminacy, so there is no rotation problem. To indicate this difference, we shall refer to PARAFAC-CANDECOMP and INDSCAL as methods which have the *intrinsic axis property*.

Strictly speaking, it is not correct to say that the intrinsic axis methods do not have the fundamental indeterminacy. Instead, they have an indeterminacy which corresponds to the fundamental indeterminacy but is quite different from it, and which does not create a problem. As we have seen, the fundamental indeterminacy associated with the models $\mathbf{B}_0$, $\mathbf{B}_1$, $\mathbf{B}_2$ and $\mathbf{B}_3$ involves the possible multiplication of part of the solution, $\mathbf{A}$, by a matrix $\mathbf{T}_1$ to form $\mathbf{A}_1 = \mathbf{A}\mathbf{T}_1$ where $\mathbf{T}_1$ is a nonsingular matrix. Although we have not discussed this, multidimensional scaling has a corresponding indeterminacy, which is slightly different because $\mathbf{T}_1$ is an *orthogonal* matrix (instead of any nonsingular matrix). For both PARAFAC-CANDECOMP and INDSCAL, there is a superficially similar indeterminacy. Here, however, $\mathbf{T}_1$ is the *product* of a *permutation matrix* and a nonsingular *diagonal matrix*. Multiplication by such a matrix may permute the axes and may stretch them, but the axes remain oriented in the same position relative to the configuration, so the corresponding "rotation problem" effectively disappears.

Of course, 3-way data is much richer than 2-way data. It is not surprising that a method which makes use of richer data should be able to get a better determined solution than one that uses poorer data. How to use the extra richness for this purpose, however, is not so obvious. Hayashi's e_{ijkl} quantification uses still richer 4-way data, and does not have the intrinsic axis property. The intrinsic axis methods succeed in using the extra richness of the 3-way array to determine the position of the coordinate axes, and do so without the need for any supplementary principle (such as "simple structure").

The model $\mathbf{T}_0$ is the basis for PARAFAC-CANDECOMP, just as model $\mathbf{B}_0$ is the basis for factor analysis and principal components analysis. (Models for PARAFAC are touched on briefly in appendix 2–2.) The observed data for $\mathbf{T}_0$ form a 3-way array which we shall call x_{ijk}, where $i = 1$ to I, $j = 1$ to J, and $k = 1$ to K. For example, stimuli might correspond to i, rating scales to j, subjects to k, and x_{ijk} might be rating of stimulus i on rating scale j as provided by subject k. The variables to be determined are a_{ir}, b_{jr}, c_{kr}, where $r = 1$ to R. Of course, here as before, R is the number of factors (or dimensions). We present the model $\mathbf{T}_0$ in several different forms:

$$\mathbf{T}_0: \quad x_{ijk} \cong a_{i1}b_{j1}c_{k1} + \ldots + a_{iR}b_{jR}c_{kR} ,$$

$$x_{ijk} \cong \sum_{r=1}^{R} a_{ir}b_{jr}c_{kr} ,$$

$$\mathbf{X} \cong [\mathbf{A}, \mathbf{B}, \mathbf{C}],$$ where $\mathbf{X}$ is I by J by K, $\mathbf{A}$ is I by R, $\mathbf{B}$ is J by R, $\mathbf{C}$ is K by R.

In the last equation, the square brackets indicate a new operation, called the *triple product* of matrices, which is defined by the expression in the second equation.

In earlier descriptions of PARAFAC and in all descriptions of CANDECOMP, it was assumed (implicitly or explicitly) that the data would be preprocessed by subtracting means and dividing by standard deviations in some suitable way, prior to applying model $\mathbf{T}_0$, just as such simple preprocessing is assumed in principal components analysis and is made part of (many types of) factor analysis. Such preprocessing is just as important in the 3-way case as in the 2-way case. However, the choice of preprocessing turns out to be surprisingly subtle, and the associated calculations surprisingly complicated in the 3-way case. Harshman made considerable progress in clarifying the preprocessing, and incorporated appropriate methods of preprocessing into PARAFAC during 1977–1979. I discuss the preprocessing in appendix 2–2. With these improvements, which are part of what distinguish PARAFAC from CANDECOMP, the PARAFAC method has changed in my opinion from an interesting idea to a major advance in geometrical data analysis. Note, however, that CANDECOMP has its own distinguishing feature. From the very beginning, CANDECOMP was defined in a more general way so as to be an n-linear model for n-way data, where n is not only 3 but also includes larger values.

AN APPLICATION

There have been over 10 nice applications of PARAFAC in less than a year, since the improvements were included. However, the method is so new that these applications have not been previously published. Consequently, to illustrate the method, I will use the one successful application which was published before the improvements, during the 10 years from first publication of PARAFAC-CANDECOMP to 1980.

Figure 2–3. X-Ray Movie Procedure

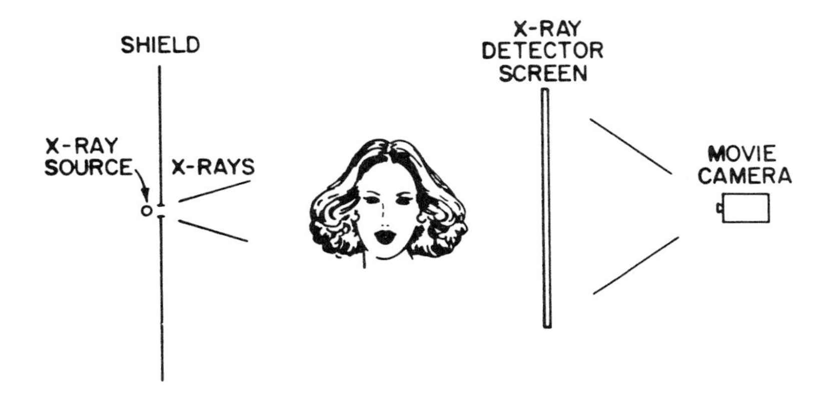

This application is taken from Harshman, Ladefoged, and Goldstein (1977) and concerns the shape and position of the tongue while a speaker is pronouncing vowels. As shown in Figure 2–3, a beam of X-rays was projected sideways through the head of the speaker, and X-ray movies were taken during speech. During the central portion of a vowel, one frame of the movie was chosen, and the position of the tongue measured. As shown in Figure 2–4, lines were drawn on the X-ray photograph, perpendicular to the vocal tract at carefully chosen positions, and

Figure 2–4. X-Ray Photograph of Vocal Tract

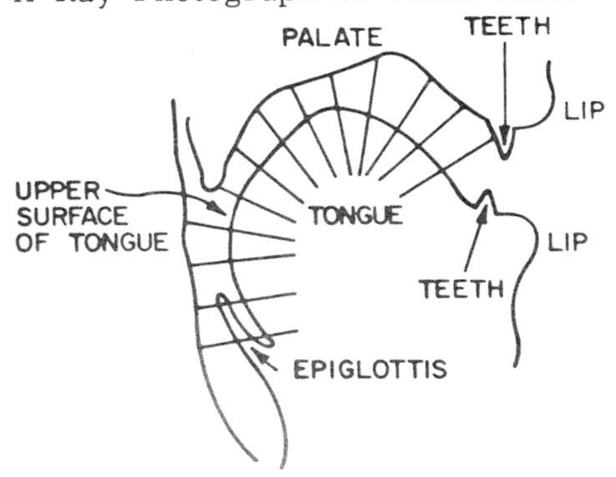

Source: Harshman, Ladefoged, and Goldstein (1977)

the height of the upper surface of the tongue (above a zero-point) was measured along those lines. Although the zero-point along each line was carefully selected, mean values of the tongue height are removed in the following analysis (details given below), so choice of the zero-point is not important. The data selected for analysis are the tongue heights for 10 American-English vowels, measured at 13 positions along the tongue, and spoken by 5 speakers:

x_{ijk} = height of tongue	for vowel i	(10 vowels)
	at position j	(13 positions)
	for speaker k	(5 speakers).

The 10 by 13 by 5 array is shown in Table 2–1.

As a first step, these data were preprocessed by removing mean tongue height for each (j,k) (each combination of position and speaker):

$$\bar{x}_{.jk} = \frac{1}{10} \sum_{i=1}^{10} x_{ijk} ,$$

$$x^*_{ijk} = x_{ijk} - \bar{x}_{.jk} .$$

After this, the model $\mathbf{T}_0(R)$ was fit to the preprocessed data x^*_{ijk}. To decide how many factors to use, that is, to decide on the value of R, several different values of R were tried and the results compared. Unfortunately, there is not enough space here to discuss the methods of comparison, but R was chosen to be 2.

The main result of the analysis consists of three matrices. The first matrix a_{ir} is 10 by 2 and corresponds to the 10 vowels. The second matrix b_{jr} is 13 by 2, and corresponds to the 13 positions along the tongue. The third matrix c_{kr} is 5 by 2, and corresponds to the 5 speakers. A familiar display for these matrices shows each row as a point in R-dimensional space: the vowels are 10 points in 2-dimensional space, the tongue positions are 13 points in 2-dimensional space, and the speakers are 5 points in 2-dimensional space. Because the 13 positions along the tongue form a natural sequence, another useful display for b_{jr} shows each column as a curve made of 13 points (j,b_{jr}). In different applications, interest focuses on different parts of the output. In this application, the most interesting parts of the output are the a_{ir} and the b_{jr}.

Here we discuss only a_{ir}, which is displayed in Figure 2–5. Next to each of the 10 points is the standard phonetic symbol. Also displayed are two classical representations of American-English vowels. One, the tongue hump diagram, is based on impressionistic values for the height and front-back position of the tongue hump used for that vowel. The other, a "formant diagram," is based on the frequencies of the first two resonances (formants) in the sound wave of vowels. The points in each representation are connected in a conventional order. The most striking fact about the main diagram is that it corresponds rather well to the two classical diagrams. At first, this may give the impression that the method has merely reproduced the obvious.

Figure 2–5. Ten Vowels. Method: PARAFAC-CANDECOMP

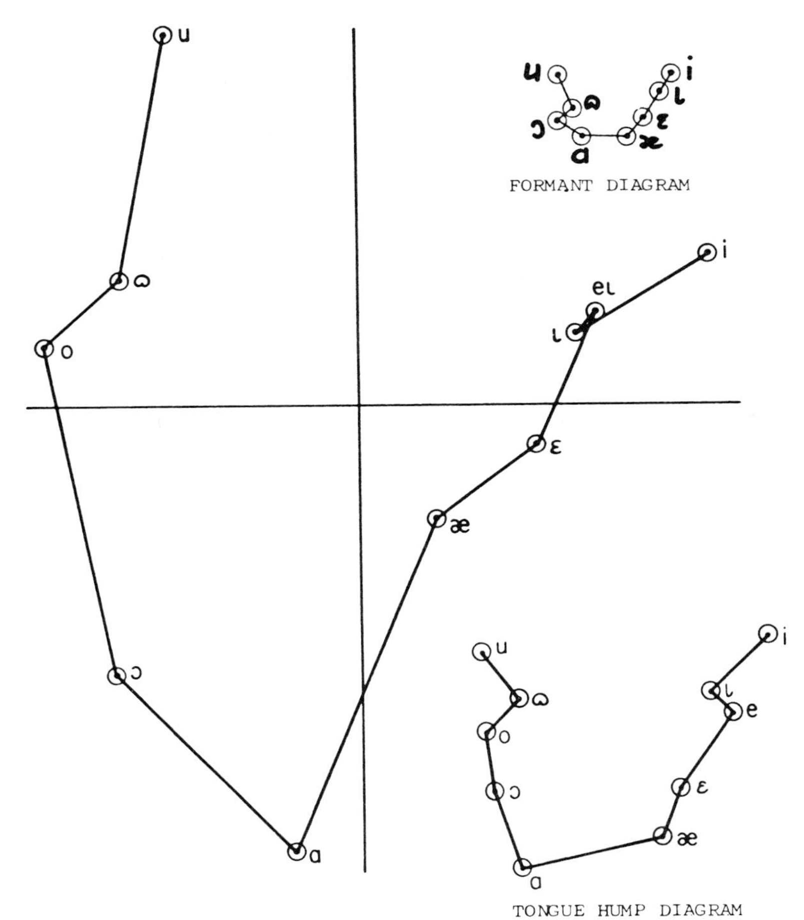

Source: Harshman, Ladefoged, and Goldstein (1977)

The diagram, however, is unique in two ways and is the first of its kind. First, though it is based on tongue shape and position like the tongue hump diagram, its coordinates are quantitative, not merely impressionistic. Second, it relies on variation among speakers to provide a well-defined rotational position, so that it becomes more meaningful to ask whether the orientation of the diagram differs from that in formant diagrams, which are based on the sound wave.

There are subtle differences from the tongue hump diagram, which depend in part on the rotational orientation of the configuration. These differences, together with evidence from other sources, led Ladefoged to propose a modified method of describing vowels. As a partial check on the reality of these differences, and as a check on the reproducibility and stability of various features of this configuration, Harshman (in unpublished work)

TABLE 2-1. Three-Way Array of Data

5 Speakers 10 Vowels k

	Speaker		Vowel	1	2	3	4	5	6	7	8	9	10	11	12	13
				13 Positions Along the Tongue j												
	1		1	2.45	2.40	2.40	2.50	2.45	2.05	1.65	1.00	0.45	0.40	0.55	0.55	0.95
	1		2	2.55	2.05	1.95	1.90	1.80	1.60	1.30	0.95	0.55	0.65	0.90	0.90	1.05
	1		3	2.35	1.90	1.80	1.80	1.80	1.55	1.30	0.95	0.65	0.70	0.90	0.85	1.05
	1		4	2.50	2.05	1.65	1.65	1.55	1.45	1.25	1.05	0.70	1.05	1.20	1.15	1.10
	1	i	5	2.05	1.50	1.35	1.50	1.55	1.45	1.30	1.15	0.95	1.40	1.55	1.40	1.25
	1		6	1.55	1.00	0.85	1.00	1.25	1.35	1.50	2.00	2.10	2.55	2.65	2.35	2.00
	1		7	1.65	0.90	0.65	0.65	0.75	0.95	1.40	1.90	2.15	2.60	2.70	2.60	2.40
	1		8	1.90	1.30	0.95	0.85	0.75	0.75	0.95	1.30	1.65	2.15	2.30	2.25	2.30
	1		9	2.40	1.60	1.45	1.25	1.00	0.95	0.80	0.85	1.10	1.50	2.10	2.00	1.65
	1		10	2.70	1.95	1.50	1.30	0.90	0.70	0.55	0.55	0.95	1.45	1.80	1.90	2.00
	2		1	2.95	2.70	2.75	2.75	2.70	2.60	2.25	1.00	0.35	0.15	0.30	0.60	1.15
	2		2	2.40	2.20	2.25	2.20	2.25	2.15	1.85	1.25	0.75	0.75	0.90	1.05	1.10
	2		3	2.25	2.45	2.65	2.65	2.40	2.20	2.05	1.55	0.95	0.85	1.10	1.40	1.65
	2		4	2.00	1.75	1.90	2.30	2.40	2.20	2.00	1.45	1.00	1.05	1.40	1.75	1.80
	2	i	5	1.25	1.15	1.30	1.65	1.95	1.90	1.80	1.65	1.40	1.70	2.15	2.45	2.60
	2		6	0.45	0.25	0.30	0.40	1.15	1.70	1.95	2.30	2.60	2.95	3.30	3.15	2.60
	2		7	0.40	0.20	0.20	0.30	0.60	1.05	1.35	1.65	2.60	3.05	3.45	3.60	3.40
	2		8	1.00	0.55	0.55	0.45	0.65	0.80	1.15	1.55	2.25	2.75	3.20	3.35	3.25
	2		9	1.30	0.70	0.65	0.45	0.65	0.90	1.20	1.45	1.90	2.40	2.85	2.80	2.45
	2		10	2.15	1.80	1.50	1.05	0.65	0.55	0.65	0.80	0.95	1.55	2.10	2.35	2.60
	3		1	2.10	2.00	2.15	2.05	1.95	1.80	1.45	1.10	0.75	0.65	0.75	0.80	0.90
	3		2	2.00	1.70	1.90	1.95	1.90	1.75	1.35	1.15	0.95	1.00	1.10	0.90	0.65
	3		3	1.95	1.80	1.80	1.95	1.95	1.95	1.65	1.25	0.90	0.85	1.05	0.95	0.90
	3		4	1.55	1.40	1.50	1.70	1.85	1.80	1.90	1.80	1.75	1.70	1.70	1.40	1.10

5 Speakers
10 Vowels
k

3	i	5	1.65	1.25	1.40	1.70	1.90	1.95	2.05	2.10	1.95	1.95	2.15	2.10	1.70
3		6	0.95	0.55	0.70	1.15	1.65	2.20	2.65	2.95	3.05	3.20	3.35	2.95	1.90
3		7	1.20	0.65	0.45	0.65	0.75	1.00	1.45	2.10	2.40	2.65	2.80	2.55	1.95
3		8	1.55	1.45	1.05	1.15	1.05	1.00	1.15	1.45	1.90	2.40	2.70	2.65	1.85
3		9	1.80	1.05	1.05	1.05	1.00	1.00	1.15	1.40	1.65	1.95	2.15	1.85	1.50
3		10	2.00	1.70	1.40	1.20	1.00	0.85	0.95	1.00	1.10	1.55	1.80	1.70	1.25
4		1	2.70	2.60	2.55	2.50	2.45	2.40	1.80	1.35	0.70	0.55	0.75	0.85	1.85
4		2	2.25	1.90	1.85	1.90	2.15	2.05	1.85	1.65	1.35	1.40	1.50	1.90	1.80
4		3	2.25	2.20	2.30	2.25	2.30	2.20	1.70	1.45	0.90	0.90	1.10	1.25	1.85
4		4	1.90	1.50	1.40	1.40	1.65	1.75	1.75	1.85	1.60	1.80	1.90	1.65	1.50
4	i	5	1.70	1.20	1.05	1.05	1.55	1.70	1.80	1.90	1.85	2.10	2.35	2.40	2.25
4		6	1.05	0.90	0.45	0.60	1.45	2.05	2.90	2.90	3.00	3.20	3.35	2.95	2.15
4		7	0.90	0.40	0.45	0.55	1.30	1.80	2.30	2.80	3.10	3.40	3.45	3.00	2.40
4		8	2.00	1.30	1.05	0.90	0.95	0.90	1.25	1.65	1.80	2.30	2.60	2.60	1.90
4		9	2.15	1.70	1.40	1.30	1.30	1.25	1.20	1.35	1.45	1.95	2.20	2.25	1.95
4		10	2.95	2.30	2.05	1.80	1.70	1.45	1.00	0.80	0.80	1.15	1.55	1.90	1.40
5		1	3.00	2.45	2.30	2.20	2.10	1.45	1.15	0.80	0.40	0.60	0.45	0.40	0.85
5		2	2.40	2.10	1.95	1.90	1.80	1.45	1.10	0.90	0.70	0.95	0.95	0.75	1.10
5		3	2.50	2.40	2.20	2.05	2.05	1.70	1.30	0.95	0.65	0.95	1.00	0.85	1.20
5		4	2.25	2.10	1.95	1.90	1.90	1.55	1.15	1.00	0.90	1.10	1.05	0.90	1.25
5	i	5	1.70	1.95	2.05	2.10	1.95	1.50	1.15	1.15	1.10	1.30	1.30	1.20	1.45
5		6	1.40	0.85	1.05	1.30	1.55	1.55	1.65	2.00	2.40	2.75	2.80	2.60	2.35
5		7	1.10	0.70	0.70	0.90	1.15	1.00	1.20	1.80	2.40	2.75	2.80	2.35	2.05
5		8	1.80	1.05	0.75	0.70	0.70	0.55	0.60	1.20	1.85	2.40	2.45	2.25	2.40
5		9	1.90	1.25	1.05	0.90	0.95	0.65	0.65	1.25	1.85	2.35	2.35	2.05	2.30
5		10	2.70	2.05	1.65	1.40	1.15	0.60	0.40	0.50	0.60	1.15	1.40	1.60	1.65

Source: Harshman, Ladefoged, and Goldstein (1977)

has obtained PARAFAC analyses of various subsets of the data set, and compared the configurations they yield with the configuration in Figure 2–5. Some of these analyses used only a subset of the 10 vowels. Other analyses used all the vowels, but used only half the positions along the tongue (e.g., the 1st, 3rd, 5th, . . . , positions along the tongue). Still other analyses used all the positions along the tongue, but used only 3 subjects or 2 subjects (instead of all 5). While it was not possible in this way to verify the reality of all the features of interest, these studies did provide significant partial verification.

APPENDIX 2–1: RELATIONSHIP AMONG SOME MODELS

Correctly speaking, $\mathbf{B}_0$ or $\mathbf{B}_0(R)$ is a one-parameter family of models, where the parameter is R, the number of factors. The same is true of $\mathbf{B}_1$, $\mathbf{B}_1'$, and $\mathbf{B}_2$. $\mathbf{B}_3$ however is parameterized by $\mathbf{U}$ and $\mathbf{V}$ as well, so this family of models would be more correctly referred to as $\mathbf{B}_3(R,\mathbf{U},\mathbf{V})$. The other models mentioned here can each be obtained from $\mathbf{B}_3$ by choosing $\mathbf{U}$ and $\mathbf{V}$ appropriately.

Before describing how to do this, it is necessary to discuss $\mathbf{B}_2$. This model has redundancy which is usually removed by using two restrictions, namely,

$$\textstyle\sum_i \; a_{i0} = 0 \qquad \text{and}$$

$$\textstyle\sum_j \; b_{j0} = 0.$$

To obtain $\mathbf{B}_2$ from $\mathbf{B}_3$, however, it is necessary to remove some of the redundancy from $\mathbf{B}_2$ in a different manner, by dropping the term m, which results in the following mathematically equivalent version of $\mathbf{B}_2$,

$$x_{ij} \cong a_{i0} + b_{j0} + \textstyle\sum \; a_{ir}b_{jr} \; .$$

This form still has some redundancy, which can be removed by using either one (but not both) of the restrictions just described.

Note that $\mathbf{A}_0$ and $\mathbf{V}$ always have I rows and $\mathbf{U}'$ and $\mathbf{B}_0'$ always have J columns. We can obtain the other models from $\mathbf{B}_3$ as follows:

- for $\mathbf{B}_0$, $\mathbf{A}_0$ and $\mathbf{U}'$ are vacuous (have 0 columns and 0 rows respectively), and $\mathbf{V}$ and $\mathbf{B}_0'$ are also vacuous;
- for $\mathbf{B}_1$, $\mathbf{A}_0$ and $\mathbf{U}'$ are vacuous, while $\mathbf{V}$ is a single column of 1's and $\mathbf{B}_0'$ is a single row of variables to be determined;
- for $\mathbf{B}_1'$, $\mathbf{V}$ and $\mathbf{B}_0'$ are vacuous, while $\mathbf{A}_0$ is a single column of variables to be determined and $\mathbf{U}'$ is a single column of 1's;
- for $\mathbf{B}_2$ (in the alternate form), $\mathbf{A}_0$ is a single column of variables to be determined, $\mathbf{U}'$ is a single row of 1's, $\mathbf{V}$ is a single column of 1's, and $\mathbf{B}_0'$ is a single row of variables to be determined.

The model

$$x_{ij} \cong m + \textstyle\sum \; a_{ir}b_{jr}$$

may appear to be a special case of $\mathbf{B}_2$, obtained by dropping the terms a_{i0} and b_{j0}, or setting them to be 0. However, dropping terms, or setting variables-to-be-determined to 0, is a significant operation which changes the model, and does not yield a special case. Setting parameters of the family (like **U** and **V** in $\mathbf{B}_3$), to particular values is very different from setting variables-to-be-determined to particular values.

APPENDIX 2–2: PARAFAC PREPROCESSING

The preprocessing discussed here only includes centering (removal of sample means) and standardization (used here to refer to division by scale factors). For data subject to so-called "object-variation," PARAFAC uses a further preprocessing step. Roughly speaking, this step is conversion to a 3-way array of covariances, but it is not covered here.

Centering, which is often thought of as subtraction of the sample mean from every element, achieves this result: the mean value of the centered data values is 0. We shall focus on the result, and think of centering as *subtracting a constant term from every element so that the resulting data values have mean 0.* The purpose of centering will be touched on lightly below.

For a 3-way array, three different types of centering are possible. One is centering of 1-way arrays, called 1-way centering, e.g.,

$$x^*_{ijk} = x_{ijk} - m_{jk}, \qquad \text{where } m_{jk} \text{ is chosen so that}$$

$$\frac{1}{I} \sum_i x^*_{ijk} = 0 \qquad \text{for every } j,\ k.$$

Another is centering of 2-way arrays, called 2-way centering (NOT double-centering), e.g.,

$$x^*_{ijk} = x_{ijk} - m_k, \qquad \text{where } m_k \text{ is chosen so that}$$

$$\frac{1}{I} \frac{1}{J} \sum_i \sum_j x^*_{ijk} = 0 \qquad \text{for every } k.$$

Another is centering of the entire 3-way array, called 3-way centering (NOT triple-centering):

$$x^*_{ijk} = x_{ijk} - m, \qquad \text{where } m \text{ is chosen so that}$$

$$\frac{1}{I} \frac{1}{J} \frac{1}{K} \sum_i \sum_j \sum_k x^*_{ijk} = 0 .$$

Of course, 1-way centering can be done in three directions: centering across i (as illustrated above), centering across j, and centering across k. Similarly, 2-way centering can be done in three directions: centering across i and j (as illustrated above), centering across i and k, and centering across j and k. Thus we already have seven different methods of centering: three 1-way methods, three 2-way methods, and one 3-way method. In addition, certain combinations of these methods appear sensible, e.g.,

1-way centering across i and also 1-way centering across j, or 1-way centering across i and also 2-way centering across j and k. (When a combination of centerings is performed, the result is the same regardless of the order in which they are performed.)

The earlier work on PARAFAC and CANDECOMP tended to approach centering very vaguely. While it was often assumed that centering was an important step, little attention was paid to what method of centering should be used. Subsequently, in unpublished work, Harshman and Kruskal independently considered the question of centering, and reached the same conclusion: 1-way centering is appropriate, but 2-way and 3-way centering have some undesirable properties. Furthermore, any combination of 1-way centerings is appropriate, but combinations involving other centerings have the same undesirable properties. Thus there are seven appropriate combinations (made from 1-way centerings), and PARAFAC permits them all:

> across i only; across j only; across k only;
> across i and across j; across i and across k; across j and across k;
> across i and across j and across k.

The first three are sometimes called single-centering; the next three are sometimes called double-centering; and the last is sometimes called triple-centering. Which combination to use with a given set of data (or whether not to center at all, though centering is almost always desirable) should be decided partly by intuition as to the nature of the data, and partly by trying different methods and comparing the results.

Unfortunately, there is not enough space here to explain the basis for these conclusions. However, some insight can be gained by comparing the following two approaches.

1. Center the array by subtracting certain terms, say $m_{jk}^{(1)}$, and $m_{ik}^{(2)}$, and then fit $\mathbf{T}_0(R)$,

$$x_{ijk}^{*} = x_{ijk} - (m_{jk}^{(1)} + m_{ik}^{(2)}) \cong \sum a_{ir}b_{jr}c_{kr} \ .$$

2. Include the corresponding terms in the model,

$$x_{ijk} \cong m_{jk}^{(1)} + m_{ik}^{(2)} + \sum a_{ir}b_{jr}c_{kr} \ ,$$

where the m's are not defined as in (1) but are variables to be determined along with the a's, the b's, and the c's. Note that this mixed trilinear model bears roughly the same relationship to $\mathbf{T}_0$ as mixed bilinear models like $\mathbf{B}_1$, $\mathbf{B}_2$, and $\mathbf{B}_3$ bear to $\mathbf{B}_0$.

By the way, one reason for centering is that a model like (2) presumably approximates the truth better than model $\mathbf{T}_0$. I believe that appropriate centering terms often achieve a considerable improvement in the reality of the model at low cost.

For 1-way centerings and combinations of them, it can be proved that the same values which are subtracted in approach (1) are true least-squares estimates for the model in approach (2).

For the other centering methods discussed, the same statement is not true, and examples which show it is not true are easy to find.

Standardization means dividing every element by a constant term, so as to achieve this result: the "scale" of the resulting data values have some fixed value (often chosen to be 1). "Scale" generally refers to some measure of variability, most often to the standard deviation. PARAFAC uses the root-mean-square instead of standard deviation, because the data are already centered by a separate operation, and for an additional reason too complex to describe here. (The relationship between centering and standardizing, will be discussed below.) The purpose of standardizing will be touched on lightly below. For a 3-way array, three different types of standardizing are possible (based on the root-mean-square measure of scale):

One is standardizing of 1-way arrays, called 1-way standardizing, e.g.,

$$x^*_{ijk} = x_{ijk}/s_{jk}\ , \qquad \text{where } s_{jk} \text{ is chosen so that}$$

$$\left[\frac{1}{I}\sum_i (x^*_{ijk})^2\right]^{1/2} = 1 \qquad \text{for every } j,\ k.$$

Another is standardizing of 2-way arrays, called 2-way standardizing, e.g.,

$$x^*_{ijk} = x_{ijk}/s_k\ , \qquad \text{where } s_k \text{ is chosen so that}$$

$$\left[\frac{1}{I}\frac{1}{J}\sum_i\sum_j (x^*_{ijk})^2\right]^{1/2} = 1 \quad \text{for every } k.$$

Another is standardizing of the entire 3-way array, called 3-way standardizing,

$$x^*_{ijk} = x_{ijk}/s\ , \qquad \text{where } s \text{ is chosen so that}$$

$$\left[\frac{1}{I}\frac{1}{J}\frac{1}{K}\sum_i\sum_j\sum_k (x^*_{ijk})^2\right]^{1/2} = 1.$$

Of course, 1-way and 2-way standardization can each be done in three directions, and in addition certain combinations of standardization may appear sensible. (When a combination of standardizations is performed, the order in which they are performed strongly affects the result. Order will be discussed below.)

2-way and 3-way standardization have an important property (described below) which 1-way standardization lacks, so 1-way standardization is seen as undesirable. The desirability of one or more 2-way standardizations depends on the characteristics of the data. 3-way standardization doesn't matter (unless the analyses of several different sets of data are to be compared). To see the important property, suppose that x_{ijk} satisfies the model $\mathbf{T}_0$ (exactly), i.e., suppose that

$$x_{ijk} = \sum a_{ir}b_{jr}c_{kr}\ .$$

Suppose we define

$$x^*_{ijk} = x_{ijk}/s_k \ ,$$

as in 2-way standardization, using arbitrary numbers s_k. Then x^*_{ijk} also satisfies the model $\mathbf{T}_0$ because

$$x^*_{ijk} = \sum a_{ir} b_{jr} (c_{kr}/s_k) \ .$$

Thus 2-way standardization *does not destroy agreement with the model.* The same is not true for 1-way standardization.

As a fuller way of referring to (say)

"standardizing across i and j"

(which is illustrated above), it is natural to use the phrase

"standardizing *across* i and j *within* (a fixed value of) k."

Harshman often finds it convenient to use a shorter phrase like

"standardizing within k."

There are seven combinations of 2-way standardization:

within i only; within j only; within k only;
within i and within j; within i and within k; within j and within k;
within i and within j and within k.

Which to use with a given set of data should be decided partly by intuition as to the nature of the data, and partly by trying different methods and comparing the results.

To discuss the purpose of standardizing, we introduce an explicit error term into the model $\mathbf{T}_0$,

$$x_{ijk} = \sum a_{ir} b_{jr} c_{kr} + \varepsilon_{ijk} \ ,$$

where ε_{ijk} is a random variable with mean 0. Because PARAFAC uses least-squares fitting, it implicitly assumes that the standard deviations of the error terms, $sd(\varepsilon_{ijk})$, are all equal. As with many other least-squares methods, the fitted values yielded by PARAFAC are not very sensitive to deviations from this assumption, but severe deviations may cause problems. If we knew $sd(\varepsilon_{ijk})$, presumably we would modify x_{ijk} by multipliers as in 2-way standardization, seeking to make the error terms after modification as nearly constant as possible. In practice, we don't know $sd(\varepsilon_{ijk})$ in most cases, so we fall back on the idea frequently used in factor analysis and in principal components analysis, namely, seeking to make the *total* standard deviations such as $sd(x_{..k})$ equal instead of seeking to make the *error* standard deviations equal. This approach has a long tradition in the bilinear methods, and is presumably as reasonable for trilinear models as for bilinear models, though a satisfying rationale for it is not known. As an option, PARAFAC may then make the further modification of not subtracting the mean before taking the root-mean-square of the values, i.e., using root-mean-square instead of standard deviation. In many cases, the centering used will make the standard deviation the same as the root-mean-square. For example, if we center across i, then standardizing within j and standardizing within k both use root-mean squares which are the same as standard deviations.

Suppose some data x_{ijk} have *already* been centered across i, so that

$$\frac{1}{I} \sum_i x_{ijk} = 0 \qquad \text{for all } j,\ k.$$

Suppose we now standardize within i, $x^*_{ijk} = x_{ijk}/s_i$. This operation will usually destroy the centering, i.e., x^*_{ijk} will *not* be centered across i. Similarly, if some data have already been standardized, then centering or another standardization will usually destroy the standardization. In general, one operation destroys the result of another operation. The only exceptions are that one centering does not destroy another, and that standardizing within one subscript does not destroy centering across another subscript.

For this reason, it is often not an elementary process to achieve the desired combination of centering and standardization. PARAFAC handles such cases by an iterative procedure. It performs whatever centerings are desired, in a fixed order, then the standardizations desired in a fixed order, then repeats the centerings followed by the standardizations, and continues repeating until adequate convergence to the desired conditions has been achieved. Ordinarily, this requires 4 or 5 repetitions. Though it has not been proved mathematically, we believe that this procedure will always converge and that the order of operations within each major iteration will not affect the final result.*

Finally, it is possible to summarize a rationale for the various types of centering by giving the following model as underlying PARAFAC:

$$x_{ijk} = m_{jk}^{(1)} + m_{ik}^{(2)} + m_{ij}^{(3)} + s_i^{(1)} s_j^{(2)} s_k^{(3)} \left(\sum_{r=1}^{R} a_{ir} b_{jr} c_{kr} + e_{ijk} \right)$$

where the e_{ijk} are independent random errors which all have mean 0 and the same standard deviation. Actually, this is not a single model, but a class of models, since the term $m_{jk}^{(1)}$ is included only if the corresponding centering is done, and the multiplier $s_i^{(1)}$ is included only if the corresponding standardization is done, and similarly for the other additive terms and the other multipliers. Then the rationale for the centerings and standardizations is to remove the effect of additive terms and the multipliers.

---

*Unfortunately, it has subsequently been discovered by Harshman that the final result does depend on the order of operations as well as other details. Thus, much mathematical clarification is still needed.

REFERENCES

Bishop, Y. M. M., S. E. Fienberg, and P. W. Holland. 1975. *Discrete multivariate analysis.* Cambridge, Mass.: The MIT Press.

Carroll, J. D., and J. J. Chang. 1970. Analysis of individual differences in multidimensional scaling via an N-way generalization of "Eckart-Young" decomposition. *Psychometrika* 35:283–319.

Carroll, J. D., and J. B. Kruskal. 1978. Scaling, multidimensional. In *International encyclopedia of statistics,* ed. W. H. Kruskal and J. M. Tanur. New York: The Free Press.

Carroll, J. D., S. Pruzansky, and J. B. Kruskal. 1980. CANDELINC: A general approach to multidimensional analysis of many-way arrays with linear constraints on parameters. *Psychometrika* 45:3–24.

Chino, N. 1978. A graphical technique for representing the asymmetric relationships between N-objects. *Behaviormetrika* 5:23–40.

Gnanadesikan, R., ed. 1983. *Proceedings of symposia in applied mathematics. Vol. 28. Statistical data analysis.* Providence, R.I.: American Mathematical Society.

Gollob, H. F. 1968. A statistical model which combines features of factor analytic and analysis of variance techniques. *Psychometrika* 33:73–115.

Harman, H. H. 1960, 1967, 1976. *Modern factor analysis.* Chicago: University of Chicago Press.

Harshman, R. A. 1970. Foundations of the PARAFAC procedure: Models and conditions for an "explanatory" multi-modal factor analysis. *UCLA Working Papers in Phonetics* 16:1–84 (University Microfilms No. 10,085).

Harshman, R. A., P. Ladefoged, and L. Goldstein. 1977. Factor analysis of tongue shapes. *Journal of the Acoustical Society of America* 62:693–707.

Hayashi, C. 1972. Two dimensional quantification based on the measure of dissimilarity among three elements. *Annals of the Institute of Statistical Mathematics* 24:251–57.

———. 1980. Personal communication.

Kruskal, J. B. 1978. Factor analysis and principal components: Bilinear methods. In *International encyclopedia of statistics,* ed. W. H. Kruskal and J. M. Tanur. New York: Free Press.

Kruskal, J. B., and M. Wish. 1978. *Multidimensional scaling.* Beverly Hills: Sage.

Saito, T. 1980. *Multidimensional scaling.* Tokyo: Asakura Shoten.

Thurstone, L. L. 1947. *Multiple-factor analysis: A development and expansion of the vectors of mind.* Chicago: University of Chicago Press.

PART II
Factor-Analytic Tradition

3

Three-Mode Principal Component Analysis: Illustrated with an Example from Attachment Theory

Pieter M. Kroonenberg

The three-mode principal component model—here referred to as the *Tucker3* model—was first formulated within the context of the behavioral sciences by Ledyard Tucker (1963). In subsequent articles, Tucker extended the mathematical description and its programming aspects (1964, 1966). In the context of multidimensional scaling, references to his model occur frequently (Carroll and Chang 1972; Takane, Young, and de Leeuw 1977; Jennrich 1972), as the Tucker3 is the general model comprising many other individual differences models. A discussion of the relation between multidimensional scaling and three-mode principal component analysis can be found in Tucker (1972), Carroll and Wish (1974), Takane, Young, and de Leeuw (1977), and Carroll and Arabie (1980). Other approaches to three-mode analysis include three-mode common factor analysis within the context of linear structural equation models (Bloxom 1968; Bentler and Lee 1978, 1979; Law and Snyder 1981). Sands and Young (1980) presented a restricted form of three-mode principal component analysis in the spirit of Harshman's PARAFAC2 model (1972), but they included an optimal scaling phase in their algorithm to accommodate data with lower measurement levels, missing data, and different data conditionalities (see also Young 1981).

In this chapter, we first present the three-mode principal component model. on a conceptual level by providing various informal ways of looking at it. Secondly, we provide an outline of some technical aspects connected with analyzing this type of model. Finally, an example treating data from attachment theory

---

I am particularly grateful to Frits Goossens for allowing me to use his data even before publication of his thesis. The conclusions with respect to the example should be seen as part of his work rather than mine. I would also like to thank Rien van IJzendoorn and other members of the "Attachment project" for their assistance in the preparation of the manuscript.

is used to illustrate some of the major aspects and possibilities of analyzing three-mode data with the three-mode principal component model.* (The method to solve the estimation of the model used and described here has been treated in full by Kroonenberg and de Leeuw [1980].)

THEORY

Informal Descriptions

In this section, we present three more or less different ways of looking at three-mode principal component analysis. First, we start with questions a researcher might ask about three-mode data and discuss the way in which these questions fit into the framework of a three-mode principal component model. Next, we take a structural point of view, postulating some structural relationships and investigating how real data might be described by a combination of structural parameters. Third, we will take a methodological point of view and demonstrate how three-mode principal component analysis is a generalization of standard principal component analysis and so-called singular value decomposition.

Research Questions Arising from Three-Mode Data

After collecting information from a number of subjects on a large number of variables, one often wants to know whether the observed scores could be described as combinations of a smaller number of more basic variables or so-called *latent variables*. As a first approximation, one generally looks for linear combinations of such underlying variables, which either account for the larger part of the variation—*principal components*—or reproduce the covariation matrix—*factors*.

As an example, one could imagine that the scores on a set of variables are largely determined by linear combinations of such latent variables as the arithmetic and verbal content. The latent variables—arithmetic and verbal content—can be found by a standard principal component analysis.

Suppose now, in the same example, that the researcher has administered the variables a number of times under various conditions of stress and time limitations. The data are now classified by three different types of quantities or *modes* of the data: subjects, variables, and conditions. First of all, the researcher is again interested in the components of the variables that explain a larger part of the variation in the data. Second, he wants to know if general characteristics can be defined for subjects as well. To put it differently, the researcher wants to know if it is possible to see the subjects as linear combinations of "idealized" subjects. In the example, we could suppose that the subjects are linear combinations of an exclusively mathematically gifted person

---

*This chapter aims to be comprehensible for the relatively uninitiated. A basic working knowledge of standard principal component analysis is, however, essential, as is an insight into eigenvalue-eigenvector problems.

and an exclusively verbally gifted person. Such persons are clearly "ideal" types. Finally, a similar question could arise with respect to conditions: Can the conditions be characterized by a set of "idealized" or "prototype" conditions?

Each of these questions can be answered by performing principal component analyses for each type of quantity or mode. In fact, the same variation present in the data is analyzed in three different ways; therefore, the components extracted must in some way be related. The question is, of course, how? In order to avoid confusion in answering this question, we will call the variable components *latent variables,* the subject components *idealized subjects,* and the condition components *prototype conditions.*

Considering the relationship between the components of the three modes, one could ask: Do idealized subject 1 and idealized subject 2 react differently to latent variable 2 in prototype condition 1? Or, is the relation between the idealized subjects and the latent variables different under the various prototype conditions?

By performing three separate component analyses, such questions are not immediately answerable, as one does not know how to relate the various components. The three-mode principal component model, however, specifies explicitly how the relations between the components can be determined. The three-mode matrix that embodies these relations is called the *core matrix,* as it is assumed to contain the essential characteristics of the data.

Structure: Raw Scores Derived from Idealized Quantities

It is often useful to look at three-mode principal components starting from the other end—the core matrix. For example, we pretend to know how an exclusively mathematically gifted person scores on a latent variable that has only mathematical content and on a latent variable that has only verbal content. Furthermore, we pretend to know these scores under a variety of prototype conditions. In other words, we pretend to know how idealized subjects react to latent variables under prototype conditions. However, in reality, we deal with real subjects, variables, and conditions. Thus, we have to find some way to construct the actual from the idealized world. A reasonable way to do this is to suppose that a real subject is a mixture of the idealized individuals and then make an analogous assumption for variables and conditions; the real scores can then be thought of as combinations of mixtures of idealized entities.

What is still lacking is some rule that indicates how the idealized quantities can be combined into real values. One of the simplest ways to do this is to weight each and then add the weighted contributions. (For instance, each latent variable could be weighted according to its average contribution over all subjects and conditions.) In more technical terms, each real variable is a linear combination of the latent variables.

We will show how to construct the score of an individual i on a test j under condition k from known idealized quantities. Suppose we have at our disposal 2 idealized persons (p_1, p_2), 2 latent variables (q_1, q_2), and 2 prototype conditions (r_1, r_2). We also know the scores of:

subject p_1 on variable q_1 under condition r_1:
$c_{p_1 q_1 r_1}$ or c_{111};

subject p_1 on variable q_1 under condition r_2:
$c_{p_1 q_1 r_2}$ or c_{112};

subject p_1 on variable q_2 under condition r_1:
$c_{p_1 q_2 r_1}$ or c_{121}; and

subject p_1 on variable q_2 under condition r_2:
$c_{p_1 q_2 r_2}$ or c_{122}.

Similarly we know the scores of subject p_2:
c_{211}, c_{212}, c_{221}, and c_{222}.

In other words, we know all the elements of the core matrix. As mentioned above, we want to construct the score of real subject i on a real variable j under a real condition k. We will do this sequentially and assemble all of the findings at the end.

We start with the observation that the score of a real subject i on the latent variable q_1 under a prototype condition r_1 is a linear combination of the scores of the idealized persons p_1 and p_2, using weights g_{ip_1} and g_{ip_2}:

$$s_{iq_1r_1} = g_{ip_1}c_{q_1r_1} + g_{ip_2}c_{p_2q_1r_1} \quad \text{or}$$

$$s_{i11} = g_{i1}c_{111} + g_{i2}c_{211} \; .$$

Similarly, for variable q_2 under condition r_1:

$$s_{i21} = g_{ip_1}c_{p_1q_2r_1} + g_{ip_2}c_{p_2q_2r_1}$$

$$= g_{i1}c_{121} + g_{i2}c_{221} \; ,$$

and the other variable-condition combinations:

$$s_{i12} = g_{i1}c_{112} + g_{i2}c_{212} \; ,$$

$$s_{i22} = g_{i1}c_{122} + g_{i2}c_{222} \; .$$

The weights g_{i1} and g_{i2} thus indicate to what extent the idealized subjects p_1 and p_2 determine the real subject i. The assumption in this approach is that these g_{i1} and g_{i2} are independent of the test and the conditions under which the subject is measured. All interrelationships between subjects, variables, and conditions are the consequence of interrelationships between the idealized entities, as reflected in the core matrix (see later section on interpretations of core matrices).

Our next step is to construct the scores for subject i on a real variable j instead of on the latent variables q_1 and q_2, as was done in the above procedure.

The score for subject i on real variable j under prototype condition r_1 is:

$$v_{ij1} = h_{jq_1}s_{iq_1r_1} + h_{jq_2}s_{iq_2r_1}$$
$$= h_{j1}s_{i11} + h_{j2}s_{i21} \ .$$

Similarly, on real variable j under prototype condition r_2, we have

$$v_{ij2} = h_{j1}s_{i12} + h_{j2}s_{i22} \ ,$$

where the weights h_{j1} and h_{j2} indicate to what extent the latent variables determine the real variable j.

Finally, we combine the idealized conditions. Subject i's score on test j under condition k may be written as

$$z_{ijk} = e_{k1}v_{ij1} + e_{k2}v_{ij2} \ ,$$

where the weights e_{k1} and e_{k2} indicate to what extent each idealized condition determines the real condition k.

Assembling the results from the three steps we get:

$$z_{ijk} = \sum_{r=1}^{2} e_{kr}v_{ijr} = \sum_{r=1}^{2} e_{kr}\{\sum_{q=1}^{2} h_{jq}s_{iqr}\}$$

which can be compactly written as

$$z_{ijk} = \sum_{r=1}^{2} e_{kr}\{\sum_{q=1}^{2} h_{jq}(\sum_{p=1}^{2} g_{ip}c_{pqr})\} \ ,$$

where

$\sum_{p=1}^{2} g_{ip}c_{pqr}$ is the linear combination of subjects p_1 and p_2;

$\sum_{q=1}^{2} h_{jq}(\sum_{p=1}^{2} g_{i}c_{pqr})$ is the linear combination of variables q_1 and q_2; and

$\sum_{r=1}^{2} e_{kr}\{\sum_{q=1}^{2} h_{jq}(\sum_{p=1}^{2} {}_{ip}c_{pqr})\}$ is the linear combination of conditions r_1 and r_2

or

$$z_{ijk} = \sum_{p=1}^{2}\sum_{q=1}^{2}\sum_{r=1}^{2} g_{ip}h_{jq}e_{kr}c_{pqr} \ ,$$

as it is usually written.

As can be seen in the next section—Formal Descriptions—this is the definition of the three-mode principal component model. In Bloxom (chapter 4), the nested form of the three-mode model is described as well, but there, the model is developed as an example of a third-order factor analysis model, in which the s are the second order and the v the third order factors.

Methodology: Extending Standard Principal Component Analysis

From a methodological point of view, three-mode principal component analysis is a generalization of standard principal component analysis, or rather, of *singular value decomposition*. Figure 3–1 schematically shows the relationship between standard principal component analysis and singular value decomposition. In essence, singular value decomposition—(or *two-mode principal component analysis*)—is a simultaneous analysis of both the individuals and the variables, in which the relationship between the components of the variables and the subjects is represented by the core matrix **C**. In Figure 3–1, the core matrix is diagonal with s diagonal elements c_{pp} ($p = 1, . . . , s$). These c_{pp} are equal to the square roots of the eigenvalues associated with the pth components of both the variables and the subjects. When **G** and **C** are combined to form **A**, as shown in Figure 3–1, we have the standard principal component solution; and when **H** and **C** are combined, we have what could be called (in Cattell's terms [1966]) "**Q**"-principal component analysis. Figure 3–2 shows the decomposition of a three-mode matrix according to the three-mode principal component model. Comparison of Figure 3–1 and Figure 3–2 shows the analogy between the singular value decomposition and three-mode principal component analysis. The core matrix now has three modes, and the relationships between the singular values or elements of the core matrix and the eigenvalues of the various modes are less simple than in the two-mode case (see later section on interpretations of core matrices).

Examples of Applications

In this section, we present some examples of the types of problems that can be handled successfully by three-mode principal component analysis.

Semantic Differential Data

A classical example of three-way classified data can be found in the work of Osgood and associates (Osgood, Suci, and Tannenbaum 1957). In the development and application of semantic differential scaling, subjects have to judge various concepts using bipolar scales of adjectives. Such data used to be analyzed after averaging over subjects, but the advent of three-mode principal component analysis and similar techniques has made it possible to analyze the subject mode as well in order to detect individual differences with regard to semantic organization of the relations between the scales and the concepts. Examples of such studies can be found in Snyder and Wiggins (1970) and Kroonenberg (1983a).

Similarity Data

Three-way similarity data—consisting of stimuli by stimuli by subjects—are generally analyzed with individual differences scaling programs, such as INDSCAL (Carroll and Chang 1970) and ALSCAL (Takane, Young, and de Leeuw 1977). However, when

Figure 3-1. Singular Value Decomposition and Principal Component Analysis

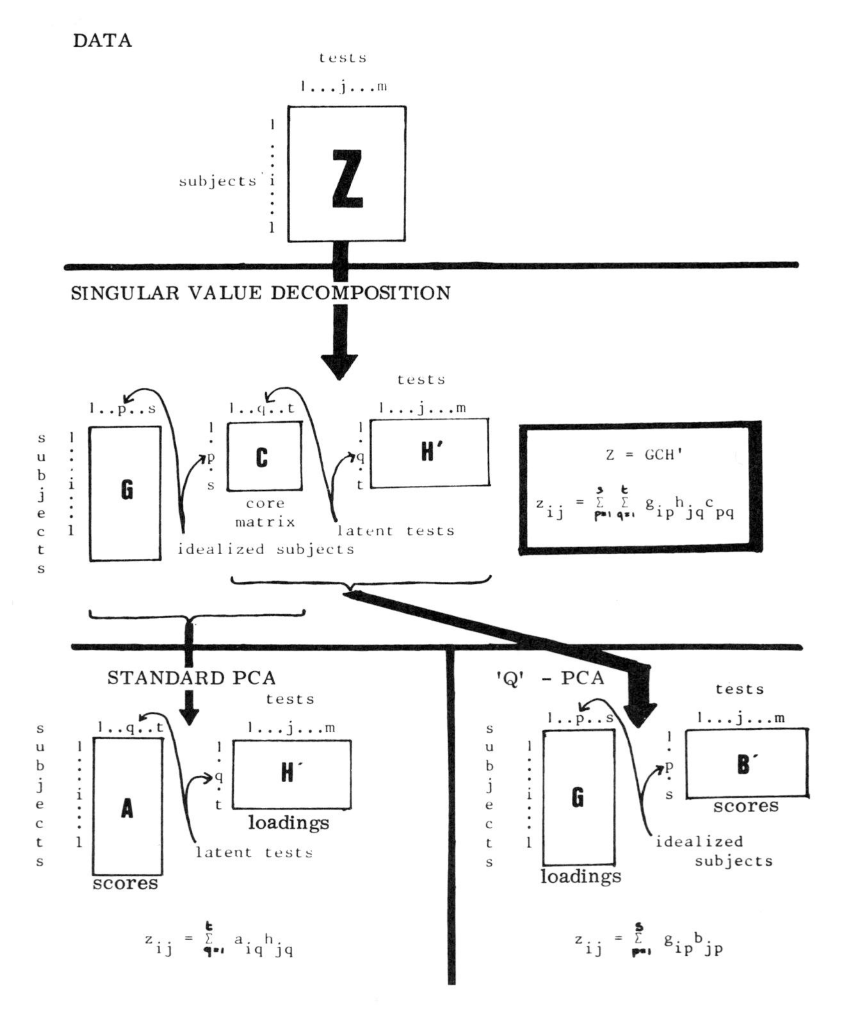

the data are asymmetric and/or a more general model is required, three-mode principal component analysis can provide useful insight. (See Carroll and Wish [1974] for details on individual differences scaling and its relation to three-mode component analysis.)

Asymmetric Similarity Data

Van der Kloot and Van den Boogaard (1978) collected data from 60 subjects who rated 31 stimulus persons on 11 personality trait scales. In the original report, the data—which can be considered asymmetric similarity data—were analyzed by canonical discriminant analysis using the stimulus persons as groups. This analy-

Figure 3–2. Three-Mode Principal Component Analysis

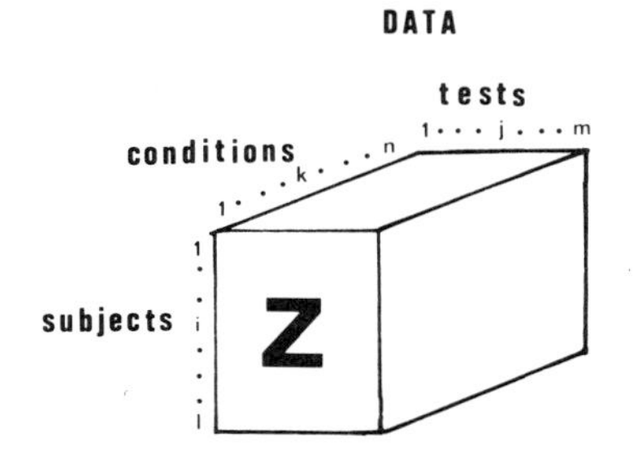

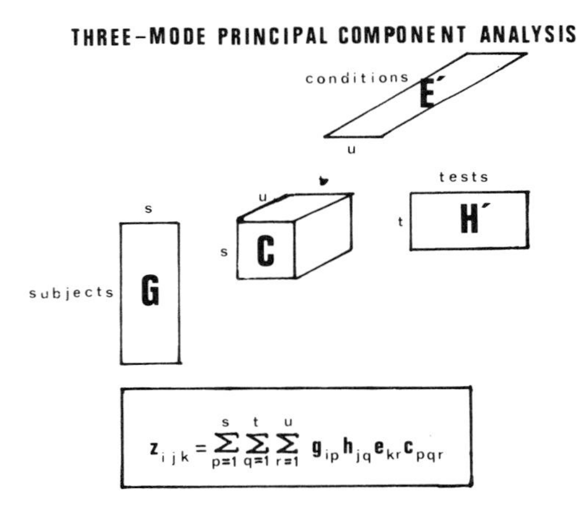

$$z_{ijk} = \sum_{p=1}^{s} \sum_{q=1}^{t} \sum_{r=1}^{u} g_{ip} h_{jq} e_{kr} c_{pqr}$$

sis yielded a circular configuration for the scales and a similar configuration for the stimuli. Van der Kloot and Kroonenberg (1982) used three-mode principal component analysis on the original data, recovering essentially the same configurations for the scales and the stimuli. However, in their analysis, it was possible to show that the two spaces were in fact identical. In addition, it was possible to assess the individual differences between subjects. These differences manifested themselves primarily in the *size* rather than the *shape* of their configurations. These differences can be explained as differences in response style (extreme versus nonextreme), rather than differences in judgmental processes.

Multivariate Longitudinal Data

In the social sciences, multivariate longitudinal data pose problems for many standard techniques. There are often too few observations and/or too many points in time for the analysis of covariance approach (Jöreskog and Sörbom 1976), or too few points in time and/or too many variables for multivariate time series analysis by some kind of ARIMA model (see, for example, Glass, Willson, and Gottman 1975; Cook and Campbell 1979, ch. 6). In such situations, three-mode principal component analysis can be very useful, especially for exploratory purposes.

Lammers (1974) presented an example of longitudinal data with a relatively large number of variables and only a limited number of points in time. Data were available for 188 hospitals measured on 22 variables in 11 consecutive years. The aim of this study was to determine if various kinds of hospitals showed different patterns or rates of growth. The general results of this study revealed that, over the years, large hospitals stayed large in relation to the initially small ones and that all hospitals grew roughly in the same manner. There were, however, a small number of hospitals that showed a specific growth pattern in a special group of variables. A reanalysis of these data can be found in Kroonenberg (1983a). (A complete survey of applications of three-mode factor and principal component analysis can be found in Kroonenberg [1983b].)

Formal Descriptions

In this section, we present a rather superficial description of the Tucker3 and Tucker2 models since our purpose here is to provide just enough detail for understanding the main principles involved. A more detailed treatment can be found in Kroonenberg and de Leeuw (1980).

Tucker3 Model

The general three-mode principal component model, or Tucker3 model, can be formulated as the factorization of the three-mode data matrix $\mathbf{Z} = \{z_{ijk}\}$, such that:

$$z_{ijk} = \sum_{p=1}^{s} \sum_{q=1}^{t} \sum_{r=1}^{u} g_{ip} h_{jq} e_{kr} c_{pqr} ,$$

$$\text{for } i = 1, \ldots, l;\ j = 1, \ldots, m;\ k = 1, \ldots, n.$$

The coefficients g_{ip}, h_{jq}, and e_{kr} are the entries of the component matrices $\mathbf{G}(l \times s)$, $\mathbf{H}(m \times t)$, and $\mathbf{E}(n \times u)$; l, m, and n are the number of elements (= rows), and s, t, and u are the number of components of the first, second, and third mode, respectively. We will *always* assume that $\mathbf{G}$, $\mathbf{H}$, and $\mathbf{E}$ are *columnwise orthonormal* real matrices, with the number of rows larger than or equal to the number of columns. The c_{pqr} are the elements of the three-mode core matrix $\mathbf{C}(s \times t \times u)$.

In practice, the three-mode data matrix is not decomposed into all its components, since one is usually only interested in the first few. Therefore, one seeks an approximate decomposition $\hat{\mathbf{Z}}$ that is minimal according to a least-squared loss function. Specifically, one solves for a $\hat{\mathbf{Z}}$ such that:

$$\sum_{i=1}^{l} \sum_{j=1}^{m} \sum_{k=1}^{n} (z_{ijk} - \hat{z}_{ijk})^2 \qquad \text{with}$$

$$\hat{z}_{ijk} = \sum_{p=1}^{s} \sum_{q=1}^{t} \sum_{r=1}^{u} g_{ip} h_{jq} e_{kr} c_{pqr}$$

attains a minimum. The algorithm to solve this minimization problem is implemented in the program TUCKALS3 (Kroonenberg 1981a). (Details about the existence and uniqueness of a minimum, the algorithm itself, and its implementation can be found in Kroonenberg and de Leeuw [1980].)

Tucker2 Model

An important restriction of the general Tucker3 model can be obtained by equating the component matrix **E** with the identity matrix. We will refer to this model as the Tucker2 model; it has also been called the generalized subjective metrics model.
The Tucker2 model can be written as

$$z_{ijk} = \sum_{p=1}^{s} \sum_{q=1}^{t} g_{ip} h_{jq} c_{pqk} ,$$

or in matrix notation

$$\mathbf{Z}_k = \mathbf{G}\mathbf{C}_k\mathbf{H}' \quad (k = 1, \ldots, n) ,$$

where $\mathbf{Z}_k (l \times m)$ is the kth frontal plane or slice of the data matrix, and $\mathbf{C}_k(s \times t)$ is the extended core matrix, respectively.

The core matrix is called "extended" because the dimension of the third mode is equal to the number of conditions in the third mode rather than to the number of components, as is the case in the Tucker3 model. The Tucker2 model only specifies principal components for the l subjects and m variables but not for the n conditions. The relationships between the components of the subjects and the variables can be investigated for all conditions together, as well as for each condition separately.

The loss function for the Tucker2 model has the form

$$\sum_{i=1}^{l} \sum_{j=1}^{m} \sum_{k=1}^{n} (z_{ijk} - \hat{z}_{ijk})^2 , \qquad \text{with}$$

$$\hat{z}_{ijk} = \sum_{p=1}^{s} \sum_{q=1}^{t} g_{ip} h_{jq} c_{pqk} .$$

The algorithm to solve this minimization problem is implemented in the program TUCKALS2 (Kroonenberg 1981b).

One important advantage of the methods discussed in this chapter over the standard procedures outlined by Tucker (1966, 297 ff) is that the estimates of the parameters are least-squares rather than estimates with ill-defined properties. Another advantage of the definition of loss functions is that it becomes possible to look at residuals (see later discussion on interpretation of residuals). A third advantage is that there exists a direct relationship between the eigenvalues of the configurations and the size of the elements in the core matrix (see introductory sections of this chapter).

Miscellaneous Topics

Various kinds of auxiliary information can be useful for the interpretation of results from a three-mode principal component analysis. Some of the most important ones will be presented here, including joint plots, component scores, use of residuals, scaling of input data, and rotations. Various ways to interpret core matrices will be discussed later in this chapter.

Joint Plots

After the components have been computed, the core matrix will provide the information about the relations between these components. It is very instructive to investigate the component loadings of the subjects jointly with the component loadings of, say, the variables, by projecting them together into one space, as it then becomes possible to specify what they have in common. The plot of the common space is called a *joint plot.*

Such a joint plot of every pair of component matrices for each of the components of the third mode—such as $\mathbf{E}$, in the TUCKALS3 case—and for the average core plane in the TUCKALS2 case, is constructed in such a way that $g_i(i = 1, \ldots, s)$ and $h_j(j = 1, \ldots, t)$ (the columns of $\mathbf{G}$ and $\mathbf{H}$, respectively) are close to each other. Closeness is measured as the sum of all $s \times t$ squared distances $d^2(g_i, h_j)$ over all i and j.

The plots are constructed as follows. For each component r of $\mathbf{E}$, the components $\mathbf{G}$ and $\mathbf{H}$ are scaled by dividing the core plane associated with that component, $\mathbf{C}_r$, between them (by using singular value decomposition), and then weighting the scaled $\mathbf{G}$ and $\mathbf{H}$ by the relative number of elements in the modes to make the distances comparable:

$$\mathbf{D}_r = \mathbf{G}\mathbf{C}_r\mathbf{H}' = \mathbf{G}(\mathbf{U}_r\Lambda_r\mathbf{V}_r')\mathbf{H}'$$

$$= \left(\frac{l}{m}\right)^{1/4}(\mathbf{G}\mathbf{U}_r\Lambda_r^{1/2})\left(\frac{m}{l}\right)^{1/4}(\mathbf{H}\mathbf{V}_r\Lambda_r^{1/2})' = \tilde{\mathbf{G}}_r\tilde{\mathbf{H}}_r' ,$$

with

$$\tilde{\mathbf{G}}_r = \left(\frac{l}{m}\right)^{1/4}\mathbf{G}\mathbf{U}_r\Lambda_r^{1/2} \quad \text{and} \quad \tilde{\mathbf{H}}_r = \left(\frac{m}{l}\right)^{1/4}\mathbf{H}\mathbf{V}_r\Lambda_r^{1/2} ,$$

$$r = 1, \ldots, u .$$

When $\mathbf{C}_r$ is not square, only the first $\min(s,t)$ components can be used. The procedure can be interpreted as rotating the component matrices by an orthonormal matrix, followed by a stretching or shrinking of the rotated components. Similar procedures for plotting two sets of vectors into one figure have been developed by Schiffman and Falkenberg (1968). (See also Schiffman et al. 1981, ch. 14; Gabriel 1971, biplot; Carroll 1972, MDPREF; Benzecri 1973, correspondence analysis; Gifi 1981, ch. 4.)

Component Scores

In some applications, it is useful to inspect the scores of all combinations of the elements of two modes on the components of the third mode. For instance, for longitudinal data the scores of each subject-time combination (or *ik*-combination) on the variable (j) components can be used to inspect the development of an individual's score on the latent variable over time. In the example presented here, these component scores in fact turn out to be the most successful summary of the relationships involved.

The component scores on the rth component of the third mode have the form

$$d_{ijr} = \sum_{p=1}^{s} \sum_{q=1}^{t} g_{ip} h_{jq} c_{pqr} \qquad \text{or} \qquad \mathbf{D}_r = \mathbf{G}\,\mathbf{C}_r\,\mathbf{H}' .$$

But by using other combinations of component matrices, three different sets of scores can be calculated. In general, only a few of these will be useful in a particular application.

One of the interesting aspects of the component scores d_{ijr} is that they are at the same time the inner products,

$$\sum_{p=1}^{\min(s,t)} \tilde{g}_{ip}^{r} \tilde{h}_{jp}^{r} ,$$

thus expressing the closeness of the elements from different modes in the joint plot.

Residuals

Kroonenberg (1983a) shows that for both the Tucker3 and the Tucker2 models the following is true:

$$\sum_{i=1}^{l} \sum_{j=1}^{m} \sum_{k=1}^{n} z_{ijk}^2 = \sum_{i=1}^{l} \sum_{j=1}^{m} \sum_{k=1}^{n} \hat{z}_{ijk}^2 + \sum_{i=1}^{l} \sum_{j=1}^{m} \sum_{k=1}^{n} (z_{ijk} - \hat{z}_{ijk})^2 ,$$

where the $\hat{z}_{ijk}$ values are the data "reconstructed" from the estimated parameters. This is, of course, a standard result in least-squares analyses. Less numerically, this may be written as

SS(Data) = SS(Fit) + SS(Residual).

In addition, it is shown that for each element f of a mode,

$$SS(Data_f) = SS(Fit_f) + SS(Residual_f).$$

By comparing the fitted sum of squares and the residual sum of squares for the *f*th element, one can gauge the correspondence of the *f*th element's configuration with the overall configuration. Large residual sums of squares indicate that a particular element does not fit very well into the structure defined by the other quantities.

Clearly, the size of the SS(Residual) depends on the SS(Total). Therefore, one should focus on the *relative residual sum of squares* (or *relative residual,* for short), which is equal to SS(Residual)/SS(Total) when assessing the role of a particular element in the final solution. Similarly, one could look at the *relative fit* (= SS(Fit)/SS(Total)). Of course, these two quantities convey essentially the same information.

The SS(Res) and the SS(Fit), as well as their relationships, can be shown directly in a so-called *sums-of-squares-plot,* which is explained and illustrated in the section on fit of the scales, episodes, and children.

Scaling of Input Data

In standard principal component analysis, the input data are often transformed into standard scores without much thought about the consequences. In other words, correlation matrices are generally analyzed with principal component analysis rather than cross-product matrices or covariance matrices. In three-mode analysis, the question of scaling the input data must be approached with more care, as there are many ways to standardize or center the data.

Two basic rules can be formulated with regard to the scaling of input data: (a) those means should be eliminated (set equal to zero) that cannot be interpreted or that are incomparable within a mode; and (b) those variances should be eliminated (set equal to one) that are based on arbitrary units of measurement or that are incomparable within a mode. If all quantities are measured in the same (possibly arbitrary) units, it is not necessary to eliminate the variances and perhaps is not even desirable.

Common scaling procedures include: (a) centering or standardizing the variables over all subject-condition combinations (*j-centering*), so that the grand mean of a variable over all subjects and conditions is zero and/or its total variance over all subjects and conditions is one; (b) centering or standardizing the variables over all subjects for each condition separately (*jk-centering*); and (c) double-centering, or centering per condition over both variables and subjects (*jk-, ik-centering*). (As before, subjects, variables, and conditions here indicate first, second, and third mode quantities, respectively.)

The decision as to which centering or standardization method is appropriate in any particular data set depends on the researcher's assessment of the origin of the variability of his data. In other words, one must assess which means and variances can be meaningfully interpreted. Harshman (chapters 5 and 6) and Kruskal (chapter 2) discuss these issues in greater detail. (For a somewhat different perspective, see Kroonenberg [forthcoming].)

Rotation of Components and Core Matrix

In standard principal component analysis and factor analysis, it is customary to rotate the solution of the variables to some kind of "simple structure," mainly by Kaiser's (1958) varimax procedure. This and other rotational procedures have been extensively applied in three-mode principal component analysis (see Kroonenberg 1983b). Various authors have advocated some particular rotation for a specific type of data. Lohmoller (1981), for instance, recommends rotation of time components to orthogonal polynomials, a proposal also put forward by Van de Geer (1974). Subject modes tend to be transformed in such a way that the axes coincide with centroids of clusters of individuals. Tucker discusses several of the above possibilities and advocates that the "first priority for these transformations should be given to establishing meaningful dimensions for the object space (variables)" (1972, 10–12).

The emphasis in the literature on rotating the component matrices first is clearly a consequence of the familiarity with such procedures in standard principal component analysis. In three-mode analysis, the core matrix is the most difficult to interpret, due to its trivariate character (see later section on interpretation of core matrices). This leads to the recommendation to concentrate on the simplicity of the core matrix rather than that of the component matrices. Simplicity here means a large number of zeros or very small values in the core matrix, preferably in the off-diagonal elements. The most simple structure would be a core matrix (for the Tucker3 model), with only non-zero elements on the body diagonal ($c_{pqr} \neq 0$, if $p = q = r$). In such a case, each component of a mode is exclusively linked to one component of another mode, so that they can be equated or at least be given the same interpretation. This model is then the orthonormal version of the PARAFAC/CANDECOMP model discussed in Harshman and Lundy (chapter 5).

An interesting observation drawn from a large number of applications is that the principal component solution already seems to produce such simple structures if they are present in the data and if they are compatible with the model employed. At present, this is just an empirical finding, but it is conjectured that it can be shown to be true for at least a number of specific cases. On the basis of this conjecture, it appears that rotating the component matrices to some kind of structure in fact destroys the simplicity of the core matrix and thus introduces unnecessary complications in its interpretation.

AN EXAMPLE FROM ATTACHMENT THEORY

Design and Data Description

To familiarize the reader with some practical aspects of three-mode analysis and to illustrate the main points of the previous sections, we will analyze data collected by Goossens (forthcoming) on the reactions of two-year-old children to a stranger and to their mothers in an unfamiliar environment within the context of a standardized observation procedure called the "Strange Situation," from *Patterns of Attachment* (Ainsworth et al. 1978). The practical aspects and theoretical considerations that form the foundation of the strange situation are covered in many publications (including the above) as the measurement procedure has become a standard one in developmental psychology. Our main purpose here is to illustrate three-mode principal component analysis rather than to dwell in detail on the strange situation itself. We will, therefore, treat its aspects only insofar as it is necessary to understand the data and the analysis.

In the course of the strange situation, the child is subjected to increasingly stressful circumstances (such as the arrival of a stranger, leaving of the mother, and being left alone) in order to elicit "attachment behaviors." *Attachment* itself is defined as "the affectional bond or tie that an infant forms between himself and his mother figure—a bond that tends to be enduring and independent of specific situations." *Attachment behaviors* are defined as "the class of behaviors that share the usual or predictable outcome of maintaining a desired degree of proximity to the mother figure" (Ainsworth et al. 1978, 302).

As Ainsworth et al. point out, the sequence of episodes was very powerful both in eliciting the expected behaviors and in highlighting individual differences (1978, 33). The major purpose of the procedure is to assess the quality of the attachment relationship of a child to its mother-figure. (A summary of the procedure is given in Table 3–1.) The major types of attachment are *secure* attachment (*B*-children), *anxiously resistant* attachment (*C*-children), and *anxiously avoidant* attachment (*A*-children). Ainsworth et al. (1978, ch. 3) have developed a more detailed classification system, which is presented in Table 3–2. The classifications of the children are made by trained judges on the basis of the children's scores on so-called *interactive scales*, which range from 1 to 7 (see Table 3–3). The child's behavior corresponding to each of the 7 categories has been explicitly defined and can be summarized as going from 1 (virtually nonexistent) to 7 (very often, very intense). The scores are awarded by trained observers while viewing videotapes of the strange situation. In the present analysis, the following scales were used: *proximity seeking* (*PROX*), *contact maintaining* (*CM*), *resistance* (*RES*), *avoidance* (*AVOI*), and *distance interaction* (*DI*).

The data of the present study consisted of observations on 65 two-year-old children on the 5 interactive scales during 4 episodes (*S*4, *M*5, *S*7, *M*8), where *S* indicates the presence of the stranger and *M* that of the mother. Details on the data and the reasons for discarding the earlier episodes can be found in Goossens (forthcoming). One might argue that a three-mode analysis is not a proper technique for these data; for instance, proximity seeking toward the stranger might not be the same variable as proximity seeking toward the mother. Also, the relationships between the scales in the stranger episodes might be different from those in the mother episodes. However, since the basic purpose of the strange situation is to assess children on the

TABLE 3-1. Description of Strange Situation

Situation Number	Persons Involved	Duration of Situation	Brief Description of Action
1	mother, child, observer	30 secs.	Observer introduces mother and baby to experimental room, then leaves.
2	mother, child	3 min.	Mother is nonparticipant while child explores; if necessary, play is stimulated after two minutes
3	stranger, mother, child	3 min.	Stranger enters. First minute: stranger silent. Second minute: stranger converses with mother. Third minute: stranger approaches child. After three minutes mother leaves unobstrusively.
4	stranger, child (S4)	3 min. or less 1)	First separation episode. Stranger's behavior is geared to that of the child.
5	mother, child (M5)	3 min. or more 2)	First reunion episode. Mother greets and/or comforts child, then tries to settle it again in play. Stranger leaves unobtrusively in the meantime. Mother leaves saying "bye bye."
6	child alone	3 min. or less 1)	Second separation episode.
7	stranger, child (S7)	3 min. or less 1)	Continuation of second separation. Stranger enters and gears behavior to that of the child.
8	mother, child (M8)	3 min.	Second reunion episode. Mother enters, greets child, then picks it up. Meanwhle stranger leaves unobtrusively.

Note: The episode is curtailed if the child is unduly distressed, and the episode is prolonged if more time is required for the child to become reinvolved in play.

Source: Ainsworth et al. (1978, 37).

basis of their reactions to the entire strange situation—and not to specific parts of it—it seems justified to treat a scale as the same variable regardless of the adult to which the behavior is directed.

Before analysis, the overall scale means were removed; specifically, the scales were centered over all children-episode combinations (j-centering). (See previous section on Scaling of Input Data.) No equalization of variances was performed. This decision was based on the consideration that the individual differ-

TABLE 3-2. Ainsworth Classification System

	Behavior toward the mother					Most salient	Behavior toward
	PROX	CM	RES	AVOI	DI	feature	stranger
A1	-	-	-	++	-	disinterested	Treatment more or less
A2	+(+)	-	(+)	++	-	mixed feelings	like mother
B1	(+)	-	-	-	++	secure	Friendly toward stranger
B2	+(+)	(+)	-	(+)	+(+)	secure	but mother is clearly
B3	++	++	-	-	-/++	very secure	preferred and sought after
B4	++	++	(+)	-	-	secure	
C1	++	++	++	-	-	angry, ambivalent	Treatment more or less
C2	(+)	(+)	++	-	(+)	passive	like mother

- = low; (+) = low to moderate; + = moderate; +(+) = moderate to high; ++ = high.

Source: Ainsworth et al. (1978, 59-63); Sroufe and Waters (1977).

TABLE 3-3. Interactive Scales

Proximity (or contact) seeking	(PROX)	A measure for the degree of active initiative a child shows in seeking physical contact with or proximity to an adult.
Contact maintaining	(CM)	A measure for the degree of active initiative a child exerts in order to maintain physical contact with a person, once such contact is achieved.
Resistance	(RES)	A measure for the degree of angry and/or resistant behavior to an adult. It is shown by physically rejecting an adult who tries to come into contact or initiate interaction with the child.
Avoidance	(AVOI)	A measure for the degree of avoiding proximity and interaction with an adult, for instance by ignoring or looking away.
Distance interaction	(DI)	A measure for the degree in which a child interacts with an adult from a distance, for instance, by showing toys and talking.

ences between children were of more interest than the overall scoring levels of the children on the interactive scales. This centering ensures that the meaningful differences in scoring levels between episodes that carry important information are retained. However, a disadvantage of using the mean values for generalization is that they are sample-dependent. For more extensive studies, some standard norm for centering scales should be devised. (It should be mentioned that this centering is not recommended by Harshman [chapter 6] or Kruskal [chapter 2], but their starting point is different from ours [see Kroonenberg, forthcoming].)

Analyses and Fit

Analyses

The main analysis reported here is a Tucker3 (T3) analysis with two components each for the first mode (*episodes*), second mode (*interactive scales*), and third mode (*children*). It will be referred to as the *2 × 2 × 2-solution,* and it will be compared with a *3 × 3 × 3-solution* using the same data. We will also refer to a Tucker2 (T2) analysis with two components for the first two modes, or the *2 × 2-solution*. (It is, by the way, not necessary to have equal numbers of components, but it is often more convenient.)

Fit

Table 3–4 shows that the fit increases with an increasing number of components but that the increase in fit in going from the 2 × 2 × 2-solution (fit = .59) to the 3 × 3 × 3-solution (fit = .68) involves estimating an additional 93 parameters. At least 60% of the variation in the (*j*-centered) data is accounted for by the three-mode model. Considering the relative difficulty of reliably measuring children's behavior and the variability inherent in it, this seems quite satisfactory.

When using the Tucker2 model—computing only components for episodes and interactive scales—a better overall fit is possible than with the Tucker3 model using the same number of components (.67 for the 2 × 2-solution versus .59 for the 2 × 2 × 2-

TABLE 3-4. Characteristics of the Solution

	T3	T3	T2
	2×2×2	3×3×3	2×2
Standardized total sum of squares--SS(Total)	1.00	1.00	1.00
Approximation of SS(Fit) from separate PCA			
on mode 1	.77	.91	.77
on mode 2	.83	.92	.83
on mode 3	.63	.71	--
Fitted sum of squares from simultaneous estimation--SS(Fit)	.59	.68	.67
Residual sum of squares from simultaneous estimation--SS(Res)	.41	.32	.33
Improvement in fit compared to initial configuration	.03	.01	.001
Parameters to be estimated	156	249	278

solution). But due to leaving the third mode uncondensed, there are more parameters in the former case (278 versus 156). Comparing the two T3-solutions, it is difficult to decide which is the "best" solution to look at in detail. No goodness-of-fit tests are available; furthermore, it seems largely a content-specific problem as to how much detail one wants to go into in describing the relations.

The "approximate fit" from the initial configuration for each of the modes, which are derived from the standard Tucker (1966) Method I solution, are upper bounds for the SS(Fit) of the simultaneous solution. Obviously, the smallest of the three is the least upper bound, which, in this case, is the one based on the third mode (.63). The initial configurations are used as starting points for the main TUCKALS algorithms. The improvement in fit indicates how much the iterative process improves the simultaneous solution over the starting solution. In this case, the improvement is not large—in other words, we might have settled for the Tucker method as far as fit is concerned. This does not mean, however, that the changes in the component matrices **G**, **H**, and **E** are also negligible.

Another point worth mentioning is the ratio of components to variables. In standard principal component analysis, it seems ill-advised to attempt to extract, for instance, three components for the four episodes. However, due to the presence of another mode, the order of the solutions in three-mode analysis may be larger than in the standard situation (see also Kruskal 1976, 1977).

Configurations of the Three Modes

One of the advantages of three-mode principal component analysis over separate analyses for each episode or each interactive scale is that one common space can be found for all episodes together, for instance, instead of one for each. The common component spaces for each mode are given in Tables 3–5, 3–6, and 3–7, respectively. In Figure 3–3, the components for scales and episodes are plotted and in Figure 3–4 those for the children are plotted. In Figure 3–3, but not Figure 3–4, the components have been multiplied by the square root of their component weights so that the plots reflect the relative importance of the axes.

The general remark can be made that the choice of a particular solution is not very crucial with respect to interactive scales and episodes. The first two components of both the scale space and the episode space are the same within reasonable bounds (roughly ± .05 and the order is preserved in all but two cases). The differences illustrate, by the way, that the solutions are *not* nested.

A point that should be made at the outset of the interpretation is that it is rather difficult to link the details of our results to those in Ainsworth et al. (1978); the latter refer mainly to one-year olds and Goossens' study deals with two-years olds. Previous research (summarized in Ainsworth et al. 1978) shows that the reaction of older children in the strange situation is different from that of one-year-olds for whom it has been validated (see also Goossens et al. 1982).

TABLE 3-5. Component Spaces--Episodes (Mode 1)

			T3: 2×2×2		T3: 3×3×3			T2: 2×2	
nr.	adult		E1	E2	E1	E2	E3	E1	E2
4	stranger	S4	.26	-.44	.25	-.37	.45	.26	-.45
5	mother	M5	.47	.25	.52	.28	.68	.48	.27
7	stranger	S7	.38	-.77	.41	-.80	-.23	.44	-.73
8	mother	M8	.75	.39	.71	.38	-.53	.71	.43
component weight (λ_p)			.37	.22	.41	.21	.07	.42	.25

Note: Labels for components: E1 = stress of situation; E2 = mother versus stranger; E3 = early versus late.

TABLE 3-6. Component Spaces--Interactive Scales (Mode 2)

		T3: 2×2×2		T3: 3×3×3			T2: 2×2	
Scales		S1	S2	S1	S2	S3	S1	S2
Proximity seeking	PROX	.32	.69	.37	.68	.04	.35	.67
Contact maintaining	CM	.26	.35	.26	.34	.14	.28	.34
Resistance	RES	.33	-.41	.30	-.39	.85	.30	-.39
Avoidance	AVOI	.27	-.48	.25	-.50	-.46	.25	-.53
Distance interaction	DI	-.81	.07	-.80	.12	.24	-.80	.10
Component weight (μ_q)		.37	.22	.43	.24	.02	.40	.27

Note: Labels for components: S1 = intensity of reaction; S2 = security seeking; S3 = interest in adult.

One of the aims of the present analysis, with regard to the content of the research, is to investigate how individual differences between the children can be traced back to their different behavior in the various episodes based solely on the interactive scales. These results will then be compared with the classification subcategories resulting from the scoring instructions in Ainsworth et al. (1978) (see Goossens, in preparation).

One qualification should be made in advance, as the research project from which these data have been derived is not yet finished. The results presented here should be seen as preliminary and not yet definitive; such final results will be published elsewhere at a later date.

TABLE 3-7. Component Spaces--Children (Mode 3)

Number	ACC	C1	C2	Number	ACC	C1	C2
55	B4	.34	.08	52	B3	-.03	.14
39	B4	.33	.12	32	B3	-.02	.14
38	B4	.30	.07	40	B3	-.05	.14
18	B4	.28	.09	19	B3	-.07	.14
62	B4	.27	-.03	7	B3	-.07	.14
20	B4	.25	-.03	33	B3	-.06	.13
48	B4	.22	.14	64	B3	-.06	.13
61	B4	.22	-.02	42	B3	-.05	.13
24	B4	.20	.05	59	B3	-.08	.12
3	B4	.19	.08	60	B3	-.08	.11
44	B3	.19	-.02	17	B2	-.04	.11
2	B4	.18	.06	47	B3	-.06	.10
41	C1	.18	-.01	30	B2	-.02	.09
11	B3/4	.15	.07	56	B3	-.04	.09
13	B3	.14	.14	16	B3	-.05	.09
34	B3	.08	.10	29	B2	-.09	.08
				43	B3	-.09	.07
14	B3	.08	.26	26	B1	-.03	.07
57	B3	-.04	.21	6	B1	-.00	.06
4	B3	.01	.20	63	B2	-.03	.05
12	B3	-.01	.19	15	B2	-.05	.04
27	B3	.04	.19	21	B3	-.07	.04
22	B3	-.01	.18	10	B2	-.04	.00
50	B3	-.04	.18	31	B3	.04	-.01
65	B3	.01	.18	35	B1	-.02	-.02
28	B3	.02	.17	8	B1	-.02	-.04
9	B3	-.00	.17	49	?	.07	-.06
25	B3	-.09	.16	53	B2	-.04	-.08
5	B3	-.09	.16	51	B2	-.01	-.09
58	B3	-.07	.16				
46	B3	-.03	.15	54	A1	-.03	-.17
1	B3	-.07	.15	37	A1	.08	-.21
36	B3	-.08	.14				
23	B3	-.10	.15	component weight		.50	.09
45	B3	-.00	.15	(ν_r)			

Note: ACC = Ainsworth's classification category; ? = unclassified; B3/4 = B3 or B4.

Episodes

With just four episodes, there is really no need to label the axes, but for further reference we will try to name them anyway. The first axis (*E*1) reflects the overall variability of the scores in the episodes, and it does not seem unreasonable to associate increasing variability with greater *stress* placed on the child. The second axis (*E*2) contrasts the behavior toward the *mother* with

Figure 3–3. Component Spaces (Scaled): Episodes and Interactive Scales

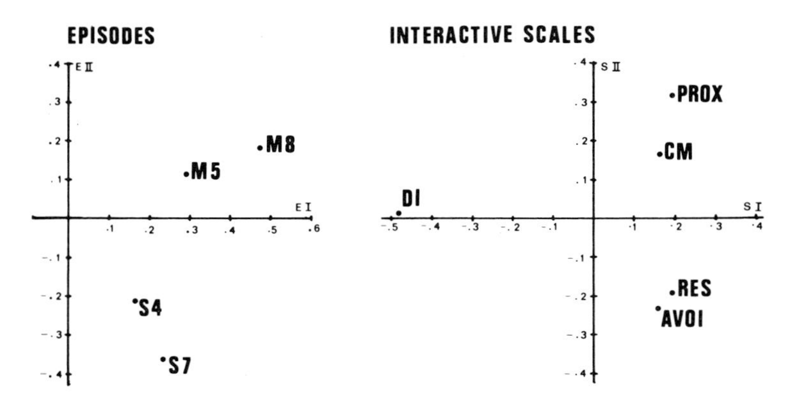

Figure 3–4. Child Space (Unscaled)

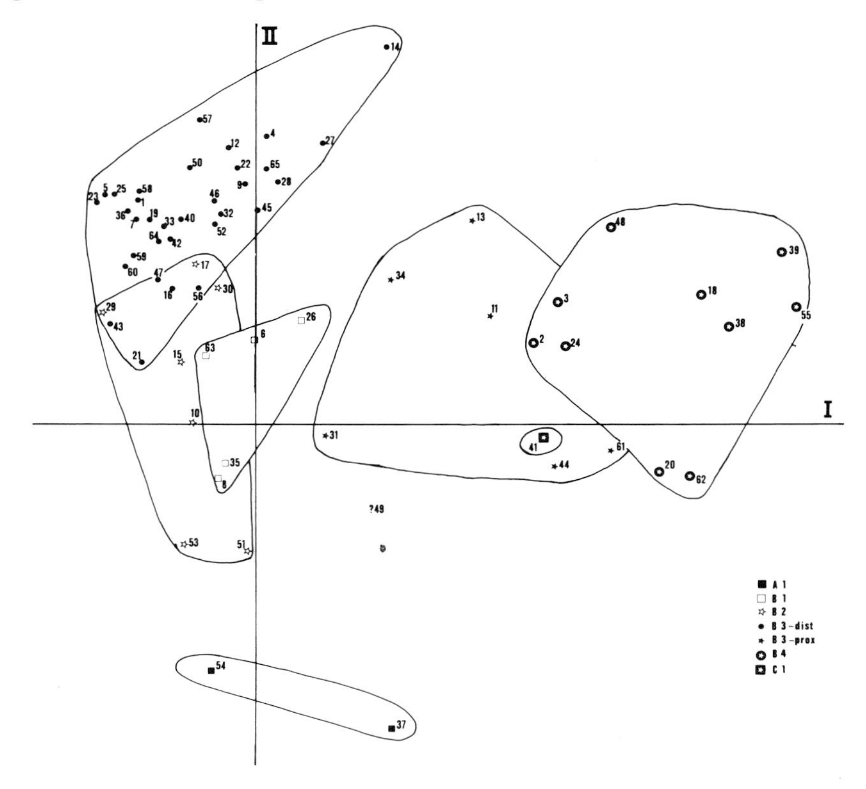

that toward a *stranger*. Finally, the third axis ($E3$) contrasts the *early* and *late* episodes, namely, those episodes before and after episode 6, in which the child has been left alone.

Interactive Scales

The first axis (S1) reflects the overall variability of the children-episode combinations around the overall scale mean. This variability is approximately equal for *PROX, CM, RES,* and *AVOI* and considerably larger for *DI*. It is clear that high scores on distance interaction reflect an opposite reaction compared to high scores on the other scales and that the same holds for low scores. This is, of course, to be expected as proximity seeking more or less precludes distance interaction and vice versa. The special position of distance interaction has been noted before; therefore, a number of researchers do not include it in their analyses (for example, see Waters 1977; Grossman et al, 1981). In Ainsworth et al. (1978), for instance, it is noted that for one-year-olds, distance interaction is a low-stress behavior of low intensity and that it differentiates less among the classification subcategories (Ainsworth et al. 1978, 246). Whether this is true for two-year-olds is still a matter for investigation; we will come back to this point later. An acceptable label for the first scale component therefore seems to be *intensity* of the reaction.

The second component (S2) distinguishes between attachment behaviors, proximity seeking and contact maintaining, and behaviors antithetical to attachment, including avoidance and resistance. It might be labeled as *security seeking*. We will not discuss the third axis (S3), due to the small amount of variation explained by it (2%), even though it shows a theoretically important contrast between resistance and avoidance.

Children

Table 3–7 and Figure 3–4 show the two-dimensional child space for the 2 × 2 × 2-solution. The children have been labeled both by a sequence number and their Ainsworth classification subcategory (see Table 3–2). These classifications are based on the same interactive scales as those in this analysis. However, it is primarily the behavior toward the mother that is taken into account, instead of that toward both the mother and the stranger, as is the case in our analysis. The classification instructions are contained in Ainsworth et al. (1978, 59–62; see also Swaan and Goossens 1982) and require extensive training. One of the aims of applying three-mode principal component analysis to these data is to assess the adequacy of the scoring instructions. For instance, it is known from psychological and medical research that people do not necessarily combine multivariate information in a very reliable way (see Sawyer 1966; Einhorn 1972).

With respect to these data, we will try to answer two questions: (a) whether the classification system is consistent; namely, whether the children who occupy the same region in the child space have the same Ainsworth classification; and (b) whether the same scales are responsible for the grouping of the children to the same extent, as is specified in the scoring instructions; it could be that the observed grouping in our analysis is the result of different combinations of scores. In other words, the present analysis is an attempt to validate the classification rules.

Ainsworth et al. (1978, ch. 6) applied discriminant analysis to check the adequacy of the classification system, but this involved the interactive scales twice: once to make the classification and then again to evaluate this classification by using the interactive scales as predictors in the discriminant functions. Here we use the interactive scales to group the children and to assess their contribution to this grouping simultaneously; only after that do we check the grouping against the classification. This provides a more adequate check of the appropriateness of the classification procedure.

The first general impression is that a reasonable separation is possible between the *B*-subcategories, although on the basis of our analysis alone the divisions could not have been made. In addition, the two *A*1-children are in their proper places, as their score pattern on the interactive scales should be the mirror-image of the *B*3-children (see Table 3–2). Furthermore, the one *C*1-child does not occupy a separate place. Finally, there are some *B*3-children seemingly belonging to the *B*4-children, and they have been labeled "*B*3-prox" for reasons to be discussed in the last section of this chapter, where we will also try to provide the answers to the above questions. In the meantime, we will use the Ainsworth classification to label the children, pretending we have already established its appropriateness.

Interpretations of Core Matrices

In this section, we will discuss the interpretational possibilities of the core matrices of the Tucker3 and Tucker2 models, both in general and within the context of the example. Three ways to interpret the values in the core matrix are given: (a) percentage of explained variation; (b) three-mode interactions; and (c) scores of idealized (latent) quantities.

Explained Variation

The core matrix indicates how the various components of the three modes relate to one another. For instance, the element c_{111} (= 19.9) of the T3 core matrix (Table 3–8) indicates the strength of the relation between the first components of the three modes, and c_{221} (= 13.5) indicates the strength of the relation between the second components of the first and second modes in combination with the first of the third mode. The interpretation of the elements of the core matrix is facilitated if one knows that the sum over all squared elements of the core matrix is equal to the unstandardized SS(Fit). In other words, c^2_{pqr} indicates how much the combination of the *p*th component of the first mode, the *q*th component of the second mode, and the *r*th component of the third mode contributes to the overall fit of the model, or how much of the total variation is accounted for by this particular combination of components. Thus, as seen in Table 3–8, 30% of the SS(Total) is accounted for by the combination of the first components of the three modes, another 14% by c^2_{221} and 3% each by c^2_{121} and c^2_{211}. Together the contributions of the elements of the first frontal plane add up to 50%, which is equal to the standardized weight of the first component of the third mode, as

TABLE 3-8. Core Matrices--TUCKALS3-Solution

2×2×2-solution (frontal planes)

Child component	Components of episodes		S1 Intensity of reaction	S2 Security seeking	Proportion variation explained			
			Components of interactive scales					
(C1): B4 versus REST	Stress of situation	E1	19.9	5.8	.30	.03	c_{111}	c_{121}
	Mother versus stranger	E2	-5.8	13.5	.03	.14	c_{211}	c_{221}
					$\nu_1 = .50$			
(C2): B3-dist versus A1	Stress of situation	E1	-6.7	3.0	.03	.01	c_{112}	c_{122}
	Mother versus stranger	E2	-2.1	7.7	.00	.05	c_{212}	c_{222}
					$\nu_2 = .09$			

3×3×3-solution (frontal planes)

C1	S1	S2	S3	C2	S1	S2	S3	C3	S1	S2	S3
E1	20.1	4.5	.6	E1	-6.5	2.8	2.6	E1	1.1	-6.7	-2.7
E2	-4.8	13.7	-2.2	E2	-2.0	6.8	0.1	E2	-1.7	-0.8	0.0
E3	-2.0	-2.3	-0.5	E3	-7.1	-0.3	0.6	E3	-4.9	-1.3	0.0
	$\nu_1 = .50$				$\nu_2 = .12$				$\nu_3 = .06$		

it should be. The core matrix thus breaks up the SS(Fit) into small parts, through which the complex relations between the components can be analyzed. It is in this way that we can interpret the core matrix as the generalization of eigenvalues or of the singular values of the singular value decomposition. It constitutes a further partitioning of the explained variation, as is indicated by the eigenvalues of standard principal component analysis.

In the present example, we see that the differences between the children on the first component (*C*1) explain half of the fitted variation. This 50% can be partitioned as follows:

a. due to c_{111} (30%): intensity of reaction (*S*1) due to the stress of situation (*E*1) for *B*4-children versus *REST* (*C*1);
b. due to c_{221} (14%): security seeking (*S*2) with the mother versus stranger (*E*2) for *B*4-children versus *REST* (*C*1);
c. due to c_{121} (3%): security seeking (*S*2) with stress of situation (*E*1) for *B*4-children versus *REST* (*C*1); and
d. due to c_{211} (3%): intensity of reaction (*S*1) with mother versus stranger (*E*2) for *B*4-children versus *REST* (*C*1).

The differences between the children on the second component (*C*2) contribute the remaining 9% explained variation, which can be divided as follows:

e. due to c_{112} (3%): intensity of reaction (*S*1) due to the stress of the situation (*E*1) for *B*3(dist)-children versus *A*1-children (*C*2);
f. due to c_{222} (5%): level of attachment (*S*2) with mother-stranger (*E*2) for *B*3(dist)-children versus *A*1-children (*C*2); and
g. due to c_{122} (1%): security seeking (*S*2) with stress of the situation (*E*1) for *B*3(dist)-children versus *A*1-children (*C*2).

Three-Mode Interactions

The percentages of explained variation only point to the important combinations but do not indicate the direction of the relationship. This information can be found in the original, not-squared core matrix. The problem is, however, what the c_{pqr} themselves represent. Their squares are variation explained; and the c_{pqr} themselves refer to what we call *three-mode interactions*.

To illustrate this three-mode interaction between loadings on components, we will look at c_{111} (= +19.9). The plus sign indicates that:

a. positive loadings on *C*1, *S*1, and *E*1 occur together the more *B*4-like children are, the more intensely they react (= the higher their scores are above average on all scales except *DI*) in more stressful situations (= *M*5/*S*7 and *M*8);
b. negative loadings on *C*1 and *S*1, occur together with positive loadings on *E*1: the more negative a child loads on *C*1 the less intensely it reacts (= scores below average on all scales except *DI*) in more stressful situations (= *M*5/*S*7 and *M*8);
c. positive loadings on *C*1 and negative loadings on *S*1 and *E*1

go together; and

d. negative loadings on $C1$ and $E1$ go together with positive scores on $S1$.

However, these combinations (c and d) do not occur in practice as all episodes load positively on $E1$.

Similarly, for each element c_{pqr} of the core matrix, such a set of statements can be made. Clearly, having four statements to explain each element of the core matrix is not particularly easy to comprehend. The situation can be simplified by omitting statements about negligible elements in the core matrix (here, c_{212} and c_{122}) and by making "conditional statements." For instance, a, b, c, and d can be simplified as:

a. for $B4$-children (with positive loadings on $C1$), intensity of the reaction ($S1$) and stress of the situation ($E1$) are positively related; and
b. for children with negative loadings on $C1$, intensity of the reaction and stress of the situation are negatively related.

Another way to gain insight in the three-mode interactions is to try and produce plots of these relationships. Making plots "conditional" upon one of the modes turns out to be singularly effective. Both the *joint plots* (Figure 3–5, Parts A and B) and the plots of the *component scores* (Figure 3–6, Parts A and B) are examples of this approach. Here these plots are made conditional on the child components. In general, the subject matter and the way the data have been generated will determine which mode can be best used for conditioning.

Scores of Idealized Quantities

This interpretation was the basis for the earlier explanation of the model at the beginning of this chapter. Each element of the core matrix represents the score of a "pure" or "ideal" child on a latent interactive scale in a prototype episode. For Goossens' data, this means that an ideal $B4$-child reacts intensely in stressful situations ($c_{111} = 19.9$), seeks much security with its mother-figure ($c_{221} = 13.5$), seeks moderate security in stressful situations ($c_{121} = 5.8$), and reacts with moderately low intensity to the mother-figure ($c_{211} = -5.8$).

The difference with the interpretation in the previous section is that there the interpretations were based on relationships between loadings on components, and here we construct interpretations in terms of the components themselves. In some applications, the former method will be easier to handle and in other applications, the latter. In the present example, using very few elements in the episode and scale modes, the naming of components is somewhat uncertain and the former approach seems more helpful. In other cases, especially when the labeling of the components as continuous variables is more adequately defined, the latter approach will be easier to use.

Figure 3–5, Parts A and B. Joint Plots of Episodes and Interactive Scales

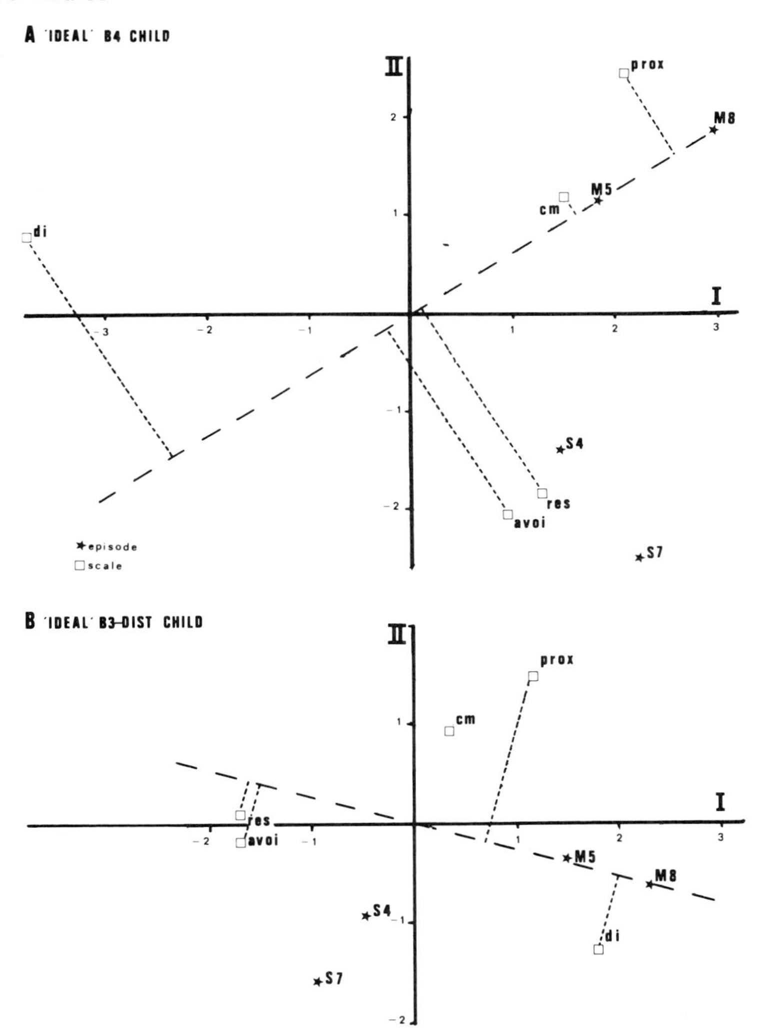

Extended Core Matrix of Tucker2 Model

So far we have only looked at the interpretation of the core matrix of the Tucker3 model. The extended core matrix can be interpreted in essentially the same way as the T3 core matrix, in terms of the amount of explained variation. Again the sum of the squared elements equals the fitted sum of squares, but now the sum of the squared elements of a frontal plane, $\mathbf{C}_k$, equals the contribution of the kth element (child) to this fitted sum of squares.

We already noted the near equality of the components for the interactive scales and the episodes in the 2 × 2-solution and

Figure 3–6, Parts A and B. Component Scores for Episode-Scale Combinations

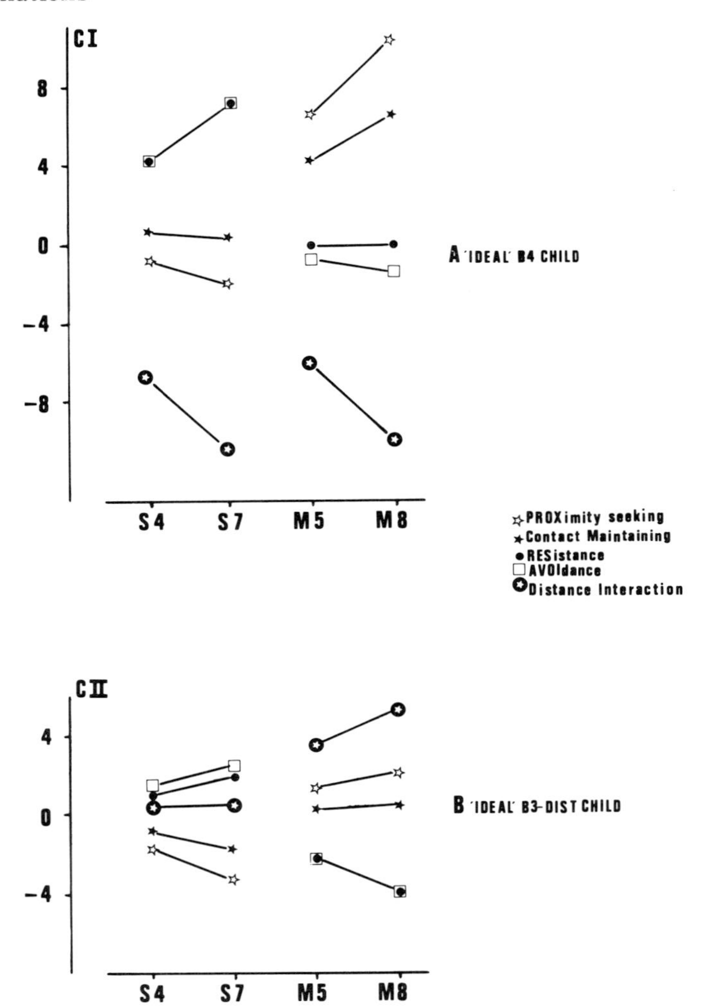

2 × 2 × 2-solution in connection with Table 3–5 and Table 3–6, and thus interpretations of those spaces are the same as before. The relationships between these components, as embodied in the frontal planes of the T2 core matrix, are given for a few selected children in Table 3–9. Four of the children were chosen because they are relatively close to one of the axes in the child space (namely, 38, 57, 29, 37) and can thus be considered "ideal individuals" (Tucker and Messick 1963).

Thus, the frontal planes indicate how the axes of the common space are related for each child, as was the case in the Tucker3 model for "ideal" children. For instance, for child 38 (a *B*4-child), intensity of reaction (*S*1) and stress of the situation (*E*1)

are positively related (see Table 3–9), as are security seeking (*S*2) and the mother versus stranger distinction (*E*2); the other combinations are immaterial. In comparison, for child 35 (a *B*1-child), none of the relationships seem very relevant (see the last section for a discussion of this phenomenon). Note also that the two *A*1-children (37 and 54) have very different patterns of relationships, despite their similar position in the child space (Figure 3–4).

Basically, one can conclude that children on the first child dimension (*C*1) weight the intensity-stress (*E*1,*S*1) combination and the mother versus stranger-security seeking (*E*2,*S*2) combination with a ratio similar to that of c_{111} to c_{221} in the T3 analysis. The overall size of the elements determines their position on the *C*1 component. High, positive numbers on the diagonal of the T2 core plane (for child 41 and child 38) lead to highly positive loadings on *C*1 and moderately negative numbers (for child 29) lead to moderately negative loadings. On the negative side of the second child component (*C*2), there are children who emphasize the (*E*1,*S*1) combination but not the (*E*2,*S*2) combination (37). On the positive side of *C*2 (57), the situation is reversed: (*E*2, *S*2) is high and (*E*1,*S*1) is low. This distinction corresponds with the opposite signs in the second frontal plane of the T3 analysis.

Direction Cosines

In those cases in which two modes are equal or the components define the same space, an additional interpretation of the core matrix is possible. For instance, within the context of multidimensional scaling of individual differences, the input similarity

TABLE 3-9. Core Matrices--TUCKALS2 Core Planes for Selected Children

	B4 (38)		B3 (57)		B2 (29)		B1 (35)	
	S1	S2	S1	S2	S1	S2	S1	S2
E1	5.7	0.9	-2.1	-0.7	-2.4	-0.2	-0.2	-1.0
E2	-0.5	5.1	-0.7	1.4	0.4	-0.6	1.1	0.1
*)	.30	.07	-.04	.21	-.09	.08	-.02	-.02
	A1 (54)		**A1 (37)**		**C1 (41)**		**B2 (51)**	
	S1	S2	S1	S2	S1	S2	S1	S2
E1	0.6	-2.2	3.0	-3.2	3.7	-.07	0.5	-0.0
E2	2.3	-0.8	-0.1	0.4	-1.0	2.9	0.2	-0.9
*)	-0.3	-.17	.08	-.21	.18	-.01	-.01	-.09

Note: B4 (38): child number 38--Ainsworth classification category B4; S1 (S2): first (second) scale component; E1 (E2): first (second) episode component; *): T3 component loadings (see Table 3-7).

matrices satisfy these conditions. Within this field, an interpretation has been developed in terms of correlations and direction cosines of the axes of the spaces common to two (generally the first and second) modes (see Tucker 1972, 7; Carroll and Wish 1974, 91).

In such situations, it makes sense to speak about the angle between the first and second component of the common space. This angle can be derived from the off-diagonal elements of the core planes, as they can be looked upon as a direction cosine or correlation between component p and component q, provided c_{pqr} is scaled by dividing it by $c_{ppk}^{1/2}$ and $c_{qqk}^{1/2}$, and that the components are standardized. The direction cosine indicates the angle under which the kth condition "sees" the axes or components of the common space. In the present example, the approach is not applicable, but in Kroonenberg (1981a) and Van der Kloot and Kroonenberg (1982), the method has been successfully used.

Joint Plots

The approach in the previous section toward the core matrices was in the spirit of Tucker's three-mode scaling (1972) and Harshman's PARAFAC2 (1972), as indicated in Carroll and Wish (1974) and Dunn and Harshman (1982). The joint plots, on the other hand, are more similar to Carroll and Chang's (1970, 1972) approach to treating the core matrix, in which the extended core matrix is decomposed by either eigenvalue-eigenvector or singular value decompositions. As pointed out earlier, the purpose of the joint plots is that elements of two modes can be plotted in one figure in order to express the relationships between the components in terms of the original variables, such as scales and episodes in our example. An advantage is that the components are now automatically scaled in accordance with their relative importance via the core plane.

With the joint plots, we can examine in some detail the relationships between the interactive scales and the episodes for each ideal-type child or child component. In Figure 3–5, Parts A and B, we present the joint plots for the two child components. The following characterization for the children loading on the positive side of the first component $C1$ can now be made.

a. They have high scores on proximity seeking and contact maintaining toward the mother (in episodes $M5$, $M8$), and they score about twice as high in $M8$ as in $M5$. With a high score, we mean relatively to the overall scale means, as we have removed these means for all interactive scales.
b. They have high scores on resistance and avoidance toward the stranger (in $S4$ and $S7$), nearly twice as high in $S7$ as in $S4$.
c. They show roughly average resistant and avoidant behavior toward the mother in $M5$ and $M8$ and even somewhat below average behavior on avoidance. Similarly, proximity seeking and contact maintaining toward the stranger have average values.
d. The scores on distance interaction do not discriminate between

the mother and the stranger and they are below average. There is less distance interaction in the later episodes.

These interpretations are derived from the fact that the scales can be seen as points and the episodes as vectors or directions in the common space, and vice versa. In this case, the former approach is to be preferred because the episodes are fixed; specifically, they are elements of the design. The relative importance of the various scales at any episode can then be assessed from their perpendicular projections on the vectors as is shown for *M*5 and *M*8 combined. The values of the projections are contained in the matrix of the scaled inner products $\mathbf{D}_r = \tilde{\mathbf{G}}_r \tilde{\mathbf{H}}_r'$ $= \mathbf{G}\mathbf{C}_r\mathbf{H}'$ (see earlier discussion of component scores).

For the positive scores on the second child component—the *B*3(dist)-children—the characterization is (see Figure 3–5, Part B):

a. low scores on resistance and avoidance toward the mother, coupled with average contact maintaining and proximity seeking; high distance interaction increasing further in *M*8;
b. low scores on proximity seeking and contact maintaining toward the stranger, with lower scores on proximity seeking; average resistance, avoidance, and distance interaction with a slight increase in the avoidance measures in *S*7.

For 37, an *A*1-child, the mirror image of the above observations is true as he lies on the negative side of the second child component (*C*2).

Component Scores

As remarked earlier in this chapter, the values of the inner products on which the above observations were made are at the same time the component scores on the child component in question; thus, they can serve as an intermediate level of condensation between the raw data and the three-mode model.

As we are looking here for characterizations of "ideal" children, it is not very useful to display the component scores in the two-dimensional child space. It is far more useful to plot the component scores of the interactive scales for each episode, as is done in Figure 3–6. In fact, for the present data—together with that of the children's loadings—these plots are the best summary of the results. Certainly they are easier to read than the joint plots with their projections on vectors. It has been noted that in longitudinal data, these plots have a similar appeal when the points in time are placed on the horizontal axis (see Kroonenberg 1983a).

Fit of the Scales, Episodes, and Children

In essence, the analysis could stop with the above interpretations. All that the technique has to offer toward breaking down complex relationships into small intelligible pieces is contained in the analysis so far. However, it is beneficial to have some additional information available to assess if there are no irregularities in the data, such as outliers, unduly influential points, and points that are not sufficiently accounted for. A useful way to investigate such questions is to inspect the residual sums of squares in conjunction with the fitted sums of squares (see earlier discussion of residuals). Whereas the core matrix informs us about the contributions of the components and their interrelationships, the sums of squares broken down by the elements or variables of the modes inform us about the contributions of these elements to the solutions.

In Table 3–10 and Table 3–11, the sums of squares for the scales and episodes are shown, respectively. From the SS(Total)s for episodes, we see that the variability as expressed by the sums of squares increases with the later episodes, as children deviate more from the scale means or perhaps show more variation among themselves. Which of the two is more important cannot be unequivocally determined from the present analysis and should be assessed separately. With respect to the scales, we see that contact maintaining has relatively little variability, while distance interaction has considerably more variation. From the residual sums of squares we note that the scales fit more or less equally well, irrespective of their total sum of squares, but that the configurations derived and discussed above are for a large part determined by the last two episodes. The structure described is therefore more representative of the later behaviors than the early ones. This explains, for instance, why an added third episode component shows an early versus late character; primarily the earlier episodes will then be fitted better.

Figure 3–7 is a so-called *sums-of-squares-plot,* which shows the residual sums of squares versus the fitted sums of squares for the children from the 2 × 2 × 2-solution. By plotting the sums of squares directly, rather than taking the relative sums of squares, the total sums of squares are also contained in the plot, and unusually large elements can be spotted directly. Moreover, it can be seen if the larger SS(Fit)s resulted only from larger total sums of squares, as is to be expected from least-squares procedures. Furthermore, whether the variations of the elements

TABLE 3-10. Sums of Squares for Episodes (Mode 1)

		2×2×2-solution				3×3×3-solution	
	SS(Total)	SS(Fit)		SS(Res)		SS(Res)	
Episode	STD	STD	REL	STD	REL	STD	REL
S4	.16	.07	.40	.10	.60	.10	.59
M5	.21	.09	.44	.12	.56	.06	.27
S7	.29	.18	.63	.11	.37	.09	.30
M8	.33	.24	.74	.09	.26	.08	.23
Overall	1.00	.59		.41		.32	

Note: STD = standardized or divided by the overall SS(Total); REL = relative sum of squares, which is defined as:

$$\text{relative SS(Res) of episode S4} = \frac{\text{SS(Residual) of episode S4}}{\text{SS(Total) of episode S4}}$$

TABLE 3-11. Sums of Squares for Interactive Scales (Mode 2)

	SS(Total)	2x2x2-solution				3x3x3-solution	
		SS(Fit)		SS(Res)		SS(Res)	
Scale	STD	STD	REL	STD	REL	STD	REL
PROX	.23	.14	.61	.09	.39	.06	.27
CM	.10	.05	.54	.05	.46	.04	.41
RES	.15	.08	.52	.07	.48	.06	.41
AVOI	.17	.08	.44	.09	.56	.08	.46
DI	.35	.24	.68	.11	.32	.07	.21
Overall	1.00	.59		.41		.32	

Note: STD = standardized or divided by the overall SS(Total); REL = relative sum of squares, which is defined as:

$$\text{relative SS(Res) of PROX} = \frac{\text{SS(Residual) of PROX}}{\text{SS(Total) of PROX}}$$

have been equalized is evident from the arrangement of the elements on a line at an angle of -45° to the positive x-axis. Note that because the axes represent sums of squares, the total sums of squares are obtained by directly adding the x-value and the y-value (according to the city-block metric).

Another interesting feature of these plots is that they show which elements have equal residual sums of squares with different total sums of squares. In other words, it becomes possible to separate points that have large residual sums of squares (because they do not fit well) from the points that have a large SS(Res) (because they have a large total sum of squares). Without a residual analysis, it is uncertain whether a point in the middle of a configuration on the first principal components is an ill-fitting point or just a point with little overall variation.

Finally, by drawing the line through (0,0) and (av.SS(Fit), av.SS(Res)), and appropriate similar lines above and below it, something similar to confidence bands can be constructed around the former line to assess the extremity of certain elements. The lines are the loci of points with equal relative residual sums of squares. A guide line for what is "appropriate" in this case—for instance, how much the individual element may deviate in relative residual sum of squares from the overall sum of squares—has not been developed yet.

A number of features should be noted for the present data. The $B4$-children fit well, have large sums of squares, and dominate the solution. (For a more detailed discussion of these $B4$-children and the relation to the other B-groups, see Van IJzendoorn et al. 1983.) Furthermore, there is a large group of $B3$-children that have small total sums of squares; thus they score about average on all scales and most of their variation is fitted well. Conversely, none of the $B1$- and $B2$-children fit very well into the overall pattern, but we have to remember that there are only few of them. Their total sums of squares are not very large, but their relative residual sums of squares are.

Figure 3–7. Sums-of-Squares Plot for Children

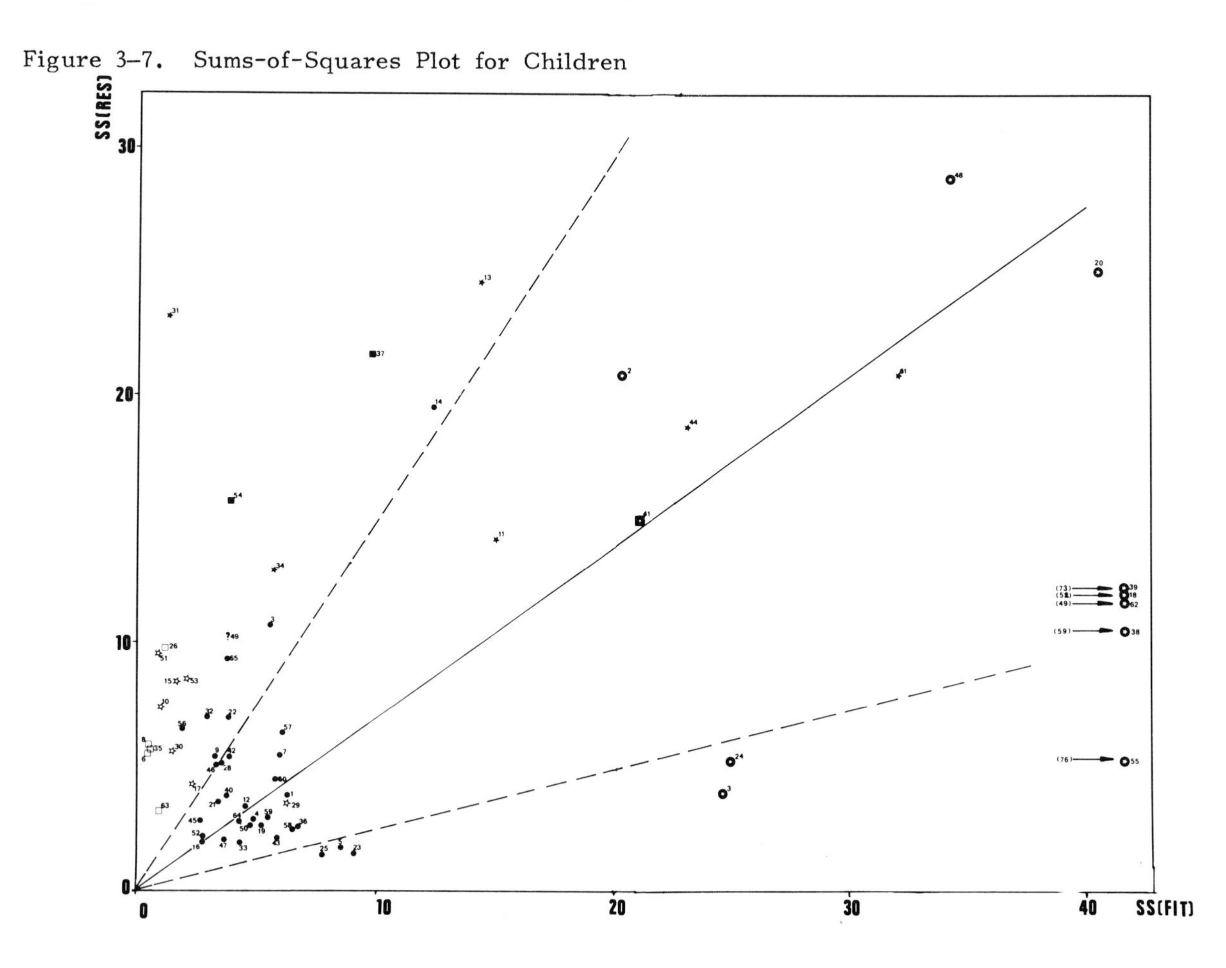

Finally, there is a number of peculiar children, and one should hesitate to draw definitive conclusions about them without further analysis. They couple considerable sums of squares with little fit, indicative of either another organization of the scale and episode relationships or of large amounts of random variation. In fact, the two *A*1-children (37 and 54) belong to this group.

Discussion

Keeping in mind the preliminary character of the data, there are some conclusions that can be drawn with respect to the example. In the first place, we note that three-mode principal component analysis succeeded in showing individual differences between the children and characterizing the kind and degree of these differences. Furthermore, the analysis presented here supports to a large degree the *consistency* of the classification procedures, as described by Ainsworth et al. (1978), especially for the *B*-children. The consistency follows from the grouping of children belonging to the same category. The presence of only two *A*-children and a single *C*-child precludes any serious statements about these classification categories, apart from the observation that their position in the child space (Figure 3–4) agrees with what one would expect, but this might be accidental.

We noted earlier the presence of two groups of *B*3-children. In Figure 3–4, they were labeled *B*3-*prox* and *B*3-*dist*. The classification instructions in Ainsworth et al. (1978, 61) for *B*3-children (see also Swaan and Goossens 1982) also suggest that there are two types of *B*3-children: those who actively seek physical contact with their mothers (*B*3-prox) and those who seem especially "secure" in their relationship with their mother and are thus content with mere interaction from a distance without seeking to be held (*B*3-dist). It is possibly due to the greater capabilities of communicating at a distance with the mother by two-year-olds that there are more children in the *B*3-dist than in the *B*3-prox group in Goossens' sample. For one-year-olds, the reverse seems to be true. (See Goossens, forthcoming, for further details.)

In Table 3–12, the characterizations of the children (derived from Figure 3–6) occupying the extremes of the axes in Figure 3–4 (child space) are presented. A comparison of this table with Table 3–2 shows global agreement and disagreement in detail. The most conspicuous differences are related to resistance and distance interaction. The comparison for resistance is probably biased by the absence of extremely resistant *C*-children; "high resistance" in Goossens' sample might be average when compared to the resistant behavior of *C*-children. The differences between distal behaviors are, of course, related to the age differences.

A number of problems remain. One is the low number of *A*-children compared to the number found in samples of one-year-old children. One possible explanation for this might be the less avoidant behavior of two-year-old children (see Goossens, forthcoming).

Another potential problem is the ill-fitting *B*1- and *B*2-children. Two reasons may explain this situation. First of all, these children have approximately average scores on all scales,

TABLE 3-12. Comparison of Ainsworth and TUCKALS Classifications (behavior toward the mother)

	Ainsworth*					TUCKALS**				
	PROX	CM	RES	AVOI	DI	PROX	CM	RES	AVOI	DI
A1	-	-	-	++	-	o	o	H	H	L
B3-dist						o	o	L	L	H
B3-prox	++	++	-	-	-	H	H	o	o	L
B4	++	++	(+)	-	-	HH	HH	o	o	LL

*For the Ainsworth classification: - = low; (+) = low to moderate; + = moderate; +(+) = moderate to high; and ++ = high.

**For the TUCKALS classification: LL = low; L = low to average; o = average; H = average to high; and HH = high.

so we are trying to fit their individual error rather than any meaningful variation. Another reason may be that their way of reacting to the Strange situation cannot be fitted very well together with the other children. The small number of these children may preclude finding a separate dimension especially for them. Clearly these conjectures are topics for further investigation.

In discussing three-mode principal component analysis, we have attempted to present as full and detailed an account as possible within the limited context of one chapter. In Kroonenberg (1983a), a more detailed account has been given of many of the points raised here. Although we have concentrated on the programs TUCKALS2 and TUCKALS3, which were developed by Kroonenberg and de Leeuw (1980) to illustrate the technique, many of the issues raised have wider relevance. Crucial to the whole approach, however, is the least-squares formulation of the method to solve the estimation of the parameters in the model. It is this formulation that allows the explanation of the core matrix in terms of amount of explained variation and makes the separation of the fitted and residual sums of squares possible.

REFERENCES

Ainsworth, M. D. S., M. C. Blehar, E. Waters, and S. Wall. 1979. *Patterns of attachment. A psychological study of the strange situation.* Hillsdale, N.J.: Lawrence Erlbaum Associates.

Bentler, P. M., and S-Y. Lee. 1978. Statistical aspects of a three-mode factor analysis model. *Psychometrika* 43:343–52.

———. 1979. A statistical development of three-mode factor analysis. *British Journal of Mathematical and Statistical Psychology* 32:87–104.

Benzecri, J. P., et al. 1973. *Analyse des donnees. II Analyse des correspondences.* Paris: Dounod.

Bloxom, B. 1968. *Individual differences in multidimensional scaling (ETS Research Bulletin 68–45).* Princeton, N.J.: Educational Testing Service.

Carroll, J. D. 1972. Individual differences and multidimensional scaling. In *Multidimensional scaling: Theory and applications in the behavioral sciences,* vol. 1, ed. R. N. Shepard, A. K. Romney, and S. Nerlove. New York: Seminar Press.

Carroll, J. D., and P. Arabie. 1980. Multidimensional scaling. *Annual Review of Psychology* 31:607–49.

Carroll, J. D., and J. J. Chang. 1970. Analysis of individual differences in multidimensional scaling via an N-way generalization of "Eckart-Young" decomposition. *Psychometrika* 35:283–319.

————. 1972. IDIOSCAL (individual differences in orientation scaling): A generalization of INDSCAL allowing idiosyncratic reference systems as well as an analytic approximation to INDSCAL. Paper presented at the spring meeting of the Psychometric Society, Princeton, N.J., March, 1972.

Carroll, J. D., and M. Wish. 1974. Models and methods for three-way multidimensional scaling. In *Contemporary developments in mathematical psychology,* vol. 2, ed. D. H. Krantz, R. C. Atkinson, R. D. Luce, and P. Suppes. San Francisco: W. H. Freeman.

Cattell, R. B. 1966. The data box: Its ordering of total resources in terms of possible relational systems. In *Handbook of multivariate experimental psychology,* ed. R. B. Cattell. Chicago: Rand McNally.

Cook, T. D., and D. T. Campbell. 1979. *Quasi-experimentation. Design and analysis issues for field settings.* Chicago: Rand McNally.

Dunn, T. R., and R. A. Harshman. 1982. A multidimensional scaling model for the size-weight illusion. *Psychometrika* 47:25–45.

Einhorn, H. J. 1972. Expert measurement and mechanical combination. *Organizational Behavior and Human Performance* 7:86–106.

Gabriel, K. R. 1971. The biplot graphical display of matrices with applications to principal components. *Biometrika* 58:453–67.

Gifi, A. 1981. *Non-linear multivariate analysis,* preliminary edition. Leiden, The Netherlands: University of Leiden, Department of Data Theory.

Glass, G. V., V. L. Willson, and J. M. Gottman. 1975. *Design and analysis of time-series experiments.* Boulder, Colo.: Colorado Associated University Press.

Goossens, F. A. Forthcoming. The quality of the attachment relationship for two-year-old children of working and non-working mothers and some associated factors. Dissertation in preparation, University of Leiden.

Goossens, F. A., J. Swaan, L. W. C. Tavecchio, M. M. Vergeer, and M. H. Van IJzendoorn. 1982. *The quality of attachment assessed (WEP-Reeks, WR 82-23-LE).* Leiden, The Netherlands: Rijksuniversiteit Leiden.

Grossman, K. E., K. Grossman, F. Huber, and U. Wartner. 1981. German children's behavior towards their mother at 12 months and their father at 18 months in Ainsworth's strange situation. *International Journal of Behavioral Development* 4:157–81.

Harshman, R. A. 1972. PARAFAC2: Mathematical and technical

notes. *UCLA Working Papers in Phonetics* 22:30–44 (University Microfilms No. 10,085).

Jennrich, R. A. 1972. A generalization of the multidimensional scaling model of Carroll & Chang. *UCLA Working Papers in Phonetics* 22:45–47.

Jöreskog, K. G., and D. Sörbom. 1976. Some models and estimation methods for analysis of longitudinal data. In *Latent variables in socioeconomic models,* ed. D. J. Aigner and A. S. Goldberger. Amsterdam, The Netherlands: North Holland.

Kaiser, H. F. 1958. The varimax rotation for analytic rotation in factor analysis. *Psychometrika* 23:187–200.

Kroonenberg, P. M. Forthcoming. Scaling of input data for three-mode principal component analysis. Manuscript in preparation.

———. 1981a. *User's guide to TUCKALS3. A program for three-mode principal component analysis (WEP-Reeks, WR-81-6-RP).* Leiden, the Netherlands: Rijksuniversiteit Leiden.

———. 1981b. *User's guide to TUCKALS2. A program for three-mode principal component analysis with extended core matrix (WEP-Reeks, WR-81-35-RP).* Leiden, the Netherlands: Rijksuniversiteit Leiden.

———. 1983a. *Three-mode principal component analysis: Theory and applications.* Leiden: DSWO Press.

———. 1983b. Annotated bibliography of three-mode factor analysis. *British Journal of Mathematical and Statistical Psychology* 36:81–113.

Kroonenberg, P. M., and J. de Leeuw. 1980. Principal component analysis of three-mode data by means of alternating least-squares algorithms. *Psychometrika* 45:69–97.

Kruskal, J. B. 1976. More factors than subjects, tests and treatments: An indeterminacy theorem for canonical decomposition and individual differences scaling. *Psychometrika* 41: 281–93.

———. 1977. Three-way arrays: Rank and uniqueness of trilinear decompositions, with application to arithmetic complexity and statistics. *Linear Algebra and its Applications* 18:95–138.

Lammers, C. J. 1974. *Groei en ontwikkeling van de ziekenhuisorganisaties in Nederland.* Leiden, The Netherlands: Rijksuniversiteit Leiden, Sociologisch Instituut.

Law, H. G., and C. W. Snyder, Jr. 1981. An introduction to the analysis of covariance structures: A general model for data analysis. In *Proceedings of a seminar on measuring social behavior in road research,* ed. J. M. Morris. Vermont South, Australia: Australian Road Research Board.

Lohmöller, J. B. 1981. Stabilität und Kontinuität in Längsschnittdaten, analysiert durch T- und trimodale Faktorenanalyse. Technical Report. Fachbereich Pädagogik, Hochschule der Bundegwehr München, Neubiberg, FRG.

Osgood, C. E., G. J. Suci, and P. H. Tannenbaum. 1957. *The measurement of meaning.* Urbana: University of Illinois Press.

Sands, R., and F. W. Young. 1980. Component models for three-way data: An alternating least-squares algorithm with optimal scaling features. *Psychometrika* 45:39–67.

Sawyer, J. 1966. Measurement *and* prediction, clinical *and* statistical. *Psychological Bulletin* 66:178–200.

Schiffman, H., and P. Falkenberg. 1968. The organization of stimuli and sensory neurons. *Physiology and Behavior* 3:197–201.

Schiffman, S. S., M. L. Reynolds, and F. W. Young. 1981. *Introduction to multidimensional scaling: Theory, methods, and applications*. New York: Academic Press.

Snyder, F. W., and N. Wiggins. 1970. Affective meaning systems: A multivariate approach. *Multivariate Behavioral Research* 5:453–68.

Sroufe, L. A., and E. Waters. 1977. Attachment as an organizational construct. *Child Development* 48:1184–99.

Swaan, J., and F. A. Goossens. 1982. *Handleiding bij de strange situation (WEP-Reeks, WR 82-24-EX)*. Leiden, The Netherlands: Rijksuniversiteit Leiden.

Takane, Y., F. W. Young, and J. de Leeuw. 1977. Non-metric individual differences multidimensional scaling: An alternating least-squares method with optimal scaling features. *Psychometrika* 42:7–67.

Tucker, L. R. 1963. Implications of factor analysis of three-way matrices for measurement of change. In *Problems in measuring change,* ed. C. W. Harris. Madison, Wis.: University of Wisconsin Press.

———. 1964. The extension of factor analysis to three-dimensional matrices. In *Contributions to mathematical psychology,* ed. N. Frederiksen and H. Gulliksen. New York: Holt, Rinehart and Winston.

———. 1966. Some mathematical notes on three-mode factor analysis. *Psychometrika* 31:279–311.

———. 1972. Relations between multidimensional scaling and three-mode factor analysis. *Psychometrika* 37:3–27.

Tucker, L. R., and S. Messick. 1963. An individual differences model for multidimensional scaling. *Psychometrika* 28:333–67.

Van de Geer, J. P. 1974. Toepassing van drieweg-analyse voor de analyse van multiple tijdreeksen. In *Groei en ontwikkeling van de ziekenhuis organisaties in Nederland,* ed. C. J. Lammers. Leiden, The Netherlands: Rijksuniversiteit Leiden, Sociologisch Instituut.

Van der Kloot, W. A., and P. M. Kroonenberg. 1982. Group and individual implicit theories of personality. An application of three-mode principal component analysis. *Multivariate Behavioral Research*. 17:471–92.

Van der Kloot, W. A., and T. Van den Boogaard. 1978. *Weights of information in impression formation*. (Report MT 001–78): Leiden, The Netherlands.

Van IJzendoorn, M. H., L. W. C. Tavecchio, F. A. Goossens, M. M. Vergeer, and J. Swaan. 1983. How B is B4? Attachment and security of Dutch children in Ainsworth's strange situation and at home. *Psychological Reports* 52:683–91.

Waters, E. 1978. The reliability and stability of individual differences in infant-mother attachment. *Child Development* 49:483–94.

Young, F. W. 1981. Quantitative analysis of qualitative data. *Psychometrika* 46:357–88.

4

Tucker's Three-Mode Factor Analysis Model

Bruce Bloxom

There are two important problems that often face a data analyst when confronted by a data matrix of responses of K subjects to L measures: (a) to assess what variables are latent in the measures, and (b) to assess the nature of the association among those latent variables.

An early approach to these problems is found in the development of factor analysis. In that analysis, the association among and much of the variance of the L measures is attributed to their linear regression on a relatively small number ($R < L$) of common factors. When the $L \times R$ matrix of regression coefficients (the factor pattern matrix) exhibits simple structure, the common factors may be interpreted in terms of latent psychological response variables. Then, the direction and magnitude of linear association among those variables may be inferred from correlations among the common factors.

The problem of studying latent variables becomes more complicated if the L measures have a crossed design; for instance, if I ratings of K subjects are made by J raters or sources on M occasions under each of N sets of instruction, L then equals $I \times J \times M \times N$. In this situation, the data matrix can have many more modes or facets than the two modes that are handled by factor analysis. Latent variables can then exist with respect to each mode separately and in combination with other modes. Simply to regress all L measures directly on a small number of common factors does not necessarily indicate the nature of the latent variables connected with each of the modes.

A number of early approaches to this problem with multimode matrices can be found in Harris (1963). One of those approaches, Tucker's three-mode factor analysis (1963, 1964, 1966), has generated a considerable amount of research in psychometrics, as is evidenced in part by the contents and citations of other chapters in this book. For this reason, this chapter focuses on Tucker's approach, although references will be made to some other important approaches as well.

Tucker's approach was originally formulated for a data matrix with three modes—I ratings of K subjects by J sources, resulting in $L = I \times J$ measures of the K subjects. Although the three-mode analysis can be generalized to accommodate a greater number of modes, the present discussion will be restricted to the three-mode case. Points that are made here can be extended to apply when more than three modes are present.

When discussing Tucker's approach, it is particularly convenient to think of it as a model or composition rule that is a special case of the composition rule used in hierarchical factor analysis (Harman 1967, 334). In hierarchical factor analysis, correlation among the common factors is attributed to the regression of those factors on a smaller number of higher-order (second-order) factors; furthermore, correlation among those second-order factors may be attributed to their regression on an even smaller number of third-order factors. One way to use hierarchical factor analysis in the analysis of three-mode data is to place particular constraints on the factor pattern matrices such that first-order common factors include latent variables associated with one mode, second-order common factors include latent variables associated with a second mode, and third-order common factors include latent variables associated with a third mode. Tucker's approach is a special case of this hierarchical approach in which additional constraints are placed on the first- and second-order factor pattern matrices. The following sections of this chapter illustrate why this is so and discuss its implications for the use of Tucker's approach.

It should be noted that this use of hierarchical factor analysis in the exposition of Tucker's three-mode factor analysis is not the same expository approach used by Tucker. The particular expository approach used here was chosen because it leads to a specification of a set of constraints that are implicit in the use of Tucker's formulation. Also, it provides a conceptual framework for suggesting alternatives to that formulation.

The following sections will focus on that aspect of Tucker's composition rule that combines the components associated with each of the three modes of the data matrix. An important problem that will not be discussed here is the treatment of residuals, those parts of the L measures that are in addition to components generated by the common factors. If the residuals are assumed to be independent of each other and of the common factors, Tucker's approach yields a composition rule that, like the common factor analysis model, can have a statistically falsifiable covariance structure (Bentler and Lee 1979). If, however, one is interested in minimizing the sum of squared residuals (see Kroonenberg and de Leeuw 1980; Sands and Young 1980), Tucker's approach leads to a kind of three-way component analysis. These and other treatments of residuals, as well as methods of parameter estimation, are discussed in other chapters of this book.

EXPOSITION OF TUCKER'S COMPOSITION RULE

We will begin by considering an artificial example from personality assessment. Imagine $J = 4$ Likert scale ratings (using the descriptors "good-natured," "outgoing," "energetic," and "resourceful") of $K = 20$ persons by $I = 3$ sources (a supervisor at work, a peer at work, and a neighbor). One way to arrange the data would be to place it in a parallelopiped in which I matrices of J columns and K rows are stacked back to back. However, a more convenient arrangement of the data for a hierarchical analysis of the kind to be discussed here is to place the data in a two-way matrix of K column supervectors, with each column containing IJ-element subvectors. The notation for one of the K data supervectors is $\mathbf{y}_k$. The first part of Table 4–1 indicates one way in which the $J = 4$ ratings by the $I = 3$ sources could be arranged to form that supervector.

With each subject's data arranged in such a supervector, we can construct a set of three hierarchical factor pattern matrices, each of which is associated with one of the modes of the data. The first part of Table 4–1 shows a factor pattern matrix, $\mathbf{B}^*$, which regresses the $I \times J = 12$ measures, $\mathbf{y}_k$, on a set of $I \times Q = 6$ first-order factors, $\mathbf{z}_k$, associated with the rating mode. The important feature of $\mathbf{B}^*$ is that for each element of one mode (the source of the ratings), there is a pattern submatrix, $\mathbf{B}_i$, which regresses the J elements of the other mode (the ratings) on a set of $Q = 2$ common factors (traits) for that second mode. All other submatrices of $\mathbf{B}^*$ are constrained to be null matrices.

A useful but unnecessary feature of the submatrices, $\mathbf{B}_i$, in Table 4–1 is that they contain elements equal to zero, which provide simple structure in the $\mathbf{B}_i$ and hence an easier interpretation of the first-order factors. If each $\mathbf{B}_i$ has the same number of columns and if the pattern of zero elements in each $\mathbf{B}_i$ is invariant across all $\mathbf{B}_i$, then simple structure is invariant across i (the sources). This in turn implies that the rating factors (traits) could have the same interpretation across sources.

The second part of Table 4–1 also shows a factor pattern matrix, $\mathbf{A}^*$, which regresses the $I \times Q = 6$ first-order factors, $\mathbf{z}_k$, on a set of $P \times Q = 4$ second-order factors, $\mathbf{w}_k$, associated with the source mode. The important feature of $\mathbf{A}^*$ is that for each of the Q common factors already associated with one mode (the ratings), there is a set of factor pattern coefficients, which regress the I elements of the other mode (the sources) on a set of $P = 2$ common factors (interpersonal contexts) for that second mode. These pattern coefficients are the diagonal elements of the submatrices $\mathbf{A}_{ip}$ in $\mathbf{A}^*$. The first diagonal element in $\mathbf{A}_{ip}$ is the pattern coefficient for regressing source i on context p for the first trait; for example, the first element of $\mathbf{A}_{11}$ is the weight given to the *supervisor's* ($i = 1$) *perception* of the *subject's friendliness* ($q = 1$) at work ($p = 1$). All off-diagonal elements of the submatrices, $\mathbf{A}_{ip}$ are constrained to be zero.

A useful but unnecessary feature of the $\mathbf{A}_{ip}$ matrices in Table 4–1 is that some of their diagonal elements are zero. The pattern of these zeros provides simple structure and hence can facilitate the interpretation of the second-order factors by showing where a source gives weight to a given context when evaluating subjects on a given trait. If the patterns of zero elements in the diagonals of the $\mathbf{A}_{ip}$ matrices are invariant within those submatrices, as is also the case here, there is an invariant simple structure across the $Q = 2$ traits. This, in turn, implies that the source factors could have the same interpretation across the traits (they are being used similarly across the traits).

Finally, the third part of Table 4–1 shows a factor pattern matrix, $\mathbf{G}$, which regresses the $P \times Q = 4$ second-order factors, $\mathbf{w}_k$, on a set of $R = 2$ third-order factors, $\mathbf{x}_k$. The third-order factors are those that are associated with the third mode (subjects) and in this example can be called *disposition factors.* No elements of $\mathbf{G}$ are constrained to be equal to zero. However, if some elements of $\mathbf{G}$ are zero, as is the case in Table 4–1, that matrix can have a simple structure, which provides an easier interpretation of the third-order factors by showing where weight is given to each disposition (subject factor) in determining the traits (rating factors) in each context (source factor). As an illustration, the pattern of nonzero coefficients in $\mathbf{G}$ in Table 4–1 shows that the first disposition, $\mathbf{x}_{1k}$, is given weight in predicting the tendency to be seen as friendly and diligent at work as well as diligent but not necessarily friendly by peers. A person with such a tendency might be considered to be oriented toward vocational achievement.

Tucker's composition rule for three-mode data is a special case of the hierarchical factor composition rule just described. In addition to constraining particular elements of the pattern matrices, $\mathbf{A}^*$ and $\mathbf{B}^*$, to be equal to zero, it places equality constraints on the nonzero elements of those matrices. These constraints are illustrated in Table 4–2.

The equality constraint placed on $\mathbf{B}^*$ is that each of the $\mathbf{B}_i$ submatrices equals each of the other $\mathbf{B}_i$ submatrices. In other words, the regression of elements of one mode (ratings) on the common factors of that mode (traits) is invariant across the elements of the other mode (sources). By using this constraint, we can represent $\mathbf{B}^*$, as the right Kronecker product of a I-dimensional identity matrix and the invariant submatrix $\mathbf{B}$, which equals $\mathbf{B}_i$ for all i. This is illustrated in the first part of Table 4–2.

The equality constraint placed on $\mathbf{A}^*$ is that the Q diagonal elements are equal in each diagonal submatrix, $\mathbf{A}^*_{ip}$. In other words, the regression of elements of one mode (sources) on the common factors of that mode (contexts) is invariant across the common factors (traits) of the other mode. By using this constraint, we can write each $\mathbf{A}_{ip}$ as a scalar multiple of a Q-dimensional identity matrix. Then, $\mathbf{A}^*$ is the right Kronecker product of a $I \times P$ pattern matrix $\mathbf{A} = \{a_{ip}\}$ and a Q-dimensional identity matrix. This is illustrated in the second part of Table 4–2.

When these equality constraints are placed on $\mathbf{A}^*$ and $\mathbf{B}^*$, the third-order factor pattern matrix, $\mathbf{G}$, becomes what Tucker termed the *core* matrix. Because $\mathbf{G}$ is a matrix of regression coefficients, only under certain conditions can it be interpreted as indicating the correlations between the second-order factors (traits in contexts) and the third-order factors (dispositions). Specifically, the elements of $\mathbf{G}$ are correlation coefficients if the second- and third-order factors have means equal to zero and variances equal to 1.0 and if the third-order factors are uncorrelated with each other. An implication of this point will be indicated in the next section.

TABLE 4-1. A Hierarchical Composition Rule for Three-Mode Data

Estimated Rating			Source: Trait	
By supervisor:	good-natured	$\hat{y}_{11k}$	Supervisor:	friendly
	outgoing	$\hat{y}_{12k}$		diligent
	energetic	$\hat{y}_{13k}$	Peer at work:	friendly
	resourceful	$\hat{y}_{14k}$		diligent
By peer at work:	good-natured	$\hat{y}_{21k}$	Neighbor:	friendly
	outgoing	$\hat{y}_{22k}$		diligent
	energetic	$\hat{y}_{23k}$		
	resourceful	$\hat{y}_{24k}$		
By neighbor:	good-natured	$\hat{y}_{31k}$		
	outgoing	$\hat{y}_{32k}$		
	energetic	$\hat{y}_{33k}$		
	resourceful	$\hat{y}_{34k}$		

$$
\begin{bmatrix} \hat{y}_{11k} \\ \hat{y}_{12k} \\ \hat{y}_{13k} \\ \hat{y}_{14k} \\ \hat{y}_{21k} \\ \hat{y}_{22k} \\ \hat{y}_{23k} \\ \hat{y}_{24k} \\ \hat{y}_{31k} \\ \hat{y}_{32k} \\ \hat{y}_{33k} \\ \hat{y}_{34k} \end{bmatrix}
=
\left[\begin{array}{cc|cc|cc}
.9 & .0 & & & & \\
.6 & .0 & \multicolumn{2}{c|}{\underline{0}} & \multicolumn{2}{c}{\underline{0}} \\
.0 & .7 & & & & \\
.0 & .8 & & & & \\
\hline
 & & .8 & .0 & & \\
\multicolumn{2}{c|}{\underline{0}} & .7 & .0 & \multicolumn{2}{c}{\underline{0}} \\
 & & .0 & .9 & & \\
 & & .0 & .6 & & \\
\hline
 & & & & .6 & .0 \\
\multicolumn{2}{c|}{\underline{0}} & \multicolumn{2}{c|}{\underline{0}} & .8 & .0 \\
 & & & & .0 & .7 \\
 & & & & .0 & .9
\end{array}\right]
\begin{bmatrix} z_{11k} \\ z_{12k} \\ z_{21k} \\ z_{22k} \\ z_{31k} \\ z_{32k} \end{bmatrix}
$$

$$
\hat{\underline{y}}_k = \begin{bmatrix} \underline{B}_1 & \underline{0} & \underline{0} \\ \underline{0} & \underline{B}_2 & \underline{0} \\ \underline{0} & \underline{0} & \underline{B}_3 \end{bmatrix} \underline{z}_k = \underline{B}^* \underline{z}_k
$$

where the $\underline{B}_i$ have elements b_{ijk}

Source: Trait	
Supervisor:	friendly
	diligent
Peer at work:	friendly
	diligent
Neighbor:	friendly
	diligent

$$\begin{bmatrix} z_{11k} \\ z_{12k} \\ z_{21k} \\ z_{22k} \\ z_{31k} \\ z_{32k} \end{bmatrix} = \left[\begin{array}{cc|cc} .7 & .0 & .0 & .0 \\ .0 & .9 & .0 & .0 \\ \hline .8 & .0 & .6 & .0 \\ .0 & .7 & .0 & .7 \\ \hline .0 & .0 & .9 & .0 \\ .0 & .0 & .0 & .6 \end{array}\right] \begin{bmatrix} w_{11k} \\ w_{12k} \\ w_{21k} \\ w_{22k} \end{bmatrix}$$

Context: Trait	
At work:	friendly
	diligent
With peers:	friendly
	diligent

$$\underline{z}_k = \begin{bmatrix} \underline{A}_{11} & \underline{A}_{12} \\ \underline{A}_{21} & \underline{A}_{22} \\ \underline{A}_{31} & \underline{A}_{32} \end{bmatrix} \underline{w}_k = \underline{A}^* \underline{w}_k$$

where all $\underline{A}_{ip}$ are diagonal with elements a_{ipq}

Context: Trait	
At work:	friendly
	diligent
With peers:	friendly
	diligent

$$\begin{bmatrix} w_{11k} \\ w_{12k} \\ w_{21k} \\ w_{22k} \end{bmatrix} = \begin{bmatrix} .5 & .8 \\ .7 & .0 \\ .0 & .8 \\ .6 & .7 \end{bmatrix} \begin{bmatrix} x_{1k} \\ x_{2k} \end{bmatrix}$$

Disposition
Vocational-achievement oriented
Community-service oriented

$$\underline{w}_k = \underline{G}\ \underline{x}_k$$

Summary of Composition Rule

$$\underline{y}_k = \hat{\underline{y}}_k + \underline{e}_k = \underline{B}^* \underline{A}^* \underline{G}\ \underline{x}_k + \underline{e}_k$$

$$y_{ijk} = \sum_q b_{ijq} \sum_p a_{ipq} \sum_r x_{kr}\ g_{pqr} + e_{ijk}$$

TABLE 4-2. Tucker's Composition Rule for Three-Mode Data

Estimated Rating:

$$\hat{\underline{y}}_k = \left[\begin{array}{cc|cc|cc} .9 & .0 & & & & \\ .7 & .0 & \multicolumn{2}{c|}{\underline{0}} & \multicolumn{2}{c}{\underline{0}} \\ .0 & .6 & & & & \\ .0 & .8 & & & & \\ \hline & & .9 & .0 & & \\ \multicolumn{2}{c|}{\underline{0}} & .7 & .0 & \multicolumn{2}{c}{\underline{0}} \\ & & .0 & .6 & & \\ & & .0 & .8 & & \\ \hline & & & & .9 & .0 \\ \multicolumn{2}{c|}{\underline{0}} & \multicolumn{2}{c|}{\underline{0}} & .7 & .0 \\ & & & & .0 & .6 \\ & & & & .0 & .8 \end{array}\right] \underline{z}_k \quad : \quad \underline{\text{Source: Trait}}$$

$$\hat{\underline{y}}_k = \begin{bmatrix} \underline{B} & \underline{0} & \underline{0} \\ \underline{0} & \underline{B} & \underline{0} \\ \underline{0} & \underline{0} & \underline{B} \end{bmatrix} \underline{z}_k = \underline{B}^* \underline{z}_k$$

$$\hat{\underline{y}}_k = \left[\begin{pmatrix} 1.0 & .0 & .0 \\ .0 & 1.0 & .0 \\ .0 & .0 & 1.0 \end{pmatrix} \otimes \underline{B} \right] \underline{z}_k$$

$$= [\underline{I}_{(3)} \otimes \underline{B}]\, \underline{z}_k$$

Source: Trait:

$$\underline{z}_k = \left[\begin{array}{cc|cc} .8 & .0 & .0 & .0 \\ .0 & .8 & .0 & .0 \\ \hline .5 & .0 & .7 & .0 \\ .0 & .5 & .0 & .7 \\ \hline .0 & .0 & .6 & .0 \\ .0 & .0 & .0 & .6 \end{array}\right] \underline{w}_k \quad : \quad \underline{\text{Context: Trait}}$$

$$\underline{z}_k = \begin{bmatrix} a_{11} \begin{pmatrix} 1.0 & .0 \\ .0 & 1.0 \end{pmatrix} & a_{12} \begin{pmatrix} 1.0 & .0 \\ .0 & 1.0 \end{pmatrix} \\ a_{21} \begin{pmatrix} 1.0 & .0 \\ .0 & 1.0 \end{pmatrix} & a_{22} \begin{pmatrix} 1.0 & .0 \\ .0 & 1.0 \end{pmatrix} \\ a_{31} \begin{pmatrix} 1.0 & .0 \\ .0 & 1.0 \end{pmatrix} & a_{32} \begin{pmatrix} 1.0 & .0 \\ .0 & 1.0 \end{pmatrix} \end{bmatrix} \underline{w}_k$$

$$= [\underline{A} \otimes \underline{I}_{(2)}]\, \underline{w}_k = \underline{A}^* \underline{w}_k$$

Summary of Composition Rule

$$\begin{aligned} \underline{y}_k = \hat{\underline{y}}_k + \underline{e}_k &= [\underline{I}_{(3)} \otimes \underline{B}]\, [\underline{A} \otimes \underline{I}_{(2)}]\, \underline{G}\, \underline{x}_k + \underline{e}_k \\ &= [(\underline{I}_{(3)}\, \underline{A}) \otimes (\underline{B}\, \underline{I}_{(2)})]\, \underline{G}\, \underline{x}_k + \underline{e}_k \\ &= [\underline{A} \otimes \underline{B}]\, \underline{G}\, \underline{x}_k + \underline{e}_k \\ y_{ijk} &= \sum_p \sum_q \sum_r a_{ip}\, b_{jq}\, x_{kr}\, g_{pqr} + e_{kjk} \end{aligned}$$

A useful technical consequence of placing these equality constraints on $\mathbf{A}^*$ and $\mathbf{B}^*$ is that the algebra of the hierarchical composition rule in the last part of Table 4–1 becomes simpler. The first- and second-order pattern matrices can be combined into a single matrix, which is the Kronecker product of $\mathbf{A}$ and $\mathbf{B}$, as is indicated in Table 4–2. This result follows from a property of Kronecker products: $[\mathbf{A} \otimes \mathbf{C}]\ [\mathbf{D} \otimes \mathbf{B}] = [(\mathbf{AD}) \otimes (\mathbf{CB})]$, assuming that $\mathbf{A}$ has as many columns as $\mathbf{D}$ has rows and that $\mathbf{C}$ has as many columns as $\mathbf{B}$ has rows.

THE USE OF TUCKER'S COMPOSITION RULE

Implications of the Equality Constraints

The equality constraints that result in the Kronecker product representation of the first- and second-order pattern matrices are important constraints because when they are utilized, the priority of modes in the first two stages of the factor hierarchy becomes arbitrary. To illustrate this point, we begin by considering Table 4–3, which shows a set of hierarchical factor matrices regressing the sources of ratings on the source factors first ($\mathbf{A}^*$) and regressing the ratings on the rating factors second ($\mathbf{B}^*$). This is opposite to the priority of the rating factors and source factors in Table 4–1. Now, if the same equality constraints are placed on $\mathbf{A}^{**}$ and $\mathbf{B}^{**}$ in Table 4–3 as were placed on $\mathbf{A}^*$ and $\mathbf{B}^*$ in Table 4–1 to produce Table 4–2, the result is the set of pattern matrices shown in Table 4–4. The product of those pattern matrices, $[\mathbf{A} \otimes \mathbf{B}]$, is the same as the product of pattern matrices obtained in Table 4–2. This is a nontrivial result because Tables 4–1 and 4–3—in which the equality constraints are not placed on the nonzero pattern coefficients—depict factor pattern matrices that are very different from each other. In Table 4–1, $\mathbf{A}^*$ regresses sources on source factors for each rating factor, but in Table 4–3, $\mathbf{A}^{**}$ regresses sources on source factors for each rating per se. In Table 4–1, $\mathbf{B}^*$ regresses ratings on rating factors for each source per se, but in Table 4–3, $\mathbf{B}^{**}$ regresses ratings on rating factors for each source factor. Such distinctions are inconsequential in Table 4–2 and Table 4–4, in which the equality constraints are used.

This arbitrariness of hierarchical ordering in Tucker's approach suggests an advantage this method has over a method in which the equality constraints are not used. The interchangeability of modes associated with the first- and second-order factors means that the elements of $\mathbf{A}$, $\mathbf{B}$, and $\mathbf{G}$ are independent of the mode associated with each of these two levels of factoring. In contrast, if the equality constraints are not used, such independence is not guaranteed. Furthermore, the decision about which mode should be associated with each level of factoring may be difficult to judge from the substantive nature of the measures being analyzed. For example, with the measures used in Table 4–1 and Table 4–3, it is not clear how one would deduce whether the source factors should be defined for each rating factor (trait) or for each rating per se.

TABLE 4-3. An Alternative Hierarchical Composition Rule

Estimated Rating:

$$\underline{\ell}_k = \left[\begin{array}{cccc|cccc} .7 & .0 & .0 & .0 & .0 & .0 & .0 & .0 \\ .0 & .8 & .0 & .0 & .0 & .0 & .0 & .0 \\ .0 & .0 & .9 & .0 & .0 & .0 & .0 & .0 \\ .0 & .0 & .0 & .8 & .0 & .0 & .0 & .0 \\ \hline .8 & .0 & .0 & .0 & .6 & .0 & .0 & .0 \\ .0 & .6 & .0 & .0 & .0 & .7 & .0 & .0 \\ .0 & .0 & .9 & .0 & .0 & .0 & .7 & .0 \\ .0 & .0 & .0 & .7 & .0 & .0 & .0 & .6 \\ \hline .0 & .0 & .0 & .0 & .9 & .0 & .0 & .0 \\ .0 & .0 & .0 & .0 & .0 & .7 & .0 & .0 \\ .0 & .0 & .0 & .0 & .0 & .0 & .6 & .0 \\ .0 & .0 & .0 & .0 & .0 & .0 & .0 & .6 \end{array}\right] \begin{bmatrix} z_{11k} \\ z_{12k} \\ z_{13k} \\ z_{14k} \\ z_{21k} \\ z_{22k} \\ z_{23k} \\ z_{24k} \end{bmatrix}$$

Context: Rating

At work:	good-natured
	outgoing
	energetic
	resourceful
With peers:	good-natured
	outgoing
	energetic
	resourceful

$$\underline{\ell}_k = \underline{A}^{**}\, \underline{z}_k$$

Context: Rating

$$\underline{z}_k = \left[\begin{array}{cc|cc} .9 & .0 & & \\ .6 & .0 & \multicolumn{2}{c}{\underline{0}} \\ .0 & .7 & & \\ .0 & .8 & & \\ \hline & & .6 & .0 \\ \multicolumn{2}{c|}{\underline{0}} & .7 & .0 \\ & & .0 & .9 \\ & & .0 & .9 \end{array}\right] \begin{bmatrix} w_{11k} \\ w_{12k} \\ w_{21k} \\ w_{22k} \end{bmatrix}$$

Context: Trait

At work:	friendly
	diligent
With peers:	friendly
	diligent

$$\underline{z}_k = \underline{B}^{**}\, \underline{w}_k$$

Summary of Composition Rule

$$\underline{y}_k = \underline{\ell}_k + \underline{e}_k = \underline{A}^{**}\, \underline{B}^{**}\, \underline{G}\, \underline{x}_k + \underline{e}_k$$

$$y_{ijk} = \sum_p a_{ijp} \sum_q b_{jpq} \sum_r x_{kr}\, g_{pqr} + e_{ijk}$$

Although the preceding discussion indicates how using the invariance constraints implicit in Tucker's model enables one to arbitrarily interchange two modes of the hierarchy, the interchangeability of modes is even more general than is indicated in Table 4–2 and Table 4–4. Using scalar notation to represent Tucker's composition rule, as is done at the bottom of Table 4–2, one can define six possible and interchangeable hierarchies of factors of the three modes, as is indicated in Table 4–5. This is because of the permutation of summation operations that is allowed under the subscripts employed in the composition rule. Thus, the choice among these six hierarchical expressions is arbitrary.

In spite of this theoretical interchangeability of hierarchies, the choice among hierarchies is not totally arbitrary in some

TABLE 4-4. An Alternative Arrangement of Tucker's Composition Rule

---

Estimated Rating:

$$\hat{\underline{y}}_k = \left[\begin{array}{cccc|cccc}
.8 & .0 & .0 & .0 & .0 & .0 & .0 & .0 \\
.0 & .8 & .0 & .0 & .0 & .0 & .0 & .0 \\
.0 & .0 & .8 & .0 & .0 & .0 & .0 & .0 \\
.0 & .0 & .0 & .8 & .0 & .0 & .0 & .0 \\
\hline
.5 & .0 & .0 & .0 & .7 & .0 & .0 & .0 \\
.0 & .5 & .0 & .0 & .0 & .7 & .0 & .0 \\
.0 & .0 & .5 & .0 & .0 & .0 & .7 & .0 \\
.0 & .0 & .0 & .5 & .0 & .0 & .0 & .7 \\
\hline
.0 & .0 & .0 & .0 & .6 & .0 & .0 & .0 \\
.0 & .0 & .0 & .0 & .0 & .6 & .0 & .0 \\
.0 & .0 & .0 & .0 & .0 & .0 & .6 & .0 \\
.0 & .0 & .0 & .0 & .0 & .0 & .0 & .6
\end{array}\right] \underline{z}_k \quad : \text{Context: Rating}$$

$$\hat{\underline{y}}_k = [\underline{A} \otimes \underline{I}_{(4)}]\, \underline{z}_k$$

Context: Rating:

$$\underline{z}_k = \left[\begin{array}{cc|cc}
.9 & .0 & & \\
.7 & .0 & \multicolumn{2}{c}{\underline{0}} \\
.0 & .6 & & \\
.0 & .8 & & \\
\hline
 & & .9 & .0 \\
\multicolumn{2}{c|}{\underline{0}} & .7 & .0 \\
 & & .0 & .6 \\
 & & .0 & .8
\end{array}\right] \underline{w}_k \quad : \text{Context: Trait}$$

$$\underline{z}_k = [\underline{I}_{(2)} \otimes \underline{B}]\, \underline{w}_k$$

Summary of Composition Rule

$$\begin{aligned}
\underline{y}_k &= \hat{\underline{y}}_k + \underline{e}_k = [\underline{A} \otimes \underline{I}_{(4)}]\, [\underline{I}_{(2)} \otimes \underline{B}]\, \underline{G}\, \underline{x}_k + \underline{e}_k \\
&= [(\underline{A}\, \underline{I}_{(2)}) \otimes (\underline{I}_{(4)}\, \underline{B})]\, \underline{G}\, \underline{x}_k + \underline{e}_k \\
&= [\underline{A} \otimes \underline{B}]\, \underline{G}\, \underline{x}_k + \underline{e}_k
\end{aligned}$$

---

applications of the model. One such application is where one of the modes—the subject mode in the present example—has elements that are chosen at random from a population of elements and where a covariance structure for the population of elements is then being estimated. In such a circumstance, it is notationally convenient to let the component that is associated with that mode form the third-order factor scores and to let the components associated with the other two modes form the factor pattern matrices in the Kronecker products shown in Table 4–2. This approach to defining the model is illustrated in papers by Bloxom (1968) and Bentler and Lee (1979).

A second kind of application in which the choice among the six hierarchies in Table 4–5 is not arbitrary is where the nonzero elements of **G**, the core matrix, are correlation coefficients. In such an application, it needs to be noted that the magnitude of

TABLE 4-5. Six Hierarchical Arrangements of Tucker's Composition Rule

1.*	$y_{ijk} = \sum_p a_{ip} \sum_q b_{jq} \sum_r x_{kr}\, g_{pqr} + e_{ijk}$
2.	$y_{ijk} = \sum_p a_{ip} \sum_r x_{kr} \sum_q b_{jq}\, g_{pqr} + e_{ijk}$
3.**	$y_{ijk} = \sum_q b_{jq} \sum_p a_{ip} \sum_r x_{kr}\, g_{pqr} + e_{ijk}$
4.	$y_{ijk} = \sum_q b_{jq} \sum_r x_{kr} \sum_p a_{ip}\, g_{pqr} + e_{ijk}$
5.	$y_{ijk} = \sum_r x_{kr} \sum_p a_{ip} \sum_q b_{jq}\, g_{pqr} + e_{ijk}$
6.	$y_{ijk} = \sum_r x_{kr} \sum_q b_{jq} \sum_p a_{ip}\, g_{pqr} + e_{ijk}$

Note: *See example in Table 4-4; **see example in Table 4-2.

the coefficients depends on which of the three components—x_{kr}, a_{ip}, or b_{jq}—forms the third-order factor scores. That component is the one that is being correlated with the second-order factor scores. The other two components are functioning as regression coefficients and not as variables being correlated. Thus, interpreting the elements of **G** as correlation coefficients requires an a priori determination of the mode of the data over which dispersion is being defined. As noted earlier, the second- and third-order factor scores need to have means equal to zero and variances equal to one, and the third-order factor scores need to be uncorrelated.

Once a factor hierarchy has been selected, albeit arbitrarily, the constraints in Tucker's model shown in Table 4–2 should not be accepted uncritically, without considering their appropriateness for the application at hand. One concern is that the constraints require invariance in the number of factors (P and Q) associated with each of the first two modes in the hierarchy and in the simple structures of the pattern matrices, **A** and **B**, associated with those modes. These two constraints may be difficult to justify or to support empirically in some applications. For example, invariance of simple structure would not hold in the present example if supervisors' ratings had a halo-effect factor (weighted for all four ratings) and a diligence factor (weighted for only two ratings) while the ratings by peers at work and by neighbors had friendliness and diligence factors (weighted for two ratings each, as in Table 4–2).

Another concern about the appropriateness of Tucker's model for some applications is the stringency of the equality constraints, or the exact equalities of the nonzero elements of the $\mathbf{B}_i$ matrices and of the diagonal elements of the $\mathbf{A}_{ip}$ matrices. At the very least, as Tucker (1966) noted, careful attention needs to be given to the metric of the elements of each mode as they are measured across the elements of the other two modes. This is important because the pattern matrices, **A** and **B**, contain regression coefficients, the invariance of which depends on the use of the same units of measurement. In the present example, if different rating scales were used by the different raters, or if the units of measurement of those scales were altered differently across raters, the use of the equality constraints would no longer be clearly defensible. This concern would be particularly germane if the $I \times J = 12$ ratings were rescaled to give them all the same variance, as would be done when computing a multitrait-multimethod correlation matrix (Campbell and Fiske 1959).

In addition to attending to the units of measurement, conditions that are still unknown may need to be considered to fully justify the use of the equality constraints used in Table 4–2. Previous work on factorial invariance (Meredith 1964; Bloxom 1972) described sets of sufficient conditions for the equality of factor pattern matrices across nonoverlapping, uncorrelated subpopulations (for instance, across different groups of subjects). There has not been a development of corresponding sets of sufficient conditions for the analysis of multimode data in which observations are correlated across subpopulations (sources as well as within subpopulations). Thus, finding sufficient conditions for equality constraints in this situation requires further work.

Alternatives to Tucker's Composition Rule

In instances in which the constraints in Tucker's model are too stringent, it may still be possible to conduct an analysis using some subset of those constraints. For example, the equality constraints may still be appropriate in $\mathbf{A}^*$ or $\mathbf{B}^*$, even when they are not appropriate for both of those pattern matrices. This could occur when the measures have been rescaled differently across the elements of one of the modes, so that the equality constraints would not hold in the first-order factor pattern matrix ($\mathbf{B}^*$ in Table 4–1) but could still hold in the second-order factor pattern matrix ($\mathbf{A}^*$ in Table 4–2).

If the equality constraints are too stringent for both $\mathbf{A}^*$ and $\mathbf{B}^*$, a hierarchical composition such as that shown in Table 4–1 or Table 4–3 can still be considered, with or without invariance in the number of factors (the number of columns of $\mathbf{B}_i$) or invariance of simple structure (zero elements in the $\mathbf{B}_i$ matrices). Furthermore, if there is not a small number of latent variables underlying one of the modes (sources), the hierarchical composition rule can be simplified by eliminating one of the pattern matrices (such as $\mathbf{A}^*$) by assuming that it is an identity matrix. Such a composition rule is implicit in Horst's oblique maximum variance method for factoring a number of sets of measures (1965, 512).

In the course of considering the hierarchical composition rules in Tables 4–1, 4–2, 4–3, and 4–4, it is important not to overlook the possible use of more parsimonious composition rules. One such rule is a first-order factor composition rule (Jöreskog 1974), in which a congeneric test theory model is used to define common factors associated with one mode in addition to common factors associated with another mode. A composition rule of this type for the present example is illustrated in the first pattern matrix, **C** *, in Table 4–6. This example shows one common factor for each element of each of two modes. Jöreskog suggested that such a composition rule could be very useful for analyzing a multitrait-multimethod correlation matrix.

Two points should be noted about the use of such an alternative composition rule. First, factors associated with different modes can be either linearly independent or correlated. This stands in contrast to the hierarchical approach, in which factors associated with one mode are necessarily linearly dependent on factors associated with other modes. A second point is that Jöreskog's (1974) first-order composition rule is not restricted to the case in which there is one common factor for each element of each of two modes. An example of an alternative to **C*** in Table 4–6 is provided by **C**** in Table 4–7.

Instead of using Jöreskog's approach as an alternative to Tucker's, one could postulate a first-order factor composition rule, as suggested by Jöreskog, plus a hierarchical factor composition rule. Such a combination of composition rules is shown at the bottom of Table 4–7. If the first-order factor composition rule were based on the pattern matrix **C*** in Table 4–6, it should be noted that the overall composition rule would contain three-mode specific factors, which were suggested by Tucker (1963). In that paper, he noted that a specific factor in a conventional

TABLE 4-6. A First-Order Factor Composition Rule

Estimated Rating:

$$\hat{\underline{y}}_k = \left[\begin{array}{cccc|ccc} .9 & .0 & .0 & .0 & .7 & .0 & .0 \\ .0 & .7 & .0 & .0 & .9 & .0 & .0 \\ .0 & .0 & .8 & .0 & .7 & .0 & .0 \\ .0 & .0 & .0 & .6 & .8 & .0 & .0 \\ \hline .7 & .0 & .0 & .0 & .0 & .8 & .0 \\ .0 & .8 & .0 & .0 & .0 & .6 & .0 \\ .0 & .0 & .9 & .0 & .0 & .9 & .0 \\ .0 & .0 & .0 & .6 & .0 & .8 & .0 \\ \hline .6 & .0 & .0 & .0 & .0 & .0 & .9 \\ .0 & .7 & .0 & .0 & .0 & .0 & .6 \\ .0 & .0 & .7 & .0 & .0 & .0 & .7 \\ .0 & .0 & .0 & .8 & .0 & .0 & .7 \end{array}\right] \left[\begin{array}{c} u_{1k} \\ u_{2k} \\ u_{3k} \\ u_{4k} \\ u_{5k} \\ u_{6k} \\ u_{7k} \end{array}\right]$$

	Factor
u_{1k}	Rated good-natured by all sources
u_{2k}	Rated outgoing by all sources
u_{3k}	Rated energetic by all sources
u_{4k}	Rated resourceful by all sources
u_{5k}	Halo effect in ratings by supervisor
u_{6k}	Halo effect in ratings by peer at work
u_{7k}	Halo effect in ratings by neighbor

$$\hat{\underline{y}}_k = \underline{C}^* \underline{u}_k$$

Summary of Composition Rule

$$\underline{y}_k = \hat{\underline{y}}_k + \underline{e}_k = \underline{C}^* \underline{u}_k + \underline{e}_k$$

TABLE 4-7. An Alternative First-Order Factor Composition Rule

Estimated Rating:

$$\hat{\underline{y}}_k = \begin{bmatrix} .9 & .0 & .7 & .0 \\ .7 & .0 & .9 & .0 \\ .0 & .8 & .7 & .0 \\ .0 & .6 & .8 & .0 \\ \hline .7 & .0 & .8 & .7 \\ .8 & .0 & .6 & .8 \\ .0 & .9 & .6 & .9 \\ .0 & .6 & .7 & .8 \\ \hline .6 & .0 & .0 & .9 \\ .7 & .0 & .0 & .6 \\ .0 & .7 & .0 & .7 \\ .0 & .8 & .0 & .7 \end{bmatrix} \begin{bmatrix} u_{1k} \\ u_{2k} \\ u_{3k} \\ u_{4k} \end{bmatrix}$$

Factor
- Perceived friendly by all sources
- Perceived diligent by all sources
- Halo effect at work
- Halo effect with peers

$$\hat{\underline{y}}_k = \underline{C}^{**} \underline{u}_k$$

Summary of Composition Rule

$$\underline{y}_k = \hat{\underline{y}}_k + \underline{e}_k = \underline{C}^{**} \underline{u}_k + \underline{e}_k$$

A Combined First-Order and Hierarchical Composition Rule

$$\underline{y}_k = \hat{\underline{y}}_k + \underline{e}_k = [\underline{C}^* \underline{u}_k + \underline{B}^* \underline{A}^* \underline{G} \underline{x}_k] + \underline{e}_k$$

two-mode factor analysis may be a common factor in a three-mode analysis. The factor pattern matrix $\mathbf{C}^*$ has a pattern that would be obtained if there were one common factor specific to each of the J rating scales across all sources and one common factor specific to each of the I sources across all rating scales.

Simple Structure Transformations

As discussed earlier, the interpretation of the factor pattern matrices in Tables 4–1, 4–2, 4–3, and 4–4 is easier if the matrices exhibit simple structure. Whether one is using the hierarchical approach (shown in Tables 4–1 and 4–3) or Tucker's approach (shown in Tables 4–2 and 4–4), it is important to know the effect of a simple structure transformation of one pattern matrix on other pattern matrices.

An appealing feature of Tucker's approach is that a simple structure transformation of either the first- or second-order factor pattern matrices can be made without affecting any other matrix besides the core matrix, $\mathbf{G}$ (Tucker 1966). This is not generally the case with the hierarchical approach illustrated in Table 4–1 and Table 4–3. Table 4–8 shows how a $P \times P$ non-singular matrix $\mathbf{T}$ (and its inverse $\mathbf{T}^{-1}$) can be used to transform $\mathbf{A}$ and, therefore, the first-order factor pattern matrix $[\mathbf{A} \otimes \mathbf{I}_{(J)}]$ in Table 4–4 with the inverse affecting only $\mathbf{G}$. The matrix $\mathbf{A}^+$ is $\mathbf{A}$ after it has been transformed to simple structure. The matrix $\mathbf{G}^+$ is $\mathbf{G}$ after it has been premultiplied by $[\mathbf{T}^{-1} \otimes \mathbf{I}_{(J)}]$. Note

TABLE 4-8. A Simple Structure Transformation with Tucker's Composition Rule

$$
\begin{aligned}
\hat{\underline{\Sigma}}_k &= [\underline{A} \otimes \underline{I}_{(J)}]\, [\underline{I}_{(P)} \otimes \underline{B}]\, \underline{G}\, \underline{x}_k \\
&= [\underline{A} \otimes \underline{I}_{(J)}]\, [\underline{T} \otimes \underline{I}_{(J)}]\, [\underline{T}^{-1} \otimes \underline{I}_{(J)}]\, [\underline{I}_{(P)} \otimes \underline{B}]\, \underline{G}\, \underline{x}_k \\
&= [\underline{AT} \otimes \underline{I}_{(J)}]\, [\underline{T}^{-1} \otimes \underline{B}]\, \underline{G}\, \underline{x}_k \\
&= [\underline{AT} \otimes \underline{I}_{(J)}]\, [\underline{I}_{(P)} \otimes \underline{B}]\, [\underline{T}^{-1} \otimes \underline{I}_{(Q)}]\, \underline{G}\, \underline{x}_k \\
&= [\underline{A}^{+} \otimes \underline{I}_{(J)}]\, [\underline{I}_{(P)} \otimes \underline{B}]\, \underline{G}^{+}\, \underline{x}_k
\end{aligned}
$$

that the second-order factor pattern matrix $[\mathbf{I}_{(P)} \otimes \mathbf{B}]$ in the last equation is the same as it is in the first equation—it is not affected by the transformation of $\mathbf{A}$. Also note that commuting the multiplication of $[\mathbf{T}^{-1} \otimes \mathbf{I}_{(J)}]$ and the second-order factor pattern matrix is possible under the composition rule shown in Table 4–4 but not under the composition rule shown in Table 4–3, because the pattern matrix in Table 4–3 does not have a Kronecker product (or invariance) structure, which is necessary for commutative multiplication. Thus, transforming the first-order pattern matrix and producing no effect on the second-order pattern matrix is attributable to the invariance constraints on the pattern matrices in Table 4–4.

It is easier to see how a simple structure transformation of the second-order pattern matrix $[\mathbf{I}_{(P)} \otimes \mathbf{B}]$ can be made without affecting the first-order pattern matrix $[\mathbf{A} \otimes \mathbf{I}_{(J)}]$. This is done by postmultiplying $[\mathbf{I}_{(P)} \otimes \mathbf{B}]$ by $[\mathbf{I}_{(P)} \otimes \mathbf{S}]$, where $\mathbf{S}$ is a $Q \times Q$ nonsingular simple structure transformation postmultiplying $\mathbf{B}$. The inverse of the transformation, $[\mathbf{I}_{(P)} \otimes \mathbf{S}^{-1}]$, is then premultiplied into $\mathbf{G}$.

After $\mathbf{T}$ and $\mathbf{S}$ are used to transform $\mathbf{A}$ and $\mathbf{B}$, respectively, to simple structure and after the inverses of those transformations are premultiplied into $\mathbf{G}$, as was just indicated, a $R \times R$ nonsingular matrix, $\mathbf{U}$, may be used on the right of $\mathbf{G}$ to attempt to obtain simple structure in that pattern matrix. To compensate for this transformation, the third-order factor scores, $\mathbf{x}_k$, are premultiplied by $\mathbf{U}^{-1}$. Tucker reported that, in some applications, this approach "met with moderate success" in providing simple structure in $\mathbf{G}$ (1966, 325). He also reported that the alternative of transforming the factor scores, $\mathbf{x}_k$, to simple structure was not "markedly successful" (1966, 309). Furthermore, Bloxom suggested using the approach of transforming $\mathbf{A}$, $\mathbf{B}$, and $\mathbf{G}$ to simple structure (1968) as described here, when the K vectors, $\mathbf{x}_k$, are sampled from a subpopulation defined by multivariate selection. When $\mathbf{x}_k$ is so defined, $\mathbf{A}$, $\mathbf{B}$, and $\mathbf{G}$, but not $\mathbf{x}_k$ or its covariance structure, can be assumed to be invariant under that selection.

More recently, Bentler and Lee (1979) suggested that the transformation $\mathbf{U}$ might be chosen to provide simple structure in the matrix product $[\mathbf{A} \otimes \mathbf{B}]\mathbf{G}$ instead of in $\mathbf{G}$ alone. In considering this suggestion it is important to note that strong simple structure in **A**, **B**, and **G** does not necessarily imply a very interpretable simple structure for the product $[\mathbf{A} \otimes \mathbf{B}]\mathbf{G}$, as can be seen from the pattern matrices and their product given in Table 4–9. Neither should it be assumed that a transformation that provides simple structure in $[\mathbf{A} \otimes \mathbf{B}]\mathbf{G}$ would necessarily provide it in **G** alone, even if **A** and **B** already exhibit such a structure. Thus, choosing **U** to transform $[\mathbf{A} \otimes \mathbf{B}]\mathbf{G}$ to simple structure may result in a very different transformed **G** matrix than if **R** is chosen to transform only **G** to simple structure.

Using **U** to transform the product $[\mathbf{A} \otimes \mathbf{B}]\mathbf{G}$ to simple structure is clearly worth considering if **A** and **B** do not have interpretable patterns, because **G** would not be interpretable even if it exhibited simple structure. However, if **A** and **B** do have interpretable patterns so that the rows of **G** can be interpreted, then the decision between transforming $[\mathbf{A} \otimes \mathbf{B}]\mathbf{G}$ or **G** alone to simple structure is more difficult. In making this decision, it may be useful to consider how the third-order factors, $\mathbf{x}_k$, are to be interpreted. If they are to be interpreted in terms of their regression on the second-order factors, as suggested earlier in this chapter, then trying to obtain simple structure in **G** may facilitate that interpretation. However, if they are to be interpreted in terms of their regression on the original measures, $\mathbf{y}_k$, then trying to obtain simple structure in the product $[\mathbf{A} \otimes \mathbf{B}]\mathbf{G}$ may facilitate that interpretation.

SUMMARY

As an approach to the analysis of three-mode data matrices, Tucker's composition rule has a number of useful features as well as some potential limitations when it is viewed as a special case of

TABLE 4-9. Tucker's Factor Pattern Matrices

$$\underline{A} = \begin{bmatrix} .8 & .0 \\ .5 & .7 \\ .0 & .6 \end{bmatrix} \qquad \underline{B} = \begin{bmatrix} .9 & .0 \\ .7 & .0 \\ .0 & .6 \\ .0 & .8 \end{bmatrix} \qquad \underline{G} = \begin{bmatrix} .5 & .8 \\ .7 & .0 \\ .0 & .8 \end{bmatrix}$$

$[\underline{A} \otimes \underline{B}]\ \underline{G}$ =

		Source:	Rating
.36	.58	Supervisor:	good-natured
.28	.45		outgoing
.34	.00		energetic
.45	.00		resourceful
.23	.86	Peer at work:	good-natured
.18	.67		outgoing
.46	.29		energetic
.62	.39		resourceful
.00	.43	Neighbor:	good-natured
.00	.34		outgoing
.22	.25		energetic
.29	.34		resourceful

a hierarchical factor composition rule. The first-, second-, and third-order factors can successively indicate latent variables that are associated with the three modes of the data matrix. At the same time, strong invariance constraints are placed on the first- and second-order pattern matrices and may not be defensible in some empirical situations. However, when these constraints are justifiable, Tucker's composition rule has marked advantages over a less restricted hierarchical approach. One advantage is that the decision about which modes are to be associated with the first-, second-, and third-order factors may be made arbitrarily, except under specified circumstances. A second advantage is that simple structure transformations of either the first- or second-order factor pattern matrices can be made without affecting the one remaining matrix that is not being transformed.

Whether Tucker's composition rule or an alternative hierarchical rule is used, the rule can be altered to include additional common factors that do not have a hierarchical structure. Such factors can include latent variables that are specific to elements of one mode across all elements of another mode.

No attempt has been made here to discuss alternative treatments of residuals (that portion of a score that is not generated by regression on common factors). Alternative approaches to that subject and to the related subject of parameter estimation are discussed in other chapters of this book. The purpose in this chapter has been to focus on those features of Tucker's model (or composition rule) that are important regardless of how residuals are treated.

REFERENCES

Bentler, P. M., and S-Y. Lee. 1979. A statistical development of three-mode factor analysis. *British Journal of Mathematical and Statistical Psychology* 32:87–104.

Bloxom, B. A note on invariance in three-mode factor analysis. *Psychometrika* 33:347–50.

———. 1972. Alternative approaches to factorial invariance. *Psychometrika* 37:425–40.

Campbell, D. T., and D. W. Fiske. 1959. Convergent and discriminant validation by the multi-method matrix. *Psychological Bulletin* 56:81–105.

Harman, H. H. 1968. *Modern factor analysis.* 2nd ed. Chicago: University of Chicago Press.

Harris, C. W., ed. 1963. *Problems in measuring change.* Madison, Wis: University of Wisconsin Press.

Horst, P. 1965. *Factor analysis of data matrices.* New York: Holt, Rinehart and Winston.

Jöreskog, K. G. 1974. Analyzing psychological data by structural analysis of covariance matrices. In *Measurement, psychophysics and neural information processing,* vol. 2, ed. D. H. Krantz, R. C. Atkinson, R. D. Luce, and P. Suppes. San Francisco: W. H. Freeman and Co.

Kroonenberg, P. M., and J. de Leeuw. 1980. Principal components analysis of three-mode data by means of alternating least-squares algorithms. *Psychometrika* 45:69–97.

Meredith, W. 1964. Notes on factorial invariance. *Psycho-*

metrika 29:177–85.

Sands, R., and F. W. Young. 1980. Component models for three-way data: An alternating least squares algorithm with optimal scaling features. *Psychometrika* 45:39–67.

Tucker, L. R. 1963. Implications of factor analysis of three-way matrices for measurement of change. In *Problems in measuring change,* ed. C. W. Harris. Madison, Wis.: University of Wisconsin Press.

———. 1964. The extension of factor analysis to three-dimensional matrices. In *Contributions to mathematical psychology,* ed. N. Frederiksen and H. Gulliksen. New York: Holt, Rinehart and Winston.

———. 1966. Some mathematical notes on three-mode factor analysis. *Psychometrika* 31:279–311.

5

The PARAFAC Model for Three-Way Factor Analysis and Multidimensional Scaling

Richard A. Harshman and Margaret E. Lundy

INTRODUCTION AND OVERVIEW

Motivation for PARAFAC

One of the oldest and most troublesome issues in factor analysis is the "rotation problem." Because the two-way factor model is inadequately constrained by the data, there are an infinite number of possible solutions consistent with any given data set. These alternatives correspond to different rotations (or more general transformations) of the coordinate axes in the factor space. To overcome this rotational indeterminacy, factor analysts have appealed to special criteria, such as "simple structure," to guide the selection of a preferred solution. But these rotation criteria are not part of the factor model itself, and the additional assumptions involved often seem hard to defend on empirical grounds. A further difficulty arises because of disagreement as

---

We would like to express our appreciation to the many colleagues (especially J. D. Carroll and J. B. Kruskal) whose stimulating comments over the last few years have contributed toward the development of the ideas presented in this chapter and the one that follows. We are particularly indebted to Henry G. Law, who (along with the other editors of this volume) initially stimulated the writing of this article, encouraged us as it was progressing, and showed extraordinary patience and forbearance in seeing it through to its final completion.

Portions of this research were supported by the Natural Sciences and Engineering Research Council of Canada, Grant #A7896. Additional funds were provided by intramural grants from the Faculty of Social Sciences, University of Western Ontario.

to which additional assumptions—or which rotation criteria—are most appropriate for determining the orientation of factors. Since alternative factor rotations sometimes lead to different and conflicting empirical generalizations, the problem of rotation is a serious one for those who seek to apply factor analysis to scientific problems.

PARAFAC is a three-way factor analysis/multidimensional scaling procedure that was developed in order to overcome the rotation problem, at least for certain classes of data. It uses three-way data in order to obtain richer information about the underlying factors than is possible with two-way data, thus providing an empirical basis for determining the "true" factor axes. The method generalizes Cattell's idea of parallel proportional profiles (1944), which is described below, into a particular strong three-way model of factor variation. (This model is closely related to the CANDECOMP model of Carroll and Chang 1970, which is also described below.) When the PARAFAC model is fit to data, a unique orientation of the factor axes is obtained as a direct consequence of determining the best-fitting factors for a given data set. Thus, if the model is appropriate, it eliminates the need for an additional factor rotation process based on controversial factor orientation criteria; the location of axes is an *intrinsic characteristic* of the factor solution itself.

Theoretical arguments (to be discussed below) indicate that intrinsic axis solutions provide a stronger basis for discovery of empirically meaningful factors than is typically provided by other criteria, such as principal components orientation, rotation to simple structure, and so forth. These arguments seem to be supported by results obtained when the model is applied to real data; the unrotated PARAFAC intrinsic axis factors are often quite meaningful (for instance, Gandour and Harshman 1978; Haan 1981; Harshman, Ladefoged, and Goldstein 1977; Harshman and Papcun 1976; Kettenring 1983; Meyer 1980; Snyder, Walsh, and Pamment 1983; Terbeek 1977; Dawson 1982; Harshman and Reddon 1983; Sentis, Harshman, and Stangor 1983; Trick 1983; Weinberg and Harshman 1980). Thus, it appears that PARAFAC can indeed provide a solution to the rotation problem in many situations.

In addition to providing meaningful intrinsic axes, PARAFAC has the extra advantage of directly decomposing the three-way array without requiring that it be "collapsed" into a two-way version. This is not a distinctive feature, of course, since PARAFAC shares this advantage with all the other three-way analysis procedures discussed in this volume. But the PARAFAC-CANDECOMP model does provide a particularly simple description of the structure of a three-way array, involving only one set of factors that are common to all three modes. In some cases, however, a more complex description is needed, and Tucker's three-mode model or one of the other models discussed in this volume might be preferable.

Limitations of PARAFAC and Attempts to Reduce Them

It is important to stress that the PARAFAC model for three-way arrays is not a completely general one. As we will discuss below,

Tucker's three-mode model is more general and can thus provide solutions for data that cannot be adequately analyzed by PARAFAC. The reason for this is that Tucker's model is both more complex and weaker in its underlying assumptions. This creates the advantage of giving Tucker's model greater generality; however, it also creates a disadvantage, as it causes the model to be underdetermined by the data. Consequently, the solution obtained with Tucker's model does not have the intrinsic axis property; hence, after fitting the model, one must perform a series of factor rotations to obtain interpretable axes. This requires using additional rotational criteria to select a preferred solution, and one must face the same controversies concerning rotation that plague two-way factor analysis. As will be argued below, one would like to have intrinsic axis solutions whenever feasible. Therefore, one might adopt the following strategy: first, try to obtain an intrinsic axis solution by fitting PARAFAC; then, if additional structure seems present, or if the intrinsic axis solution is uninterpretable, proceed to apply the more general models, rotating axes to aid interpretation.

In the last five or six years, considerable effort has gone into broadening the applicability of PARAFAC, and several useful findings have been made:

a. By means of data preprocessing, the model has been extended to embrace a wider class of three-way data, while still retaining the important intrinsic axis property.
b. In some situations, however, meaningful solutions are not obtained, even with careful preprocessing, indicating that the extended PARAFAC model is apparently not general enough. For such cases, specially constrained methods of PARAFAC analysis have been developed (for instance, requiring orthogonality of axes in one mode) that often allow an interesting *subset* of the variance to be fit; useful intrinsic axis solutions can then be obtained, although additional systematic variance remains and should be analyzed with a more general model.
c. Diagnostic procedures have been developed that will alert the investigators to problems with the data or the analysis procedures (see appendix C in this volume).
d. And finally, still under development are even more general models that will allow the intrinsic axis property and other advantages of PARAFAC to be obtained with still wider classes of data.

Overview

Before discussing these more recent developments (see chapter 6), we will first consider the basic forms of the PARAFAC three-way factor analysis model: (a) the raw data or profile data form; (b) the form for analysis of covariance matrices; and (c) the form for multidimensional scaling (essentially equivalent to the INDSCAL model). Next, we will examine the intrinsic axis or "uniqueness" property of the model, briefly discussing its limitations and our interpretation of its significance. We will then consider the relationship between PARAFAC and some other models, including two-way factor analysis and Tucker's three-mode factor analysis. Finally, in chapter 6 we will consider the more recent PARAFAC extensions and other developments mentioned above.

THE BASIC PARAFAC MODEL

PARAFAC1 Applied to Raw Score or Profile Data

Generalizing the Two-Way Factor Model

The Traditional Factor Model. Let us begin by recalling the basic two-way factor-analytic model. The factor structure underlying a particular data point x_{ij} can be represented in scalar form as

$$x_{ij} = \sum_{r=1}^{q} (a_{ir} f_{jr}) + e_{ij} \qquad (5\text{–}1)$$

and in matrix form as

$$\mathbf{X} = \mathbf{AF'} + \mathbf{E} \ , \qquad (5\text{–}2)$$

where $\mathbf{X}$ is a two-way data array with n rows and m columns. Although $\mathbf{X}$ might be any kind of data, let us assume, for convenience of discussion, that the n rows represent attributes (variables), and the m columns represent entities (cases, such as persons); thus, x_{ij} represents the value of the ith variable obtained for the jth case (person). For a model in terms of q factors, $\mathbf{A}$ is an n by q matrix of factor weights on variables, often called factor "loadings," and $\mathbf{F}$ is an m by q matrix of factor weights on cases or persons, often called factor "scores." The term a_{ir} represents the weight of the ith variable on the rth factor; similarly, f_{jr} represents the weight of the jth case on the rth factor. The weights represent the degree to which the factor is expressed in the particular variable or case. The size of the contribution of factor r to x_{ij} is thus a function of its importance both for that variable and for that person. $\mathbf{E}$ represents a matrix of random error terms (and "specific factor" contributions).

The model (5–1) and (5–2) can be considered either a principal components model or a common factor model, depending on our assumptions about $\mathbf{E}$. Current usage among some psychometricians assigns the terms "factor" and "factor score" specific meanings, which precludes their application to dimensions obtained by fitting a principal-components-like model. However, this convention is not followed by other factor analysts, such as Horst (1965), and to simplify exposition it will not be strictly followed here. We will often use "factor" in a general sense as a latent component of the kind described by (5–1), regardless of the assumptions about the error terms. We will also use the more general terminology of Kruskal (1978, 1981, and chapter 2), who calls (5–1) a *bilinear model* because its nonrandom or structural part is a bilinear expression. (The structural part of [5–1] is bilinear because it would be linear in the a_{ir} coefficients if the f_{jr} coefficients were considered fixed or given and vice versa.) Our use of the term "factor" in the more general sense and our adoption of Kruskal's classification scheme both allow us to focus on the characteristics of the structural part of the model independent of the particular assumptions made about the stochastic (random error) part. (For more details on Kruskal's terminology, see Kruskal, chapter 2, in this volume.)

The PARAFAC Three-Way Generalization. The PARAFAC generalization is based on Cattell's (1944) idea of parallel proportional profiles, which we will discuss in more detail in a later section. We simply note here that it describes the relationship between the factor loadings on two different occasions in which the factors change their relative influence. It specifies that if on the second occasion the influence of a given factor is increased by some amount, then on that occasion all the loadings for that factor should be increased by the same proportion. Thus, if the influence of factor 2 were increased by 20%, then all the a_{i2} coefficients should be 20% larger.

One easy way to express proportional changes without rewriting all the a coefficients is to introduce a third set of loadings, or o_{kr} coefficients, corresponding to occasions. Just as the a_{ir} coefficients proportionally increase or decrease the contributions of factor r from variable to variable, the o_{kr} coefficients proportionally increase or decrease the contributions of factor r from occasion to occasion. If for occasion k the coefficient represents a 20% increase in the effect of factor r relative to some baseline, then the revised loadings might be thought of as $(o_{kr}a_{ir})$, where $o_{kr} = 1.2$. An equivalent perspective is to consider the loadings to be fixed but the factor scores increased by 1.2 on the particular occasion. Thus, we might write the factor score for occasion k as $o_{kr}f_{jr}$.

Although Cattell conceptualized his principle of parallel proportional profiles as a means of relating the factors obtained in two different factor analyses of two different data sets, we generalize his notion to apply to the simultaneous factor analysis of many different occasions (factor analysis of a three-way data array). Thus, instead of x_{ij}, we now consider the triply subscripted data entry x_{ijk}, an element from a data array organized in terms of three different ways or (in Tucker's terminology) "modes" of classification. To extend the earlier example, the three modes of the data array could correspond to variables, persons, and occasions. The three modes could also refer to stimuli, rating scales, and individuals doing the rating, or any other such classifications that would define a data cell in a three-way array.

To maintain generality, we will often refer to the three modes of a data array as "Mode A," "Mode B," and "Mode C." From the point of view of PARAFAC, there is nothing distinctive about the mathematical properties of any mode as compared to any other. Because PARAFAC will be applied to many different kinds of three-way arrays, we no longer have any particular reason for treating the expression of factors in one of the three modes differently from the expression of factors in another. Therefore, we drop the distinction between factor "loadings" in one mode and factor "scores" in another. We consider factors to have weights or loadings in all three modes, and these loadings are of the same kind in all modes, although we might choose to scale them differently in one mode or another (see below). We thus change notation, from f_{jr} for a factor score to b_{jr} for a Mode B factor loading, and from o_{kr} to c_{kr} for a Mode C factor loading. The PARAFAC generalization of the two-way factor model of (5–1) takes the following simple form:

$$x_{ijk} = \sum_{r=1}^{q} (a_{ir} b_{jr} c_{kr}) + e_{ijk} \; . \qquad (5\text{–}3)$$

Here, x_{ijk} is an entry in a three-way data array; it might, for example, be the score on the ith variable obtained by the jth person on the kth testing session. The a_{ir} term represents the loading of factor r on the ith level of Mode A; in terms of our example, it would stand for the importance or size of contribution of the rth factor to the ith variable. Similarly, b_{jr} represents the loading of factor r on the jth level of Mode B and so stands for the importance or size of the contributions of the rth factor for person j. Finally, the c_{kr} coefficient stands for the loading of factor r on the kth level of Mode C and in our example data set, would represent the importance of factor r on occasion k. Kruskal (1981, 1983, chapter 2) describes (5–3) as a *trilinear model,* since there are now three different sets of loadings, and the model is linear in each set if the other two are considered fixed. (Quadrilinear and higher-order generalizations have also been envisioned [Harshman 1970, 22] but have not been implemented in the PARAFAC program. However, the CANDECOMP procedure implemented by Carroll and Chang [1970] will analyze up to seven-way tables.)

The matrix representation of the PARAFAC model is in some ways less elegant than the scalar one. The evenhanded treatment of all three modes by PARAFAC results in a three-way symmetry of the model that is easy to represent in scalar terms but harder to carry over into matrix notation.[1] To facilitate the use of conventional matrix notation, we must divide up the three-way array into a stack of two-way matrices. The direction of division is arbitrary—we could divide the array in any of several different ways. Let us adopt the convention that the three-way array is "sliced" into matrices corresponding to different levels of Mode C. Thus, the array is represented as a stack of two-way matrices, each matrix being n by m and corresponding to a Mode A by Mode B set of observations; there would be p such matrices, one for each level of Mode C. We then write an expression representing all "slices" in the array by means of describing its arbitrary kth slice, just as we wrote a scalar expression describing all the elements in the array by describing its arbitrary ijth element.

If we let $\mathbf{X}_k$ represent that n by m matrix, which is the kth slice of the n by m by p three-way array, we can then write the expression

$$\mathbf{X}_k = \mathbf{A}\,\mathbf{D}_k\,\mathbf{B}' + \mathbf{E}_k \; , \qquad (5\text{–}4)$$

where $\mathbf{A}$ is an n by q factor-loading matrix for Mode $\mathbf{A}$, B is an m by q factor-loading matrix for Mode B, and $\mathbf{D}_k$ is a q by q

diagonal matrix, with diagonal elements taken from the kth row of **C**, a p by q factor-loading matrix for Mode C. Thus, the diagonal matrix $\mathbf{D}_k$ provides weights for the kth occasion, which step up or down the sizes of the columns of **A** (or, equivalently, the rows of $\mathbf{B}'$). The r diagonal elements of $\mathbf{D}_k$ thus represent the effect of the changes in the relative importance or influence of the r factors on occasion k.

Scaling and Interpretation of A, B, and C Matrices. As we have already noted, traditional two-way factor analysis has developed different names for the Mode A versus Mode B weights. In the example described above, if we considered only one occasion, the Mode A weights would be called factor loadings and the Mode B weights would be called factor scores. In addition to distinctive names, the two sets of weights have become associated with different scaling conventions and frameworks of interpretation. These conventions enhance the meaning of the weights by allowing one set (the loadings) to be interpreted as standardized regression coefficients—beta weights—and sometimes as correlations between variables and factors, while the other set (the scores) can be interpreted as z-scores describing the amount of each factor attributed to each case.

Since PARAFAC views the three modes as being interchangeable, it seeks to treat them evenhandedly. Thus, the basic model is worked out in a more general fashion in which the weights need have no special interpretation other than describing the linear composite of factors that would predict the data. This is analogous to the interpretation assigned to regression weights when neither the predictors nor the predicted variable are standardized in any way. It also means that with appropriate scaling of the output, PARAFAC-CANDECOMP can be viewed as a three-way generalization of the singular value decomposition of a two-way matrix. (For a discussion of the singular value decomposition, see Green 1978; Kruskal, chapter 2. For an application of PARAFAC to provide the singular value decomposition, see Reddon, Marceau, and Jackson 1982.)

However, by adopting special conventions for standardizing both the input data and the output **A**, **B**, and **C** matrices, it is possible to produce PARAFAC weights that are strictly analogous to the loadings and scores of traditional factor analysis. PARAFAC solutions obtained in this way will have one mode that can be given the same special interpretations that traditional two-way loadings are given (for instance, as factor-variable correlations in the orthogonal case), while the other two modes can be interpreted as typical z-scores analogous to estimates of factor scores. Other standardizations that provide more general interpretations are also possible. The issues involved in standardization of data and loadings to achieve different meanings require some knowledge of the effects of different kinds of preprocessing (discussed in chapter 6), and so the relevant conclusions cannot be proven here. A more detailed discussion and proofs are provided in appendix 5-1. To show what is possible, however, we present in the following few paragraphs a summary of some of the most useful results.

Suppose, for example, that we are planning a PARAFAC analysis of a three-way set of profile data consisting of a group of variables measured on a set of cases under several different conditions. Furthermore, suppose that we wish Mode *A*, the variable weights, to be interpretable as loadings in the traditional sense, with Modes *B* and *C* weights interpretable as factor or component scores for the cases and conditions. To permit this we must standardize the data in a certain way before the analysis, and then we must standardize the loadings in a certain way after the analysis. Before analysis, the data must be mean-standardized or "centered" (that is, have means removed according to the procedures described in chapter 6) across levels of Modes *B* and/or *C*. In addition, the data must be size-standardized within levels of Mode *A* so that each level of Mode *A* has a variance of 1.0. After the analysis, the output loadings must also be standardized. The mean-squared loading for Modes *B* and *C* must be set to 1.0 so that Mode *A* loadings will reflect the scale of the data.

If these things are done, the Mode *B* and/or *C* loading matrices will be composed of *z*-scores, with each column having zero-mean and unit variance. This leads to two possible interpretations. In one, the Mode *B* and/or *C* weights are themselves factor scores of the traditional kind.[2] In the other, they represent *average* factor scores, while the actual factor scores are taken to be estimated by the $b_{jr}c_{kr}$ products. For simplicity of discussion, we only consider the second interpretation in this chapter. From this perspective, each PARAFAC weight (such as b_{jr}) corresponds to an average (that is, root-mean-square) factor score. For example, b_{jr} will give the root-mean-square score for factor *r* across all observations at level *j* of Mode *B*. This root-mean-square average will be equal to the standard deviation of the factor scores at that level, if the other mode is centered (for instance, b_{jr} will be the standard deviation of the factor scores at level *j* of Mode *B* if Mode *C* is centered, and vice versa). For each factor, the outer product of its Mode *B* and *C* PARAFAC weights will give a table of traditional *z*-score factor scores, with rows corresponding to levels of Mode *B*, columns to levels of Mode *C*, and entries ($b_{jr}c_{kr}$) corresponding to the individual factor score of case *j* in condition *k* for factor *r*.

If the entries in the factor score table for factor *r* are uncorrelated with the corresponding entries in the table for factor r^*, then factors *r* and r^* are "orthogonal" in the traditional sense (assuming zero factor score means as before). In terms of our example, this will happen whenever the two factors have orthogonal loadings in Mode *B* or *C* (or both). Thus, to have an "orthogonal solution," in which every table of factor scores is orthogonal to every other, it is sufficient that either the Mode *B* or *C* PARAFAC loading matrix have mutually orthogonal columns. (In fact, this latter condition is sufficient to produce orthogonal factor scores even when the factor score means are not zero.)

In a suitably standardized orthogonal PARAFAC solution, Mode *A* loadings can be interpreted as correlations between variables and underlying factors, just as in an orthogonal two-way solution. Conversely, if the tables of factor scores have correlated entries, the PARAFAC Mode *A* loadings will be interpretable as standardized regression coefficients—beta weights—in the same way as the traditional factor loadings that appear in the factor pattern matrix of an oblique two-way solution.

By applying the same preprocessing and standardization procedures to other combinations of modes, it is possible to interpret either Mode *B* or *C* PARAFAC loadings as factor loadings of the traditional kind. And even if the data are not centered in the necessary way, it is still possible to generate interpretations of a more general kind (in terms of mean-squares contributed by factors, and so forth). (These various combinations of input data processing and output loading standardization may seem complicated to select and cumbersome to employ in any actual analysis, but experience shows that this is not the case. They are available as defaults and/or easily specified options in the recent versions of the PARAFAC program; hence, they can be applied and the effects of different standardizations can be compared with little or no effort.) For further discussion of the different types of preprocessing and output standardization, including mathematical proofs of some of the assertions made above, see chapter 6 and appendix 5–1.

The Conceptual Model Underlying PARAFAC

Behind any *mathematical* model created for data analysis there must lie a *conceptual* model. The conceptual model provides a framework for logical or semantic interpretation of the terms in the mathematical model and thus determines the conditions under which its application to data can be deemed reasonable. In order to fully understand the PARAFAC generalization of two-way factor analysis, we need to consider the way in which it generalizes the conceptual as well as the mathematical two-way model. This task is not as straightforward as it might first seem. As has been pointed out previously (Harshman 1970, 19–25), the two-way *mathematical* model of factor analysis is an ambiguous representation of several fundamentally different *conceptual* models. And although these conceptual models have equivalent algebraic representations in the two-way case, their three-way generalizations are quite different. For example, one gives rise to PARAFAC, another to Tucker's three-mode model, and a third to a representation that cannot be fit by any current model of three-way data arrays and so must be fit indirectly.

System versus Object Variation. We can illustrate the implications of different conceptual models by comparing two models (discussed previously in Harshman 1970) that involve a differing "locus of action" for the factors. In the first model, the *system variation model,* the factors reside in the system under study and through the system affect the particular objects; the factor influences exhibited by particular objects would thus vary in a synchronous manner across a third mode such as occasions. In the second model, the *object variation model,* separate instances of the factors can be found in each of the objects, and these within-object factors would not be expected to show synchronous variation across levels of a third mode, such as occasions.

Consider, for example, an economist doing factor analysis of several production measurements on ten industries across a number of different years. He might believe that any factors which he uncovers—such as "raw material costs" or "energy costs"—reflect aspects of the economic *system* under study. Thus, he might find it natural to assume that when a factor increases its impact on some occasion, it does so proportionally for all measures and industries through the system. This would be a case of *system variation*.

In contrast, consider a psychologist studying personality change across time. He might think that the factors he uncovers —such as introversion-extroversion—refer to a *type* of influence rather than a single *source* of variation, with the actual effect of introversion-extroversion occurring at innumerable independent loci. Indeed, each individual under study would presumably be a source of his own introversion-extroversion variation. In this case, the factors would be considered resident in the objects under study (the persons) rather than in some single system. Thus, the variations across time (in introversion-extroversion) would show a different pattern for each individual, corresponding to his idiosyncratic experiences and life history. This would be a case of *object variation*.

Mathematical Expression of the Two Models. These two different conceptions of "factor" imply different three-way factor models. To compare them, first consider this very general expression for factor variation in a three-way array, which we call the *unconstrained object variation model*:

$$x_{ijk} = \sum_r (a_{ir} f_{jkr}) + e_{ijk} \; . \qquad (5\text{–}5)$$

Because the factor score component in (5–5) is triply subscripted, there is no constraint placed on the pattern of variation of factor scores across Modes B and C of the data. (Some may question whether this is really a three-way model at all, since it is easy to fit [5–5] by conventional two-way methods: Simply place all the two-way slices of the three-way array end to end, thus "stringing-out" the data into a large two-way matrix that is i by jk, and then perform a two-way analysis on this matrix. Nonetheless, [5–5] is a reasonable model for a three-way array. Such ambiguities of classification are not uncommon when one considers extreme limiting cases of particular ideas.)

The system variation concept of factor variation would imply that

$$f_{jkr} = b_{jr} c_{kr} \; , \qquad (5\text{–}6)$$

which says that if we consider object j on occasion k, we find that the influence or expression of factor r is stepped up or down by an amount c_{kr}. This shrinkage or expansion in the importance of factor r occurs in parallel for all objects, but the effect for any particular object is proportional to the basic sensitivity of object j to effects of factor r (as designated by b_{jr}). Thus, the change in the cost of energy would affect the different measures and different industries in proportion to their energy dependency.

Another example of an appropriate application of the system variation model is semantic differential or similar data, in which each of several stimuli are rated on each of several scales by a number of different judges. It is reasonable to assume that the set of rating scales taps a smaller set of underlying attributes (factors), and that these attributes are found across all the different stimuli and are used to greater or lesser degree by most of the judges. Moreover, suppose we assume that: (a) underlying attributes change in their relative value from one stimulus to the next (consistently across scales and judges); (b) that each judge is more or less sensitive to each attribute factor (consistently across stimuli and scales); and (c) that each scale can be described as measuring some combination of these basic attributes (more or less consistently across the stimuli and judges). Then we can justify the assumption that the influence of a given factor goes up or down proportionally as one goes from one stimulus to the next, one judge to the next, and one attribute or rating scale to the next. The rating scale example of system variation is particularly useful in that it demonstrates the actual symmetry of the PARAFAC system variation model; the requirement is that factors change influence proportionally across the levels of *each* mode. We need not focus on a particular mode as representing factor loadings and the others as representing factor scores, except to facilitate discussion. It also shows how the "system" involved in the system variation model might be quite abstract. In the case of the ratings, the system is presumably the cognitive/semantic system in which the dimensions of meaning are defined; this allows a given dimension to be expressed to a greater or lesser degree by a given rater, a given stimulus, and/or a given scale.

In contrast to the system variation model, we might consider the object variation model to be represented as follows:

$$x_{ijk} = a_{ir}(b_{jr} + v_{jkr}) + e_{ijk} \, . \tag{5–7}$$

Here, the basic factor score for object j on factor r is modified on occasion k by a variation v_{jkr}; this variation is not necessarily similar to the variation of any other object.

Other Types of Data. Data that requires the generality of the unconstrained object variation model (5–5) or the object variation model (5–7) cannot be adequately described by more restricted models such as PARAFAC or Tucker's three-mode model. However, there are intermediate forms of variation in which these more restricted models might be useful. For example, many longitudinal data sets may contain a combination of object and system variation. To analyze such data, one might fit the system variation part and treat the object variation as "error." One might instead choose to fit the data initially with a model that does not distinguish the two parts (for instance, [5–12], below; also see the discussion in Harshman and Berenbaum 1981 and application in Haan 1981).

There are also forms of systematic three-mode variation that are less general than (5–7) but more general than (5–3). Some of these will be captured by the extended PARAFAC model, to be discussed below. Others require three-way models with a more complex structure, such as Tucker's three-mode model, but may be captured in part by specially constrained PARAFAC analyses (see chapter 6). Tucker's three-mode model is based on a conceptualization of "factor" that is different from either the system variation or object variation models. For Tucker, a factor does not represent a distinct additive source of variation in the data; rather, it represents an idealized aspect or pattern of variation in a given mode, which generates the data by interaction with different idealized aspects of other modes. Tucker's model thus conceptualizes a separate set of factors for each mode of the data (indeed, the number of factors in one mode need not match the number of factors in another). This model will be considered further when PARAFAC is compared with other three-way factor analysis procedures.

Because the application of PARAFAC to raw or profile data presupposes the system variation model, it is important to consider the type of variation expected in one's data before performing an analysis. If one believes that object variation is likely to be the main kind of factor variation found, then direct application of PARAFAC is not warranted. However, an indirect application of PARAFAC analysis is still possible, provided the data is preprocessed by conversion to covariances before analysis. We will show in the next section that this data transformation allows one to fit a version of the PARAFAC model appropriate for object variation data.

PARAFAC1 Applied to Covariance Data

Importance of Covariance Analysis

Because factor analysis provides a structural model for one's original observations ([5–1] and [5–2]), it would seem natural to obtain factor loadings and factor scores by directly fitting this model to a set of data. Some investigators (including Horst 1965; Kruskal 1978) have favored this approach, pointing out that it is the most mathematically and logically straightforward method of estimation. Historically, however, factor analysis has focused on analysis of correlations rather than the raw data, and this perspective has maintained its dominance to the present day, perhaps because of the reduction in computational effort that it permits. Consequently, factor analysis is usually performed by what Kruskal (1978) calls the *indirect fitting* approach: The data are first transformed into a set of correlations among variables (or cases), and then a factor model (derived from [5–1]) is fit to the correlations. It turns out that in the two-way case, these approaches are equivalent. That is, the indirect fitting of the data based on a least-squares fit of the derived factor model to the correlation matrix yields the same loadings as the direct least-squares fit of (5–1) to the z-score matrix. Thus, the distinction between fitting methods has not been given much attention (aside from Kruskal 1978 and McDonald 1979).

With three-way data, the equivalence between direct and indirect fitting no longer holds. The results of direct and indirect PARAFAC analysis differ both at the *statistical* level, in terms of the precise values of the loadings and residual errors that are obtained when fitting the system variation model to a given data set, and at a deeper *structural* level, in terms of the patterns of data variation that can be implicitly represented by the solution. The different structural implications of the two approaches were pointed out earlier (Harshman 1972b), but the differences in statistical characteristics are only now becoming known. Because we are still investigating these statistical differences, we will only briefly summarize some current findings and will focus the bulk of our discussion on the different structural models that can be fit by the two methods.

Statistical Nonequivalence. In the three-way case, direct and indirect fitting will provide identical loadings only in very special circumstances that would not be realized with real data (for instance, when factors are orthogonal in Mode *B* and the data are fit perfectly, or when all the variance is not fit by the extracted dimensions but the unextracted dimensions—systematic *and error*—are perfectly orthogonal to the extracted dimensions in both Modes *A* and *B*). With realistic cases involving fallible data and less than orthogonal dimensions, the loadings obtained by indirect fitting may show varying degrees of resemblance to those obtained by direct fitting. Generally, the resemblance is fairly close, but in a few of these cases, the differences can be large enough to substantially affect interpretation.

Currently, our interpretation is that the general equivalence of two-mode solutions obtained by least-squares direct and indirect fits depends in part on the arbitrary axis orientation that is possible in the two-way case but not in the PARAFAC three-way case. In the two-way case, it is always possible to obtain successive best fitting dimensions that are orthogonal to all previous ones in both modes. As a result, successive solutions are "nested," in the sense that the best fitting two-dimensional solution forms the first two dimensions of the best fitting three-dimensional solution, and so on. In the three-way case, the best fitting dimensions are not generally orthogonal to one another in both modes, and they cannot be made orthogonal because of the intrinsic axis property. Thus, the nesting of solutions does not generally occur. And since in the indirect solution Mode *A* is represented twice, Mode *C* loadings are squared, and Mode *B* is not represented at all (see below), the degree and pattern of effective variance overlap of dimensions will be different from the direct fitting solutions. Hence, reduced-rank approximations (for example, the two-dimensional approximation of a three-dimensional data set) obtained by direct versus indirect fitting methods will involve different compromises and so will not generally be the same. The only circumstance in which the compromises would be the same is when the nonarbitrary PARAFAC orientation of axes happens to provide orthogonal dimensions in both Modes *A* and *B*.

Nonequivalent Structural Implications. As we shall see below, both direct and indirect methods can be used to fit the system variation model to three-way data. However, indirect fitting does not make use of the same information in the three-way array, and it involves different assumptions about the patterns of three-way data variation. As a result, it can also be used to fit non-system-variation versions of the PARAFAC structural models. For example, the indirect fitting approach permits PARAFAC (and Tucker's three-model model) to handle object variation data and thus provides a means of fitting three-way models that have much greater generality. At the same time, the simplest PARAFAC implementation of the indirect fitting approach requires assumptions about factor orthogonality that are not required by the direct fitting approach. These structural differences will be considered in detail in the following sections.

The differences in statistical and structural characteristics of direct and indirect fitting in the three-way case make it important to consider in some detail the implications of analysis of covariances as opposed to raw or preprocessed profile data. Unless the investigator understands the different implications of the two approaches, he may not be able to properly determine which is most appropriate for a given problem.

Approach and Terminology

In the discussion that follows, we focus on analysis of summed cross-products and covariances rather than analysis of correlations. This gives our discussion greater generality, since correlations are a special case of covariances in which the data for each variable are scaled to have unit variance, and covariances are in turn a special case of summed cross-products, in which the data for each variable have zero-mean. There is also a more serious reason for avoiding correlations, however. As we will show below, computing correlation matrices for each of several occasions would generally impose separate scalings on each variable for each occasion, which would complicate any expression for the size of contributions of factors to a given variable across occasions. Thus, we will see that analysis of correlations is not in general appropriate for either PARAFAC, Tucker's three-mode, or other current three-way factor analysis models.

The analysis of covariances by PARAFAC can be thought of in either of two ways: (a) as fitting to covariances the same model used for raw data ([5–3] and [5–4]); or (b) as fitting to the covariances a model *derived* from (5–4), one which represents the structure that would underlie covariance matrices if they were computed from raw data with structure described by (5–4). From the first perspective, the model is unchanged but the solution is said to take on a "special form" (for example, Mode A loadings equal Mode B loadings). From the second perspective, the special form of the solution corresponds to the special form of the derived model. In either case, the same computational algorithm can be used to perform the analysis, since the special form emerges as a result of the data rather than as a result of any constraints imposed by the analysis procedure. Thus, we consider the two perspectives to be interchangeable and so will sometimes talk of the "PARAFAC1 model for covariances" and at other times talk of the "PARAFAC1 model" in a more general way that is meant to include its application to raw data, covariances, and even multidimensional scaling applications.

We refer to "PARAFAC1" rather than simply "PARAFAC" in order to distinguish the model of (5–3) and (5–4) and its derivatives (such as [5–10], below) from generalizations such as (5–9), which is called "PARAFAC2" (Harshman 1972b), and (T1–7) (see Table 5–1), which is called "PARAFAC3." These latter generalizations require different computer algorithms and can no longer be considered trilinear models (for instance, PARAFAC2 is quintilinear, at least from one perspective).

Deriving the PARAFAC1 Covariance Model from the Raw-Score Model

A General Expression for Cross-Products. Suppose we begin with data that have PARAFAC latent structure, as described in (5–4), and compute cross-products among variables for each occasion, as follows:

$$\mathbf{C}_k = (\mathbf{X}_k \mathbf{X}'_k) \; . \qquad (5\text{–}8)$$

By substituting the PARAFAC representation of $\mathbf{X}_k$ from (5–4) into (5–8), we get

$$\mathbf{C}_k = (\mathbf{A}\,\mathbf{D}_k \mathbf{B}' + \mathbf{E}_k)\;(\mathbf{A}\,\mathbf{D}_k \mathbf{B}' + \mathbf{E}_k)' \; ,$$

and if we assume that the error is orthogonal to the systematic part, the cross-products of error and systematic terms drop out, leaving

$$\mathbf{C}_k = (\mathbf{A}\,\mathbf{D}_k \mathbf{B}')\;(\mathbf{A}\,\mathbf{D}_k \mathbf{B}')' + (\mathbf{E}_k \mathbf{E}'_k) \; .$$

By taking transposes and regrouping terms, we obtain

$$\mathbf{C}_k = \mathbf{A}\,\mathbf{D}_k (\mathbf{B}'\mathbf{B})\,\mathbf{D}_k\,\mathbf{A}' + \mathbf{E}_k \mathbf{E}'_k \; ,$$

and if we let $\mathbf{W} = (\mathbf{B}'\mathbf{B})$, the matrix of cross-products among Mode B factor loadings, we get the general PARAFAC model for cross-products:

$$\mathbf{C}_k = \mathbf{A}\,\mathbf{D}_k \mathbf{W}\,\mathbf{D}_k \mathbf{A}' + \mathbf{E}_k \mathbf{E}'_k \; . \qquad (5\text{–}9)$$

For maximum generality, we have developed (5–9) in terms of summed cross-products. Several special cases should be noted. If the $\mathbf{X}_k$ matrices are individually row-centered (so that the mean for each variable in each matrix is zero), then $\mathbf{C}_k$ is the deviation sums of squares and cross-products matrix for occasion k. By taking the further step of dividing each entry in $\mathbf{C}_k$ by m, the number of entries in each row of $\mathbf{X}_k$, then $\mathbf{C}_k$ represents the covariances among variables on occasion k and (5–9) gives the general PARAFAC representation of the structure underlying a three-way array of covariance matrices.

PARAFAC2. The model (5–9) corresponds to the PARAFAC2 generalization of PARAFAC (Harshman 1972b). PARAFAC2 is a nonorthogonal factor model for summed cross-product and covariance matrices and will be discussed briefly later in this chapter. The standard algorithm used to fit (5–4) to data cannot be used to fit this model to data.

The $\mathbf{W}$ matrix, the only new component to the PARAFAC2 model, can be interpreted as describing obliqueness or nonindependence among the factors in the mode over which covariances were computed (Mode B). When the $\mathbf{X}_k$ are row-centered and multiplied by $1/m$ so that the $\mathbf{C}_k$ become covariance matrices, the

$\mathbf{B}'$ matrix in the underlying PARAFAC representation of $\mathbf{X}_k$ is also row-centered. If we absorb the $1/m$ multiplier of $\mathbf{X}_k$ into $\mathbf{B}$, then the $\mathbf{B}'\mathbf{B}$ matrix becomes a matrix of covariances among the Mode B factor loadings.

If the columns of $\mathbf{B}$ (rows of $\mathbf{B}'$) are further scaled so as to have mean-square of 1.0, with compensatory rescaling of the $\mathbf{D}_k$ (which is always possible; see appendix 5–1), then $\mathbf{W}$ becomes a matrix of factor intercorrelations, and (5–9) corresponds to a three-way generalization of the two-way oblique factor analysis model. The $\mathbf{A}$ matrix is the same n by q matrix of factor loadings that appears in (5–4), $\mathbf{D}_k$ is the same q by q matrix that gives the weights of the factors on occasion k, and $\mathbf{W}$ is a q by q matrix that in this case gives the cosines of the angles among the factors. For this special case, $\mathbf{W}$ could be designated Φ to correspond with traditional notation for the factor intercorrelation matrix. Finally, $\mathbf{E}_k\mathbf{E}_k'$ gives the residual covariances not fit by the model.

PARAFAC1. If we make the additional assumption that $\mathbf{W} = \mathbf{I}$ (that is, that the factors are orthogonal in Mode B), then the $\mathbf{W}$ matrix in (5–9) disappears, and the model can be written as follows:

$$\mathbf{C}_k = \mathbf{A}\mathbf{D}_k^2\mathbf{A}' + \mathbf{E}_k\mathbf{E}_k' \ . \tag{5–10}$$

This is called the "PARAFAC1 model for covariances" (or other cross-product data). The numeral "1" designates that it corresponds to the simplest PARAFAC model and can be fit by the same algorithm as is used to fit (5–4) to a raw data matrix. Thus, as noted earlier, it can be considered an alternative application of the original PARAFAC procedure, one in which the solution takes on a special form. When this model is fit to covariance data by the same general three-way PARAFAC algorithm used to fit (5–4) to profile data, the first and second tables of factor loadings (which in [5–4] would correspond to Mode A and Mode B loadings, respectively) start out containing different values but end up at convergence to be identical to one another; this is a natural consequence of the symmetry of Modes A and B that occurs when the input data are covariances. Furthermore, the Mode C matrix obtained by analyzing the $\mathbf{C}_k$ will contain the *squares* of the factor weights for occasion k, namely, the squares of entries that would be obtained if the raw $\mathbf{X}_k$ matrices were analyzed. This occurs because each covariance is a mean of cross-products, and in each cross-product, the Mode C weight occurs twice and hence is squared.

Consequences of Indirect versus Direct Fitting

If the raw data can be assumed to have a structure similar to (5–4) and the factors are orthogonal in the mode over which covariances are computed (in the example above, if $\mathbf{B}'\mathbf{B}$ is diagonal), then the PARAFAC1 model for covariances (5–10) is appropriate and the factor loading matrix $\mathbf{A}$ obtained by indirect fitting will usually be quite similar to the matrix obtained by direct factoring of the raw data. However, indirect fitting does not optimize the same fit criterion as direct fitting: One is least-squares to the covariances, the other least-squares to the profile data. As noted earlier, these criteria are not strictly equivalent in the three-way case and can sometimes lead to differences in the resulting solution that are large enough to affect interpretation. Thus, some thought should be given to selection of the approach that would seem most appropriate for any given problem.

The solution obtained by indirect fitting will entail some initial loss of information, because loadings for Mode *B* of the raw data do not occur. (However, they can be estimated after the analysis by regression methods.) On the other hand, there may be some economy of computation in the indirect fitting procedure if there are many more levels to Mode *B* than to Mode *A*. These are not usually the main considerations, however. When the direct fitting model is deemed appropriate and gives interpretable results, it would often seem more straightforward and thus preferable.

There is one respect in which the indirect fitting model might sometimes seem too restrictive. If the factors are *not* orthogonal in the mode over which covariances are computed, then the PARAFAC1 model for covariances is not strictly appropriate and the solution will be distorted to some extent. If the divergence from orthogonality is not great, then the distortion will be quite small and will not affect the interpretation of the factors. If, however, the underlying factors are actually quite oblique in the mode over which covariances were computed, then the solution will be considerably distorted and a PARAFAC1 analysis may be misleading or uninterpretable. Thus, the necessity of assuming that the factors are orthogonal in the mode over which covariances are computed can occasionally be a serious disadvantage of the indirect fitting approach. Note that direct fitting of (5–4) to the raw data does not make any assumptions concerning the orthogonality of factor-loading matrix $\mathbf{B}$. Thus, it would usually seem preferable to use direct fitting when other considerations are roughly equal.

Sometimes, however, the restriction on the solution imposed by the implicit orthogonality assumption in the covariance model (5–10) is a help rather than a hindrance. Certain difficult data sets—in which the patterns of factor variation do not closely approximate those of any PARAFAC model (see chapter 6)—tend to produce degenerate solutions when the raw data is analyzed directly using the PARAFAC1 raw data model (5–4); very highly correlated factors occur and interpretation is not feasible. These degenerate solutions appear to be less likely when indirect fitting using (5–10) is employed, in part, perhaps, because of the stability resulting from the implicit orthogonality assumption, which tends to block the appearance of highly correlated factors.

The most important advantage of indirect fitting has not yet been discussed, because is it not apparent from our derivations thus far: By analyzing cross-products or correlations rather than raw data, one can estimate the parameters for models more general than (5–4), while still retaining the desirable intrinsic axis property. In particular, one can obtain meaningful intrinsic axis PARAFAC solutions when the raw data follows the object variation model, and it is even possible to fit data sets in which a different sample of cases is measured at each level of Mode *C*, so that there is no direct correspondence whatsoever between the factor score for a given case on one occasion and the score on the next.

Deriving the PARAFAC1 Model for Covariances from More General Assumptions

Sampling from Several Populations. Let $\mathbf{X}_k$ be an n by m_k data matrix, consisting of measurements of n variables on m_k cases. Assume that there are p such matrices obtained by sampling from p different populations, and that while the number of cases in the samples may differ, all samples are measured on the same n variables. If we postulate that these variables tap the same set of factors in all the samples, but expect that the relative importance of the factors may change across populations and therefore samples, then we might represent the underlying structure of such data as follows:

$$\mathbf{X}_k = \mathbf{A}\,\mathbf{D}_k\,\mathbf{B}'_k + \mathbf{E}_k \; . \qquad (5\text{–}11)$$

Here, $\mathbf{A}$ is a common n by q factor-loading matrix for all samples; $\mathbf{D}_k$ is a diagonal q by q weight matrix that gives the relative importance of the q factors on the kth occasion; and $\mathbf{B}_k$ is an m_k by q matrix of person weights or factor scores for sample k. $\mathbf{E}_k$ is an n by m_k matrix of error terms for occasion k.

In its raw data form, (5–11) cannot be fit by the PARAFAC model. One could perform a two-way analysis, by "stringing out" or concatenating all the matrices, to obtain a single matrix with n rows and as many columns as cases in all samples combined. Of course, the two-way analysis of this array will not have the intrinsic axis property. However, the data can be *indirectly* fit by PARAFAC if they are first converted to cross-products (covariances if the $\mathbf{X}_k$ were all row-centered). This analysis *will* have the intrinsic axis property.

Let $\mathbf{C}_k$ be the kth such cross-product or covariance matrix, defined as in (5–8). By substituting the expression $\mathbf{X}_k$ from (5–11) into (5–8) and then applying the same steps that follow (5–8), we obtain:

$$\mathbf{C}_k = \mathbf{A}\,\mathbf{D}_k\,(\mathbf{B}'_k\,\mathbf{B}_k)\,\mathbf{D}_k\,\mathbf{A}' + (\mathbf{E}_k\,\mathbf{E}'_k) \; .$$

If we let $\mathbf{W}_k = \mathbf{B}'_k\,\mathbf{B}_k$, then

$$\mathbf{C}_k = \mathbf{A}\,\mathbf{D}_k\,\mathbf{W}_k\,\mathbf{D}_k\,\mathbf{A}' + (\mathbf{E}_k\,\mathbf{E}'_k) \; . \qquad (5\text{–}12)$$

This is a general expression for the factor structure of the kth matrix of summed cross-products. If the data arrays had been row-centered, then this would be an expression for deviation cross-products, and if $\mathbf{C}_k$ is divided by m_k, then $\mathbf{C}_k$ is a covariance matrix, and $\mathbf{W}_k$ can be standardized to represent correlations or cosines among the factors in Mode B for occasion k. This then becomes a three-way generalization of the oblique factor model. In this generalization, the angles among the factors are not expected to remain constant across different levels of Mode C. Hence, this is more general than the PARAFAC2 model derived earlier by assuming system variation.

The model (5–12) is similar to Carroll and Chang's IDIOSCAL model (1972). It is more general than Tucker's three-mode model, although it is a special case of a variant of Tucker's model that Kroonenberg and de Leeuw (1980) call the "Tucker2" model. The representation of (5–12) was also discussed by Meredith (1964). The relations among such models will be considered in a later section.

In many cases, it will be reasonable to assume that the factors maintain roughly the same intercorrelations across the different levels of Mode *C*; hence, $\mathbf{W}_k = \mathbf{W}$, a common matrix of angles among factors for all covariance matrices. In this case, (5–12) reduces to (5–9), the PARAFAC2 model.

If we assume that in all samples the factors are orthogonal in Mode *B*, then $\mathbf{W}_k = \mathbf{I}$ and (5–12) reduces to (5–10), the PARAFAC1 model for covariance matrices. Thus, the PARAFAC1 model for covariances can be derived from assumptions far less restrictive than the system variation model from which it was originally obtained. Consequently, PARAFAC1 can be used to estimate the factors underlying raw data of the very general form (5–11), provided that indirect fitting is used.

Object Variation in One Population. In the case where all $\mathbf{B}_k$ matrices are measured on the same subjects, and so are all the same size, then (5–11) can be taken as a representation of a three-way data array with object variation. Thus, PARAFAC1 can also be used to analyze object variation data, by means of indirect fitting.

However, in all these applications of PARAFAC1, there is the need to adopt an orthogonality assumption that was not part of the system variation model: We must assume that the factors are orthogonal in the mode over which the covariances were computed (in these examples, Mode *B*).

Issues of Scaling and Interpretation

Scaling of Loadings and Data. As noted earlier, nonstandardized loadings can be interpreted as describing relative importance of different variables for a given factor by simply comparing their sizes within a column of the loadings matrix, provided the data sum of squares is similar across variables. However, if one wants to make additional interpretations of loadings in terms of correlations between factors and data and so forth, and if one wants PARAFAC loadings to be directly comparable to the loadings that are obtained in two-way factor analysis of correlation matrices, then special standardization of the data and of the output loadings is required. In essence, the covariance matrices must be scaled so that the average covariance matrix would have unit diagonals and therefore be interpreted as a correlation matrix. The loadings must be scaled so that the identical Mode *A* and Mode *B* tables jointly reflect the scale of the data, with the Mode *C* table having column means of 1.0. (This arrangement is discussed in appendix 5–1.) With this rescaling, factor loadings of the traditional kind are obtained. It is important to note, however, that one *cannot* simply proceed by converting the covariance matrices individually to correlation matrices and then analyzing the correlation matrices.

Inappropriateness of Correlations. If, in the above derivations, we had proceeded to rescale the covariances to obtain correlation matrices, no simple PARAFAC representation would have been possible. Such rescaling destroys the comparability of the **A** matrix across levels of Mode *C*. To see this, let us write the expression for a correlation matrix $\dot{\mathbf{C}}_k$. Let $\dot{\mathbf{D}}_k$ be an n by n diagonal matrix with diagonal entries equal to the reciprocal of the square roots of the diagonals of $\mathbf{C}_k$. We can write an expression for the kth correlation matrix as:

$$\dot{\mathbf{C}}_k = \dot{\mathbf{D}}_k \ \mathbf{A} \ \mathbf{D}_k \ \mathbf{W} \ \mathbf{D}_k \ \mathbf{A}' \ \dot{\mathbf{D}}_k + \dot{\mathbf{D}}_k \ \mathbf{E}_k \ \mathbf{E}_k' \ \dot{\mathbf{D}}_k \quad . \tag{5–13}$$

Thus, $\dot{\mathbf{D}}_k$ rescales the entries in $\mathbf{C}_k$ from covariances to correlations. However, this causes the factor-loading matrix **A** for each occasion to be row-rescaled. In order to analyze correlations, we would need a model with a separate $\mathbf{A}_k$ matrix for each occasion, permitting $\mathbf{A}_k = \dot{\mathbf{D}}_k \mathbf{A}$, or with special row-rescaling $\dot{\mathbf{D}}_k$ parameters in the model to modify the factor loading matrix for each occasion. None of the existing models for three-way data—including PARAFAC and Tucker's model—incorporate such parameters. They do not really need to, however, when use of covariances avoids the problem completely. It should be clear from the above, then, that correlations are not only inappropriate as input for PARAFAC—they are inappropriate for *any* currently existing three-way factor analysis procedures. Similarly, Jöreskog (1971) points out that correlations are not appropriate input for his methods of factor analysis in multiple populations.

One of the motivations for use of correlation coefficients is to remove arbitrary differences in the scale of variables by setting all variables to a constant unit mean-square. This objective can be accomplished within the context of covariance analysis by standardizing the total variance of each variable across all levels of Mode *C*, rather than the separate variances within each level. Thus, we would use an unsubscripted $\dot{\mathbf{D}}$ matrix in (5–13), for which the ith diagonal entry would be:

$$\dot{d}_{ii} = \left(\frac{1}{p} \sum_{k=1}^{p} c_{iik} \right)^{-1/2} .$$

In fact, this is a built-in PARAFAC option and has become a standard procedure to use for PARAFAC analysis of covariance matrices; it is called "equal average diagonal standardization."

Principal Components versus Common Factor Models

A further difference between direct and indirect fitting is that indirect fitting allows a three-way generalization of either the common factor or principal components models to be fit to the data, whereas direct fitting only allows the three-way principal components model to be fit. In this chapter, we have not stressed the distinction between the *principal components* and the *common factor* models for factor analysis. In general, we agree with Harris (1975) that from a data-analytic standpoint, the distinction is not critical, since the two different factor estimation methods usually give very similar sets of factor loadings.

From a theoretical standpoint, however, it seems clear that the principal components model ignores an obvious source of bias that is corrected by the common factor model. Consider any of our expressions for the structure underlying a covariance matrix, such as equation (5–10). The off-diagonal elements of the error covariance matrix $\mathbf{E}_k\mathbf{E}'_k$ will tend to be small and randomly distributed about zero, due to the small covariances that arise by chance among finite samples of random errors. They will not bias the solution. The diagonals, however, will be systematically larger and all positive, since each diagonal represents the covariance of an error component with itself, or, in other words, the error variance for a variable. There is a good argument, then, for ignoring the diagonal elements of a covariance data matrix, since they are likely to contain larger error components and be systematically biased upward from their "true" values (defined as the contribution of the factors before the error component is considered).

In some factor analysis procedures, it is not easy to fit off-diagonal cells and ignore the diagonal cells. In such cases, an alternative approach is sometimes used: The diagonal values are modified before the analysis to more closely represent an unbiased estimate of what they would equal if it were not for the inflating effect of the error. These communality estimates may be based on the squared multiple correlation between the variable and all the other variables in the data set or on some other estimation procedure. One of the more common approaches is to iterate on the diagonals. This method uses the results of a factor analysis to estimate the size that the diagonal cells would have if they only contained common factor variance; it replaces the diagonal with these estimates and then refactors the modified matrix. This procedure is supposed to be iterated until it converges, although tests have demonstrated that actual convergence usually takes a larger number of iterations than have commonly been applied in the past. In PARAFAC, the method used for ignoring the diagonals—and any other cells identified as containing missing data —can be viewed as a modified version of this iterative estimation procedure (see below).

In practice, the bias introduced by including the inflated diagonal values in the data set often has very little effect on the factor loadings (as Harris [1975] demonstrates). In some Monte Carlo studies (including Velicer, Peacock, and Jackson 1982), the principal components solutions have actually recovered underlying structure as well or better than common factor solutions, perhaps because of greater stability arising from the component model's fewer parameters.

Whether there will be an appreciable difference between factors determined using the principal component versus common factor models will depend on the characteristics of the data being analyzed. When one is analyzing large matrices, the diagonal cells constitute such a small proportion of the total data that the systematic bias introduced by including them in the analysis has a minimal effect on the final solution. Furthermore, unless the communality estimates are widely divergent, the pattern of loadings is not noticeably changed, even with smaller matrices (Harris 1975). Nonetheless, there are certain circumstances in which a difference between the two methods may be noticeable. If the covariance matrix is small—for instance, less than 15 or 20 variables—and if the estimated communalities are quite different from one another, so that relations among diagonal elements would be altered by communality estimation, then the common factor model might be expected to give an appreciably more accurate solution.

A method for fitting the common factor model by direct decomposition of the raw data array has been suggested for two-way data by McDonald (1979) but has not yet been generalized to the three-way case. However, when doing indirect fitting by means of covariances, the common factor model is easily fit in the three- as well as the two-way case, by simply ignoring the diagonal and/or placing communality estimates in the diagonals of the covariance matrices.

The PARAFAC computer program allows either model to be fit to covariances. If the user decides to fit the common factor model, the program takes the same approach as Harman and Jones' (1966) MINRES procedure; namely, parameters of the PARAFAC model are estimated by fitting only the off-diagonal elements of the covariance matrices. One of the program's analysis options allows the user to ignore the diagonal cells of each covariance matrix using the same technique that PARAFAC uses to ignore cells with missing data (by continuous iterative reestimation within the alternating least-squares algorithm; see Harshman 1972b). The method appears to work quite well. In fact, our experience suggests that this method may work better than more traditional iteration methods commonly used with two-way arrays, converging more rapidly to accurate estimates of the appropriate values (as tested by Monte Carlo experiments). This may in part be due to the fact that reestimation occurs repeatedly during the factor estimation phase, not just at the end. It may also be in part attributable to the stronger information provided by the three-way data.

Initial applications of the common factor model to real data problems have confirmed our theoretical expectations. With a large problem (Haan 1981), including or ignoring the diagonals made very little difference in the final **A** matrix—at most, a small change in the second decimal place of a factor loading (based on comparisons conducted in our laboratory but not reported in Haan's article). This was to be expected, since the data (which was from a longitudinal study of personality) consisted of 12 matrices, each of which was 86 by 86; thus, less than 2% of the data cells were ignored when the common factor model was estimated. Furthermore, examination of the communality estimates showed that most were of comparable size. Conversely, the common factor model provided a noticeably more interpretable solution in a three-way analysis of the WAIS and WAIS-R intelligence test standardization data (Harshman and Reddon 1983). This latter data set had those characteristics that might lead one to expect a difference; the 9 covariance matrices were small—only 11 by 11—and the communality estimates for the 11 variables turned out to differ considerably in size both within and across occasions.

The PARAFAC1 Model Applied to Multidimensional Scaling of Proximity Data

The Relationship between Factor Analysis and Metric MDS

Multidimensional Scaling (MDS) is sometimes used in a broad sense to mean any procedure that represents a large matrix of observations or relationships in terms of a small set of underlying dimensions. PARAFAC is classed as a multidimensional scaling procedure in this broad sense by Carroll and Arabie (1980). We shall use "multidimensional scaling" in a more restricted sense, however, meaning a procedure that derives a dimensional representation of relationships among entities from data on their proximity (such as pairwise similarity or dissimilarity). From this perspective, MDS is distinguished by the fact that it takes as input a type of data that is not directly appropriate for factor or component analysis: interpoint "distances," dissimilarities, or other distancelike quantities. Distances are not suitable for factor analysis because a distance cannot be directly decomposed into the additive contributions of dimensions. As the Pythagorean theorem demonstrates, it is the *squared* distances along dimensions that add together to produce the *squared* distance between two points in a space.

Nonetheless, factor analysis and multidimensional scaling are very closely related. In fact, metric multidimensional scaling can be accomplished by performing a factor (or principal component) analysis of distancelike data that has been appropriately preprocessed. The preprocessing turns dissimilarities into squared interpoint distances and then converts these into scalar products, which are essentially equivalent to covariances. Factor analysis of the scalar products then proceeds in the same way as analysis of covariance matrices. Thus, the application of factor or component analysis to MDS problems involves a further example of indirect fitting: an MDS distance model is indirectly fit to the original distancelike data by directly fitting a factor or component model to the covariancelike matrices obtained after the data is preprocessed.

The conversion of dissimilarities to scalar products has been described elsewhere (Kruskal and Wish 1978; Spence 1977; Torgerson 1958) and will therefore only be briefly recounted here. Three steps are involved for each subject: (a) an additive constant is estimated which converts that subject's dissimilarities into distances (that is, from interval-scale to ratio-scale measures of subjective distance between stimuli); (b) each entry in the distance matrix is then squared; and (c) the matrix of squared distances is then double-centered (row means are removed and then column means are removed from the residuals, so that the resulting matrix has both rows and columns summing to zero). This last step removes undesired constants that entered at the squaring stage and also adjusts the resulting scalar products so that their origin is at the centroid of the configuration of stimulus points in the space. The resulting values are all multiplied by -.5 for geometric reasons, as explained in Torgerson (1958).

When PARAFAC is used to perform three-way MDS, the mathematical model that is being indirectly fit to the data differs from those that we have discussed previously. It is called the "weighted Euclidean model." If, for example, we considered x_{ijk} to represent the distance between stimulus i and j as judged by the kth individual, the weighted Euclidean model would represent that distance in terms of a generalization of the Euclidean distance formula:

$$x_{ijk} = \left(\sum_r (w_{kr} (a_{ir} - a_{jr}))^2 \right)^{1/2} + e_{ijk} \ . \qquad (5\text{–}14)$$

In this formula, a_{ir} and a_{jr} represent the stimulus projection of the ith and jth stimuli onto dimension r, and w_{kr} represents the weight that subject k places on dimension r, or, in other words, the salience of that dimension for that subject.

Use of the PARAFAC program to indirectly fit the weighted Euclidean model is essentially equivalent to use of Carroll and Chang's (1970) INDSCAL procedure. Both PARAFAC and INDSCAL follow the same steps: First, transform the input data from similarities or dissimilarities to distances; then, convert the distances to scalar products; and finally, fit to the scalar products a trilinear model (5–3), which, because of the covariancelike structure of the data, takes the special form of (5–10). In the resulting representation, the $\mathbf{A}$ matrix gives the projections of stimuli onto dimensions, and the $\mathbf{D}_k^2$ matrix gives the squares of the dimension weights or saliences for the kth subject.

It is interesting that although INDSCAL and PARAFAC were developed independently and from somewhat different perspectives, both approaches led to the same trilinear model, based on related reasoning about systematic variations of dimensions underlying levels of a three-way array. We call (5–3) the "PARAFAC" model, alluding to Cattell's parallel proportional profiles idea, from which it was derived. But Carroll and Chang developed the model from somewhat different perspectives related to Horan's (1969) findings concerning individual differences in MDS; they call (5–3) the "CANDECOMP" (for CANonical DECOMPosition) model. Because of this parallelism, (5–3) is sometimes referred to as the "PARAFAC-CANDECOMP" (Kruskal 1981) or "CANDECOMP-PARAFAC" model (Carroll and Arabie 1980, 635).

Properties of Three-Way MDS Solutions

Advantages. When the trilinear model (5–3) is applied to scalar products in order to accomplish multidimensional scaling, the solutions have the same important intrinsic axis property as when it is applied to raw profile or covariance data to accomplish three-way factor analysis. This intrinsic axis property is perhaps the most important reason INDSCAL has been such a popular MDS procedure, with more than 75 successful applications to date. The success of INDSCAL in multidimensional scaling strengthens arguments for the potential value of the intrinsic axis property in factor analysis, as well.

However, MDS applications are in some respects easier than factor-analytic ones. For one thing, subtle preprocessing issues that arise in three-way factor analysis (which will be discussed in chapter 6) are avoided in MDS. Also, there are implied orthogonality constraints resulting from the indirect fitting of the MDS application that can increase the stability of the solution and help to prevent oblique axis problems that sometimes trouble factor-analytic applications. Finally, it appears from our accumulating experience that the patterns of dimensional variation in three-way profile data are sometimes more complex than provided for by the system variation model; when this occurs, the recovery of sensible solutions requires special constraints on the form of the loadings matrices. For all of these reasons, the general application of the trilinear PARAFAC-CANDECOMP model to profile data has been a more challenging problem than the application of this model to MDS data. Thus, it has taken longer to develop all the necessary theory, as well as preprocessing and other techniques, to make three-way factor analysis relatively trouble-free.

Orthogonality Assumptions. As we have pointed out previously (Harshman 1972b; Terbeek and Harshman 1972; see also Carroll and Wish 1974; McCallum 1976), the MDS application of the PARAFAC-CANDECOMP trilinear model (5–3) involves implicit orthogonality assumptions that are not required in two-way multidimensional scaling. Since fitting PARAFAC to scalar products is equivalent to analyzing covariances, it requires the same kind of assumption of orthogonality (discussed above). However, the interpretation of this assumption is slightly different in the case of MDS. When we analyze scalar products derived from dissimilarities, there is no "Mode *B*" over which we must assume uncorrelated variations in factor influence. Instead, the diagonal form of the matrix **W** reflects the special orthogonal form of the Euclidean distance formula. In the more general case, in which the axes in a space are not orthogonal, (5–10) will not describe the scalar products. Instead, a more general Euclidean formula must be used, thus corresponding to (5–9) (see Harshman 1972b).

There is some evidence that oblique perceptual dimensions may indeed arise when two or more properties by which people perceive a set of stimuli are strongly associated in the minds of those making the judgments. In an earlier unpublished study (Harshman 1973), subjects judging the similarity of objects differing only in size and weight appeared to treat the dimensions of size and weight as if they were oblique in their perceptual space. Tucker (1972) also obtained MDS results that suggested oblique dimensions.

Several three-way MDS models have been formulated to deal with the case of nonorthogonal perceptual dimensions. If we apply the PARAFAC2 model (5–9) to scalar product matrices, we can obtain a three-way MDS solution with oblique perceptual dimensions but involving the assumption of a common pattern of obliqueness across all subjects. This model may share the PARAFAC1 intrinsic axis property, although the question of uniqueness is more complex. (This will be pursued later in our discussion of PARAFAC2.) An even more general model would allow different patterns of obliquely related dimensions for each subject, as in the very general covariance model (5–12). This model corresponds to what Carroll and Chang (1972) call the "IDIOSCAL" model and Kroonenberg and de Leeuw (1980) call the "Tucker2" or "T2" model. A version of this model that puts restrictions on the different $\mathbf{W}_k$ is the Tucker application of three-mode factor analysis to multidimensional scaling (Tucker 1972). Neither IDIOSCAL, T2, nor Tucker's three-mode scaling provide intrinsic axis solutions.

Diagonal Cells in MDS. It often occurs that the distances or dissimilarities used for MDS input are obtained in such a way that the diagonal cells may not be directly comparable to the other cells or indeed may be set to zero by assumption and not even measured. For example, when subjects are asked to judge the pairwise dissimilarity of elements from some stimulus set, they are often not presented with pairs consisting of a stimulus matched with itself. Such stimuli are usually going to be judged to have infinite similarity or zero dissimilarity by an alert subject; hence, these trivial judgments are omitted to save time, even though they might serve a valuable "anchor" function in the experiment. When similarities are obtained by other means, such as stimulus confusions, it still may often be the case that the diagonal cells of the data matrices are subject to special considerations that make them not directly comparable to the off-diagonal cells.

The question thus arises: Would it be advisable to ignore the diagonal cells of MDS, much as is done when fitting the common factor model in three-way factor analysis? As in factor analysis, if the stimulus matrix is large and/or the subjective diagonal values that might be estimated by the dimensions (that is, the analogs of communality estimates) would be fairly uniform across stimuli, then the solution would probably not be noticeably affected by ignoring the diagonals. But if the stimulus set were small—as is common in MDS applications in which all pairwise comparisons are needed—and the subjective diagonals somehow divergent, then ignoring the diagonal might be useful. Exploration of the effects of such an approach would easily be accomplished by means of the PARAFAC option to ignore diagonal cells, but no systematic study of the effects of such a procedure has yet been attempted.

UNIQUENESS PROPERTIES OF PARAFAC-CANDECOMP

Why is the Intrinsic Axis Property Important?

Viewed geometrically, a factor analysis or multidimensional scaling solution provides two basic types of information: (a) a configuration of points in a low-dimensional space; and (b) a set of axes spanning that space. The configuration of points provides a compact description of the observed relationships among the things represented by the points. Such a description may clarify our insight by simplifying the patterns we are trying to understand, but it does not present us with anything essentially new. The axes, on the other hand, can potentially take us beyond observed relationships into new inferred ones. In the appropriate conditions, the projections of points onto correctly oriented axes could be taken to indicate relationships between observed surface variables and unobserved, but empirically real, latent variables. If a factor analysis can indicate which set of axes is most likely to approximate empirically real processes, it can provide us with genuinely novel information. But if the choice of a particular axis orientation is somewhat arbitrary, or if it is based on a criterion that is hard to defend empirically, the orientation of factor axes will be less informative and inferences from factor axes to theoretical constructs will seem dubious.

As noted in the introduction, PARAFAC was developed primarily to provide a stronger theoretical/empirical basis for determining orientation of factor axes. This advantage is important to scientists who would like to use factor analysis inferentially, either inductively, as an aid to development of theoretical constructs in a poorly understood domain, or deductively, as a means of testing previously hypothesized constructs with new data. The intrinsic axis property of the PARAFAC model provides a plausible theoretical basis for such inferences when the solution can be replicated across samples. In this section, we explain the intrinsic axis property, discuss its mathematical basis and its limitations, and show why it often provides stronger grounds for attributing empirical reality to factors than the grounds provided by simple structure or other commonly used criteria.

Is there a "Most Valid" Rotation?

Some investigators have claimed that the particular orientation of axes in a factor-analytic solution is not important. Different rotations of a factor-analytic solution are said to correspond to different yet equally valid perspectives on the same complex phenomenon (for example, see Thurstone 1947, 332).

Such an attitude may be appropriate when one only uses factor analysis to provide a condensed description of the particular data set. However, if one uses it to obtain clues to the empirical processes responsible for the observed patterns, then alternative rotations cannot be considered "equally valid." They will lead to competing hypotheses concerning the underlying empirical reality, and these hypotheses will, in turn, lead to different and competing predictions of what would be expected in novel nonequivalent situations. It is this difference in implied predictions that gives empirical meaning to the claim that there is a "most valid" axis orientation.

A search for the most valid rotation or orientation of factor axes can be viewed as a part of the scientist's search for theoretical constructs that will prove the most generalizable (namely, those constructs not only successful at accounting for patterns of variation within the given data set, but those able to be used in explaining patterns in new and systematically different data sets, as well). Proportional profiles rotation seeks to maximize this sort of generalizability.

The argument can also be made on less pragmatic grounds: We believe that there are particular component processes that actually exist in the empirical situation and that they are likely to be better approximated by some factorial descriptions than others. If we could identify those axes that most closely correspond to the underlying component processes (which provide the best description of them), we would know which axis orientations should be considered most valid. Because of their better correspondence with empirical reality, we expect that such axes will lead to the most successful scientific constructs or theories, theories that will make the most accurate predictions in the widest range of situations.

Consider, as one example, the classic debate concerning the dimensions of intelligence. This is in part a debate over correct orientation of axes in the factor space of intellectual tests. In one position, we obtain a general factor and a set of uncorrelated group factors corresponding to specific intellectual abilities. In another orientation, we obtain several correlated broad intellectual abilities. These alternative solutions are not just different perspectives, since they lead to different predictions concerning the effect of drugs or brain damage on intellectual performance. Empirical findings can provide support for one set of factors as against the other. In fact, Bock (1973) has argued that recent discoveries about the functional specialization of the two hemispheres (based in part on effects of brain damage on cognitive abilities) have helped to confirm those factor-analytic theories that distinguish a broad class of verbal abilities from another broad class of spatial abilities.

To take another example, consider a recent disagreement between Eysenck and Guilford over the "real factor of Extroversion-Introversion" (Eysenck 1977; Guilford 1977). Guilford (1977) points out explicitly that the disagreement hinges on choice of axis orientation. In order to resolve such disputes, appeal must be made to empirical data beyond the particular two-way factor analysis, in the hope that different predictions of the two different factor-based constructs can be somehow tested empirically.

Limitations of Traditional Criteria for Orienting Axes

Since the two-way factor model provides a family of possible solutions—all of which fit the data equally well—investigators have had to devise additional criteria external to the model to aid in selecting a "best" solution from among this family. Confirmatory factor analysis uses rotation-to-target techniques and/or allows loadings to be constrained in certain ways. Exploratory factor analysis rotates to optimize some desired characteristics of the resulting factors; usually this is an index of simplicity, although other criteria have also been employed (see Comrey 1967; Eysenck 1950). In many situations, however, the factors that optimize these criteria may or may not be the "real" ones, that is, the ones that most closely approximate the empirical processes generating the observed relationships. For example, we often do not have strong empirical reasons for preferring the particular rotation of factors that maximizes simplicity, and hence we do not have strong grounds for attributing empirical validity to constructs that the simplest factors suggest. Furthermore, controversy can arise because different investigators prefer different rotation criteria and thus obtain differently oriented factors. This occurred in the Eysenck-Guilford dispute mentioned earlier.

We have discussed the limitations of traditional factor rotation procedures elsewhere (Harshman 1970, 8–14) and will therefore only briefly summarize some objections here. First of all, the likelihood of obtaining empirically valid dimensions by theoretically guided target rotation will depend on the validity of the theory used. Such rotations can provide evidence against a theory, if the target cannot be approximated by any rotation of the obtained factors. However, these rotations usually provide only weak evidence in favor of a theory; when factors can be found that are consistent with a given target, the theory is not discredited, but other rotations of the same axes may provide factors consistent with other theories. Thus, target rotation will not usually result in the discovery of theoretically novel dimensions.

A second objection is that rotation in search of meaningful interpretations lacks objectivity. In addition, it is limited by the investigator's notion of what is "meaningful"; a dimension may be difficult to interpret because it reveals an unfamiliar truth about the domain under investigation.

A final objection to current methods of factor rotation is that the likelihood of obtaining valid dimensions by rotation to simple structure (or its analytic approximations, such as Varimax) depends on one's assurance that maximizing the simplicity criterion is appropriate for the particular situation. Often, relationships between variables and underlying factors will not be maximally simple. As Comrey notes:

> If we sample at random from the entire universe of factors and use predominately factor-pure measures, simple structure will no doubt give results which are reasonably satisfactory. In many real-life factor analyses, however, where selection of variables is anything but random, and measures of considerable factor complexity sometimes predominate, one can only hope that the simple structure criterion is approximately applicable. . . . There is no particular reason why the variance of the squared factor loadings must be maximized, except that loadings are more easily interpreted if they are high or low rather than medium in absolute value. It may well be, however, that they should be in medium range, rather than high or low, depending on the data being analyzed. (Comrey 1967, 143)

The Principle of Parallel Proportional Profiles

Cattell (1944) reviewed seven principles for choice of rotation and concluded that the "most fundamental" was one that he called the "principle of parallel proportional profiles." This principle may be thought of as an application of the method of cross-validation to the problem of factor rotation, in order to determine empirically which rotation corresponds to "real functional unities in the psychological situation." However, Cattell recognized that "to require agreement in factors and factor loadings among correlation matrices derived from the same or similar test variables on the same or similar population samples, is an empty challenge. No new source of rotation determination is introduced, for such matrices will differ only by sampling errors and there will be an infinite series of possible parallel rotations in the two or more analyses." What is needed is a method for *meaningful* validation across samples, demonstrating a consistent empirical pattern that would not be found if the factors were a mere "mathematical artifact." How could this be obtained?

> The special and novel required condition is that any two matrices should contain the same factors, but *that in the second matrix each factor should be accentuated or reduced in influence by the experimental or situational design,* so that all its loadings are proportionately changed, thereby producing, from the beginning, an actual correlation matrix different from the first. (Cattell 1944, 274 [his italics])

Cattell states the principle even more clearly in a later article:

> The basic assumption is that, *if a factor corresponds to some real organic unity, then from one study to another it will retain its pattern, simultaneously raising or lowering all its loadings according to the magnitude of the role of that factor under the different experimental conditions of the second study.* No inorganic factor, a mere mathematical abstraction, would behave in this way. . . . This principle suggests that every factor analytic investigation should be carried out on at least two samples, under conditions differing in the *extent* to which the same psychological factors . . . might be expected to be involved. We could then anticipate finding the "true" factors by locating the unique rotational position (simultaneously in both studies) in which each factor in the first study is found to have loadings which are proportional to (or some simple function of) those in the second: that is to say, a position should be discoverable in which the factor in the second study will have a pattern which is the same as the first, but stepped up or down. (Cattell and Cattell 1955, 84 [their italics])

Although he was convinced of the potential importance of this idea, Cattell had problems implementing it as a practical rotation procedure. He developed an algebraic method for orthogonal proportional profiles rotation of two sets of factor loadings (long before a similar method was proposed by Schönemann [1972] for MDS), but was unable to formulate a solution that allowed oblique axes (Cattell and Cattell 1955). Even more discouraging, his attempt at demonstrating a practical application of the orthogonal method to real data was not successful. Finally, he realized that his expectation of proportionality across occasions was an oversimplification: The relationship between two solutions obtained by conventional factor analysis of correlation matrices would be more complicated. In 1955, he did not see how to overcome this problem (Cattell and Cattell 1955), but it was subsequently resolved by Meredith (1964), who pointed out that strict proportionality could be obtained by use of covariances rather than correlations. More recently, Cattell refers to continued work on the method, which he now calls "confactor rotation." He reports encouraging results with synthetic data, but has yet (as far as we know) to report a successful application to real data.[3]

PARAFAC overcomes the problems that plagued Cattell in

several ways. First, Cattell's idea is generalized from a method of comparing separate factor analyses into a true three-mode factor-analytic model. Second, the basic model is formulated in terms of the direct fitting of profile data, and so avoids the restriction to orthogonal axes implied by the indirect fitting case. Finally, the indirect fitting case is formulated in terms of summed cross-product or covariance matrices, overcoming the problem of nonproportionality, which occurs with correlation matrices. Also, there is an oblique axis generalization for indirect fitting (PARAFAC2; see Harshman 1972b), which has been applied successfully to at least one data set (Terbeek and Harshman 1972). Despite these modifications, however, the basic idea behind PARAFAC is still the insight by Cattell that a search for proportional profiles across several nonequivalent two-way data sets would determine the most empirically meaningful factor axes. Thus, the recent success of PARAFAC with many kinds of real data is a confirmation of what he has been proposing for more than thirty years.

Nature of the Intrinsic Axis Property

What is Unique?

Suppose we perform a PARAFAC-CANDECOMP three-way factor analysis (or multidimensional scaling) of a set of matrices $\{\mathbf{X}_k\}$. As described previously, we obtain a solution of the form

$$\mathbf{X}_k = \mathbf{A}\,\mathbf{D}_k\mathbf{B}' + \mathbf{E}_k . \tag{5–15}$$

We now ask: Which characteristics of this representation are unique? That is, among the alternative representations (of the same structural form in the same number of dimensions) that would fit the data as well, which characteristics must remain the same, and which characteristics are free to vary?

It has been proven elsewhere (by Jennrich, reported in Harshman 1970; and by Harshman 1972a; Kruskal 1976, 1977) that with "adequate" data, the only alternative solutions involve changing the order of columns and/or stepping all the entries in a given column up or down by a constant multiplier. ("Adequate" data will be defined below.) To express this algebraically, let an alternative solution be represented as

$$\mathbf{X}_k = \overset{*}{\mathbf{A}}\,\overset{*}{\mathbf{D}}_k\,\overset{*}{\mathbf{B}}' + \mathbf{E}_k , \tag{5–16}$$

where $\overset{*}{\mathbf{A}}$ is an alternative version of the factor-loading matrix $\mathbf{A}$, and similarly $\overset{*}{\mathbf{B}}$ and $\overset{*}{\mathbf{D}}_k$ are alternative versions of $\mathbf{B}$ and $\mathbf{D}_k$, respectively. The uniqueness or intrinsic axis property of PARAFAC-CANDECOMP insures that $\overset{*}{\mathbf{A}}$ can only differ from $\mathbf{A}$ by a rearrangement of columns (which can be represented by postmultiplying $\mathbf{A}$ by a permutation matrix $\mathbf{P}$) and/or a multiplicative rescaling of its columns (which can be represented by postmultiplication of $\mathbf{A}$ by a diagonal matrix $\mathbf{D}_a$), which is compensated for by an inverse rescaling of the columns of $\mathbf{B}$ and/or $\mathbf{C}$. Similar statements hold for the $\overset{*}{\mathbf{B}}$ and $\overset{*}{\mathbf{C}}$ matrices. (The permutation matrix $\mathbf{P}$ must both pre- and postmultiply $\mathbf{D}_k$ so that while the order of the diagonal elements is changed, it remains diagonal.) Thus, the alternative $\mathbf{A}$, $\mathbf{B}$, and $\mathbf{D}_k$ matrices must be related to the corresponding "original" versions of these matrices as follows:

$$\overset{*}{\mathbf{A}} = \mathbf{A}\,\mathring{\mathbf{D}}_a\,\mathbf{P}\ , \quad \overset{*}{\mathbf{B}} = \mathbf{B}\,\mathring{\mathbf{D}}_b\,\mathbf{P}\ , \quad \overset{*}{\mathbf{D}}_k = \mathbf{P}'\,\mathbf{D}_k\,\mathring{\mathbf{D}}_c\,\mathbf{P}\ , \qquad (5\text{–}17)$$

and since the diagonal of $\overset{*}{\mathbf{D}}_k$ is the kth row of $\overset{*}{\mathbf{C}}$, we can also write:

$$\overset{*}{\mathbf{C}} = \mathbf{C}\,\mathring{\mathbf{D}}_c\,\mathbf{P}\ . \qquad (5\text{–}18)$$

In addition, since the effects of any internal rescalings (described by the $\mathring{\mathbf{D}}$ matrices) must "cancel out," the $\mathring{\mathbf{D}}$ must conform to the requirement that

$$\mathring{\mathbf{D}}_a\,\mathring{\mathbf{D}}_b\,\mathring{\mathbf{D}}_c = \mathbf{I}\ . \qquad (5\text{–}19)$$

These indeterminacies can be considered trivial because they do not affect the interpretation of the solution. Geometrically, the effect of the permutation matrix $\mathbf{P}$ is simply to "renumber" the axes without moving them, and the effect of the rescaling $\mathring{\mathbf{D}}_a$ and so forth is simply to stretch or contract the axis in one space, with compensatory contraction or stretch of the corresponding axis in the space of another mode. None of these changes affect the *orientation* of the axes in any of the three spaces. Since it is the orientation of axes that determines their meaning (by determining the pattern of loadings or projections of points onto the axes), the intrinsic axis property insures that the characteristics crucial for *interpretation* of dimensions are uniquely determined as a consequence of simply minimizing the error of fit (provided, as mentioned earlier, that the data are "adequate").

Although they do not affect interpretation, the indeterminacies can be distracting when comparing different solutions; thus, they are often removed by adopting simple conventions. The convention of ordering the factors from the largest to the smallest variance-accounted-for will generally fix the columnar order, and the convention of scaling two of the three loading matrices so that each column has a mean-square of 1.0 will determine two of the three $\mathring{\mathbf{D}}$ matrices in (5–19). (Motivation for this second convention will be provided below when PARAFAC is compared to traditional two-way factor analysis.) The third $\mathring{\mathbf{D}}$ matrix will then be determined by the requirement that the triple product of all three internal rescalings be the identity matrix. To fix sign patterns, dimensions in Modes C and A are reflected (multiplied by -1) when needed so that their mean cubed value is positive, with compensatory reflection(s) applied to Mode B.

Why is it Unique?

There are several perspectives from which the PARAFAC-CANDECOMP intrinsic axis property can be explained: (a) in terms of the algebra; (b) in terms of the geometry; (c) in terms of the added information used; and (d) as a function of "implicit constraints" imposed.

Algebraic Perspective. The proofs of uniqueness (Jennrich, in Harshman 1970; Harshman 1972a; Kruskal 1976, 1977) are too long to recount here, but with the algebraic notation defined above, we can quickly convey some idea of what is behind the uniqueness, based on the approach taken in the Harshman (1972a) proof.

First, let us recall the algebraic basis for the indeterminacy in the two-way case. The rotation problem arises because we can start with any given factorial representation

$$\mathbf{X} = \mathbf{A}\,\mathbf{B}' + \mathbf{E} \tag{5–20}$$

and generate an alternative representation by applying an arbitrary nonsingular linear transformation $\mathbf{T}$ to the loading matrix $\mathbf{A}$, and the compensatory transformation $\mathbf{T}'^{-1}$ to the "factor score" matrix $\mathbf{B}$, as follows:

$$\overset{*}{\mathbf{A}} = \mathbf{A}\,\mathbf{T}, \qquad \overset{*}{\mathbf{B}} = \mathbf{B}\,\mathbf{T}'^{-1} . \tag{5–21}$$

In the q-factor case, $\mathbf{A}$ is an n by q matrix whose columns give the projections of the n variables onto the q factor axes. $\mathbf{T}$ is a nonsingular q by q transformation matrix, the columns of which give the projections of the original axes of $\mathbf{A}$ onto the new axes of $\overset{*}{\mathbf{A}}$, and so $\overset{*}{\mathbf{A}}$ is an n by q matrix whose columns give the projections of the variables onto the new axes.

It is easy to show that the resulting alternative representation of $\mathbf{X}$, that is,

$$\mathbf{X} = \overset{*}{\mathbf{A}}\,\overset{*}{\mathbf{B}}' + \mathbf{E} , \tag{5–22}$$

is equivalent to our original representation (5–20). Simply substitute from (5–21) into (5–22) to obtain:

$$\begin{aligned} \mathbf{X} &= (\mathbf{A}\,\mathbf{T})\,(\mathbf{B}\,\mathbf{T}'^{-1})' + \mathbf{E} = \mathbf{A}(\mathbf{T}\,\mathbf{T}^{-1})\mathbf{B}' + \mathbf{E} \\ &= \mathbf{A}\,\mathbf{B}' + \mathbf{E} . \end{aligned} \tag{5–23}$$

In the three-way case, we can use a similar approach to examine the question of permissible transformations of the $\mathbf{A}$ matrices. Let us consider two alternative representations of a given set of $\mathbf{X}_k$ matrices, such as (5–15) and (5–16), again defining the transformation generating the second set of factor loadings and factor scores in general fashion, as in (5–21). (We will allow our alternative occasion-weight matrices, $\overset{*}{\mathbf{D}}_k$, to be any nonsingular diagonal matrices.)

By substituting from (5–21) into (5–16), we obtain the expression for $\mathbf{X}$ in terms of our original $\mathbf{A}$ and $\mathbf{B}$ matrices, and then determine the conditions under which it is equivalent to (5–15). The substitution of (5–21) into (5–16) yields

$$\mathbf{X}_k = (\mathbf{A}\,\mathbf{T})\,\overset{*}{\mathbf{D}}_k(\mathbf{B}\,\mathbf{T}'^{-1})' + \mathbf{E}\,,$$

which, in general, simplifies to

$$\mathbf{X}_k = \mathbf{A}(\mathbf{T}\,\overset{*}{\mathbf{D}}_k\,\mathbf{T}^{-1})\mathbf{B}' + \mathbf{E}\,.$$

If we define

$$\mathbf{H}_k = \mathbf{T}\,\overset{*}{\mathbf{D}}_k\,\mathbf{T}^{-1}\,, \qquad (5\text{–}24)$$

our expression becomes

$$\mathbf{X}_k = \mathbf{A}\,\mathbf{H}_k\mathbf{B}' + \mathbf{E}\,. \qquad (5\text{–}25)$$

The permissible $\mathbf{T}$ matrices—that is, the ones that produce alternative PARAFAC representations—are restricted to those for which $\mathbf{H}_k$ is diagonal; otherwise, (5–25) will not have the appropriate PARAFAC form. Furthermore, $\mathbf{H}_k$ must equal $\mathbf{D}_k$ if (5–25) is to be valid.

Unlike the two-way case, we find that not all axis transformations $\mathbf{T}$ will work. If $\mathbf{T}$ is a general orthogonal or oblique transformation, then $\mathbf{H}_k$ will not in general remain diagonal for different $\overset{*}{\mathbf{D}}_k$, as required by the PARAFAC model. In fact, it can be proven that this will occur only if $\mathbf{T}$ is a diagonal or a permutation matrix, or some product of these (Harshman 1972a). This corresponds to the case in which $\mathbf{T} = \mathring{\mathbf{D}}_a\mathbf{P}$, so that the relationship between $\mathbf{A}$ and $\overset{*}{\mathbf{A}}$ is as stated in (5–17).

It is easy to verify that when $\mathbf{T}$ has this form, the $\mathbf{X}_k$ represented by the alternative solution is equal to the original $\mathbf{X}_k$. By substituting from (5–17) into the right-hand side of (5–16), we obtain:

$$\overset{*}{\mathbf{A}}\;\overset{*}{\mathbf{D}}_k\;\overset{*}{\mathbf{B}}' = (\mathbf{A}\,\mathring{\mathbf{D}}_a\mathbf{P})(\mathbf{P}'\,\mathbf{D}_k\,\mathring{\mathbf{D}}_c\,\mathbf{P})(\mathbf{B}\,\mathring{\mathbf{D}}_b\mathbf{P})' \qquad (5\text{–}26)$$

$$= \mathbf{A}\,\mathring{\mathbf{D}}_a\,(\mathbf{P}\,\mathbf{P}')(\mathbf{D}_k\,\mathring{\mathbf{D}}_c)(\mathbf{P}\,\mathbf{P}')(\mathring{\mathbf{D}}_b\,\mathbf{B}')\,. \qquad (5\text{–}27)$$

And because $\mathbf{P}\mathbf{P}' = \mathbf{I}$ for any permutation matrix $\mathbf{P}$, these terms in (5–27) vanish. Further, since the diagonal matrices commute, $\mathring{\mathbf{D}}_a\mathbf{D}_k = \mathbf{D}_k\,\mathring{\mathbf{D}}_a$, and by using (5–19), we obtain:

$$\overset{*}{\mathbf{A}}\,\overset{*}{\mathbf{D}}_k\,\overset{*}{\mathbf{B}} = \mathbf{A}\,\mathbf{D}_k\,(\mathring{\mathbf{D}}_a\mathring{\mathbf{D}}_c\mathring{\mathbf{D}}_b)\mathbf{B}' = \mathbf{A}\,\mathbf{D}_k\,\mathbf{B}' = \mathbf{X}_k\,. \qquad (5\text{–}28)$$

If the foregoing discussion seems involved, it might be useful to consider a simple algebraic analogy that makes the uniqueness seem less surprising and which will also be useful when we consider data "adequacy" below. PARAFAC-CANDECOMP performs what might be thought of as the analog of using simultaneous equations to obtain a unique solution: It performs simultaneous factor analysis of several two-way matrices. While the equations underlying a single two-way matrix are not sufficiently constrained to provide a unique solution, the PARAFAC-CANDECOMP equations used for the simultaneous factoring of a set of two-way arrays do constitute a sufficiently constrained set of relations to provide a unique solution, when the data are "adequate." To borrow the example given in Harshman, Ladefoged, and Goldstein (1977), the equation $x + y = 20$ has no unique solution, since an infinite number of x,y pairs will satisfy the weak constraints imposed by the specified relationship. Similarly, the matrix equation $\mathbf{A}\mathbf{B}' = \mathbf{X}$ has no unique solution. However, if we require that our values of x and y also satisfy a second equation, such as $2x + 3y = 55$, then a unique solution is obtained. Likewise, if we require that the same loading matrix $\mathbf{A}$ satisfy two different equations, $\mathbf{A}\mathbf{D}_1\mathbf{B}' = \mathbf{X}_1$ and $\mathbf{A}\mathbf{D}_2\mathbf{B}' = \mathbf{X}_2$, then with adequate data the $\mathbf{A}$ (and $\mathbf{B}$, $\mathbf{D}_k$ matrices) become determined uniquely, except for the trivial indeterminacies of columnar order and scale noted above.

Geometric Perspective. Suppose we plotted the point configurations for each of the two-way slices $\mathbf{X}_k$ of our three-way array, but left out any axes. (These plots are easiest to visualize in a two-dimensional case, so we might imagine that our plots are based on the first two principal components of each slice.) When the PARAFAC-CANDECOMP model is appropriate, we would observe a very similar configuration of points across the various slices. However, there would be certain systematic differences between the plots. When we compared one plot to another, all the points on one might be displaced outward from the origin along a certain line by an amount proportional to their distance from the origin along that line. This would correspond to a "stretching" of the space in a certain direction. In other directions, the space might be "contracted." By comparing the several spaces, one can detect such stretching and contraction and identify the directions along which the space is recurrently stretched or contracted.

Now, to explain these systematic variations in the configurations in terms of simple variations of a common set of factors, we must select an orientation of factor axes in the spaces that "line up with" the directions of stretch or contraction. Then, we can account for the systematic displacements of the point locations (the "stretches" or the configurations) as simply due to the increase or decrease in the importance of particular factors. This geometric interpretation bears a simple relation to the algebraic representation: The $\mathbf{A}$ (or $\mathbf{B}$) matrix would describe the common pattern of projections of points onto factor axes, and the $\mathbf{D}_k$ matrix would give the stretches or contractions in the kth space; that is, the rth diagonal cell of the $\mathbf{D}_k$ matrix would describe the proportion by which factor r was stretched on occasion k.

When one imagines looking at plots of two or more such spaces, each resembling the others but stretched or contracted in certain directions, it is easy to see how lining up the factor axes with the directions of stretch will force a particular unique orientation for the axes. Furthermore, to the extent that the axes deviate from proper alignment with these directions of stretch, the resulting factor solution will deviate from reproducing the actual stretches in the configurations across different slices of the three-way array, and so the goodness-of-fit will drop from its maximum value. Thus, by simply maximizing the fit of the PARA-

FAC-CANDECOMP model, one necessarily determines the axis orientation, without any recourse to outside criteria such as simple structure.

Use of Richer Information. From both the geometric and algebraic explanations, it becomes apparent that the PARAFAC-CANDECOMP model takes advantage of the extra information present in a three-way array to determine the "true" factor axis orientations. It is not surprising that proper use of this extra information can resolve the rotational indeterminacy of two-way factor analysis. Yet other three-way models, with access to the same type of information, do not provide a unique orientation for the factor axes because they do not use the information in the same way. Tucker's three-way model, for example, allows representations in which the axes can be oriented in any position, and the Mode *A* axes can be oriented independently of the Mode *B* axes, which in turn can be oriented independently of those in Mode *C*. The model compensates for the effects of these variations by adopting a more complex representation of the relationships among dimensions. Not only are there contributions from the Mode *A* loadings of dimension one multiplied by the Mode *B* and Mode *C* loadings for that dimension, but also contributions from the Mode *A* loadings of dimension one multiplied by the Mode *B* loadings of dimension two, times the Mode *C* loadings of dimension three, and so on for all possible combinations of dimensions in the different modes. This model produces a family of more complex representations that can generate the same patterns of changes across the three-mode data set as the PARAFAC-CANDECOMP model (and which can generate more complex patterns, as well). However, when the Tucker and PARAFAC models have approximately the same goodness-of-fit, the PARAFAC representation would probably be preferred because of its simplicity and straightforward empirical interpretation. In fact, some implementations of Tucker's model have methods of transforming the solution so as to approximate PARAFAC-CANDECOMP-type representations (see Kroonenberg and de Leeuw 1980).

Use of "Implicit Constraints." It might be argued that fitting the PARAFAC-CANDECOMP model is equivalent to imposing extra constraints on the more general Tucker-type of representation. (The exact form of the "constraints" that make the two models equivalent will be discussed below.) From this perspective, three-way analysis with a model "constrained" to optimize proportional profile form is like two-way analysis with the solution "constrained" to optimize simple structure or some other special form. Using this analogy, one might argue that it is these extra constraints of factor proportionality across levels of Mode *C* that remove the rotational indeterminacy, just as "simple structure" constraints remove the indeterminacy in the two-way case.

Even if one adopts this perspective, however, it is interesting to note that the PARAFAC "constraints" do not specify anything about the *form* of the factor-loading matrices **A** , **B** , or **C** . The factors may be simple or complex in structure, and they may be orthogonal or oblique (in the direct-fit case). PARAFAC-CANDECOMP does not specify preferred loading relationships *within* a mode, as simple structure does, but rather constrains the way a factor varies *across* different slices of the three-way array

($A_k = AD_k$). We will argue below that the PARAFAC-CANDECOMP-type of "constraint" differs from other two- and three-way constraints in that it is intrinsic to a plausible model of factor variation, and that comparison of split-half solutions can be used to test its appropriateness. Furthermore, if the model is found to be appropriate, this result can be used to provide empirical evidence in support of particular theoretical constructs being inferred from, or tested by means of, the given solution.

When is it Unique?

Not all sets of three-way data contain the information necessary to uniquely determine the orientation of factor axes; the data set must be "adequate." If we think of three-way data as composed of several two-way "slices," then the conditions of data adequacy can be briefly summarized as follows: The slices must involve the same set of factors *and* differ in the relative importance or contribution of these factors.

Same Factors across Slices. All current three-way factor analysis models (with the exception of Corballis and Traub 1970; and Swaminathan, chapter 8) assume that there is a common set of factors that generates data at all different levels of the three-way array. The models only differ in terms of how the weighting or pattern of combination of the factors is allowed to vary across the different levels in order to account for systematic differences between successive two-way arrays.

This assumption of a common set of factors may not always be strictly appropriate. In a long-term longitudinal study, for example, the pattern of loadings on a given factor may shift somewhat, because the meaning of test or personality items may not be the same when administered to children as when administered to adults, even if the underlying trait or process is the same (Harshman and Berenbaum 1981). It is also possible that the empirical entity represented by a particular factor will change somewhat in "quality" or nature across time.

To take another example, suppose we are analyzing an array of stimulus ratings, where each of several stimuli are rated on each of several scales by several raters. The cognitive or perceptual dimensions underlying the ratings may not have exactly the same quality from one rater to another, even if they are constant within a given rater across stimuli and rating scales. Suppose two raters both use the dimension of "valuable-worthless." One rater may think of the dimension in a slightly different way than another, which would imply that one person's pattern of factor loadings might differ from the other by more than a simple proportional reweighting.

Nonetheless, the assumption of common factors but different weighting or combination rules across the levels of the array is a useful approximation that greatly reduces the number of parameters in the solution and usually leads to an interpretable summary account of what is going on. (For a successful application of PARAFAC to lifespan longitudinal data, see Haan 1981.) Indeed, the same kind of common factor assumption is basic to the use of two-way factor analysis or multidimensional scaling.

The PARAFAC-CANDECOMP model actually allows considerable flexibility in the dimensional representation across levels, because it does not insist that every factor be involved at every level of the data. It is possible that some factors have zero loadings on some variables, or on some people, or on some occasions. In fact, one can imagine a complete change in the factor structure of the slices over successive occasions, as one set of factors comes to take on zero loadings and another begins to take on nonzero loadings. The only "sameness" that must be assumed is the qualitative sameness of a given factor across slices; that is, if a factor is present on two or more occasions, it is assumed to have the same pattern of factor loadings on those occasions, except that it may be stepped up or down by the occasion weights.

Differences in Factor Importance across Slices. As Cattell noted in the quote cited earlier, comparison of two-way arrays that are merely replications of one another—differing only by random sampling—will provide no added information over a single two-way array; hence, it will not reduce the indeterminacy of the solution. There must be systematic differences between at least some of the slices of the three-way array. In particular, the slices must differ in terms of relationships among variables, and these differences must be attributable to shifts in the relative importance of underlying factors. Furthermore, the pattern of shifts must be distinct for each factor; if two factors shift in importance in the same way across levels, they will not be uniquely determined. In other words, each factor that is to be uniquely recovered must have a pattern of effects in each mode that is distinct from all other factors.

We can gain an intuitive understanding of the necessary conditions for uniqueness by referring back to the same algebraic and geometric analogies that we employed earlier.

Algebraic Interpretation. We have compared the simultaneous factorization of several two-way arrays to the solution of simultaneous equations. As a simple example, we took the equations $x + y = 20$ and $2x + 3y = 55$; in this case, there was only one value of x and y that would satisfy both equations (namely, $x = 5$, $y = 15$). However, if our second equation were $2x + 2y = 40$, there would be no unique solution. In order for x and y to be determined uniquely, the ratio of the x coefficients in the two equations must be different from the ratio of the y coefficients. Similarly, for two factors to be distinguished and their loadings determined uniquely, there must be at least two levels of Mode *C* for which the ratio of the coefficients for the first factor differs from the ratio for the second factor. If two factors always change by the same percentage across levels of Mode *C*, then in the solution these two factors will not be uniquely determined.

Because all three modes have the same status in the basic PARAFAC model (5–3), the requirement that factors have distinct patterns of variation applies in each mode. If two factors have proportional loadings in any one of the three modes, but show distinct loading patterns in the other two, then there will be a family of possible solutions in which different linear combinations of these two factors occur (for example, see Harshman 1970, 41). (We note in passing that if *two* factors have proportional loadings in *two* modes, they cease to be distinct factors; their contribution is represented by a single factor that has loadings proportional to theirs in the two modes where they were proportional, and has loadings equal to a weighted sum of their loadings in the third mode. The same thing happens if the two factors have proportional loadings in all three modes.)

Geometric Interpretation. As we pointed out earlier, the orientation of factor axes can be established geometrically by comparing the configuration of points derived from different two-way slices and determining the directions along which one configuration of points is stretched relative to another. But such comparisons will only be informative if the factors are stretched by different amounts. If two axes were each stretched by the same amount, we would observe only a uniform dilation of the plane in which those axes lie; all interpoint distances—that is, their projections in that plane—would be increased to the same degree. Since such a uniform dilation would be produced by an equal stretch of any set of axes in that plane, regardless of orientation, we could not determine the correct orientation of the two axes in that plane from a comparison of these two configurations. In order to distinguish and uniquely orient these two factors, we would have to compare other spaces (using other slices of our three-way array) until we found two spaces for which the factors in that plane did not increase or decrease by the same proportion.

Partial Uniqueness. The geometric interpretation of the necessary conditions for uniqueness permits us to easily grasp how the uniqueness need not break down entirely if the conditions are not met for all factors. Those factors that have distinct patterns of changes across three modes will be uniquely determined, but those that do not will show rotational indeterminacy within the subspace that they span. For a concrete example, consider the comparison of two three-dimensional spaces in which factors 1 and 2 each increase by 50%, but factor 3 increases by 75%. The second space will look like one that is uniformly expanded by 50% and then stretched by an additional 25% along one direction. This extra stretch will allow us to identify the axis orientation for factor 3. But the plane in which factors 1 and 2 reside is uniformly expanded, and so the axes can be rotated to any arbitrary position within this plane and still be able to reproduce the relationship between the two spaces.

Theoretically, a partial breakdown of uniqueness should only create ambiguities of interpretation within the subspace(s) spanned by the inadequately distinct factors. In practice, however, if major factors are nonunique, it may seem as though no interpretable solution can be obtained with a given data set. This is because PARAFAC solutions are usually obtained in low dimensionalities first, to determine if any sensible dimensions can be extracted, before proceeding to higher dimensionalities. If nonunique factors account for a substantial proportion of the variance, these factors may emerge in the lower-dimensional solutions, producing uninterpretable results that discourage attempts at further analysis.

When several factors are nonunique, it is commonly because they do not change across one of the three modes. Such factors can be removed from the data by appropriate PARAFAC preprocessing (by centering across the mode in which they are constant). This allows the remaining factors to be analyzed successfully.

Minimum Conditions for Uniqueness. Successive proofs of uniqueness have established successively less stringent requirements for determination of unique axes. Jennrich's original proof (in Harshman 1970) showed that q factors could be determined uniquely if the **A**, **B**, and **C** matrices all have rank q. But empirical results (Harshman 1970, 39–44) suggested that this condition was stronger than necessary. Also, Cattell and Cattell (1955) had presented an algorithm for orthogonal rotation of two factor-pattern matrices to proportional profiles that would provide a solution for any number of factors. In 1972, a PARAFAC uniqueness proof was formulated that showed that if the factor-loading matrices for two modes had rank q, the third mode need only have two levels (and hence a maximum rank of 2), even if there were more than two factors in the solution (Harshman 1972a). Uniqueness was obtained so long as each factor had a distinct ratio of change across the two levels. Although it was not stressed at the time, the proof also implied that even easier conditions obtained when there were more than two levels. In that case, not all factors need show distinct amounts of change across any particular pair of levels; some factors might be distinguished by means of distinct changes between levels 1 and 2, others by distinct changes between levels 2 and 4, and so on.

Some surprising uniqueness properties of three-way arrays have been discovered empirically, one of which was reported in the original PARAFAC monograph (Harshman 1970, 44). It was found that more factors could be uniquely determined by a three-way array than would seem possible from a conventional two-way perspective. In particular, it was found that more factors could be determined uniquely than there were levels in any of the three modes of the data array! In the example cited, 10 factors with random loadings were used to construct an $8 \times 8 \times 8$ data array, and PARAFAC correctly and uniquely recovered these 10 factors upon analysis of the array. Since no two-way slice of such an array could have more than rank 8, the unique recovery of 10 factors showed the remarkable degree to which the extra richness of relationships in a three-way array permits recovery of information that would not be possible in a two-way array.

Kruskal, intrigued by this empirical finding, undertook a rigorous investigation of the mathematical issues of uniqueness and generalized rank of three-way arrays. His results (Kruskal 1976, 1977) constitute by far the most complete treatment of this question to date. Among other things, he shows mathematically why 10 and even 11 PARAFAC factors can be uniquely determined by an $8 \times 8 \times 8$ array. He proves a number of important theorems concerning uniqueness and generalized rank. However, in this chapter, we will sketch only one of his basic results concerning uniqueness.

To lay the groundwork for Kruskal's result, we first restate a familiar fact concerning rank and then, in parallel terms, explain a stronger property related to rank. Consider the $n \times q$ matrix **A**. The rank of **A** is equal to r_a if the columns are linearly independent in at least one set of r_a columns from **A**, and if there is no set of $(r_a + 1)$ columns from **A** that consists of linearly independent columns. Now let us consider a stronger property, which Kruskal defines but does not name, but which we shall call *k*-rank ("Kruskal-rank"). **A** has *k*-rank of k_a if the columns are linearly independent in every set of k_a columns from **A**, and if there is at least one set of $(k_a + 1)$ columns of **A** that includes linearly dependent columns. For example, if **A** has six columns, all of which are linearly independent except one (which is a linear combination of three others), then **A** has rank of five but *k*-rank of three.

A PARAFAC solution consists of three factor-loading matrices, each with its corresponding *k*-rank. We have used k_a to represent the *k*-rank of **A**; in a similar fashion, we can let k_b represent the *k*-rank of **B** and k_c the *k*-rank of **C**. Kruskal shows that if the factor-loading matrices **A**, **B**, and **C** provide an exact *q*-factor PARAFAC representation of the three-way array **X**, this representation is unique (up to trivial permutations and rescalings, as discussed earlier) whenever

$$(k_a + k_b + k_c) \geqq (2q + 2) \ . \tag{5–29}$$

(Most often, what we called **X** will in fact be $\hat{\mathbf{X}}$, the *q*-dimensional fitted part of a data array of higher dimensionality.)

Not only does Kruskal's result prove that it is possible to extract more factors than variables and still obtain a unique solution, but it also confirms another surprising property that was indicated by earlier results obtained with synthetic data: PARAFAC can uniquely recover factors that are linearly dependent on other factors. In unpublished research, we have found that if a factor is a linear combination of three other factors in two modes, but has linearly independent patterns of change in a third mode, this factor can be uniquely recovered by PARAFAC analysis. However, if it is only a linear combination of two other factors, then it will not be uniquely recovered. While Kruskal's articles do not make this consequence explicit, his result implies that this will happen and further implies an even stronger result. It follows from Kruskal's theorems that a factor can be a linear combination of other factors in *all three modes* and still be uniquely recovered, provided that it is a combination of a sufficient number of other factors. (A factor that is a linear combination of three other factors in all three modes cannot be uniquely recovered, but one that is a linear combination of four other factors can be uniquely recovered.) In further synthetic data tests, we have recently verified that uniqueness holds under these conditions as well.

Aside from their mathematical interest, such results have potential bearing on the question of "higher-order" factors, which are sometimes thought of as linear combinations of "lower-order" factors. They are also relevant to the question of an oblique axis generalization of the indirect fitting model (PARAFAC2), since data computed from oblique axes can be reformulated as data generated by orthogonal axes plus appropriate linearly dependent extra dimensions (Harshman 1973). The surprising strength of the intrinsic axis property suggests that all the interesting characteristics and potentials of PARAFAC-CANDECOMP trilinear representation of three-way arrays have yet to be uncovered.

Empirical/Inferential Significance of the Intrinsic Axis Property

We have argued that the intrinsic axis property of PARAFAC-CANDECOMP gives the factors obtained by this method a special empirical significance. But some authors have questioned this, pointing out that other matrix decomposition methods, such as principal components, can also be interpreted as providing a unique set of dimensions (for instance, see Schönemann 1972). In fact, even two-way factor extraction followed by analytic rotation (using Varimax or some similar criterion) can be interpreted as a two-step procedure that provides a unique solution. Why, then, should the "unique axes" provided by PARAFAC-CANDECOMP be any more meaningful than the "unique axes" provided by these other methods? We suggest that there are two basic reasons: (a) empirical plausibility; and (b) empirical confirmability.

Plausibility of Assumptions

The first basic consideration is the empirical plausibility of the assumptions required to obtain a unique solution. With intrinsic axis methods, one makes no additional assumptions beyond the appropriateness of a particular three-way factor model. Since for this model different axis orientations yield different fit values, the "correct" orientation of factor axes is established as an intrinsic part of fitting the model to data. With the other methods, in addition to the plausibility of the factor model, one must also consider the plausibility of the additional assumptions involved in selecting axis orientation; as noted earlier, these assumptions have little or no empirical rationale in many applications.

In principal components analyses, for example, the axis locations are fixed by the additional requirement that the first dimension explain the maximum amount of variance possible and similarly that each successive dimension explain as much remaining variance as possible. But if there are several factors underlying a given data set, maximizing the variance explained by the first dimension usually makes the first dimension represent a *combination* of the underlying influences operating in the situation, rather than any one of them. It is not likely that real influences or processes will correspond to axes which have the successive variance-maximizing characteristics assumed by principal components analysis. Likewise, as noted earlier, the assumptions underlying simple structure rotation are often debatable, particularly in the many common applications of factor analysis in which the variables are not selected with an eye to eventual factor rotation but are determined by other characteristics of the study. With such data, the variables are likely to be factorially complex and concentrated in a particular domain in a way inconsistent with simple structure.

It would be misleading, however, to say that PARAFAC involves no new assumptions. The assumptions that determine axis orientation are external to the factor model in the two-way case and intrinsic to it in the PARAFAC-CANDECOMP three-way case. As was pointed out earlier, the direct application of PARAFAC to profile data requires the assumption of system variation, although indirect fitting by means of covariance analysis allows weaker assumptions. We argue that these assumptions are "natural" or empirically plausible for many types of three-way data (so much so that they constitute a straightforward generalization of the factor model to three-way arrays). They are also, in a certain sense, less restricting; there are no assumptions about the pattern of factor loadings *within* a given occasion but only assumptions about the form of factor variation *between* occasions. Thus, the factors can have simple or complex structure and can be orthogonal or oblique (in the profile data case), whichever best fits the data.

Confirmability

Since PARAFAC-CANDECOMP is not a completely general three-way model, we need to make an argument for its appropriateness when applying it to a given data set. When a PARAFAC solution is to be used as a starting point for scientific inferences, this argument needs to be particularly persuasive. Hopefully, it would consist of more than pointing out the conceptual plausibility of the model in a particular application. It would consist of empirical *evidence* that two essential preconditions are fulfilled: (a) the PARAFAC-CANDECOMP model is appropriate for these data (at least as a first approximation); and (b) the pattern of factor variation within the data is adequate to determine reliably the orientation of axes. Only then can the orientation of axes in the solution be taken as substantial evidence for or against specific empirical hypotheses. (In the same sense, factors determined by rotation of a two-way analysis to some target orientation or to the position that maximizes some "simplicity" function do not provide substantial evidence for any empirical hypothesis until one is persuaded of the likely appropriateness of the rotation criterion.)

Fortunately, one can test whether the two essential preconditions are fulfilled. It is an important consequence of the intrinsic axis property that analyses can be performed in such a way as to provide *confirmatory evidence* for both the appropriateness (or partial appropriateness) of the model and the adequacy of the data to determine a unique axis orientation. As far as we are aware, it is not generally possible to obtain comparable internal evidence for the validity of axes when their orientation is determined by other methods, such as simple structure or principal components.

Empirical Tests of PARAFAC Axes. The simplest and most common approach to testing a given PARAFAC solution has been to use a split-half or double-split-half technique. Suppose we wish to demonstrate that the pattern of factor variation across all three modes of a given data set is appropriate and adequate to determine a stable orientation of factor axes. If we split our data set into two halves (for example, by randomly assigning the subjects to two different subsamples) and perform a separate PARAFAC analysis of each half, we can compare the solutions to assess whether or not the necessary pattern of three-way variation was present in the data. If the factors did not show adequate proportional changes in relative importance across levels of Mode *C* (as well as *A* and *B*), then the same set of axes will not be obtained in the two solutions. Instead, arbitrary orientations of factor axes will be obtained, and these will differ from one split-half to the other. If certain factors are well determined, but other are not, this too will be revealed. *Thus, the observation of similar factors in two different split-halves provides confirmatory evidence that the data contains enough systematic variance of the appropriate kind to determine a unique, stable set of factor axes.*

The split-half procedure is an approximate test of stability, but the results obtained will depend to some extent on the particular random split that is used for the test. More sophisticated procedures allow one to minimize this dependence on a particular split. The simplest approach is to use a double-split-half procedure. "Orthogonal" splits can be generated by dividing the data into four subsamples—call these S_1, S_2, S_3, and S_4—and defining the two split-half comparisons as follows: Split One is $(S_1 + S_2)$ versus $(S_3 + S_4)$, and Split Two is $(S_1 + S_3)$ versus $(S_2 + S_4)$. With such a procedure, an accidental inequality between the halves of one split is likely to be rectified in the other orthogonal split. Validation of a given dimension in either of the two orthogonal splits is sufficient to demonstrate its "reality."

More sophisticated tests are possible using resampling methods, such as "jackknifing" (Mosteller and Tukey 1977; chapter 8) and "bootstrapping" (Efron 1982). In these methods, one takes multiple repeated samples from the original data set and performs independent analyses of these samples. By examining the variation in the solutions that result, inferences can be made about the reliability of different aspects of the solution. A brief discussion of these techniques can be found in Gifi (1981, section 12) and a very simple introduction to bootstrapping is given in Diaconis and Efron (1983). Weinberg, Carroll, and Cohen (1983) have recently applied these methods successfully to INDSCAL solutions.

It is also possible to perform significance tests for particular structural characteristics in three-way data by applying the methodology of permutation and randomization tests (Edgington 1969, 1980). For example, one can test whether there is significant "system variation" in a three-way array by proceeding as follows: Fit a PARAFAC model in q dimensions and note the goodness-of-fit value (Stress or r-squared). Then randomly permute the data observations across levels of Mode *C* but within the same levels of Modes *A* and *B* (permuting the values of subscript k, but keeping the i and j subscripts unchanged). By applying a different random permutation within each Mode *C* "tube" of the data (a different permutation of the k for each i,j pair), we can "scramble" the systematic variation across the third mode while preserving the systematic variation across Modes *A* and *B*. The resulting data will still contain q factors, but they will not show the systematic variation across Mode *C* necessary to determine a reliable unique solution. If the null hypothesis of no Mode *C* system variation is true, then the original data should not differ systematically from these permuted data sets. If the alternate hypothesis of system variation is true, the fit of the PARAFAC model to the original data should be better than to the permuted data.

Suppose we perform the permutation process 19 times and fit the *q*-factor PARAFAC model to each of the permuted data sets. We then rank the 20 fit values (19 permuted plus one nonpermuted). If there is no system variation, the fit for the unpermuted data is equally likely to be at any of the 20 possible ranks. Thus, if the fit value obtained for all 19 permuted data sets is less than that for the original nonpermuted data, we have observed a ranking that would happen only one time out of 20 under the null hypothesis. Thus, we can reject at the .05 level the null hypothesis of no system variation; there appears to be systematic variation across Mode *C* that is fit by the PARAFAC model. (Similar applications of permutation tests to multilinear models are described in Harshman, Green, Wind, and Lundy 1982; Harshman and Reddon 1983; Hubert 1983).

No Similar Tests for Two-Way Solutions. Note the contrast between intrinsic axis solutions and principal components or simple structure solutions. With either of these latter methods, the finding of consistent axis orientations across two split-halves (or by other means, such as the bootstrap) does *not* constitute evidence for their empirical validity. Similar axis orientations necessarily occur, so long as the configurations of points in the factor spaces are similar across split-halves, since these two-way methods determine axis orientation by finding directions in the configuration that maximize some simplicity or variance criterion. Indeed, so long as the split-half configurations are similar, any arbitrary rotation principle based on relations of axes to points in the space (for instance, a "most-complex-structure" criterion) would show similar results in the two split-halves. Obviously, no evidence for the correctness of a particular rotation criterion is provided by such a replication. For intrinsic axis methods, however, a consistent configuration is not sufficient to ensure consistent axis orientations. In each split-half there must be systematic stretches and contractions of the configuration as one proceeds across levels of the third mode, and these stretches must be in consistent directions in the two split-halves. Thus, replication of an intrinsic axis solution demonstrates the reliable presence of those characteristics of the data postulated by the three-way model and required to orient factor axes. Hence, in some sense, it validates the criterion used for orienting axes.

Cattell (1978) and others have sometimes argued for the confirmability of simple structure rotation. They have suggested that the occurrence of clear simple structure in the rotated solution will provide evidence for the "reality" of the hyperplanes and thus the appropriateness of seeking simple structure in the first place. While there is some logic to this argument, the conditions under which it might actually be persuasive seem quite limited. The presence of "clear hyperplanes," with many near-zero loadings on each factor, may often be an artifact of how the variables were selected (for example, if they were selected in clusters of related items) and may not indicate any structure in nature. But even if we could obtain "neutral" or representative selections of variables, there is little distributional or Monte Carlo data on the likelihood of obtaining various degrees of simplicity "by chance" when the variables are selected "randomly." (However, see Cattell 1978, appendix A.6.) Furthermore, it is not clear how to go about getting better data of this kind, since such Monte Carlo work would very likely have to be based on questionable assumptions concerning distributions of variables and of factor loadings on variables. Thus, although there might be a theoretical argument for the confirmability of simple structure, based on clarity of hyperplanes when the variables are selected in a certain way, this approach is much more difficult to apply in practice than the one for confirmability of intrinsic axis solutions by split-half or related methods. Consequently, the evidence provided by split-half confirmation of an intrinsic axis solution is usually much stronger.

Empirical Implications of Confirmation

Split-half or bootstrap confirmation of intrinsic axis factors has empirical significance beyond simply verifying that the model was (at least partly) appropriate and the data adequate. By demonstrating the occurrence of a particular kind of systematic variation within the data, it presents us with a fact that is hard to explain, except in terms of the conceptual framework that underlies the proportional profiles idea (that is, in terms of variation of contributions of "empirical unities" of the sort conceptualized by the PARAFAC three-way factor model).

To put it geometrically, suppose separate factor analyses of individual slices of a three-way array reveal a series of parallel configurations in which variables show basically the same relationships, but in which the points in some configurations are displaced outward from the origin in a particular direction by an amount proportional to their distance from the origin and contracted inward in other directions in a similar proportional way. Suppose, in other words, that the configurations show the sort of coordinated shifts of points describable as systematic stretches and contractions, as postulated by the PARAFAC-CANDECOMP model. This systematic pattern of relationships constitutes a striking empirical fact about the data that is very hard to explain except as the result of variations in the strength of a few underlying influences or processes that affect the relationships among variables. The systematic displacement of points in a configuration along a certain direction seems to imply some common influence on the several variables that get displaced, a common influence that has increased or decreased in magnitude from one configuration to the next. In other words, there is an empirical reality that corresponds (at least to some extent) to the abstract mathematical "factor" oriented along the direction of stretch.

Inductive versus Deductive Use of Factor Analysis. PARAFAC and related intrinsic axis methods significantly strengthen the inductive or hypothesis generating ability of factor analysis. When split-half testing (or similar methods, such as bootstrapping) confirms a set of dimensions, one can proceed with considerably stronger confidence to the construction of empirical hypotheses based on these dimensions. However, the significance of the split-half evidence can be both inductive and deductive. In one sense, the appeal to split-half methods is itself a test of a hypothesis; that is, in each half, one generates a hypothesis about the orientation of axes and tests it by replication in the other half. However, PARAFAC can also be used in a purely deductive mode to test hypotheses generated by other means, such as theory concerning configurations and axis orientations. In this mode, it will usually differ from other "deductive" or hypothesis-testing applications of factor analysis in that the test would not involve rotation-to-target to see how close a given matrix can be approximated; instead, the solution would be compared to the theoretical matrix without rotation (but possibly with rescaling of columns) to see how close both the configurational part of the hypothesis and the axis orientation part of the hypothesis agree with the data.

The configuration and axis orientation aspects of theory can be tested separately, however. If the fit value does not decrease much when the PARAFAC solution is rotated from the observed position to the position closest to the theoretical hypothesis, and if the configuration provides a good match to theory, then the evidence against the axes predicted by theory might be considered weak. However, if a PARAFAC solution in several different data sets shows an axis orientation that diverges from theory in a consistent manner, then the theory would be cast into doubt even if the theoretical orientation of axes had only slightly poorer fit. This is because the theory would presumably be unable to account for the consistency of the divergent PARAFAC results.

Limitations on Empirical Interpretation. Replication of intrinsic axis solutions—across split-halves or different experiments—implies that something meaningful and systematic is going on that can, at least in part, be captured by the PARAFAC-CANDECOMP model. But we must remember, of course, that this evidence does not imply that the model is perfectly appropriate, just that it captures enough variance to be a generalizable approximation to a more complex world. As noted earlier, for example, the model's assumption that common factors are present at the different levels of the data may not always be correct. However, the nature of the model is such that the axes established by such an approximation are likely to retain empirical meaning, even when the reality is considerably more complex.

When the reality is too complex for such an approximation to capture the bulk of the systematic variance, analysis of both real and simulated data suggests that the PARAFAC model will be much more likely to give degenerate solutions, with uninterpretable highly correlated dimensions, than misleading interpretable ones. And when these degenerate solutions arise, there are ways of constraining the PARAFAC solution so that a meaningful subset of the variance can often be captured and interpreted. With much more complex data, one might want to go to a more general model, such as Tucker's three-mode model, but one then loses the intrinsic axis property. To compensate for this loss, a constrained PARAFAC solution might be attractive as a supplement; it could provide an intrinsic axis reference point to aid in interpretation and rotation of the Tucker solution. Another intermediate approach is to rotate Tucker's core matrix to approximate diagonal form (McCallum 1976; Cohen 1974; Kroonenberg 1981b).

COMPARISON WITH TUCKER'S AND OTHER MODELS

Comparison with Tucker's Three-Mode Model

Tucker's Model "T3"

Tucker was the first to extend factor analysis to three-way arrays (Tucker 1963, 1964, 1966), and his model is still the basic reference point to which all other three-way procedures are compared. It is a very general model and incorporates most of the other three-way models as special cases. Kroonenberg and de Leeuw (1980) distinguish two versions of the model: (a) Tucker's original version, which reduces dimensionality of all three modes; and (b) a variant, which only reduces dimensionality in two modes (that is, it does not reduce the dimensionality of Mode C). They call the first model "T3" and the second "T2." We will adopt this terminology and concentrate our initial discussion on T3.

Our objective in this discussion is to compare and contrast the general T3 model and PARAFAC-CANDECOMP. For more details on T3 and its applications, the reader is referred to the works by Tucker (cited above) and to other chapters in this volume, including those by Bloxom and Kroonenberg.

Tucker's T3 in Scalar Notation. If x_{ijk} is an entry in a three-way data array, the model T3 can be written in scalar notation as follows:

$$x_{ijk} = \sum_{r=1}^{q_a} \sum_{s=1}^{q_b} \sum_{t=1}^{q_c} a_{ir} b_{js} c_{kt} g_{rst} \quad . \qquad (5\text{–}30)$$

Here the a, b, and c coefficients have the same meaning as in the PARAFAC model (equation [5–3]); they are the factor loadings or weights for Modes A, B, and C, respectively. Tucker calls the small three-way array of g_{rst} coefficients the "core matrix." The g coefficients describe the sizes of interactions among factors across modes, as will be explained below. Instead of a single summation, there is a triple summation that runs over all combinations of entries in the three modes. The number of factors in one mode does not necessarily equal the number in another, and so in (5–30), q_a is used to represent the number of factors in Mode A, q_b the number in Mode B, and q_c the number in Mode C.

Since we have already discussed PARAFAC in some detail, we begin our discussion of T3 by noting the basic characteristics that the two models share. First, they are both designed primarily for direct fitting of the data matrix rather than indirect fitting of covariances, although, like PARAFAC-CANDECOMP, T3 can be applied to covariances to perform indirect factor analysis or to scalar products in order to perform multidimensional scaling (for example, see Tucker 1972). Second, both T3 and PARAFAC-CANDECOMP extend the idea of proportional changes in factor contributions to include the third mode; the Mode *C* weights are analogous to Mode *A* and *B* weights in that they simply multiply the contributions of a given factor by a coefficient representing the importance of that factor at a particular level of the given mode. Third, both T3 and PARAFAC-CANDECOMP are even-handed with respect to the role of different modes; all modes have the same formal properties—none has a preferred or special status.

There are, however, five main differences between Tucker's model and PARAFAC. The first is conceptual; the other four are algebraic and follow from the first one:

a. T3 is based on a different conception of "factor" than PARAFAC, although reconciliation is possible, as we shall see below.
b. In T3, all between-mode combinations of loadings are permitted, not only $a_{i1}b_{j1}c_{k1}$ but $a_{i1}b_{j3}c_{k2}$, and so forth; hence, each factor in Mode *A* potentially "interacts with" every one of the factors in Modes *B* and *C*, in all possible combinations; similarly, all factor combinations are possible for the other modes.
c. T3 incorporates a fourth set of coefficients, the g_{rst}, which describe the size of the factor "interactions" across modes; hence, T3 is a quadrilinear rather than trilinear model.
d. T3 allows the number of factors in Mode *A* (the number of columns of **A**) to differ from the number of factors in Mode *B*, which in turn may differ from the number in Mode *C*.
e. T3 does not have the intrinsic axis property but instead has a greater transformational indeterminacy than two-way factor analysis.

Differences in the Conceptual Model. As noted above, the Tucker T3 model was derived from a different basic conceptual idea of "factor" than that used by PARAFAC. Briefly, the difference is that a PARAFAC factor is more global than a T3 factor. For PARAFAC, a factor represents an empirical entity, process, or influence that is observed as a result of the situation being measured, rather than as a result of any particular mode of data classification or measurement. Thus, a single PARAFAC factor would be expected to have influence across the levels of all three modes of measurement. In other words, a PARAFAC factor is not "in a mode" but in a situation itself, with influences on or relationships with each of the three modes. In contrast, a T3 factor is conceptualized as an aspect, characteristic, or idealized type of level of a given mode. From Tucker's perspective, it is not the factors by themselves that generate variation in the data but rather the *interaction* of factors, that is, the interaction of particular aspects (factors) of Mode *A* with particular aspects (factors) of Modes *B* and *C*.

The distinction may be clarified by an example. Suppose we have a set of "semantic differential" data, a 25 × 30 × 200 array consisting of ratings of 25 stimuli on 30 rating scales, as made by 200 raters. From the PARAFAC perspective, the factors that we are seeking are global "dimensions of affective meaning." In the terms used by Osgood, Suci, and Tannenbaum (1957), these semantic dimensions might correspond to "Evaluation," "Potency," and "Activity." Such dimensions are not "in" any of the modes but are thought to be basic to the nature of affective meaning and are thus a natural part of the data-generating situation; through the particular stimulus-rating procedure, these global dimensions have patterns of relative influence or importance (given by loadings) for the several levels of each of the data modes. Thus, the first column of the Mode *A* loading matrix might describe the relationship between "Evaluation" and each of the stimuli (that is, which stimuli were perceived as strongly "good" or "bad"). The first column of **B** would then describe the relationship between "Evaluation" and the rating scales (that is, which rating scales were strongly expressive of "goodness" or "badness"). Finally, the Mode *C* loadings would show which raters stressed evaluative aspects strongly in making their ratings.

From Tucker's perspective, the semantic differential rating scale data is described differently. Just as each original data value is the result of the interaction of a particular rater with a particular stimulus and a particular rating scale, so the underlying patterns in the data are described as the result of interactions of features of stimuli, aspects of rating scales, and traits of raters. The objective of three-mode factor analysis is to discover those few different *features* or *types* of the stimuli (the Mode *A* factors), *kinds of components* of the rating scales (the Mode *B* factors), and *traits* or *types* of raters (the Mode *C* factors) that permit an adequate description of the patterns underlying the observed data variation. Tucker would call these factors "idealized" stimuli, "idealized" rating scales, and "idealized" raters. The core matrix **G** gives the pattern of interactions among these idealized entities and hence might be thought of as a miniature idealized image of the original data set.

A T3 factor-loading matrix for stimuli would then describe how the patterns for each actual stimulus can be approximated by some linear combination of the patterns for these idealized stimuli, implicitly telling how much the stimulus partakes of each of the idealized stimulus features. Likewise, the factor loadings for scales describe how the pattern of scores for each rating scale can be represented as some weighted linear combination of the idealized scale characteristics; similarly, each rater would be represented as some combination of the idealized raters or traits. When a given rater makes a rating, each of his idealized traits reacts to the various features (idealized stimulus components) in the particular stimulus and to the characteristics of the particular rating scale (the mixture of idealized scales in the actual scale) to produce the rating.

Tucker's conception of "factor" is not the same as the conventional notion most commonly used in two-way factor analysis. Thus, while PARAFAC reduces to two-way factor analysis when there is only one level to Mode *C*, this is not true of T3, unless the core is an identity matrix. This point is made particularly clear when one reads the Levin (1965) explication of the Tucker model.

Resulting Algebraic Differences. It is easy to see how these two contrasting notions of "factor" lead to the algebraic differences between T3 and PARAFAC. For T3, the Mode *A* factors are different in kind from the Mode *B* and Mode *C* factors, each arising from a different source; thus, there is no reason to require the same number of factors in each mode. However, for PARAFAC, if a factor is present in the situation, it is reflected in all three modes; hence, the dimensionality of the three modes should be the same (except when a factor fails to vary or to vary independently in a given mode—see below). For T3, interactions between all factors of one mode and all factors of another is what generates the data variance. Therefore, computation of all combinations of loadings across modes and the use of a core matrix to weight these combinations is natural and essential. But for PARAFAC-type factors, such interaction between modes does not have any meaningful interpretation and hence is not a part of the model. (However, an interpretation of the core matrix in slightly different terms is possible, as will be explained below.)

Lack of Uniqueness. Because of the complexity of relations permitted by the core matrix, the Tucker three-mode model permits free rotation or linear transformation of any three of its four sets of component parameters, with no loss of fit to the data (provided the compensatory transformation is applied to the fourth set of parameters). This means, for example, that each of the three loading matrices can be subjected to an arbitrary nonsingular linear transformation, and the inverses of these transformations can all be absorbed into the core matrix. Even system-variation data, which has only one PARAFAC representation, can be represented by T3 in terms of many alternative sets of factors, because each alternative set also has a different pattern of interactions between modes.

Since the T3 model allows a far greater family of possible solutions than were available for two-way factor analysis, it places even stronger demands on the user for intelligent application of rotation or transformation criteria; one might, for example, use simple structure when appropriate, or target rotation, or even rotation to transform T3 into a simpler model, perhaps based on some external theory (as in Dunn and Harshman 1982). Such "freedom of rotation" can prove useful for exploration purposes, but many investigators regret the loss of the empirical information which would have been provided by the intrinsic axis property. As a compromise, some programs that fit the Tucker model include options to approximate the simpler PARAFAC-CANDECOMP model by a special rotation to approximately "diagonalize" the core matrix and thus perhaps allow some of the information about factors provided by intrinsic axis methods to be recovered (see below).

Both PARAFAC and T3 models seem to make sense. Both can be applied to the same data and yield useful insights. Are they really, then, as different as they seem? As we shall see, the two models can be interpreted in such a way as to make them much more closely related. We will first see how the PARAFAC model can be considered as a special case of the Tucker T3 model and how in this case the interpretations can also be parallel. We will then look at the reverse relationship, constructing the T3 model as a special case of PARAFAC-CANDECOMP. Finally, a particular formulation of the T3 model in matrix terms permits it to be interpreted in terms of "global" or PARAFAC-like factors that vary in a more complex fashion than provided for in the PARAFAC-CANDECOMP model.

PARAFAC-CANDECOMP as a Special Case of T3 and Vice Versa

Embedding PARAFAC-CANDECOMP in T3. If we start with the T3 model but eliminate multiple interactions of factors across modes, we obtain a special case of T3 that is structurally equivalent to PARAFAC-CANDECOMP. This can be accomplished by simply requiring that the core matrix elements g_{rst} be zero except when $r = s = t$, in which case we specify that they are equal to 1.0. (This equivalence is well known and has been pointed out by many authors, including Carroll and Chang 1970; Harshman 1970; Tucker 1972.) These restrictions turn the core matrix into a three-way analog of an identity matrix, with zeros in all cells except for the "superdiagonal" cells running diagonally through the body of the cube from the upper left cell of the front slice of the cube to the lower right cell in the back slice.

Now, suppose instead that we fit the more general T3 model but after suitable rotation find that the core matrix has large entries on the superdiagonal and small entries elsewhere. How would we interpret this solution in terms of Tucker's conception of "factor"? Suppose for the moment that we are analyzing the semantic differential data described earlier and that three large superdiagonal cells are identified, so the core matrix is $3 \times 3 \times 3$. A Tucker interpretation might proceed as follows: There are three different basic factors in the rating scale mode, which, under suitable rotation, correspond to "Evaluation," "Activity," and "Potency" (that is, all the rating scales employing highly evaluative terms load on the Evaluation factor, and so forth). There are also three major features of the stimuli. In the rotation that "diagonalizes" the core matrix, these three stimulus mode factors seem to group the stimuli that are highly evaluative into one group, those that are highly active into another, and those that are highly potent into a third, with stimuli showing a mixture of several of these characteristics loading on several of the factors. Thus, it might be appropriate to give the same names to the three stimulus dimensions as to the rating scale dimensions.

It is also significant, so the interpretation might proceed, that the Evaluation feature of the rating scales interacts almost exclusively with the "good-bad" feature of the stimuli and very little with the Potency or Activity aspects of the stimuli (a consequence of the superdiagonal core matrix). Likewise, the Activity and Potency factors of the scale mode interact mainly with the corresponding characteristics of the stimulus mode. Finally, the patterns of ratings of different individuals seem to vary mainly in terms of their relative emphasis on the three factors of the other two modes. In such a state of affairs, it would seem plausible to suggest that Evaluation, Activity, and Potency are "global" or general cognitive-affective dimensions, determined perhaps by culture or human psychology, but in any case more general than any particular mode of measurement by which the rating data were collected. It appears that the same cognitive-affective dimensions influenced raters' perceptions of both the rating scales and the stimuli. In this sense, we might say that the first factor of Mode *A* is the *same factor* as the first factor of Modes *B* and *C* and likewise for the second and third factors. We would conclude that a common set of three factors appears in all three modes of the Tucker analysis, thus arriving at a very PARAFAC-like interpretation of the data.

Embedding T3 in PARAFAC-CANDECOMP: Method I. It is instructive to see how the structure and interpretive framework of the two models can be made to correspond in a complementary way by representing a general T3 data structure in PARAFAC-CANDECOMP terms. In the one-dimensional case, this correspondence is trivial, since both models are directly equivalent. In the multidimensional case, however, we need some way to represent in the PARAFAC-CANDECOMP model all the interactions provided for by the T3 core matrix. The most direct but cumbersome way to do this is to provide a distinct PARAFAC factor for each additive component in the T3 model, that is, for the additive contribution corresponding to each cell in the core matrix.[4]

For simplicity, let us consider the case in which T3 has two factors in each mode—and thus a 2 × 2 × 2 core matrix. To represent data with this T3 structure in PARAFAC-CANDECOMP terms, we could construct a PARAFAC model with eight factors, one to represent each possible interaction of a T3 Mode *A* factor with a T3 Mode *B* factor and a T3 Mode *C* factor. The PARAFAC factor representing a given interaction would have loadings in each mode corresponding to the T3 factor in that mode that was involved in the interaction (see Figure 5–1). For example, the PARAFAC factor corresponding to the g_{212} term would have loadings in Mode *A* corresponding to the second Mode *A* factor of T3, loadings in Mode *B* corresponding to the first Mode *B* factor of T3, and loadings in Mode *C* corresponding to the second Mode *C* factor of T3. In order that these interactions have the correct relative sizes, one mode, say Mode *C,* should have each set of factor loadings multiplied by the corresponding core matrix g value. There would appear to be considerable redundancy in such a PARAFAC representation. In this example, the same set of Mode *A* loadings would appear four times, each time linked with a different set of Mode *B* and/or *C* loadings; similar redundancies would occur in the other two modes (Figure 5–1).

What might we make of such a solution if it were obtained with real data? Let us assume that the two factors involved are Activity and Evaluation. Our first PARAFAC factor might have Activity loadings in both Modes *A* and *B* and so we would interpret it in the normal PARAFAC fashion as a factor representing the connotative dimension of activity. But the second factor might have Activity loadings in Mode *A* and Evaluation loadings in Mode *B,* representing the contribution of the g_{121} cell of the T3 core matrix. We would be forced to conclude that somehow the raters were adding a component to their responses on evaluative scales that was determined by the activity feature of the stimuli. While at first this might seem peculiar, on reflection, we might

Figure 5–1. Carroll's Method of Transforming a Tucker Representation into the Corresponding PARAFAC-CANDECOMP Representation

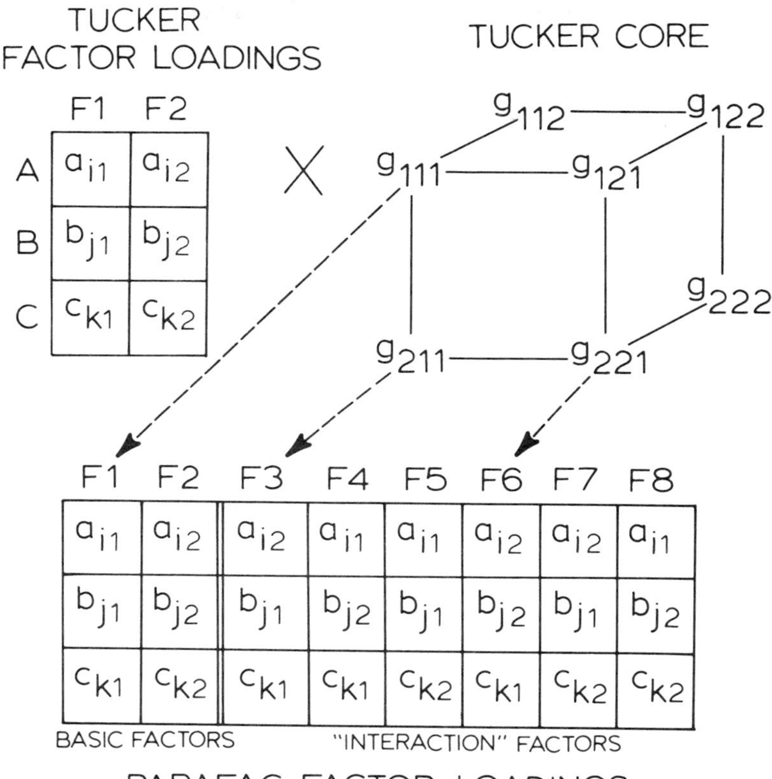

conclude that for these raters, *activity itself* might seem good. Thus, this factor, expressing an Activity-Evaluation interaction, might be said to arise from a nonindependence of the two concepts in the minds of the raters. Similar interpretations could be made about the other "interaction" factors; variations between subjects in the weights assigned to these factors would correspond to variations in overlap of Activity and Evaluation. Note that here we are still talking of "global" factors of the PARAFAC kind but interpreting the interaction components as arising from dependent effects of these global factors—a kind of nonorthogonality.

Since PARAFAC already permits nonorthogonality of factors, it might seem puzzling that additional factors would be needed to express such overlap of meaning. If activity has overtones of goodness in the minds of our raters, then this could be represented by incorporating some loadings on evaluative scales into the basic Activity dimension. Such a nonorthogonality in the basic factor, however, would produce the same pattern of correlated overtones for all subjects. The addition of interaction factors allows *individual differences* in the degree of obliqueness, as determined by the Mode *C* weights. Tucker introduces this kind of interpretation of core matrix interaction terms, for example, when discussing multidimensional scaling applications of T3 (Tucker 1972). The nondiagonal cells of the first frontal plane of the core matrix allow the model to represent one pattern of nonorthogonality between dimensions. Likewise, the nondiagonal cells of the second plane represent a different pattern of nonorthogonality. Each subject's individual pattern of obliqueness is some combination of these core patterns. It is with such reinterpretations that we can bridge the gap between the two models.

By embedding T3 in PARAFAC, it is possible to describe T3 as a "special case" of the PARAFAC-CANDECOMP model, one in which the factors do not all have distinct loading patterns within a given mode but rather repeat their loading patterns to produce all possible combinations of factors across the three modes. Unfortunately, our chance of obtaining such a recognizable pattern in a PARAFAC solution with real data is practically nil (unless we employ special rotation methods), because the axis orientation would not be unique. Recall that uniqueness requires distinct patterns of factor variation across the levels of each mode. In the PARAFAC solution that represents T3, there are two sets of four factors in each mode that have identical loading patterns (although these are different groups of four in each mode). Clearly, the requirement that factors show distinct patterns of variation is grossly violated, and so this kind of PARAFAC representation of data with general T3 structure will show as much indeterminacy as the T3 representation.

While Method I provides the most straightforward way of embedding T3 in PARAFAC-CANDECOMP, there are other ways that require fewer PARAFAC dimensions. However, the less mathematically inclined reader can skip the next two sections and go directly to the section "Reconciliation of T3 and PARAFAC Perspective" with no loss of continuity.

Simplifying the Representation: Method II. There are more compact ways of embedding T3 in PARAFAC-CANDECOMP. A first simplification allows us to reduce the number of PARAFAC dimensions needed from one per cell of the core to one per cell of the smallest face of the core. This is possible because, as noted earlier, any two factors that share the identical pattern of loadings in two of three modes can be combined into a single factor. Our initial method of embedding T3 in PARAFAC-CANDECOMP systematically creates factors with identical loading patterns in two modes. All factors, for example, that correspond to the interaction of a particular Mode *A* factor with a particular Mode *B* factor differ only in the Mode *C* loadings. Such factors can be combined so that a more compact representation is obtained, with one PARAFAC factor for each combination of Mode *A* and *B* factors from T3, rather than one for each combination of Mode *A, B,* and *C* factors.

Suppose we take two factors with identical loading patterns in Modes *A* and *B* but different patterns in Mode *C*. These factors can be combined as follows:

$$a_{i1}b_{j1}c_{k1}g_{111} + a_{i1}b_{j1}c_{k2}g_{112}$$
$$= a_{i1}b_{j1}(c_{k1}g_{111} + c_{k2}g_{112}) \ . \qquad (5\text{–}31)$$

In terms of our example, the two factors that have the Evaluation pattern of Mode *A* and *B* loadings can be combined into one, and the two that have the Activity pattern of Mode *A* and *B* loadings can also be combined into one. Similarly, the two factors with Activity loadings in Mode *A* and Evaluation loadings in Mode *B* (or vice versa) can be combined. The result is a four-dimensional PARAFAC representation, and a new set of Mode *C* weights.

In general, we can use this technique to reduce the PARAFAC representation of an arbitrary q_a by q_b by q_c T3 model from $q_a q_b q_c$ PARAFAC dimensions to $q_a q_b$ PARAFAC dimensions, or (by collapsing across Mode *A* or *B*) $q_b q_c$ or $q_a q_c$ dimensions if this product is smaller. This procedure will work regardless of the number of levels of the mode that collapses. For example, if Mode *C* has the highest dimensionality, then we can define our representation as

$$a_{iu}b_{ju}c_{ku} = a_{ir}b_{js} \ (\sum_{t=1}^{q_c} c_{kt}g_{rst}) \qquad (5\text{–}32)$$

with

$$u = r + (s - 1)(q_a) \ ,$$

where u is the index of the PARAFAC factor, r, s, and t are indices of the T3 factors in Modes *A*, *B*, and *C*, respectively, and there is a distinct u for each a,b combination. (An alternative way of demonstrating that no more than q_a times q_b dimensions are needed is given below, where a matrix formulation of the model is considered.)

Still More Compact Representations: Method III. By simplifying the core matrix before applying (5–32), it is often possible to achieve an even more condensed PARAFAC representation of a T3 structure. This simplification, interestingly enough, can be accomplished by applying PARAFAC to the core itself. To summarize briefly, by using enough PARAFAC dimensions, it is always possible to find an exact PARAFAC representation of an arbitrary core. In general this will require more PARAFAC factors than Tucker factors—and thus more PARAFAC factors than rows or columns of the core matrix (see Kruskal 1976, 1977)—but still relatively few dimensions compared to the number required by Method II. This exact-fitting PARAFAC solution provides a reexpression of the core in terms of a set of diagonal matrices that are pre- and postmultiplied by particular transformations (in other words, in terms of an equation of the form [5–4]). These transformations are then absorbed into the Tucker *A* and *B* matrices to produce the new T3 representation. Because of its intrinsic interest, as well as its relationship to previous proposals for simplifying the T3 core (McCallum 1976; Carroll and Pruzansky 1979; Cohen 1974), this procedure is described in more detail in appendix 5–2.

Method III often compresses our PARAFAC representation of a given T3 structure into considerably fewer dimensions than even

Method II requires. For example, it is possible to represent two-dimensional T3 data by means of a 2- or 3-factor PARAFAC model.[5] In the case of larger core matrices, our Monte Carlo tests indicate that the number of PARAFAC dimensions needed to embed the T3 model is considerably less than the $q_a q_b$ upper bound give by Method II. In the $3 \times 3 \times 3$ case, it appears that the PARAFAC model can be reduced from 9 to 5 factors. In the $2 \times 3 \times 4$ case, the representation can be reduced from 6 to 4 factors. Examples such as this demonstrate that often many of the cells in the core matrix are "nonessential" in the sense that they could be constrained to be zero with no loss in fit.

Because one can transform a T3 representation into a PARAFAC representation by applying PARAFAC to the core, as described in appendix 5–2, the question of how many PARAFAC dimensions are required to embed a given T3 model reduces to the question of how many PARAFAC dimensions are required to represent its core matrix. In the most general case, the core matrix is simply an arbitrary three-way array, and so the number of PARAFAC dimensions needed to represent T3 data that has q_a, q_b, and q_c dimensions in Modes A, B, and C, respectively, is formally equivalent to the generalized or trilinear rank of a q_a by q_b by q_c three-way array. The issue of trilinear rank is considered in some detail by Kruskal (1976, 1977).[6]

Reconciliation of T3 and PARAFAC Perspective

T3 in Matrix Notation. A useful alternative perspective on T3 can be gained by considering the following matrix representation of the model

$$\mathbf{X}_k = \mathbf{A}\,\mathbf{H}_k\,\mathbf{B}' + \mathbf{E}_k , \tag{5–33}$$

where

$$\mathbf{H}_k = \sum_{t=1}^{q_c} c_{kt}\,\mathbf{G}_t . \tag{5–34}$$

In (5–33), $\mathbf{X}_k$ is the kth slice of the three-way data array, $\mathbf{A}$ is an n by q_a matrix of Mode A factor loadings, $\mathbf{B}$ is an m by q_b matrix of Mode B factor loadings, and $\mathbf{H}_k$ is a q_a by q_b matrix giving the relationships between the Mode A and Mode B factors for the kth level of Mode C (for instance, for the kth person if Mode C represents individuals). For each k, the $\mathbf{H}_k$ matrix is obtained by taking a weighted sum of q_c slices of the core matrix $\mathbf{G}$, with the weights given by the kth row of the Mode C factor-loading matrix.[7]

Let us return once again to our example of the rating scale data. On the one hand, (5–33) can be interpreted in terms of "factors" in the sense meant by Tucker—as aspects, features, or idealized levels of a given mode. Treating our semantic differential example from this perspective, $\mathbf{H}_k$ gives the kth individual's pattern of interactions of the stimulus features with the rating scale attributes. On the other hand, it is possible to reinterpret (5–33) in terms of "factors" in the sense meant by PARAFAC—as global influences acting in the situation and expressed in all three modes; in our example, the factors would be basic semantic or affective features, with Mode *A* giving their relation to the stimuli and Mode *B* giving their relation to the rating scales. In this second interpretation, $\mathbf{H}_k$ would describe the way that the *k*th individual modified the use of or shifted the overtones associated with these global factors.

In order to reinterpret $\mathbf{H}_k$ as describing individual differences in use of global dimensions, we need to decide how to apply it to the **A** and **B** matrices. In terms of our example, we need to decide whether $\mathbf{H}_k$ describes individual differences in the perceptions of the stimuli, in the semantic dimensions underlying the scales, or in both.

Individual Differences in Mode A. One possibility is to apply the $\mathbf{H}_k$ solely to the stimulus mode (which in our example we assume to be Mode *A*), and so conceptualize the model as follows:

$$\mathbf{X}_k = (\mathbf{A}\,\mathbf{H}_k)\,\mathbf{B}' \qquad (5\text{–}35)$$

or

$$\mathbf{X}_k = \mathbf{A}_k\,\mathbf{B}' \;. \qquad (5\text{–}36)$$

This approach attributes the variations in patterns of judgments across different raters to variations in the perceptions of the stimuli. The perceptual dimensions would be assumed constant —an unchanging pattern of rating-scale loadings would reflect a fixed notion of Evaluation, Activity, and Potency—but the degree to which these properties were perceived as characteristic of different stimuli would change from subject to subject.

Note, however, that the changes in perceived stimulus properties would only be of certain particular kinds. The model cannot represent the fact that an individual point might be the only one to shift position relative to other points in one rater's space. The $\mathbf{H}_k$ produces a linear transformation of the stimulus space and so describes coordinated changes in the positions of all the stimuli. In fact, in the T3 model (as contrasted with the T2 model, to be discussed briefly below), the patterns of individual variations are even more restricted: Every $\mathbf{H}_k$ is made up of a weighted combination of slices $\mathbf{G}_t$ of the core matrix, $\mathbf{G}$, with each slice describing a characteristic pattern of transformation of the stimulus space. Thus, individual raters differ only in the degree to which they incorporate one or another of these few basic patterns of individual variation into their personal perception of the stimuli.

As with PARAFAC-CANDECOMP, the stimulus space for subject *k* might be stretched or contracted along the axes, but in addition it might also be subjected to rotation, shear, or other linear transformations. If, for example, the space were compressed in such a way that certain "active" stimuli were located closer to the "good" ones and no longer at right angles to them, then we might interpret the solution as indicating that for person *k* those stimuli were perceived as having a mixture of activity and goodness, not simply activity alone.

Individual Differences in Mode B. Now suppose that we group terms in the opposite way and interpret the model as

$$\mathbf{X}_k = \mathbf{A}\,(\mathbf{H}_k\,\mathbf{B}') \tag{5–37}$$

or

$$\mathbf{X}_k = \mathbf{A}\,\mathbf{B}'_k \,. \tag{5–38}$$

With this grouping, the model could be interpreted as describing differences in the nature of the semantic dimensions across individuals. For example, the kth rater might have an Activity dimension (column two of his $\mathbf{B}_k$) that was a linear combination of the basic Activity dimension (described by column two of $\mathbf{B}$) and Evaluation (column one of $\mathbf{B}$); this could be interpreted as indicating that the notions of activity and goodness were not as independent for this person as they were in the group space described by $\mathbf{B}$.[8]

Individual Differences in Both Modes. The third possible interpretation of (5–33) involves splitting $\mathbf{H}_k$ into two or three pieces. Here $\mathbf{H}_k$ is considered to be the product of several matrices that represent transformations of both the $\mathbf{A}$ and $\mathbf{B}$ matrices and possibly also a diagonal matrix of weights on the dimensions. For example, we can let

$$\mathbf{H}_k = \mathbf{T}_{ak}\,\mathbf{T}'_{bk} \quad \text{or} \quad \mathbf{H}_k = \mathbf{T}_{ak}\,\mathbf{D}_k\,\mathbf{T}'_{bk}\,, \tag{5–39}$$

so that

$$\mathbf{X}_k = (\mathbf{A}\,\mathbf{T}_{ak})(\mathbf{T}'_{bk}\,\mathbf{B}') \text{ or } \mathbf{X}_k = (\mathbf{A}\,\mathbf{T}_{ak})\,\mathbf{D}_k\,(\mathbf{T}'_{bk}\,\mathbf{B}')\,.$$

Here we would interpret the differences in judgments made by rater k as due to shifts both in the perception of the stimuli and in the overtones of the fundamental semantic dimensions used in making the evaluations. The more detailed representation would insert a $\mathbf{D}_k$ matrix of dimension weights between $\mathbf{T}_{ak}$ and $\mathbf{T}'_{bk}$ to represent general dimension weights for occasion k. Alternatively, we can view the dimensions as fixed except for differences in weight or salience and the interactions among the dimensions as changing across levels of Mode C. In this case, the $\mathbf{H}_k$ matrix would be written

$$\mathbf{H}_k = \mathbf{D}_k\,\mathbf{T}_k\,\mathbf{D}_k\,, \tag{5–40}$$

so that

$$\mathbf{X}_k = (\mathbf{A}\,\mathbf{D}_k)\;\mathbf{T}_k\;(\mathbf{D}_k\,\mathbf{B}')\,.$$

This latter approach is sometimes taken when $\mathbf{X}_k$ is symmetric, such as when the data are covariances. Then, $\mathbf{T}_k$ expresses obliqueness of axes and $\mathbf{D}_k$ dimension weights or saliences.

We can see that by reinterpreting the $\mathbf{H}_k$ matrices, we can formulate an alternative conceptual framework for Tucker's three-way algebraic model, one which involves the same sort of global factors as underlie the conceptual framework for PARAFAC. That is, loadings for a given column in all three modes can be interpreted as referring to the same factor. The greater complexities that can be represented by the Tucker model are seen as referring to more complex patterns of individual variations. In particular, these complexities are seen as referring to changes across individuals in the perceived grouping of stimuli (5–36), in the overtone content of semantic dimensions (5–38), or both (5–39). Alternatively, the variations can be attributed to individual differences in the interactions or obliqueness among dimensions (5–40).

Different Dimensionality in Modes A and B. While reconciliation of Tucker and PARAFAC schemes of interpretation seems feasible if there are the same number of factors in each of the three modes, what can we make of the T3 models in which there are a different number of factors in one mode compared to another? At first this would seem to represent a serious block to application of our scheme of reinterpretation. There is, however, a simple solution: Situations in which T3 modes have differing dimensionality correspond to situations in which two or more PARAFAC factors do not show linearly independent patterns of variation in a particular mode but are independent in the other two modes. In the PARAFAC representation of this situation, all three factor-loading matrices will have the same number of columns, but the rank of the matrices would differ. With T3, however, all loading matrices are *by convention* required to have full column rank; thus, T3 would typically represent this situation by having fewer factors in some modes than others. With fallible data, true linear dependence would normally not occur, even if we filled out the T3 loading matrices for all three modes to the dimensionality of the mode with the most factors; but sometimes when this is attempted, the eigenvalues corresponding to the last few factors are very small in certain modes and so the corresponding dimensions are typically considered to make trivial contributions in those modes. By convention, the factors are eliminated from the modes to which they do not make substantial *independent* contributions.

Suppose that for a particular stimulus set, two factors happen to show a more or less identical pattern of variation across stimuli (levels of Mode *A*), although they are clearly distinguished in terms of their weights on rating scales and people. If we adopt the usual practice of requiring that all the Tucker factor-loading matrices $\mathbf{A}$, $\mathbf{B}$, and $\mathbf{C}$ have full column rank, then a Tucker solution will be obtained in which the Mode *A* loading matrix has one less dimension than Modes *B* and *C*. The "missing" linearly dependent factor loadings in Mode *A* will be recreated, however, if the rectangular $\mathbf{H}_k$ matrix is used to transform the Mode *A* loadings, as in (5–36). There will be an extra column of elements in $\mathbf{H}_k$ that will describe how to regenerate the "missing" loadings in Mode *A*; that is, $(\mathbf{AH}_k)$ or $\mathbf{A}_k$ will have as many columns as $\mathbf{B}$ and will contain the column of linearly dependent loadings needed to go with the additional column of the Mode *B* and *C* factor-loading matrices. By thinking of T3 as in (5–36), for example, we can maintain our alternative interpretation in terms of "global" PARAFAC-like dimensions, even when the dimensionality of the formal T3 model varies across modes.

Transforming the Core to Approximate PARAFAC-CANDECOMP. Once the T3 model is interpreted in a way consistent with the

PARAFAC notion of "factor," it becomes possible to use proportional profiles to help deal with the transformational indeterminacy of the T3 solution. By rotating the solution into an approximation of PARAFAC form, we might be able to simplify the interpretation of the T3 dimensions, while testing the feasibility of going to a strict PARAFAC analysis (by observing how much the best approximation still diverged from a true superdiagonal core matrix). If a T3 solution can be rotated or transformed so that the core is very close to being diagonal (or superdiagonal), then the resulting axes can be interpreted in terms of the PARAFAC-CANDECOMP model. An approximate intrinsic axis property will apply, as well. In such a case, if the deviations from "pure" PARAFAC-CANDECOMP structure are attributed to chance (the off-diagonal core cells are considered to be small deviations from zero due to fitting error in the data), then it might be appropriate to replace the T3 solution with a true least-squares PARAFAC-CANDECOMP solution. If, on the other hand, there are a few large off-diagonal cells in the core matrix, even after transformation to achieve closest approximation to diagonal form, then by studying the core, the analyst may be able to gain further insight into the patterns of individual difference in the data. (Current research on "degenerate" PARAFAC solutions, discussed in chapter 6, suggests that strict minimization of all off-diagonal core cells may not lead to the most interpretable rotation when certain kinds of T3 structure are present. It may be better to retain a few large off-diagonal cells when this will cause the others to be much closer to zero and/or the loadings matrices to be less correlated and more interpretable. Such a rotation of the T3 core might produce factor loadings that correspond to an orthogonally constrained PARAFAC analysis—see chapter 6.)

This type of cross-model comparison was originally suggested in the slightly more restricted context of three-way multidimensional scaling; Harshman (1972b) suggested transformation of IDIOSCAL (symmetric T2) to maximize agreement with PARAFAC2, and Cohen (1974) suggested transformation of T3 or T2 for comparison with INDSCAL. The first successful demonstration of this idea was by Cohen (1974), who implemented it by the ingenious method of applying CANDECOMP directly to the core matrix to obtain an approximation in the same dimensionality as the Tucker solution, then using the factor-loading matrices of the CANDECOMP solution to transform the Tucker dimensions. Similar procedures involving transformation of the Tucker core to INDSCAL form were subsequently suggested by McCallum (1976), and Carroll, Pruzansky, and Kruskal (1980). The latter authors developed an application of their linearly constrained three-way model CANDELINC, which turned out to be precisely equivalent to Cohen's procedure and suggested that the INDSCAL-like solution that they could obtain in this way would be useful as a starting position for true least-squares INDSCAL analysis (see Carroll and Pruzansky 1979). Kroonenberg and de Leeuw (1980) (see also Kroonenberg 1981b) have developed a program called TUCKALS2 for fitting the T2 model to data by the method of alternating least-squares. They have incorporated into this procedure an option for approximate diagonalization of the core matrix, based on an algorithm described by de Leeuw and Pruzansky (1978).

The "Extended Core" Tucker Model T2

Tucker's original model places fairly strong constraints on the patterns of individual differences in use of dimensions. The $\mathbf{H}_k$ for each level of Mode C must be a weighted combination of the slices of the core matrix. A more general model would be to allow each individual to have an arbitrary $\mathbf{H}_k$. This was suggested in the context of analysis of symmetric scalar product data for MDS by Carroll and Chang (1970), who subsequently called the model IDIOSCAL, for IDIOsyncratic SCALing model (Carroll and Chang 1972). The same model, given a slightly different interpretation, was independently proposed by Jennrich (1972). Tucker (1972) incorporated the possibility for this kind of generality by allowing the third dimension of the core matrix to become "extended" until it had as many levels as there were levels of Mode C of the data array itself. The extended core also was useful in interpretation of Tucker's model in the context of MDS. The Tucker model with this extended core is called T2 by Kroonenberg and de Leeuw (1980) because it provides a reduced-dimensionality representation for only two of the three modes of the data. The T2 algebraic model would be given by (5–33), in which each $\mathbf{H}_k$ would be one slice (the kth frontal plane) of the core.

Most of the points that were made above about the T3 model also apply to T2. The same alternative interpretations of "factor" can be applied, and the same reinterpretations of the model can be used to place it and PARAFAC-CANDECOMP in a common conceptual framework. Interestingly, much the same limits on dimensionality also apply. The earlier demonstration that it is possible to combine all the dimensions that have the same Mode A and Mode B loadings but different Mode C loadings applies regardless of the number of levels of Mode C. Hence, an upper bound for the number of PARAFAC dimensions needed to "embed" or represent a given T2 structure is still the product of the two smaller ways of the core matrix—in this case, $q_a q_b$—regardless of the number of distinct levels of Mode C. For example, a T2 model with two dimensions in Mode A, three in Mode B, and 80 in Mode C (for 80 subjects) would be equivalent to a six-dimensional PARAFAC model or a T3 model with a $2 \times 3 \times 6$-dimensional core.

Comparison with Other Three-Way Models

Several other three-way models have been proposed, but space limitations prevent us from discussing them in any detail here. Table 5–1 presents a summary of one family of related models for three-way profile data. Some of these models have been proposed and programmed by one or more groups of investigators, others have been proposed but not yet incorporated into a working program, and still others are included simply for comparison or theoretical interest. The models range from the most general, which allows a completely different set of dimensions for each two-way slice of the three-way array (T1–1), to the most restricted, which assumes exactly the same structure in all slices (T1–9). Tucker's T3 and T2 and the PARAFAC-CANDECOMP

TABLE 5-1. Some Three-Way Models for Profile Data

(T1-1)	$\underline{X}_k = \underline{A}_k\underline{D}_k\underline{B}'_k$	Independent spaces
(T1-2)	$\underline{X}_k = \underline{AD}_k\underline{B}'_k$ or $\underline{AB}'_k$	Unconstrained object variation (equals two-mode analysis of the "string out" data)
(T1-3)	$\underline{X}_k = \underline{A}(\underline{B} + \underline{V}_k)'$	("Continuous") Object variation
(T1-4)	$\underline{X}_k = \underline{AD}_k(\underline{B} + \underline{V}_k)'$	Mixed variation
(T1-5)	$\underline{X}_k = \underline{AH}_k\underline{B}'$	Tucker's T2
(T1-6)	$\underline{X}_k = \underline{A}(\sum_{t=1}^{q_c} c_{kt}\underline{G}_t)\underline{B}'$	Tucker's T3
(T1-7)	$\underline{X}_{kf} = \underline{AD}_k\underline{HD}_f\underline{B}'$	PARAFAC3
(T1-8)	$\underline{X}_k = \underline{AD}_k\underline{B}'$	PARAFAC1
(T1-9)	$\underline{X}_k = \underline{AB}'$	Two-mode analysis of "collapsed" (averaged) data

models will be found at intermediate positions. This table provides an idea of some of the different variations of three-way factor analysis that might be of interest. It focuses on versions of the models suitable for direct fitting of profile data and does not include special forms that arise from indirect fitting.

Corballis' Three-Way Model

A different kind of three-way model not included in Table 5–1 has been presented by Corballis (1973) as an extension of a factor-analytic model for change proposed by Corballis and Traub (1970). It is defined in terms of correlation matrices rather than profile data matrices. Even though it has been neglected in the three-mode literature, it is mentioned here because it provides an interesting complement to the more familiar models given in Table 5–1. It has the form

$$\mathbf{C}_{kk^*} = \mathbf{A}_k \mathbf{D}_{kk^*} \mathbf{A}'_{k^*} + \mathbf{E}_{kk^*} \quad , \tag{5–41}$$

where $\mathbf{C}_{kk^*}$ is the matrix of correlations or covariances between the values obtained for a set of variables when they were measured on a set of cases on occasion k and the values for the same variables on the same cases but measured on occasion k^*. $\mathbf{A}_k$ is the factor-loading matrix for occasion k; $\mathbf{A}_{k^*}$ is the corresponding factor-loading matrix for occasion k^*; and $\mathbf{D}_{kk^*}$ is a diagonal matrix of weights, the product of the diagonal scaling matrices for occasions k and k^*. Thus the model considers both the matrices of covariances *within* occasions (when $k = k^*$) and *between* occasions (when $k \neq k^*$). Note that this model differs from any we have considered previously in that the factor-loading matrix $\mathbf{A}$ is subscripted. The model incorporates the idea that the precise pattern of factor loadings and factor scores for a given factor might change between occasions (such as when a given test is not measuring exactly the same thing when administered to subjects at different ages). These changes are not assumed to be proportional or to have any other particular form; in fact, they are not, in general, representable as a linear transformation of some common $\mathbf{A}$. In this respect, the model is more flexible than any other three-way models that we have considered.

It might be thought that this model reduces to the overly general model given in the first line of Table 5–1, where there is no constraint on how the dimensions for one value of k (one occasion or individual) are related to those for another. This is not true, however, because (5–41) requires an assumption that each factor is orthogonal to all other different factors not only within occasions, but also between occasions. This requirement has the further consequence that all patterns of factor *change* across occasions must also be orthogonal.

Corballis and Traub (1970) and Corballis (1973) point out that the strong orthogonality assumptions of this model are generally sufficient to insure a *unique* solution for all the $\mathbf{A}_k$ loading matrices and for the other parameters, as well. Is this uniqueness another possible source of empirically meaningful axes? Once again, we would suggest that one's confidence in the orientation of axes provided by any procedure is a function of the plausibility of the assumptions invoked to obtain the unique solution. To many investigators, the orthogonality assumptions might seem too strong to be plausible. But it is an interesting question: How plausible are they *as an approximation,* and how distorted would the recovered factors be when the "true" factors are moderately correlated? This is presumably a problem for a Monte Carlo study. Corballis (1973) cautions us that "a potential difficulty is that the factor rotations specified by the model may not make sense, [that is], the factors may be uninterpretable." To our knowledge, there have been no applications of this model other than the demonstration application in the Corballis article (1973).

PARAFAC3 and PARAFAC2

It might be useful to note briefly some related intermediate versions of three-way factor-analytic models that form a bridge between PARAFAC1 and T3 or that have been considered elsewhere as interesting variants. However, this somewhat technical discussion is not essential for understanding the subsequent sections of this article; therefore, the less theoretically inclined reader may wish to skip to the discussion of ALSCOMP, below.

PARAFAC3 in General Form. In Table 5–1, a model called PARAFAC3 is placed between T3 and PARAFAC1. If we consider this model in its most general form, where the left-hand matrix of weights $\mathbf{D}_k$ need not be the same as the right-hand matrix $\mathbf{D}_f$, PARAFAC3 can be interpreted as a four-mode or fourway model. Applied to the analysis of cross-product or covariance matrices, it resembles the Corballis (1973) model in that it fits both the between-occasion and within-occasion cross-products. In this application, $\mathbf{X}_{kf}$ would represent the matrix $(\mathbf{X}_k' \mathbf{X}_f)$, in other words, the matrix of cross-products between data on occasions k and f. In this application, PARAFAC3 differs from Corballis' model in that it assumes constant $\mathbf{A}$ and $\mathbf{B}$ matrices rather than ones that vary across occasions. It also incorporates a central matrix $\mathbf{H}$, which allows for oblique relationships among dimensions.

The PARAFAC3 model is not restricted to analysis of cross-products or covariances; it can be applied to rectangular matrices of raw-score or profile data or to square matrices of asymmetric relationships. Such data might be obtained in some kind of four-way experiment. For example, $\mathbf{X}_{kf}$ might represent the ratings of stimulus k in context or situation f, and all $\mathbf{X}_{kf}$ matrices would contain ratings made by the same common set of raters (represented by the rows of $\mathbf{X}$) on a common set of rating scales (represented by the columns of $\mathbf{X}$). Or, it might be applied to transition matrices, such as a matrix whose rows and columns correspond to the same categories of automobiles and in which the cell x_{ij} represents the trade-in of an automobile in category i for a new one in category j. The matrix $\mathbf{X}_{kf}$ might then represent the number of people involved in such transitions who are trading in their old automobiles manufactured in year k for new automobiles manufactured in year f. In this application, the $\mathbf{H}$ matrix would represent an asymmetric pattern of transitions between categories. Such an application would be a four-way example of DEDICOM (see below).

The PARAFAC3 model has not been incorporated into a computer program, but it has been subjected to some mathematical analysis. An important reason for interest in this model is that it has been proven (Harshman 1981) that PARAFAC3 has the intrinsic axis property under reasonable conditions of data adequacy that are similar, in some respects, to those for PARAFAC1. It provides a unique solution without requiring the orthogonality assumptions of the Corballis model.

PARAFAC3 in Three-Mode Form. If we take the special case of PARAFAC3, where $k = f$, we obtain a three-mode model with intermediate generality between T3 and PARAFAC1. This version can be interpreted in the same terms as the general Tucker T3 model, in which there are "interactions" between the Mode A and Mode B dimensions; but in PARAFAC3, these interactions are fixed and described for all subjects by $\mathbf{H}$. It can also be reinterpreted in terms of more "global" dimensions by the methods noted earlier for T3. In either interpretation, individual differences between subjects (or whatever sources of variation are represented by the levels of Mode C) are of the simpler kind assumed by all the PARAFAC series models: Differences in the importance or salience of the dimensions for different levels of Mode C. The weights defining the significance or contribution of dimensions on the kth occasion are given by the diagonal elements of the diagonal matrices $\mathbf{D}_k$.

PARAFAC2 and DEDICOM. When the three-mode version of PARAFAC3 is applied to symmetric matrices of scalar products or covariances, we obtain the PARAFAC2 model. This model is useful for oblique axis indirect fitting of PARAFAC and as a multidimensional scaling model that allows individual differences in the saliences of *oblique* dimensions (Harshman 1972b). While PARAFAC2 is more general than the PARAFAC1 model for covariances, since the indirectly fit axes can be oblique, it requires a constant set of angles between dimensions across levels of Mode *C* (for instance, across occasions) and thus is more restricted than T3.

PARAFAC2 involves an **H** that is square and symmetric. (The symmetry is not required, however.) When PARAFAC3 or PARAFAC2 is applied to a set of square *n* by *n* matrices describing relationships among *n* things—such as stimulus confusion matrices (see Dawson and Harshman 1983), matrices describing the number of people making telephone calls or perhaps migrating from the row place to the column place, the number of people switching from the row product to the column product, and so forth—we obtain the three-way DEDICOM models (see Harshman, Green, Wind, and Lundy 1982; Harshman 1978). If we allow **H** to be asymmetric but require **B** = **A**, we obtain the three-way "single domain" DEDICOM model; if we relax this requirement, we obtain the three-way "dual domain" model.

Despite its intermediate position in Table 5–1, PARAFAC2 (and the three-mode case of PARAFAC3) is not easily obtained as a special case of T3. This is because PARAFAC2 requires that the individuals' $\mathbf{H}_k$ matrices be obtained by weighting a fixed cosine matrix. In other words,

$$\mathbf{H}_k = \mathbf{D}_k \mathbf{W} \mathbf{D}_k \quad , \qquad (5\text{–}42)$$

where $\mathbf{D}_k$ is a diagonal matrix giving the weights of the dimensions for the kth individual (or on the kth occasion). It is not obvious that a set of matrices of this kind can be obtained by taking different linear combinations of a few slices of any core matrix. Nonetheless, because any "extended" T2 core can be replaced by a moderate-sized diagonal or superdiagonal T3 core (as shown earlier), it is possible to represent PARAFAC2 and PARAFAC3 in terms of a T3 model. First, we set up a T2 model for which the slices of the extended core matrix have the form given in (5–42); then, we "collapse" the extended core to a T3 core with at most $q_a q_b$ slices.

Given that such a T3 representation is possible for any PARAFAC2 and PARAFAC3 structure, should we call these models special cases of T3? This may straining the sense of "special case." Even though there exist T3 solutions in moderately higher dimensionalities that would provide the same fitted values $\hat{x}_{ijk}$ as these PARAFAC models, one cannot obtain these models from T3 (or T2 or IDIOSCAL) by simply constraining certain terms to be zero, or fixed, or equal, and so forth. In order to find an "extended core" that has the special form required by PARAFAC2 and PARAFAC3, one must solve what is really the same problem as fitting these PARAFAC models to any other data array. Thus, it is perhaps better to treat PARAFAC2 and PARAFAC3 as separate models with their own interesting properties, such as intrinsic axes in certain circumstances (Harshman 1981).

The ALSCOMP Procedure

The Basic Idea of ALSCOMP: Nonmetric PARAFAC-CANDECOMP. Sands and Young (1980) have developed an algorithm and associated computer program called ALSCOMP (for Alternating Least-Squares COMPonents analysis). ALSCOMP is not really a new structural model but rather a new and more flexible estimation procedure, one that allows *nonmetric* fitting of the basic PARAFAC-CANDECOMP model (2–3) and (2–4) to a wide class of data. With their procedure, the data can be treated as ratio, interval, ordinal, or nominal scale, or some combination of these, and can be considered as "subject conditional," or "variable conditional," or the like (that is, having a measurement scale that differs from subject to subject, variable to variable, and so on).

The authors compare their procedure to other recently developed three-way procedures (including those for PARAFAC, CANDECOMP, and the Tucker three-mode model) and stress that "because all of these [other] . . . procedures place stringent requirements on the measurement characteristics of the data, none of them are applicable to most of the data types usually encountered in psychological research" (Sands and Young 1980). They suggest, therefore, that the ALSCOMP algorithm provides the only suitable method of fitting the PARAFAC-CANDECOMP model to most social sciences data.

How Essential is Nonmetric Fitting? While it is true that early applications of PARAFAC and CANDECOMP analysis often met with discouraging results, this probably occurred because the intrinsic axis property was interfered with by problems with the data characteristics and/or model appropriateness. PARAFAC has performed well in recent years, providing meaningful solutions to a wide range of data. How are we to explain this fact? The answer probably lies in the modifications and improvements to the analysis and particularly to the data preprocessing procedures that have been developed more recently. These improvements (discussed in chapter 6) have resulted in much greater flexibility in application of the model to real data; in fact, they have extended the flexibility or applicability of PARAFAC in some ways beyond that of ALSCOMP (although in other ways ALSCOMP is still more general).

Where the more sensitive intrinsic axis property is not part of the solution, metric procedures have been successful for some time. Sands and Young do not comment on the apparently successful application of Tucker's model to a number of different data sets (as reviewed elsewhere in this volume), nor do they mention the much longer history of the successful social sciences applications of conventional two-way factor analysis, which is also a metric procedure and thus involves the same "stringent" theoretical requirements on the measurement characteristics of the data.

Comparison of ALSCOMP with PARAFAC. What are the advantages and disadvantages of ALSCOMP versus PARAFAC algorithms for fitting (5–3) and (5–4)? We do not have space to discuss this issue in detail here, but there are several points that should be mentioned. PARAFAC preprocessing methods (discussed in the next section) eliminate the requirement for ratio-scale data and permit several kinds of data conditionality, so the basic criticism of Sands and Young is considerably weakened. Nonetheless, PARAFAC is still a metric analysis procedure. As we noted earlier, however, almost all factor analysis and many MDS procedures are metric; yet, they have been shown to be generally quite robust, perhaps because the solution is highly overdetermined by the data. The differences between the results obtained by metric and nonmetric fitting are typically modest, even when the data are known to be only ordinal scale, as Weeks and Bentler (1979) concluded from a Monte Carlo study using two-way data and as Harshman and Howe (1979) found in a Monte Carlo study using three-way MDS data that violated the Euclidean distance formula. Carroll and Chang (1970) also developed a "quasi-nonmetric version of INDSCAL," but in practice it gave results that were not appreciably different from the metric version. Hence, it was not considered to justify the extra analysis cost (Carroll and Chang 1970). (They did not develop a fully nonmetric version, because they doubted that it would provide any greater advantages.)

The distinction between continuous and discrete data is also not crucial for PARAFAC, as demonstrated by recent Monte Carlo results of Sentis, Harshman, and Stangor (1983), who found that PARAFAC quite successfully recovered continuous latent structure from binary data. Thus, it appears that in the vast majority of cases, nonmetric fitting may provide at best only a subtle improvement over the solution obtained by metric procedures.

Despite all the reservations listed above, there are no doubt certain cases involving systematic and extreme violations of interval-scale properties in which nonmetric procedures provide significantly better representations of the data than metric procedures. It is not clear, however, whether such extreme cases occur in real data except on rare occasions. Furthermore, in those special cases in which nonmetric analysis would make a substantial improvement, there may be special approaches to metric analysis that would provide similar results. Weeks and Bentler (1979) suggest that preprocessing the data by conversion to ranks (their "rank-linear" procedure) will provide most of the benefits of nonmetric analysis in such cases. We have not yet explored the effects of this kind of preprocessing in conjunction with PARAFAC. However, for cases in which nonlinearity is the problem, it has been demonstrated earlier (Harshman 1970, chapter 6; Terbeek and Harshman 1972) that the intrinsic axis properties of PARAFAC may often make it feasible to explicitly represent nonlinearities as extra dimensions and thus not only fit such data but quantify their nonlinear structure. Similar representations of nonlinearities are possible with intrinsic axis multidimensional scaling methods such as INDSCAL or the MDS application of PARAFAC. In fact, Chang and Carroll (1978) in an INDSCAL study of color perception, found "extra" dimensions beyond what would be theoretically expected; the form of the dimensions suggested that they might be due to nonlinear components involved in the reported similarities of colors.

Since nonmetric analysis involves fewer assumptions than metric analysis, it might seem advisable to use nonmetric approaches "just to be safe." However, one must balance potential benefits of nonmetric analysis against potential disadvantages. Sometimes the nonmetric procedures provide worse solutions than metric procedures, because they are subject to certain degeneracies—such as a number of points collapsing into a single location in the space—that the metric procedures avoid. Study of the Sands and Young article also suggests that, with some data, ALSCOMP may also be more subject to local optimum problems than metric ALS (Alternating Least-Squares) procedures such as PARAFAC. A nonmetric analysis fits far more parameters to a given data set than does a metric analysis; sometimes the data set is not large enough to determine all these parameters adequately. Also, ALSCOMP requires the user to specify the level of measurement of the data. The Monte Carlo results reported by Sands and Young (1980) indicated that incorrect specification could lead to considerably poorer recovery of the latent structure.

Finally, it should be pointed out that some of the most serious problems with fitting the PARAFAC-CANDECOMP model may be due to other things than violation of metric assumptions. Two cases, in particular, are worth noting:

a. Often the problem is that the data include certain unwanted components that interfere with intrinsic axis solutions (for instance, multiple factors constant in one mode but not others). It takes particular kinds of preprocessing to remove such contaminants and permit meaningful intrinsic axes to be defined by the part of the data that remains.
b. At other times, the data is technically inappropriate for either PARAFAC or ALSCOMP because it is generated by a process with a more complex structure (similar to the structure of Tucker's model). As we note in chapter 6, uninterpretable "degenerate" solutions with very highly correlated factors sometimes provide better fit to such data than interpretable solutions with factors resembling the "true" latent axes. The problems caused by such complex data should interfere equally with metric and nonmetric analysis. To cope with this situation, extended PARAFAC incorporates options for special analysis constraints that can block such degenerate solutions. Analysis with these constraints will permit recovery of reasonable approximations of the true dimensions, even though the fit to the data is lower than for the degenerate solution.

For these reasons, the extended PARAFAC model—made possible by the current PARAFAC procedure that incorporates special preprocessing and constraint options—provides certain kinds of generality that the ALSCOMP model does not.

In summary, then, both the ALSCOMP and PARAFAC programs have their particular strengths and weaknesses. Although there may still be situations in which the ALSCOMP nonmetric procedure could reveal more than PARAFAC (such as nominal data), there appear to be other circumstances in which ALSCOMP could not deal with the data as effectively as PARAFAC preprocessing and the extended PARAFAC model. We do not yet know the relative frequency of the two classes of situations "in the real world." Overall, it may be that PARAFAC and ALSCOMP will frequently give similar results, but such a statement is speculative, since to our knowledge no systematic comparison of the two procedures has yet been undertaken.

Other Models and Procedures. Space limitations prevent us from discussing a number of other models and procedures for three-way factor analysis that have been developed. For example, Bloxom (1968) has developed a version of Tucker's model in which the subject weights and errors are treated more specifically as random variables. Bentler and Lee (1978, 1979) have carried the statistical development of T3 even further; their approach also links three-mode factor analysis with structural equation modeling and permits confirmatory factor analysis, estimates of standard errors, and so on. Finally, the important models for the analysis of covariance structures proposed by Jöreskog (1971) and others are certainly relevant in the broader context of three-way data analysis but had to be omitted from this discussion in order to permit more detailed development of ideas within the scope of this chapter.

Comparison with Two-Way Factor Analysis

Of the various three-way models, PARAFAC is probably the one most directly related to traditional two-way factor analysis. When we apply PARAFAC to a three-mode array with only one level to the third mode (namely, the two-way special case), it reduces directly to traditional two-way factor or component analysis. In contrast, the Tucker model reduces to a less conventional but interesting representation in which there is a core matrix that has only one slice. (The two-way version of Tucker's model is discussed in some detail by Levin 1965.)

On the one hand, with appropriate preprocessing and scaling of the output loadings (as noted briefly earlier and described in more detail in appendix 5–1), PARAFAC gives results identical to those obtained with traditional programs for two-way singular value decomposition or principal component analysis. This property was exploited by Reddon, Marceau, and Jackson (1982), who used PARAFAC as an efficient way to obtain the first few vectors of the singular value decomposition of the more than five hundred items of the MMPI; their solution was then rotated by Varimax to obtain an interpretable principal axis solution. On the other hand, by fitting covariances or correlations and choosing the option to ignore the diagonal, PARAFAC can be used to perform common factor analysis; the solution that results is equivalent to that produced by the MINRES procedure (Harman and Jones 1966) or by the more common principal factor method based on iterating on the diagonal (when iteration on the diagonal is allowed to reach true convergence).

Of course, when PARAFAC is applied to two-way data, the intrinsic axis property is not obtained; the solution shows the classical rotational indeterminacy of two-way factor analysis. However, by exercising the option to require orthogonal loading matrices in both modes simultaneously, a unique solution can be obtained with axes oriented as in unrotated principal components analysis or principal factor analysis. This solution can then be rescaled and rotated by some suitable analytic or graphical method to obtain the final desired result.

The fact that PARAFAC bears this very direct relationship to two-way factor analysis facilitates interpretation of the loadings in the three-way case. We saw earlier how interpretive conventions can be carried over directly from the two-way case. For example, when doing direct fitting, weights for one of the three modes can be given a conventional interpretation as "factor loadings" and weights for the other two can be interpreted as defining "factor scores" (or "factor score estimates") of the traditional kind.

In this chapter, we have discussed the PARAFAC models for factor analysis (both direct and indirect fitting) and multidimensional scaling. We have examined the intrinsic axis property and compared PARAFAC with other models, particularly Tucker's three-mode factor analysis models.

In chapter 6, we will describe the results of our efforts at extending the domain of PARAFAC analysis to a wider range of data types by means of three-way data preprocessing and special analysis procedures.[9]

APPENDIX 5–1: SCALING AND INTERPRETATION OF PARAFAC LOADINGS

Size-Standardization of Loadings

Size Indeterminacy

There is a multiplicative indeterminacy in the scaling of factor loadings for PARAFAC, as there is with all factor-analytic procedures. The loadings for a given factor can be scaled upward or downward in one mode, provided that compensatory adjustments are made to the size of loadings in the other mode(s). For example, all the loadings for a given factor can be doubled in one mode and halved in another, and the resulting factor contributions (triple products) remain unaffected. In general, we can rescale the a_{ir}, b_{jr}, and c_{kr} loadings for factor r by any constant multipliers $k_{r(a)}$ for Mode A, $k_{r(b)}$ for Mode B, and $k_{r(c)}$ for Mode C, so long as $(k_{r(a)})*(k_{r(b)})*(k_{r(c)}) = 1$. The new loadings for factor r would be defined as follows:

$$\overset{*}{a}_{ir} = (k_{r(a)})a_{ir}\ , \quad \overset{*}{b}_{jr} = (k_{r(b)})b_{jr}\ , \quad \overset{*}{c}_{kr} = (k_{r(c)})c_{kr}\ .$$

These rescaled loadings would provide exactly the same fitted values, residuals, and so forth as the prior loadings.

Since within any mode such rescaling multiplies all the loadings for a given factor by the same constant, this size indeterminacy does not affect the pattern of relationships used for identification and interpretation of a given factor. It *can* affect interpretation, however, if loadings below a given size are considered "insignificant" and are disregarded; a factor scaled up or down will appear to have more or fewer large-sized loadings. Furthermore, if different k_r values are used for each r (that is, for each factor), then rescaling will change the relative sizes of loadings in different columns and will therefore complicate the comparison of relative contributions of factors to a given variable or occasion.

Standardization Conventions

A similar indeterminacy exists in two-way factor and principal component analysis of profile data, where it has been resolved by adopting a convention in which the factor weights for one mode are set to unit variance and called "factor scores," "component scores," or "factor score estimates." As a result, weights in the other mode reflect the scale of the data and the "absolute" size of factor contributions and are called "loadings." For example, when the data consist of variables measured on cases, the standardization is usually applied to the person weights while the size of the factor contributions is expressed in the variable weights.

We have adopted a similar convention for PARAFAC analysis of three-way profile data: Two of the three loading matrices are typically size-standardized so that the mean-squared factor loading for each factor in each mode is equal to 1.0; as a result, the size of loadings in the remaining factor-loading matrix is determined by the size of factor contributions to the data. (In contrast to the two-way case, we often use the term "loadings" to refer to weights in any of the three modes, since we treat the modes more even-handedly.) The loadings in this nonstandardized matrix take on the same units as the original data measurements; thus, they have some sort of "absolute" meaning that allows one to compare the size of loadings across columns, and to evaluate the size of a given loading against some external criterion to determine whether it reflects a substantial or trivial relationship. With standardized modes, on the other hand, loadings are evaluated in relative terms; a large loading means that the factor has a relatively strong relationship to that level of the mode, compared to other levels. Thus, the loadings are interpreted in the same way as any other z-score standardized variables.

Any one of the three matrices resulting from analysis of a given data set could be taken as "the" primary loadings matrix, provided the k_r values used to standardize the solution had been selected so that the columns of the other two matrices were set to unit mean-squares. Indeed, the results of a given analysis can be rescaled several different ways to change the mode that is interpreted as "loadings" in the sense used in two-way factor analysis. The investigator could therefore initially look at "loadings" on variables and factor or component "scores" on people and occasions but then rescale the variable and occasion entries and look at occasion "loadings" and variable and person "scores." No new analysis or estimation procedures would be required to get values in any mode that would reflect the scale of the data and provide cross-factor comparability. However, if the "loadings" were to be given additional interpretations—for example, as beta weights or variance components—then particular data-centering and size-standardization would be required, as described below. Since in some data sets these prerequisite characteristics might not be true of every mode, the special interpretations of loadings that they permit might not be possible for every mode. (Details on the required conditions are discussed in the following two parts of this appendix.)

Interpretations of Loadings

In two-way factor analysis, factor loadings are often given special interpretations, for example, as beta weights. If the factors are orthogonal, loadings are often interpreted as the correlation between a variable and a factor, and the sum of the squared loadings for a given variable is the proportion of the variance predicted by the factors, also known as the "communality." Before we can invest PARAFAC loadings with such additional meaning, certain conditions must be met. The data must be size-standardized in certain ways, and the loadings must then be scaled in a coordinated fashion. In the following discussion, we will first consider the more general interpretations of PARAFAC loadings that are possible, regardless of data standardization, and then develop the additional meanings that can be attributed to the loadings in certain special cases.

General Interpretation of Loadings as Regression Weights

The trilinear PARAFAC and bilinear factor analysis or principal component models can be thought of as equivalent to multiple regression models, except that the data is being predicted from latent factors rather than observed variables. Recall that the basic multiple regression model can be written (in the q predictor case) as

$$y_j = a_1 x_{j1} + a_2 x_{j2} + \dots + a_q x_{jq} + e_j \ , \qquad (5\text{–}43)$$

where the y_j is the value of the dependent variable for the jth case, and $x_{j1}, x_{j2}, \dots$ are the values of the independent or predictor variables for the jth case; the $a_1, a_2, \dots$ are the regression weights; and e_j is the residual or error of prediction. Typically, there is also an intercept or constant term included, but we omit this and consider regression "through the origin." The intercept term would permit the regression model to account for an additive constant in y. We consider such an additive constant as one of the h-terms in our model of interval-scale conditional-origin data (chapter 6, equation [6–3]) and assume that it has been removed by an appropriate preprocessing stage.

The regression model (5–43) can be transformed into the two-way factor-analytic model by letting $x_{j1} = b_{j1}$, that is, by taking the predictor variables to be the scores of each case on latent factors rather than on observed variables. To complete the transformation into a factor-analytic model, we simply consider n

different y-variables at the same time, so that we have n different regression equations predicting the n variables from the q latent factors. For the ith such variable, the score of the jth case would then be written y_{ij}, and the equation predicting that variable would have the regression weights a_{i1}, a_{i2}, and so forth, and an error term e_{ij}.

To generalize this to three-way factor analysis, we simply take a j^* that ranges over two modes rather than one—for example, over occasions as well as cases. If there are m cases (that is, if the maximum value of j is m) and p occasions (that is, the maximum value of k is p), then $j^* = j + (k - 1)m$, and the maximum value of j^* is (mp). To obtain a regression of the same form as (5–43), we replace x_j with x_{j^*}, where $x_{j^*} = (b_j c_k)$. Thus, we obtain a regression in which the data values are predicted from factor or component scores for each case on each occasion, and these scores are given by the product of Mode B and Mode C weights. By considering i variables simultaneously, we obtain a set of regression equations for which the ith equation is

$$y_{ij^*} = a_{i1}x_{j^*1} + a_{i2}x_{j^*2} + \ldots + a_{iq}x_{j^*q} + e_{ij^*} , \quad (5–44)$$

or

$$y_{i(jk)} = a_{i1}(b_{j1}c_{k1}) + a_{i2}(b_{j2}c_{k2}) + \ldots + a_{iq}(b_{jq}c_{kq}) + e_{i(jk)} . \quad (5–45)$$

We see that the factor loadings a_{i1}, a_{i2}, and so on are simply regression weights in a multiple regression equation predicting the y values of the ith variable from scores on the factors, as before; only now we consider the scores to be a function of the occasion as well as the case. Carroll and Chang (1980) use similar reasoning in their development of the alternating least-squares method for fitting CANDECOMP.

Interpreting Regression and Factor Weights

In the most general case, application of regression involves no standardization of predictor or predicted variables. In such applications, the B weights (as nonstandardized regression weights are often called) have no special interpretation except that they give the amount of each predictor variable needed to generate the composite that best approximates the predicted variable. Similarly, when PARAFAC is applied to raw data with no special scaling, the sizes of the factor loadings have no special interpretation except that they give the amount that must be contributed by each factor in each mode for the q factors to best predict the data array. Additional meaning can be attributed to regression weights or factor loadings only when the data and the factors have special properties, such as zero-mean, unit variance, and/or orthogonality.

For example, recall the effects of variable standardization in regression. Size-standardizing the x values permits straightforward comparison of the sizes of regression weights. Differences in weights now directly reflect differences in size of the contribution of the associated predictors to y, without being influenced by the overall size (mean-square) of each predictor variable.

Size-standardizing y permits the size of regression weights to be compared across different variables, data sets, and so forth. When both the predicted and predictor variables are standardized to be z-score variables (with mean zero and variance 1.0), then the B weights of the regression become fully standardized and are called "beta weights." In addition, if the predictor variables are mutually uncorrelated, the transformation of both x and y to z-scores permits each beta weight to be interpreted as the simple product-moment correlation between the predictor and the predicted variable.

Similarly, in PARAFAC analysis, standardization of y (the data) and x (the predictor "factor scores") allows the loadings to take on the additional interpretations possible with z-score regression. Suppose, for example, that we want Mode A weights to take on this extra meaning. We need to standardize the data and the output factor loadings so that Mode A factor weights are interpretable as the beta weight "loadings" and the products of Mode B and C weights give the predictor variable "factor scores" in z-score form. To do this, the data should be centered across Mode B and/or C, so that there is a mean of zero within each level of Mode A. (Recall that fiber-centering a given mode will slab-center the other two modes, as is apparent from inspection of Figure 6–1 in chapter 6.) The data should also be size-standardized on Mode A, so that within each level of Mode A the data have a mean-square of 1.0 (computed across all levels of the other two modes). This would transform the data at each level of A into z-scores and thus provide the necessary standardization of y in our corresponding regression equation.

In addition, the output *loadings* should be standardized so that the b and c weights have a mean-square of 1.0 for each factor (and thus the Mode A loadings reflect the scale of the data); then the x_{j*r} factor scores or factor contributions will also be z-scores. To show this, we first show that the x_{j*r} will have a mean of zero and then that they will have a variance of one.

It is easy to see why the x_{j*r} predictor variables must have a mean of zero. Since we are assuming that either Mode B or C (or both) has been centered, it follows that either the Mode B or Mode C factor-loading table is column-centered (as proven in chapter 6). As a result, the total set of "factor scores," across all levels of both Modes B and C—that is, the products $b_{jr}c_{kr}$ for all jk combinations—has the zero-mean property of z-scores. This is because

$$\sum_j \sum_k (b_{jr}c_{kr}) = \sum_j b_{jr} \left(\sum_k c_{kr} \right) = \sum_k c_{kr} \left(\sum_j b_{jr} \right) ,$$

which implies that the sum of the bc products are zero whenever either the sum of the b or of the c terms is zero.

It is also easy to establish that the bc products for any factor have unit variance whenever both the b and the c weights for that factor have unit variance, since the mean-square of the bc products is equal to the product of their respective mean-squares. In other words,

$$\frac{1}{mp} \sum_j \sum_k (b_{jr} c_{kr})^2 = \frac{1}{mp} \sum_j \sum_k b_{jr}^2 c_{kr}^2$$

$$= \frac{1}{m} \sum_j b_{jr}^2 \left(\frac{1}{p} \sum_k c_{kr}^2\right) .$$

Thus, the bc terms, the x_{j*} of equation (5–44), have unit variance and zero means and hence are z-scores.

We have now shown how to set up a PARAFAC analysis so that both the y and x terms in the corresponding regression equation (5–44) will be z-scores. When this is done, the factor loadings in the nonstandardized mode (in the example, Mode A) of the PARAFAC output can be considered beta weights. This makes them strictly comparable to loadings obtained by conventional two-way factor analysis.

If, in addition, the factors are orthogonal across Mode B or C (or simply within each B by C slice), the Mode A factor loadings can be interpreted as simple product-moment correlations between the data (at that level) and the contributions of the factor (at that level). Additional interpretations in terms of variance components are also possible, but they will be discussed later in this appendix.

Of course, when analyses are performed without imposition of an orthogonality constraint in Mode B or C, exact orthogonality of the resulting loadings will be very unlikely. However, if for each pair of factors the product of the Mode B cosine and the Mode C cosine is small, the above relationships will be closely approximated.

If the factors are not orthogonal across levels of Mode B or C (or at least within the Mode B by C slices at each level of Mode A), standardizing the data and the output to obtain a loading matrix interpretable as beta weights will produce a factor-loading matrix of the kind obtained in traditional *oblique* two-way factor analysis. In terms of the above example, the Mode A loadings matrix would be the analog of the factor pattern matrix in an obliquely rotated two-way solution. The $b_{jr} c_{kr}$ products themselves would be the estimated factor scores. Thus, the matrix of correlations between the $b_{jr} c_{kr}$ scores for the factors would be the phi matrix of "correlations among the factors," as traditionally interpreted. This correlation could be quickly computed for any pair of factors by computing the cosine of the angle between the two factors in Mode B and similarly in Mode C and then taking the product of these two cosines. (The factor cosines are computed as the inner product of the factor-loading vectors after the vectors have been scaled to unit length.)

Other interpretations of loadings from two-way factor analysis can also be carried over into the three-way domain. Thus, for those accustomed to looking at traditional factor loadings and interpreting them in particular ways, a method is available for obtaining interpretations of this kind from three-way PARAFAC analysis.

Interpretation of Loadings as Variance (or Mean-Square) Components: Orthogonal Case

If the factors are orthogonal, it is possible to interpret the loadings as describing components of the total variance (or mean-square) contributed by particular factors at particular levels of a mode, even when the data have not been size-standardized and centered in the way required for interpretation of loadings as beta weights. In the following discussion, we will prove such relationships for Mode A factor loadings only, since the arguments for Mode B and Mode C factors are parallel.

We start with an expression for MSQ, the mean-square data value at level i of Mode A (if the data has zero-mean at each level of A, MSQ will be the variance of level i):

$$\frac{1}{mp} \sum_j \sum_k (x_{ijk})^2 = MSQ_i \, . \qquad (5\text{–}46)$$

By substituting the PARAFAC model for x_{ijk}, we obtain:

$$\frac{1}{mp} \sum_j \sum_k \left(\sum_r (a_{ir} b_{jr} c_{kr}) + e_{ijk} \right)^2 = MSQ_i \, . \qquad (5\text{–}47)$$

Now, because the error part is orthogonal to the systematic factor part, we have:

$$\frac{1}{mp} \sum_j \sum_k \left(\left(\sum_r (a_{ir} b_{jr} c_{kr}) \right)^2 + (e_{ijk})^2 \right) = MSQ_i \, . \qquad (5\text{–}48)$$

If we assume that the factors are orthogonal to one another, their cross-products vanish, so we can write:

$$\frac{1}{mp} \sum_j \sum_k \left(\sum_r (a_{ir}^2 \, b_{jr}^2 \, c_{kr}^2) + e_{ijk}^2 \right) = MSQ_i \, . \qquad (5\text{–}49)$$

By rearranging the order of summation, pulling the constant a-coefficient outside the summation over j and k, and moving the constant divisors m and p inside where appropriate, we obtain:

$$\frac{1}{mp} \sum_r \left(a_{ir}^2 \left(\sum_j \sum_k (b_{jr}^2 \, c_{kr}^2) \right) \right) + \frac{1}{mp} \sum_j \sum_k e_{ijk}^2 = MSQ_i \qquad (5\text{–}50)$$

or

$$\sum_r \left(a_{ir}^2 \left(\frac{1}{m} \sum_j b_{jr}^2 \right) \left(\frac{1}{p} \sum_k c_{kr}^2 \right) \right) + \frac{1}{mp} \sum_j \sum_k e_{ijk}^2 = MSQ_i \, .$$

If we have size-standardized the PARAFAC *loadings* (as opposed to standardizing the *data*) so that they have a mean-square of 1.0 in Mode B and Mode C, we obtain:

$$\sum_r (a_{ir}^2)(1)(1) + \frac{1}{mp}\sum_j \sum_k e_{ijk}^2 = MSQ_i$$

or, subtracting the error mean-square from both sides,

$$\sum_r (a_{ir}^2) = MSQ_i - \frac{1}{mp}\sum_j \sum_k e_{ijk}^2 \quad . \qquad (5\text{–}51)$$

In other words, when the factors are orthogonal in Mode B or C (or simply across all the points in the B-C slice), the sum of squared loadings in a row of the Mode A matrix gives the mean-square accounted for by the q-factor model at that level of Mode A (namely, the total mean-square minus the error mean-square at that level of Mode A). Furthermore, it follows that: (a) each squared loading equals the mean-square contribution of a particular factor at that level; (b) by computing the average squared loading for each column of the Mode A matrix, one can obtain the mean-square contribution of each factor to the total data; and (c) the sum of these quantities across the r factors equals the total mean square predicted by the PARAFAC model at that dimensionality.

Interpretation for Nonorthogonal (or Orthogonal) Factors

When factors are not orthogonal, then the step from (5–48) to (5–49) does not follow. However, if we have size-standardized Modes B and C so that their mean-squared loading is 1.0, there is still a straightforward interpretation for the size of a_{ir} (disregarding sign) that holds whether or not factors are orthogonal: a_{ir} gives the root-mean-square average size of the contribution of factor r to level i of the data. (Recall that by a factor's "contribution" we mean the triple product $a_{ir}b_{jr}c_{kr}$, one of the q additive components of the predicted value for x_{ijk}.) The size of factor contributions is expressed in the same units as the data; for example, if the original measurements were in centimeters displacement, the factor contributions will of necessity be in terms of centimeters displacement, and so a_{ir} can be interpreted as giving the (RMS) average centimeters of displacement at level i due to factor r (see, for example, Harshman, Ladefoged, and Goldstein 1977). This property of loadings—sharing the same units as the data—is sometimes quite useful for application of the results of the factor analysis (see, for example, Ladefoged, Harshman, Goldstein, and Rice 1978).

It is easy to establish the validity of this general interpretation. When the Mode B and C loadings for each factor have a mean-square of 1.0, then, as we proved above, the bc products also have a mean-square of 1.0. But whenever a set of numbers with unit mean-square are multiplied by a constant (such as a_{ir}), their mean-square becomes the square of that constant (a_{ir}^2) and so their root-mean-square must be the absolute value of that constant ($|a_{ir}|$). The algebraic argument for this is obtained by following the same steps as those that took us from equation (5–50) to (5–51) but considering only the one-factor case, so that the sum over r vanishes.

Since the mean-square of the contributions of factor r at level i of Mode A is $(a_{ir})^2$, the average squared value in column r of A gives the mean-square contribution of factor r in the data as a whole. When the factors are not orthogonal, the sum of the column mean-squares will not equal the total mean-square predicted by the model, because some of the mean-squares will be overlapping. But as the factors become less and less oblique, individual factor mean-squares overlap less and less, and their sum approaches the mean-square of $\hat{x}_{ijk}$, the part of the data fit by the model. This is true not only in the data as a whole, but also at each level of A.

Special Cases

If the data are centered across Mode B or C and thus have a mean of zero at each level of Mode A, then the factor contribution mean-squares discussed above become equal to factor contribution *variances*, and the equations above can be interpreted in terms of *variance components*. For example, $(a_{ir})^2$ would give the variance of the contributions of factor r to level i of the data; thus, the absolute value of the loading would give the standard deviation of the factor contributions, a reasonable measure of factor influence and one which (as noted earlier) is in the same scale as the data itself. To take another useful example, (5–51) would imply that for orthogonal factors, the squared loadings in a given row of the Mode A matrix sum to the variance predicted by the factors (the "variance accounted for") at that level.

When the data are also size-standardized so that their overall mean-square is 1.0, then, for orthogonal factors, each squared factor loading can be interpreted as the *proportion* of variance contributed by that factor at that level. Thus, when the squared loadings are summed across rows, the sums can be interpreted as the proportion of data variance accounted for by the q-factor model at each level of Mode A (the "communality" of each level). When averaged down columns, the mean-squared loading of each factor can be interpreted as the proportion of the total data variance accounted for by that factor. (In this latter case, taking the column *sums* of squares, rather than mean-squares, will give the three-way equivalent of eigenvalues or squared singular values.)

If the preprocessing of a given data set had included double size-standardization—for example, both within levels of Mode A and within levels of Mode B, when the data were centered on Mode C—then it would be possible to consider two different sets of alternative standardized loadings for the data. One might scale the output so that the Mode B and C loadings have sums of squares equal to 1.0 (in order to interpret the Mode A loadings in the traditional way), and then rescale the output so that Modes A and C have mean-squares of 1.0 (to interpret the Mode B loadings in the traditional way).

Before we invest too much effort in such procedures, however, we should remind ourselves that a given set of factor loadings can be examined to determine the meaning of the factor or its pattern of effects in a given mode, regardless of the scaling of the loadings matrix. Such basic interpretation is accomplished by looking at the relative sizes of different loadings within the factor and determining what might best distinguish those items lying at one pole of the dimension from those lying at the opposite pole. Since such comparisons within a factor are not affected by columnar scale adjustment of the loadings matrix, it is only for special interpretation in terms of factor-variable correlations or variance-accounted-for that one need worry about the special standardizations discussed above.

"Factor (Component) Scores"

Give the strong parallel that can be established between one of the PARAFAC loadings matrices (in the example above, the Mode *A* matrix) and the traditional two-way "factor loadings" matrix, what can we make of the PARAFAC loadings matrices for the other two modes (in the example, the Mode *B* and *C* matrices)? As we have already noted, these matrices can be compared to the traditional factor score or component score matrices from a two-way analysis. A slightly different perspective, which we have adopted in this chapter, interprets the entries in the Mode *B* and Mode *C* loading matrices as expressing the *average* (that is, root-mean-square) size of factor or component scores for each factor, within particular levels of the mode in question. The individual factor score estimates, or principal component scores, are given by the direct or Kronecker product of the Mode *B* and Mode *C* loadings vectors for each factor (that is, the $b_{jr}c_{kr}$ double-product terms corresponding to each jk combination).

It should be noted, however, that factor scores computed in this fashion strictly conform to the "strong" restrictions imposed by the PARAFAC "system variation" model, as explained earlier. In contrast, indirect estimation of factor scores by regression methods (as described by Harshman and Berenbaum 1981) would allow estimation of variance components of the factor scores that might follow more general patterns of variation.

As we noted earlier, some contemporary psychometricians have adopted conventions regarding the use of the terms "factor" versus "component" that are more restricted than our usage here. They would refer to the ($b_{jr}c_{kr}$) products as "component scores," since they are based on loadings obtained by direct fitting. They would also use "component scores" to refer to any scores estimated after indirect fitting, so long as the diagonals of the covariance matrices used in the indirect fitting contained unaltered variances. They would reserve the term "factor score estimates" for the scores obtained (by regression or other methods) after fitting the "common factor model" (namely, when indirect fitting is performed in which the covariance matrix diagonals are replaced by estimates of common variance or "communality"; this is accomplished in PARAFAC when the diagonals are "ignored" by iterative reestimation during the ALS fitting procedure).

Loadings Standardization when Doing Indirect Fitting

If the data being analyzed are summed cross-products, or covariances, a *modified* standardization convention is required to maintain comparability between the results of three-way and two-way analysis and between direct and indirect fitting methods. Basically, the Mode *C* loadings must be scaled so that their mean is 1.0 and the scale of the data is jointly reflected in Modes *A* and *B*.

For solutions obtained by analysis of covariance matrices, the Mode *A* and Mode *B* loadings tables are identical, since the covariance data is symmetric across Modes *A* and *B*. Both sets of loadings correspond to the Mode *A* table of the direct fit solution. However, the Mode *C* table contains entries that correspond to the *squares* of the entries obtained with direct fitting. This is because the data variances and covariances are averages of cross-products, and in each cross-product, the Mode *C* loadings occur twice. For this reason, we scale the Mode *C* weights so that the average *first power* of the entries in each column is equal to one; it is not appropriate to set their average *square* to one, since they are already squared quantities. As a result, we obtain Mode *C* weights that directly equal the factor (or "factor score") variances at each level of Mode *C*; the average factor score variance over all levels of Mode *C* is 1.0, as is appropriate for *z*-scores. The scale of the data is then jointly reflected in Modes *A* and *B*. (Both Mode *A* and *B* loadings are multiplied by the square root of the scale factors that would otherwise be applied to a single mode—these scale factors are the square roots of our three-mode generalization of eigenvalues.)

When the data have been suitably size-standardized so that the average covariance matrix is a correlation matrix[10] (we call this "Equal Average Diagonal" or EAD standardization), then the resulting Mode *A* or *B* weights will have the characteristics of traditional "loadings." They can be interpreted as beta weights and have the same properties as the loadings obtained from two-way factor analysis of correlations or from factor analysis by direct fitting of *z*-score standardized data. In fact, with error-free data, when extracting the correct number of factors for perfect fit, the two kinds of loadings will be identical. That is, the Mode *A* loadings obtained by indirect fitting, when scaled in this modified fashion, are identical to the Mode *A* loadings that would be obtained by direct fitting of the profile data, when appropriately scaled and centered, as described earlier. In the more general case of fallible data with "true" and "error" factors that are not strictly orthogonal in any mode, the result of indirect fitting will be similar but not identical to the result of direct fitting.

When doing indirect fitting, one can compute measures of the fit to the covariances (for example, mean-square error, Stress, *R*-squared), but one often wants to obtain an index of the implied fit of the factors to the original data from which the covariances were computed. This is what is normally reported in two-way factor analysis. Such a fit value is easily obtained by simply averaging the squared loadings in each column of the Mode *A* matrix, to obtain the variance accounted for by that factor (or summing the squared loadings to obtain the generalized eigenvalue for each factor). By summing variance estimates across factors, we obtain the total variance accounted for by the solution. The latter value could also have been obtained by summing squared loadings in each row of **A** to get the predicted diagonal elements of all the covariance matrices fit by the analysis. The average of these diagonals gives the desired quantity, namely, the variance of the original profile data that is indirectly fit by the model; since the original diagonal elements of the covariance matrices represent the variances of the variables in the profile data, the fitted part of these diagonals represents the fitted part of the original data variance.

When EAD normalization is used, the average diagonal in the data is 1.0, and so the average predicted diagonal gives the proportion of variance accounted for by the solution, and the mean-squared loading for a given factor gives the proportion of the total variance contributed by that factor. The proportion of the common variance contributed by each factor is simply the mean-squared Mode *A* loading for that factor divided by the total of mean-squared Mode *A* loadings for all factors in the solution.

APPENDIX 5–2: A METHOD OF TRANSFORMING ANY TUCKER REPRESENTATION INTO A COMPACT PARAFAC-CANDECOMP REPRESENTATION

Suppose a three-way array has an exact Tucker representation and we wish to find the corresponding PARAFAC-CANDECOMP representation that provides perfect fit in the lowest possible dimensionality. The following is a method of deriving the desired PARAFAC-CANDECOMP representation by operations on the Tucker representation. We will describe the method in terms of operations on a T3 model; however, the same procedure will work with a T2 model, as noted at the end of this appendix.

This procedure demonstrates that the trilinear rank of an array—that is, the number of "triads" or PARAFAC dimensions needed to provide exact fit to the array (Kruskal 1977)—is determined by the trilinear rank of the core matrix, which is the number of PARAFAC dimensions needed to exactly fit the core matrix.

The T3 Starting Point

Standard Form

Let **X** be an n by m by p three-way array. For notational convenience, we conceptualize it as composed of p successive n by m "slices" or two-way arrays; the kth such slice is called $\mathbf{X}_k$. To express the T3 representation of **X** in matrix terms, we use the same convention as in (5–4); that is, we represent the three-way array by providing a general expression for the kth slice. We call the initial form of our T3 model $(\text{T3})_1$ and write it as follows:

$$(\mathrm{T3})_1: \quad \mathbf{X}_k = \mathbf{A} \left[\sum_{t=1}^{q_c} c_{kt}\mathbf{G}_t \right] \mathbf{B}' , \qquad (5\text{–}52)$$

where the matrices **A**, **B**, and **C** are factor-loading matrices for Modes *A*, *B*, and *C*, respectively. **A** is n by q_a with arbitrary element a_{ir}, **B** is m by q_b with arbitrary element b_{js}, and **C** is p by q_c with arbitrary element c_{kt}. The "core matrix" **G** is a three-way array considered for convenience as a set of slices $\mathbf{G}_t$; there are q_c such slices, each slice being q_a by q_b.

Individualized Form

If we let $\mathbf{H}_k$ represent the matrix of interactions between Mode *A* and *B* dimensions for the kth level of Mode *C* (for instance, for the kth person), then we can write an "individualized" form of the model,

$$\mathbf{X}_k = \mathbf{A}\, \mathbf{H}_k\, \mathbf{B}' , \qquad (5\text{–}53)$$

where

$$\mathbf{H}_k = \sum_{t=1}^{q_c} c_{kt}\, \mathbf{G}_t \qquad (5\text{–}54)$$

In this "individualized" form of the model, each level of Mode *C* has a matrix $\mathbf{H}_k$, which gives that level's interactions between the Mode *A* and *B* dimensions. For example, if the levels of Mode *C* represent persons, then each $\mathbf{H}_k$ gives the idiosyncratic changes in Mode *A* and/or *B* axis orientations and weights for the kth person. As shown in (5–54), each $\mathbf{H}_k$ is a weighted combination of the slices of the core matrix, with the tth slice weighted by the kth person's loading on the tth dimension of Mode *C*.

The Transformation

To find the most compact PARAFAC-CANDECOMP model that represents this same data structure, we need to perfectly "diagonalize" the slices of the core array. We can do this by applying PARAFAC to the core, although other related methods, such as that of de Leeuw and Pruzansky (1978), should also work. Note that we do not seek an approximation but rather a perfectly fitting PARAFAC representation. To find this exact representation, we take advantage of the fact that PARAFAC can fit more dimensions than there are levels to any way of the array (Harshman 1970; Kruskal 1976). For an arbitrary core that is roughly cubical, the required diagonalized equivalent will in general be slightly larger than the original core, but not by much. As Kruskal has pointed out (personal communication, April 1983), in all such cases that we have examined, the PARAFAC dimensionality q does not exceed $(q_a + q_b - 1)$, where q_a and q_b are the number of levels of the two smaller ways of the core.

Thus, we begin by applying PARAFAC to the core. To describe the PARAFAC representation of the core, we employ a matrix formulation of the same kind as used earlier in this chapter. The PARAFAC representation of the three-way core array **G** is specified by giving a general matrix representation of an arbitrary tth slice $\mathbf{G}_t$:

$$\mathbf{G}_t = \bar{\mathbf{A}}\,\bar{\mathbf{D}}_t \bar{\mathbf{B}}' , \tag{5–55}$$

where (if an exact PARAFAC solution requires q factors) $\bar{\mathbf{A}}$ would be a q_a by q matrix of PARAFAC loadings on Mode A of the core, and $\bar{\mathbf{B}}$ would be a q_b by q matrix of PARAFAC loadings on Mode B of the core, with $\bar{\mathbf{D}}_t$ being a diagonal q by q matrix, whose diagonal elements are the PARAFAC loadings on Mode C of the core. To provide an exact representation, the $\bar{\mathbf{D}}_t$ will sometimes be larger than the corresponding $\mathbf{G}_t$. For example, in the case of a $2 \times 3 \times 5$ core matrix, where each $\mathbf{G}_t$ slice is 2×3, the PARAFAC representation would require at most four dimensions to perfectly fit **G**. (For some $2 \times 3 \times 5$ cores, it would require less.) In the four-dimensional case, $\bar{\mathbf{A}}$ would be 2×4, $\bar{\mathbf{B}}$ would be 3×4 (and so $\bar{\mathbf{B}}'$ would be 4×3), and $\bar{\mathbf{C}}$ would be 5×4. Each $\mathbf{D}_t$ would be a diagonal 4×4 matrix.

We now need to find the factor loadings for the data array **X** that go along with the "diagonalized" form of its core array. These will be obtained by appropriately transforming the original Tucker loadings. Substituting our PARAFAC-CANDECOMP representation of the core (5–55) back into our basic model $(\mathrm{T3})_1$ (given by [5–52]), we obtain:

$$\mathbf{X}_k = \mathbf{A}\,[\,\sum_{t=1}^{q_c} c_{k\,t}\,(\bar{\mathbf{A}}\,\bar{\mathbf{D}}_t\bar{\mathbf{B}}')\,]\,\mathbf{B}' . \tag{5–56}$$

Since $\bar{\mathbf{A}}$ and $\bar{\mathbf{B}}$ do not change with t, we can move these constant terms outside the summation over t, and obtain:

$$\mathbf{X}_k = \mathbf{A}\,[\,\bar{\mathbf{A}}\,(\sum_{t=1}^{q_c} c_{kt}\bar{\mathbf{D}}_t)\,\bar{\mathbf{B}}'\,]\,\mathbf{B}' . \tag{5–57}$$

Regrouping terms, we obtain:

$$\mathbf{X}_k = (\mathbf{A}\,\bar{\mathbf{A}})\,(\sum_{t=1}^{q_c} c_{kt}\bar{\mathbf{D}}_t)\,(\bar{\mathbf{B}}'\,\mathbf{B}') . \tag{5–58}$$

We can interpret the $\bar{\mathbf{A}}$ and $\bar{\mathbf{B}}$ matrices as defining linear transformations of the original $(\mathrm{T3})_1$ loading matrices for Modes A and B. So, if we define the transformed loading matrices as

$$\overset{*}{\mathbf{A}} = \mathbf{A}\,\bar{\mathbf{A}} , \qquad \overset{*}{\mathbf{B}} = \mathbf{B}\,\bar{\mathbf{B}} , \tag{5–59}$$

and if we define a new "diagonalized" core matrix $\bar{\mathbf{G}}$ composed of q_c diagonal slices $\bar{\mathbf{G}}_t$ such that

$$\bar{\mathbf{G}}_t = \bar{\mathbf{D}}_t ,$$

then we can write an expression for a partially transformed T3 model, which we call $(\mathrm{T3})_2$:

$$(\mathrm{T3})_2: \quad \mathbf{X}_k = \overset{*}{\mathbf{A}} \left(\sum_{t=1}^{q_c} c_{kt} \overline{\mathbf{G}}_t \right) \overset{*}{\mathbf{B}}' \, . \tag{5–60}$$

This model has the same structural form as $(\mathrm{T3})_1$ given above in (5–52) but uses transformed component matrices $\overset{*}{\mathbf{A}}$ and $\overset{*}{\mathbf{B}}$, which are n by q and m by q, rather than $\mathbf{A}$ and $\mathbf{B}$, which are n by q_a and m by q_b. It also has a modified core $\overline{\mathbf{G}}$, which is q by q by q_c, instead of $\mathbf{G}$, which is q_a by q_b by q_c. It is not fully transformed into a PARAFAC-CANDECOMP form, however, since it has the same matrix of Mode C loadings as $(\mathrm{T3})_1$, and the core matrix is not superdiagonal.

Nonetheless, model $(\mathrm{T3})_2$ gives rise directly to a PARAFAC-CANDECOMP representation when we write it in "individualized" form, as was done for $(\mathrm{T3})_1$ in (5–53). We simply define

$$\overline{\mathbf{H}}_k = \sum_{t=1}^{q_c} c_{kt} \overline{\mathbf{G}}_t = \sum_{t=1}^{q_c} c_{kt} \overline{\mathbf{D}}_t \, , \tag{5–61}$$

which allows us to write $(\mathrm{T3})_2$ as

$$\mathbf{X}_k = \overset{*}{\mathbf{A}} \, \overline{\mathbf{H}}_k \, \overset{*}{\mathbf{B}}' \, . \tag{5–62}$$

Now it is apparent from (5–61) that all the matrices that are summed to produce $\overline{\mathbf{H}}_k$ are diagonal, and so $\overline{\mathbf{H}}_k$ is itself diagonal. Hence, (5–62) can be considered a PARAFAC model. For uniformity of notation, we can let $\overline{\mathbf{H}}_k = \overset{*}{\mathbf{D}}_k$ and rewrite (5–62) as the PARAFAC-CANDECOMP model

$$\mathbf{X}_k = \overset{*}{\mathbf{A}} \, \overset{*}{\mathbf{D}}_k \, \overset{*}{\mathbf{B}}' \, , \tag{5–63}$$

where $\overset{*}{\mathbf{A}}$ is an n by q matrix of Mode A loadings, $\overset{*}{\mathbf{B}}$ is an m by q matrix of Mode B loadings, and $\overset{*}{\mathbf{D}}_k$ is a q by q diagonal matrix whose diagonal elements constitute the kth row of $\overset{*}{\mathbf{C}}$, a p by q matrix of Mode C loadings. This gives the PARAFAC-CANDECOMP representation of the three-way array $\mathbf{X}$, which is equivalent to the Tucker representation $(\mathrm{T3})_1$.

T3 in "Superdiagonal" (PARAFAC-CANDECOMP) Form

From (5–63) we can directly obtain a T3 representation in which the core matrix is superdiagonal (that is, has nonzero entries only when $r = s = t$). We simply define our transformed Mode C loading matrix for T3 as a matrix that has the diagonals of the $\overset{*}{\mathbf{D}}$ matrices as its rows. That is, $\overset{*}{\mathbf{C}}$ is a p by q matrix for which

$$\overset{*}{c}_{kt} = \overset{*}{d}_{tt(k)} \, ,$$

where $\overset{*}{d}_{tt(k)}$ is the tth diagonal element of $\overset{*}{\mathbf{D}}_k$. With these revised Mode C loadings, we can define the core matrix as the three-mode equivalent of an identity matrix. Thus, if $\overset{*}{g}_{uut}$ is the element in the uth diagonal cell of the tth slice of the revised core matrix $\overset{*}{\mathbf{G}}$, then

$$\overset{*}{g}_{uut} = 1 \quad \text{if} \quad t = u, \quad 0 \text{ otherwise.}$$

We can now write the T3 model in explicit superdiagonal or PARAFAC-CANDECOMP form, as follows:

$$(\text{T3})_3: \quad \mathbf{X}_k = \overset{*}{\mathbf{A}} \left[\sum_{t=1}^{q} \overset{*}{c}_{kt} \overset{*}{\mathbf{G}}_t \right] \overset{*}{\mathbf{B}}' . \tag{5–64}$$

Transforming T2

The same procedure can be applied to find the PARAFAC-CANDECOMP representation for a Tucker T2 model, in which the core matrix has as many Mode C levels as the data array $\mathbf{X}$. We simply apply PARAFAC to the extended core. In the argument above, this would mean that $q_c = p$. This presents no new difficulties for the PARAFAC decomposition in (5–55), and the rest of the transformation follows as before.

The fact that corresponding dimensions across Mode C can always be "collapsed" (as described earlier in this chapter) places the absolute upper bound on the number of needed PARAFAC dimensions as $(q_a q_b)$, or the product of whichever two dimensionalities are the smallest. (Kruskal [1977] gives the same upper bound.) Consideration of parameter counts or "degrees of freedom" of the core and corresponding PARAFAC representation shows that for T2, "extended cores" that have many levels in one mode, $(q_a q_b)$ will often be both a lower and upper bound. On the other hand, for small cores in which q_a, q_b, and q_c are similar, our experience suggests that the actual upper bound is much lower, perhaps $(q_a + q_b - 1)$. In either case, the transformed representation would be relatively more compact than the original.

As noted earlier in this chapter, the idea of approximately "diagonalizing" the Tucker T2 core array has been proposed previously by several authors (the first of which, to our knowledge, was Cohen [1974]). Thus, the transformations presented here should be compared with the proposals of Carroll and Pruzansky (1979), Cohen (1974), de Leeuw and Pruzansky (1978), McCallum (1976), and the application of this idea in Kroonenberg (1981b).

NOTES

1. A matrix approach that clearly displays the three-way symmetry of the PARAFAC model can be implemented by means of the Kronecker product notation. This is the approach used by Jennrich in Harshman (1970, chapter 5). (Tucker [1966] also

employed the Kronecker product for discussion of three-mode factor analysis.) The data would be represented as the sum of q different three-way arrays, each array corresponding to the contribution of one factor. The array corresponding to the contribution of the rth factor would be represented as the Kronecker product of the rth column of $\mathbf{A}$ with the rth column of $\mathbf{B}$ and $\mathbf{C}$. If we let $\mathbf{X}$ equal the three-way array $\{x_{ijk}\}$ and $\mathbf{E}$ the three-way array of error terms, and if we use $\mathbf{a}_r$ to represent the rth column of $\mathbf{A}$, and similarly for $\mathbf{B}$ and $\mathbf{C}$, then, by using $\otimes$ to represent the Kronecker product of two vectors, we can write:

$$\mathbf{X} = \sum_r \mathbf{a}_r \otimes \mathbf{b}_r \otimes \mathbf{c}_r + \mathbf{E} .$$

While elegant, this representation would involve us with notation and mathematics less familiar to many who work in this area; hence, it has not been adopted here.

2. Some would insist on the term "component scores" rather than "factor scores," as noted earlier.

3. Cattell discusses proportional profiles in several places in a recent book (Cattell 1978), describing the "confactor rotation" method with some optimism. However, he still considers the implicit orthogonality constraint (which we saw earlier to be a consequence of the use of indirect fitting) to be an unsolved problem. He does not discuss direct fitting using PARAFAC1 or indirect fitting using the PARAFAC2 model, as possible ways around this problem.

4. The authors would like to thank J. D. Carroll for suggesting this simple method of embedding T3 in PARAFAC-CANDECOMP.

5. Kruskal (personal communication, March 1983) has proven that a $2 \times 2 \times 2$ core matrix will have a maximum rank of 3 and that a $3 \times 3 \times 3$ core will have a maximum rank of 5, consistent with our Monte Carlo results. The interesting thing is that under a wide range of plausible conditions, a $2 \times 2 \times 2$ core will have rank 2 or 3, depending on the relative size of different core elements. Kruskal provides algebraic conditions (inequalities based on products of particular elements of the core) that determine whether a $2 \times 2 \times 2$ core will have rank 3 or 2.

6. Some of our Monte Carlo results have led Kruskal to reexamine these proofs; he now cautions us (personal communication, March 1983) that the theorems 3a–3d in Kruskal (1977) require minor modification and are not correct as stated. Thus, these particular theorems concerning trilinear rank have been revised. However, theorems in Series 1, 2, and 4 of that paper still appear to be valid. Copies of corrected versions of Kruskal (1977) are available from J. Kruskal.

7. This expression for T3 leads to a simple demonstration of an upper bound on Tucker dimensionalities (closely related to the upper bound on the embedding PARAFAC representation, given in Method II and equation [5–31], above). Any Tucker representation can always be replaced by one in which the largest mode of the core has no more dimensions than the product of the dimensionalities of the other two modes. We show this as follows:

Suppose, without loss of generality, that in our model Mode *C* has the most dimensions (that is, q_c is larger than either q_a or q_b). We can construct an alternative core array and corresponding Mode *C* loading matrix that will generate the same data as the original model. First, we use the original q_a by q_b by q_c core and Mode *C* loadings to construct the *p* different $\mathbf{H}_k$ matrices, one for each level of Mode *C*, as defined by (5–34). (If we are starting with a T2 model, this first step is unnecessary.) Then we construct a new core matrix that is q_a by q_b by $q_a q_b$—that is, one that has q_a times q_b slices, rather than q_c. We construct the slices of the new core such that there is only one nonzero element in each slice, and this element occurs at a different location in each slice; furthermore, we set these nonzero elements equal to 1.0. This establishes the new core. Note that we now have one slice of the core for each cell of an $\mathbf{H}_k$ matrix, since $\mathbf{H}_k$ is q_a by q_b. This allows us to obtain the Mode *C* matrix corresponding to this new core by simply assigning to each person a set of Mode *C* loadings equal to the values in his $\mathbf{H}_k$ matrix. In particular, we obtain the *k*th person's loading for factor *t* from the cell of $\mathbf{H}_k$ that corresponds to the nonzero cell in the *t*th slice of the new core matrix.

8. The question might be raised as to whether this second interpretation (5–38) is really different from the first one (5–36). We would argue that they *could* describe different states of affairs. On the one hand, if a new set of stimuli, when judged by person *k*, failed to show the same close association between active and good objects, it would appear that person *k*'s distinct perspective was specific to particular stimuli, and model (5–35/5–36) might be more appropriate. On the other hand, if *all* new stimuli would be judged by person *k* in a way more closely linking goodness and activity, then it might be argued that model (5–37/5–38) is more appropriate. (Even here, however, it would seem that one could conceptually distinguish two different cases. In the first case, the basic *meaning* of "activity" is unaltered, but associated with this meaning is a derivative evaluation of "goodness," perhaps because person *k* believes that activity has good consequences. In the second case, the basic semantic category itself is changed; for example, person *k* might not use "activity" in its pure sense but rather as some notion of "vitality" that incorporates both aspects of activity and goodness.) This is a subtle issue, but we believe that the distinction is "real" if it can lead to different predictions. One way in which this might occur would be in a four-mode situation.

It is possible to construct four-mode models in which the indeterminacies of the Tucker model are reduced or eliminated because different ways of resolving these indeterminacies make different predictions about how the data might vary across a fourth mode (see, for example, PARAFAC3, given in Table 5–1 and discussed above under Other Models). One might imagine such a four-mode model being applied to studies in which each rater evaluates a given set of stimuli, using a standard set of rating scales, but repeats the ratings for each stimulus with respect to several different situations or contexts.

9. General acknowledgments for chapters 5 and 6 are made at the beginning of this chapter. We would also like to thank

John Reddon for helpful comments on chapter 5.

10. The covariance matrices are uniformly rescaled as follows:

$$\overset{*}{c}_{ijk} = \frac{c_{ijk}}{[(\overline{c}_{ii.})^{1/2} \ (\overline{c}_{jj.})^{1/2}]},$$

where

$$\overline{c}_{ii.} = \frac{1}{P} \sum_{k=1}^{p} c_{iik},$$

and similarly for $\overline{c}_{jj.}$. This imposes the same rescaling on all levels of Mode C and produces the same covariances as would have been obtained if the raw data for each level of Mode A had been converted to z-scores before covariances were computed.

REFERENCES

Bentler, P. M., and S.-Y. Lee. 1978. Statistical aspects of a three-mode factor analysis model. *Psychometrika* 43:343–52.

————. 1979. A statistical development of three-mode factor analysis. *British Journal of Mathematical and Statistical Psychology* 32:87–104.

Bloxom, B. 1968. A note on invariance in three-mode factor analysis. *Psychometrika* 33:347–50.

Bock, R. D. 1973. Word and image: Sources of the verbal and spatial factors in mental test scores. *Psychometrika* 38:437–57.

Carroll, J. D., and P. Arabie. 1980. Multidimensional scaling. *Annual Review of Psychology* 31:607–49.

Carroll, J. D., and J. J. Chang. 1970. Analysis of individual differences in multidimensional scaling via an N-way generalization of "Eckart-Young" decomposition. *Psychometrika* 35:283–319.

————. 1972. IDIOSCAL (Individual DIfferences in Orientation SCALing): A generalization of INDSCAL allowing idiosyncratic reference systems as well as an analytic approximation to INDSCAL. Paper presented at the spring meeting of the Psychometric Society, Princeton, New Jersey, March.

Carroll, J. D., and S. Pruzansky. 1979. Use of LINCINDS as a rational starting configuration for INDSCAL. Manuscript, Bell Laboratories, Murray Hill, New Jersey.

Carroll, J. D., and M. Wish. 1974. Multidimensional perceptual models and measurement methods. In *Handbook of perception,* vol. 2, ed. E. C. Carterette and M. P. Friedman. New York: Academic Press.

Carroll, J. D., S. Pruzansky, and J. B. Kruskal. 1980. CANDELINC: A general approach to multidimensional analysis of many-way arrays with linear constraints on parameters. *Psychometrika* 45:3–24.

Cattell, R. B. 1944. "Parallel Proportional Profiles" and other principles for determining the choice of factors by rotation. *Psychometrika* 9:267–83.

———. 1978. *The scientific use of factor analysis in behavioral and life sciences.* New York: Plenum.
Cattell, R. B., and A. K. S. Cattell. 1955. Factor rotation for proportional profiles: Analytical solution and an example. *British Journal of Statistical Psychology* 8:83–92.
Chang, J. J., and J. D. Carroll. 1978. Three are not enough: An INDSCAL analysis suggesting that color space has seven (± one) dimensions. Manuscript, Bell Laboratories, Murray Hill, New Jersey.
Cohen, H. S. 1974. Three-mode rotation to approximate INDSCAL structure (TRIAS). Paper presented at the annual meeting of the Psychometric Society, Palo Alto, California, March.
Comrey, A. L. 1967. Tandem criteria for analytic rotation in factor analysis. *Psychometrika* 32:143–53.
Corballis, M. C. 1973. A factor model for analysing change. *British Journal of Mathematical and Statistical Psychology* 26:90–97.
Corballis, M. C., and R. E. Traub. 1970. Longitudinal factor analysis. *Psychometrika* 35:79–98.
Dawson, M. 1982. Multidimensional responses to metaphor. Master's thesis, University of Western Ontario.
Dawson, M., and R. A. Harshman. Forthcoming. Applications of an asymmetric multidimensional scaling model to alphabetic confusion matrices. Manuscript, University of Western Ontario.
De Leeuw, J., and S. Pruzansky. 1978. A new computational method to fit the weighted Euclidean distance model. *Psychometrika* 43:479–90.
Diaconis, P., and B. Efron. 1983. Computer-intensive methods in statistics. *Scientific American* May:116–26, 128, 130.
Dunn, T. R., and R. A. Harshman. 1982. A multidimensional scaling model for the size-weight illusion. *Psychometrika* 47:25–45.
Edgington, E. S. 1969. Approximate randomization tests. *Journal of Psychology* 72:143–49.
———. 1980. *Randomization tests.* New York: Marcel Dekker.
Efron, B. 1982. The jackknife, the bootstrap and other resampling plans. *Society for Industrial and Applied Mathematics Monographs* 38.
Eysenck, H. J. 1950. Criterion analysis—an application of the hypothetico-deductive method to factor analysis. *Psychological Review* 57:38–53.
———. 1977. Personality and factor analysis: A reply to Guilford. *Psychological Bulletin* 84:405–11.
Gandour, J. T., and R. A. Harshman. 1978. Cross-language differences in tone perception: A multidimensional scaling investigation. *Language and Speech* 21:1–33.
Gifi, A. 1981. *Non-linear multivariate analysis.* Preliminary edition. Leiden, the Netherlands: University of Leiden, Department of Data Theory.
Green, P. E. 1978. *Analysing multivariate data.* Hinsdale, Illinois: The Dryden Press.
Guilford, J. P. 1977. Will the real factor of Extroversion-Intro-

version please stand up? A reply to Eysenck. *Psychological Bulletin* 84:412–16.

Haan, N. 1981. Common dimensions of personality development: Early adolescence to middle life. In *Present and past in middle life,* ed. D. H. Eichorn, J. A. Clausen, N. Haan, M. P. Honzik, and P. H. Mussen. New York: Academic Press.

Harman, H. H., and W. H. Jones. 1966. Factor analysis by minimizing residuals (MINRES). *Psychometrika* 31:351–68.

Harris, R. J. 1975. *A primer of multivariate statistics.* New York: Academic Press.

Harshman, R. A. 1970. Foundations of the PARAFAC procedure: Models and conditions for an "explanatory" multi-modal factor analysis. *UCLA Working Papers in Phonetics* 16:1–84 (University Microfilms No. 10,085).

————. 1972a. Determination and proof of minimum uniqueness conditions for PARAFAC1. *UCLA Working Papers in Phonetics* 22:111–17 (University Microfilms No. 10,085).

————. 1972b. PARAFAC2: Mathematical and technical notes. *UCLA Working Papers in Phonetics* 22:30–44 (University Microfilms No. 10,085).

————. 1973. Oblique coordinate systems in multidimensional scaling: Theory and experimental test. Manuscript, University of California at Los Angeles.

————. 1978. Models for analysis of asymmetrical relationships among N objects or stimuli. Paper presented at the first joint meeting of the Psychometric Society and the Society for Mathematical Psychology, Hamilton, Canada, August.

————. 1981. Generalizing INDSCAL and PARAFAC1: Uniqueness of certain representations of three-way crossproduct arrays. Manuscript, University of Western Ontario.

Harshman, R. A., and S. A. Berenbaum. 1981. Basic concepts underlying the PARAFAC-CANDECOMP three-way factor analysis model and its application to longitudinal data. In *Present and past in middle life,* ed. D. H. Eichorn, J. A. Clausen, N. Haan, M. P. Honzik, and P. H. Mussen. New York: Academic Press.

Harshman, R. A., and M. L. Howe. 1979. Systematic effects of noneuclidean input on the recovered subject and stimulus spaces in three-way MDS. Paper presented at the annual meeting of the Psychometric Society, Monterey, California, June.

Harshman, R. A., and G. Papcun. 1976. Vowel normalization by linear transformation of each speaker's acoustic space. *Journal of the Acoustical Society of America* 59:S71 (abstract).

Harshman, R. A., and J. R. Reddon. 1983. Three-way (PARAFAC) factor analysis of WAIS and WAIS-R suggests different axes than simple structure. Paper presented at annual meeting of the Classification Society (North American Branch), Philadelphia, May.

Harshman, R. A., P. E. Green, Y. Wind, and M. E. Lundy. 1982. A model for the analysis of asymmetric data in marketing research. *Marketing Science* 1:205–42.

Harshman, R. A., P. Ladefoged, and L. Goldstein. 1977. Factor analysis of tongue shapes. *Journal of the Acoustical Society of America* 62:693–707.

Horan, C. B. 1969. Multidimensional scaling: Combining observations when individuals have different perceptual structures. *Psychometrika* 34:139–65.

Horst, P. 1965. *Factor analysis of data matrices.* New York: Holt, Rinehart, and Winston.

Hubert, L. J. 1983. Inference procedures for the evaluation and comparison of proximity matrices. In *Numerical Taxonomy,* ed. J. Felsenstein. New York: Springer Verlang.

Jennrich, R. 1972. A generalization of the multidimensional scaling model of Carroll and Chang. *UCLA Working Papers in Phonetics* 22:45–47.

Jöreskog, K. G. 1971. Simultaneous factor analysis in several populations. *Psychometrika* 36:409–26.

Kettenring, J. R. 1983. A case study in data analysis. In *Statistical data analysis,* ed. R. Gnanadesikan. Providence, R.I.: American Mathematical Society.

Kroonenberg, P. M. 1981a. *User's guide to TUCKALS3. A program for three-mode principal component analysis (WEP-Reeks, WR 81-6-RP).* Leiden, The Netherlands: University of Leiden.

———. 1981b. *User's guide to TUCKALS2. A program for three-mode principal component analysis with extended core matrix (WEP-Reeks, WR 81-35-RP).* Leiden, The Netherlands: University of Leiden.

Kroonenberg, P. M., and J. de Leeuw. 1980. Principal components analysis of three-mode data by means of alternating least-squares algorithms. *Psychometrika* 5:69–97.

Kruskal, J. B. 1976. More factors than subjects, tests than treatments: An indeterminacy theorem for canonical decomposition and individual differences scaling. *Psychometrika* 41: 281–93.

———. 1977. Three-way arrays: Rank and uniqueness of trilinear decompositions, with application to arithmetic complexity and statistics. *Linear Algebra and Its Applications* 18: 95–138.

———. 1978. Factor analysis and principal components: Bilinear methods. In *International encyclopedia of statistics,* ed. W. H. Kruskal and J. M. Tanur. New York: Free Press.

———. 1981. Multilinear models for data analysis. *Behaviormetrika* 10:1–20.

———. 1983. Multilinear methods. In *Statistical data analysis,* ed. R. Gnanadesikan. Providence, R.I.: American Mathematical Society.

Kruskal, J. B., and M. Wish. 1978. *Multidimensional scaling.* Beverly Hills: Sage.

Kruskal, J. B., R. A. Harshman, and M. E. Lundy. 1983. Relationships between Tucker's three-mode factor analysis, PARAFAC-CANDECOMP and CANDELINC. Paper presented at the annual meeting of the Psychometric Society, Los Angeles, June.

Ladefoged, P., R. A. Harshman, L. Goldstein, and L. Rice. 1978. Generating vocal tract shapes from formant frequencies. *Journal of the Acoustical Society of America* 64:1027–35.

Levin, J. 1965. Three-mode factor analysis. *Psychological Bulletin* 64:442–52.

Linker, W. 1982. Articulatory and acoustic correlates of labial activity in vowels: A cross-linguistic study. *UCLA Working Papers in Phonetics* 56:1–134.

MacCallum, R. C. 1976. Transformation of a three-mode multidimensional scaling solution to INDSCAL form. *Psychometrika* 41:385–400.

McDonald, R. P. 1979. The simultaneous estimation of factor loadings and scores. *British Journal of Mathematical and Statistical Psychology* 32:212–28.

Meredith, W. 1964. Notes on factorial invariance. *Psychometrika* 29:177–85.

Meyer, J. P. 1980. Causal attribution for success and failure: A multivariate investigation of dimensionality, formation and consequences. *Journal of Personality and Social Psychology* 38:704–18.

Mosteller, F., and J. W. Tukey. 1977. *Data analysis and regression: A second course in statistics.* Reading, Mass.: Addison-Wesley.

Osgood, C. E., G. J. Suci, and P. H. Tannenbaum. 1957. *The measurement of meaning.* Urbana: University of Illinois Press.

Reddon, J. R., R. Marceau, and D. N. Jackson. 1982. An application of singular value decomposition to the factor analysis of MMPI items. *Applied Psychological Measurement* 6: 275–83.

Sands, R., and F. W. Young. 1980. Component models for three-way data: An alternating least squares algorithm with optimal scaling features. *Psychometrika* 45:39–67.

Schönemann, P. H. 1972. An algebraic solution for a class of subjective metrics models. *Psychometrika* 7:441–51.

Sentis, K. P., R. A. Harshman, and C. Stangor. Forthcoming. PARAFAC three-way factor analysis of dichotomous data: A Monte Carlo study. Manuscript, Ted Bates Advertising/New York.

Snyder, C. W., Jr., W. D. Walsh, and P. R. Pamment. 1983. Three-mode PARAFAC factor analysis in applied research. *Journal of Applied Psychology* 68:572–83.

Spence, I. 1977. Multidimensional scaling. In *Quantitative ethology,* ed. P. Colgan. New York: Wiley-Interscience.

Terbeek, D. 1977. A cross-language multidimensional scaling study of vowel perception. *UCLA Working Papers in Phonetics* 37:1–271.

Terbeek, D., and R. A. Harshman. 1972. Is vowel perception non-Euclidean? *UCLA Working Papers in Phonetics* 22:13–29.

Thurstone, L. L. 1947. *Multiple-factor analysis: A development and expansion of the vectors of mind.* Chicago, Illinois: University of Chicago Press.

Torgerson, W. S. 1958. *Theory and methods of scaling.* New York: Wiley.

Trick, L. 1983. The effect of individual differences in analogic reasoning and perceived tenor-vehicle similarity of metaphor comprehension and appreciation. Master's thesis, University of Western Ontario.

Tucker, L. R. 1963. Implications of factor analysis of three-way matrices for measurement of change. In *Problems in measuring change,* ed. C. W. Harris. Madison, Wis.: University of

Wisconsin Press.
———. 1964. The extension of factor analysis to three-dimensional matrices. In *Contributions to mathematical psychology,* ed. N. Frederiksen and H. Gulliksen. New York: Holt, Rinehart, and Winston.
———. 1966. Some mathematical notes on three-mode factor analysis. *Psychometrika* 31:279–311.
———. 1972. Relations between multidimensional scaling and three-mode factor analysis. *Psychometrika* 37:3–27.
Velicer, W. F., A. C. Peacock, and D. N. Jackson. 1982. A comparison of component and factor patterns: A Monte Carlo approach. *Multivariate Behavioral Research* 17:371–88.
Weeks, D. G., and P. M. Bentler. 1979. A comparison of linear and monotone multidimensional scaling models. *Psychological Bulletin* 86:349–54.
Weinberg, S. L., and R. A. Harshman. 1980. The structure of semantic space: A reexamination. Paper presented at the annual meeting of the National Council on Measurement and Education, San Francisco, April.
Weinberg, S. L., J. D. Carroll, and H. S. Cohen. Forthcoming. Confidence regions for INDSCAL using the jackknife and bootstrap techniques. Manuscript, Bell Laboratories, Murray Hill, New Jersey.

6

Data Preprocessing and the Extended PARAFAC Model

Richard A. Harshman and Margaret E. Lundy

DATA PREPROCESSING

Definitions and Objectives

Definitions

Two Kinds of Preprocessing. By *data preprocessing* we mean any transformations of the data values that are performed before fitting the main analysis model. For convenience, we define two broad classes of preprocessing: data *conversion* and data *adjustment.* This distinction is somewhat arbitrary, but, if not taken too seriously, it will simplify discussion.

Data conversions take one kind of data and transform it into a different kind, usually entailing a change in the form of the model that is subsequently fit to the data. In chapter 5, we discussed two types of data conversions and their implications: (a) the conversion of profile data into covariances (or sums of cross-products) prior to indirect fitting, and (b) conversion of proximity data into scalar products as part of multidimensional scaling. Data adjustments, on the other hand, take data of a given kind and transform it into an improved or more standardized version of data of the same kind. Adjustments do not require a change in the form of the model, although they may implicitly extend the model being fit, as we shall see later. In this section, we will only consider data adjustments.

There are two basic types of data adjustments used before direct fitting of profile data: (a) additive adjustments—for example, to "center" the data; and (b) multiplicative adjustments—for example, to "rescale" or "normalize" the data. Occasionally, a third type of adjustment might also be useful: (c) nonlinear adjustments —such as log transformation to "linearize" the data, or the Weeks and Bentler (1979) rank transformation to deal with nonmetric data. (We will not consider this third case in this chapter.)

This article has been reprinted with the permission of the authors, Richard A. Harshman and Margaret E. Lundy (Copyright 1983). See page 222 for acknowledgments.

Additive Adjustments. In traditional two-way factor analysis, the raw score matrix is often transformed so that the scores for each variable have zero-mean. This involves a simple additive adjustment—the mean for each variable is subtracted from each score involving that variable. At other times, score matrices are transformed so that the mean for each individual is zero, producing what is called *ipsative* data. Both kinds of additive adjustments are commonly called *centering,* presumably because they place the origin or zero-point of the measurement scale at the mean of a set of data values. There are also other kinds of additive adjustments that are sometimes used. For example, in multidimensional scaling, an *additive constant* is estimated for each subject, which transforms that subject's interval-scale dissimilarities into ratio-scale distances. Such additive adjustment does not produce data with zero-mean, in part because negative distances are not permitted. (These additive constants were mentioned briefly in chapter 5.)

In this section, we will restrict our discussion of additive adjustments to consideration of various kinds of centering. There are many ways of centering a three-way array, but we will find that only some of these are appropriate methods of preprocessing for three-way factor analysis or multidimensional scaling. In particular, we will demonstrate that, to be appropriate, centering must be performed over one-way subarrays, such as rows, columns, or "tubes."

Multiplicative Adjustments. In two-way factor analysis, the data are often adjusted so that the variance of each variable is unity. This is accomplished by multiplying all the data points for each variable by the reciprocal of the standard deviation for that variable. In other applications (such as multidimensional scaling), the data may be adjusted so that the scores for each subject have a sum of squares equal to unity. In general, such multiplicative adjustments are accomplished by considering the data in subsets and multiplying the values in each subset by a constant selected so that the size of observations in the subset will have some desired property (for example, mean-square 1.0). When used to equate the size of observations across subsets, such multiplicative adjustments are often called *standardizing* or *normalizing* (although *standardizing* is also used to refer to the combination of additive and multiplicative adjustment that transforms a set of values to *z*-scores). Multiplicative adjustments can also be used to vary the overall data size across different subsets (for example, as a means of accomplishing a weighted least-squares analysis, to be described below).

Multiplicative adjustments are often useful in three-way applications, but there are many possible ways of performing such adjustments, and, once again, we shall find that only some of these are appropriate for three-way factor analysis or multidimensional scaling. It will be demonstrated that multiplicative adjustments, unlike centering, must be performed over two-way subarrays (that is, "slices" of the three-way array).

Objectives of Preprocessing

Preprocessing plays a very important role in three-way analysis. We saw in the previous chapter how conversion of profile data to covariances or scalar products allows us to use indirect fitting, which permits modified or alternative models of the data, thus extending the power and generality of the PARAFAC procedure. Preprocessing plays an equally important role in direct fitting of profile data. Once again, it will allow us to fit extended versions of our analysis model and thus increase the generality of the three-way procedure. We will explore this extended PARAFAC perspective in more detail later in this chapter. In the present section, however, we adopt a complementary perspective, where preprocessing is used to make the data appropriate for the PARA-FAC model. We will also focus on several additional advantages of preprocessing data before analysis.

Eight Objectives. There are at least eight basic reasons for preprocessing data before performing a PARAFAC analysis: (a) to make the data appropriate for the PARAFAC model by removing unwanted constants and eliminating some kinds of *conditionality* (to be explained below); (b) to emphasize relationships among patterns of change in certain modes, rather than among baseline values; (c) to weight and thus emphasize or ignore particular subsets of the data during the analysis, or, conversely, to equate the influence that different variables, stimuli, subjects, and so forth have on the form of the final solution; (d) to equate the size of presumed error variance components across different subsets of the data; (e) to standardize the data so that comparison of loadings across levels is facilitated; (f) to standardize the overall scale so that comparisons across data sets is facilitated; (g) to standardize the data so that useful added interpretations of the factor loadings are possible (for example, as correlations, variance components, and so on); and (h) to permit fitting of an extended PARAFAC model.

Objectives (a) and (b) are accomplished by various additive adjustments, and objectives (c) and (d) are accomplished by multiplicative adjustments. Finally, objectives (e) through (h) are accomplished by means of both additive and multiplicative adjustments. Of these various objectives, (a) is probably the most general and important application of preprocessing, and it will be considered in detail in the following discussion. After that, we will consider (b) through (g). Interpretation in terms of the extended PARAFAC model (objective [h], which is closely related to objective [a]) will be taken up in the latter part of the chapter.

Making One's Data More Appropriate for the PARAFAC Model

Restrictiveness of the Formal Model

The form of the basic PARAFAC-CANDECOMP model (5–3) appears to require what Stevens (1946) would call *ratio-scale* measurements. That is, the data values must be strictly proportional to the underlying quantities being measured, so that a zero in the data represents a complete lack of the property being measured, and 4 units represents exactly twice as much of the property as 2 units. This strong level of measurement appears necessary because the factor model is formulated in terms of proportional changes in factor contributions from one level of a given mode to the next. Computing such proportional changes implies estimation of a true zero for each factor, which implies knowledge of the true zero-point on the variables from which those factors are to be extracted. Yet with much of the data encountered in the social sciences and elsewhere, one does not know the true zero; there are arbitrary origins and unknown additive constants that make the measurements interval scale at best. Furthermore, social sciences data is often "row-conditional" or "column-conditional"; that is, there may be a different measurement scale from one row to the next or from one column to the next. For example, each person may have a different subjective origin to his scale, or each variable might involve a different additive constant. Since there are no explicit terms in the model and no features in the fitting procedure designed to take care of such arbitrary origins and constants, Sands and Young (1980) have suggested that the practical application of the model may be seriously limited.

However, the model's formal measurement level is not as crucial as one might expect. While it is true that PARAFAC treats the data as ratio scale, so do all other metric factor analysis models, including Tucker's three-mode model and almost all varieties of traditional two-way factor analysis. Despite this formal limitation, factor analysts have not had trouble coping with interval-scale data. In part, this is because factor analysis was traditionally conceptualized in terms of correlation matrices. The conversion of raw data to correlations before performing factor analysis eliminated the problem of origin and so allowed investigators to avoid the issue. The first stages of such conversion implicitly perform a centering operation that transforms an interval-scale data matrix into appropriate ratio-scale form. (Such centering operations will be discussed below.) But conversion to correlations is not necessary for interval-scale data to be successfully factor-analyzed. As we shall see, additive constants can be represented as extra factors with constant loadings in one or more modes, and thus, in theory, can be easily accommodated within the factor-analytic model by slightly increasing the dimensionality of the solution. It can be argued, therefore, that the limitation of metric factor analysis to ratio-scale data is more apparent than real.

Unfortunately, however, extracting biases and constants as extra factors has drawbacks when one tries to discover the preferred *rotation* for the factor axes; hence, an alternative approach is preferred. Authors such as Horst (1965), who applied two-way factor analysis directly to raw profile data, found a simple solution to the question of origin that bypassed full consideration of measurement issues: First, convert the data to deviation score form (usually as z-scores), either by rows, columns, or both, and then apply the factor analysis to these standardized scores. This approach is usually rationalized in an ad hoc fashion, as a method of concentrating on deviations from a group mean; the mean is taken to be an alternative reference point suitable for cases in which the true origin was unknown or undefined. As we shall see, however, this standardizing procedure can be justified in terms of a more theoretically based rationale.

With three-way arrays, the question of proper centering and scale adjustment becomes much more subtle; many alternative possibilities exist, and the "easy" or "natural" approaches are not always correct. It becomes important to develop an explicit rational basis for choosing among different procedures for preprocessing. We advocate here a set of criteria based primarily on an algebraic analysis of the effects of the preprocessing on the latent structure of the data. To do this, we must begin with a simple model of the raw data. This model includes terms representing the factors we wish to extract, along with terms representing the undesirable constants and biases that obscure the true origin and make the data interval scale (and row- or column-conditional) rather than ratio scale. We then study the effects of alternative preprocessing transformations on both the factors and on the unwanted components. "Good" preprocessing will remove the undesired components while leaving the latent factors relatively unchanged.

We first apply this theoretical analysis to the question of preprocessing two-way arrays and demonstrate that it confirms the appropriateness of the methods currently used in two-way factor analysis. We then extend it to the three-way case to discover which of the many possible three-way centering and standardizing techniques are in fact appropriate for three-way factor analysis.

A Model for Interval-Scale Data

Two-Way Case. Suppose we have a data matrix $\mathbf{X}$ that fits the structural model of factor or principal component analysis, except that there are additional constants present that obscure the origin of the variables. There may be a constant term h offsetting the data as a whole from true ratio-scale properties. In addition, if the origin of the data is row- or column-conditional, then there would be additional constants h_i or h_j, which would describe how the ith row or jth column deviates from ratio scale (after adjustment by h). If we let x_{ij} be the element in the ith row and jth column of $\mathbf{X}$, our expression for the latent structure of a data point might be written as follows:

$$x_{ij} = \sum_r (a_{ir} b_{jr}) + h_i + h_j + h + e_{ij} \; . \qquad (6\text{–}1)$$

Here, as earlier, the a_{ir} and b_{jr} terms represent the factor loadings for Modes A and B, respectively, and e_{ij} represents random error.

For ease of discussion, let us interpret (6–1) in terms of a concrete example. Suppose that the data matrix $\mathbf{X}$ is a two-way array of stimulus ratings, with rows corresponding to stimuli (pictures of products) and columns corresponding to judges, and that two factors underlie the systematic ratio-scale part of the data. All judges are rating a common set of stimuli on a particular quality (such as attractiveness). Thus, the data entry described by (6–1) is the attractiveness score of stimulus i as rated by judge j. The a-loadings indicate how much each stimulus has of each type of attractiveness on a ratio scale. For example, the two types of attractiveness might be the aesthetic and economic advantages of the products. The b-loadings (which in some cases would be called *factor scores* or *factor score estimates*) are proportionality coefficients that relate the size of a given subject's ratings to the amount of each type of attractiveness possessed by the stimulus. When considered in this way, it becomes apparent that two-way factor analysis is formulated in terms of ratio-scale quantities.

In addition to the $a_{ir}b_{jr}$ terms representing the factor contributions, we have included additional h_i, h_j, and h-terms in the model, representing the unknown constant offsets for each row and column, and an overall additive constant, which disturb proportionality and thus cause the data to have conditional interval-scale rather than unconditional ratio-scale properties. Let us consider the interpretations that these extra terms might have in our hypothetical analysis of rating scales. In such a study, the true zero-point for attractiveness would be hard to define and would almost certainly differ from a zero data value. In fact, the responses would typically be made on a rating scale, with points on the scale ranging from 1 to 9 and the two ends of the scale labeled "unattractive" and "attractive." It is clear that a rating of "2" on this scale is unlikely to represent twice as much attractiveness as a rating of "1." Thus, an overall additive constant is necessary to shift the zero-point somewhere nearer the middle of the scale. This is provided by the h-term. But a single constant may not be enough. If a given rater views "unattractive" as a negative quality equal and opposite to "attractive," the neutral point may be at 4.5. But for other raters who view "unattractive" as simply a lack of attractiveness, it may be nearer the low end of the scale. Because of such subject-to-subject variation, an h_j term is also needed to specify the shifts in the baseline for the jth rater. Finally, there might also be shifts in the baseline from stimulus to stimulus, although it seems less likely in this particular example. In other cases, however, shifts from row to row might seem plausible; for example, if the data were test scores, with rows representing the particular tests and columns representing the subjects. To provide for the general case, our model includes both an h_i and h_j term.[1]

Direct factor analysis of our hypothetical ratings data, without preprocessing, would be an example of fitting a ratio-scale model to row- and column-conditional interval-scale data. If we only fit r-factors, the obtained loadings would be distorted by the presence of the constants h_i, h_j, and h in the data. However, the fact that the ratings do not properly reflect the zero-point of the attributes being measured can be easily corrected within the context of the two-way factor model. One solution would be to extract two extra dimensions. If, as in our example above, there were two "real" dimensions present, then a third and fourth dimension would be needed. The dimensions could be defined so that

$$a_{i3} = h_i \ , \ a_{i4} = 1 \qquad (6\text{–}2)$$

and

$$b_{j3} = 1 \ , \ b_{j4} = h_j + h \ .$$

(Here we have absorbed the overall bias h into the fourth dimension. There is no reason why it might not have been divided between the two dimensions or absorbed into the third dimension instead; the choice is arbitrary.) With such additional factors, one can represent conditional origin interval-scale data within the framework of a ratio-scale factor model.

For this example, there should be a four-dimensional solution that would recover the factor loadings for the first two dimensions correctly while isolating the constant biases into a third and fourth dimension. However, because of error in the data, the four factors that explain the largest proportion of the variance will not exactly correspond to the two "true" factors plus the two factors representing the additive constants. If the error is not too large, it should nonetheless be possible to find a rotation in which these factors are closely approximated.

PARAFAC Three-Way Case. Both the PARAFAC model and Tucker's three-mode model focus on components that an analysis of variance would call *three-way interactions*. That is, they describe components whose pattern of change from one level of Mode A to the next depends simultaneously on the levels of Mode B and Mode C involved, as well as the levels of Mode A. The sizes of the factor contributions normally vary as a function of all three modes, because normally the loadings for a given factor vary in size across the levels of all three modes. Thus, in order to take all the unwanted offsets into account, we need to consider the consequences of not only the global constant h and "one-way" effects (the h_i, h_j, and now h_k terms) that are constant across two modes, but also the consequences of "two-way" effects, which arise out of the interaction of a particular pair of modes and are constant over a third mode.

For simplicity, let us return to our example involving rating-scale data. Suppose that each judge now rates each stimulus on several scales; in addition to "unattractive-attractive," he also rates the stimulus pictures on "practical-impractical," "fun-dull," "appealing-unappealing," "economical-uneconomical," and so on. This would generate a three-way array whose general entry would be the rating of stimulus i by judge j using scale k. As before, we might want to have an overall baseline term and terms for biases due to particular people, stimuli, and now scales. But we can also imagine that a particular person has an idiosyncratic interpretation of a particular scale and uses that scale as if it had a different neutral point. This would be a two-way interaction of person and scale that would be constant across stimuli. We might adjust for such effects by means of a doubly subscripted constant h_{jk}. Similarly, we should consider possible two-way interactions of particular people with particular stimuli and particular scales with particular stimuli. For example, if one of the stimulus pictures included a beautiful model, the scale "attractive-unattractive" might take on a special meaning with respect to this stimulus, which would uniformly alter the neutral point of ratings for this stimulus-scale combination across all subjects. Thus, within the PARAFAC framework, we might represent three-way arrays of interval-scale data in terms of the following general form:

$$x_{ijk} = \sum_r (a_{ir}b_{jr}c_{kr}) + h_{ij} + h_{ik} + h_{jk} + h_i + h_j + h_k + h + e_{ijk} \,. \tag{6–3}$$

This representation could easily be adapted to incorporate the Tucker three-mode model or other metric three-way models by simply modifying the first term. In (6–3) the first term represents the PARAFAC-CANDECOMP factor model; by replacing it with the corresponding triple summation from Tucker's model and applying the algebraic transformations discussed below, we could prove the appropriateness or inappropriateness of different types of preprocessing for these models. It quickly becomes apparent, however, that the consequences of the transformations on the h-terms are the same regardless of the exact representation of the factors and that the consequences of the transformations on the factor component are similar for all current metric three-way models. Thus, all of the conclusions reached concerning representation of interval-scale data and the appropriateness of different centerings will apply equally to PARAFAC and Tucker's three-mode or other metric three-way models, with the exception of those considerations that involve the intrinsic axis property of the PARAFAC-CANDECOMP model.

Now, since the unsubscripted and singly subscripted h-terms can be absorbed into the doubly subscripted ones as the investigator sees fit, the general expression for a interval-scale three-way array might be written more simply:

$$x_{ijk} = \sum_r (a_{ir}b_{jr}c_{kr}) + h_{ij} + h_{ik} + h_{jk} + e_{ijk} \,. \tag{6–4}$$

Given this representation, let us consider how it can be incorporated within the ratio-scale framework of three-way factor analysis. The effects represented by the h_{ij} term, for example, cannot in general be represented as a single factor. In fact, this set of effects constitutes a two-way array (replicated identically across all levels of the third mode) which may have a complex structure, with a rank possibly equal to the number of rows or columns (whichever is less) in the h_{ij} table. Hopefully, however, the systematic part of the two-way biases will have a relatively low dimensionality; we might therefore approach the representation of these biases as a traditional two-way factor-analytic problem. We could decompose each of the two-way arrays of h-values into a set of factors in order to represent its effects in terms of our model.

Although decomposing a two-way table of, say, h_{ij} values would normally yield two-way factors, in order to incorporate these effects into our three-way factor model, we could express them in terms of three-way factors with constant loadings across a third mode. This would give us a three-way (such as PARA-

FAC) representation of the three-way array of bias components that is added to the data array by the h_{ij} term in equation (6–3). Obviously, the h_{ik} and h_{jk} terms could be represented in analogous fashion, and so we could rewrite our expression for the interval-scale data as:

$$x_{ijk} = \sum_r (a_{ir}b_{jr}c_{kr}) + \sum_s (a_{is}b_{js}c_{ks}) + \sum_t (a_{it}b_{jt}c_{kt}) + \sum_u (a_{iu}b_{ju}c_{ku}) + e_{ijk} , \qquad (6\text{–}5)$$

with the constraints that:

$$\begin{aligned} &\text{all } c_{ks} = \text{a constant,} \\ &\text{all } b_{jt} = \text{a constant, and} \\ &\text{all } a_{iu} = \text{a constant.} \end{aligned} \qquad (6\text{–}6)$$

Thus, two-way interactions can be decomposed into a set of two-way factors, which can be incorporated into the three-way model as factors that have constant loadings over a third mode. If desired, these two-way factors could further be decomposed into deviations plus *main effects* (that is, factors that are constant across two modes), to more closely parallel the longer form of (6–3). Such additional decomposition would be purely to facilitate some preferred scheme of interpretation, however, since no improvement in fit would be obtained by the extra terms and the particular manner of the decomposition would be somewhat arbitrary.

Disadvantages of Treating Constants as Extra Factors. Although (6–5) shows how the factor-analytic model can be considered appropriate for conditional-origin interval-scale data, the representation of bias and constant terms as extra factors can pose problems for the actual processes of factor estimation and rotation. As noted earlier, the presence of random error in the data normally prevents the extraction of factors with precisely constant loadings; factors often emerge with some of the constant variance mixed in with the variance of factors that are nonconstant across all modes. Even if it were possible to extract the constants unperturbed by error, existing methods for factor rotation do not automatically find the desired solution—one that has dimensions of the form specified in (6–6) or, in the two-way case, as described by the constraints of (6–2).

When fitting the PARAFAC model to three-way arrays, a further consideration is involved: We want to preserve the intrinsic axis property of the solution. As noted in chapter 5, however, it is generally necessary that factors show distinct patterns of variation across *all three modes* for their contributions to be unambiguously sorted out from other possible linear combinations with other factors (that is, in order for their axis orientation to be uniquely determined). It is apparent from (6–5) and (6–6) that the extra factors that we might incorporate to represent constants and two-way interactions do not show the necessary distinct patterns of variation across all three modes. Indeed, since they are defined by what are essentially two-way arrays (replicated across a third mode), they show the classic rotational indeterminacy of two-way factor analysis. If the sum of squares due to these extra factors is substantial, then they tend to emerge even in solutions of low dimensionality, probably mixed in with the unique three-way factors that we are trying to identify. The indeterminacy of these two-way interaction factors causes the factors of interest to appear indeterminate, and the resulting solution in general is quite hard to interpret. Thus, the approach of treating unwanted constants as extra factors seems to have too many drawbacks to recommend it for normal three-way PARAFAC analysis. There is, however, an alternative way to deal with the unwanted constants: Remove them by appropriate preprocessing.

Transforming Interval-Scale into Ratio-Scale Data

Instead of including extra factors in our model to represent any unwanted constants or biases that may be present in the data, another approach is to remove such constants by special data transformations preceding the factor analysis, or in other words, by data preprocessing. First, we will show how this is done for two-way data and then extend it to the three-way case.

Two-Way Case. As mentioned earlier, the problem of arbitrary origin was traditionally dealt with in two-way factor analysis by means of *centering,* removing means from the rows and/or columns of a data matrix. This set the new origin at the group mean on each row or column. The procedure was justified as a way of focusing attention on the deviations from the mean. But is such a justification adequate? What information is lost or distorted by this procedure? In more precise terms: How is our interpretation of the underlying "true" factors of these centered scores different from the interpretation we would have obtained could we have factored the original uncentered data without the interfering constants?

If we algebraically examine the traditional centering process using our model of interval-scale data ([6–1]), we see that centering has very desirable properties. It removes the unwanted and unknown constants that obscure the origin of the data and at the same time performs only a minor transformation on the factors in which we are interested. Since this transformation does not usually interfere with our ability to interpret the solution, we are able to get essentially the same information that we would have obtained had the original data been ratio scale in the first place. This algebraic result provides a firm basis for concluding that centering is an appropriate method of preprocessing in two-way factor analysis.

Let us adopt the convention that a dot in place of a subscript means that the term has been averaged across the levels of that subscript. For example, if there are n levels to Mode A, then

$$x_{.j} = \frac{1}{n} \sum_{i=1}^{n} (x_{ij}) \ , \qquad (6\text{–}7)$$

and similarly for other subscripted terms.

Let us return again to our example of the ratings matrix **X**, where x_{ij} represents the rating of the ith stimulus by the jth judge. Suppose the original data differs from ratio scale because of an overall constant h and an offset h_j for each judge, but there are no differential offsets for each stimulus. Thus, our data point before centering has the underlying structure

$$x_{ij} = \sum_r (a_{ir} b_{jr}) + h_j + h + e_{ij} \; . \tag{6–8}$$

We wish to determine what happens to this structure after centering. We start with the description of the effects of centering defined in terms of the observed scores. If we let $\overset{*}{x}_{ij}$ represent the data point after centering across stimuli, then we can start with the expression:

$$\overset{*}{x}_{ij} = x_{ij} - x_{.j} = x_{ij} - \frac{1}{n} \sum_r (x_{ij}) \; . \tag{6–9}$$

To determine the effects on the underlying structure, we substitute from (6–8) into (6–9) and simplify, obtaining:

$$\begin{aligned} \overset{*}{x}_{ij} = {} & \sum_r (a_{ir} b_{jr}) + h_j + h + e_{ij} \\ & - \frac{1}{n} \sum_{i=1}^{n} \Big(\sum_r (a_{ir} b_{jr}) + h_j + h + e_{ij} \Big) \; . \end{aligned} \tag{6–10}$$

By distributing the summation over i and rearranging the terms of (6–10), we obtain:

$$\begin{aligned} \overset{*}{x}_{ij} = {} & \sum_r (a_{ir} b_{jr}) + h_j + h + e_{ij} - \sum_r \Big(b_{jr} \Big(\frac{1}{n} \sum_{i=1}^{n} (a_{ir}) \Big) \Big) \\ & - \frac{1}{n} \sum_{i=1}^{n} (h_j) - \frac{1}{n} \sum_{i=1}^{n} (h) - \frac{1}{n} \sum_{i=1}^{n} (e_{ij}) \; . \end{aligned}$$

Since the average of a constant is the same constant, the sixth and seventh terms are not affected by the summation. We can use the dot notation to simplify the remaining terms and obtain:

$$\begin{aligned} \overset{*}{x}_{ij} = {} & \sum_r (a_{ir} b_{jr}) + h_j + h + e_{ij} - \sum_r (b_{jr} (a_{.r})) \\ & - h_j - h - e_{.j} \end{aligned}$$

Collecting like terms, we obtain:

$$\overset{*}{x}_{ij} = \sum_r ((a_{ir} - a_{.r}) b_{jr}) + (e_{ij} - e_{.j}) \; .$$

If we let $\overset{*}{a}_{ir} = a_{ir} - a_{.r}$ and $\overset{*}{e}_{ij} = e_{ij} - e_{.j}$, we have the following simple expression for the underlying structure of the data matrix after centering:

$$\overset{*}{x}_{ij} = \sum_r (\overset{*}{a}_{ir} b_{jr}) + \overset{*}{e}_{ij} . \tag{6–11}$$

The consequences of centering the data are simple: The factor loadings for one mode are centered, as are the error terms, and the constants h and h_j disappear. The centered data is now ratio scale and can be represented (aside from the random error) simply in terms of the underlying factors of attractiveness. We can recover these factors accurately after the centering operation, since the b-coefficients (in this example, person weights or factor score estimates) are unaffected by the centering, and the a-coefficients (or factor loadings) are the same except that the mean value of a for each factor has been subtracted out (that is, the loading matrix **A** is column-centered). Thus, the shape of the profile of factor loadings is unaffected, and only the elevation is changed. Since most of the interpretation of a factor is based on the relative values of the loadings on the different stimuli, rather than the absolute size of the loadings, the centering operation does not normally interfere with our interpretation of the factor.

Similarly, if there had been bias or offset terms that differed across stimuli but were constant across persons (h_i terms), then centering across persons would have removed these terms. In this case the factor scores or b-terms would have been centered but not the loadings or a-terms. As noted earlier, the occurrence of h_i terms might be more plausible if our data consisted of ability test scores for a group of people; this is where centering across persons is most commonly employed.

Centering both rows and columns of **X** can be accomplished by applying the second centering operation to the result obtained by the first centering. If we use $\overset{\circ}{x}_{ij}$ to represent an element of the double-centered data array $\overset{\circ}{\mathbf{X}}$, then we could compute it as follows:

$$\overset{\circ}{x}_{ij} = \overset{*}{x}_{ij} - \overset{*}{x}_{i.} .$$

Double-centering removes both h_i and h_j terms; it also column-centers both the Mode A and Mode B tables of factor loadings and double-centers the error terms. It otherwise leaves the latent structure unchanged. As has been indicated elsewhere (Horst 1965), the effects of double-centering are the same, regardless of the order of the centering operations performed; this becomes clear when one sees that double-centering can also be represented—more compactly but less transparently—as a single operation defined in terms of the original row and column means:

$$\overset{\circ}{x}_{ij} = x_{ij} - x_{i.} - x_{.j} + x_{..} .$$

When the response scale is clearly offset from psychological zero, as in the rating scale example, the most important consequence of the centering operation is often the removal of the overall constant h. This could be accomplished by centering in either direction. However, the elimination of the conditionality of origin can also be important, and this requires centering the correct mode.

Matrix Formulation of the Effects of Centering. In the preceding discussion, we used scalar notation to deduce the relationships between the form of a given factor before and after centering. We used this method because it easily generalizes to n-way arrays. However, with two-way arrays, there is a matrix representation of centering that allows one to demonstrate its effect on the latent structure most clearly. Horst (1965) defines a matrix that he calls a *centering matrix*. We will call this matrix $\mathbf{L}_n$ (for the n by n version of $\mathbf{L}$ needed for column-centering) or $\mathbf{L}_m$ (for the m by m version needed for row-centering). The matrix is defined as follows:

$$\mathbf{L}_n = \mathbf{I} - \frac{1}{n}(\mathbf{1}\mathbf{1}') \; , \qquad (6\text{–}12)$$

where $\mathbf{1}$ is an n-element column vector, all of whose entries are 1, and $\mathbf{I}$ is an n by n identity matrix. The matrix $\mathbf{L}_n$ is symmetric, idempotent, and has rank $n - 1$. Its rows and columns sum to zero. Its diagonal elements are equal to $(n - 1)/n$, and off-diagonal elements are equal to $-1/n$. The matrix

$$\overset{*}{\mathbf{X}} = \mathbf{L}_n \mathbf{X} \qquad (6\text{–}13)$$

is the column-centered version of $\mathbf{X}$; that is, the sum of each column of $\overset{*}{\mathbf{X}}$ is zero.

To determine the effects of centering on the latent structure of $\mathbf{X}$, we may replace $\mathbf{X}$ by its representation in terms of factors. So, if

$$\mathbf{X} = \mathbf{A}\mathbf{B}' + \mathbf{E} \; , \qquad (6\text{–}14)$$

where $\mathbf{A}$ is the n by r matrix of factor loadings and $\mathbf{B}$ is the m by r matrix of person loadings (factor scores or component scores), then by premultiplying both sides of (6–14) by $\mathbf{L}_n$, we obtain

$$\mathbf{L}_n \mathbf{X} = \mathbf{L}_n (\mathbf{A}\mathbf{B}' + \mathbf{E}) \qquad (6\text{–}15)$$

or

$$\mathbf{L}_n \mathbf{X} = (\mathbf{L}_n \mathbf{A})\mathbf{B}' + (\mathbf{L}_n \mathbf{E}) \; .$$

If we let $\overset{*}{\mathbf{A}} = \mathbf{L}_n \mathbf{A}$ and $\overset{*}{\mathbf{E}} = \mathbf{L}_n \mathbf{E}$, then

$$\overset{*}{\mathbf{X}} = \overset{*}{\mathbf{A}}\mathbf{B}' + \overset{*}{\mathbf{E}} \; . \qquad (6\text{–}16)$$

That is, column-centering $\mathbf{X}$ simply column-centers the $\mathbf{A}$ matrix of factor loadings and the matrix of error terms, as we determined earlier in scalar notation. Similarly, we can row-center $\mathbf{X}$ by multiplying it on the right by an m by m matrix $\mathbf{L}_m$, defined analogously to $\mathbf{L}_n$ ($\mathbf{L}_m = \mathbf{I} - (1/m)(\mathbf{1}\mathbf{1}')$ where $\mathbf{1}$ is a m-element column vector all of whose entries are 1). The result is:

$$\mathbf{X}\,\mathbf{L}_m = (\mathbf{A}\,\mathbf{B}' + \mathbf{E})\,\mathbf{L}_m\,,$$

and so

$$\mathbf{X}\,\mathbf{L}_m = \mathbf{A}\,(\mathbf{B}'\,\mathbf{L}_m) + \mathbf{E}\,\mathbf{L}_m\,. \qquad (6\text{–}17)$$

Thus, row-centering $\mathbf{X}$ simply row-centers the matrix $\mathbf{B}'$ and the error matrix $\mathbf{E}$. Since $\mathbf{B}'$ is the transpose of the person-loading matrix $\mathbf{B}$, row-centering $\mathbf{B}'$ is equivalent to column-centering $\mathbf{B}$.

Finally, as noted earlier, we can apply both row- and column-centering to a data matrix, in either order, with the same result: Both row sums and column sums will be zero. The effect of double-centering on the latent structure is:

$$\mathbf{L}_n\,\mathbf{X}\,\mathbf{L}_m = (\mathbf{L}_n\,\mathbf{A})\,(\mathbf{B}'\,\mathbf{L}_m) + \mathbf{L}_n\,\mathbf{E}\,\mathbf{L}_m = \overset{*}{\mathbf{A}}\,\overset{*}{\mathbf{B}}{}' + \overset{\circ}{\mathbf{E}}\,. \qquad (6\text{–}18)$$

Double-centering the data matrix simply column-centers both matrices of factor loadings and double-centers the error matrix.

Some Interpretations of the Effects of Centering. From these two-way results, we can state a rule that will enable us to generalize to three-way arrays: Centering across any mode of the data matrix simply centers the errors across that mode and column-centers the factor-loading matrix for that mode. For any given axis orientation, a column-centered factor matrix generally leads to the same interpretations as an uncentered factor matrix. Geometrically, after centering, the axes are parallel to the corresponding axes before centering; they have simply been translated as a unit from the original arbitrary origin to a new origin at the centroid of the configuration. Thus, centering by rows or columns can be said to *preserve the factor structure* underlying the data.[2]

The conclusion that centering the data across levels of a given mode centers the corresponding factor loadings also implies that centering removes unwanted components that are constant across that mode. We showed earlier how terms that are constant across levels of a given mode are equivalent to factors with constant loadings in that mode. Because centering the data across levels of a given mode column-centers the factor-loading matrix for that mode, those factors with constant loadings in the centered mode are transformed into factors with zero loadings in that mode and hence vanish.

The above discussion shows that the effects of different centerings are quite straightforward in the two-way case. Nonetheless, there has been considerable controversy on this point, with some authors claiming that different factors can be uncovered by different centerings. One reason that some investigators may have gotten this impression is that different centerings can give different relative *emphasis* to the several dimensions underlying a data set; the relative emphasis of dimensions before and after centering is determined by the relative sizes of the sums of squares of their original loadings compared to their column-centered loadings. This change in sums of squares will change the relative contributions of the dimensions to each unrotated factor or principal component. If some nonnegligible components are discarded to obtain a lower-dimensional solution, then different parts of the factor space might be discarded in the centered versus uncentered solutions.

Perhaps the most serious obstacle to recognizing the sameness of factors before and after centering has been the rotational indeterminacy of the two-way model. Centering either mode will change the orientation of the principal components, and centering across levels of a given mode will also change the best simple structure orientation of the factors in that mode. These artificial differences have led authors to overlook the essential sameness of the centered factors.

For example, Horst (1965) examined the effects of centering on the underlying dimensions by comparing the basic structure (similar to the unrotated principal components or singular value decomposition) of a data matrix before centering to its basic structure after centering. But the orientation of axes in the basic structure solution is determined by the directions that successively maximize variance-accounted-for. These directions will usually be different after the translation of origin that results from centering, and so comparing the two basic structures amounts to comparing factors that have not only been translated but also rotated into different orientations. Naturally, this obscures the essential sameness of the dimensions. Although Horst presented a complex procedure for deriving the centered basic structure solution from the noncentered, and vice versa, he failed to mention the very simple relation between the two structures that applies *with suitable rotation to restore the equivalence of axis orientations in the factor space.*

The straightforward effects of centering may also be obscured if the pre- and postcentered versions are rotated to simple structure using Varimax, Oblimin, graphical rotation, or some other such criterion. Column-centering a factor-loading matrix changes the number and positions of near-zero entries in the matrix and thus affects the axis orientations obtained when simple structure or related analytic rotation techniques are applied to the dimensions of the centered mode. By centering across variables, for example, and then rotating the variable loadings to simple structure, one could easily get the idea that novel dimensions have been revealed by this ipsative analysis. Once again, this is an illusion caused by looking at the same configuration from a different angle and location. Furthermore, it could be argued that the empirical rationale behind rotation in search of hyperplanes does not directly apply to a mode that has been centered; any hyperplane that results from a number of variables being unaffected by a given factor should go through the true ratio-scale origin of the space and not the centroid origin that results from centering.

It must be acknowledged, however, that the error terms are also modestly affected by centering and hence the perturbations of the true factor loadings due to error may change slightly with different centerings. If the error is random with expected value of zero, the deviations from zero-mean error in the columns or rows of any particular matrix of observed errors will usually be small and so centering would not normally have much effect on these components.

Extension to Three-Way Centering. In three-way factor analysis of raw or profile data, the question of origin becomes considerably more complicated. Thus, we must start by defining some terms. There are three types of centering that might seem natural. We have called them *fiber-, slab-,* and *grand mean- (global-)* centering; Kruskal (chapter 2) calls them *one-way, two-way,* and *three-way centering.*[3]

Fiber- or one-way centering subtracts means computed over one-way subarrays. Since there are three kinds of fibers—rows, columns, and "tubes" (see Figure 6–1)—there are three corresponding varieties of fiber-centering. Slab- or two-way centering subtracts means computed over two-way subarrays. Since there are three kinds of slabs—lateral, horizontal, and frontal (see Figure 6–1)—there are also three types of slab-centering. Grand mean- or three-way centering subtracts the mean computed over the entire three-way array (that is, the *grand mean*). There is only one variety of three-way centering.

Fiber-centering removes means computed across all levels of a given mode, holding the level of the other two modes fixed. Thus, we say that fiber-centering is done across Mode A, across

Figure 6–1. Fibers and Slabs in a Three-Way Data Array

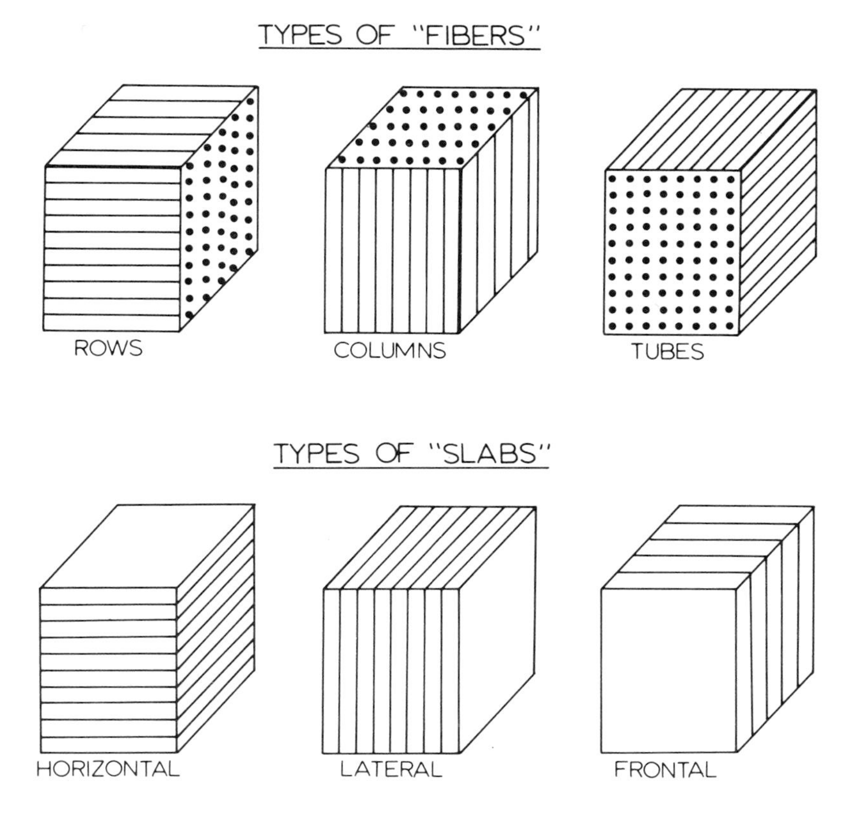

Mode B, or across Mode C. In slab-centering, means are computed across all levels of two modes, holding the third fixed. Thus, we say that slab-centering is done across Modes A and B, across Modes B and C, or across Modes A and C.

Examples of each type of centering are:

$$\overset{*}{x}_{ijk} = x_{ijk} - x_{...} \quad \text{global- (three-way) centering;} \qquad (6\text{–}19)$$

$$\overset{*}{x}_{ijk} = x_{ijk} - x_{..k} \quad \text{slab- (two-way) centering across Modes } A \text{ and } B; \qquad (6\text{–}20)$$

$$\overset{*}{x}_{ijk} = x_{ijk} - x_{.jk} \quad \text{fiber- (one-way) centering across Mode } A. \qquad (6\text{–}21)$$

(Here, again, a dot in place of a subscript indicates that a mean was computed across all levels of the missing subscript.)

To determine which of these centering procedures would provide appropriate preprocessing for three-way factor analysis, we will algebraically determine their effects on a model of three-way interval-scale data. For example, when testing different centerings as possible preprocessing for PARAFAC, we will use a model of data that incorporates the PARAFAC model of the factor component, along with our standard terms for the two-way, one-way, and overall biases. Similarly, we can see which centerings provide appropriate preprocessing for Tucker's model by applying them algebraically to a model of interval-scale data in which the T2 or T3 representation of factors replaces the PARAFAC-CANDECOMP representation. It will become apparent, however, that the same conclusions about appropriate preprocessing hold for all current metric three-way factor models.

The model of three-way interval-scale data ([6–3]) is rather long and unwieldy. To simplify our discussion, we will break it into three parts. We can consider the expression for a data point to be constructed of error plus two different systematic parts: (a) a systematic trilinear part represented by the basic factor model (initially taken to be the PARAFAC-CANDECOMP model), which we will call t_{ijk}; and (b) a systematic but troublesome part not represented by the basic PARAFAC-CANDECOMP model (except by adding factors with constant loadings in one or more modes), which we will call h_{ijk}. This latter part is a sum of the parts explicitly represented in (6–3) as two-way and one-way terms plus an overall constant. Thus,

$$x_{ijk} = t_{ijk} + h_{ijk} + e_{ijk} \quad , \qquad (6\text{–}22)$$

where

$$t_{ijk} = \sum_r (a_{ir} b_{jr} c_{kr}) \qquad (6\text{–}23)$$

and

$$h_{ijk} = h_{ij} + h_{ik} + h_{jk} + h_i + h_j + h_k + h \ . \qquad (6\text{–}24)$$

We are now able to test different types of centering to see if they eliminate the h_{ijk} component but leave the t_{ijk} component relatively unchanged (that is, changed only in a simple way that does not increase the dimensionality or obscure the interpretation). First, we note that we can investigate the effects on the two systematic components separately, since the linearity of the centering operation insures that the effects on each part are independent of the other part. For example, since

$$x_{.jk} = \frac{1}{n}\sum_i (t_{ijk} + h_{ijk} + e_{ijk})$$

$$= \frac{1}{n}\sum_i (t_{ijk}) + \frac{1}{n}\sum_i (h_{ijk}) + \frac{1}{n}\sum_i (e_{ijk}) ,$$

we can rewrite this as:

$$x_{.jk} = t_{.jk} + h_{.jk} + e_{.jk} . \quad (6\text{–}25)$$

Substituting into our general expression for one-way centering and grouping like terms, we obtain:

$$x_{ijk} - x_{.jk} = (t_{ijk} - t_{.jk}) + (h_{ijk} - h_{.jk}) + (e_{ijk} - e_{.jk}) . \quad (6\text{–}26)$$

In other words, centering the data by a given method can be thought of as independently centering each of the additive components of the data by that method.

Let us now ask what the effects of the different centerings are on the trilinear and bilinear parts of our model. We begin with fiber-centering, which happens to have desirable properties.

Centering of One-Way Arrays. Applying fiber- or one-way centering to the trilinear part of the model, we obtain:

$$t_{ijk} - t_{.jk} = \sum_r (a_{ir}b_{jr}c_{kr}) - \frac{1}{n}\sum_i (\sum_r (a_{ir}b_{jr}c_{kr})) \quad (6\text{–}27)$$

$$t_{ijk} - t_{.jk} = \sum_r (a_{ir}b_{jr}c_{kr}) - \sum_r (b_{jr}c_{kr}(\frac{1}{n}\sum_i ((a_{ir})) \quad (6\text{–}28)$$

$$t_{ijk} - t_{.jk} = \sum_r (b_{jr}c_{kr}(a_{ir} - a_{.r})) . \quad (6\text{–}29)$$

If, as before, we place a "*" above a centered component, we obtain:

$$\overset{*}{t}_{ijk} = \sum_r (\overset{*}{a}_{ir} b_{jr} c_{kr}) \ . \tag{6–30}$$

Thus, fiber-centering has a very simple effect on the trilinear part of three-way arrays, strictly analogous to the effect of traditional centering on the bilinear part of the two-way arrays: It results in the same number of dimensions and the same loadings, except that the loadings for the mode across which the centering was done are themselves column-centered. In this case, since we centered across Mode A, the Mode B and C loadings were unaffected, and the Mode A loadings for each factor had the mean Mode A loading for that factor subtracted out.

Now, to complete our examination of fiber-centering, we must test its effects on the troublesome h_{ijk} part. We proceed in the same way as before:

$$h_{ijk} - h_{.jk} = (h_{ij} + h_{ik} + h_{jk} + h_i + h_j + h_k + h) \tag{6–31}$$
$$- (h_{.j} + h_{.k} + h_{jk} + h_{.} + h_j + h_k + h)$$

$$h_{ijk} - h_{.jk} = (h_{ij} - h_{.j}) + (h_{ik} - h_{.k}) \tag{6–32}$$
$$+ (h_{jk} - h_{jk}) + (h_i - h_{.})$$
$$+ (h_j - h_j) + (h_k - h_k) + (h - h)$$

$$\overset{*}{h}_{ijk} = \overset{*}{h}_{ij} + \overset{*}{h}_{ik} + 0 + \overset{*}{h}_i + 0 + 0 + 0 \ . \tag{6–33}$$

We find that fiber-centering across Mode A causes all the h-terms that do not have an i-subscript to vanish. In other words, fiber-centering across Mode A removes all the terms that are constant across the levels of Mode A. For the other components, which vary across levels of Mode A, the original terms are replaced by the corresponding i-centered versions.

The effect of fiber-centering on the error term is simply to center the errors across that mode; that is,

$$e_{ijk} - e_{.jk} = \overset{*}{e}_{ijk} \ . \tag{6–34}$$

Because our model of three-way interval-scale data treats all three modes symmetrically, it follows that the effects of centering across Mode B and Mode C are equivalent to the effects of Mode A-centering, shown above, except that the affected modes are correspondingly changed. Thus, centering across Mode B causes the b_{jr} term of the trilinear part to be centered, the h-terms that do not have j-subscripts to vanish, and the error components to be centered across Mode B.

As with two-way arrays, a three-way array may be double-centered or triple-centered by centering across several modes in succession. The order of the operations does not matter, so long as the values input to each successive centering operation are the residuals from the prior stage of centering. Thus, if we first center over Mode A, then the doubly centered data point is computed as follows:

$$\overset{\circ}{x}_{ijk} = \overset{*}{x}_{ijk} - \frac{1}{m}\sum_{j} (\overset{*}{x}_{ijk}) \ . \tag{6–35}$$

It does *not*, in general, equal

$$x_{ijk} - x_{.jk} - x_{i.k} \ .$$

With fiber-centering, it is possible to represent the effects of centering a three-way array by matrix notation. If we consider the three-way array to be composed of a set of p two-way arrays, each n by m, then we can represent double-centering across both Modes A and B by extending (6–18) to the three-way case as follows:

$$\{x_{ijk}\} = \{\mathbf{X}_k\} = \text{the three-way array} \ , \tag{6–36}$$

$$\mathbf{L}_n \ (\mathbf{X}_k)\, \mathbf{L}_m = \mathbf{L}_n \ (\mathbf{A}\ \mathbf{D}_k\, \mathbf{B}' + \mathbf{E})\ \mathbf{L}_m \ , \text{ and} \tag{6–37}$$

$$\mathbf{L}_n \ (\mathbf{X}_k)\, \mathbf{L}_m = (\mathbf{L}_n\mathbf{A})\ \mathbf{D}_k \ (\mathbf{L}_m \mathbf{B})' + \mathbf{L}_n\, \mathbf{E}\, \mathbf{L}_m \ . \tag{6–38}$$

If we represent the doubly centered data as $\overset{\circ}{\mathbf{X}}$, then

$$\overset{\circ}{\mathbf{X}}_k = \overset{*}{\mathbf{A}}\ \mathbf{D}_k\ \overset{*}{\mathbf{B}}{}' + \overset{\circ}{\mathbf{E}} \ . \tag{6–39}$$

That is, double-centering the three-way array by means of two fiber-centerings is equivalent to double-centering each constituent $\mathbf{X}_k$ matrix. The result is that the Mode A and Mode B factor-loading tables are column-centered, and the error component is double-centered.

If we represent our unwanted h-terms as extra factors that are constant across the levels of one or more particular modes, then we can see that double-centering eliminates all the factors that are constant in either Mode A or B, since both of their factor-loading tables are column-centered. The only h-terms that remain are those that vary across both Modes A and B but are constant across levels of Mode C.

The more stages of centering that are performed, the greater the number of h-terms that are removed. Single-centering removes the overall constant h and two of the three one-way effects, along with one of the three two-way effects. Double-centering removes the remaining one-way effect and another one of the two-way effects, leaving only one of the doubly subscripted h-terms remaining. Finally, it can easily be shown that triple-centering the data array removes all of the h-terms and leaves all three of the factor-loading tables for the trilinear part column-centered.

In theory, then, triple application of fiber-centering leaves the data in ideal condition, with all the troublesome constants and one- and two-way components removed. In practice, however, triple-centering can have an additional, less desirable effect. It sometimes causes too severe a reduction in the "signal-to-noise" ratio in the data, as will be discussed below. Since double-centering removes all the unwanted components except one, it can be very effective in cleaning up even difficult data. It is often possible to choose the modes to be centered so that the two-way term remaining after double-centering is so small as to be insignificant. Thus, double-centering is often the optimum method of removing unwanted constants from one's data. Naturally, however, not all data require double-centering. Some may not require any centering at all, although this would be unusual.

Centering the data centers one or more tables of factor loadings, and this can sometimes affect one aspect of factor interpretation: It can become more difficult to distinguish unipolar factors (with high influence on a few variables and no effect on the rest) from certain kinds of bipolar factors (those with high influence on some variables and moderate opposite influence on many other variables). The problem does not arise with more clear-cut bipolar factors, with roughly equal numbers of loadings at each pole and a larger group of loadings closer to the origin. However, when the configuration of loadings does not make the bipolarity of a factor clear and the unipolar-bipolar distinction is important in a particular mode, then one might want to consider centering another mode instead.

Centering of Two-Way Subarrays. We can now compare the effects of fiber-centering versus slab-centering. Many investigators have considered slab-centering to be the most natural kind for three-way factor analysis. Yet, as we shall shortly see, it has undesirable effects. It does not remove all the h-terms and, most seriously, it distorts the trilinear part of the data, increasing its dimensionality.

In slab- or two-way centering, the mean is computed for all the points at a given level of a particular mode (namely, for all the points in that two-way subarray or slice of the data) and then subtracted from all the points at that level. For example, means might be obtained for each stimulus by averaging across persons and scales; the points are then centered as follows:

$$\overset{*}{x}_{ijk} = x_{ijk} - x_{i..} = x_{ijk} - \frac{1}{mp} \sum_j \left(\sum_k (x_{ijk}) \right) . \qquad (6\text{–}40)$$

Now it is easily shown (by arguments parallel to those made above for fiber-centering) that slab-centering the data is equivalent to slab-centering each additive part in our representation of the data. Hence, we once again can simplify our discussion by considering the effects of centering on each of the three components of the data ([6–22]) separately. First, let us consider its effects on the trilinear part, which after slab-centering would have the form:

$$t_{ijk} - t_{i..} = \sum_r (a_{ir} b_{jr} c_{kr}) - \frac{1}{mp} \sum_r \left(\sum_j \left(\sum_k (a_{ir} b_{jr} c_{kr}) \right) \right) . \qquad (6\text{–}41)$$

Rearranging the terms, we obtain:

$$t_{ijk} - t_{i..} = \sum_r (a_{ir}\{b_{jr}c_{kr} - \frac{1}{mp}(\sum_j [\sum_k (b_{jr}c_{kr})])\}) \qquad (6\text{–}42)$$

Unfortunately, no further simplification of (6–42) is possible. Obviously, the effect of slab-centering on the trilinear part of the data is not as simple as the effect of fiber-centering, since there is no direct representation in terms of changes in the Mode B or C loadings.

This example shows that slab-centering across Modes B and C removes the mean $b_{jr}c_{kr}$ product from each $b_{jr}c_{kr}$ part before it is multiplied by the a_{ir} coefficient. A more straightforward statement of the effects of slab-centering on the trilinear part is given by (6–41). Slab-centering across Modes B and C adds an artifactual dimension, whose loadings in Mode A are -1.0 times the mean value in each Mode A slice, and whose loadings in Modes B and C are all equal to 1.0. Thus, the constant that is added to center the data becomes another constant factor, rather than simplify the solution.

Let us now look at the effects of slab-centering on the h_{ijk} part of the data. We begin with:

$$h_{ijk} - h_{i..} = (h_{ij} + h_{jk} + h_{ik} + h_i + h_j + h_k + h) - (h_{i.} + h_{..} + h_{i.} + h_i + h_{.} + h_{.} + h) \qquad (6\text{–}43)$$

$$h_{ijk} - h_{i..} = (h_{ij} - h_{i.}) + (h_{jk} - h_{..}) + (h_{ik} - h_{i.}) + (h_i - h_i) + (h_j - h_{.}) + (h_k - h_{.}) + (h - h) \qquad (6\text{–}44)$$

The effect of centering across two-way subarrays is to eliminate any h-terms that are constant in those subarrays. In our example, we centered across the bc subarray for each level of i. This eliminated the h-terms that were constant across both b and c (that is, constant across all the cells in each bc slice). Thus, a single application of two-way centering will remove any overall constant h and a single one-way constant (for example, h_i). The preprocessing also alters the two-way h-terms. In our example, the h_{ij} and h_{ik} arrays were each single-centered, and the h_{jk} two-way array was globally centered (that is, the overall or grand mean for that array was removed).

The first important fact to notice about the effects of slab-centering on the h-terms is that far fewer terms vanish than with fiber-centering. The second point is that none of the two-way interaction terms, the doubly subscripted h-terms, vanish. Instead, these two-way arrays are partially centered. It is clear that a single application of slab-centering is less effective than a single application of fiber-centering in removing h-terms.

By applying the same algebraic analysis, it is easy to show that successively applying slab-centering across two or three modes will still not remove all the unwanted h-terms. Triple slab-centering—for instance, by first performing bc-centering, then ac-centering, then ab-centering—will remove all the h-terms that are constant over two modes. Thus, it will remove all the one-way effects in the h-component of the data (reintroducing such effects from distortion of the t-component, however; see below). It will double-center all the two-way h-arrays but will not in general cause any of them to vanish.

Even though slab-centering removes a singly subscripted h-component, it reintroduces an unwanted constant of the same form because of its distortion of the t-component of the data. Thus, for example, any advantage to be gained from bc-centering because of the removal of h_i might be cancelled out by the addition of the extra dimension needed to fit the $t_{i..}$ term. This extra dimension has exactly the same form as h_i; that is, it varies across Mode A and is constant across Modes B and C. If this extra dimension is smaller in mean-square contribution than the constant that is removed, the bc-centering might provide a net gain in interpretability of the solution. This could occur if the underlying factor structure of the data happened to be such that the products of the Mode B and C loadings had means of approximately zero. But, on the other hand, if the h_i bias in the data were small and the mean product of Mode B and C loadings were large for some factors, then the slab-centering could prove clearly detrimental.

Thus, we find that slab-centering is undesirable for several reasons: (a) it introduces unwanted constants because it transforms the trilinear or t-component of the data in a manner inconsistent with the factor model; (b) it does not remove two-way interaction terms; and (c) it removes only one of the one-way constants. If the basic requirements for good preprocessing are that it remove much of the unwanted part of the data and preserve the wanted part, then two-way centering is theoretically inadequate on both counts. The practical consequences of these inadequacies will depend on the particular structure of the data but will sometimes be quite substantial.

Three-Way Centering (removing the grand mean). The results for global-centering follow closely those for slab-centering and can be proven by very similar arguments. Thus, they will only be stated briefly without proof. Removal of the grand mean has little effect on the structure of the data. Its effect on the trilinear t-component of the data is again an inappropriate shift of origin, which introduces a spurious factor; in this case, the factor is constant across all modes, with contributions equal to $t_{...}$. All of the individual h-terms are globally centered; consequently, the only h-term to vanish is the overall constant h. The benefit versus the harm of global-centering thus depends on the relative size of $h_{...}$ compared to $t_{...}$, which in turn depends on the unknown values of the h-terms compared to the $a_{ir}b_{jr}c_{kr}$ loadings.

Appropriate Centering Independent of One's Model of the Biases. If one believed the data had the structure

$$x_{ijk} = \sum (a_{ir}b_{jr}c_{kr}) + h + e_{ijk} , \qquad (6\text{–}45)$$

then one might be tempted to apply global-centering; that is, remove the grand mean, in order to get rid of the h-term. One might reason that a centering that treats all points uniformly would be appropriate since the conceptualized disturbance h affects all points uniformly. This would overlook the fact that global-centering also removes the mean from the trilinear part of the data, and this disturbs its ratio-scale properties. The proper constant to add to data with structure as in (6–45) is $-h$. However, since the size of h is unknown, then there is no obvious global adjustment that can be performed. Subtracting out the grand mean subtracts the quantity $(h + t_{...})$, which can distort the data a little or a lot, depending on the size of $t_{...}$.

A similar argument demonstrates why slab-centering is inappropriate, even when one is convinced that the only unwanted biases present are terms that are constant within two-way slices of the data (that is, when one is convinced that the only h-terms present are singly subscripted ones). Slab-centering will remove these h-terms but introduce equivalent singly subscripted offsets due to its effect on the trilinear part. For example, if bc-centering is used, the old h_i values will be eliminated, but a new set of h_i will be introduced which equal $-(t_{i..})$. And these spurious offsets will appear regardless of whether or not the original data had such offsets. Indeed one can start with synthetic data that is perfectly appropriate for the trilinear model and which is fit exactly in r dimensions and, by triple application of slab-centering, distort the data so that it takes $r + 3$ dimensions to fit it exactly. In Monte Carlo experiments that we have performed, the fit in r dimensions after slab-centering has been as low as two-thirds of the fit that would have been obtained with correct centering.

Summary. Surely, a minimum requirement of "appropriate" preprocessing should be that it preserves the appropriateness of any factor structure underlying the data. In particular, if we start out with error-free r-dimensional data appropriate to the model, the preprocessing should not make the data inappropriate to the model or create spurious factors and thus inflate the dimensionality. Fiber-centering preserves appropriateness, but slab- and global-centering do not.

Since the main objective of centering is to turn interval-scale data into ratio-scale data and eliminate conditionality of origin, appropriate centering methods should remove additive constants and one- and two-way effects that would otherwise interfere with the analysis. There is no single centering that removes all these effects, but application of one-way or fiber-centering to all three modes can do so. (However, application to two rather than three modes removes all but one of the possible biases and often provides preferable results.) Two-way or slab-centering does not generally remove two-way interaction components, and it replaces any preexisting one-way bias effects with new one-way biases. One must guess about the factor structure and the preexisting biases in order to decide whether the bias removed is likely to be more or less severe than the one introduced. Fortunately, however, one need not guess about the loading patterns of the latent factors in order to justify fiber-centering, since it does not introduce spurious components.

Multiplicative Adjustments: Reweighting and/or Equating Aspects of the Data

Principles Behind Reweighting

Not for Improving Appropriateness. The objectives of multiplicative adjustment are somewhat different from those of additive adjustment. PARAFAC does not use data rescaling to make data more consistent with the factor model. If there is size-conditionality in the data, the differences in scale usually vary across levels of a mode (that is, they are slab-conditional) and so are easily accommodated within the PARAFAC model. For example, if one subject has a more extreme response style than another and hence tends to use the endpoints of his rating scales, this is captured within the model by assigning the subject larger weights on all factors. Consequently, all entries in the row of the subject-weight matrix corresponding to that subject will be proportionally increased in order to reflect the larger size of his ratings. Similar effects occur in the variable-weight matrix if one variable is measured in different units than another. Thus, despite such slab-conditional variations in scale size, the dimensionality of the solution will not be affected nor will the structure of the factor-loading matrices (except as differences in the overall sizes of entries in different rows).

If however, there are *fiber*-conditional differences in size, the data will not be strictly appropriate for the PARAFAC model. This might happen, for example, if a subject arbitrarily uses extreme ratings when rating some stimuli and very moderate ratings when rating other stimuli, simply because of fluctuations in his response style that were independent of his perceptions of the stimuli. PARAFAC preprocessing does not have a satisfactory general method of making such fiber-conditional data appropriate for the PARAFAC model. However, the success of PARAFAC with real data suggests that such fiber-conditionality of scale size is usually either absent or insignificant and so does not seriously interfere with the data analysis process. The importance and effects of fiber-conditionality of size have yet to be thoroughly investigated.

While slab-conditionality of scale does not make the data inappropriate for the PARAFAC model, it does have other undesirable consequences that often make multiplicative adjustment desirable. It can complicate comparison of factor loadings between rows of a factor matrix and can cause some subjects or variables to have much more influence on the form of the final solution than others. To cope with such difficulties, PARAFAC incorporates size-rescaling options as a standard part of preprocessing. Here we discuss how size-rescaling can be used to change the relative influence of different parts of the data; this application is called *reweighting*. We then discuss how size-rescaling can be used to make factor loadings more directly comparable across slabs, as well as to facilitate the comparison of fit values and so forth across data sets; this second application is called *size-standardization*.

Need for Reweighting. With real data, the analysis model is never expected to fit perfectly; any solution is a compromise.

Hopefully, however, it is a compromise that most clearly resolves those aspects of the data of greatest interest or importance, even though it may lose some information concerning other less important aspects. To make such a beneficial compromise more likely, we will often need to influence how the compromise is to be resolved. Thus, with any data set, we should consider questions of relative sensitivity of the analysis to different aspects of the data. We may wish to modify the relative influence on the solution that can be exerted by different aspects—such as baseline versus deviations—or different segments of the data—such as one group of variables versus another. It may be appropriate to equalize the influence of different parts of the data (for instance, data from different subjects) or, conversely, to weight certain parts of the data more than others. In the following discussion, we consider how preprocessing can be used to accomplish such changes.

Definition of "Influence." What does it mean when we say that one part of the data has "more influence" on the form of the final solution than another? Basically, we mean that the structure underlying the influential part of the data has a stronger effect on the factor loadings. Suppose, for example, we compare the effects on the solution of randomly permuting or scrambling the elements within different subsets of the data. If one subset has more influence than another, then scrambling the elements in that subset will cause a larger change (on the average) in the factor loadings obtained by PARAFAC analysis than would be caused by scrambling the elements within the other subset. Alternatively, we could compare the effect on the loadings of split-half analyses when we exchange different subsets of the data between split-halves.

Reweighting to Modify "Influence." For least-squares fitting procedures, the influence of a given part of the data is determined by the amount that the error sum of squares can be reduced by fitting that part. Thus, to change the relative influence of particular data segments, the relative proportion of the total squared errors contributed by these parts must be changed. In some analysis procedures, this is accomplished by means of weights applied within the error computation algorithm; each error component is multiplied by the weight chosen for that part of the data before it is added to the error sum of squares. This provides a weighted least-squares solution, which allows the investigator to modify the influence of different parts of the data. However, the PARAFAC fitting procedure does not *directly* incorporate any method of assigning different weights to the errors contributed by various parts of the data; it simply minimizes the error sum of squares over the entire data set. Nonetheless, one can *indirectly* weight different sources of error—change the relative error sum of squares contributed by different parts of the data—by changing the actual size of different parts of the data itself.

Surprisingly, one useful type of multiplicative reweighting can be accomplished by additive adjustment of the data. A set of data points can be thought of as having two underlying aspects—baselines and deviations from baselines—and additive adjustment preprocessing alters their relative influence. In particular, centering emphasizes deviations or change across levels of the centered mode by effectively multiplying baselines by zero and deviations by 1.0 in that mode.

In contrast, multiplicative adjustment alters the relative influence of one set of data points as compared to another. If all the data points within a particular subset of the data are increased in size, then the points in this subset become more influential in determining the form of the final solution. The opposite happens if they are decreased in size. There are two reasons why one might want to change the relative importance of different subsets of the data. The first is to deliberately make the influence of certain segments of the data unequal; one might want to reduce the relative importance or impact of data from certain segments of the three-way array (such as certain variables, subjects, stimuli, and so on) because these are considered unreliable or unimportant. The second reason is the converse of the first, that is, to equalize the weight given to different aspects of the data. For example, certain segments of the data (such as certain variables) may happen to contain much larger numbers than other segments because of arbitrary differences in scale of measurement, yet we may not want them to influence the form of the final solution more than any other variables. Reweighting can equate the influence of the segments.

Weighted Least-Squares

Weighted Least-Squares via Preprocessing. By multiplying all the data points for a particular variable by a constant k, we accomplish exactly the same effect as if we weighted all the errors involved in fitting that variable by the constant k (and also multiplied the factor loadings for that variable by k). Thus, we can obtain a weighted least-squares solution by ordinary least-squares fitting, if it is preceded by appropriate preprocessing and followed by compensatory rescaling of the resulting factor loadings.

For example, suppose we wished to include a particular variable in our analysis in order to see how the factors would load on it, but we suspected that the data for that variable might not have the same structure as the other data in the analysis or might otherwise be inappropriate for the PARAFAC model determined by the rest of the data. Thus, we might want the data for that variable to have very little effect on the final solution. We could accomplish this by dividing all the values for that variable by some constant, such as 10^{+3}, so that they would be very much smaller than all the other data in the array. Two things would happen as a result. First, the factor loadings for that variable would become very small—10^{-3} times the size they would otherwise have had. Second, both the fitted parts and the *residuals* for the data points involving that variable would be reduced by a factor of 10^{-3}. As a consequence of this second fact, there would be no appreciable contribution to the error sum of squares from that variable. The least-squares fitting criterion would determine the factor loadings for the rest of the data without being influenced by any need to reduce the errors contributed by that variable, since they would already be so small. When the final solution is obtained, we could recover the appropriately scaled factor loadings for the variable in question by multiplying its factor loadings by 10^{+3}.

We can algebraically develop the general rationale for this preprocessing approach to weighted least-squares. In an unweighted PARAFAC analysis of a three-way array, we seek to minimize the quantity

$$\sum_i \left(\sum_j \left(\sum_k \left([e_{ijk}]^2 \right) \right) \right) . \qquad (6\text{–}46)$$

Since e_{ijk} is equal to the difference between the observed x_{ijk} and the PARAFAC estimate of x_{ijk}, (6–46) is equivalent to:

$$\sum_i \left(\sum_j \left(\sum_k \left([x_{ijk} - \sum_r (a_{ir} b_{jr} c_{kr})]^2 \right) \right) \right) .$$

With the standard PARAFAC fitting procedure, all squared errors of a given size count equally toward the total error sum that is to be minimized. Now suppose, for example, that we wanted to weight the errors for some variables more than others. If variables corresponded to levels of Mode A, we could accomplish this weighting by multiplying each error from level i of Mode A by a weight factor w_i. We now seek to minimize

$$\sum_i \left(\sum_j \left(\sum_k \left([w_i e_{ijk}]^2 \right) \right) \right) , \qquad (6\text{–}47)$$

or

$$\sum_i \left(\sum_j \left(\sum_k \left([w_i \{x_{ijk} - \sum_r (a_{ir} b_{jr} c_{kr})\}]^2 \right) \right) \right) ,$$

which is equal to:

$$\sum_i \left(\sum_j \left(\sum_k \left([w_i x_{ijk} - \sum_r (w_i a_{ir} b_{jr} c_{kr})]^2 \right) \right) \right) . \qquad (6\text{–}48)$$

Now, suppose we consider a set of weighted raw data points, $\overset{*}{x}_{ijk}$, such that $\overset{*}{x}_{ijk} = w_i x_{ijk}$, and likewise consider a row-weighted Mode A factor-loading matrix $\overset{*}{\mathbf{A}}$, such that $\overset{*}{a}_{ir} = w_i a_{ir}$. Then, we could rewrite (6–48) as:

$$\sum_i \left(\sum_j \left(\sum_k \left([\overset{*}{x}_{ijk} - \sum_r (\overset{*}{a}_{ir} b_{jr} c_{kr})]^2 \right) \right) \right) . \qquad (6\text{–}49)$$

To find the $\overset{*}{a}$, b, and c loadings that will minimize (6–49), we merely need to perform an ordinary least-squares PARAFAC analysis of the weighted data matrix $\overset{*}{\mathbf{X}}$. But minimizing (6–49) is in fact equivalent to minimizing (6–47). Thus, we see that an ordinary least-squares analysis of the weighted data yields the required weighted least-squares solution. However, since the $\overset{*}{a}_{ir}$ values are equal to a $w_i a_{ir}$, they must be multiplied by the inverse of the weights applied to the data to give the a_{ir} loadings that are used to predict unweighted $\mathbf{X}$.

We can readily see how weighting the variables will result in much closer fit to the data values that have large weights than to those that have small ones. We need only represent this weighting as premultiplication of $\mathbf{X}_k$ by a diagonal matrix $\mathbf{W}$ whose ith diagonal element is the weight for the ith variable. Now, if

$$\overset{*}{\mathbf{X}}_k = \mathbf{W}\,\mathbf{X}_k = \overset{*}{\mathbf{A}}\,\mathbf{D}_k\,\mathbf{B}' + \overset{*}{\mathbf{E}}_k\,,$$

then

$$\mathbf{X}_k = \mathbf{W}^{-1}\,\overset{*}{\mathbf{X}}_k = \mathbf{W}^{-1}\,\overset{*}{\mathbf{A}}\,\mathbf{D}_k\,\mathbf{B}' + \mathbf{W}^{-1}\,\overset{*}{\mathbf{E}}_k\,. \qquad (6\text{–}50)$$

In this representation of $\mathbf{X}_k$, the rows of the factor loading matrix $\overset{*}{\mathbf{A}}$ are multiplied by the inverse of the weights applied to the data, and the rows of the error matrix are multiplied by the inverse weights, as well. Now $\overset{*}{\mathbf{E}}$ was an ordinary least-squares error matrix for the estimation of $\overset{*}{\mathbf{X}}$, and so most of the rows and columns of this matrix will usually have errors of comparable size. However, the rows of $\mathbf{W}^{-1}\mathbf{E}_k$ in (6–50) will usually be very unequal in size. Where the diagonal weight matrix $\mathbf{W}$ had a large diagonal entry, the corresponding row of the error matrix $\mathbf{W}^{-1}\mathbf{E}_k$ will have small entries, and vice versa. Therefore, those rows of $\mathbf{X}$ that were made large by $\mathbf{W}$ are associated with small errors of fit in (6–50), while those rows that were made small are associated with large errors of fit. On the other hand, if we thought that there were big imbalances in the sizes of the errors in $\mathbf{X}$ before reweighting, then reweighting could be used to make the errors more equal.

Two Bases for Reweighting. One approach to justifying the reweighting of variables in this way is to assume that we are making the variance of the error terms uniform throughout the data array. This then justifies a statistical model with fixed error variance, as mentioned in Kruskal (1981; see also chapter 2) and discussed later in this chapter. Another approach, however, is based on the realization that in addition to *random* error, there is likely to be *specification* error for each component of the data. That is, there is likely to be systematic error because the model is only an approximation to a more complex systematic structure. From such a perspective, we would often like to reweight data to adjust the trade-offs between different sources of systematic error that arise as part of this approximation process. In essence, we may want to equate the sensitivity of the analysis to the specification error in the different parts of the data or to deliberately make the sensitivity unequal for theoretical reasons. Such alteration of the sensitivity of the model to different parts of the data is also sometimes done to obtain "robust" statistical procedures.

Appropriateness of Different Rescaling Methods

When dealing with three-way data, we see that there are several different ways in which rescaling can be performed. We could apply multiplicative adjustments to fibers or one-way subarrays, to slabs or two-way subarrays, or to the entire three-way array. To determine the appropriateness of different rescaling schemes, we proceed in a manner similar to our investigation of centering. We start with a model of the structure underlying the data and then consider the effects of the rescaling on this structure. If the structure is preserved or only modified in a simple way that does not adversely affect interpretation, then the rescaling method is said to be *appropriate*. If, at the same time, the objectives of rescaling are achieved, then the method is said to be *effective*.

For investigation of the effects of rescaling, we can consider a simplified trilinear model of the structure underlying the data. We could assume that one- and two-way components have been removed by prior centering operations. Even if we do not want to make this assumption, the simple trilinear model is appropriate since, as we have shown earlier, the one- and two-way components can be embedded in it by simply considering them to be extra factors.

One-Way or Fiber-Rescaling. In fiber- or one-way rescaling, multiplicative adjustments are made to one-way subarrays—to rows, columns, or "tubes." It does not matter what determines the size of these adjustments, but for simplicity we can assume that they adjust the fibers so that they each have mean-square of 1.0. If these scale factors are designated by the letter s followed by subscripts, we have three different directions of rescaling that are possible: rescaling of each row of each slab by the coefficient s_{ik}, rescaling of each column of each slab by the coefficient s_{jk}, and rescaling of each tube by the coefficient s_{ij}. Let us consider, as an example, rescaling of columns. Our representation of the rescaled data is:

$$\overset{*}{x}_{ijk} = s_{jk}x_{ijk} \;, \qquad (6\text{–}51)$$

or

$$\overset{*}{x}_{ijk} = s_{jk} \sum_r (a_{ir}b_{jr}c_{kr}) + s_{jk}e_{ijk} \;, \qquad (6\text{–}52)$$

or

$$\overset{*}{x}_{ijk} = \sum_r (a_{ir}(s_{jk}b_{jr}c_{kr})) + s_{jk}e_{ijk} \;. \qquad (6\text{–}53)$$

The expression (6–53) cannot be represented by any simple modification of the three-way structure that existed before standardization. The factor contributions to each column (each jk pair) are subject to a potentially different rescaling, but no such column-specific rescaling is feasible within the PARAFAC or Tucker model. Hence, the number of PARAFAC factors required to fit the data after such preprocessing will in general be higher than the number of factors required before such preprocessing. Empirical tests with error-free synthetic data show that fiber-rescaling of just one mode will add more than three extra dimensions to the data structure and decrease the fit value at the original true dimensionality by as much as 20–40%.

We conclude that fiber-rescaling of the data points is not appropriate for the PARAFAC model. Similarly, if we carry through the same reasoning with the Tucker model, we obtain similar results at each step and end up with:

$$\overset{*}{x}_{ijk} = \sum_r (a_{ir} \sum_s (\sum_t (s_{jk} b_{js} c_{kt} g_{rst})) + s_{jk} e_{ijk} . \tag{6–54}$$

Thus, it is apparent that fiber-rescaling is also inappropriate for the Tucker model.

Two-Way or Slab-Rescaling. When performing slab-rescaling, we multiply all the elements in a given two-way subarray of the three-way array by the same constant (for size-standardization, we use the constant that gives a mean-squared data value of 1.0 in the rescaled slab). For example, if we rescale Mode B, we multiply the entries in each level j of Mode B by some multiplier s_j. The algebraic representation of the effects of this transformation is:

$$\overset{*}{x}_{ijk} = s_j \sum_r (a_{ir} b_{jr} c_{kr}) + s_j e_{ijk} , \tag{6–55}$$

which equals

$$\overset{*}{x}_{ijk} = \sum_r (a_{ir} (s_j b_{jr}) c_{kr})) + s_j e_{ijk} . \tag{6–56}$$

Thus, slab-rescaling has a simple representation in terms of the PARAFAC or Tucker model: It just multiplies each row of the Mode B loading matrix by the corresponding constant s_j. Similar results hold for slab-rescaling of Mode A or C. In all cases, the dimensionality of the data remains the same, and the loadings for two of the three modes are unaffected.

Slab-rescaling can also be represented in matrix terms as multiplication of the rescaled mode by a diagonal matrix with diagonal entries equal to the rescaling coefficients. For example, rescaling of Mode B would be accomplished by means of an m by m diagonal scaling matrix $\mathbf{S}_b$, and its effects on the underlying structure would be written as follows:

$$\begin{aligned} \overset{*}{\mathbf{X}}_k = \mathbf{X}_k \mathbf{S}_b &= \mathbf{A} \mathbf{D}_k \mathbf{B}' \mathbf{S}_b + \mathbf{E}_k \mathbf{S}_b \\ &= \mathbf{A} \mathbf{D}_k \overset{*}{\mathbf{B}}{}' + \overset{*}{\mathbf{E}}_k , \end{aligned} \tag{6–57}$$

where $\overset{*}{\mathbf{B}} = \mathbf{S}_b \mathbf{B}$ and $\overset{*}{\mathbf{E}}_k = \mathbf{E}_k \mathbf{S}_b$. Since $\mathbf{S}_b$ is diagonal, each row j of $\overset{*}{\mathbf{B}}$ is equal to the corresponding row of $\mathbf{B}$ multiplied by a constant s_j.

In the affected mode, the transformation can be visualized geometrically as moving each point along its radius from the origin, so that its distance from the origin is s_j times its prior distance. When the s_j are selected so that all the levels of the rescaled mode have equal mean-square, then the points are projected out onto the surface of a hypersphere of unit radius. However, if the model does not fit the data perfectly, then the points will fall short of the surface of the hypersphere.

Now that we have seen the simple effects of rescaling a single mode, it is easy to show that multiple rescalings—such as rescaling both the levels of Mode A and Mode B—can be accomplished without changing the dimensionality of the solution and without changing the patterns within rows of the factor loading matrices. In general, if

$$\overset{*}{x}_{ijk} = s_i s_j s_k \left(\sum_r (a_{ir} b_{jr} c_{kr}) + e_{ijk} \right) , \tag{6–58}$$

then

$$\overset{*}{x}_{ijk} = \sum_r ((s_i a_{ir})\ (s_j b_{jr})\ (s_k c_{kr})) + s_i s_j s_k e_{ijk} . \tag{6–59}$$

We conclude that two-way rescaling is appropriate for PARAFAC; similar arguments show that it is also appropriate for Tucker's three-mode factor analysis.

Slab-Rescaling of Several Modes. While the effect of multiple rescalings on the factor loadings is straightforward, the effect on the slab mean-squares is somewhat more complicated. If rescaling of Mode A is followed by rescaling of Mode B, the effects of the second rescaling will modify the results of the first one. For example, if we size-standardize within the levels of Mode A so that each level has a mean-square of 1.0 and then size standardize within the levels of Mode B, the result will generally be that the levels of Mode A no longer have a mean-square of 1.0. However, if one wants to size-standardize both Mode A and B simultaneously, this can be accomplished by iteratively applying standardization to first one mode, and then the other, and then the first again, and so on until the results converge on a solution in which both modes have a mean-square arbitrarily close to 1.0 for all levels. Likewise, such iterative procedures can be used to size-standardize all three modes simultaneously. Iterative rescaling is an option of the PARAFAC program, and experience with this procedure demonstrates that it usually converges in a small number of iterations (usually 3–6, seldom more than 20).

The purpose of iterative rescaling is to discover the multiplications that will jointly standardize several modes simultaneously. By representing the process of iterative rescaling in matrix terms, we can see that the final result could be accomplished by a single rescaling matrix $\tilde{\mathbf{S}}$ applied to each rescaled mode. If we use $\mathbf{S}_{a1}$ to represent the rescaling matrix applied to Mode A on iteration 1, and similarly $\mathbf{S}_{b1}$ for rescaling Mode B on iteration 1, and so forth, we can represent the process of iterative rescaling Modes A and B as follows:

$$\overset{*}{\mathbf{X}}_k = \ldots \mathbf{S}_{a4}\mathbf{S}_{a3}\mathbf{S}_{a2}\mathbf{S}_{a1}\mathbf{X}_k\mathbf{S}_{b1}\mathbf{S}_{b2}\mathbf{S}_{b3}\mathbf{S}_{b4}\ldots \qquad (6\text{–}60)$$

$$= \tilde{\mathbf{S}}_a \mathbf{X}_k \tilde{\mathbf{S}}_b \quad .$$

Iterative rescaling of several modes to obtain a mean-square of 1.0 in each level of each rescaled mode produces the same effect on the data regardless of the order in which the modes are rescaled. For example, iteratively standardizing the modes in the order *ABABA* . . . produces the same converged result as is obtained by starting the sequence with Mode B (that is, using the order *BABA* . . .). This mode-order independence is a highly desirable property, which obtains with PARAFAC preprocessing but not with other plausible methods. Rescaling based on variance rather than mean-square does not have this property (unless the two criteria are equivalent, due to prior centering that causes slab means in each standardized mode to be zero), nor does rescaling in which each iteration involves centering *and* rescaling a given mode before the next mode is preprocessed.

Note that we have been consistently referring to standardization of the *mean-square* (or, equivalently, *root-mean-square*) of the levels of a mode, rather than standardization of the variances. There are several reasons for this. We feel that this is a more general criterion, one that applies equally well both to cases in which we also center the data and thus remove baselines (hence, in which a mean-square of 1.0 is equivalent to a variance of 1.0) and to cases in which we do not center because we consider the baselines of interest (hence, in which equating mean-squares does not generally equate variances). In the latter case, it would seem illogical to standardize the data on the basis of mean-squared deviations (variance) when the quantities being minimized—and which determine the influence of the various data points—are not deviations but rather the mean-squared data values themselves (or, more precisely, their residuals).

The most conclusive reason, however, for size-standardizing the mean-square (or root-mean-square) rather than the variance is that use of the mean-square makes standardization of multiple modes possible. It is generally not possible to standardize several modes to unit variance simultaneously. For example, when we iteratively rescaled the slabs of an uncentered three-way array such that on each step unit variances were produced within the levels of a rescaled mode, the procedure converged on a solution that only had unit variances in the slabs of the last mode rescaled. The slab variances in the other rescaled mode or modes were uniform but were not equal to 1.0. In one case, for example, the variances were close to 1.2 in one mode and 1.1 in another. After a few cycles, level variances in all modes remained unchanged by successive iterations.

The PARAFAC iterative rescaling procedure is closely related to the iterative proportional fitting procedure used to equate the marginal frequencies in contingency tables. The convergence and other properties of iterative proportional fitting are discussed by Bishop, Fienberg, and Holland (1975), among others.

Three-Way Rescaling. For completeness, we mention the case in which all entries in a three-way array are multiplied by the same constant **s**. If the three-way array constitutes the entire data set under analysis, then three-way rescaling will have no effect on the solution other than to multiply all the coefficients by **s** in the mode selected to represent the scale of the data. For example, if Modes *A* and *B* were standardized and Mode *C* represented the scale of the data, then all the Mode *C* loadings would be multiplied by **s**. Those fit measures that are scale-dependent (such as mean-square error) will be multiplied by $\mathbf{s}^2$ or **s**, depending on the measure.

There are occasions when one might want to multiply only a part of the total three-way array by some constant. As noted earlier, this technique can be used to perform a weighted least-squares analysis. For example, one might wish to fit a model determined primarily by variables 1–15 and have the solution only modestly affected by the characteristics of variables 16–30. A simple approach would be to scale variables 1–15 so that their mean-squares were 1.0 but scale variables 16–30 so that their mean-squares equaled some smaller constant, such as .2. The resulting solution will, in some sense, be five times more sensitive to the characteristics of variables 1–15 than to variables 16–30. In terms of squared errors, the solution will be 25 times more sensitive to variables 1–15. In addition, the two groups of variables will have different sized loadings; all loadings for variables 16–30 will have been effectively multiplied by .2, and so these loadings should be multiplied by 5 to restore them to equivalent scale before they are compared to the loadings of variables 1–15.

Multiplicative Adjustment for Data Standardization

In the preceding section, we discussed multiplicative adjustments used to alter the relative *influence* of different parts of the data. In this section, we consider multiplicative adjustments for a different purpose: To facilitate comparisons of estimated parameters either within or between solutions. One may want to standardize the size of different parts of a data set—such as different variables, subjects, and so forth—to facilitate comparison of the loadings across these parts or to standardize the size of the data as a whole to permit comparison of loadings, mean-square error values, and likewise across several data sets. Finally, size-standardization is used in conjunction with centering so that useful added interpretations of the factor loadings (as correlations) becomes possible.

Standardization of Levels within a Data Set

It often happens that different variables included in a three-way data set might have different and incomparable units of measurement. It would be meaningless to compare, for instance, the size of tremor in millimeters with reaction time in milliseconds. In such cases, the differences in size of measurements for these variables should generally be removed by size-standardization;

otherwise, they will act as a confounding source of size differences in factor loadings. They are typically removed in two-way factor analysis.

Sometimes, differences in overall size between levels arise from the same sources as differences within levels. In the Harshman, Ladefoged, and Goldstein (1977) analysis of variations in tongue shapes, certain locations on the vocal tract showed small variations in tongue position, whereas others showed much larger variations. Yet it was not considered appropriate to use standardization to remove these overall size differences, because they were interpreted as being a part of the same overall linguistic pattern that generated the within-location variations in position. The differences in mean-squares between levels of that mode were interpreted as a natural consequence of the differences in effects of the underlying tongue factors.

There are times, however, when one should consider size-standardization within levels of a mode, even when all the levels of that mode are measured in the same nonarbitrary units. This can happen when there are large, naturally occurring differences between the size of the values corresponding to different levels of a given mode. For example, in one study, each of 24 persons was videotaped while performing 7 tasks involving manipulation of blocks to solve verbal or spatial puzzles (Hampson forthcoming). The videotapes were then scored for the frequencies of 38 different behaviors made during performance of each task. In the resulting 24 × 7 × 38 three-way array, some classes of behaviors had much smaller overall frequencies than others. Consequently, the factor loadings for these behaviors were much smaller than the loadings for other behaviors, even when the low-frequency behaviors were in some sense crucially related to the given factor. This interfered with the interpretation of the factor-loading matrix for the behavior mode. Therefore, in some analyses, Hampson size-standardized the levels of the mode corresponding to behavior type. This made the sizes of loadings for the different behaviors directly comparable. In general, after such size-standardization, a small loading means that there is little relation between the factor and that level of the mode (for example, that class of behavior); a large loading means a strong relationship.

However, there is sometimes a disadvantage to standardizing the size of the different behaviors. This problem arises in the Hampson example. Some cells in the data array correspond to behaviors with very small overall frequencies and are thus based on such small samples of that type of behavior that they are unreliable estimates of the relative population frequencies for those behaviors. When slabs containing many such cells are increased in size to make the size of their factor loadings directly comparable to those of different behaviors, there is a substantial increase in the proportion of error in the overall data set. The most unreliable low-frequency data values are increased in influence, and the most reliable high-frequency data values are decreased in influence. Consequently, the solution changes. If the reweighting is severe, there could be a deterioration of the stability and thus the interpretability of the solution. In general, the investigator employing reweighting should be alert to the danger of overinflation of error.

An alternative method of making the data values directly comparable is to standardize the loadings but not rescale the data. For example, one could divide each row of the factor-loading matrix for a given mode by the standard deviation of the data at that level of that mode. This provides standardized loadings without altering the relative weighting of the data values during the analysis, an approach that is discussed in more detail later. When deemed useful, this approach could be combined with judicious data reweighting.

Standardization of Overall Size across Data Sets

As noted earlier, when PARAFAC is used to perform direct fitting of profile data, the output loadings are usually standardized in such a way that factor loadings for one of the three modes *reflect the scale of the data*—namely, they are in the same units as the data and of the appropriate size to predict the observed data values. Similarly, when cross-product or covariance data is analyzed, the output loadings are usually standardized such that the two identical modes jointly reflect the scale of the data. To facilitate comparison of such loadings between data sets, it is often useful to standardize the overall size of the data values. In two-way factor analysis, this is usually accomplished by conversion of the data on each variable to z-scores. With PARAFAC, similar standardization options are available to set the mean-square for each level of one or more modes to 1.0 before the analysis and thus equate the overall size of the data. When no slab-standardization was desired, one could simply divide all of the data points in the array by the root-mean-square value of the data.

Standardization to give Added Meaning to Factor Loadings

It was noted in chapter 5 (and in more detail in appendix 5–1) that particular combinations of centering and size-standardization would permit factor loadings to take on additional meaning. For example, if the data were size-standardized within levels of Mode A (so that the mean-square for each variable is 1.0) and centered across the levels of Mode B and/or C (so that the mean for each variable is 0), then the data consists of z-scores for each variable. If the output is also standardized so that Mode A reflects the scale of the data, then Mode B and C loadings can be interpreted as estimates of factor scores with Mode A as factor loadings of the traditional kind (factor pattern coefficients in the oblique case, and correlations between factors and variables in the orthogonal case). However, the joint application of both centering and size rescaling requires more complex iterative preprocessing, as will be discussed below.

Combined Additive and Multiplicative Adjustments

Often, an investigator would like to accomplish several of the above listed objectives of preprocessing in the same data set. Centering one or more modes is almost always advisable in order to remove unwanted constant components and two-way interactions, as well as to emphasize interesting variations rather than baselines. At the same time, equalization of the influence of different variables and subjects is also usually desirable, along with the standardization of the data that facilitates interpretation of the resulting factor loadings. However, the joint use of centering and size-standardization raises new issues and complications, because the two types of preprocessing interact.

Interaction among Preprocessing Operations

The joint application of both size-standardization and centering requires more complex iterative preprocessing than either alone, since some multiplicative adjustments disturb previous additive adjustments, and vice versa. For example, if the data are first fiber-centered on Mode *B* and then slab size-standardized on Mode *B*, the standardization disturbs the effects of the centering; consequently, the data is standardized but is no longer centered. If we recenter Mode *B*, then the standardization of Mode *B* is disturbed, although the mode is closer to equal mean-squares within levels than it was initially. To jointly accomplish both the centering and the size-standardization of a given mode requires repeated iteration of the preprocessing operations.

In general, the interactions of the multiplicative and additive preprocessing steps can be described as follows: Size-standardization of any mode disturbs prior centering on that mode but not on the other two modes; centering on a given mode not only disturbs prior standardization of that mode but of the other two modes as well. Thus, there are only a restricted number of combinations of additive and multiplicative preprocessing that can be performed without iteration between the additive and multiplicative stages. Pure centering requires no iteration whatsoever. Pure standardization of more than one mode requires iteration within the standardization step (as described in [6–60]). Iteration between centering and standardization steps is only required if both centering and size-standardization are being performed on the same mode. (Here we always refer to fiber-centering and slab-standardization, since we established earlier that these are the appropriate forms of these operations.)

Matrix Representation

The results of iterative application of joint additive and multiplicative preprocessing can be represented in compact matrix terms. Suppose that both centering and size-standardization are requested for Modes *A* and *B* of a three-way array. The same centering and standardization properties would thus be required across all levels of Mode *C*. If $\mathbf{X}_k$ represents the slice of that array at the *k*th level of Mode *C*, we can represent the resulting operations as follows:

$$\mathbf{X}_k = \ldots (\tilde{\mathbf{S}}_{a2}\,(\mathbf{L}_n(\tilde{\mathbf{S}}_{a1}(\mathbf{L}_n\,\mathbf{X}_k\,\mathbf{L}_m)\tilde{\mathbf{S}}_{b1})\mathbf{L}_m)\tilde{\mathbf{S}}_{b2}) \ldots \qquad (6\text{–}61)$$

$$= \mathbf{M}_a\,\mathbf{X}_k\,\mathbf{M}'_b \ .$$

That is, all the operations performed on the left-hand side of $\mathbf{X}_k$ can be represented by a single matrix $\mathbf{M}_a$; similarly, the operations on the right-hand side can be represented by $\mathbf{M}'_b$. Note that each size-standardization step in (6–61) is the result of an iterative standardization, since, in this example, two different modes were being standardized, and the standardization of the first would disturb the standardization of the second. Thus, in (6–61), the $\tilde{\mathbf{S}}$ matrices each represent the compressed product of a series of individual iterations, so that, for example, $\tilde{\mathbf{S}}_{a1}$ in (6–61) is equivalent to the whole string of diagonal rescalings summarized in (6–60) as $\tilde{\mathbf{S}}_a$. Note also that the rescaling and centering matrices are not specific to the kth slice but rather are applied uniformly across all slices of the three-way array.

The properties of the summary standardization matrices $\mathbf{M}_a$ and $\mathbf{M}_b$ are still under study. It can be shown, for example, that they are necessarily column-centered. (Note, however, that the right-hand transformation in [6–61] is written $\mathbf{M}'_b$ and hence is row-centered as it occurs there.) When the initial step in the iterative sequence is centering, they are necessarily double-centered. More importantly, it is known that when $\mathbf{M}_a$ and $\mathbf{M}_b$ are determined by iterative preprocessing, their form is not unique for a given $\mathbf{X}$ and a particular set of centering and/or standardizing conditions but depends also on the sequence of operations used, and, in particular, on whether the initial operation is centering or size-standardization. As a consequence, the preprocessed data that results are also not unique. Fortunately, however, with real data, the different solutions obtained by starting the iterative preprocessing sequence with centering versus standardization seem to lead to similar results, at least in the cases we have tried to date.

To simplify the relationship between the raw and the preprocessed data, we now sometimes employ only part of the iterative sequence, equivalent to the initial centering followed by the standardization operations shown with subscript 1 in (6–61). This *one cycle* preprocessing gives results similar to those obtained when iterative preprocessing is carried to convergence but simplifies the extended PARAFAC model associated with the analysis. (The relationship between preprocessing issues and the extended PARAFAC model will be discussed briefly later in this chapter.)

General Linear Preprocessing

More General Preprocessing Transformations

An Example Involving First-Differences. Suppose that we are studying economic time-series data and that each successive level of Mode *C* represents the same variables, measured for the same industries, but in successive years. For some purposes, we might want to focus on the rate of changes in such data and ignore the overall trends. This could be done by taking first-difference scores, which would be the discrete analog of the first derivative of the variables across time. To prepare a data set for factor analysis, we could construct a preprocessed data array in which each Mode *C* slab represents the difference between two successive levels of the raw data array. But would such a preprocessed array be appropriate as input for a PARAFAC analysis? If the initial structure of the raw data were appropriate for the model, would the transformed data also be appropriate?

This example prompts us to inquire into a wider range of possible data preprocessing operations and to ask how we can decide whether these are "appropriate" in the sense used earlier. For example, does the taking of first-differences of successive levels of a mode constitute appropriate preprocessing? If we start with a model of our data and carry through the algebraic representation of this preprocessing, we find that it does. With error-free data, the solution remains unchanged except that the Mode *C* loadings for the preprocessed data are equal to the first-differences of the adjacent levels of the original Mode *C* factor-loading matrix. Thus,

$$\overset{*}{x}_{ijk} = x_{ijk} - x_{ij(k-1)}$$

$$= \sum_r (a_{ir} b_{jr} (c_{kr} - c_{(k-1)r})) + (e_{ijk} - e_{ij(k-1)}) \ .$$

This result shows that the taking of first-differences has the same desirable properties that we noted above for fiber-centering and slab-preprocessing. That is, it does not inflate the dimensionality of the array, and it has straightforward consequences on the factor-loading matrices. As before, the effect of the preprocessing on the factor loadings mirrors its effect on the data.

Matrix Formulation. These parallels are not coincidental but rather can be seen as part of a more general pattern. To understand this pattern, it is useful to formulate the preprocessing in matrix terms. For our example of taking differences of successive years, we can represent the preprocessing transformation in matrix terms by slicing our three-way array into lateral slabs, which are variables by occasions, rather than frontal slabs, which are variables by industries. We then represent the arbitrary *n* by *p* slab for the *j*th level of Mode *B* (for the *j*th industry) as follows:

$$\overset{*}{\mathbf{X}}_j = \mathbf{X}_j \mathbf{T}_{\text{diff}} \quad ,$$

where $\mathbf{T}_{\text{diff}}$ is a *p* by (*p* - 1) matrix that has the form:

$$\begin{matrix}
-1 & 0 & 0 & 0 & 0 & \cdots & 0 \\
1 & -1 & 0 & 0 & 0 & \cdots & 0 \\
0 & 1 & -1 & 0 & 0 & \cdots & 0 \\
0 & 0 & 1 & -1 & 0 & \cdots & 0 \\
0 & 0 & 0 & 1 & -1 & \cdots & 0 \\
\cdot & \cdot & \cdot & \cdot & \cdot & & \cdot \\
\cdot & \cdot & \cdot & \cdot & \cdot & & \cdot \\
\cdot & \cdot & \cdot & \cdot & \cdot & & \cdot \\
0 & 0 & 0 & 0 & 0 & \cdots & -1 \\
0 & 0 & 0 & 0 & 0 & \cdots & 1
\end{matrix}$$

The resulting matrix $\overset{*}{\mathbf{X}}_j$ is an n by $(p - 1)$ matrix whose rows represent variables and whose columns represent first-differences of the value of those variables over successive pairs of years.

Linear Transformations Provide "Appropriate" Preprocessing. More generally, a very broad family of preprocessing operations can be written as linear transformations or recombinations of the levels of a mode. For example, we commented earlier that the result of combined fiber-centering and slab-standardization applied to several modes requires iteration to determine the correct transformations but once determined can be written in terms of single linear transformation matrices for each mode (namely, $\mathbf{M}_a$, $\mathbf{M}_b$, and $\mathbf{M}_c$), as defined in (6–61). It can easily be shown, in fact, that any kind of preprocessing that can be expressed as a linear recombination of the levels of a given mode (and thus in matrix terms as multiplication of slices of $\mathbf{X}$ by some common matrix $\mathbf{T}$) will be appropriate preprocessing in the sense meant in this article.

Consider any preprocessing representable by multiplying all the slices of the matrix $\mathbf{X}$ by the transformation $\mathbf{T}$. We obtain the following simple expression for the effect of this preprocessing on the latent structure of the preprocessed data $\overset{*}{\mathbf{X}}$:

$$\overset{*}{\mathbf{X}}_k = \mathbf{X}_k\,\mathbf{T} = \mathbf{A}\,\mathbf{D}_k\,(\mathbf{B}'\,\mathbf{T}) + (\mathbf{E}_k\,\mathbf{T}) \qquad (6\text{–}62)$$

$$= \mathbf{A}\,\mathbf{D}_k\,\overset{*}{\mathbf{B}}{}' + \overset{*}{\mathbf{E}}_k\,,$$

or, for Tucker's model T3 we would write:

$$\overset{*}{\mathbf{X}}_k = \mathbf{X}_k\,\mathbf{T} = \mathbf{A}\,\mathbf{H}_k\,(\mathbf{B}'\,\mathbf{T}) + (\mathbf{E}_k\,\mathbf{T}) \qquad (6\text{–}63)$$

$$= \mathbf{A}\,\mathbf{H}_k\,\overset{*}{\mathbf{B}}{}' + \overset{*}{\mathbf{E}}_k\,,$$

where the constituent matrices are as defined in chapter 5. The effects on the latent structure can be interpreted as simply an application of the linear transformation $\mathbf{T}$ to the factor-loading matrix for the transformed mode and to the corresponding mode of the three-way error matrix. Hence, the preprocessing is appropriate since (a) the dimensionality of the data is not inflated, and (b) the post-preprocessing structure bears a straightforward relationship to the original structure. As in all the other cases considered earlier, the transformation of the affected factor loadings is of the same kind as the transformation applied to the preprocessed mode and should thus be easily intelligible to the investigator. The preprocessing is effective because when it is applied to data that is inappropriate (in ways anticipated by a plausible model of our raw data), many of the inappropriate characteristics of the raw data disappear and new spurious components are not introduced. In the case of first differences, we might have in mind a model of the raw data in which there are large but irrelevant components that change in unpredictable ways but only slowly so that they are very similar from one occasion to the next. First-difference preprocessing would greatly reduce these irrelevant components and bring out the changes of interest. It would also remove two of the three singly subscripted and one of the three doubly subscripted h-terms in our model of conditional-origin interval-scale data.

In this particular example, we apply the preprocessing to Mode B. However, since all three modes have equivalent status, it can be shown that the same properties hold when the preprocessing is applied to any of the three modes. Furthermore, since such a transformation of one mode does not interfere with the factor structure of any other mode, the effects of double and triple preprocessing can be obtained by simply applying the several preprocessing transformation matrices to their respective factor-loading tables.

The underlying unity of all these appropriate transformations can be seen most clearly by treating them all as instances of a generalized kind of matrix multiplication.

Appropriate Preprocessing as Array-Matrix Multiplication

Multiplying a Three-Way Array by a Two-Way Matrix. Mathematicians have generalized the operation of matrix multiplication to include the multiplication of a three-way array by a two-way matrix (for instance, see Kruskal 1977). The logic of this generalization is straightforward. Consider the matrix product

$$\mathbf{Y} = \mathbf{X}\,\mathbf{T}\ .$$

Suppose we initially take $\mathbf{X}$ to be an n by m two-way array and postmultiply it by an m by u matrix $\mathbf{T}$ to obtain an n by u product matrix $\mathbf{Y}$. The columns of $\mathbf{Y}$ are simply linear recombinations of the columns of the original $\mathbf{X}$. Here, the weights contained in the jth column of $\mathbf{T}$ are applied to the *columns* of $\mathbf{X}$, which are then summed to produce the jth column of $\mathbf{Y}$.

Now suppose that $\mathbf{X}$ is not an n by m two-way matrix but an n by m by p three-way array. Instead of thinking of $\mathbf{X}$ as composed of m vertical *columns,* we now consider it to be made up of m vertical *slabs*. Each slab is n by p. The product $\mathbf{Y}$ is now an n by u by p three-way array, made up of u vertical slabs. In this case, the weights contained in the jth column of $\mathbf{T}$ are applied to the *vertical slabs* of $\mathbf{X}$ to produce the jth vertical slab of $\mathbf{Y}$. An alternative way of looking at the multiplication of a three-way array by a matrix is to consider it as a set of parallel two-way matrix multiplications. If the n by m by p array $\mathbf{X}$ is postmultiplied by an m by u matrix $\mathbf{T}$, the result obtained is the same as if each n by m slice of the three-way array is postmultiplied by the m by u matrix $\mathbf{T}$.

In a similar fashion, for premultiplication of the three-way array $\mathbf{X}$ by a u by n matrix $\tilde{\mathbf{T}}$, the n elements in the ith row of $\mathbf{T}$ are the weights applied to the horizontal slabs of $\mathbf{X}$ so that their sum will equal the ith horizontal slab of $\mathbf{Y}$. Finally, if a p by u matrix $\mathbf{T}$ is applied to the third mode, the p elements in the kth column of $\mathbf{T}$ are the weights applied to the frontal slabs of $\mathbf{X}$ to obtain the kth frontal slab of $\mathbf{Y}$. (For a picture of the three kinds of slabs, see Figure 6–1.)

Appropriate Preprocessing is Array-Matrix Multiplication. From the definition of array-matrix multiplication just given, it should be apparent that all examples of appropriate preprocessing that we have discussed can be considered instances of multiplication of a three-way array by one or more matrices. In fact, all examples of appropriate preprocessing of which we are presently aware fall into this class. Such array-matrix multiplication generates the class of operations that can be represented as linear transformations of a mode, as in (6–62), and thus have the desirable properties that make them appropriate as preprocessing for three-way factor analysis.

This unified characterization of appropriate preprocessing methods resolves the apparent anomaly that emerged earlier in which the appropriate additive adjustment seemed to require a different kind of operation (fiber transformation) than was required by the appropriate multiplicative adjustment (slab transformation). We can now interpret both kinds of preprocessing, along with their various iterative combinations, as special cases of a single kind of operation: multiplication of a three-way array by a matrix. At the same time, we find that the inappropriate preprocessing methods (slab-centering and fiber-standardization) cannot be interpreted in this way.

Of course, not all matrices constitute effective preprocessing transformations, even when correctly applied to the data via array-matrix multiplication. In order to determine which family of preprocessing transformations is most appropriate for a given kind of data set, it is still essential to consider a mathematical model of the structure of the data and examine the effect of proposed preprocessing transformations on that model, as was done above. In particular, the terms representing the inappropriate part of the data should vanish, and those representing the appropriate part should retain their simple form. And in any given analysis, one must further select the optimal member or members of the family of appropriate preprocessing transformations, based on the particular characteristics of the data set at hand.

Some Practical Guidelines for Application of Preprocessing

To conclude this section, we should like to give the reader some brief summary guidelines for selection of preprocessing procedures, based partly on general principles and partly on experience:

1. It is almost always desirable to center across at least one mode. This removes unwanted constants and provides an approximation to ratio-scale data. Centering two modes is often optimal but rarely is centering three.

2. If there are meaningless differences in origin across the levels of a given mode—for example, due to different arbitrary or incomparable measurement scales—it is usually advisable to center to remove these differences.

3. In selecting which modes to center, consider where there are likely to be effects that do not vary across levels of a mode. In other words, identify likely singly subscripted and doubly subscripted h-terms, and center those modes in which the

unwanted effects are constant. For example, each subject's overall biases, which are constant across the ratings of all stimuli, can be removed by centering across stimuli.

4. For reasons not yet well understood (but probably related to the complex individual differences in relationships that lead to degenerate solutions, as described later), it is usually undesirable to center across the person mode. This result is surprising, since centering across persons is a traditional method of preprocessing in two-way factor analysis. However, in three-way intrinsic axis methods, such centering sometimes seems to emphasize those individual differences that are inconsistent with the model and that cause problems for the determination of axis orientations. Centering across persons may not be problematic if used as preprocessing for Tucker's T2 or T3 model, in which axis orientation is determined by reference to outside criteria.

5. If there are meaningless arbitrary differences in scale size across the levels of a given mode—perhaps due to differences in the unit of measurement—it is generally advisable to remove these differences by standardizing within the levels of that mode.

6. One may sometimes want to remove overall differences in mean-square across levels of a mode, even when these differences are meaningful and result from genuine differences in the process being measured (such as different frequencies of movements, in the example given in the section on Multiplicative Adjustment for Data Standardization), to equalize the influence of different levels on the solution.

7. Reweighting other than size-standardizing (such as for weighted least-squares analysis) may sometimes seem advisable, as described in the section on Reweighting and/or Equating Aspects of the Data.

8. With three-way rating-scale data—for example, stimuli by scales by persons or judges—good results have been repeatedly obtained when the scales and the stimuli are centered and the scales and the persons are size-standardized.

9. More generally, it has often been found useful to standardize two modes and center two modes, but not the same two modes, so that one mode is simply centered, one mode is simply standardized, and one mode is both centered and standardized.

10. It is often necessary to try several different preprocessing methods and compare their effects on the form of the factor-analytic solutions that result. Select the preprocessing that produces the most interpretable solution.

11. Technically inappropriate centerings can sometimes work out fairly well in practice if the distortions they introduce are

smaller than the distortions they remove. Nonetheless, there seems no reason to use an inappropriate method when appropriate ones exist that work at least as well and do not risk introducing serious distortions into the solution. Many examples of both appropriate and inappropriate preprocessing are given by Kroonenberg (1983).[4]

THE EXTENDED PARAFAC MODELS

PARAFAC Preprocessing as a Basis for More General Models

Extension to Conditional-Origin Interval-Scale Data

Two Perspectives. In our earlier treatment of the topic, we interpreted centering as a method of making a wide class of data appropriate for a more restricted model. However, the opposite perspective is also possible. We can also consider centering to be a way of making the PARAFAC model less restricted and hence appropriate for more general types of data.[5] In any PARAFAC analysis in which preprocessing is employed, the combined preprocessing-factor extraction procedure can be thought of as fitting an extended PARAFAC model. There is a family of such models, and the particular model that is fit in any particular analysis is determined by the specific preprocessing options selected (for instance, which modes are centered for that analysis). Similar extensions of Tucker's model can also be formulated.

Extension via Centering. Since PARAFAC centering is based on an explicit model of conditional-origin interval-scale data, this model ([6–3] or [6–4]) can be considered part of an extended PARAFAC model. From this perspective, the means that are removed by centering are considered additional terms in the PARAFAC representation.

Suppose, for example, that we center across all three modes before performing a PARAFAC analysis. This centering process can be thought of as estimation and then removal of the first three components of the following model:

$$x_{ijk} = h_{ij} + h_{jk} + h_{ik} + \sum_r (a_{ir} b_{jr} c_{kr}) + e_{ijk} \,. \qquad (6\text{–}64)$$

In addition to the ratio-scale components that were part of the basic PARAFAC model (5–3), this extended model directly incorporates the offsets that make the data interval scale and fiber-conditional in origin.

While the form of (6–64) is fairly compact, we should keep in mind that it implicitly incorporates the additional constant terms given in (6–3). The doubly subscripted h-terms of (6–64) can be considered to have absorbed and can thus also represent the effects of the unsubscripted additive constant h, the singly subscripted main effects h_i, h_j, and h_k, as well as the two-way interactions. If desired, the *grand mean* and *main effects* (to use analysis-of-variance terminology) can be extracted from the two-way h_{ij}, h_{jk}, and h_{ik} arrays by conventional methods. The resulting model would be closely related to those proposed by Gollob (1968) for the two-way case and by Gower (1977) for the three-way case.

As noted earlier, one could also subject these h-arrays to two-way factor analysis to bring out further structure. For example, we saw how h_{jk} two-way interactions could be due to one or more factors that are constant across Mode A. Two-way factor analysis of this array could uncover these dimensions. The factors would, of course, be rotationally indeterminate and would therefore need to be rotated to simple structure or to congruence with factors extracted by PARAFAC from the three-way interactions (see below). These additional factors could then be incorporated into the analysis model by giving them constant loadings in the third mode, much as in (6–5) and (6–6).

To provide the most complete example, the extended model that results from triple-centering was represented in (6–64). As noted earlier, however, triple-centering often reduces the size of the systematic part of the data too much without appreciably reducing the size of the random error; thus, it impairs the signal-to-noise ratio of the data. On occasion, we have recovered meaningful factors from triple-centered data, but we usually obtain better results with double-centering. The mode left uncentered, if carefully chosen, can be one that has minimal constant components not already removed by the other centerings.

The extended model that would result from double-centering would have two of the three h-terms in (6–64). For example, if Mode C were not centered, the h_{ij} term would not be part of the extended model being fit in that particular analysis.

One-Stage and Two-Stage Least-Squares. From the extended model perspective, PARAFAC can be considered to be a two-stage least-squares fitting procedure. Since the mean is a least-squares estimate, computation of means as part of centering is the first least-squares stage of the procedure. Least-squares fitting of the factor model to the residuals is the second stage.

Unpublished theorems by Kruskal (1977) show that when the centering is across *fibers,* this two-stage least-squares procedure is equivalent to a one-stage least-squares fit of an extended model (such as [6–64]). That is, the procedure of first estimating and extracting fiber means and then performing PARAFAC analysis of the residuals produces exactly the same model parameters, fitted values, and residuals as would be produced by using a more general least-squares procedure (such as a hill-climbing procedure) to directly solve for the least-squares fit of all the means and factor loadings at the same time.

Kruskal's theorems also show that this equivalence of two-stage and one-stage least-squares procedures does *not* hold if slab-centering or global- (grand mean-) centering is used (Kruskal 1977). These other kinds of centering do not as neatly allow one to treat the combined centering-factoring procedure as the fitting of an extended model. Thus, although Kruskal's results were not known at the time PARAFAC centering was being developed, they provide an additional reason for preferring fiber-centering over other kinds of centering for three-way factor analysis.

Further Extensions: Additive Factor Change

Additive Change in One Mode. The system variation model ([5–3]) permits only multiplicative changes in factors across levels of a mode. That is, the factor weights (for instance, factor or component scores) can change from one occasion to the next only by being multiplied by some coefficient, as shown in (5–6). However, we might want to consider a more general expression for the way that component scores are related across occasions, one that incorporates additive shifts in the factor baseline, as well as multiplicative shifts in the factor size or variance.

We might expect additive shifts in the factor score baseline values in, for example, a study of animal stress and adaptation in novel environments.[6] Suppose we measure animals' behavior—including number of vocalizations, exploratory movements, attempts to escape, and so on—during several successive exposures to the novel environment. We could then apply PARAFAC to these data in an attempt to identify a few factors—such as general activity, stress, and so forth—that would explain the behavioral differences among animals and across occasions.

Besides proportional changes in these factor scores across occasions, we might also anticipate additive shifts in their baseline values due to the animals' adaptation to the new environment. Hence, we would want a more general model than (5–3) to incorporate these baseline shifts. With the desired extended model, the expression for f_{jkr} (the factor score of the jth animal on the kth occasion for factor r) would not simply be $b_{jr}c_{kr}$ but instead would have the general linear form

$$f_{jkr} = (b_{jr}c_{kr} + d_{kr(b)}) \ , \qquad (6\text{–}65)$$

where the b- and c-terms are defined as usual, and the $d_{kr(b)}$ term represents the additive shift in factor r on occasion k that is constant across levels of Mode B (namely, animals). This model is more general than that for system variation ([5–6]) but less general than the one for object variation ([5–7]); here, the additive shifts are the same for all animals.

Expression (6–65) can be viewed as a linear regression equation that predicts f_{jkr}, the factor score of the jth animal on occasion k from the animal's basic level of factor r. The prediction equation multiplies the base level of the factor, b_{jr}, by a slope term for occasion k, c_{kr}, and adds an intercept term for occasion k, d_{kr}. Thus, the full PARAFAC representation becomes:

$$x_{ijk} = \sum_{r=1}^{q} (a_{ir}(b_{jr}c_{kr} + d_{kr(b)})) + e_{ijk} \ , \qquad (6\text{–}66)$$

which can be rewritten:

$$x_{ijk} = \sum_{r=1}^{q} (a_{ir}b_{jr}c_{kr}) + \sum_{r=1}^{q} (a_{ir}d_{kr(b)}) + e_{ijk} \ . \qquad (6\text{–}67)$$

Thus, we obtain a PARAFAC model with up to q extra factors. There is one extra factor for each dimension that shows nonzero additive shifts across occasions.

The extra factors in (6–67) have two distinguishing characteristics: (a) they are constant in Mode B; and (b) factor r in the second set has the same Mode A loadings as factor r in the first set. These two characteristics make it possible to fit models like (6–67) as another extension of the original PARAFAC analysis procedure.

The extra factors generated by the baseline shifts in the r original factors appear as two-way interactions constant over Mode B and would thus contribute to the h_{ik} terms of the extended PARAFAC model. By centering on Mode B, these extra components are extracted during the first stage of fitting and so are eliminated from the second stage. Thus, the presence of factor change of this more general kind does not interfere with extraction of the correct form of the Mode A factor loadings, nor does it interfere with the estimation of each Mode B base level for each factor (the b_{jr} terms).

To identify the extra factors contributed by the additive shifts, we could rotate factors extracted by two-way analysis of the h_{ik} terms into a position where the Mode A loadings resemble as closely as possible the Mode A loadings obtained from the PARAFAC analysis of the centered data. Alternatively, one could simply use regression to estimate weights for these factors. Starting with the assumption that there are dimensions in the h_{ik} table that have the same Mode A factor loadings as the dimensions subsequently extracted from the centered data, one could simply estimate their weights on occasion k by using regression to predict the raw data from the Mode A loadings.

Additive Change in Several Modes. In a completely general PARAFAC model that treats all three modes even-handedly, one would want to allow for additive shifts of factor baselines in any of the three modes. For a given factor r, there are six possible two-way additive contribution factors: two with the same Mode A loadings (one of these two is constant in Mode B and the other is constant in Mode C), two with the same Mode B loadings (constant in Mode A or C), and two with the same Mode C loadings (constant in Mode A or B). The full model would have the form:

$$x_{ijk} = \sum_r (a_{ir}b_{jr}c_{kr} + a_{ir}d_{kr(b)} + a_{ir}d_{jr(c)} + b_{jr}d_{kr(a)} + b_{jr}d_{ir(c)} + c_{kr}d_{jr(a)} + c_{kr}d_{ir(b)}) \quad (6\text{–}68)$$

The simultaneous interpretation of all the shifts in such a model is not straightforward; indeed, it is not clear that all these shift terms would ever be needed simultaneously. Nonetheless, (6–68) displays all the varieties that might be needed in one or another situation. (We will not pursue this model further in this chapter.)

Perhaps the most important point to be made about additive factor shifts is that they need not interfere with the identification of the basic q three-way factors when proper centering is employed. Furthermore, the shift scores can in fact be estimated after the central PARAFAC analysis by factor score estimation methods.

Extensions via Reweighting and Size-Standardization

The PARAFAC model can also be extended by incorporation of the effects of multiplicative preprocessing. Suppose, for example, that we had rescaled data values within levels of all three modes before performing a PARAFAC analysis, perhaps to size-standardize within the levels of all three modes. As a result, we analyze a data array whose entries were equal to $s_i s_j s_k x_{ijk}$, where s_i is the constant multiplier for the ith level of Mode A (the ith diagonal element of the rescaling matrix $\tilde{\mathbf{S}}_a$, as defined earlier in this chapter), s_j is the multiplier for the jth level of Mode B, and s_k is the multiplier for the kth level of Mode C. We might then consider our solution to represent the fitting of the extended model

$$x_{ijk} = \frac{1}{s_i} \frac{1}{s_j} \frac{1}{s_k} \left(\sum_r (a_{ir} b_{jr} c_{kr}) + e_{ijk} \right) , \qquad (6\text{–}69)$$

or equivalently,

$$x_{ijk} = \sum_r (\overset{*}{a}_{ir} \overset{*}{b}_{jr} \overset{*}{c}_{kr}) + \frac{1}{s_i} \frac{1}{s_j} \frac{1}{s_k} (e_{ijk}) , \qquad (6\text{–}70)$$

where $\overset{*}{a}_{ir} = a_{ir}/s_i$, $\overset{*}{b}_{jr} = b_{jr}/s_j$, and $\overset{*}{c}_{kr} = c_{kr}/s_k$. Our interpretation of (6–69) and (6–70) depends on our motivation for applying the rescalings. We will mention two basic motivations: (a) an attempt to achieve uniform error standard deviations, and (b) a desire to differentially weight particular subsets of the errors.

Making Error Standard Deviations Uniform. We could adopt the perspective advocated by Kruskal (chapter 2) and assert that the rescaling was applied because we believed that the original $\mathbf{X}$ array had entries with greater error in some levels of each mode than others and that our goal in rescaling was to make $\mathbf{X}$ appropriate for a model in which all the e_{ijk} in (6–69) are drawn from distributions with the same standard deviation. The scaled errors in (6–70) then represent our estimates of how the size of the error in $\mathbf{X}$ actually varies, and so the terms s_i and so forth are an extension of the model to explicitly represent error standard deviations.

Of course, as Kruskal points out (see chapter 2), we cannot actually observe the error, and so the rescalings are determined on the basis of our beliefs about the errors. Usually, its size is assumed to be approximately proportional to the size of the data itself. There are several justifications for such an assumption. For example, there are many psychological situations in which this proportionality is known to hold, such as when subjects are asked to make judgments of the properties of stimuli (for instance, the weight of objects). More generally, it is often thought that the variables in a given data set are roughly equal in reliability so that the random component will be approximately the same proportion of the total variance across all variables. Of course, this will not be exactly true, but in the absence of other knowledge, it can often be considered a best approximation; it would be unusual to have very precise and reliable observations in the same data set with very imprecise and unreliable ones. Furthermore, if we knew that we had such a combination of reliable and unreliable observations, then we would presumably not rescale the data uniformly, or at least we would not adopt this uniform-error justification for our rescaling.

If a substantial part of the error is thought to be *specification* error—systematic error introduced because the model is not strictly appropriate for the data (for example, when the model is too simple to represent all the complexities of the data)—then it is often reasonable to consider this specification error roughly proportional to the size of the data. The assumption is that the model has roughly the same degree of inappropriateness for all the different observations (for all variables, subjects, and so on). Again, in the absence of other knowledge, this is often a good working assumption. It implies that those levels of a mode in which the mean-square is larger probably have error components that are also larger by roughly the same proportion.

However, if we expected that certain parts of the data would have greater specification error than others, we might choose rescaling coefficients that were not based on size-standardization of the data itself but instead were based on our external estimates of the error components in different parts of the data. In this case, the preprocessing would probably be a noniterative multiplication of the levels of particular modes by particular values.

Weighting Errors Differentially. Alternatively, we could view (6–70) as a weighted least-squares model. In this case, the s_i, s_j, and s_k values might be determined by other means than data-standardization. When the weighting of different parts of the data is based on our estimate of the reliability of those parts or on our beliefs about the relative inappropriateness of the PARAFAC model for different parts of the data, this weighted least-squares approach does not differ much from the approach that aims to equate error terms across the data. However, when our weights are based on assessment of theoretical importance of different parts of the data or on the relative "cost" associated with not fitting different parts of the data (in some practical application), then the weighted least-squares approach introduces a novel rationale for (6–70).

Rescaling Loadings but Not Data. Recall from our earlier treatment of multiplicative adjustment that the second reason for rescaling aspects of the data was to facilitate comparison of loadings through size-standardization within levels of a particular mode. This method of achieving comparability of levels of a mode is appropriate if one also wants to equalize the influence of the levels of that mode on the solution, since such influence equalization is a necessary consequence of size-standardization.

However, suppose one does *not* think that the levels of a particular mode should have an equal influence, perhaps because those levels with small values are based on fewer observations and hence are less reliable (as in the example given in the section on Multiplicative Adjustments for Data Standardization). Then, a different method of obtaining comparable loadings is called for. In an earlier section, we briefly mentioned an alternative approach that involves rescaling the loadings but not the data. Here, we explain the extended model that this implies.

The size of a factor loading is influenced by two things: (a) the strength of the relationship between factor and variable—such as how much the factor influences a particular variable, subject, occasion, and so on; and (b) the overall size of the data values that the factor loading is supposed to predict—such as whether the variable is measured in large units or small ones. For example, the larger the data values at the ith level of Mode A (in comparison to other levels of that mode), the larger the corresponding row of factor loadings will have to be (in comparison to other rows of the Mode A factor-loading matrix). When these variations in overall size are deemed extraneous, we need to disentangle their effects from loading differences that are due to variations in strength of factor-variable relationship so that the two influences can be interpreted separately.

We can construct an extended version of the PARAFAC model in which these two different influences on factor-loading size are represented by separate parameters in the model. As a first approximation, we could divide each row of the factor-loading matrices A, B, and C by the root-mean-square of the corresponding level of the data, obtaining row-standardized loading matrices $\overset{*}{\mathbf{A}}$, $\overset{*}{\mathbf{B}}$, and $\overset{*}{\mathbf{C}}$. To maintain accurate representation of the data, we would compensate for these changes in the loadings by incorporating new $\dot{s}$-coefficients that explicitly represent the differences in overall size or scale for the levels of each mode. This extended model would be written as follows:

$$x_{ijk} = \dot{s}_i \dot{s}_j \dot{s}_k \sum_r (\overset{*}{a}_{ir} \overset{*}{b}_{jr} \overset{*}{c}_{kr}) + (e_{ijk}) \ , \qquad (6\text{–}71)$$

where $\dot{s}_i$ is the scale coefficient (in this case taken to be the root-mean-square of the ith level of Mode A), and $\overset{*}{a}_{ir}$ is the size-standardized factor loading defined as $a_{ir}/\dot{s}_i$. This model is equivalent to (5–3) except that the expression for the systematic part has been expanded slightly; the effects of differences in data size across levels are now represented by separate terms, leaving the factor-loading matrices to express unambiguously the strength of relationship between a factor and a variable, person, and so forth. Note particularly that in both (5–3) and (6–71), the error term is unweighted. The size-standardized factor loadings of (6–71) can be obtained by an unweighted analysis. Hence, the potential problems of the weighted model in (6–70), such as inflation of error in levels with small mean-squares, do not arise.

Issues Concerning Estimation of Size-Coefficients. In the foregoing discussion, we suggested estimating the $\dot{s}$-terms in (6–71) by simply taking the root-mean-square of the corresponding levels of the data. In some situations, however, this might not represent the effects of data size as completely or correctly as desired. That is, a data matrix multiplied by the inverse of these $\dot{s}$-values would not generally have all the effects of data scale removed. This is because the effects of the several standardizations on level sizes would interact; the resulting data array would thus have modest variations from a mean-square of 1.0 across the different levels of two modes. (The last mode standardized, of course, would not have this problem.) For some purposes, it would be more desirable to use $\dot{s}$-values determined by iterative size-standardization of a "spare copy" of the data (that is, define the $\dot{s}$ to be the diagonal elements of the $\tilde{\mathbf{S}}$ matrices defined above).

The issue of which $\dot{s}$-coefficients to use in (6–71) would seem to hinge on the following question: What is the most appropriate weighting of the elements within a given slab when computing the relative size of that slab compared to other slabs in the same mode? One approach is to use the "raw" slab root-mean-squares. This implies that each data point in the slice is taken as it is, independent of what position it has in other modes, and with no consideration of how the weighting coefficients of other modes might reduce or enlarge it to accomplish removal of scale effects from the other modes. Another approach mentioned above is to use the $\dot{s}$-values that would be determined by an iterative size-standardization of the data. By this method, the investigator equates the relative size of the slabs in a given mode *after removing* size differences due to variations across levels of other modes. This implies that each data point in a given slice is weighted by the $\dot{s}$-values determined for the other two modes before the $\dot{s}$-value for that mode is computed. We are not experienced enough yet to know when each perspective is more appropriate; indeed, the choice may depend on the particular data application and perspective of the investigator. We speculate, however, that the latter approach might seem more appropriate when two ways of the data correspond to the same entities; an example of this as part of a two-way DEDICOM analysis of frequencies of different automobile trade-ins can be found in Harshman, Green, Wind, and Lundy (1982), along with a more detailed discussion of rationale and comparison of results with those obtained by reweighting the data before analysis.

Even when one or two (but not all three) modes of the data have been rescaled to reweight the influence of different parts, it still might be deemed desirable to incorporate data size-coefficients into the final model. One would then have a mixed model, which would combine an attempt to separate factor importance from data size effects and which would also have a weighted error term. We have not yet had any experience with application of these more complex adjustments for data size differences and thus do not know whether they have important advantages over the simpler models.

Complications Arising from Combined Additive and Multiplicative Preprocessing

As noted earlier, combined additive and multiplicative preprocessing is accomplished by iterative procedures. This approach was adopted on pragmatic grounds: (a) it is a method for discovering the transformations that accomplish several different standardization objectives simultaneously (for instance, set fiber means across two modes to zero and simultaneously set slab levels to mean-square of 1.0 in two modes); and (b) data preprocessed in this way often reveals meaningful structure that was not accessible before preprocessing. Often, when meaningless degenerate solutions are obtained with a given data set, the right combination of iterative preprocessing options will allow the investigator to recover meaningful, interpretable solutions.

However, when we consider the effects of iterative combined centering and size-standardization in the light of possible extended models of the data, we realize that such procedures introduce unanticipated subtleties and complications of interpretation. More than one way of preprocessing will accomplish a given set of standardization objectives, but the different methods result in different standardized data sets, which can sometimes yield dissimilar solutions. One is then faced with the question of how to determine if a particular approach is "optimal." Even more perplexing is how to relate the solution for iteratively preprocessed data to an extended model of the original data. We are still investigating these issues of data preprocessing and cannot yet present a fully satisfactory solution; however, we mention them here to introduce the reader to our current thinking on them.

Dependence on Order of Operations. While the results obtained from iterative combinations of additive and multiplicative preprocessing depend on the order in which the preprocessing operations are performed, things could have been much worse. The nature of this order dependence is less serious with PARAFAC preprocessing than with any other method of iteratively combining multiplicative and additive preprocessing of which we are aware.

It is useful to distinguish two kinds of order dependence that iterative preprocessing schemes might be subject to: (a) *mode-order* dependence; and (b) *operation-order* dependence. Mode-order dependence occurs when the result of iterative preprocessing depends on the order in which the modes are preprocessed. In contrast, operation-order dependence occurs when both centering and standardization are involved in iterative preprocessing; the converged result depends on whether the process begins with centering or with standardization.

Let us first consider an example of mode-order dependence. Suppose we want to center and size-standardize both Mode *A* and Mode *B* of a two-way array. If we do this by first centering and standardizing Mode *A*, then centering and standardizing Mode *B*, and then repeating *A*, *B*, *A*, and so on until the process converges, we end up with a different final result than if we start by centering and size-standardizing Mode *B*, then *A*, then *B*, *A*, and so forth until convergence. Both converged matrices are doubly standardized; that is, they have both variances of 1.0 and row and column means of zero. But their particular data entries —and more importantly their factor structures—will differ quite substantially. Thus, this procedure has the undesirable property that the result obtained by preprocessing the transpose of a matrix is not the transpose of the result obtained by preprocessing the original matrix itself.

This type of order dependence was first noted by Cattell (1966), who experimented with the method of iteratively centering *and* size-standardizing one mode before proceeding to the other, as described above. Cattell's discouraging findings caused some investigators (such as Kroonenberg 1981a) to take a relatively gloomy view of attempts to iteratively standardize several modes. However, as we pointed out earlier, pure size-standardization of several modes, if iterated and carried to convergence, is independent of the order in which the modes are standardized. Thus, PARAFAC size-standardization (like PARAFAC centering) is not subject to mode-order dependence.

When performing combinations of additive and multiplicative preprocessing that require iteration, PARAFAC uses a different sequence of operation from that employed for two-way arrays by Cattell. It first centers all requested modes, then iteratively standardizes all requested modes, then returns to update the centering, and so on. Because in the most recent versions of PARAFAC full iterative size-standardization is employed between each centering step, the combined procedure is a sequence of individual parts, each of which is free of mode-order dependence; therefore, the total preprocessing is also free of mode-order dependence. With PARAFAC, the transpose of a preprocessed matrix or three-way array is thus equal to the result that would be obtained by preprocessing the transpose of the original matrix or array, provided the requested preprocessing objectives are transposed appropriately.

While PARAFAC preprocessing procedures are free from mode-order dependence, they are subject to operation-order dependence. With some synthetic data sets, when we compare results obtained by starting with centering versus starting with size-standardization, we find that we have two different outcomes, both of which fulfill the preprocessing objectives but have quite different factor structures in the preprocessed modes (although the same number of dimensions as the original data). How, then, is one to decide which operation to use first? And how does one argue for the optimality of a given solution when a different result could have been obtained by a different order of preprocessing operations? We currently employ the following tentative principles for preprocessing.

When, as is usually the case, we suspect that major components of the raw data are constant in one or more modes, and we think that removing them would make the data more appropriate for the model, we consider it preferable to center as the first stage of iterative preprocessing. This eliminates these unwanted components and lets successive preprocessing and analysis proceed on the parts of the data of greatest interest. When the raw data contains such constants, standardization as a first stage of preprocessing would disturb the constants and presumably make their extraction more difficult.

If, on the other hand, we suspect that the major constant biases and two-way interactions have been obscured by different scales of measurement, subject response styles, or other influences on the sizes of entries in different levels of particular modes, then it might be necessary to standardize these modes to make these artifactual components appropriately constant before centering. When there is a serious question as to whether it is more appropriate to size-standardize first or instead to center —or whether to do only a partial size-standardization, then center, then a fuller iterative size-standardization and centering—it is necessary to try out the different procedures and compare the interpretability of the resulting solutions.

To date, when we have used real data to compare the results of initial centering versus initial standardization, we found very little difference between the PARAFAC solutions obtained with the two kinds of iterative preprocessing. For example, both preprocessing methods were used for analysis of the semantic-differential ratings of automobiles and celebrities described by Harshman and De Sarbo (appendix C). To our surprise, the correlations and factor congruence measures for the corresponding factors in the two solutions were usually .99; none was below .98. Similar results have been obtained with other data sets of the same general kind. But at the same time, parallel comparisons with certain synthetic data sets have resulted in factor correlations more in the range of .7. We do not yet understand what properties of the real data sets make them so relatively immune to preprocessing order-dependence. It may be that the problem of order-dependence is of greater theoretical than practical concern, but it might instead be that we have not checked out the degree of order-dependence on a sufficiently wide range of data types.

The Relation between Original and Preprocessed Data. Earlier in this chapter, we demonstrated how fiber-centering preserves a very simple relationship between the structure of the original data and that of the preprocessed data. In the error-free case, the factor-loading tables of the centered modes are themselves column-centered. Similarly, we determined that slab-rescaling also preserves a simple relation between the original and preprocessed solutions; namely, the rows of the factor-loading table for the rescaled mode or modes are multiplied by the same rescaling coefficients as were applied to the corresponding levels of the data (again, in the error-free case). This was found to be true even when several modes were iteratively rescaled. But when we considered the combined iterative application of centering and size-standardization, we were not able to describe the effect on the loadings in any simple straightforward way. The transformation of the factor loadings is described by the $\mathbf{M}_a$ (or $\mathbf{M}_b$ or $\mathbf{M}_c$) matrix, whose general properties are not well understood.

The possible complexity of the transformation imposed by iteratively centering and standardizing a given mode becomes clearer when we consider the effects geometrically. The centering operation shifts the origin of the axes to the centroid of the points. Then the size-standardization operation shifts these points along their radii from this origin out toward the surface of the unit hypersphere. If we consider the points in the full dimensionality that provides perfect fit, then the points are in fact moved to the surface of a hypersphere of unit radius in this high dimensional space. This shift redistributes the points, and so their centroid is no longer at the origin. The successive centering operation shifts the origin to the new centroid and the process continues until the points are distributed in such a way across the surface of the hypersphere that their centroid is at its origin.

Algebraically, we can describe the consequences of the successive operations applied to the data in terms of the equivalent operations applied to the factor-loading matrix. That is, the factor-loading matrix is sequentially column-centered and then row-rescaled until the process converges.

One theoretical consequence of this apparent complexity is that we are not able to write a simple extended PARAFAC model. The best we can do—at least until we understand the properties of the **M** matrices better—is to write an opaque expression, such as:

$$\mathbf{X}_k = (\mathbf{M}_a^{-1}\,\mathbf{A})\,\mathbf{D}_k\,(\mathbf{M}_b^{-1}\,\mathbf{B})' + \mathbf{M}_a^{-1}\,\mathbf{E}_k\,\mathbf{M}_b'^{-1}\,. \qquad (6\text{–}72)$$

This example represents the model resulting from iterative preprocessing of Modes A and B but not C. Similar models would result from other combinations of preprocessing.

Perhaps when using fully iterative preprocessing we would do better to talk of the extended PARAFAC *procedure,* rather than an extended model for the original data. The objective of such an extended procedure is to reveal certain structure that can be found in the data after other unwanted effects are removed.

"One-Cycle" Preprocessing. One method of simplifying the relationship between the original data structure and the preprocessed data structure is to truncate the iterative process after only one full iteration. Thus, preprocessing would consist of centering all requested modes, followed by iterative size-standardization of all requested modes (or vice-versa, depending on whether initial centering or initial size-standardization was used). This "one-cycle" preprocessing allows us to write a simple extended PARAFAC model for the data; the preprocessing matrix **M** has the simple form $\tilde{\mathbf{S}}\mathbf{L}$ or $\mathbf{L}\tilde{\mathbf{S}}$. The scalar form of the extended PARAFAC model can be written as:

$$x_{ijk} = \frac{1}{s_i}\,\frac{1}{s_j}\,\frac{1}{s_k}\,[h_{ij} + h_{jk} + h_{ik} \qquad (6\text{–}73)$$

$$+ \sum (a_{ir} b_{jr} c_{kr}) + e_{ijk}]\ ,$$

if initial centering is used, and in a slightly different form—with the h-terms outside of the parentheses and the s-terms inside—if initial size-standardization is used.

Because of the usual rapid convergence of the iterative proprocessing procedure, one-cycle preprocessing has produced results quite similar to full iterative preprocessing. The interpretation of the factors has been virtually identical. However, all the preprocessing objectives are not fully met at the end of one cycle. If, for example, initial centering is used, then the subsequent standardization usually disturbs the zero-means in the preprocessed modes. Thus, the h-terms of the extended model do not represent the entire two-way mean components in the data. Also, the trilinear part of the model generates fitted values whose means are not exactly zero. While this is different from the familiar roles played by such terms in two-way models, we see nothing wrong with such a representation.

The *h*-terms of the three-way model are defined by their distinctive functional roles (having to do with offsets arising from conditional origins, factors constant across one mode, plus part of the two-way effects in the trilinear component), rather than by the fact that they express all the two-way components of the fitted part of the data.

Other Approaches. Kettenring (1983) discusses ways of extending the information obtained in analysis by combining PARAFAC and analysis of variance. Models that combine the two perspectives have also been proposed by Gollob (1968), Gower (1977), and Lohmoller (1979, as noted in Kroonenberg [1983]). (The reader is referred to Kroonenberg [1983, 135–41] for further discussion of these models.)

An Orthogonally Constrained Model: Degenerate Solutions and Their Remedies

We have seen several extensions of PARAFAC's applicability that result from implicit *generalizations* of the algebraic model. We now discuss an important extension of applicability that results, surprisingly, from placing a *restriction* on the model. The restriction consists of orthogonality or zero correlation constraints on the factor-loading patterns in one or more modes of the solution. For some three-way data sets it appears that the restricted model has a "robustness" that allows it to provide more meaningful solutions, despite its formal inappropriateness.

While the usefulness of an orthogonality constraint has been demonstrated repeatedly in the last few years, we are only now beginning to understand why it works. The following account should be considered simply as a brief progress report of ongoing research.

The Classical Patterns of Degeneracy

During the initial period after development of PARAFAC, its primary applications were indirect fitting to accomplish multidimensional scaling (see Gandour and Harshman 1978; Terbeek 1977; Terbeek and Harshman 1972) and direct fitting of physiological/acoustic profile data (see Harshman, Ladefoged, and Goldstein 1977; Harshman and Papcun 1976; Lindau, Harshman, and Ladefoged 1971). It generally worked well in these applications. Eventually, however, PARAFAC began to be applied to a wider range of data types, particularly profile data of more psychological origin, such as semantic differential ratings. With these data, we would frequently observe certain characteristic but uninterpretable patterns of factor loadings that we have come to call *degenerate solutions*. We saw that the occurrence of these degenerate solutions was one of the most serious problems interfering with wide application of PARAFAC to profile data.

Degenerate solutions are characterized by factors with very high correlations (.80–.98) in Modes *A* and *B*, and usually Mode *C*, as well. However, Mode *C* is typically used for person weights; thus, correlations are sometimes lower (.50–.80) in this mode, even in degenerate cases, perhaps because of the lower reliability of person weights that are based on only one subject's data. Mode *C* is also typically the one mode left uncentered, so it shows high cross-products even when it does not show such high correlations.

The loading patterns on these factors are usually quite difficult to interpret, and in any case, because of the high correlations, there would be only one interpretation for the several dimensions in the solution. In the classic pattern of "hard-core" degeneracy, the correlations between a pair of factors will have a triple product across modes that is negative; that is, their Mode *A* correlation times their Mode *B* correlation times their Mode *C* correlation will be large and negative. Therefore, a high negative correlation between factors will be found in either one or in all three modes.

Circumstances Yielding High Factor Correlations. We now distinguish three or four conditions in which highly correlated factors can occur. The first is not considered evidence of degeneracy, whereas the other three represent successively more extreme types of degenerate solutions.

The first condition in which high correlations might be observed occurs when more factors are extracted from a data set than are actually present in the data set or can be supported by the data set. It can be demonstrated with synthetic data that highly correlated factors will sometimes occur in this situation, but several features distinguish this case from one in which a true degeneracy is present. The highly correlated factors emerge after the extraction of several meaningful nondegenerate factors. The triple product of factor correlations for these extra factors might sometimes be high and negative but at other times might be high and positive and would more often be lower in magnitude. The high correlations, when found, presumably are due to chance collinearity as a result of the random axis orientation in the subspace spanned by the redundant dimensions. Thus, such correlated dimensions were not necessarily interpreted as a symptom of degeneracy but rather as an indication that we had exceeded the correct dimensionality for a given data set.

With error-free synthetic data, the highly correlated dimensions resulting from extracting too many factors will vary across different starting positions. With real data containing error, the solution may or may not be unique, since the fitting of error might cause one orientation to be strongly preferred. However, the arbitrary nature of the axis orientation for these extra dimensions should be revealed by comparing solutions obtained in two split-halves of the data set, in which the random error would not be the same.

The second occasion on which high correlations are observed occurs when the data have not been adequately preprocessed. If the dimensionality is low and yet we observe the "classical" degenerate pattern of very high factor correlations (with one or three of them negative), we may suspect a degenerate solution, particularly if consideration of other information (such as fit values, number of dimensions extracted, replicability of the degenerate solution across split-halves, theoretical expectations, and so on) suggests that there should be more systematic dimensions present. If, as is often the case, better preprocessing allows additional meaningful dimensions to be uncovered, then we know that we were dealing with a "soft" or easily overcome degeneracy.

The third situation is degeneracy that persists even after several attempts at carefully chosen preprocessing. Then we might be confronted with hard-core degeneracy. This would be most strongly confirmed if the pattern of degenerate dimensions was replicable across split-halves of a data set and if imposition of orthogonality constraints caused a noticeable drop in the fit value yet revealed much more meaningful dimensions (which, sample size permitting, should be replicable across split-halves).

There is a fourth situation that could develop, although we have not yet observed it except with synthetic data. A hard-core degenerate solution might occur that could not be overcome, even by imposition of orthogonality constraints. In such a case, imposition of orthogonality constraints would presumably produce solutions that were not interpretable and perhaps not replicable across split-halves.

"Soft" (easily overcome) Degeneracies. Sometimes highly correlated and uninterpretable dimensions even occur in a two-dimensional analysis of a given data set. Yet examination of the fit-versus-dimensionality curve for these data sets often suggests that more than one dimension is present. Our Monte Carlo studies have shown that highly correlated dimensions can sometimes be obtained as one consequence of analyzing data that does not have the independent patterns of variation for each factor needed to determine a unique solution (for instance, when several factors are constant in a given mode) or when artificial additive constants (h-terms) have been added to the data. We originally thought that all degenerate solutions were caused by such extraneous components or conditionality in the data, and PARAFAC preprocessing methods were developed primarily so that meaningful solutions could be recovered with data of this kind. For many data sets, the preprocessing does just that.

As an example, consider the analysis discussed by Harshman and De Sarbo (appendix C). The data consisted of ratings of 25 stimuli (celebrities and automobiles) on 39 rating scales by 34 individuals. (Appendix C, Table C–1 gives the loadings obtained before data preprocessing.) Note that the first factor is constant in two of the three modes; its Mode A loadings are close to 1.0 and its Mode C loadings are close to 3.5, which is the center of the 7-point rating scale. Its Mode B loadings show a pattern of variation across scales that represents a strong positive evaluation, indicating that subjects took the zero-point for these stimuli to be of very high positive evaluation. We might suppose that this dimension is adjusting for the arbitrary displacement of the origin from true zero, that is, the h-term in (6–3). However, it shows variations in the baseline across stimuli and so would more accurately be described as representing $h + h_j$ in (6–3). The explanation is not as simple as this, however. The second and third dimensions do not represent independent patterns of influence freed from the baseline that the first factor represented. Instead, they are very highly intercorrelated and also resemble the first dimension in Mode *B* but show more complex patterns in Modes *A* and *C*. The correlations among dimensions (appendix C, Table C–4) reveal the classic high negative triple product across modes. However, there is a dramatic difference between this solution and the one obtained after iterative preprocessing: The highly correlated dimensions have disappeared, and the solution is much more interpretable (see appendix C, Tables C–5 and C–6, and Figures C–1 through C–3).

Difficult or "Hard-Core" Degeneracies. With the development of PARAFAC preprocessing procedures, we were frequently able to prevent or overcome degenerate solutions. However, there remained some solutions that were not improved, even after the most careful and extensive preprocessing efforts. Even when attempts were made to encourage noncorrelated dimensions (for instance, by using an orthogonal approximation to the solution as a starting position for further iterations), the solution would quickly return to the degenerate form. Furthermore, this degenerate form was consistent across different random starting positions and replicable across split-halves of the data. And so it would seem that the highly correlated dimensions were strongly determined by the data itself. Indeed, any less correlated solutions obtained by subsequent application of orthogonality constraints usually had noticeably poorer fit. This indicated that despite the high correlations, successive dimensions in these degenerate solutions did not represent redundant information.

The pattern of factor correlations obtained in these cases always showed a characteristic "signature": The triple product of the Mode *A* times Mode *B* times Mode *C* correlation between a pair of factors not only was quite high, but invariably was negative in sign. Yet both theoretical arguments and empirical tests with synthetic data indicated that if the problem was simply one of too many dimensions or inadequate determination of axis orientation by the data, then only some high correlations would be expected, and for these, the triple product of correlations in the three modes would be positive at least as often as it was negative. It seemed that something additional was causing the high negative correlations, that some unknown phenomenon was forcing a special form on the solution. We soon found a practical method of overcoming the problem, but only now are we beginning to understand theoretically what causes it.

Application of Orthogonality Constraints

In an attempt to block the classical pattern of degeneracy, a special option was incorporated into the PARAFAC program that allowed the user to constrain the factors in one or more modes to be orthogonal (or uncorrelated, which would be more appropriate if the mode contained all positive loadings). We reasoned that if highly correlated factors were prohibited by means of such a constraint, then interpretable dimensions might emerge.

The constraint worked surprisingly well. In most cases, it was sufficient to apply the constraint to only one of the three modes. This prevented the degenerate dimensions from emerging in all modes and provided dimensions with quite distinct interpretations. Once again the systematic nature of the degenerate solutions was emphasized. Apparently high correlations in *all three* modes (with either one or three negative correlations) were necessary for the fit value to benefit from the degeneracy. If high correlations were prevented in one mode, then they did not form in the other two. In the case of three-way rating-scale data, the constraint was generally applied to the scales mode, although sometimes the stimulus mode was used with success (Dawson 1982; also see below). Interestingly, the constraint did not act as effectively when applied to the subject mode.

Not only were the dimensions obtained after application of the constraint almost always interpretable, but we have begun to accumulate evidence that they are in some sense empirically valid. One example of such evidence is Harshman and De Sarbo (appendix C). In this study, preprocessing fixed the initial "soft" degeneracy, and the classic pattern of "hard core" degeneracy did not emerge until four dimensions were extracted, at which point only two dimensions participated in the degeneracy. When split-half analyses of the data set were performed, however, highly correlated dimensions emerged in the three-dimensional solutions (due perhaps to the small sample size [17] in each half). But by applying orthogonality constraints to the scales mode, we were able to obtain interpretable three-dimensional split-half solutions. In this case, we knew what the valid three-dimensional solution should look like, based on the interpretable full-sample solution. Thus, it is noteworthy that the solutions that emerged from the constrained split-half analyses very closely resembled the interpretable three-dimensional unconstrained solution obtained with the total sample.

In another recent PARAFAC application, Dawson (1982) performed an analysis of judges' rating-scale reactions to a set of metaphors. His unconstrained two-dimensional analysis yielded highly degenerate solutions. By application of the orthogonality constraint in the stimulus mode, however, he obtained a two-dimensional solution that was not only highly interpretable but congruent with theoretical predictions.

In certain special cases, it is useful to apply orthogonality constraints to two or all three modes to obtain a solution whose axes are oriented in the directions more closely analogous to unrotated principal components. In Snyder, Walsh, and Pamment (1983), this approach was justified by special theoretical considerations and provided a satisfying solution. In general, however, it is desirable to apply the orthogonality constraint to only one mode. This allows the orientation of axes to be established by the proportional profiles criterion without undue influence from the orthogonality constraints. Constraints applied to only one mode would allow the intrinsic axis property to determine the choice among the infinite set of solutions that are orthogonal in one mode; however, double or triple constraints would force the solution to be some compromise between the unique orientation determined by proportional profiles and that determined by principal-components-type criteria.

As we gain experience with these constraints, we are becoming increasingly convinced that the distortions that they impose on a solution generally have relatively minimal effects on the interpretation, as can be demonstrated by applying the constraints to synthetic data with oblique axes in all three modes. In exchange for these minor distortions, the constrained solution is robust in the sense that it usually has a much more meaningful interpretation than the unconstrained degenerate solution does. Sometimes it is useful to apply these constraints even when actual degeneracies are not present in the unconstrained solution. Moderately high correlations and dimensions that are difficult to interpret might also justify use of the constraint.

Searching for the Cause of "Hard-Core" Degeneracies

As we mentioned earlier, we initially attributed the problem of degenerate solutions to data peculiarities, such as extra constants, two-way interactions, and so on. This explanation was at least partly correct, since the problems with many data sets disappeared after application of adequate preprocessing. However, there were degeneracies that were not corrected in this fashion. They consistently showed a high negative product of factor correlations for the three modes. Also, we found that orthogonally constrained solutions sometimes had substantially poorer fit values than unconstrained solutions in the same dimensionality, even though the unconstrained solutions appeared much more redundant than the constrained ones. (However, careful examination of the fitted data produced by a two-dimensional degenerate solution suggested that the two highly correlated dimensions were not as redundant as they might seem. Because of the high negative correlation between the dimensions, their common part was largely cancelled out when they were added together to fit data, and the subtle differences between them became magnified to produce different and more interpretable dimensions.) Thus, we had to seek some other special characteristic of the data that could account for these properties of the degenerate solutions.

Hypothesis Testing by Data Synthesis. Since the most striking characteristic of the degenerate solutions is the fact that successive dimensions look like only slightly modified versions of previous ones, we began with the premise that the degenerate solutions were an attempt to fit individual variations in the patterns of the loadings. If each subject had points that were idiosyncratically displaced relative to those of other subjects, then the PARAFAC program might extract two or more similar dimensions to more closely approximate the subjects' alternative versions of a given dimension.

Therefore we generated "perturbed" subject spaces; that is, we synthesized profile-type data from two underlying "true" factors that had been additively "perturbed," with a different random perturbation used for each subject. We then performed a two-dimensional PARAFAC analysis of the data to see if the solution would be degenerate. The perturbations themselves were uniform random numbers selected from the range (-1,1), a different one for each loading that was to be perturbed. Before being added to the loadings, they were rescaled so that the mean-squared perturbation value for a factor in a given mode was equal to some proportion of the mean-squared true loading value for that factor; for example, sometimes we used perturbations that had a mean-squared value equal to half of the mean-squared loading value. Since we took Mode *C* to represent person weights, we never distorted the true loadings in Mode *C* but selectively perturbed the factors in one or both of the other two modes.

In addition to varying the size of the perturbations, we also varied the amount of additional error (using 0%, 5%, and 20%) and the relative size of the underlying factors in the data. Regardless of the combination of perturbation and error in the data, however, we were not able to produce fully degenerate solutions. We frequently obtained factors that were highly correlated (that is, 0.77 or above) in one mode; occasionally we saw correlations above 0.7 in two modes, but never in all three. The solution that was most like a degenerate one had factor correlations of 0.68, -0.99, and 0.73 in Modes *A*, *B*, and *C*, respectively; it was obtained by analyzing error-free 18 × 18 × 18 data with one underlying factor that was almost three times the size of the other and with the larger factor equally perturbed in both Modes *A* and *B*. On the basis of these tests, therefore, we concluded that perturbations of the underlying factors were not an adequate explanation of the degenerate solutions that we had seen. However, idiosyncratic variations in the location of points in each subject's space might contribute to some instances of highly correlated factors in one or perhaps two modes.

Hypothesis Testing by "Filtering" the Data. Finally, to test the assumption that the problem was due to some pattern of individual perturbations in the loadings of each dimension, we decided to remove all such perturbations (while retaining stretches and other linear transformations) and see if the degeneracies disappeared. For this test, we used real data rather than synthetic data. We applied a "filter" to the data for each person, making sure that each row of that person's data fit into the common row space for all subjects and similarly that each column fit into the common column space.

To accomplish this "filtering," we first determined an arbitrary set of basis vectors for the common space by averaging the data across persons and then performing a component analysis or singular value decomposition of the matrix of means. Suppose we represent the average subject matrix as $\mathbf{X}$; then we can write this decomposition as follows:

$$\frac{1}{p}\sum_{k}^{p} \mathbf{X}_k = \overline{\mathbf{X}} = \mathbf{P}\,\mathbf{D}\,\mathbf{Q}' \ .$$

To test the cause of degeneracies in a u-dimensional PARAFAC solution, we used u-dimensional filters; that is, we used the orthogonal matrices $\mathbf{P}_u$ and $\mathbf{Q}_u$, consisting of the u columns of $\mathbf{P}$ and $\mathbf{Q}$, respectively, to construct orthogonal projection matrices that projected each person's data into the u-dimensional subspace of the common space.

The filtered matrix for subject k is called $\mathbf{F}_k$ and has the property that each column of $\mathbf{F}_k$ is some linear combination of the columns of $\mathbf{P}_u$ and each row is some linear combination of the columns of $\mathbf{Q}_u$. This is accomplished with the orthogonal projection matrices $\mathbf{P}_u \mathbf{P}'_u$ and $\mathbf{Q}_u \mathbf{Q}'_u$, as follows:

$$\mathbf{F}_k = \mathbf{P}_u \mathbf{P}'_u \mathbf{X}_k \mathbf{Q}_u \mathbf{Q}'_u \ . \qquad (6\text{–}74)$$

We then submitted this filtered version of the data to PARAFAC analysis. The results were clear cut. Although the fit value increased dramatically—from R-squared of .44 to .94 in one case and from .28 to .89 in another—the loadings obtained in Modes A and B were almost identical to those obtained by analysis of raw data. The solution was still degenerate, although Mode C loadings were sometimes much less correlated. This demonstrated to us that it was not some idiosyncratic fluctuations or perturbations of individual loadings that were the cause of the degenerate solutions, since these had all been removed. Instead, the source of the degeneracies had to lie in the linear transformations that related each subject's space to the common space. Indeed, differences in the subjects' linear transformations of the common space constituted the only possible differences left after the filtering operation, yet the degeneracies still appeared.

Such individual differences, where each person's space was a different linear transformation of a common space, had to go beyond individual differences in the amount of stretch of the factor axes, since such differences could readily be represented by PARAFAC without degeneracies. There must also have been differences in either the obliqueness of dimensions or the directions of stretch (that is, individual rotations of axes before stretching or contracting them), or perhaps both. These are just the kind of more general transformations permitted by the Tucker three-mode model. In other words, it appeared that in the hard-core degenerate cases, the problems may have been caused by the fact that the data showed the kind of variations that could only be compactly represented by the Tucker T3 or T2 models; it appeared that the patterns in the data were too complex for the PARAFAC model.

Our hypotheses thus became the following: The hard-core degenerate solutions were attempts by PARAFAC to approximate the more general patterns of variation consistent with Tucker's model, using a bizarre combination of the more restricted PARAFAC factors; when we applied the orthogonality constraint, we somehow prevented this approximation procedure and forced PARAFAC to find a solution that had lower overall fit but presented the underlying dimensions in a more straightforward, interpretable manner. In other words, we hypothesized that constrained PARAFAC provided a meaningful fit to a *subset* of the Tucker structure that had a simple PARAFAC representation.

Successful Synthesis of Data Causing Degeneracies. As a first test of this hypothesis, we generated data using the Tucker T3 model and analyzed it with PARAFAC to see if we could find T3 structures that produced the characteristic signature of degenerate solutions found with real data. We were successful. Furthermore, we began to see which features of the Tucker structures would give rise to the degeneracy. For the same pattern of off-diagonal cells, we synthesized data sets in which we varied the size of the superdiagonal elements of the Tucker core matrix (the elements for which $r = s = t$). When these were large compared to the other elements of the core (such as 2 or 4 versus -.5 to .9) then the resulting data had a structure dominated by aspects for which the PARAFAC model was appropriate, and thus the solution was not degenerate. However, when we synthesized data with superdiagonals that were only moderately larger than the other entries (such as 1.0 to 1.5 versus -.5 to .9), we obtained degenerate solutions. The sign pattern for the off-diagonal cells of the core matrix also seemed to affect the likelihood of obtaining degenerate solutions. Some patterns (for instance, negative entries in certain locations) increased the likelihood of obtaining degenerate solutions, but other patterns of core matrix entries did not produce degeneracies at all. However, since these investigations are still in the preliminary stages, further generalizations would be premature.

When the degenerate PARAFAC dimensions obtained upon analysis of the synthetic data were correlated with the "true" Tucker dimensions used to generate it we found a very interesting pattern. A particular dimension (for instance, dimension 1) would be well recovered in one mode, but another (for instance, dimension 2) would be well recovered in the other mode. This surprising pattern was similar to a phenomenon that was observed in the Harshman and De Sarbo (appendix C) analysis. When the "true" dimensions of the three-dimensional solution—that is, those obtained with the total sample—were correlated with the degenerate dimensions of the three-dimensional solutions (found when analyzing the split-half samples), we sometimes found that the "redundant" dimension of the degenerate solution resembled the second true dimension in Mode *A* but the third true dimension in Mode *B*. This similarity in the behavior of our synthetic and real data analyses further indicated that we were "on the right track" with our hypothesis about the cause of degenerate solutions.

Testing Recovery via the Orthogonality Constraint. As a next step, we tried applying the orthogonality constraint to the synthetic T3 data to see if it aided in recovering the "true" underlying dimensions. (By "true" dimensions, we now mean the ones associated with large superdiagonal cells in the T3 core matrix used to generate the data.) The success depended, once again, on the size of the superdiagonal cells versus the off-diagonal cells. For all but the most severe cases, however, the orthogonality constraint worked quite well, providing solutions in which the PARAFAC dimensions correlated 0.84 to 0.99 with the "true" Tucker dimensions. Furthermore, the intrinsic axis property of the PARAFAC solution allowed the "true" T3 dimensions associated with large superdiagonals to be recovered and interpreted *without rotation*.

Current State of this Research

Currently, we are collaborating with J. B. Kruskal of Bell Laboratories on a series of mathematical and empirical investigations of how and why the T3 model can give rise to such patterns of degeneracy. We are also exploring psychological models of patterns of individual differences (and thus particular core matrix structures) that might result in degeneracies and are beginning to examine real data that causes degeneracies to determine the core matrix that would be estimated by suitable Tucker analysis of such data.

As part of this work, Kruskal has been able to confirm mathematically that T3 data in which the elements of the core matrix have certain sizes will lead to degeneracy when fit by a PARAFAC model. For example, he can define relations among the core matrix elements that determine whether a $2 \times 2 \times 2$ core matrix has trilinear rank of three or two. When the core is rank 2 and far above the boundary to the rank 3 region, the PARAFAC solution will have nondegenerate form and should recover the "true" dimensions well. When the core is rank 3, however, unusual things can happen, including degenerate two-dimensional PARAFAC solutions. Some preliminary results were reported in two papers presented at the 1983 Psychometric Society meetings (Harshman, Kruskal, and Lundy 1983; Kruskal, Harshman, and Lundy 1983).

To summarize, it seems likely that the latent structure of many three-way profile data sets is well approximated by the PARAFAC-CANDECOMP model (after preprocessing); these do not yield degenerate solutions. And for others that may not conform quite as closely to the model, preprocessing may often make them suitable for PARAFAC analysis by changing the size of different elements in the core matrix so that the preprocessed data will yield acceptable solutions even without orthogonality constraints; this might account for many cases of "soft" degeneracies. For still other data sets, however, the T3 or T2 model may be the only sufficiently general representation, even after preprocessing; these data would give rise to "hard core" degeneracies.

Nonetheless, by use of orthogonality constraints, many of these latter cases might be usefully subjected to PARAFAC analysis. The orthogonality constraint apparently forces PARAFAC to fit a *subset* of the Tucker variations, those that can be simply and meaningfully expressed in terms of orthogonal PARAFAC dimensions.

To understand the full complexity of such data, of course, one needs to apply the Tucker model(s), perhaps by means of Kroonenberg and de Leeuw's (1980) TUCKALS programs (see also Kroonenberg 1981b, 1981c). However, it might be useful to use them in conjunction with the orthogonally constrained extended PARAFAC model, since it has the intrinsic axis property and so might help to determine an empirically meaningful rotation of the Tucker solution. In the same vein, it might be useful to add some Tucker-related options to PARAFAC, such as a method of estimating the core matrix associated with a given set of PARAFAC axes. There would seem to be a convergence of the two perspectives, and methods of combining the Tucker and PARA-

FAC-CANDECOMP approaches would seem to provide a valuable direction for future research.

More Basic Generalizations

In these chapters, we have considered PARAFAC1 and its extensions. We have only noted in passing that there are also fundamental generalizations of the model, such as PARAFAC2 (mentioned in chapter 5). These would allow the proportional profiles and intrinsic axis methods to be applied to data that currently cannot be properly fit, even by the extended versions of the PARAFAC1 model. For example, PARAFAC2 (Harshman 1972) allows oblique axis analysis of cross-product matrices, and PARAFAC3 allows even more general structures to be investigated. DEDICOM (Harshman 1978; Harshman, Green, Wind, and Lundy 1982) allows a type of factor analysis or multidimensional scaling of a matrix of relationships that are not necessarily symmetric (for instance, the number of telephone calls between different towns, alphabetic confusions, international trade balances, and so on). This model has both two- and three-way versions. There is also a "linked-mode" version of PARAFAC1, in which two or more different three-way data sets that share a common mode (for example, if all were based on the same stimuli) can be analyzed with the constraint that the dimensions for the common or linked mode be the same. Discussion of generalizations such as these is beyond the scope of this chapter. We merely mention that work on such models is being undertaken and that they may further extend the domain of data to which three-way factor-analytic and multidimensional scaling methodologies can be usefully applied.

NOTES

1. We should note in passing that less subjective measures can also have the problem of an uncertain origin. Interval-scale data can arise even when the measures have a superficial true zero (for example, counts of the number of observations that fulfilled certain criteria, or the number of people in a certain location at a certain time) if there is a threshold before nonzero counts begin to be observed. Even though such counts represent a ratio-scale measure of the directly observed quantities (for example, the number of observations fulfilling a particular criterion), they may represent only an interval- or ordinal-scale estimate of the underlying *influence* that one is trying to get at by means of those quantities. Horst (1965) gives the example of items designed to measure intelligence. A count of the number of test items passed successfully might seem like a ratio-scale value with a true zero, but if the items were all of sufficient difficulty, then younger children or less-bright adults might fail all the items and yet still have considerable intelligence.

2. Some might object that the effect of centering cannot always be this simple, since when $\mathbf{X}$ is full column rank (rank m for an n by m matrix), row-centering will reduce it to rank $m - 1$. But even this is just the effect of a translation of axes.

For example, consider a 3 × 3 square matrix representing the position of 3 points in a three-dimensional space. Centering this matrix shifts the points so that the origin of the space is at the centroid of the points. This makes all three points lie on a plane through the origin and hence reduces the dimensionality of their space from three to two, simply by translation of axes.

3. Others have informally called them "spaghetti," "lasagna," and "meatloaf" centering.

4. We do not agree with some of Kroonenberg's (1983) preprocessing recommendations, for reasons that should be apparent after reading the preceding section of this chapter. His discussion of preprocessing and many examples are most instructive, however, and his book is a valuable source of information about three-mode factor analysis in general. Unfortunately, we did not receive it in time to discuss it elsewhere in this chapter.

5. This perspective was first suggested to us by Kruskal (personal communication, 1980) and is explicitly mentioned in Kruskal (1983), reprinted in this volume. We have also been recently informed by Kroonenberg (personal communication, 1983) that Lohmoller (1979) has considered three-way models with some additional additive terms. We have not yet seen this work, however, and so cannot comment on it.

6. This example is based on a study being done at the University of Western Ontario by P. Ossenkopp, L. Sorensen, and D. Mazmanian.

REFERENCES

Bishop, Y. M. M., S. E. Fienberg, and P. W. Holland. 1975. *Discrete multivariate analysis: Theory and practice.* Cambridge, Mass.: The MIT Press.

Cattell, R. B. 1966. The data box: Its ordering of total resources in terms of possible relational systems. In *Handbook of multivariate experimental psychology,* ed. R. B. Cattell. Chicago: Rand McNally.

Dawson, M. 1982. Multidimensional responses to metaphor. Master's thesis, University of Western Ontario.

Gandour, J. T., and R. A. Harshman. 1978. Cross-language differences in tone perception: A multidimensional scaling investigation. *Language and Speech* 21:1–33.

Gollob, H. F. 1968. A statistical model which combines features of factor analysis and analysis of variance techniques. *Psychometrika* 33:73–116.

Gower, J. C. 1977. The analysis of three-way grids. In *The measurement of intrapersonal space: Dimensions of intrapersonal space,* vol. 2, ed. P. Slater. London: John Wiley and Sons.

Hampson, E. Forthcoming. Three-way factor analysis of hand movement patterns during verbal and nonverbal tasks. University of Western Ontario.

Harshman, R. A. 1972. PARAFAC2: Mathematical and technical notes. *UCLA Working Papers in Phonetics* 22:30–44 (University Microfilms International, Ann Arbor, Michigan, Order No. 10,085).

———. 1978. Models for analysis of asymmetrical relationships among N-objects or stimuli. Paper presented at the first joint meeting of the Psychometric Society and the Society for Mathematical Psychology, Hamilton, Canada, August.

Harshman, R. A., and G. Papcun. 1976. Vowel normalization by linear transformation of each speaker's acoustic space. *Journal of the Acoustical Society of America.* 59:S71 (abstract).

Harshman, R. A., P. E. Green, Y. Wind, and M. E. Lundy. 1982. A model for the analysis of asymmetric data in marketing research. *Marketing Science* 1:205–42.

Harshman, R. A., J. B. Kruskal, and M. E. Lundy. 1983. Some theoretical and practical relationships between Tucker's three-mode factor analysis and PARAFAC-CANDECOMP analysis. Paper presented at the annual meeting of the Psychometric Society, Los Angeles, June.

Harshman, R. A., P. Ladefoged, and L. Goldstein. 1977. Factor analysis of tongue shapes. *Journal of the Acoustical Society of America* 62:693–707.

Horst, P. 1965. *Factor analysis of data matrices.* New York: Holt, Rinehart and Winston.

Kettenring, J. R. 1983. A case study in data analysis. In *Statistical data analysis,* ed. R. Gnanadesikan. Providence, R.I.: American Mathematical Society.

Kroonenberg, P. M. 1981a. Scaling of input data for three-mode principal component analysis. Rijksuniversiteit Leiden.

———. 1981b. *User's guide to TUCKALS3. A program for three-mode principal component analysis (WEP-Reeks, WR 81-6-RP).* Leiden, the Netherlands: University of Leiden.

———. 1981c. *User's guide to TUCKALS2. A program for three-mode principal component analysis with extended core matrix (WEP-Reeks, WR 81-35-RP).* Leiden, The Netherlands: University of Leiden.

———. 1983. *Three-mode principal component analysis: Theory and applications.* Leiden, The Netherlands: DSWO Press.

Kroonenberg, P. M., and J. de Leeuw. 1980. Principal components analysis of three-mode data by means of alternating least-squares algorithms. *Psychometrika* 45:69–97.

Kruskal, J. B. 1977. Some least-squares theorems for matrices and N-way arrays. Manuscript, Bell Laboratories, Murray Hill, New Jersey.

———. 1983. Multilinear methods. In *Statistical data analysis,* ed. R. Gnanadesikan. Providence, R.I.: American Mathematical Society.

Kruskal, J. B., R. A. Harshman, and M. E. Lundy. 1983. Relationships between Tucker's three-mode factor analysis, PARAFAC-CANDECOMP and CANDELINC. Paper presented at the annual meeting of the Psychometric Society, Los Angeles.

Lindau, M., R. A. Harshman, and P. Ladefoged. 1971. Factor analysis of formant frequencies of vowels. *UCLA Working Papers in Phonetics* 19:17–25 (University Microfilms International, Ann Arbor, Michigan, Order No. 10,085).

Lohmoller, J. B. 1979. Die tridmodale faktorenanalyse von Tucker: Skalierungen, rotationen, andere modelle. *Archiv fur Psychologie* 131:137–66.

Sands, R., and F. W. Young. 1980. Component models for three-way data: An alternating least squares algorithm with optimal scaling features. *Psychometrika* 45:39–67.

Snyder, C. W., Jr., W. D. Walsh, and P. R. Pamment. 1983. Three-mode PARAFAC factor analysis in applied research. *Journal of Applied Psychology* 68:572–83.

Stevens, S. S. 1946. On the theory of scales of measurement. *Science* 103:677–80.

Terbeek, D. 1977. A cross-language multidimensional scaling study of vowel perception. *UCLA Working Papers in Phonetics* 37:1–271 (University Microfilms International, Ann Arbor, Michigan, Order No. 10,085).

Terbeek, D., and R. A. Harshman. 1972. Is vowel perception non-Euclidean? *UCLA Working Papers in Phonetics* 22:13–29.

Weeks, D. G., and P. M. Bentler. 1979. A comparison of linear and monotone multidimensional scaling models. *Psychological Bulletin* 86:349–54.

7

The Invariant Factors Model for Multimode Data

Roderick P. McDonald

In this chapter, we further our analysis of multimode observations using the *invariant factors model*. This model is an extension of one proposed by McDonald (1980). Its basic purpose is to extend the classical common factor or linear latent trait model to multimode data sets in as simple and straightforward a manner possible. In contrast to most models for such data, the modes are not treated with mathematical or conceptual symmetry. If, for example, a sample consists of q tests measured on p occasions (or under p test conditions) on N examinees, only the q tests are thought of as yielding common factors, while the p occasions (or test conditions) and the N examinees do not. Rather, the N examinees are treated as being randomly drawn from an infinite population, and the p occasions or test conditions are treated as having an effect only on the distribution of the common factor scores of that population.

Indeed, the basic notion of the model is to carry over to the repeated measures multivariate design the fundamental axiom of common factor analysis (latent trait theory) in the independent samples design. This fundamental axiom states that with suitable scaling, the regression curves—and in particular the factor loadings—of measures obtained from experimental units (usually test or item scores from examinees) are invariant across populations of experimental units and that only the distributions of the factor scores (latent traits) may possibly change from one population to another (see Lord and Novick 1968; McDonald 1982). Here we suppose, analogously, that the regression curves—and in particular the factor loadings—of the measures on a set of experimental units in a repeated measures design are invariant over conditions of measurement and that only the distributions of the factor scores are possibly altered by the conditions of measurement. This presupposes that we have a clear distinction between *what is measured* on an experimental unit (examinee) and a *condition of measurement* (an occasion or situation). We refer to what is measured on an experimental unit as a *test score* of an examinee and to a condition of measurement briefly as a *condition*.

It is both a possible advantage and a possible limitation of this approach that we can expect to need only the classical interpretive devices of the common factor model to understand a substantive analysis. That is, we interpret each common factor of the tests as an abstractive attribute of the examinees that the high-loading tests measure in common (see McDonald 1981), but we do not expect to find and name factors of the conditions of measurement as abstractive attributes (of the examinees or of the conditions) that the conditions in some sense have in common. Briefly, the position taken here is that the tests have common factors, with interpretable columns of loadings, and the conditions do not. (Consider the difficulty, in using longitudinal data, of naming two common factors, associated respectively with two lists of ages of examinees, that are not necessarily sequential.) Nevertheless, there may exist multimode data sets in which the distinction between measures and conditions is not clear or in which the measures are concatenations of two or more measurement operations. In the former case, it may be difficult or inappropriate to apply the invariant factors model; in the latter case, it would possibly require a further generalization.

The invariant factors model is defined by three assumptions: (a) that q measures repeated under p conditions fit the common factor model with a factor-loading matrix that is invariant across the p conditions; (b) residuals of distinct measures about their regressions on the common factors are uncorrelated both within and between conditions; and (c) the factor-score covariance matrices are transformations of some basic factor-score covariance matrix.

The next section provides a somewhat more technical description of the model. This description is followed by a treatment of the matters of identifiability, scaling, and estimation. Finally, in the section on relations with other models, it is shown that, in a sense, the model is a special case of Tucker's (1966) model for three-mode data and that it in turn contains Harshman's PARAFAC as a special case.

THE INVARIANT FACTORS MODEL

Let $\mathbf{y}' = [\mathbf{y}'_1, \ldots, \mathbf{y}'_p]$ be a $pq \times 1$ random vector partitioned into p subvectors, $\mathbf{y}_k$, each having q components. In typical applications of the theory, the components of $\mathbf{y}$ will represent scores on q tests given to an examinee under p conditions of measurement. The p conditions may be distinct situations and/or replications over a series of occasions and may consist of some or all of $p_1 \times p_2 \times \ldots$ situations $\times$ occasions $\times \ldots$. That is, any number of modes of classification of the components of $\mathbf{y}$ may be supposed to be included in the set of p conditions under which the q tests are administered.

The basic notion of the model is that in each of the p conditions, the q tests fit the common factor model, with r common factors and a factor-loading matrix that is invariant over all conditions. Accordingly, the *invariant factors multimode model* may be written as:

$$\mathbf{y}_k = \mathbf{B}\mathbf{x}_k + \delta_k + \mathbf{e}_k , \quad k = 1, \ldots, p , \tag{7–1}$$

where $\mathbf{B}$ is a $q \times r$ matrix of factor loadings, δ_k is a vector of constants, $\mathbf{x}_k$ is an $r \times 1$ vector of factor scores, and $\mathbf{e}_k$ is a $q \times 1$ vector of residuals of $\mathbf{y}_k$ about its regression on $\mathbf{x}_k$. We also write:

$$\mu_k = E\{\mathbf{y}_k\} \tag{7–2}$$

and

$$\nu_k = E\{\mathbf{x}_k\} , \quad k = 1, \ldots, p , \tag{7–3}$$

for the mean vectors of the tests and factor scores under each condition, and

$$\Phi_{kl} = \text{Cov}\ \{\mathbf{x}_k, \mathbf{x}_l\} , \quad k = 1, \ldots, p , \quad l = 1, \ldots, p , \tag{7–4}$$

for the cross-covariance matrices of factor scores between conditions.

If the subvectors of $\mu' = [\mu'_1, \ldots, \mu'_p]$ are unconstrained, it follows that the parameters ν_k, δ_k are then not jointly identifiable, since

$$\mu_k = \mathbf{B}\mathbf{v}_k + \delta_k , \tag{7–5}$$

and for any choice of ν_k, we may always write:

$$\delta_k = \mu_k - \mathbf{B}\nu_k . \tag{7–6}$$

In such a case, we may reparameterize the model by writing:

$$\nu_k = \mathbf{0} \tag{7–7}$$

and

$$\delta_k = \mu_k , \quad k = 1, \ldots, p . \tag{7–8}$$

Alternatively, we might suppose that any departure of the mean of $\mathbf{y}_k$ from some $q \times 1$ mean vector δ is due only to changes in the mean factor scores from condition to condition. That is, we might suppose that:

$$\mu_k = \mathbf{B}\ \nu_k + \delta . \tag{7–9}$$

We write:

$$\mathbf{U}_{kl} = \text{Cov}\ \{\mathbf{e}_k, \mathbf{e}_l\} , \quad k = 1, \ldots, p , \quad l = 1, \ldots, p , \tag{7–10}$$

for the $q \times q$ residual cross-covariance matrices between the kth and lth conditions, and

$$\mathbf{U} = [\mathbf{U}_{kl}] \tag{7–11}$$

for the entire $pq \times pq$ residual covariance matrix. To obtain a testable, falsifiable model, we require some form of restriction on the residual covariance matrix $\mathbf{U}$. The basic notion of a common factor model applied to the same subjects under p conditions implies that in any condition, the residuals should be uncorrelated; that is, $\mathbf{U}_{kk}$ should be a diagonal matrix, $k = 1, \ldots, p$. We would thus assume a fortiori that the residuals of distinct tests are uncorrelated between distinct conditions as well as within any condition. On the other hand, in general, we would wish to allow each test to have a specific component that is stable from condition to condition so the residuals of each test may be correlated between conditions. We therefore follow McDonald's (1969) treatment of a multimode model and assume that the residual covariance matrix $\mathbf{U}$ consists of $p \times p$ submatrices $\mathbf{U}_{kl}$ that are each diagonal. If the rows and columns of $\mathbf{U}$ are permuted to yield the alternative natural order of the variables in which successive groups of p variables represent each test measured in the p conditions, then the resulting residual covariance matrix is diagonal block and yields a Cholesky (triangular) factor that is also diagonal block. It follows, on repermuting, that we may write:

$$\mathbf{U} = \mathbf{T}\mathbf{T}' , \qquad (7\text{–}12)$$

where $\mathbf{T}$ is a lower-triangular matrix, each of whose $q \times q$ submatrices T_{kl} is diagonal, $k = 1, \ldots, p$: $l = 1, \ldots, p$. We can therefore fit the model with $\mathbf{U}$ given by (7–12) and $\mathbf{T}$ a patterned matrix of real numbers and thus avoid obtaining a generalized Heywood case, that is, an improper solution in which $\mathbf{U}$ is not a possible covariance matrix (at least nonnegative definite).

We might count the case obtained by supposing that $\mathbf{U}$ is a diagonal (at least nonnegative definite) matrix as a special case of the model. This amounts to supposing that the tests possess no specific component that is stable from one condition to another. While such an assumption may appear unrealistic, it may yield a satisfactory fit to some data sets.

We may also obtain a counterpart of component analysis by setting $\mathbf{U} = \mathbf{0}$ in the covariance structure resulting from the model and fitting it by least-squares. This amounts to refusing to allow structured error within the model and leaving the entire account of the residuals to the misfit between a sample data set and the model. In doing so, we sacrifice falsifiability. However, since this is a sacrifice that some users of multivariate analysis still seem willing to make, we will regard it as an option in the model.

Next, we adopt a convenient assumption for changes in the factor-score covariance matrix across conditions. Without loss of generality, the covariance matrix Φ_{kk} of the factor scores in condition k may be written:

$$\Phi_{kk} = \mathbf{A}_k \Phi \mathbf{A}_k' , \quad k = 1, \ldots, p , \qquad (7\text{–}13)$$

where Φ is a nonsingular $r \times r$ matrix, and $\mathbf{A}_k$ is lower triangular. Given Φ and Φ_{kk}, $\mathbf{A}_k$ is unique. This follows from the fact that Φ and Φ_{kk} each have unique lower-triangular factorizations:

$$\Phi = \mathbf{L}\mathbf{L}' \tag{7–14}$$

and

$$\Phi_{kk} = \mathbf{L}_k \mathbf{L}'_k \ , \tag{7–15}$$

so that, uniquely,

$$\mathbf{A}_k = \mathbf{L}_k \mathbf{L}^{-1} \ , \tag{7–16}$$

is lower triangular.

It then seems convenient to assume that the cross-covariance matrix Φ_{kl} of the factor scores in conditions k and l is given by:

$$\Phi_{kl} = \mathbf{A}_k \Phi \mathbf{A}'_l \ , \qquad k = 1, \ldots, p \ , \quad l = 1, \ldots, p \ . \tag{7–17}$$

This is equivalent to supposing that the factor-score vector $\mathbf{x}_k$ in condition k is given by an $r \times r$ transformation $\mathbf{A}_k$ of some basic factor-score vector $\mathbf{x}$. That is,

$$\mathbf{x}_k = \mathbf{A}_k \mathbf{x} \ , \qquad k = 1, \ldots, p \ , \tag{7–18}$$

where

$$\text{Cov}\{\mathbf{x}\} = \Phi \ . \tag{7–19}$$

The model (7–1) can be written as:

$$\mathbf{y}_k - \delta_k = \mathbf{B}\mathbf{A}_k \mathbf{x} + \mathbf{e}_k \ , \qquad k = 1, \ldots, p \ , \tag{7–20}$$

or, alternatively, as

$$\mathbf{y} - \delta^* = \mathbf{B}^* \mathbf{A}^* \mathbf{x} + \mathbf{e} \ , \tag{7–21}$$

where

$$\mathbf{B}^* = \begin{bmatrix} \mathbf{B} & & & & \\ & \mathbf{B} & & & \\ & & \cdot & & \\ & & & \cdot & \\ & & & & \cdot \\ & & & & \mathbf{B} \end{bmatrix} , \text{ a } pq \times pr \text{ matrix}, \tag{7–22}$$

and

$$\mathbf{A}^* = \begin{bmatrix} \mathbf{A}_1 \\ \cdot \\ \cdot \\ \cdot \\ \mathbf{A}_p \end{bmatrix} , \text{ a } pr \times r \text{ matrix}, \tag{7–23}$$

with

$$\delta^* = \begin{bmatrix} \delta_1 \\ \cdot \\ \cdot \\ \cdot \\ \delta_p \end{bmatrix} . \qquad (7\text{–}24)$$

The model yields the covariance structure

$$\mathbf{C} = \mathrm{Cov}\,\{\mathbf{y}\} = \mathbf{B}^*\mathbf{A}^*\Phi\mathbf{A}^{*\prime}\mathbf{B}^{*\prime} + \mathbf{U} . \qquad (7\text{–}25)$$

It is important to note that while (7–25) can be written as

$$\mathbf{C} = \mathbf{B}^*\Phi^*\mathbf{B}^{*\prime} + \mathbf{U} , \qquad (7\text{–}26)$$

where

$$\Phi^* = \begin{bmatrix} \Phi_{11} & \cdots & \Phi_{1p} \\ \cdot & & \\ \cdot & & \\ \cdot & & \\ \Phi_{p1} & & \Phi_{pp} \end{bmatrix} \qquad (7\text{–}27)$$

it is not the same as (7–26) with the matrices Φ_{kl} unrestricted, since by (7–17), the $pr(pr + 1)/2$ distinct elements of these matrices are restricted to be functions of the $pr(r + 1)/2$ elements in the lower-triangular submatrices of $\mathbf{A}^*$ and the $r(r + 1)/2$ distinct elements of Φ. By the assumption (7–17), the rank of $\mathbf{C} - \mathbf{U}$ is r. Without this assumption, the rank of $\mathbf{C} - \mathbf{U}$ is possibly as great as pr.

A special case of the model may be obtained by supposing that the factor-score distribution may possibly change with respect to the variances of the factor scores, but their correlations remain unchanged; that is, we suppose that each of the matrices $\mathbf{A}_1, \ldots, \mathbf{A}_p$ in (7–17) is diagonal. This particular case of the model was described by McDonald (1980); the invariant factors multimode model is a further development of it.

The general form of the proposed model is given by (7–20) with resulting covariance structure (7–25) and mean (first-moment) structure (7–5). We distinguish $2 \times 2 \times 3 = 12$ variants on the model, obtained by the following choices:

a. between mean vectors μ_k that are either unconstrained from condition to condition or altered only as a result of changes in mean factor scores, as indicated by (7–9);
b. between residual covariance matrices $\mathbf{U}$ that are either patterned with every submatrix diagonal (to allow for stable specific components), or diagonal (implying that all residual variance is error-of-measurement), or null (so that all residuals are treated as misfit between the model and sample data, as in component analysis and many treatments of multidimensional scaling); and
c. between transformation matrices $\mathbf{A}_k$ for the factor-score covariance matrices that are either lower triangular or diagonal.

IDENTIFIABILITY, ESTIMATION, AND HYPOTHESIS TESTING

The general covariance structure (7–25) of the invariant factors multimode model may be written in complete form as:

$$\mathbf{C} = \begin{bmatrix} \mathbf{B} & & & \\ & \mathbf{B} & & \\ & & \ddots & \\ & & & \mathbf{B} \end{bmatrix} \begin{bmatrix} \mathbf{A}_1 \\ \vdots \\ \mathbf{A}_p \end{bmatrix} \Phi [\mathbf{A}'_1 \; . \; . \; . \; \mathbf{A}'_p] \begin{bmatrix} \mathbf{B}' & & & \\ & \mathbf{B}' & & \\ & & \ddots & \\ & & & \mathbf{B}' \end{bmatrix} \qquad (7\text{–}28)$$

$$+ \begin{bmatrix} \mathbf{U}_{11} & . \; . \; . & \mathbf{U}_{1p} \\ \vdots & & \\ \mathbf{U}_{p1} & . \; . \; . & \mathbf{U}_{pp} \end{bmatrix}$$

It is obvious that this model is not in general identified. We can at least postmultiply $\mathbf{B}$ by an arbitrary lower-triangular matrix $\mathbf{T}$ and premultiply each $\mathbf{A}_k$ by $\mathbf{T}^{-1}$ and then apply a similar operation to each $\mathbf{A}_k$ and to Φ. A convenient way to eliminate these indeterminacies is to choose one of the p conditions—say the first—and set its submatrix of $\mathbf{A}^*$—say $\mathbf{A}_1$—equal to the identity matrix. Then, either impose the restriction that $\Phi = I_r$ and $\mathbf{B}$ has echelon form (for exploratory work) or apply the usual restrictions to $\mathbf{B}$ for oblique restricted common factor analysis, fixing elements of $\mathbf{B}$ equal to zero to yield prescribed simple structure and setting $\text{diag}\{\Phi\} = \mathbf{I}_r$ to fix scale.

In practice, we could choose a particular condition and have its submatrix of $\mathbf{A}^*$ set equal to the identity matrix, thus clearly defining it as a *standard, base line, control,* or *initial* condition. If such a choice is not indicated by the nature of the investigation, an arbitrary designation of a condition as the *standard* condition can always be made. Although notions of mathematical symmetry might, in such a case, lead us to impose constraints on some average representation, such as

$$\frac{1}{p} \sum_{k=1}^{p} \mathbf{A}_k \qquad \text{or} \qquad \frac{1}{p} \sum_{k=1}^{p} \mathbf{A}_k \Phi \mathbf{A}'_k \, ,$$

it is neither easy nor advantageous to do so.

If care is taken to ensure sufficient tests with nonzero loadings defining each factor (see McDonald and Krane 1977, 1979), we can reasonably expect the parameters in the structure (7–25) to be identified with the other restrictions recommended. This follows from the fact that, in the standard condition, the parameters in $\mathbf{B}$ and Φ are identified under the classical conditions of restricted (confirmatory) and unrestricted (exploratory) factor analysis. If these are identified, the lower-triangular transformation matrices $\mathbf{A}_k$, $k = 2, \ldots, p$, yielding the factor-score cross-covariance matrices Φ_{kl} are, in turn, uniquely determined by the argument of (7–14) through (7–16). (This also assumes that the covariance matrix of the factor scores is nonsingular in the standard condition, that is, that no common factor that possesses nonzero variance in some other condition has its variance reduced to zero in the standard condition.)

In the special case of the model in which we restrict each $\mathbf{A}_k$ to be diagonal, corresponding to an invariant correlation matrix of the factor scores, we can obtain identifiability just by setting, for instance, $\mathbf{A}_1 = \mathbf{I}_r$ and $\text{diag}\{\Phi\} = \mathbf{I}_r$, to fix scale, without placing any restrictions on $\mathbf{B}$. This is because two or more $p \times r$ matrices of factor loadings are fixed in respect of rotation by the condition that their sth columns are related by a constant of proportionality, $s = 1, \ldots, r$. (Of course, this has been well known since Cattell's work on *parallel proportional profiles*.)

If we use the more general model in an exploratory analysis, with $\mathbf{B}$ conveniently assumed to have echelon form, we are free to apply an $r \times r$ transformation (*rotation*) matrix to the estimate $\hat{\mathbf{B}}$ of $\mathbf{B}$, perhaps seeking an approximation to orthogonal or oblique simple structure. If oblique simple structure is approximated, the inverse transformation may be applied in the usual way to each estimated cross-covariance matrix of factor scores. That is, if

$$\hat{\mathbf{B}}^{(t)} = \hat{\mathbf{B}}\,\mathbf{T} \tag{7–29}$$

is the resulting factor-loading matrix when $\hat{\mathbf{B}}$ is transformed by $\mathbf{T}$ to approximate simple structure, then the cross-covariance matrix of the corresponding transformed factor scores in conditions k and l is given by:

$$\Phi_{kl}^{(t)} = \mathbf{T}^{-1}\,\Phi_{kl}\,\mathbf{T}'^{-1} . \tag{7–30}$$

It is important to recognize that the elimination of rotational indeterminacy by choosing the special case of the model (in which the $\mathbf{A}_k$ matrices are diagonal) is in fact obtained at the expense of the strong assumption that the factor scores have an invariant correlation matrix across conditions. By making this assumption, we sacrifice the possibility of obtaining prescribed or approximate simple structure and thus lose the most widely accepted basis for reflecting the design of a set of tests in our analysis and for interpreting the substantive results obtained.

It does not seem possible to set up the invariant factors model as a scale-invariant model for the analysis of correlation structures and fit it to a sample correlation matrix, with the standard deviations of the variables estimated by their sample values (see Krane and McDonald 1978). Indeed, since the model is intended to account, in part, for changes in test variances and, in the first-moment structure, for changes in test means, it would seem undesirable to obtain a counterpart scale-invariant model for correlations even if this were mathematically straightforward. We might question, more generally, whether models for multimode data should ever be fitted to sample correlation matrices, although in the past this has been a common practice.

In the usual applications to tests whose origins and scales are arbitrary, we are free to apply, across all conditions, a change of origin and scale that is chosen to be convenient in one of the conditions. In applications, it may prove desirable to standardize the tests in the standard condition and to apply the same rescaling to them in the nonstandard conditions (which will not then result in standardized scores). The estimated factor loadings are then standardized regression coefficients in the standard condition, and the conventional notions of traditional common factor analysis can be applied to the interpretation of their magnitudes.

The invariant factors multimode model is a fairly simple linear structural model for multivariate data and may be fitted by a general purpose program for the analysis of covariance structures. Here, we will show how the model may be fitted by the COSAN program, written by Fraser to fit McDonald's model (1978, 1980). The only technical question that arises concerns the application of COSAN to a model with structure on the means and estimation by maximum likelihood. We first prove a general result on this question.

If $\mathbf{y}$ $(n \times 1)$ has a multivariate normal distribution with mean vector μ and covariance matrix $\mathbf{C}$, and $\mathbf{C}$ is a function of a vector of parameters Θ while μ is unrestricted, it is well known that the ML estimator of Θ is the minimum point of the function of likelihood (scaled log-likelihood ratio):

$$\phi(\Theta) = \log |\mathbf{C}(\Theta)| - \log |\mathbf{S}| + \mathrm{Tr}\,\{\mathbf{S}\mathbf{C}^{-1}(\Theta)\} - n \,, \qquad (7\text{–}31)$$

where $\mathbf{S}$ is a sample covariance matrix. The function ϕ is a loss function; that is, it is nonnegative everywhere and zero if and only if we can find Θ such that $\mathbf{C}(\Theta) = \mathbf{S}$. The expression (7–31) is a specialization of the loss function

$$\phi^*(\Theta) = \log |\mathbf{C}(\Theta)| - \log |\mathbf{S}| \qquad (7\text{–}32)$$

$$+ \mathrm{Tr}\,\{[\mathbf{S} + (\overline{\mathbf{y}} - \mu(\Theta)(\mathbf{y} - \mu(\Theta))']\mathbf{C}(\Theta)\}^{-1} - n \,,$$

where $\overline{\mathbf{y}}$ is a vector of sample means, which is a more general scaled log-likelihood ratio for estimating Θ if both $\mathbf{C}$ and μ are functions of Θ. Again, $\phi^*(\Theta)$ is a loss function because it is nonnegative everywhere and zero if and only if we may find Θ such that $\mathbf{C}(\Theta) = \mathbf{S}$ and $\mu(\Theta) = \overline{\mathbf{y}}$. (If μ is unrestricted, then the estimate of μ is given by $\overline{\mathbf{y}}$ and [7–32] reduces to [7–31].)

We define

$$\mathbf{G} = \left[\begin{array}{c|c} 1 & \overline{\mathbf{y}}' \\ \hline \overline{\mathbf{y}} & \mathbf{S} + \overline{\mathbf{y}}\overline{\mathbf{y}}' \end{array}\right] = \left[\begin{array}{c|c} 1 & \overline{\mathbf{y}}' \\ \hline \overline{\mathbf{y}} & \mathbf{S}^+ \end{array}\right] , \qquad (7\text{–}33)$$

where, as before, $\mathbf{S}$ is the sample covariance matrix and $\mathbf{S}^+ = \mathbf{S} + \overline{\mathbf{y}}\overline{\mathbf{y}}'$ is the sample raw product-moment matrix. We also define

$$\mathbf{H} = E\{G\} = \left[\begin{array}{c|c} 1 & \mu' \\ \hline \mu' & \mathbf{C} + \mu\mu' \end{array}\right] = \left[\begin{array}{c|c} 1 & \mu' \\ \hline \mu & \mathbf{C}^+ \end{array}\right] , \qquad (7\text{–}34)$$

where, as before, $\mathbf{C}$ is the covariance matrix of $\mathbf{y}$ and $\mathbf{C}^+ = E\{\mathbf{yy}'\}$. We then find that

$$\phi^+ (\Theta) = \log |\mathbf{H}| - \log |\mathbf{G}| + \text{Tr}\,\{\mathbf{G}\mathbf{H}^{-1}\} - n - 1 \qquad (7\text{–}35)$$

is the same as the function (7–32). That is, we can now use COSAN to simultaneously estimate a structure for the covariance matrix and a structure for the mean vector by fitting the $(n + 1) \times (n + 1)$ matrix $\mathbf{H}(\Theta)$, with the structure indicated, to the $(n + 1) \times (n + 1)$ sample matrix $\mathbf{G}$ containing sample means and the raw sample product-moment matrix.

Proof of this statement follows from the fact that

$$|\mathbf{H}| = |\mathbf{C}| \, , \qquad (7\text{–}36)$$

$$|\mathbf{G}| = |\mathbf{S}| \, , \qquad (7\text{–}37)$$

and

$$\left[\begin{array}{c|c} 1 & \mu' \\ \hline \mu & \mathbf{C} + \mu\mu' \end{array}\right]^{-1} = \left[\begin{array}{c|c} 1 + \mu'\mathbf{C}^{-1}\mu & -\mu'\mathbf{C}^{-1} \\ \hline -\mathbf{C}^{-1}\mu & \mathbf{C}^{-1} \end{array}\right] \qquad (7\text{–}38)$$

so

$$\text{Tr}\,\{\mathbf{G}\mathbf{H}^{-1}\} = \text{Tr}\,\left\{ \left[\begin{array}{c|c} 1 & \bar{\mathbf{y}}' \\ \hline \bar{\mathbf{y}} & \mathbf{S} + \mathbf{yy}' \end{array}\right] \left[\begin{array}{c|c} 1 + \mu'\mathbf{C}^{-1}\mu & -\mu'\mathbf{C}^{-1} \\ \hline -\mathbf{C}^{-1}\mu & \mathbf{C}^{-1} \end{array}\right] \right\} \qquad (7\text{–}39)$$

$$= \text{Tr}\,\{\mathbf{S}\mathbf{C}^{-1} + 1 + \mu'\mathbf{C}^{-1}\mu - 2\mu'\mathbf{C}^{-1}\bar{\mathbf{y}}' + \bar{\mathbf{y}}'\mathbf{C}^{-1}\bar{\mathbf{y}}\} \, ,$$

which gives (7–32) from (7–35) on rearranging terms (see Graybill [1969], theorem 8.2.1). We now apply this result to the invariant factors multimode model.

The parameter matrices of the model are $\mathbf{B}^*$, defined in (7–22); $\mathbf{A}^*$, defined in (7–23); $\mathbf{T}$, defined in (7–12); δ^*, defined in (7–24); and ν^*, defined by $\nu^{*\prime} = [\nu'_1, \ldots, \nu'_p]$, where ν is defined by (7–3). If the mean structure (7–9) is assumed, then, to fit the model by COSAN, we define the matrices

$$\mathbf{F}_1 = \left[\begin{array}{c|c|c|c} 1 & & & \\ \hline & \mathbf{B}^* & \mathbf{I}_{pq} & \mathbf{I}_{pq} \end{array}\right] , \qquad (7\text{–}40)$$

a $(pq + 1) \times (2pq + pr + 1)$ matrix,

$$\mathbf{F}_2 = \left[\begin{array}{c|c|c} 1 & & \\ \hline \nu^* & \mathbf{I}_{pr} & \\ \hline \delta^* & & \\ \hline & & \mathbf{I}_{pr} \end{array}\right], \tag{7–41}$$

a $(2pq + pr + 1) \times (pq + pr + 1)$ matrix in which $\delta^{*\prime} = [\delta^\prime, \ldots, \delta^\prime]$, a vector of pq components in which the subvector δ is repeated p times,

$$\mathbf{F}_3 = \left[\begin{array}{c|c|c} 1 & & \\ \hline & \mathbf{A}^* & \\ \hline & & \mathbf{T} \end{array}\right], \tag{7–42}$$

a $(pq + pr + 1) \times (pq + r + 1)$ matrix, and

$$\mathbf{P} = \left[\begin{array}{c|c|c} 1 & & \\ \hline & \Phi & \\ \hline & & \mathbf{I}_{pq} \end{array}\right], \tag{7–43}$$

a $(pq + r + 1) \times (pq + r + 1)$ symmetric matrix. Then we may fit $\mathbf{H}$, a $(pq + 1) \times (pq + 1)$ matrix, given by

$$\mathbf{H} = \mathbf{F}_1 \mathbf{F}_2 \mathbf{F}_3 \mathbf{P} \mathbf{F}^\prime_3 \mathbf{F}^\prime_2 \mathbf{F}^\prime_1 , \tag{7–44}$$

to the sample matrix of raw product-moments and means, given by

$$\mathbf{G} = \left[\begin{array}{c|c} 1 & \bar{\mathbf{y}}^\prime \\ \hline \bar{\mathbf{y}} & \mathbf{S}^+ \end{array}\right], \tag{7–45}$$

where, writing $\mathbf{Y} = [y_{ji}]$ for the $pq \times N$ raw score matrix, we have

$$\bar{\mathbf{y}} = \frac{1}{N} \mathbf{Y} \mathbf{1}_N , \tag{7–46}$$

the $n \times 1$ vector of sample means, and

$$\mathbf{S}^+ = \frac{1}{N} \mathbf{Y} \mathbf{Y}^\prime , \tag{7–47}$$

the $pq \times pq$ sample raw product-moment matrix (without correction for means). Minimizing (7–35) then yields ML estimates of $\mathbf{B}^*$, $\mathbf{A}^*$, Φ^*, ν^*, δ^*, $\mathbf{T}$, and an asymptotic chi-square test for the entire first- and second-moment structure as standard output from COSAN.

If the mean vector μ is unrestricted, then by (7–7) and (7–8) we reparameterize the model to eliminate first-moment structure and fit a simpler model to the sample covariance matrix:

$$\mathbf{S} = \frac{1}{N}\mathbf{YY}' - \overline{\mathbf{y}}\overline{\mathbf{y}}' , \qquad (7\text{–}48)$$

namely,

$$\mathbf{C} = \mathbf{F}_1\mathbf{F}_2\mathbf{P}\mathbf{F}'_2\mathbf{F}'_1 , \qquad (7\text{–}49)$$

where

$$\mathbf{F}_1 = [\mathbf{B}^* : \mathbf{I}_{pq}] , \qquad (7\text{–}50)$$

a $pq \times (pq + pr)$ matrix,

$$\mathbf{F}_2 = \left[\begin{array}{c|c} \mathbf{A}^* & \\ \hline & \mathbf{T} \end{array}\right] , \qquad (7\text{–}51)$$

a $(pq + pr) \times (pq + r)$ matrix, and

$$\mathbf{P} = \left[\begin{array}{c|c} \Phi & \\ \hline & \mathbf{I}_{pq} \end{array}\right] , \qquad (7\text{–}52)$$

a $(pq + r) \times (pq + r)$ symmetric matrix.

Good starting points for the minimization procedure in program COSAN can be found by fitting the restricted or unrestricted classical common factor model to the q variables in the standard situation to give approximations to $\mathbf{B}$ and the diagonal elements of ν. It may be a good strategy to fit the models with an unrestricted mean vector first and to use the fitted factor-loading matrix $\hat{\mathbf{B}}$ to get a good estimate of ν^* for the purpose of fitting the model with mean structure (7–9). It is easy to show by the usual least-squares theory that, conditional on an estimate $\hat{\mathbf{B}}$ of $\mathbf{B}$, the least-squares estimate of ν_k for each of the nonstandard situations $k = 2, \ldots, p$ is given by

$$\nu_k = (\hat{\mathbf{B}}'\hat{\mathbf{B}})^{-1} \hat{\mathbf{B}}'(\overline{\mathbf{x}}_k - \overline{\mathbf{x}}_1) . \qquad (7\text{–}53)$$

The model and method will be illustrated by an analysis of a data set originating in the work of Meyer and Bendig (1961), who collected data on five of Thurstone's tests of primary mental abilities—*V*erbal ability, *S*patial ability, *R*easoning ability,

*N*umerical ability, and *W*ord fluency—given at age 8 and at age 11 to a sample of 49 boys and 51 girls. Meyer and Bendig did not publish their entire correlation matrix, but they supplied it for reanalysis to Harris, who did then publish it (1963). We can reconstruct the covariance matrix from Harris' published correlation matrix and from the means and t-values published by Meyer and Bendig for the comparison of test scores for both boys and girls on each test at each age. It is easy to solve the expression for t in terms of composite sample sums of squares.

Table 7–1 contains the raw product-moment matrix, bordered by means as required to form matrix $\mathbf{G}$ in equation (7–33), as reconstructed from the original sources. The data have been in effect rescaled by subtracting the means on the first occasion. Accordingly, in the submatrix of the first occasion, the cross-products are covariances, while in the submatrix for the second occasion, they are covariances plus products of mean differences.

Four models were fitted to this data set, namely those obtained by choosing $\mathbf{A}_2$ diagonal versus $\mathbf{A}_2$ general (lower-triangular) and by choosing an unrestricted mean vector versus a structured mean vector (structured as in [7–9]). In all cases, the residual covariance matrix was patterned with every submatrix diagonal to allow the specific components to be correlated between occasions, and two common factors were prescribed, with an exploratory analysis in which just one factor loading was restricted to be zero, to fix a basis in common factor space. In a previous analysis of the correlation matrix by McDonald (1980), in which the model was used with a patterned residual covariance matrix and a diagonal transformation matrix (but described rather differently in terms of the Khatri-Rao matrix product), the loading of the fifth test on the first factor was very close to zero. Accordingly, to aid comparison with the earlier analysis, $\mathbf{B}$ was restricted to have its element $\mathbf{b}_{51} = 0$.

The results of the four analyses were fairly similar, so it is not necessary to report all of them in detail. With the problem stated as in (7–48) through (7–52), the model was fitted with unrestricted means by program COSAN. Then, a good starting point was found for ν_2 by using (7–53) with $\hat{\mathbf{B}}$ from the first run, namely $\nu'_2 = [-.602 \; 1.953]$. The problem was then rearranged as in (7–40) through (7–47) and again run on COSAN. This gave a chi-square of 49.2915 on 32 degrees of freedom (df) and a corresponding probability value of .0261. The residual matrix, given in Table 7–2, may seem at first to be quite unacceptable. If we rescale the original raw product-moment matrix to have unit diagonal elements and apply the same rescaling to the residuals, typically dividing them by a number in the order of a hundred, we recognize that the residuals are in fact of the order of smallness that would correspond to a satisfactory approximation to the data by a common factor analysis of a sample of this size.

The matrix of factor loadings $\hat{\mathbf{B}}$ is given in Table 7–3 in three formats: (a) as fitted; (b) as rescaled by the standard deviation of the variables on the first occasion to yield standardized factor loadings; and (c), for comparison, as obtained by McDonald (1980). The various transformation matrices are shown in Table 7–4: (a) the fitted general (lower-triangular) transformation matrix $\mathbf{A}_2$; (b) the diagonal transformation matrix from the first

TABLE 7-1. Meyer-Bendig Augmented Raw Product-Moment Matrix

		V	S	R	N	W	V	S	R	N	W
	1.00										
V	0.00	49.14									
S	0.00	25.25	92.74								
R	0.00	17.67	19.13	36.00							
N	0.00	28.89	10.47	17.83	61.15						
W	0.00	29.17	10.83	13.23	20.03	116.86					
V	11.32	57.56	33.75	29.63	46.13	34.83	230.15				
S	9.86	25.32	64.91	12.53	-2.90	12.68	147.16	203.10			
R	5.16	17.75	18.49	26.93	21.40	17.19	86.10	61.81	62.63		
N	5.51	25.41	11.81	21.21	51.23	18.19	112.54	60.04	57.54	110.46	
W	8.29	20.05	17.27	12.02	13.75	54.60	122.01	100.19	56.63	62.67	206.08

TABLE 7-2. Residual Matrix

		V	S	R	N	W	V	S	R	N	W
	0.000										
V	0.160	-0.966									
S	-0.406	-9.437	-15.729								
R	-0.156	-3.848	5.605	-2.092							
N	0.035	0.293	-2.607	0.783	1.425						
W	-0.366	-0.621	-9.353	0.189	0.732	1.080					
V	-0.524	7.616	-5.603	3.035	8.113	-2.189	2.731				
S	0.670	-10.853	-18.364	-3.982	-19.003	-12.620	0.064	-4.124			
R	0.138	-4.040	4.205	-3.068	-2.161	0.721	-1.525	1.251	-2.251		
N	0.053	-2.464	1.674	-0.438	-0.632	-4.952	3.145	-2.680	-2.087	-3.098	
W	0.902	-10.432	-5.107	-4.610	-12.334	-9.577	-2.355	10.434	1.018	-10.496	-6.635

TABLE 7-3. Estimated Factor Loadings

	(a) Fitted		(b) Rescaled		(c) McDonald*	
	I	II	I	II	I	II
V	-.809	7.024	-.115	1.002	.178	.899
S	-2.026	4.714	-.210	.490	-.051	.442
R	.672	3.144	.112	.524	.266	.503
N	4.133	4.544	.529	.581	.593	.433
W	.000	4.250	.000	.393	.050	.366

Source: McDonald 1980.

TABLE 7-4. Transformation Matrices

(a) General transformation	(b) Diagonal transformation	(c) Factor-score covariance--General transformation	(d) Factor-score covariance--Diagonal transformation
$\begin{bmatrix} 1.536 & \\ .298 & 1.095 \end{bmatrix}$	$\begin{bmatrix} 1.886 & \\ & 1.036 \end{bmatrix}$	$\begin{bmatrix} 2.359 & .458 \\ .458 & 1.286 \end{bmatrix}$	$\begin{bmatrix} 3.557 & \\ & 1.073 \end{bmatrix}$

analysis; and the resulting factor-score covariance matrices, both (c) general transformation and (d) diagonal transformation.

The final estimate of ν_2 was $\nu'_2 = [-.490\ 1.652]$ and $\delta' = [-.160\ .406\ .156\ -.035\ .366]$. Table 7–5 gives the fitted variances and covariances of the residuals, from $\hat{\mathbf{U}}$, rearranged for convenience into a set of five 2×2 matrices, containing the covariance matrix of the two occasions for each test in turn. Table 7–5 also gives the correlations of the residuals between occasions, calculated from these. Numerical ability has contributed a generalized Heywood case and verbal ability has almost done so. These are possibly due to the fact that the first common factor is very ill-defined, since only the numerical ability test has a reasonably large loading on it, and it is known that we need at least three variables with large loadings on each factor to obtain identifiability.

TABLE 7-5. Variances and Covariances of Residuals--Correlations between Occasions

Verbal

$$\begin{bmatrix} .089 & -1.500 \\ -1.500 & 27.275 \end{bmatrix} \qquad \rho = -.963$$

Spatial

$$\begin{bmatrix} 81.975 & 51.744 \\ 51.744 & 93.205 \end{bmatrix} \qquad \rho = .592$$

Reasoning

$$\begin{bmatrix} 27.731 & 17.067 \\ 17.067 & 23.929 \end{bmatrix} \qquad \rho = .665$$

Numerical

$$\begin{bmatrix} 22.165 & -2.100 \\ -2.100 & .199 \end{bmatrix} \qquad \rho = -1.000$$

Word Fluency

$$\begin{bmatrix} 78.837 & 37.478 \\ 37.478 & 117.051 \end{bmatrix} \qquad \rho = .390$$

The model with a restricted mean vector and a diagonal transformation matrix gives a chi-square of 56.8205 on 33 df, hence a probability of .0061. This model is nested in the previous one, so we might reasonably regard the difference in chi-square of 7.5290 on 1 df, $p < .01$, as suggesting that we should reject the hypothesis that the two occasions differ only by a change in factor variances while the factor correlations remain unchanged. As shown in the next section, this amounts to rejecting Harshman's PARAFAC model (with a patterned residual matrix) in favor of a more general one.

With an unrestricted mean vector, the general transformation matrix gives a chi-square of 36.2145 on 28 df, hence a probability of .1372, while the diagonal transformation gives a chi-square of 43.4053 on 29 df. (In this run, however, convergence was not obtained, and the true chi-square may be lower.) On the numbers obtained, the differences in chi-square suggest that we should reject the structure 7–9) for the mean vector, as well as the diagonal transformation matrix. Not surprisingly, the best-fitting model is the least restrictive one, namely the model with the unrestricted mean vector and the general transformation matrix. This model yields the estimate of the factor loadings and the transformation matrix and the resulting factor-score covariance matrix shown in Table 7–6.

It would have been possible to retain the restrictions on the transformation matrix and the mean vector and improve the fit of the model by postulating three common factors, but five variables could hardly serve to define three interpretable common factors, and we already have a Heywood case and an ill-defined factor when only two are fitted.

RELATIONS WITH OTHER MODELS

It is interesting to observe that we may rewrite the covariance structure (7–25) in the form

$$\mathbf{C} = (\mathbf{A} \otimes \mathbf{B})\, \mathbf{F}\Phi\mathbf{F}'\, (\mathbf{A}' \otimes \mathbf{B}') + \mathbf{U}\ , \qquad (7\text{–}54)$$

where $\mathbf{B}$, $p \times r$, is as before, where

$$\mathbf{A} = \begin{bmatrix} (\mathrm{Vec}\{\mathbf{A}'_1\})' \\ (\mathrm{Vec}\{\mathbf{A}'_2\})' \\ \cdot \\ \cdot \\ \cdot \\ (\mathrm{Vec}\{\mathbf{A}'_{\dot{p}}\})' \end{bmatrix}, \qquad (7\text{–}55)$$

with the notation Vec{**M**} indicating the rearrangement of an $s \times t$ matrix **M** as an $st \times 1$ vector obtained by arranging the s rows of

TABLE 7-6. Factor Loadings, Transformation Matrix, and Factor-Score Covariance Matrix

	Factor Loadings I	Factor Loadings II	Transformation Matrix		Factor-Score Covariance Matrix	
V	-1.392	6.876	1.581		2.500	.539
S	-1.683	2.858	.341	1.209	.539	1.578
R	.609	2.529				
N	3.601	4.576				
W	.000	3.114				

M successively as columns to form its subvectors and where **F** is a diagonal block matrix, each of whose diagonal blocks is the $r^2 \times 1$ vector Vec$\{I_r\}$. Matrix **A** is of order $p \times r^2$ and **F** is $r^3 \times r$. The proof of the identity of (7–54) with (7–25) is by a straightforward but tedious scalar expansion of the two expressions, as in analogous identities in the McDonald-Swaminathan matrix calculus (McDonald and Swaminathan 1973), from which the identity was conjectured.

For example, if $q = 4$, $r = 2$, and $p = 2$,

$$\mathbf{B}^*\mathbf{A}^* = \begin{bmatrix} b_{11} & b_{12} & & \\ b_{21} & b_{22} & & \\ b_{31} & b_{32} & & \\ b_{41} & b_{42} & & \\ & & b_{11} & b_{12} \\ & & b_{21} & b_{22} \\ & & b_{31} & b_{32} \\ & & b_{41} & b_{42} \end{bmatrix} \begin{bmatrix} a_{11}^{(1)} & \\ a_{21}^{(1)} & a_{22}^{(1)} \\ a_{11}^{(2)} & \\ a_{21}^{(2)} & a_{22}^{(2)} \end{bmatrix},$$

from which (7–25) may be completed, and

$$(\mathbf{A} \times \mathbf{B})\mathbf{F} = \left(\begin{bmatrix} a_{11}^{(1)} & a_{21}^{(1)} & 0 & a_{22}^{(1)} \\ a_{11}^{(2)} & a_{21}^{(2)} & 0 & a_{22}^{(2)} \end{bmatrix} \times \begin{bmatrix} b_{11} & b_{12} \\ b_{21} & b_{22} \\ b_{31} & b_{32} \\ b_{41} & b_{42} \end{bmatrix} \right) \begin{bmatrix} 1 & 0 \\ 0 & 0 \\ 0 & 0 \\ 1 & 0 \\ 0 & 1 \\ 0 & 0 \\ 0 & 0 \\ 0 & 1 \end{bmatrix},$$

from which (7–54) may be completed. In the form (7–54), the model can be regarded as a special case of Tucker's model for three-mode data, as described by Bloxom (1968; chapter 4), except for possible choices with respect to the structure of the residual covariance matrix.

In Tucker's interpretation of his model, matrix **B** is a matrix of *factor loadings,* possibly in the usual sense of regression weights of the q test variables in their regressions on r common factors or latent traits, and matrix **A** is interpreted as a *factor-loading matrix,* of the p conditions on what in this case would be r^2 factors. Matrix **F** in the Tucker model has also been interpreted as, in some sense, a factor-loading matrix. It seems, however, that such an interpretation is rather questionable in the present case. Here, the matrix **B** is interpretable as a matrix of regression coefficients of the q observed variables on r common factors, invariant over p conditions. Each of the p rows of **A** contains, as a number string, the elements of an $r \times r$ matrix that transforms the factor-score covariance matrix appropriately for each of the p conditions. Further, **F**, an $r^3 \times r$ matrix of unities and zeros, is just an operator enabling the rewriting of (7–25) as a model of the Tucker-Bloxom type.

It is intuitively evident that, in general, we could not hope to obtain the model (7–54) by transformation (rotation) of the matrices fitted in an exploratory three-mode analysis, either of the type suggested by Tucker or of the type recommended by Bentler and Lee (1979). Indeed, we may commonly have $r^2 > p$, so **A**, which is $p \times r^2$, will not in general be of full column rank, a necessary condition for the results of an exploratory three-mode analysis to be transformable into the invariant factors model.

While the invariant factors multimode model is a special case of Tucker's model, in the sense just described, it in turn contains Harshman's PARAFAC (1970; chapter 5) as a special case. If we assume that the mean vectors are unrestricted, that $\mathbf{U} = \mathbf{0}$, and that $\mathbf{A}_k$ is diagonal for all k, the model becomes

$$y_j^{(k)} = \sum_{s=1}^{r} a_{ks} b_{js} x_s \,, \tag{7–56}$$

where $y_j^{(k)}$ is the jth component of $\mathbf{y}_k$, a_{ks} is the sth diagonal element of $\mathbf{A}_k$, and b_{js} is the (j,s)th element of **B**. This is Harshman's PARAFAC with (in practice) the equality replaced by an approximate equality.

The invariant factors model for multimode data can thus be seen to occupy an intermediate level of generality between Harshman's PARAFAC and Tucker's three-mode method. As we have seen in the previous section, conditions for identifiability in the general invariant factors model can be obtained by a natural extension of those in the classical common factor model, namely, by imposing appropriate restrictions on the $q \times r$ factor-loading matrix **B**, usually in accordance with notions of simple structure, to eliminate the classical problem of rotational indeterminacy. Alternatively, if we restrict the transformation matrices $\mathbf{A}_k$ to be diagonal, we also eliminate rotational indeterminacy and obtain an identified factor-loading matrix **B** which, in general, will not possess simple structure. It is this severely restrictive assumption—that factor scores do not change their correlational structure from condition to condition (but only their variances)—that frees three-mode analysis of the PARAFAC-type from the classical problem of rotational indeterminacy. Perhaps only a great deal of practical experience will determine whether it is a better strategy to place restrictions on **A** (as in PARAFAC) or on **B** (as in classical factor analysis) in order to obtain identified, interpretable, and otherwise acceptable results.

SUMMARY AND DISCUSSION

The invariant factors multimode model is a simple extension of the classical common factor model to repeated-measures designs. Its basic assumption, an analogue of the fundamental axiom of latent trait theory, states that the regressions of the measures on a set of common factors are invariant across conditions of measurement.

Twelve variants on the model have been described, depending on choices between a structured and an unstructured mean vector, a residual covariance matrix that is patterned, diagonal, or null, and a transformation of the factor-score covariance matrices that leaves their correlations invariant or allows them to change.

In the section on identifiability, estimation, and hypothesis testing, it was shown that in the general case the model is identified on applying the conditions employed in classical exploratory and confirmatory factor analysis to the invariant factor-loading matrix and the basic factor-score covariance matrix. That is, we may employ the classical procedures for eliminating rotational indeterminacy. In the special case in which we require the factor scores to have an invariant correlation matrix, the theory of Cattell's *parallel proportional profiles* eliminates rotational indeterminacy in the well-known fashion, apparently without the need for a supplementary principle such as simple structure. Since this special case is a variant on Harshman's PARAFAC method of analysis (as is shown in the section on relations with other models), it clarifies a conjecture by Kruskal (1981) that PARAFAC eliminates rotational indeterminacy because of the comparative *richness,* in some sense, of three-mode data compared to two-mode data. Essentially, in the invariant factors model for three-mode data, we can choose between the supplementary principle of simple structure and the supplementary principle of invariant factor-score correlations, which from one point of view defines PARAFAC. It would be misleading, however, to regard the first choice as introducing a supplementary principle and the second choice as not doing so, as we might easily have supposed from Kruskal's account of PARAFAC.

It was also shown in this section that the invariant factors model should not be treated as a scale-invariant model for the analysis of correlation structures, and recommendations are made on choices of scale to facilitate interpretation. It is shown how the model may be fitted, with or without a structured mean vector, by Fraser's COSAN program for McDonald's model for the analysis of covariance structures.

In treating relations with other models, it was shown that the invariant factors model is a special case of Tucker's three-mode model, as well as containing Harshman's PARAFAC. The relation of the model to Tucker's model raises questions for further research about the latter. At the very least, we must conclude that the interpretive principles suggested for his model by Tucker (1966) and the somewhat modified interpretive principles for it developed by Bloxom (chapter 4) do not seem appropriate in this special case and therefore cannot be said to be appropriate in all cases of the model. This suggests, as a matter for further inquiry, that Tucker's model may contain a variety of special cases with quite distinct conceptual properties, requiring distinct and incompatible approaches to their interpretation. It may prove more fruitful, however, to continue to build more specialized models for repeated-measures designs for multivariate data and to ask whether any such model is contained in Tucker's model, rather than attempt to classify specializations of his model according to their apparent conceptual principles. Harshman's PARAFAC, for example, as a specialization of Tucker's model, shares with it the property of mathematical symmetry between measurements and conditions. But if we view it as a specialization of the invariant factors model, we impose a conceptual asymmetry between the factor-loading matrix of the measures and the factor-score transformation matrix of the conditions. Mathematical form is not a sure guide to substantive interpretation in a psychometric model.

There are at least two directions in which the invariant factors model can be extended in further research. The first, which is quite obvious, is to allow the cross-covariance matrices Φ_{kl} in Φ^* in (7–26) to be free from the constraint (7–17) or possibly follow a more general law. The second is to do what has been deliberately avoided in the basic model presented here and impose some structure on the residual covariance matrix beyond the requirement that residuals of distinct tests be uncorrelated within and between conditions. It is convenient to imagine the variables permuted to the alternative natural order in which we have q subvectors of order $p \times 1$, corresponding successively to the score on each test as recorded in each of the p conditions. The residual covariance matrix is then diagonal block. It might be reasonable in some cases to apply a common factor model to account for the residual covariances of the measures from each test between the p conditions, generally with factor loadings and numbers of factors varying from test to test. It might even be possible to interpret the numbers so obtained. However, the basic principles from which the invariant factors model was developed suggest that the structure of the relations between repetitions of a measure on a test under different conditions should be determined by other well-known principles governing univariate repeated-measures designs, rather than by the mechanical application of common factor analysis to the q covariance matrices. If, for example, the p conditions correspond to a systematic multifactorial design, this can be reflected easily in the structure of the mean vector, and we should also be able to model corresponding systematic changes in the factor-score covariances. If, in particular, we have longitudinal data in which the p conditions constitute a series of occasions in time, we would naturally introduce a trend line or growth curve governing changes in mean factor scores and a time series model of the well-known Box-Jenkins ARIMA- (Autoregressive Integrated Moving Average) type for the residual covariances. Program COSAN is readily adapted to many such modifications of the model, since it is designed to allow the user to introduce prescribed, possibly nonlinear functions, such as growth curves, by supplementary programming (see McDonald 1980).

REFERENCES

Bentler, P. M., and S-Y Lee. 1979. A statistical development of three-mode factor analysis. *British Journal of Mathematical and Statistical Psychology* 32:87–104.

Bloxom, B. 1968. A note on invariance in three-mode factor analysis. *Psychometrika* 33:347–50.

Graybill, F. A. 1969. *Matrix algebra with statistical applications.* Belmont, Calif.: Wadsworth.

Harris, C. W. 1963. Canonical factor models for the description of change. In *Problems in measuring change,* ed. C. W. Harris. Madison, Wis.: University of Wisconsin Press.

Harshman, R. A. 1970. Foundations of the PARAFAC procedure: Models and conditions for an "explanatory" multi-modal factor analysis. *UCLA Working Papers in Phonetics* 16:1–84 (University Microfilms No. 10,085).

Krane, W. R., and R. P. McDonald. 1978. Scale invariance and the factor analysis of correlation matrices. *British Journal of Mathematical and Statistical Psychology* 31:218–28.

Kruskal, J. B. 1981. Multilinear models for data analysis. *Behaviormetrika* 10:1–20.

Lord, F. M., and M. R. Novick. 1968. *Statistical theories of mental test scores.* Reading, Mass.: Addison-Wesley.

McDonald, R. P. 1969. A generalized common factor analysis based on residual covariance matrices of prescribed structure. *British Journal of Mathematical and Statistical Psychology* 22:149–63.

———. 1978. A simple comprehensive model for the analysis of covariance structures. *British Journal of Mathematical and Statistical Psychology* 31:59–72.

———. 1980. A simple comprehensive model for the analysis of covariance structures: Some remarks on applications. *British Journal of Mathematical and Statistical Psychology* 33:161–83.

———. 1981. The dimensionality of tests and items. *British Journal of Mathematical and Statistical Psychology* 34:100–17.

———. 1982. Linear versus nonlinear models in item response theory. *Applied Psychological Measurement* 6:379–96.

McDonald, R. P., and W. R. Krane. 1977. A note on local identifiability and degrees of freedom in the asymptotic likelihood ratio test. *British Journal of Mathematical and Statistical Psychology* 30:198–203.

———. 1979. A Monte Carlo study of local identifiability and degrees of freedom in the asymptotic likelihood ratio test. *British Journal of Mathematical and Statistical Psychology* 32:121–32.

McDonald, R. P., and H. Swaminathan. 1973. A simple matrix calculus with applications to multivariate analysis. *General Systems* 18:37–54.

Tucker, L. R. 1966. Experiments in multimode factor analysis. In *Testing problems in perspective,* ed. A. Anastasi. Washington, D.C.: American Council on Education.

8

Factor Analysis of Longitudinal Data

Hariharan Swaminathan

The study of change and the procedures that enable an investigator to assess change are not only of fundamental importance to psychologists but are also of great importance to the educational researcher who is concerned with the evaluation of instructional programs. In the evaluation of such programs, the researcher is primarily interested in establishing the effectiveness or the ineffectiveness of the program, which is most readily accomplished by studying the progress of the students in the program and the change the program has effected on them.

It is often not clear what the researcher means when he asks the question: Did a change occur? Traditionally, *change* has been interpreted as the change in *elevation,* namely, the change in the mean score of a group of students. The problems connected with the assessment of change in elevation are well-documented; thus, it is not necessary to pursue the subject here.

The question dealing with changes in the *pattern* or *profile* of a group of students is less frequently asked. Nevertheless, this appears to be a more interesting and important question. Suppose that we administer a battery of tests to a group of students being introduced to a particular instructional program; then, at the conclusion of the program, we repeat the tests. Questions dealing with changes in pattern or profile would then be represented by a question of whether or not the abilities or the achievements represented in the battery tend to remain invariant, tend to become integrated, or tend to become more specific (Harris 1963).

The study of change in the pattern or profile would involve, in the language of factor analysis, isolating factors that are stable over time, studying correlates of the factors at the various times, determining the changes over time in factors of interest, and observing the effects of the intervening experimental treatments on these factors. Hence, it is clear that there is a need for methods of factor-analyzing data obtained in longitudinal studies.

A number of factor-analytic techniques have been made available recently for the analysis of data obtained in longitudinal studies. One of the first of these, the *P*-technique, was given by Cattell (Cattell 1952, 1963). The *P*-technique, in essence, involves obtaining repeated measurements of one person on a set of psychological variables, correlating them, and factor-analyzing the resulting correlation or dispersion matrix. Although this technique is potentially useful for studying change, it has several drawbacks. In order to factor-analyze the data, it has to be assumed that the unique scores of the individual on two or more occasions are uncorrelated, an assumption that has been questioned by Anderson (1963). (A complete discussion of the statistical problems involved is given by Anderson [1963] and, in reply, by Cattell [1963].)

The other techniques that are available for factor-analyzing longitudinal data can be thought of arising from two distinct statistical models, one of which assumes that factor scores remain invariant over occasions (Harris 1963; Rao 1965) and the other which holds the factor pattern constant over time (Tucker 1963).

The assumption that factor scores remain invariant seems to be highly restrictive. Corballis and Traub (1970) have argued that it is unreasonable to assume that factor scores do not change while the test scores change. In addition, the researcher who is interested in studying how the constructs or factors (and hence the factor scores) change over time will find his other interests in conflict with the assumption of constant factor scores.

The assumption of an invariant factor pattern, although more realistic, is again restrictive. Changes in factor pattern would indicate how the factors change and whether or not the tests are measuring different abilities on different occasions. In fact, more often than not, the researcher may want to *establish* that the tests are measuring the same factors, and in order to do this, it is necessary that the factor pattern be allowed to vary.

The model developed by Corballis and Traub (1970) was the result of an attempt to solve the above problems. The important feature of this model, particularly in comparison to its predecessors (with the exception of the model of Cattell), is that both factor scores and factor pattern are permitted to vary from occasion to occasion. However, like the models developed by Harris (1963) and Rao (1965), it is restricted to two occasions and is therefore unsuitable when there are more than two. Corballis (1973) solved this problem by providing a multioccasion extension of the Corballis-Traub model. However, the relationship postulated among the factor-score vectors over occasions was ignored in the estimation phase. In addition, the two-stage least-squares procedure adopted for the estimation of the parameters does not necessarily yield the true least-squares estimators. Furthermore, like their predecessors, these researchers have not provided any statistical procedures for either evaluating the goodness of fit of the model or for testing the various hypotheses of interest.

Jöreskog (1970a) has outlined a general procedure for the analysis of multitest, multioccasion data. Although this procedure is potentially the most useful for factor-analyzing longitudinal data, it has a major drawback: The unique scores are assumed to be uncorrelated over occasions. The works of Jöreskog (1979), Jöreskog and Sörbom (1977), and Roskam (1976) address the notions of changing factor scores, factor pattern, and unique scores through a variety of approaches without considering these issues simultaneously. It should be pointed out, however, that the models provided by these authors can be adapted to address these issues. A major drawback is that the hypotheses of interest in longitudinal analysis have not been discussed comprehensively by these authors.

The aim of this chapter is to: (a) formulate a factor model for k occasions; (b) provide a statistically efficient procedure, at least for large samples, for the estimation of parameters; (c) develop tests of significance for testing various hypotheses; and (d) extend the notion of factorial invariance to the multioccasion situation.

THE MODEL

Suppose n tests are administered to the same N individuals on k separate occasions. We shall assume that the tests have the same p common factors on each occasion. Within each occasion, we assume that

$$\mathbf{y}_i = \mathbf{F}_i \mathbf{x}_i + \mathbf{e}_i \ , \qquad i = 1, \ldots, k \ , \tag{8–1}$$

where

$\mathbf{y}_i$ is the $(n \times 1)$ random vector of observed scores for an individual on the ith occasion,

$\mathbf{F}_i$ is the $(n \times p)$ matrix of factor loadings,

$\mathbf{x}_i$ is the $(p \times 1)$ random vector of factor scores, and

$\mathbf{e}_i$ is the $(n \times 1)$ random vector of unique scores.

Without loss of generality, we can assume that

$$E(\mathbf{y}_i) = \mathbf{0} \ , \tag{8–2}$$

$$E(\mathbf{e}_i) = \mathbf{0} \ , \tag{8–3}$$

and

$$E(\mathbf{x}_i) = \mathbf{0} \ , \tag{8–4}$$

where $E(\ )$ is the expectation operator.

We also require that for $i, j = 1, 2, \ldots, k$,

$$E(\mathbf{x}_i \mathbf{x}_i') = \mathbf{I} \ , \tag{8–5}$$

$$E(\mathbf{x}_i \mathbf{e}_i') = \mathbf{0} \ , \tag{8–6}$$

$$E(\mathbf{x}_i \mathbf{e}_j') = \mathbf{0} \ , \tag{8–7}$$

and

$$E(\mathbf{e}_i\ \mathbf{e}_j') = \mathbf{U}_{ij} = \mathbf{U}_{ji}\ , \quad \text{diagonal.} \tag{8–8}$$

Equation (8–5) is simply the assumption that the factors are uncorrelated within each occasion. Conditions (8–6) and (8–7) indicate that the factor scores are uncorrelated with the specific scores, within as well as between occasions. It is assumed that within occasions, the unique scores are uncorrelated, and that between occasions, unique scores on the same test are correlated but unique scores on different tests are uncorrelated. Hence, the requirement that $\mathbf{U}_{ij}$ (= $\mathbf{U}_{ji}$) be a diagonal matrix.

In order to formulate a model for changing factor scores, we consider the regression of $\mathbf{x}_i$, the factor scores on the ith occasion, on the factor scores $\mathbf{x}_{i-1}, \mathbf{x}_{i-2}, \ldots, \mathbf{x}_1$. We will assume that the factor scores $\mathbf{x}_i$ can be generated by a first-order autoregressive series:

$$\mathbf{x}_i = \mathbf{D}_i\ \mathbf{x}_{i-1} + \delta_i\ , \qquad (i = 2, 3, \ldots, k)\ , \tag{8–9}$$

that is, that the factors satisfy Guttman's Simplex model (Anderson 1960). We assume that $\mathbf{D}_i$ is a $(p \times p)$ diagonal matrix of constants. The change scores, given by δ_i, are assumed to satisfy the following assumptions:

$$E(\delta_i) = \mathbf{0}\ , \tag{8–10}$$

$$E(\delta_i\ \mathbf{x}_{i-1}') = \mathbf{0}\ , \tag{8–11}$$

and

$$E(\delta_i\ \delta_j') = \mathbf{0}\ , \qquad (i \neq j)\ . \tag{8–12}$$

Assumption (8–11) implies that the change scores δ_i are uncorrelated with the initial factor scores $\mathbf{x}_{i-1}$. Assumption (8–12) states that the change scores on any two different occasions are uncorrelated. It immediately follows that, if the elements of $\mathbf{D}_i$ are nonzero, then

$$E(\mathbf{x}_j\ \delta_i') = \mathbf{0}\ , \qquad (j = 1, 2, \ldots, i\text{-}1)\ . \tag{8–13}$$

We shall define

$$E(\delta_i\ \delta_i') = \Psi_i^2\ , \qquad (i = 2, \ldots, k) \tag{8–14}$$

and require that Ψ_i^2, the $(p \times p)$ dispersion matrix of the change scores within each occasion, be a diagonal matrix.

It immediately follows that

$$\Psi_i^2 + \mathbf{D}_i^2 = \mathbf{I}\ , \qquad (i = 2, \ldots, k), \tag{8–15}$$

where $\mathbf{I}$ is the $(n \times n)$ identity matrix, and

$$E(\mathbf{x}_j\ \mathbf{x}_i') = \mathbf{D}_{j+1}\ \mathbf{D}_{j+2} \cdots \mathbf{D}_i\ , \quad (j = 1, \ldots, i\text{-}1). \tag{8–16}$$

We shall denote the collection of the observed scores, factor scores, and unique scores, respectively, by

$$\mathbf{y}' = [\mathbf{y}_1' \ \mathbf{y}_2' \ \ldots \ \mathbf{y}_k'] \ , \tag{8–17}$$

$$\mathbf{x}' = [\mathbf{x}_1' \ \mathbf{x}_2' \ \ldots \ \mathbf{x}_k'] \ , \tag{8–18}$$

and

$$\mathbf{e}' = [\mathbf{e}_1' \ \mathbf{e}_2' \ \ldots \ \mathbf{e}_k'] \ . \tag{8–19}$$

Rewriting (8–1) in terms of the collection of vectors (8–17), (8–18), and (8–19), we obtain

$$\begin{bmatrix} \mathbf{y}_1 \\ \mathbf{y}_2 \\ \cdot \\ \cdot \\ \cdot \\ \mathbf{y}_k \end{bmatrix} = \begin{bmatrix} \mathbf{F}_1 & \mathbf{0} & \mathbf{0} \ldots \mathbf{0} \\ \mathbf{0} & \mathbf{F}_2 & \ldots\ldots \mathbf{0} \\ \cdot & & \\ \cdot & & \\ \cdot & & \\ \mathbf{0} & \ldots\ldots & \ldots \mathbf{F}_k \end{bmatrix} \begin{bmatrix} \mathbf{x}_1 \\ \mathbf{x}_2 \\ \cdot \\ \cdot \\ \cdot \\ \mathbf{x}_k \end{bmatrix} + \begin{bmatrix} \mathbf{e}_1 \\ \mathbf{e}_2 \\ \cdot \\ \cdot \\ \cdot \\ \mathbf{e}_k \end{bmatrix} , \tag{8–20}$$

which can be written compactly as:

$$\mathbf{y} = \mathbf{F} \ \mathbf{x} + \mathbf{e} \ . \tag{8–21}$$

THE STRUCTURAL MODEL

It follows from (8–21), as well as the assumptions made, that the population dispersion matrix Σ of the observed scores has the structure

$$\Sigma = E(\mathbf{y} \ \mathbf{y}') = \mathbf{F} \ E(\mathbf{x} \ \mathbf{x}') \ \mathbf{F}' + E(\mathbf{e} \ \mathbf{e}') \ , \tag{8–22}$$

which can be written as:

$$\Sigma = \mathbf{F} \ \Phi \ \mathbf{F}' + \mathbf{U} \ . \tag{8–23}$$

The matrix Σ, the supermatrix of interoccasion dispersion matrices, is given by

$$\Sigma = \begin{bmatrix} \Sigma_{11} & \Sigma_{12} \ldots\ldots & \Sigma_{1k} \\ \cdot & & \\ \cdot & & \\ \cdot & & \\ \cdot & & \\ \cdot & & \\ \Sigma_{k1} \cdot \ldots\ldots & & \Sigma_{kk} \end{bmatrix} , \tag{8–24}$$

F, the interoccasion matrix of factor patterns is given by

$$\mathbf{F} = \begin{bmatrix} \mathbf{F}_1 & \mathbf{0} & \cdots & \mathbf{0} \\ \mathbf{0} & \mathbf{F}_2 & \cdots & \mathbf{0} \\ \vdots & & & \\ \mathbf{0} & \cdots & \cdots & \mathbf{F}_k \end{bmatrix} , \quad (8\text{–}25)$$

Φ, the interoccasion matrix of the factor score correlations, is given by

$$\Phi = \begin{bmatrix} \Phi_{11} & \Phi_{12} & \cdots & \Phi_{1k} \\ \vdots & & & \\ \Phi_{k1} & \cdots & \cdots & \Phi_{kk} \end{bmatrix} , \quad (8\text{–}26)$$

and finally, **U**, the dispersion matrix of unique scores, is given by

$$\mathbf{U} = \begin{bmatrix} \mathbf{U}_{11} & \mathbf{U}_{12} & \cdots & \mathbf{U}_{1k} \\ \vdots & & & \\ \mathbf{U}_{k1} & \cdots & \cdots & \mathbf{U}_{kk} \end{bmatrix} . \quad (8\text{–}27)$$

It should be noted that Σ, Φ, and **U** are partitioned or supermatrices. Thus, the submatrix element Σ_{ij} of Σ is $(n \times n)$, $\mathbf{F}_i$ of **F** is a $(n \times p)$ matrix, Φ_{ij} of Φ is a diagonal matrix of order $(p \times p)$ with $\Phi_{ij} = \Phi_{ji}$, and $\mathbf{U}_{ij}$ of **U** is a diagonal matrix of order $(n \times n)$ with $\mathbf{U}_{ij} = \mathbf{U}_{ji}$.

It follows from (8–5) and (8–16) that

$$\Phi_{ii} \equiv E(\mathbf{x}_i \, \mathbf{x}_i') = \mathbf{I} , \quad (8\text{–}28)$$

and, if $j < i$,

$$\Phi_{ji} \equiv E(\mathbf{x}_j \, \mathbf{x}_i') = \mathbf{D}_{j+1} \, \mathbf{D}_{j+2} \cdots \mathbf{D}_i , \quad (8\text{–}29)$$
$$= \Phi_{ij} .$$

Hence,

$$\Phi = \begin{bmatrix} \mathbf{I} & \mathbf{D}_2 & \mathbf{D}_2\mathbf{D}_3 & . & \mathbf{D}_2\mathbf{D}_3\ldots\mathbf{D}_i & . & \mathbf{D}_2\mathbf{D}_3\ldots\mathbf{D}_k \\ . & \mathbf{I} & \mathbf{D}_3 & . & \mathbf{D}_3\ldots\mathbf{D}_i & . & \mathbf{D}_3\ldots\mathbf{D}_k \\ . & & & & & & \\ . & . & . & . & \mathbf{I} & . & \mathbf{D}_{i+1}\ldots\mathbf{D}_k \\ . & & & & & & \\ . & . & . & . & . & . & \mathbf{I} \end{bmatrix} . \qquad (8\text{–}30)$$

Since Φ has this special structure, it is not possible to treat (8–23) as the conventional factor model. In order to estimate the parameters, the matrix Φ has to be transformed to a more amenable form. It is easily verified that

$$\Phi = \Delta\ \mathbf{T}\ \Delta^{-1}\ \Psi\ \Delta^{-1}\ \mathbf{T}'\ \Delta\ , \qquad (8\text{–}31)$$

where

$$\Delta = \begin{bmatrix} \mathbf{I} & \mathbf{0} & . & . & . & . & \mathbf{0} \\ \mathbf{0} & \mathbf{D}_2 & . & . & . & . & \mathbf{0} \\ \mathbf{0} & \mathbf{0} & \mathbf{D}_2\mathbf{D}_3 & . & . & . & \mathbf{0} \\ . & & & & & & \\ \mathbf{0} & . & . & . & . & . & \mathbf{D}_2\mathbf{D}_3\ldots\mathbf{D}_k \end{bmatrix} , \qquad (8\text{–}32)$$

$$\mathbf{T} = \begin{bmatrix} \mathbf{I} & \mathbf{0} & . & . & . & . & \mathbf{0} \\ \mathbf{I} & \mathbf{I} & \mathbf{0} & . & . & . & \mathbf{0} \\ \mathbf{I} & \mathbf{I} & \mathbf{I} & \mathbf{0} & . & . & \mathbf{0} \\ . & & & & & & \\ \mathbf{I} & \mathbf{I} & \mathbf{I} & . & . & . & \mathbf{I} \end{bmatrix} , \qquad (8\text{–}33)$$

and

$$\Psi = \begin{bmatrix} \mathbf{I} & \mathbf{0} & . & . & . & . & \mathbf{0} \\ \mathbf{0} & \Psi_2^2 & \mathbf{0} & . & . & . & \mathbf{0} \\ . & & & & & & \\ . & & & & & & \\ \mathbf{0} & . & . & . & . & . & \Psi_k^2 \end{bmatrix} , \qquad (8\text{–}34)$$

with

$$\mathbf{I} - \mathbf{D}_k^2 = \Psi_k^2 . \qquad (8\text{–}35)$$

Thus, the structural model (8–21) takes the form:

$$\Sigma = \mathbf{F} \, \Delta \, \mathbf{T} \, \Delta^{-1} \, \Psi \, \Delta^{-1} \, \mathbf{T}' \, \Delta \, \mathbf{F}' + \mathbf{U} . \qquad (8\text{–}36)$$

The parameter matrices to be estimated are $\mathbf{F}$, Δ, and $\mathbf{U}$. The matrix $\mathbf{T}$ is a matrix of constants. Since Ψ_k is a function of $\mathbf{D}_k$, the matrix Ψ should not be estimated separately. Overall, the number of parameters to be estimated is: npk from the k factor-pattern matrices, $(k - 1)p$ from the matrix Δ, and the $n. \frac{1}{2}k(k + 1)$ uniqueness parameters.

In the specification of the model, Corballis (1973) treated submatrices of (8–30) as independent, an assumption that is clearly not warranted. Jöreskog (1970a) and Jöreskog and Sörbom (1977) ignored the interdependence of Ψ and Δ in (8–31) and, in the sequel, the structure of Σ given by (8–36). Since these assumptions have implications for the proper estimation of parameters, we shall take into account the special nature of the structure of Σ in the estimation and hypothesis testing phases.

STRUCTURAL PROPERTIES OF THE MODEL

The most striking feature of the structural model (8–23) and hence (8–36) is that once a solution matrix $\mathbf{F}$ is obtained, it cannot be rotated further to yield another acceptable solution. This is because it is impossible to find a matrix $\mathbf{L}_i$ such that

$$\mathbf{L}_i \, \mathbf{L}_i' = \mathbf{I}$$

and also such that

$$\mathbf{L}_i \, \Phi_{ii} \, \mathbf{L}_i'$$

is diagonal.

Although statistically the above feature is desirable, some authors have indicated that this may lead to uninterpretable solutions (Corballis and Traub 1970). Apparently, the only way this problem can be overcome is by formulating a model that permits oblique solutions.

The absence of the rotation problem does not imply that all the parameters are uniquely specified. Returning to the structural equations given in (8–36), we see that the total number of equations is $\frac{1}{2}kn\,(kn + 1)$—this is the number of independent elements in Σ. The number of parameters to be estimated is

$$knp + (k - 1)p + \tfrac{1}{2}nk\,(k + 1)\ .$$

Hence, the number of equations minus the number of unknowns is

$$\nu = \tfrac{1}{2}kn\,(kn + 1) - knp - (k - 1)p - \tfrac{1}{2}nk\,(k + 1)\ . \qquad (8\text{–}37)$$

If $\nu \leq 0$, an algebraic solution possibly exists. However, if $\nu > 0$, an algebraic solution is not possible in general; this seems to be a minimal requirement for the solution to be identified. This is made clear by the fact that ν represents the degrees of freedom for the likelihood ratio test for testing the hypothesis that the dispersion matrix Σ has the structure given by (8–36).

ESTIMATION OF PARAMETERS

Since maximum likelihood estimates have desirable properties (at least asymptotically) and since maximum likelihood estimates are scale invariant, we will choose to estimate the parameters of (8–36) by the method of maximum likelihood.

We will assume that the random vector of observations $\mathbf{y}$ are independently and normally distributed with dispersion matrix Σ; that is,

$$\mathbf{y} \sim N\,(\mathbf{0}, \Sigma)\ , \qquad (8\text{–}38)$$

where $N(\mu, \Sigma)$ is the multivariate normal density function with mean vector μ and dispersion matrix Σ.

The maximum likelihood estimates of $\mathbf{F}$, Δ, and $\mathbf{U}$ are those values of the parameters that maximize the likelihood function of the observations $\mathbf{y}$. The logarithm of the likelihood function is given by

$$\log L = -\tfrac{1}{2}Nnk \log 2\pi - \tfrac{1}{2}N \log |\Sigma| - \tfrac{1}{2}N\ \mathrm{Tr}\ (\mathbf{S}\,\Sigma^{-1}). \qquad (8\text{–}39)$$

Maximizing (8–39) is equivalent to minimizing the following variant of the likelihood function (Jöreskog 1970b):

$$f = \log |\Sigma| + \mathrm{Tr}\ (\mathbf{S}\,\Sigma^{-1}) - \log |\mathbf{S}| - nk\ . \qquad (8\text{–}40)$$

Here $\mathbf{S}$ is the sample dispersion matrix given by

$$\mathbf{S} = \frac{1}{N} \sum_{i=1}^{N} (\mathbf{y}_i - \bar{\mathbf{y}})\,(\mathbf{y}_i - \bar{\mathbf{y}})'\ .$$

It is well known that $(N/N - 1)\mathbf{S}$ is the unbiased estimate of Σ.

The maximum likelihood estimates are thus to be found among the solutions of the system of likelihood equations:

$$\partial f/\partial \mathbf{F} = \mathbf{0} \ ,$$

$$\partial f/\partial \Delta = \mathbf{0} \ , \tag{8–41}$$

and

$$\partial f/\partial \mathbf{U} = \mathbf{0}$$

Since $\mathbf{F}$ and $\mathbf{U}$ are block matrices, it is advantageous to obtain the derivatives of the likelihood function with respect to only these diagonal blocks. In the sequel, we shall denote the vector of the diagonal elements of a matrix $\mathbf{Z}$ as $\mathbf{Z}_{\text{diag}}$, where

$$\mathbf{Z}'_{\text{diag}} = \{z_{11} \ z_{22} \ . \ . \ . \ z_{nn}\} \ .$$

The derivatives in (8–41) can be obtained in a number of ways. See, for example, McDonald and Swaminathan (1973) and Swaminathan (1976). It can be shown that

$$\partial f/\partial \mathbf{F}_{ii} = 2\{(\mathbf{W}\ \mathbf{F}\ \Phi)_{ii}\} \tag{8–42}$$

and

$$\partial f/\partial \mathbf{U}_{ij\,\text{diag}} = k\,\mathbf{W}_{ij\,\text{diag}} \ , \tag{8–43}$$

where

$$k = \begin{cases} 1 \text{ if } i = j \\ 2 \text{ if } i \neq j \end{cases}$$

and

$$\mathbf{W} = \Sigma^{-1} - \Sigma^{-1}\mathbf{S}\,\Sigma^{-1} \ . \tag{8–44}$$

The derivatives with respect to the elements, $\mathbf{D}_2, \ . \ . \ . \ , \mathbf{D}_k$ are more complicated to obtain in view of the functional dependencies that exist in the matrices Δ and Ψ. From the chain rule and the McDonald-Swaminathan system for matrix derivatives, it follows that

$$\partial f/\partial \mathbf{D} = (\partial \Delta/\partial \mathbf{D})(\partial f/\partial \Delta) + (\partial \Psi/\partial \mathbf{D})(\partial f/\partial \Psi) \ . \tag{8–45}$$

Now

$$\partial f/\partial \Delta = \text{vec}\ [\mathbf{F}'\mathbf{W}\mathbf{F}\Phi\Delta^{-1} + (\mathbf{F}'\mathbf{W}\mathbf{F}\Phi\Delta^{-1})' \tag{8–46}$$

$$- \Delta^{-1}\mathbf{T}'\Delta\mathbf{F}'\mathbf{W}\mathbf{F}\Delta\mathbf{T}\Delta^{-1}\Psi\Delta^{-1} - (\Delta^{-1}\mathbf{T}'\Delta\mathbf{F}'\mathbf{W}\mathbf{F}\Delta\mathbf{T}\Delta^{-1}\Psi\Delta^{-1})']$$

and

$$\partial f/\partial \Psi = \text{vec}\ [\Delta^{-1}\mathbf{T}'\Delta\mathbf{F}'\mathbf{W}\mathbf{F}\Delta\mathbf{T}\Delta^{-1}] \ . \tag{8–47}$$

Here, vec $\mathbf{A}$ is a column vector made up of the rows of the matrix $\mathbf{A}$ (McDonald and Swaminathan 1973). In evaluating the derivative $\partial f/\partial \mathbf{D}$, we let

$$\mathbf{D} = \begin{bmatrix} \mathbf{D}_2 & \mathbf{0} & \mathbf{0} & \mathbf{0} \\ \mathbf{0} & \mathbf{D}_3 & \mathbf{0} & \mathbf{0} \\ \mathbf{0} & & & \mathbf{D}_k \end{bmatrix}, \qquad (8\text{–}48)$$

and

$$\partial f/\partial \Delta = \text{vec}\,\mathbf{Q} .$$

In terms of these matrices, it can be shown, after some reduction, that

$$(\partial \Delta/\partial \mathbf{D}_2)\ (\partial f/\partial \Delta) = \mathbf{Q}_{22} + \ldots + \mathbf{Q}_{kk}\mathbf{D}_k\mathbf{D}_{k-1}\mathbf{D}_3,$$

$$(\partial \Delta/\partial \mathbf{D}_3)\ (\partial f/\partial \Delta) = \mathbf{D}_2\mathbf{Q}_{33} + \ldots + \mathbf{D}_2\mathbf{Q}_{kk}\mathbf{D}_k\mathbf{D}_{k-1} \ldots \mathbf{D}_4,$$

or, in general,

$$(\partial \Delta/\partial \mathbf{D}_{i\ \text{diag}})\ (\partial f/\partial \Delta) = [\sum_{j=1}^{k} \mathbf{D}_2\mathbf{D}_3 \ldots \mathbf{D}_{i-1}\mathbf{Q}_{jj}\mathbf{D}_j\mathbf{D}_{j-1} \qquad (8\text{–}49)$$

$$\ldots \mathbf{D}_{i+1}]_{\text{diag}} .$$

Here, the $(p \times p)$ matrix $\mathbf{Q}_{ii}$ is the ith diagonal block of the matrix $\mathbf{Q}$.

Similarly, it can be shown that

$$(\partial \Psi/\partial \mathbf{D}_{i\ \text{diag}})\ (\partial f/\partial \Psi) = -2\{\mathbf{D}_i\ (\Delta^{-1}\mathbf{T}'\Delta\mathbf{F}'\mathbf{W}\mathbf{F}\Delta\mathbf{T}\Delta^{-1})_{ii}\}_{\text{diag}} . \qquad (8\text{–}50)$$

From these two equations, $\partial f/\partial \mathbf{D}_{i\,\text{diag}}$ can be readily calculated.

It does not seem that the solution of the likelihood equations given by (8–41) is available in a closed form and thus numerical procedures have to be employed. There are several algorithms available that are potentially useful for the solution of these equations, or equivalently, for minimizing the function (8–40). The most well known is the Fletcher-Powell algorithm made popular by Jöreskog (1970b). The algorithm requires the input of the derivatives given by (8–42) through (8–50).

TESTS OF HYPOTHESES

As mentioned earlier, maximum likelihood estimates enjoy certain properties. Asymptotically, maximum likelihood estimates are consistent, efficient, and sufficient. In addition, if

$$\hat{\theta}' = [\hat{\theta}_1, \hat{\theta}_2 \ldots \hat{\theta}_t]$$

is the maximum likelihood estimate of

$$\theta_o = [\theta_{1o}, \theta_{2o} \ldots \theta_{to}] ,$$

it is known that, asymptotically,

$$\hat{\theta} \sim N(\theta_o, (2/N)\mathcal{I}^{-1}) ,$$

that is, θ is asymptotically multinormal with mean vector θ_o and dispersion matrix $(2/N)\mathcal{I}^{-1}$. The information matrix, $\mathcal{I}$, is given by

$$\mathcal{I} = E\{\mathbf{H}(\mathbf{F}, \Delta, \mathbf{U})\} , \qquad (8\text{–}51)$$

$$= \mathbf{E}\begin{bmatrix} \partial^2 f/\partial \mathbf{F}^2 & & \\ \partial^2 f/\partial \mathbf{D}\partial \mathbf{F} & \partial^2 f/\partial \mathbf{D}^2 & \\ \partial^2 f/\partial \mathbf{U}\partial \mathbf{F} & \partial^2 f/\partial \mathbf{U}\partial \mathbf{D} & \partial^2 f/\partial \mathbf{U}^2 \end{bmatrix} ,$$

where **H**, the Hessian, is the matrix of second derivatives of the likelihood function with respect to **F**, Δ, and **U**. Its elements are easily obtained when the form of the Hessian is known and the actual information matrix to be used in constructing confidence intervals is obtained by evaluating the information matrix at the solution point of the likelihood equations. However, as a result of the complexity of the expression, they are not given here.

If the convergence to normality is uniform in compact intervals of θ_o, then confidence intervals for θ_{io} can be constructed readily. For any parameter θ_{io} with MLE $\hat{\theta}_i$ and variance estimate $(2/N)\mathcal{I}_{ii}^{-1}$, the $(1 - 2\alpha)$th percentile confidence interval is

$$\hat{\theta}_i - Z_\alpha \sqrt{\{(2/N)\ \mathcal{I}_{ii}^{-1}\}} \leq \theta_i \leq \hat{\theta}_i + Z_\alpha \sqrt{\{(2/N)\ \mathcal{I}_{ii}^{-1}\}} . \qquad (8\text{–}52)$$

In this way, it is possible to test the hypothesis that a specified element in one or more parameter matrices has a given value.

Testing joint hypotheses about the parameters or any combinations of them is also possible. For instance, if the hypothesis of interest is

$$H_o : \mathbf{C}\theta = \mathbf{C}\theta^*$$

against

$$H_1 : \mathbf{C}\theta \neq \mathbf{C}\theta^* ,$$

where C is a matrix of constants, the appropriate statistic is

$$(\hat{\theta} - \theta^*)'\mathbf{C}'(\mathbf{C}\ \mathcal{I}^{-1}\mathbf{C}')^{-1}\mathbf{C}(\hat{\theta} - \theta^*) . \qquad (8\text{–}53)$$

It is well known that this quantity has a χ^2 distribution with r degrees of freedom, where r is the rank of **C**.

Instead of testing hypotheses regarding specific parameters, it may often be necessary to test hypotheses concerning the parametric structure of the model against an alternate hypothesis. Let H_o be the hypothesis concerning the parametric structure of H_o, and let H_1 be any alternate hypothesis, that is,

$$H_o : \Sigma \in \omega \tag{8–54}$$

versus

$$H_1 : \Sigma \in \Omega , \tag{8–55}$$

where $\omega \subseteq \Omega$.

Let L_ω be the maximum of log L defined in (5–2) under H_o and L_Ω the maximum of log L under H_1. Then, obviously,

$$L_\omega \leq L_\Omega ,$$

since the maximum of a function in a restricted space, ω in this case, cannot be greater than the maximum of the function over the entire space, Ω in this case. Thus, the likelihood ratio criterion

$$\log \lambda = L_\omega - L_\Omega \tag{8–56}$$

is negative; that is, $o < \lambda \leq 1$.

It has been proven that, under H_o, $-2 \log \lambda$ is asymptotically distributed as χ^2, with degrees of freedom equal to the difference in the number of parameters estimated under H_1 and H_o.

In general, this requires the computation of the solution both under H_o and H_1. However, for most useful alternatives H_1, the solution is known and the value of L_Ω can be computed quite readily. We shall consider one such alternative.

Suppose that

$$H_o : \Sigma = \mathbf{F} \, \Phi \, \mathbf{F}' + \mathbf{U} \tag{8–57}$$

against the alternative

$$H_1 : \Sigma \text{ is any positive definite matrix.} \tag{8–58}$$

This can be regarded as an omnibus test that determines whether or not the dispersion matrix has the hypothesized structure given by (8–36). The maximum L_Ω under H_1 is obtained when

$$\Sigma = \mathbf{S} ,$$

where $\mathbf{S}$ is the sample dispersion matrix. Hence,

$$L_\Omega = -(N/2) \{\log |\mathbf{S}| + nk\} + \text{constant} , \tag{8–59}$$

and

$$L_{\omega} = -(N/2)\ \{\log |\hat{\Sigma}| + \mathrm{Tr}\ (\mathbf{S}\ \hat{\Sigma}^{-1})\} + \text{constant}\ . \qquad (8\text{–}60)$$

Here, $\hat{\Sigma}$ denotes the estimate of Σ under H_0 obtained by substituting the estimated values of the parameter matrices, $\hat{\mathbf{F}}$, $\hat{\Delta}$, $\hat{\Psi}$, and $\hat{\mathbf{U}}$ in (8–36); that is,

$$\hat{\Sigma} = \hat{\mathbf{F}}\ \hat{\Phi}\ \hat{\mathbf{F}}' + \hat{\mathbf{U}}\ . \qquad (8\text{–}61)$$

Hence,

$$-2 \log \lambda = 2\ (L_{\Omega} - L_{\omega}) \qquad (8\text{–}62)$$

$$= N\ \{\log |\hat{\Sigma}| + \mathrm{Tr}\ (\mathbf{S}\ \hat{\Sigma}^{-1}) - \log |\mathbf{S}| - nk\}\ , \qquad (8\text{–}63)$$

which simplifies to

$$-2 \log \lambda = \{N\ (\text{minimum of } f)\}\ . \qquad (8\text{–}64)$$

If N is large, the criterion $-2 \log \lambda$ has the χ^2 distribution with degrees of freedom ν, where ν is given by (8–37):

$$\nu = \tfrac{1}{2}\ kn\ (kn + 1) - knp - (k - 1)p - \tfrac{1}{2}\ nk\ (k + 1)\ .$$

Other hypotheses of interest may be tested similarly. We will discuss several of the more important hypotheses in the next section.

FACTORIAL INVARIANCE

The notion of factorial invariance arises naturally when one is concerned with the study of similarities and dissimilarities in the factorial composition between different groups. Several procedures are available in such situations and are well documented by Evans (1971). Jöreskog (1971) has outlined a statistical solution to this problem and, in one sense, this work can be considered as the most significant contribution in this area.

The situation is rather different when a group of students are given a battery of tests at different times. The works of Evans (1967), Tucker (1958, 1963), and McDonald (1969, 1970) are potentially useful for the study of factorial invariance over time. In this section, we will attempt to provide a statistical procedure for the study of factorial invariance. Following Jöreskog (1971), we will examine the problem of factorial invariance in the framework of a sequence of tests of hypotheses.

We begin by testing the invariance of the dispersion matrices over the occasions:

$$H_\Sigma : \Sigma_{11} = \Sigma_{22} = \ldots = \Sigma_{kk} = \Sigma_o \; . \tag{8–65}$$

This hypothesis closely resembles the well-known hypothesis of equal dispersion found in work by Anderson (1958). Unfortunately, the present situation is not that straightforward. Since the same groups of individuals are involved, we cannot assume that the observations are independent.

Hypothesis (8–65) implies the equality of the diagonal submatrices in the supermatrix, Σ, and hence that Σ has the form

$$\Sigma = \begin{bmatrix} \Sigma_o & \Sigma_{12} & \cdots & \Sigma_{1k} \\ \vdots & & & \\ \Sigma_{21} & \Sigma_o & \cdots & \Sigma_{2k} \\ \vdots & & & \\ \Sigma_{k1} & \cdots & \cdots & \Sigma_o \end{bmatrix} .$$

The test statistic for testing (7–1) can be shown to be (McDonald 1974),

$$-2 \log \lambda = N \{\log |\hat{\Sigma}| + \mathrm{Tr}\,(\mathbf{S}\,\Sigma^{-1}) - \log |\mathbf{S}| - nk\} \; , \tag{8–66}$$

where $\hat{\Sigma}$ is the solution of the system

$$\partial f / \partial \Sigma = \mathbf{0} \; . \tag{8–67}$$

It can be shown that

$$\partial f / \partial \Sigma_o = \sum_{i=1}^{k} \{2\mathbf{W}_{ii} - \mathrm{Diag}\,\mathbf{W}_{ii}\} \tag{8–68}$$

and

$$\partial f / \partial \Sigma_{ij} = 2\mathbf{W}_{ij} \; , \tag{8–69}$$

where

$$\mathbf{W} = \Sigma^{-1} - \Sigma^{-1}\mathbf{S}\Sigma^{-1} \; .$$

If the null hypothesis is true, $-2 \log \lambda$ is distributed as χ^2 with degrees of freedom, d_Σ, given by:

$$d = \tfrac{1}{2}(k - 1)\, n(n + 1) \; . \tag{8–70}$$

If (7–1) is not rejected and if the structure is identified, we can then conclude that $\mathbf{F}_1 = \mathbf{F}_2 = \ldots = \mathbf{F}_k$, $\mathbf{D}_2 = \ldots = \mathbf{D}_k = \mathbf{I}$, and $\mathbf{U}_{11} = \mathbf{U}_{22} = \ldots = \mathbf{U}_{kk}$. Hence, the characteristics common to the group over the occasions can be obtained from the pooled dispersion matrix. However, this does not imply that a complete factoral invariance is established. For example, this does not establish the invariance of *factor scores* but that of their *joint distribution*. When (8–65) is rejected, the following hypotheses may be tested.

In order to establish the invariance of parameters, we first assume that the factor model is valid and begin by testing the hypothesis of *invariant factor pattern;* namely, we test the hypothesis

$$H_F : \mathbf{F}_1 = \mathbf{F}_2 = \ldots = \mathbf{F}_k = \mathbf{F}_o \tag{8–71}$$

against the alternative,

$$H_1 : \mathbf{F}_i \neq \mathbf{F}_j \qquad \text{for some } i,\ j.$$

In order to test the hypothesis, we obtain the minimum value of the function

$$f(\mathbf{F}) = \log |\Sigma| + \mathrm{Tr}\ (\mathbf{S}\,\Sigma^{-1}) - \log |\mathbf{S}| - nk \tag{8–72}$$

under the null hypothesis and also under the alternate hypothesis. We shall denote the minimum value of the function under the null hypothesis by $f_{\min}(\mathbf{F})$ and the minimum value of the function under the alternate hypothesis by $f_{\min}$. Then, the statistic for testing the hypothesis of invariant factor pattern, $\chi^2(\mathbf{F})$, is given by:

$$\chi^2(\mathbf{F}) = N\{f_{\min}(\mathbf{F}) - f_{\min}\}\ . \tag{8–73}$$

The quantity $\chi^2(\mathbf{F})$ is asymptotically distributed as a χ^2 variable with degrees of freedom

$$\nu(\mathbf{F}) = (k - 1)np\ . \tag{8–74}$$

The minimum value of the function $f_{\min}$ under the alternate hypothesis is immediately obtained as indicated in the previous section. In order to determine the minimum value, $f_{\min}(\mathbf{F})$, of the function under the null hypothesis, we solve the likelihood equations

$$\partial f / \partial \mathbf{F} = \mathbf{0}\ ,$$

$$\partial f / \partial \mathbf{D} = \mathbf{0}\ ,$$

and

$$\partial f / \partial \mathbf{U} = \mathbf{0}\ .$$

The expressions for the derivatives $\partial f/\partial \mathbf{D}$ and $\partial f/\partial \mathbf{U}$ are given by (8–49), (8–50), and (8–43). However, the expression for $\partial f/\partial \mathbf{F}$ has to be obtained, keeping in mind that, under the null hypothesis, the matrix $\mathbf{F}$ has the special form

$$\mathbf{F} = \begin{bmatrix} \mathbf{F}_o & \mathbf{0} & \mathbf{0} \ldots \mathbf{0} \\ \mathbf{0} & \mathbf{F}_o & \mathbf{0} \quad \mathbf{0} \\ \mathbf{0} \ldots\ldots\ldots & & \mathbf{F}_o \end{bmatrix}.$$

It can be shown that, under the null hypothesis,

$$\partial f/\partial \mathbf{F}_o = 2 \sum_{i=1}^{k} (\mathbf{WF\Phi})_{ii} \ . \tag{8–75}$$

The matrices $\hat{\mathbf{F}}_o$, $\hat{\Phi}$, and $\hat{\mathbf{U}}$, which are obtained as the solutions of the likelihood equations when substituted in (8–72) yield the minimum value of f, $f_{\min}(\mathbf{F})$.

The next hypothesis of interest is the hypothesis of *invariant factor scores*. Since the factor scores are given by the relation

$$\mathbf{x}_i = \mathbf{D}_i \mathbf{x}_{i-1} + \delta_i \qquad (i = 2, \ldots, k) \ ,$$

the condition that the factor scores remain invariant implies that

$$\delta_2 = \delta_3 = \ldots = \delta_k = \mathbf{0} \tag{8–76}$$

and

$$\mathbf{D}_2 = \mathbf{D}_3 \quad \ldots = \mathbf{D}_k = \mathbf{I} \ . \tag{8–77}$$

This, in turn, implies that the structural model (8–32) has the form

$$\Sigma = \mathbf{F} \, \Phi_o \, \mathbf{F}' + \mathbf{U} \ , \tag{8–78}$$

where $\mathbf{F}$ and $\mathbf{U}$ are unrestricted but the matrix Φ_o is now a supermatrix, all of whose submatrix elements are identity matrices.

In order to test the hypothesis of invariant factor scores, we thus test the hypothesis

$$H_x \ : \mathbf{D}_2 = \mathbf{D}_3 = \ldots = \mathbf{D}_k = \mathbf{I} \tag{8–79}$$

against the alternative,

$$H_1 \ : \mathbf{D}_i \neq \mathbf{I} \qquad \text{for some } i.$$

It should be noted that, although the condition of invariant factor scores implies (8–79), the converse is not necessarily true. The hypothesis (8–79) does imply that the variances of the change scores, Ψ_i^2, are zero; that is,

$$\Psi_2^2 = \Psi_3^2 = \ldots = \Psi_k^2 = \mathbf{0} \ . \tag{8–80}$$

This would obviously be true if

$$\delta_2 = \delta_3 = \ldots = \delta_k = \mathbf{0} \; . \qquad (8\text{–}81)$$

However, this would also be true if the change scores were nonzero constants. Therefore, this test should be interpreted with caution.

The statistic for testing the hypothesis of invariant factor scores, $\chi^2(x)$, is given by

$$\chi^2(x) = N\{f_{\min}(x) - f_{\min}\} \; . \qquad (8\text{–}82)$$

Here, $f_{\min}(x)$ is the minimum value of the function (8–72) under the null hypothesis (8–79). The quantity $f_{\min}$ is the minimum under the alternate hypothesis, and it has the same value as the $f_{\min}$ discussed earlier. The statistic $\chi^2(x)$ is asymptotically distributed as a χ^2 variable with degrees of freedom

$$\nu(x) = (k - 1)p \; . \qquad (8\text{–}83)$$

In order to obtain the minimum of f under the null hypothesis, we solve the likelihood equations

$$\partial f/\partial \mathbf{F} = \mathbf{0}$$

and

$$\partial f/\partial \mathbf{U} = \mathbf{0},$$

where the expressions for the derivatives $\partial f/\partial \mathbf{F}$ and $\partial f/\partial \mathbf{U}$ are given by (8–42) and (8–43), respectively. The solution matrices $\hat{\mathbf{F}}$ and $\hat{\mathbf{U}}$, when substituted in (8–72) yield $f_{\min}(x)$.

Finally, the researcher may want to establish the *invariance* of the *unique scores*. This involves testing the hypothesis

$$H_U : \mathbf{U}_{ij} = \mathbf{U}_o \; , \qquad i, j = 1, \ldots, k, \qquad (8\text{–}84)$$

against the alternative,

$$H_1 : \mathbf{U}_{ij} \neq \mathbf{U}_o \qquad \text{for some } i, j \; . \qquad (8\text{–}85)$$

The test statistic, $\chi^2(\mathbf{U})$, is

$$\chi^2(\mathbf{U}) = N\{f_{\min}(\mathbf{U}) - f_{\min}\} \qquad (8\text{–}86)$$

and is asymptotically distributed as χ^2 with degrees of freedom

$$\nu(\mathbf{U}) = \tfrac{1}{2}n(k - 1)(k + 2) \; . \qquad (8\text{–}87)$$

In order to obtain $f_{\min}(\mathbf{U})$, we have to solve the likelihood equations

$$\partial f/\partial \mathbf{F} = \mathbf{0} ,$$

$$\partial f/\partial \mathbf{D} = \mathbf{0} ,$$

and

$$\partial f / \partial \mathbf{U}_{o\,\text{diag}} = \mathbf{0} .$$

It can be shown that

$$\partial f / \partial \mathbf{U}_{o\,\text{diag}} = \sum_{i=1}^{k} \sum_{j=1}^{k} \mathbf{W}_{ij\,\text{diag}} \quad . \tag{8–88}$$

The solutions of these equations are then substituted in (8–72) to yield $f_{\min}(\mathbf{U})$.

We have outlined tests for each of the hypotheses concerning the invariance of parameters separately. It may be more meaningful to test these hypotheses in sequence. For example, in order to establish factorial invariance, we could begin by testing the hypothesis

$$H_F\text{:}\;\; \mathbf{F}_1 = \mathbf{F}_2 = \ldots = \mathbf{F}_k \tag{8–89}$$

against

$$H_{F1}\text{:}\; \mathbf{F}_i \neq \mathbf{F}_j \qquad \text{for some } i,\, j \;. \tag{8–90}$$

If H_F is tenable, we may proceed to test the hypothesis

$$H_{Fx}\text{:}\; \mathbf{F}_1 = \mathbf{F}_2 = \ldots = \mathbf{F}_k = \mathbf{F}_o\,; \quad \mathbf{D}_2 = \mathbf{D}_3 = \ldots = \mathbf{D}_k = \mathbf{I} \tag{8–91}$$

against

$$H_F\text{:}\;\; \mathbf{F}_1 = \mathbf{F}_2 = \ldots = \mathbf{F}_k = \mathbf{F}_o \;. \tag{8–92}$$

If (8–90) is tenable, we test the hypothesis

$$H_{FxU}\text{:}\;\; \mathbf{F}_1 = \mathbf{F}_2 = \ldots = \mathbf{F}_k = \mathbf{F}_o\,; \quad \mathbf{D}_2 = \mathbf{D}_3 = \ldots = \mathbf{D}_k = \mathbf{I}\,; \quad \mathbf{U}_{11} = \mathbf{U}_{12} = \ldots = \mathbf{U}_{kk} = \mathbf{U}_o \tag{8–93}$$

against

$$H_{Fx}\text{:}\; \mathbf{F}_1 = \mathbf{F}_2 = \ldots = \mathbf{F}_k = \mathbf{F}_o\,; \quad \mathbf{D}_2 = \mathbf{D}_3 = \ldots = \mathbf{D}_k = \mathbf{I} \;. \tag{8–94}$$

If (8–93) is tenable, then we can conclude that a very strict form of factorial invariance exists.

The statistic for testing the hypothesis of invariant factor pattern is $\chi^2(\mathbf{F})$ with degrees of freedom $\nu(\mathbf{F})$. The statistic for testing H_{Fx} against H_F, denoted by $\chi^2(x\,|\mathbf{F})$, is

$$\chi^2(x\,|\mathbf{F}) = N\{f_{\min}(\mathbf{F}x) - f_{\min}(\mathbf{F})\} \;, \tag{8–95}$$

with degrees of freedom

$$\nu(x|\mathbf{F}) = \nu(\mathbf{F}x) - \nu(\mathbf{F}) \ . \qquad (8\text{–}96)$$

In order to obtain $\chi^2(\mathbf{F}x)$ we solve the likelihood equations

$$\partial f/\partial \mathbf{F}_o = \mathbf{0} \qquad (8\text{–}97)$$

and

$$\partial f/\partial \mathbf{U}_{ij\ \text{diag}} = \mathbf{0} \ , \qquad (8\text{–}98)$$

where $\partial f/\partial \mathbf{F}_o$ is given by (8–75) and $\partial f/\partial \mathbf{U}_{ij\ \text{diag}}$ is given by (8–43). Substitution of these solution matrices in (8–72) yields $f_{\min}(\mathbf{F}x)$. The degrees of freedom, $\nu(\mathbf{F}x)$ is given by

$$\nu(\mathbf{F}x) = (k - 1)np + 2(k - 1)p \ .$$

Thus,

$$\nu(x|\mathbf{F}) = (k - 1)p \ . \qquad (8\text{–}99)$$

The statistic for testing H_{FxU} against H_{Fx} is given by

$$\chi^2(\mathbf{U}|\mathbf{F}x) = N\{f_{\min}(\mathbf{F}x\mathbf{U}) - f_{\min}(\mathbf{F}x)\} \qquad (8\text{–}100)$$

with degrees of freedom

$$\nu(\mathbf{U}|\mathbf{F}x) = \nu(\mathbf{F}x\mathbf{U}) - \nu(\mathbf{F}x) \ . \qquad (8\text{–}101)$$

In order to obtain $f_{\min}(\mathbf{F}x\mathbf{U})$, we solve the likelihood equations

$$\partial f/\partial \mathbf{F}_o = \mathbf{0} \qquad (8\text{–}102)$$

and

$$\partial f/\partial \mathbf{U}_{o\ \text{diag}} = \mathbf{0} \ , \qquad (8\text{–}103)$$

where $\partial f/\partial \mathbf{F}_o$ is given by (8–75) and $\partial f/\partial \mathbf{U}_{o\ \text{diag}}$ is given by (8–88). Substitution of these solution matrices in (8–72) yields $f_{\min}(\mathbf{F}x\mathbf{U})$. The degrees of freedom $\nu(\mathbf{F}x\mathbf{U})$ is given by

$$\nu(\mathbf{F}xu) = (k - 1)np + 2(k - 1)p + \tfrac{1}{2}n(k - 1)(k + 2). \qquad (8\text{–}104)$$

Hence,

$$\nu(\mathbf{U}|\mathbf{F}x) = \tfrac{1}{2}n(k - 1)(k + 2) \ . \qquad (8\text{–}105)$$

In order to illustrate the above procedure, artificial data were generated with six variables ($p = 6$) and for three occasions ($k = 3$) for 150 subjects ($N = 150$). The data were analyzed using a modification of a program developed by Leong (1975).

The analysis revealed that two factors adequately describe the data and that the structure implied by the model (8–36) is valid. For the sake of interest, the model given by McDonald and Swaminathan (1972) and Corballis (1973) was fitted to the data. According to this model,

$$Ho : \Sigma_{ij} = \mathbf{F}_i \; \Phi_{ij} \; \mathbf{F}_j' + \mathbf{U}_{ij} \; , \tag{8–106}$$

with submatrices ϕ_{ij} and ϕ_{rs} being independent and diagonal. We test this hypothesis against the alternative $H_1 : \Sigma_{ij}$ is unrestricted. In addition, estimates of the parameters are obtained by solving the likelihood equations

$$\partial f / \partial \mathbf{F}_i = \mathbf{0} \, ,$$

$$\partial f / \partial \phi_{ij} = \mathbf{0} \, ,$$

and

$$\partial f / \partial \mathbf{U}_{ij} = \mathbf{0} \, .$$

These derivatives—except for $\partial f / \partial \phi_{ij}$—are given by (8–42) and (8–43). It can be shown that

$$\partial f / \partial \phi_{ij \, \mathrm{diag}} = -2 (\mathbf{W}_{ij})_{\mathrm{diag}} \; .$$

The likelihood ratio Lo for testing this hypothesis against the alternative has a χ^2 distribution with degrees of freedom ν_o given by

$$\nu_o = \tfrac{1}{2} kn(kn + 1) - knp - \tfrac{1}{2} pk(k - 1) - \tfrac{1}{2} pk(k + 1) \; .$$

This hypothesis was not rejected. Thus, the general structural model given by (8–36) and the model given above fitted the data.

(For the purpose of illustration, the model given by (8–36) was assumed and the hypotheses relating to factoral invariance were tested. The results are summarized in Table 8–1.)

A GENERAL MODEL FOR LONGITUDINAL ANALYSIS

Although it is sufficiently general, the model developed here suffers from several drawbacks. The orthogonality of factors simplifies the model and as an added bonus removes the indeterminacy due to rotation. However, at the same time, this may result in uninterpretable solutions (Corballis and Traub 1970). The obvious solution is to relax the assumptions of orthogonal factors, but this has implications regarding the nature of the autoregressive parameter matrices, $\mathbf{D}_i$.

In extending the basic model given by (8–23), it is only necessary to change the structure of $\mathbf{x}_i$ and the attendant parameters. We shall assume, instead of (8–9), that:

$$\mathbf{x}_i = \mathbf{D}_i \, \mathbf{x}_{i-1} + \delta_i \qquad (i = 2, \ldots, k) \, , \tag{8–107}$$

TABLE 8-1. Summary of Analyses for Hypothetical Data

Hypothesis	Degrees of freedom	χ^2
Hoa: Σ in (8-36)	119	148.4
Hob: Σ in (8-106)	117	151.8
H_F	24	35.3
H_{FX}	4	20.2
H_{FXU}	30	44.3

Note: N = 150, n = 6, k = 3, p = 2.

where the matrix $\mathbf{D}_i$ is a $(p \times p)$ nondiagonal matrix of regression coefficients.

As before, we assume that

$$E(\delta_i) = \mathbf{0},$$

$$E(\mathbf{x}_j \delta_i') = \mathbf{0},$$

$$E(\delta_i \delta_j') = \mathbf{0},$$

and define

$$E(\delta_i \delta_i') = \Psi_i$$

as nondiagonal.

Further defining

$$E(\mathbf{x}_1 \mathbf{x}_1') = \Phi_{11} ,$$

we obtain

$$E(\mathbf{x}\ \mathbf{x}') = \Phi = \Delta\ \mathbf{T}\ \Delta^{-1}\ \Psi(\Delta^{-1}\ \mathbf{T}\ \Delta)^{1} ,$$

where Δ has the same form as (8–32) except that the $\mathbf{D}$'s are not diagonal, and

$$\Psi = \begin{bmatrix} \Phi_{11} & \mathbf{0} & & \mathbf{0} \\ \mathbf{0} & \psi_1 & & \mathbf{0} \\ \mathbf{0} & \mathbf{0} & \psi_2 & \mathbf{0} \\ & & & \\ \mathbf{0} & & & \psi_k \end{bmatrix} .$$

Since in this case, the elements of Ψ and the matrices $\mathbf{D}$ are not directly related according to (8–35), it is convenient to make a one-to-one transformation of the type

$$\xi_2 = \mathbf{D}_2 ,$$

$$\xi_3 = \mathbf{D}_2\mathbf{D}_3 ,$$

and

$$\xi_k = \mathbf{D}_2\mathbf{D}_3 \ldots \mathbf{D}_k$$

and estimate $\xi_2, \xi_3, \ldots, \xi_k$ instead of $\mathbf{D}_2, \mathbf{D}_3, \ldots, \mathbf{D}_k$. This transformation considerably simplifies the evaluation of first and second derivatives of the likelihood function.

The following transformations leave the structure invariant:

$$\mathbf{F}^* = \mathbf{F}\,\mathbf{P} ,$$

$$\Delta^* = \mathbf{P}^{-1} ,$$

and

$$\psi^* = \mathbf{P}^{-1}\,(\mathbf{P}^{-1})' .$$

Hence, at least p^2 restrictions have to be imposed to identify the model. The total number of equations in this case is

$$\tfrac{1}{2}nk\,(nk + 1) + p^2 ,$$

while the number of parameters is

$$npk + kp^2 + (k - 1)p^2 + \tfrac{1}{2}nk\,(k + 1) .$$

Thus, for identification,

$$\tfrac{1}{2}nk\,(nk + 1) + 2p^2 - npk - 2kp^2 - \tfrac{1}{2}nk\,(k + 1) > 0 .$$

Once this condition is met, the estimation may proceed in the manner indicated for the orthogonal model.

CONCLUSION

In this chapter, we have formulated a general model for the factor analysis of longitudinal data, provided a statistical procedure for the estimation of parameters and for testing hypotheses of interest, and outlined a procedure for studying the important problem of factorial invariance over occasions.

This does not imply that we have solved the many problems involved in factor-analyzing longitudinal data. For instance, the problem of estimating the factor scores and, in particular, change scores remain to be solved. Furthermore, the model we have formulated is rather restrictive in the sense that the factor scores are orthogonal. While an attempt was made to indicate how the model can be generalized, further work is clearly needed with oblique models. It may also be more appropriate to assume that the unique scores, like the factor scores, follow an autoregressive scheme. The above considerations may have to be built in the model if the test for the goodness-of-fit of the model yields significant χ^2 values. In all, we can only conclude that a great deal of further research is necessary in the area of longitudinal factor analysis.

REFERENCES

Anderson, T. W. 1958. *An introduction to multivariate statistical analysis.* New York: Wiley.

———. 1960. Some stochastic process models for intelligence test scores. In *Mathematical methods in the social sciences,* ed. K. W. Arrow et al. Stanford: Stanford University Press.

———. 1963. The use of factor analysis in the statistical analysis of multiple time series. *Psychometrika* 28:1–25.

Cattell, R. B. 1952. The three basic factor-analytic research designs—Their interpretations and derivatives. *Psychological Bulletin* 49:499–520.

———. 1963. The structuring of change by P-technique and incremental R-technique. In *Problems in measuring change,* ed. C. W. Harris. Madison: University of Wisconsin Press.

Corballis, M. C. 1973. A factor model for analysing change. *British Journal of Mathematical and Statistical Psychology* 26:90–97.

Corballis, M. C., and R. E. Traub. 1970. Longitudinal factor analysis. *Psychometrika* 35:79–98.

Evans, G. T. 1967. Factor analytic treatment of growth data. *Multivariate Behavioral Research* 2:109–34.

———. 1971. Transformation of factor matrices to achieve congruence. *British Journal of Mathematical and Statistical Psychology* 24:22–48.

Harris, C. W. 1963. Canonical factor models for the description of change. In *Problems in measuring change,* ed. C. W. Harris. Madison: University of Wisconsin Press.

Jöreskog, K. G. 1970a. Factoring the multitest-multioccasion correlation matrix. In *Current problems and techniques in multivariate psychology. Proceedings of a conference honoring Paul Horst,* ed. C. E. Lunnborg. Seattle: University of Washington.

———. 1970b. A general method for the analysis of covariance structures. *Biometrika* 57:239–52.

———. 1971. Simultaneous factor analysis in several populations. *Psychometrika* 36:409–26.

———. 1979. Statistical estimation of structural models in longitudinal developmental investigations. In *Longitudinal research in the study of behavior and development,* ed. J. R. Nesselroade and P. B. Baltes. New York: Academic Press.

Jöreskog, K. G., and D. Sörbom. 1977. Statistical models and methods for the analysis of longitudinal data. In *Latent variables in socioeconomic models,* ed. D. J. Aigner and A. S. Goldberger. Amsterdam: North Holland.

Leong, K. S. 1975. A comprehensive model for covariance structure analysis. Doctoral dissertation, University of Toronto.

McDonald, R. P. 1969. A generalized common factor analysis based on residual covariance matrices of prescribed structure. *British Journal of Mathematical and Statistical Psychology* 22:149–63.

————. 1970. Three common factor models for groups of variables. *Psychometrika* 35:111–28.

————. 1974. Testing pattern hypotheses for covariance matrices. *Psychometrika* 39:189–201.

McDonald, R. P., and H. Swaminathan. 1972. *Structural analysis of dispersion matrices based on a very general model.* Toronto: Ontario Institute for Studies in Education.

————. 1973. A simple matrix calculus with applications to multivariate analysis. *General Systems* 18:37–54.

Rao, C. R. 1965. *Linear statistical inference and its applications.* New York: Wiley.

Roskam, E. E. 1976. Multivariate analysis of change and growth: Critical review and perspectives. In *Advances in psychological and educational measurement,* ed. D. N. M. Gruijter and L. J. T. Van der Kamp. New York: Wiley.

Swaminathan, H. 1976. Matrix calculus for functions of partitioned matrices. *General Systems* 21:95–99.

Tucker, L. R. 1958. An inter-battery method of factor analysis. *Psychometrika* 23:111–36.

————. 1963. Implications of factor analysis of three-way matrices for measurement of change. In *Problems in measuring change,* ed. C. W. Harris. Madison: University of Wisconsin Press.

9

N-Way Factor Analysis for Obtaining Personality-Environment-Test Contribution Systems to Any Response: A Plasmode Illustration

Raymond B. Cattell, Daniel D. Blaine, and Jerry M. Brennan

THREE EXISTING APPROACHES AND TWO POSSIBLE MODELS FOR REPRESENTING ENVIRONMENT IN THE BEHAVIORAL EQUATION

The present contribution can be regarded by statisticians as a first mathematical-statistical illustration of *N*-way factoring as previously given in algebraic outline (Cattell 1980). But to the psychologist it is something more, namely, an answer to the problem raised in various ways over the last 30 years of integrating personality traits, specific behaviors and environmental situations in a single predictive system. Indeed, it was in the psychological context that a behavior specification equation was first proposed by Cattell (1978, 1979) combining the *factorial properties* and *values* of person, stimulus, response form, and ambient situation.

Technical discussion of the need to represent environment in personality theory has historically taken several courses, only three highlights of which need be mentioned here. First, a simple concern for an adequate classification and mapping of a naturalistic *taxonomy* of environment—the definition of situations—was pursued from various angles by Barker (1968), Cattell (1963), Sells (1963) and others. Secondly, a statistical analysis of behavior variance in terms of relative contributions of person differences and situation differences, over various stimuli, was begun by Endler and Hunt (1966), Magnusson and Endler (1977) and others. Thirdly, a more finely conceptually-structured way of breaking down the behavioral variance was proposed by Cattell (1963, 1973, 1979) using one vector for person properties and another for situation properties. This accepted the ordinary

---

This article has been reprinted with the permission of the authors (Cattell, Blaine, and Brennan). It will also be included in a forthcoming book from Plenum Press.

factorial behavioral specification equation as the most sophisticated existing form of the old and fundamental S-O-R model, and developed the meaning of the loadings—the behavioral indices—as expressions of the particular environment and response form. For one must note that these b terms in the behavioral equation, stating how much each personality factor is invoked and involved in the behavior being predicted, are functions both of the stimulus situation *and* the form of response. However, as seen below, the proposed new model (Cattell 1979) breaks down b into p, e and s terms, separating the involvement of the situation in the final effect from that of the response form.

While recognizing that there is still much of psychological importance to be researched in the first of the three above approaches, i.e., mapping a quantitative taxonomy of man's cultural and natural physical environment, the present chapter nevertheless concentrates on further development of the breakdown of the magnitude of behavior into effects from several sources. For, strategically, this breakdown must be cleanly accomplished before a taxonomy of situations, tests, etc. can be made by applying pattern similarity coefficients among the various vectors.[1]

Within the behavioral equation model two distinct, but operationally related, environment treatments have been suggested by Cattell (1978) called the *personality centered* and the *additive id* (or additive attribute) models.

1. The *personality centered model* uses the factor specification equation just mentioned.

$$a_{ij} = b_{j1}T_{1i} + b_{j2}T_{2i} + \ldots + b_{jp}T_{pi} \; , \qquad (9\text{–}1)$$

in which the vector of T's gives the individual, i's, trait pattern, and the vector of behavioral indices, b's, gives the extent of the determination of the behavior a_j through these traits operating in the given situation and response forms. In this formulation the combined situation and response derives its values and meaning from the human traits involved (as we are reminded by there being as many b's as T's), whence the description "person-centered."

As mentioned above, the b values can be further split, but the model remains a person-centered one. The splitting—the experimental basis of which is discussed elsewhere (Cattell 1979)—is determined by the five-coordinate form of the *basic data relation matrix* (Cattell 1966) into persons, stimuli, responses, situations and observers. The last means "error," but if we wish to include it we shall have beyond p, e, s, and T, an r term as follows:

$$a_{hijko} = \sum p_{hx} e_{jx} s_{kx} r_{ox} T_{xi} \; . \qquad (9\text{–}2)$$

Here a is the measure of the magnitude of the response act j, to stimulus h, by individual i, in ambient[2] situation k, recorded (with error) by observer o. The coefficients p_h, e_j, s_k, and r_o determine the several variance fractions of a due to a given trait T. This equation is comprehensive because there are only five sets of different ids in the data box defining any psychological event.

2. The *additive id attribute* model, upon the development of which the present article concentrates, does not hinge so centrally on personality, but rests on the addition of the properties (factor dimensions) of each of the five possible ids[3] entering the psychological act. This is different from using the personality factor scores as the central vectors on which other vectors hinge in meaning.

To see this in the full perspective of possibilities one must refer back to the principles of the psychologist's *basic data relation matrix,* or *data box* (Cattell 1946, 1966) which defines any psychological event as a point fixed by five Cartesian coordinates of persons (*i*'s), stimuli (*h*'s), responses (*j*'s), ambient situations (*k*'s) and observers (*o*'s) (collectively called ids). It helps discussion on *N*-way factoring to use the terms *facet, face, grid,* and *box* (Cattell 1966) to describe the various score matrices that can be used (for ANOVA or CORAN purposes). Thus, Figures 9–1 and 9–2 show that a *facet* is a "slab," one id thick, e.g., a persons × tests score matrix for *one* situation of testing. Further they show *face* as a summation and averaging of several facets, e.g., a person × tests matrix averaged over many situa-

Figure 9–1. Score Matrix for a Three-Way Factor Analysis

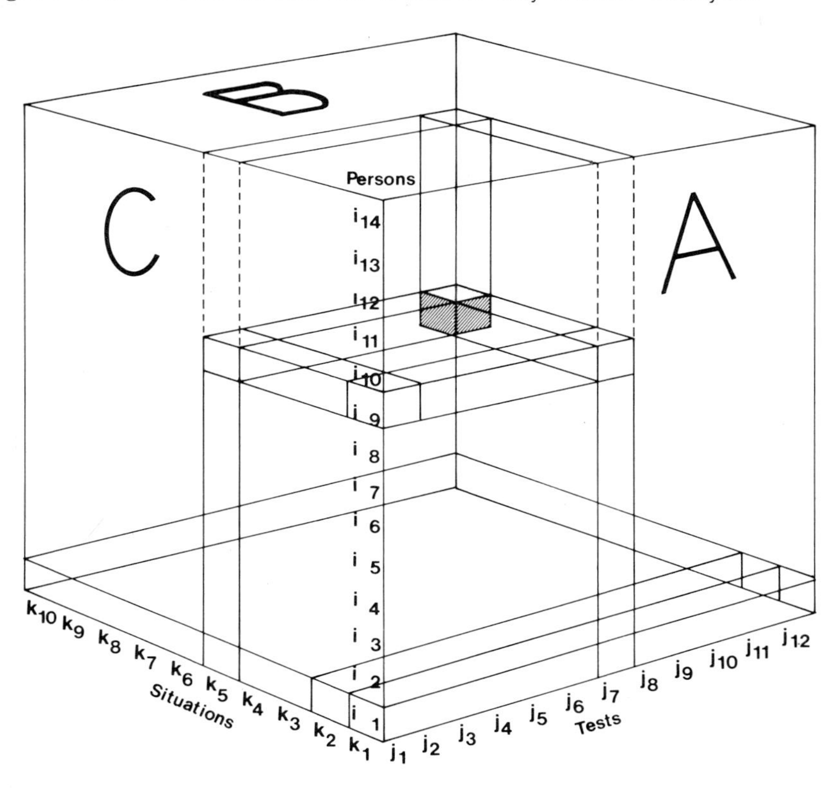

tions. Figure 9–2 shows in relation to these a grid, the properties of which are more fully discussed in Cattell (1966).

The most complete of possible psychological data relation matrices—the 5 dimensional box—will provide factor score matrices

Figure 9–2. Facets, Faces, and Files as Derivative Matrices from the Data Box

Deriving a Face from Several Facets

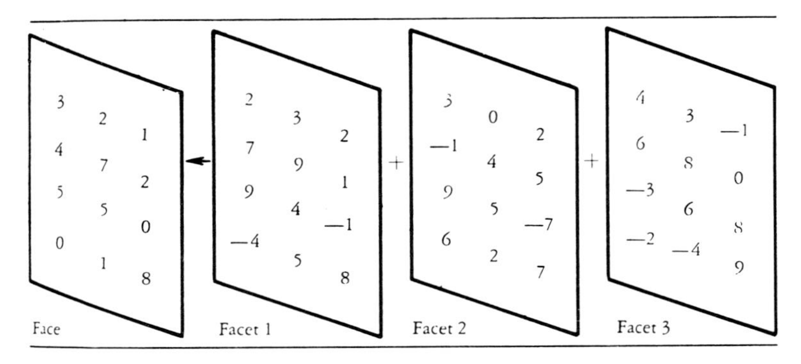

Grid from Unfolding a Three-Stimulus, Four-Environment, Five-Person Frame

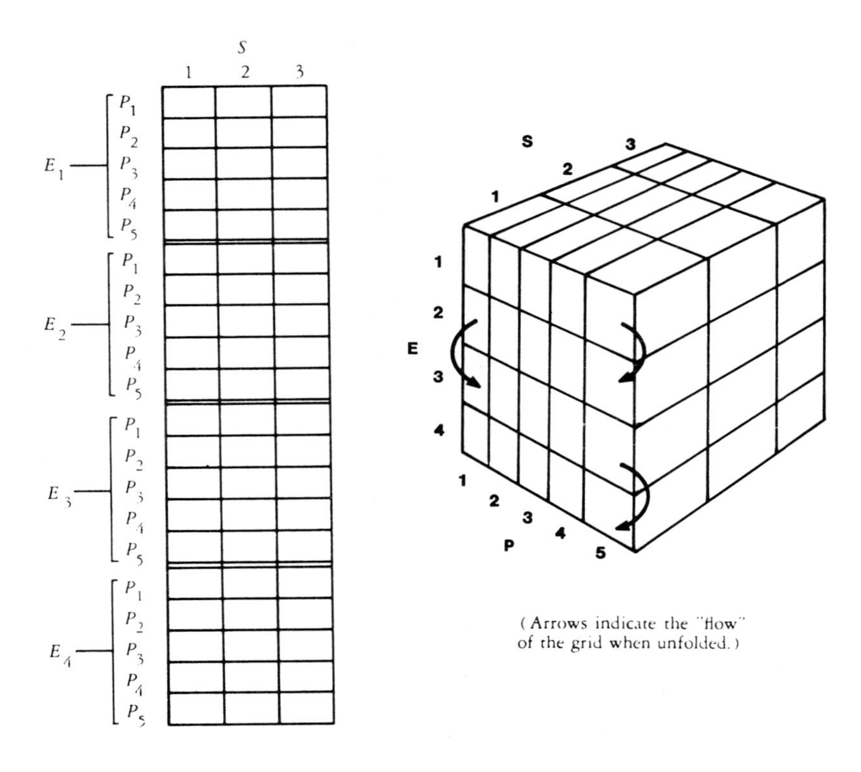

(Arrows indicate the "flow" of the grid when unfolded.)

consisting of no fewer than 10 faces (or grids) (20 if we consider transposed factorings). Thus 10 (or 20 with transposes) would be the limit of N-way factoring contributions. In this article we propose to illustrate the general problem by taking a box of only 3 dimensions (leaving the extension to 5 to whomsoever will pursue matters further!). An initial glance at the objective in a 5-way factoring is shown schematically in equation (9–3), where the magnitude of the recorded response is an addition of (independent) factors representing the properties of each of five different *sets* of *ids*. As we shall see later, this use of five matrices of factors from five different sets of ids is a simplification, which we shall hope to justify, since one could, mathematically, extract factors from pairing each of the five sets of relatives with each of the remaining four sets of referees:

$$a_{hijko} = \sum vT_h + \sum wT_i + \sum xT_j + \sum yT_k + \sum zT_o \,. \qquad (9\text{–}3)$$

Here v, w, x, y, and z are factor loadings, respectively for factors in stimuli, people, forms of response, ambient situations, and observers. T_h, T_i, T_j, etc. are similarly factor trait scores estimated respectively from stimuli, persons, etc., to give magnitudes belonging to a particular id (in common psychometric work a person).

It should be pointed out that the procedures to be discussed here are separate and distinct from Tucker's (1963, 1964, 1966) ingenious N-mode analysis. Tucker's procedure involves the determination of three sets of factors, one from each of the modes and an "inner core" matrix that "gives the interrelations that connect the three sets of factors . . . " (Levin 1965, 445). "The inner core can be interpreted in terms of psychological types. Each type consists of a matrix which can be explained in terms of interaction among the factors pertaining to the other two modes" (Levin 1965, 451). The current development does not deal with such an inner core but rather is concerned with the identification of the three factor sets directly from two-way matrices conditioned by summing across the unconsidered "ways" or "modes."

THE UTILITY OF ALTERNATIVE BREAKDOWNS IN CONTRIBUTING FACTOR SYSTEMS

As implied in the above, the numbers of different kinds of factor sources of variance one chooses to take into account can be varied according to how much one chooses to have uncontrolled and unrecorded in "error." Psychologists are accustomed to taking factors in "tests" (locked combinations, each of a kind of stimulus and a kind of response) the scores on these factors being attached to people. If scores are allowed to vary additionally over, say, situations and observers, the part of that variance associated with situations and with observers can also be represented in factors.

It helps from the beginning to think in terms of the data box (or *Basic Data Relation Matrix,* BDRM [Cattell 1946, 1966]). Therein a score at any point in any face can be estimated, along analysis of variance lines, from a mean value for the row and a mean value from the column corresponding to the point—if there is significant variance of such means. The same can be done from factoring the face, when the column (say the test) is now represented by a particular combination of loadings on test factors, and the particular row, say the person, is represented by the set of factor scores for that person. In principle it does not matter, therefore, whether the face is represented by scores in columns of tests and rows of persons, or estimates of those scores in the face from factors—except that the latter will lack completeness of calculation because of unknown specifics.

In a specification equation for any given score, a_{hijko}, one can proceed from two, three, four, or five pairs of sets. But if one proceeds from two one will estimate factorially only, say a_{hi}, at a fixed j and k and o (in a facet) or a mean $\bar{j}$, $\bar{k}$ and $\bar{o}$ in a face.

In what follows we shall use for convenience and precision the terms developed elsewhere for such matrices in the data box. The concepts are: *relatives* (variables to be correlated); *referees* (the referent ids over whom measures are taken); *facet,* a matrix of scores of referees on relatives on *one* "occasion," i.e., referent to only one id on any one of the remaining three id sets (dimensions); *face* as a sum (or mean) of values across all parallel congruent facets, i.e., along a *file* orthogonal to the intersection of a *row* and a column; an *id* as an entity in any and all of the five sets (coordinates) and a *grid* as a matrix of scores with one set of relatives at the top and all combinations of the other sets as rows in it. These are fully defined and illustrated originally in the *Handbook of Multivariate Experimental Psychology* (Cattell et al. 1966) and a brief diagrammatic reminder is given in Figure 9–2.

Now our general aim in N-way factoring, as just indicated, has been to re-construct a score value in any cell in the data box (typically denoted with five subscripts, as in a_{hijko}) from factors organically present in each of the five kinds of ids, i.e., from factoring each and all *faces*. However, if one considers faces in the full data box it is at once evident that each species of set has factors with respect to the remaining four different other sets in the role of relatives. For example, if we use responses as relatives (to take the most familiar case—tests as relatives—now known in practice) they will have factors as patterns in their a_j magnitudes that reside in (and come from) four kinds of referees: people, stimulus variations, situations, and observers. Incidentally here and later we must distinguish between *in* and *attached to*. In ordinary people-test, R-technique analysis the factors, scored on such dimensions as surgency, intelligence, etc. are *in test variables*. That is to say, they are defined in loading patterns in test variables, but the *scores* on the factors *are attached to and belong to people*. This rule holds overall possible combinations of the five id sets, in their changing roles as relatives and referees.

The statement that four kinds of score matrices and factors exist, i.e., as relatives, to describe factors in, say, people as referees, is part of the general statement above that the number of combinations in the full (5 set) data box is 20, i.e., 5 × 4.

However, what we have elsewhere (Cattell 1978) called the "transpose problem" arises here. For there is redundancy in those because they actually consist of 10 faces, each in a "direct" and a "transpose" form. For example, the factors we get from a score matrix of people × responses (averaged over situations, stimuli and observers, and thus constituting a face) invoke most of the same relations as the transpose, responses by people, in which people are correlated.

Here we encounter the old problem of the relations of factors from direct and transpose matrix correlations, i.e., of *R*- and *Q*-technique. If the factors from the two were psychologically the same there would be no problem. We should have to deal with only 10 matrices to complete prediction from all sources. Using *h, i, j, k* and *o* consistently as elsewhere for stimulus, individual, response form, ambient situation and observer we should have 10 score matrices denotable as *hi, hj, ij, ik, jk, jo, ko, kh, oh,* and *oi*. Without duplicating any matrix combination this would give two kinds of factors for each species of id and would theoretically (if direct and transpose factorings take care of the same relations, i.e., covariances) account for all but the specific variances in the $n_h \times n_i \times n_j \times n_k \times n_o$ entries in the data box (n_h being the number of *h*'s, and so on).

Without going far afield on this digression let us note that the arguments about the equivalence of direct (*R*-technique) and transpose (*Q*-technique) factorings go back to the debates between Stephenson (1953) and many others who believed *Q*-technique gave a new domain of information, while Burt (1937), Ross (1963) and Cattell (1966) concluded that *Q*-technique mostly gave the old information re-arranged. Space forbids going over this old debate, now settled—though admittedly there are minor odd corners still unresolved. Burt's statement (1937) had the caution that *R* and *Q* came out with the same factors if the matrix was first *double standardized*. That is to say the *R*-technique factor of say, intelligence, would be recognized by high loadings on intelligence subtests; but with appropriate rotation in *Q*-technique factor analysis the same factor of intelligence would duly appear, now recognizable by high loadings on intelligent persons. The present senior author showed it was difficult, except by long iterations, actually to *get* any matrix double-standardized (as distinct from double-centered). He pointed out that though the identical relations necessarily exist in a score matrix uprighted on its side (transposed to be exact) what one extracts from it by correlation or covariance factoring is another matter. Briefly, the story is that the first extracted factor in *R*-technique is lost in *Q*-technique, and vice versa. This follows from the correlation coefficient snowing out the differences of means in the variables it correlates. Except by extreme coincidence the loss of the first principal component is, however, unlikely to be the same as the loss of a subsequent simple-structure-rotated (psychologically meaningful) factor. Consequently when meaningful rotated factors are reached *R* and *Q* results are unlikely to have a different roster of factors. But the same factors will have different variances. This has been demonstrated for psychologists most lucidly by Ross (1963).

The answer therefore unfortunately seems to be that direct

and transposed factorings only convey the same information up to a point, and that 20 rather than 10 factor systems are necessary. This raises the difficulty, however, that we shall then have perhaps about a 4/5ths redundancy when we add the contributions from the 20 faces in a single specification equation. The choice that seems best is to lose a fraction of the possible information rather than face the obscure problems of allowing for redundancy. This alternative (10 factor systems) permits and requires some choices among the faces used. Each set of relatives is now factored over two sets of referees and there is now no "overlap" of any two faces in the ten chosen. Rather than examine this problem within the large table that would be necessary for 5-way, we have bowed to convenience by representing, in Table 9–1, of reasonable size, the 12 possible faces in a 4-way, and the 6 faces that could be chosen for final use of six kinds of factors in the specification equation.

Turning now to the actual plasmode example that we shall work out, we note that with three dimensions there are now three set combinations, each constituting a face. The value of any *variable* in each face will here be restored (as far as is possible by the common factors, of course) by common factor scores in tests (over people), in people (over situations) and in situations (over tests). From such factor-estimated variable values in the faces (or from the original variable true values including the specifics) any cell value in the box, i.e., any actual, empirical a_{kij} score from an experiment, will be obtainable by our model, as the sum of the corresponding values in the three faces. This is shown diagrammatically in Figure 9–1 by the perpendiculars dropped on the three surfaces (faces) from the given cell. Algebraically this is simply:

$$a_{kij} = a_{ki} + a_{kj} + a_{ij} \quad . \tag{9–4}$$

This procedure thus involves in the model estimating a_{ki}, a_{kj}, and a_{ij} from the factors in the respective faces and then combining them by 4 into an estimate of the expected experimental value.

In an actual experiment, however, we are given the a_{kij} values in the box cells and have to add them across—along files—to get the values in the faces that we shall then factor. It turns out, however, that there are some difficulties in exactly obtaining what should be the face values, a_{ki}, a_{kj}, etc. The next section will be devoted to illustrating these difficulties and possible solutions. However, before doing so we should at least pause briefly to mention that there exists a totally different approach possible from that of faces. This consists in analyzing instead by *grids*. A grid is most easily conceived, visually, as the slicing of the box by successive facets and laying them like dominoes end-to-end (Cattell 1966, 122). From a full five box there are five such matrices, each with one of the five sets as relatives. The referees are combinations of the remaining ids. For example, if we took kinds of response—*i*'s—as relatives, the entries of referees would be a_{hiko}'s and there would $n_h \times n_i \times n_k \times n_o$ of them. The factors from such a grid would be factors in kinds of responses, and the factor scores would attach to the class of en-

TABLE 9-1. Listing of 12 Possible Face Factorings in 4-Way Analysis, and a Selected 6 for Final Use

(1) Three Coordinate "Faces" (submatrices) mathematically possible:

Face 1	Face 2	Face 3	Face 4
PRS(E)*	RPE(S)	SRE(P)	PES(R)

(2) Two coordinate faces, suitable for factoring, in which averaging will be done over two coordinate sets:

PR(S) (E)	RP(E) (S)	SR(E) (P)	PE(S) (R)
RS(P) (E)	PE(R) (S)	RE(S) (P)	ES(P) (R)
SP(R) (E)	ER(P) (S)	ES(R) (P)	SP(R) (E)

(3) Re-ordering of (2) to show grouping by referees to whom the factors will belong:

PR(SE)	RP(ES)	SR(PE)	EP(SR)
PS(RE)	RE(SP)	SE(RP)	ES(PR)
PE(RS)	RS(EP)	SP(ER)	ER(PS)
Factors (T's) scored for people	T's scored for responses	T's scored for stimuli	T's scored for situations

(4) Alternative re-ordering of (2) by relatives to show domains of relatives in which factor patterns will appear:

RP(ES)	PR(SE)	PS(RE)	SE(RP)
SP(ER)	SR(PE)	RS(EP)	PE(RS)
EP(SR)	ER(PS)	ES(PR)	RE(SP)
T's defined in people	T's defined in responses	T's defined in stimuli	T's defined in situations

(5) Matrices chosen to give factors without redundancy:

PR(SE)	SP(ER)	RS(EP)	ER(PS)	SE(RP)	PE(RS)

Note: The fact that P and S appear twice as references and E and R only once is not an imbalance, since the factors connected with latter appeared also in SE and PR.

*The sets of ids over which an averaging is done, removing (the bulk of) their variance, are given in each case in parentheses. As in matrix representations generally the first symbol refers to the rows.

(In a score matrix, as the text explains, the relatives are the variables to be related [correlated] and the rows are the referees, i.e., the individual ids to which the relatives refer.)

tries a_{hiko}. There are no calculation problems here—indeed the great excess of referees ($n_h \times n_i \times n_k \times n_o$) over relatives avoids problems in the face factoring—and clearly one has factors from each of the five sources of covariance that could be meaningful. The combining of them—since the population of a_{hiko}'s is different from, say, that of a_{hijo}'s—however, presents problems, and since we henceforth concentrate on the face factoring we shall not pursue them here.

THE PROBLEM OF SIDE-EFFECTS IN ANALYSIS BY FACES

It should be kept in mind that the information in a given face can be represented either by *variable* scores or by *factor* scores plus a known factor pattern from which those scores (with specifics known) can be derived as shown in equation 9–3 and Figure 9–1. As we have seen the derivation of the value in a particular cell of the box can be obtained as the sum of the values in the face cells to which it corresponds—or, ultimately, from the sum of the factor-derived contributions of the contributory faces. As indicated, instead of considering the 10 independent contributory faces we have recognized to exist in the full data box we propose here to work on the case, both theoretically and in a plasmode, of just three dimensions, which is capable, in less bewildering abundance, of settling the general principles. As in Figure 9–1, the three-dimensional data box could concretely be thought of as bounded by sets of people (i's), of tests (actually tests are combinations of h's and j's but are symbolized here as j's) and of situations (k's). As one can infer from Table 9–1 this will operate with only three factor analyses.

From an ANOVA standpoint we ought first to ask if the model is correct in considering the box values restorable by simple addition. The algebraic analysis is relegated to a footnote.[4] In ANOVA terms it must unfortunately be concluded that 3-way factoring (or any N-way factoring) by faces (but not by grids, below) is likely in some kinds of data to be incapable of completely predicting the individual person-situation-response event. It will fall short of doing so according to the ratio of the magnitude of the three-way interaction term to the lower order terms. Since in most empirical studies known to us this ratio is quite low we feel one should not be deterred from using the model (in all but exceptionally distorted cases) by this loss. Thus if we adopt a model in which a score in the box, i.e., any literal experimental score of a person, reacting to a test, in a situation, is considered accounted for by three sets of factors, phenomenally real factorial influences, the accuracy of prediction will fall short of perfection from two causes so far recognized for summarizing: (1) the three-way interaction, and (2) the fact that in practice one will have to depend on common (broad) factor scores without possible knowledge of specific factor scores. In short, N-way factoring means an estimation not an exact calculation of cell (empirical) values.

A third approximation not familiar like (1) and (2) in ANOVA × FACTAN must finally be recognized. It arises in the process that would have to be followed in actual experiment of calculating the variable scores in the *faces* from those in the data box. (The model itself makes the box values *dependent* variables from the *independent* face variables: but actual experiment reverses this giving us box values to start with and asking us to derive first the face values and then the factors.)

This new problem which we may contingently briefly call that of "side effects" can best be introduced and recognized by looking back at Figure 9–1. As we consider the *person* × *test* face (*A*) it will be seen that as we add in the values from the "roof" or "floor" (*situations* × *tests*) (*B*), to the box values they will add the same amount to all terms down one column (a test at one situation) but will bring in different values from test column to test column. This, if true, will alter the *mean* score of each test in the (*A*) facet or face, but the rank order of persons on any test will remain the same. Thus, when we correlate test columns in the (*A*) face this intrusion will have no effect on the resulting correlation matrix, or the factors derived from it.

On the other hand, as the sum of each raw score on the *situation* × *person* face is added to the box data, to give the totals for the *person* × *test* face it will make a constant addition to one person on all tests, *but a different constant from person to person*. This *will* alter the correlations among tests, though not radically. It will have the effect of making all their correlations more positive and in effect producing one more common factor positive in all of them. (Note the scree test on the obtained principal components from the actual plasmode tended to show such an increase in number of factors above that of the factors fed in as contributors.)

What we need, of course, is some way of operating on the given data in the plasmode in such a way as to eliminate from each face this "side effect" as we may briefly call it. It might seem that since we have, from adding through the data box, the values in each face we could (keeping discussion to the *person* × *test* face), subtract across each *row* (person) the value which we know has intruded from the *person* × *situation* face. The magnitude of this intrusion could be obtained from a single vector, from adding rows and averaging each in that face. Subtracting this from each column in the person-test face would at first seem to bring precisely the restoration that we want to reach the original undistorted values. However, this fails to recognize that the values in the *person* × *situation* face have *themselves* been distorted by additions from the original values in the *person* × *test* face. (Again the *situation* × *test* face will have added only a constant to each of the *person* × *situation* columns, all the way down each, and so will not have brought any further uneveness to what is added by the *person* × *situation* rows to the person test rows.)

As far as the factoring outcomes of the *person* × *situation* and *person* × *test* faces are concerned, the effect of the *situation* × *test* face can be ignored. For, as seen above, it is not a systematic affector of correlations, being only an addition of constants. But the remaining systematic contamination discussed in the last paragraph has to be faced. Although we are discussing a three-dimensional (and by implication a four- and five-dimensional) case, this part of the problem can be handled in a two-dimensional abstraction. In fact, the problem essentially becomes that of solving for the *boundary* (*face,* in more dimensions) values in a two-dimensional matrix. The development of the problem, however, is rather lengthy and somewhat digressive and is thus included as an appendix. The upshot of these considerations is that there is indeed a systematic error unless certain assumptions are made about the relative variances contributed by situations, people, tests, etc. Since one would prefer not to make meretricious assumptions, but, in fact, to *discover* the relative contributions, as part of the solution, we shall not pursue these channels. Instead, since we have argued that with larger samples the side effects will be reduced, we propose by an actual example (plasmode) to find out how intrusive they may continue to be in a moderate sized experiment.

NUMERICAL PLASMODE ILLUSTRATION

This illustration hopefully will at the same time illuminate the algebraic treatment, as concrete examples often do, and give us some idea of the magnitude of samples needed for an acceptable approximate solution.

To keep close to current practical interests and also to restrict to the more tractable 3-way case so far pursued, we shall collapse the stimulus and response coordinates into one axis, i.e., to one set of ids, namely tests. In a test a given style of response is locked by agreement to a given stimulus. In fact this is the original simple data box—the "covariation chart" (Cattell 1946) —containing persons, tests and situations.

Our intention here is to construct a plasmode making a reasonable simulation of real data on all counts. Thus it should yield fewer factors than variables in all three faces, i.e., we wish to end with factors for persons, for situations and for tests, and it should admit experimental error. A *face* matrix is, as stated above, a sum or average across all possible *facet* matrices bounded by the given pair of sets (Figures 9–1 and 9–2). For example, one face would give the mean scores of people on tests averaged across all facets, each facet representing a person-test score made on a single situation. Since error, and poorness of estimates, become serious in any real factor analysis with few relatives and still more with fewness of referees we shall finally "sprinkle" error randomly into the box values obtained from summing the faces, even in illustration, into tolerable sample sizes, namely 45 tests, 50 situations, and 55 people, giving in the box matrix 45 × 50 × 55 = 123,750 scores corresponding (theoretically) to actual empirical measures in an experiment.

Our aim in the illustration (plasmode) is thus to begin with a known, designed factor structure underlying each of the three faces; to generate scores on variables therefrom in each of the three faces, and then to add the given person, test, and situation scores on variables in the faces, to get the 123,750 scores in the box that would correspond to what an experimenter actually initially observes. In Figure 9–1 this is illustrated for one box value (cell value) a_{iqj7k5} with its projections from the faces. This sum will necessarily be constituted as a raw score, because if the faces themselves were put in standard scores the box scores would still not be yielded as standard scores. It seems best therefore to work in raw scores throughout, for in any case the box values would first come in an experiment as raw scores.

Having thus proceeded from known "back stage" determiners to the actual score box matrix figures, we propose to work our way backward, using only the resources an experimenter would actually have—namely the box scores. Thus we shall proceed to face scores and then by factor analysis (with simple structure rotation), to see how well the original, known factor structure of the three sources can be restored in the obtained matrices.

If we define a "file" as in Figure 9–1, i.e., as a vector of values orthogonal to the intersection of row and column, it will be seen at once that when we add back along a file to get (when converted to an average) what we will call the "experimentally obtained" face (henceforth the "obtained face") this variable value will be different, due to some intrusions, from that in the "contributory" face. The reasons for this have been set out in the preceding algebraic analysis and we shall now see them operating numerically.

Let us first describe the construction, on as "naturalistic" (experiment resembling) a basis as possible, by four steps, as follows.

1. *Setting up an a priori typical simple structure factor matrix for each of the three faces.* This matrix, different for each face, defines what we should hope to retrieve, if our hypothesis and model are correct, perhaps slightly biased by the side effect intrusions. To match typical realities it should be simple structure and oblique. The matrices constructed will be deposited at NAPS. They were not made oblique at the beginning but would become so through sampling errors by the time we reach Table 9–2 which is what we shall actually work with. The original 3 factor matrices are available from the authors.
2. *Deriving from the given factor matrices the raw scores on variables in what are to be the contributory faces.* Random normal deviates were assigned to the referees (normally people) in each relative column, for both the broad and the specific factors set out in the tables referred to above. From these, scores on variables were derived by:

$$\underset{(N \times n)}{\mathbf{S}_v} = \underset{(N \times (k+n))}{\mathbf{S}_f} \quad \underset{((k+n) \times n)}{\mathbf{V}_{fp}'} \qquad (9\text{–}5)$$

(N = number of referees; n of relatives; k of a broad common factor).

From the standard scores in $\mathbf{S}_v$ a correlation matrix $\mathbf{R}_v = \mathbf{S}_v' \mathbf{S}_v / N$ was derived, and factored. It was found to yield, by scree test, the exact number of factors we had deliberately put in each face. On rotating for simple structure, it gave the values shown in Table 9–2. Here the factors are now slightly oblique in spite of the orthogonal restriction. Table 9–2 is in the first place simply a check on the matrices entered from the NAPS model, but more importantly, as the *actual* matrices (slightly different through sampling and

TABLE 9-2. The Actual Contributory Factor Matrices for the Faces

	Factors					
	1	2	3	4	5	6
Face A						
1	-0.02	0.09	-0.08	-0.06	0.68	0.17
2	-0.02	0.03	-0.05	-0.02	0.58	0.05
3	0.34	0.12	0.02	-0.10	-0.61	0.11
4	0.08	0.00	0.37	0.08	0.56	0.12
5	0.26	-0.01	-0.01	-0.02	0.76	0.06
6	0.04	0.30	0.09	-0.19	-0.43	0.01
7	-0.04	0.34	-0.16	-0.16	0.52	0.15
8	0.01	-0.21	0.03	0.11	0.69	0.02
9	-0.14	0.04	0.08	-0.10	-0.69	0.20
10	0.08	0.11	0.15	-0.01	0.20	0.58
11	0.12	-0.03	0.15	-0.08	0.28	-0.46
12	0.22	0.48	0.02	0.06	0.03	0.04
13	0.06	-0.07	0.27	-0.03	0.24	-0.74
14	0.29	-0.06	-0.43	0.09	-0.13	0.56
15	-0.03	0.12	0.12	-0.04	0.12	0.80
16	0.06	-0.13	0.27	0.10	0.17	0.51
17	0.03	-0.06	-0.03	-0.10	0.07	0.91
18	0.16	0.30	-0.07	0.12	-0.07	-0.46
19	0.12	0.06	0.07	0.84	0.04	-0.11
20	0.09	0.11	0.15	0.75	-0.18	-0.02
21	0.05	-0.11	0.11	-0.62	-0.12	0.20
22	-0.07	-0.05	-0.13	0.69	-0.26	0.03
23	0.04	0.01	-0.00	0.77	0.09	0.01
24	0.16	0.15	0.07	-0.54	-0.21	-0.18
25	-0.03	-0.14	-0.01	0.42	0.12	0.02
26	-0.06	0.12	-0.03	0.64	0.12	0.06
27	0.73	0.03	-0.04	0.01	-0.01	0.12
28	-0.60	-0.22	-0.12	-0.06	-0.06	-0.27
29	0.46	0.24	-0.21	0.04	0.23	-0.11
30	0.60	-0.15	0.11	0.12	-0.13	0.06
31	-0.85	-0.13	-0.01	0.13	-0.09	0.07
32	0.72	-0.12	-0.05	-0.09	0.06	-0.10
33	0.68	-0.17	-0.02	0.05	0.04	-0.08
34	-0.53	0.35	0.08	0.17	0.28	0.16
35	-0.13	0.49	-0.38	-0.07	0.18	-0.10
36	0.14	-0.53	-0.04	-0.04	0.03	0.01
37	-0.18	0.60	0.11	0.01	-0.21	0.01
38	0.00	0.92	-0.04	0.01	-0.03	-0.12
39	0.11	-0.54	0.03	-0.12	-0.17	-0.07
40	0.02	-0.89	-0.03	-0.00	-0.02	0.09
41	-0.00	0.62	0.12	-0.03	-0.17	0.12
42	0.21	0.77	0.00	-0.02	-0.13	-0.01
43	0.05	-0.02	-0.44	0.17	-0.36	0.18
44	0.07	-0.20	-0.50	-0.07	0.08	-0.18
45	-0.08	-0.08	0.76	0.03	-0.02	0.12
46	0.12	0.06	0.70	0.19	-0.07	-0.23
47	0.01	-0.07	0.73	-0.14	-0.08	0.04
48	-0.23	-0.05	-0.53	-0.05	-0.16	-0.13

TABLE 9-2. Continued

	Factors					
	1	2	3	4	5	6
49	-0.10	0.13	0.67	-0.21	0.02	-0.03
50	0.04	0.02	-0.79	-0.02	0.15	0.08
Factor 1	1.00	0.07	-0.06	-0.04	0.20	-0.01
Factor 2	0.07	1.00	0.01	-0.01	0.11	0.17
Factor 3	-0.06	0.01	1.00	-0.06	-0.06	-0.07
Factor 4	-0.04	-0.01	-0.06	1.00	0.06	-0.05
Factor 5	0.20	0.11	-0.06	0.06	1.00	0.11
Factor 6	-0.01	0.17	-0.07	-0.05	0.11	1.00
Face B						
1	0.58	0.30	0.37	0.37	-0.21	
2	-0.67	0.10	-0.01	0.15	0.09	
3	0.81	0.02	-0.22	-0.15	-0.02	
4	0.54	-0.10	-0.04	0.10	-0.19	
5	-0.71	0.08	-0.08	-0.04	-0.09	
6	0.88	0.09	0.10	-0.04	-0.06	
7	0.51	0.36	0.00	-0.03	-0.16	
8	0.60	-0.13	-0.30	0.09	0.00	
9	-0.58	-0.07	0.45	-0.08	-0.15	
10	-0.00	0.71	-0.00	-0.20	-0.17	
11	0.07	-0.79	0.16	-0.13	-0.14	
12	-0.28	0.44	0.08	0.05	0.28	
13	-0.02	0.88	0.00	0.01	0.06	
14	-0.02	-0.83	0.04	-0.02	0.04	
15	-0.06	0.76	0.10	-0.09	-0.32	
16	0.17	0.70	-0.25	0.25	0.08	
17	0.01	-0.57	0.10	0.24	0.14	
18	0.10	0.58	0.33	-0.15	0.16	
19	0.53	0.14	-0.03	-0.44	0.32	
20	-0.18	-0.05	-0.02	0.68	0.02	
21	0.12	-0.04	0.06	0.61	0.26	
22	0.08	0.09	0.03	0.52	-0.24	
23	0.16	0.09	0.11	-0.60	0.15	
24	0.09	0.17	-0.04	0.73	0.24	
25	-0.04	-0.04	0.04	0.78	-0.06	
26	0.07	-0.17	-0.06	0.36	0.58	
27	-0.02	0.06	-0.05	-0.81	0.11	
28	-0.36	0.25	-0.07	0.11	-0.61	
29	0.04	-0.10	-0.19	-0.19	0.61	
30	0.05	0.13	0.18	-0.03	0.69	
31	-0.21	-0.01	0.06	-0.19	0.82	
32	0.31	-0.12	0.06	-0.27	-0.50	
33	-0.30	-0.09	-0.10	0.11	0.68	
34	-0.21	0.01	0.16	0.11	0.67	
35	0.12	-0.11	0.30	-0.00	0.40	
36	0.09	-0.16	-0.01	0.08	-0.42	
37	0.14	0.14	0.75	0.11	0.15	
38	0.12	-0.27	0.84	0.05	-0.10	

TABLE 9-2. Continued

	Factors					
	1	2	3	4	5	6
39	-0.20	0.07	0.74	-0.04	-0.01	
40	0.09	-0.06	-0.61	0.30	-0.13	
41	-0.19	0.14	0.59	-0.08	0.04	
42	-0.03	-0.11	0.59	0.24	-0.08	
43	0.01	0.11	-0.72	0.09	-0.01	
44	-0.15	0.35	-0.67	-0.00	0.01	
45	-0.14	0.16	0.69	0.08	0.19	
Factor 1	1.00	-0.03	-0.13	-0.13	-0.29	
Factor 2	-0.03	1.00	0.08	-0.10	-0.01	
Factor 3	-0.13	0.08	1.00	-0.10	0.05	
Factor 4	-0.13	-0.10	-0.10	1.00	0.08	
Factor 5	-0.29	-0.01	0.05	0.08	1.00	
Face C						
1	0.03	0.13	-0.71	0.17		
2	0.15	-0.03	0.60	0.04		
3	-0.15	-0.15	-0.68	-0.02		
4	-0.11	-0.02	0.62	-0.33		
5	0.05	-0.06	0.85	0.25		
6	-0.08	-0.08	0.90	0.04		
7	-0.02	0.06	-0.77	-0.13		
8	0.21	0.29	0.42	0.12		
9	0.05	0.17	0.57	0.14		
10	-0.03	0.00	0.64	0.14		
11	0.10	-0.10	-0.72	0.09		
12	-0.00	0.01	0.57	-0.27		
13	-0.20	-0.18	0.37	0.07		
14	-0.12	-0.03	0.08	-0.75		
15	-0.01	-0.11	-0.08	-0.81		
16	0.12	0.03	-0.28	0.65		
17	-0.09	-0.11	-0.02	0.74		
18	0.07	0.27	-0.16	-0.71		
19	0.03	-0.13	0.04	0.49		
20	0.19	-0.28	-0.04	0.64		
21	-0.34	0.15	-0.04	0.73		
22	-0.01	-0.00	-0.07	-0.72		
23	0.04	0.22	0.10	0.69		
24	0.11	0.04	0.12	0.52		
25	0.38	0.15	0.06	0.29		
26	0.16	-0.12	0.06	-0.71		
27	0.25	-0.66	-0.12	-0.09		
28	0.01	-0.59	-0.11	0.01		
29	0.00	0.68	0.09	0.16		
30	-0.10	0.52	-0.25	0.22		
31	0.23	-0.47	-0.08	-0.42		
32	-0.08	-0.48	0.22	0.11		
33	-0.10	0.58	-0.13	0.27		
34	0.16	0.79	-0.25	0.03		

TABLE 9-2. Continued

	Factors					
	1	2	3	4	5	6
35	0.17	0.72	0.19	-0.45		
36	-0.06	0.77	0.25	0.03		
37	0.06	-0.80	-0.03	0.32		
38	0.07	0.94	0.05	0.04		
39	0.02	0.53	-0.26	-0.18		
40	0.03	-0.70	0.04	0.11		
41	-0.58	0.34	-0.01	-0.20		
42	0.62	0.19	-0.22	0.13		
43	-0.43	-0.12	-0.27	0.36		
44	0.78	-0.06	0.19	-0.03		
45	0.55	0.04	0.32	-0.08		
46	-0.09	0.04	-0.26	-0.29		
47	0.71	0.06	0.06	0.04		
48	0.40	-0.29	0.05	0.15		
49	0.75	-0.16	-0.01	0.13		
50	-0.70	0.13	0.04	0.03		
51	0.69	0.05	0.09	-0.04		
52	0.67	0.03	0.00	-0.04		
53	-0.62	-0.05	0.28	-0.07		
54	0.63	-0.09	0.04	-0.14		
55	0.61	0.04	-0.06	0.14		
Factor 1	1.00	-0.26	0.31	0.14		
Factor 2	-0.26	1.00	-0.03	0.22		
Factor 3	0.31	-0.03	1.00	0.05		
Factor 4	0.14	0.22	0.05	1.00		

obliquity from that original) constituting the real basis of all final plasmode construction and analysis.

3. *Converting the standard scores in the variable faces to raw scores.* To match what an experimenter would usually meet in the "box scores" it is necessary to convert from standard to raw scores by multiplying by a raw sigma and adding a mean, to each column. The "raw score faces" resulting are not reproduced here but may be obtained from the authors.
4. *Adding the three contributory matrices to get the raw scores in the data box.* The 123,750 scores in the box from adding the three faces are on tape and may be obtained from Dr. J. Brennan. From this point we begin "retracing our steps" to see if we can reach the factors with which the plasmode began. This was done in four steps as follows.

1. *Adding and averaging the values in every file in the box* (in three directions) to give the face values in the three faces. These, except for the "side effects" we have noted above, should be the same as we put in. At a casual glance they in

fact seem only mildly affected, but the real test remained to be applied in the factor analyses.

2. *Correlating and factoring within the obtained face score matrices.* The factoring followed standard procedures (Cattell 1978). The scree test for number of factors tended to show in each case some indication of an increase in number above the contributory matrices, by one or two, as shown in Figure 9–3, Parts A, B, and C. After iterating communalities to the indicated number, rotation to simple structure was carried out, with results discussed below.
3. *Correlating and factoring after experimental error is added to true scores in the data box.* In any real case the values which we find in the data box would not be true, as assumed up to this point, but would contain random measurement error. We simulated a typical psychological experiment, and one, indeed, with appreciable error of measurement with a reliability of only .75, in effect we scattered in the box error additions as random normal deviates from a distribution having a standard deviation corresponding to a vari-

Figure 9–3, Part A. Face *A* Scree Plot

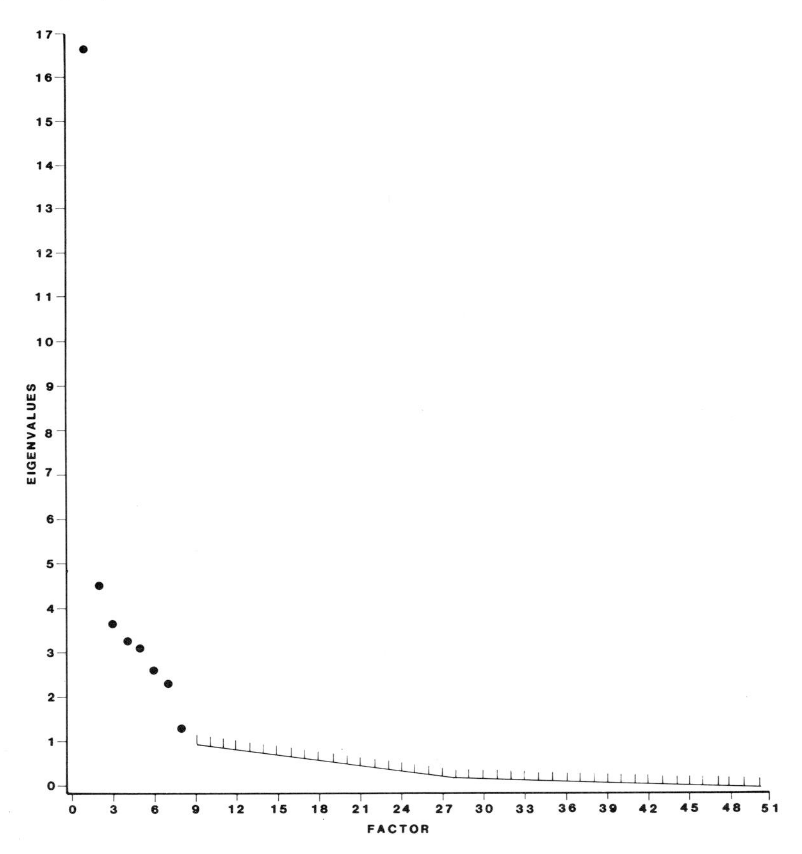

Figure 9–3, Part B. Face *B* Scree Plot

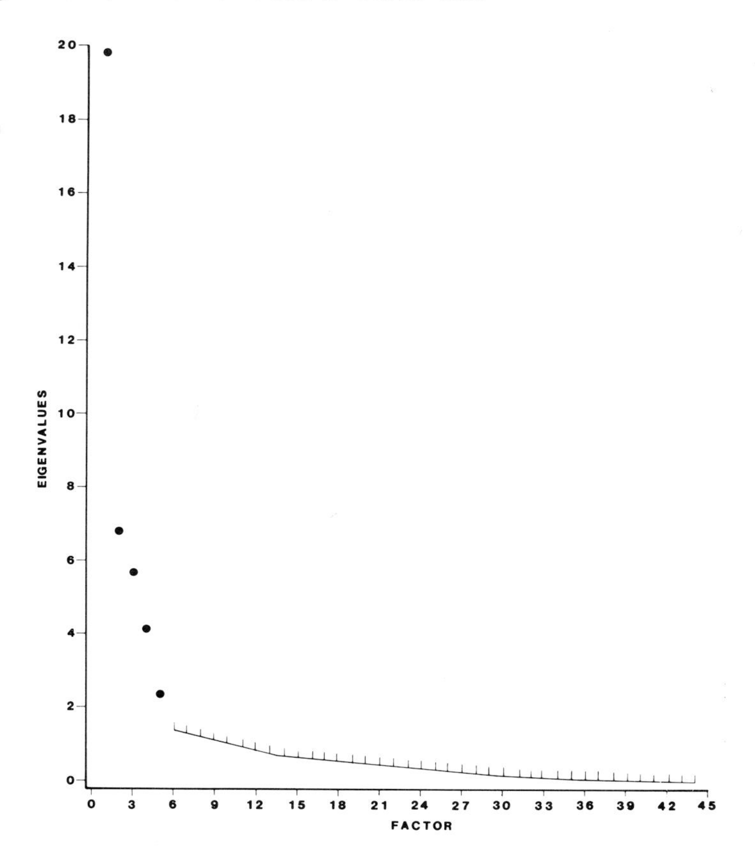

ance of 1/3d of the true variance we found existing there. The effect of this was calculated directly in the faces. If we call the above standard error σ_e, then in a face with files of n entries random deviates from a distribution whose $\sigma = \sigma_e/\sqrt{n}$ were scattered over the face values.

It was found that the correlations of the columns in the faces altered surprisingly little through this random error as such. And since, in any case, the factoring of **R**'s with r's corrected for attenuation is known to produce a factor structure very little different *in form* (Cattell 1978) from that for the uncorrected r's (except for raised communality) we were not surprised to find congruence coefficients close to 1.0 between factors from the true and the error-injected data. This reduction of error in 3-way analysis to trivial levels is worthy of comment.

4. *Ascertaining capacity of the procedure to retrieve the original contributory matrices.* It is shown (Table 9–3) that error of measurement even in a relatively high degree (dependability coefficient of .75) has a quite trivial effect so that any

Figure 9–3, Part C. Face *C* Scree Plot

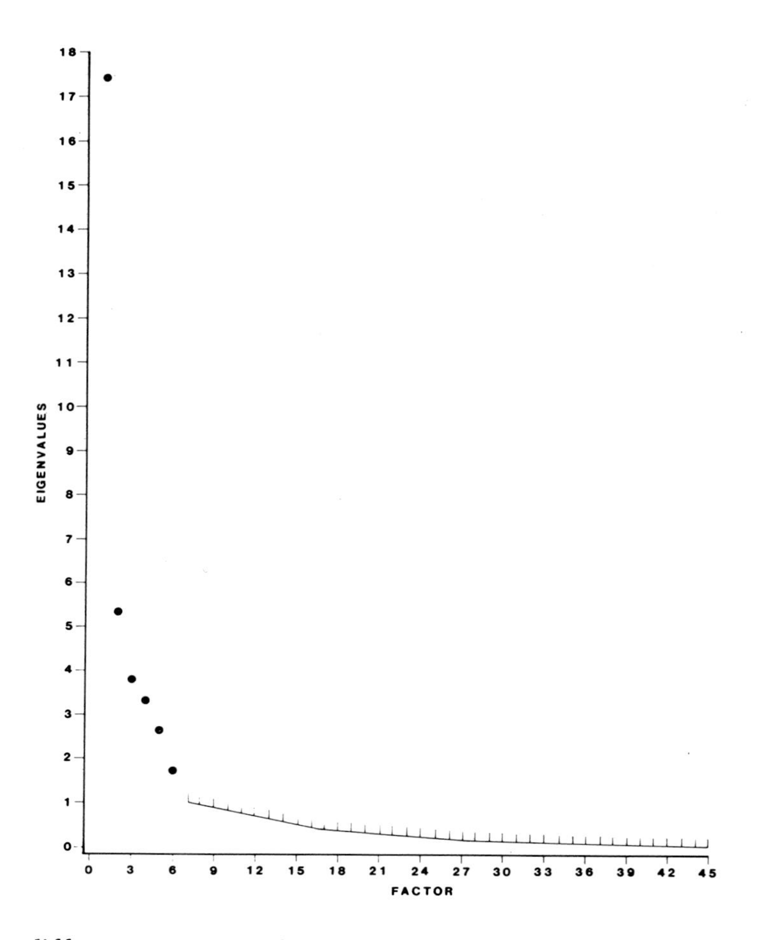

difference we now find must be due to the unavoidable systematic "side effect" we have anticipated. Thus we now approach the crucial test by factoring the scores (*with* error) in the three faces. This factoring was carried out with the objectivity and just the available resources that any experimenter unaware of the background would have in this situation. That is to say we proceeded through (i) a test for number of factors, (ii) iteration of communalities to that number, and (iii) rotation to maximum simple structure by Promax (as an experimenter would usually have to do, having no "target matrix" to use).

Figure 9–3, Parts A, B, and C, on scree test shows that the *obtained* matrices have a slight tendency to run to more factors than in the *contributory* factor matrices. Already in many earlier studies this has been recognized as due to producing one or two random error factors by the error we lately introduced. But it might also be due to more systematic

TABLE 9-3. Congruences of Errorless and Error-affected Factor Patterns comparing obtained 0 percent error to obtained 25 percent error

	Congruence Coefficients							
	1	2	3	4	5	6	7	8
Face A								
1	1.00	0.11	0.05	-0.00	0.04	0.04	0.01	0.05
2	0.11	1.00	0.06	0.00	0.01	0.06	0.06	-0.22
3	0.05	0.06	1.00	0.01	0.05	0.17	0.09	0.06
4	-0.00	0.00	0.01	1.00	0.04	0.07	0.09	-0.00
5	0.04	0.01	0.05	0.04	1.00	0.05	0.06	0.19
6	0.04	0.06	0.17	0.07	0.05	1.00	0.08	0.10
7	0.01	0.06	0.09	0.09	0.06	0.09	1.00	0.10
8	0.04	-0.21	0.05	-0.00	0.18	0.10	0.09	1.00
Face B								
1	1.00	0.01	0.07	0.11	0.21	0.08		
2	0.01	1.00	0.00	0.10	0.03	0.08		
3	0.07	0.00	1.00	0.01	0.15	0.10		
4	0.11	0.10	0.01	1.00	0.10	-0.04		
5	0.21	0.03	0.15	0.10	1.00	0.08		
6	0.08	0.08	0.10	-0.04	0.08	1.00		
Face C								
1	1.00	0.07	0.06	0.14	0.07			
2	0.07	1.00	0.12	0.09	0.13			
3	0.06	0.12	1.00	0.09	0.15			
4	0.14	0.09	0.09	1.00	0.23			
5	0.07	0.12	0.15	0.23	1.00			

error as described in the next section.

The outcome to compare with the contributory matrices is given for all three sets in Table 9–4. The remaining sets with random error on top of intrusive error are deposited at NAPS, but the congruences are given in Table 9–5 for all sets.

At this point the conclusion, on empirical grounds, from our plasmode, with data very typical of experiment, seems to be: (1) That random error, with reliabilities common in psychometric work, has a quite negligible effect on the factoring of the resultant face matrices and yields congruences of 1.0, and (2) That the effect of the systematic intrusion of the contributions from faces other than that being analyzed is to produce departures from the true values that are not negligible but not sufficient to obscure the essence of the true factor structures when operating with about 45 or more ids in each of the three sets.

TABLE 9-4. Factor Matrices from the Obtained (Retrieved) Faces (Errorless)

	Factors							
	1	2	3	4	5	6	7	8
Face A								
1	0.19	0.68	0.08	-0.05	-0.07	0.13	0.18	-0.08
2	0.03	0.68	0.01	-0.04	-0.16	0.06	0.00	0.02
3	0.08	-0.36	0.20	-0.02	0.22	0.11	0.04	0.54
4	0.12	0.56	0.44	0.11	0.02	0.02	-0.18	-0.03
5	0.10	0.84	-0.07	-0.04	0.16	-0.04	-0.02	-0.09
6	0.03	-0.14	0.43	-0.06	-0.03	0.31	0.16	0.35
7	0.17	0.42	-0.18	-0.17	-0.02	0.49	0.02	-0.26
8	0.05	0.63	0.13	0.09	-0.00	-0.13	0.23	-0.23
9	0.18	-0.61	0.40	-0.03	-0.03	0.18	0.21	0.12
10	0.54	0.27	0.30	0.06	0.03	0.10	0.01	0.05
11	-0.40	0.37	0.35	-0.04	0.06	0.01	0.07	0.08
12	0.06	0.17	0.18	0.11	0.10	0.44	0.04	0.19
13	-0.61	0.29	0.50	0.02	0.05	0.01	0.12	0.00
14	0.52	0.01	-0.41	0.11	0.17	0.01	0.32	0.34
15	0.76	0.18	0.21	0.00	-0.06	0.10	-0.10	0.04
16	0.51	0.04	0.32	0.11	0.21	-0.04	-0.18	-0.30
17	0.91	0.14	0.08	-0.09	-0.00	-0.06	0.08	0.03
18	-0.31	-0.06	0.16	0.14	0.19	0.44	0.31	-0.07
19	-0.07	0.06	-0.05	0.91	0.14	-0.00	-0.07	-0.02
20	-0.03	-0.19	0.17	0.76	0.16	0.16	-0.07	-0.07
21	0.23	-0.09	0.48	-0.61	0.13	0.06	0.26	-0.00
22	0.03	-0.24	-0.13	0.76	-0.02	-0.02	0.21	0.02
23	0.04	0.15	0.20	0.56	0.02	0.07	0.28	0.06
24	-0.06	0.01	0.49	-0.24	0.09	0.31	0.35	0.16
25	0.04	0.15	0.16	0.42	-0.04	-0.05	0.30	-0.02
26	0.08	0.16	-0.10	0.70	-0.09	0.05	0.02	0.03
27	0.10	0.10	0.02	0.02	0.59	0.16	0.09	0.17
28	-0.23	-0.05	0.32	-0.01	-0.47	-0.05	0.63	-0.05
29	-0.02	0.19	0.09	0.08	0.32	0.36	0.35	-0.02
30	0.07	-0.19	0.14	0.14	0.72	0.01	0.02	-0.12
31	0.10	-0.07	0.41	0.17	-0.63	-0.05	0.50	-0.08
32	-0.08	0.28	0.11	-0.04	0.50	-0.03	0.17	0.35
33	-0.01	0.11	0.30	0.11	0.40	0.12	0.34	0.05
34	0.16	0.22	0.19	0.19	-0.48	0.38	0.01	-0.19
35	0.00	0.04	-0.14	-0.03	0.02	0.68	0.50	-0.31
36	0.05	-0.10	0.26	0.01	0.36	-0.19	0.54	-0.32
37	0.04	0.00	0.43	0.07	-0.11	0.46	0.23	0.15
38	-0.09	0.03	0.11	0.06	-0.04	0.90	-0.04	0.08
39	0.02	-0.00	0.47	-0.00	0.10	-0.13	0.58	0.05
40	0.08	0.04	0.33	0.02	0.80	-0.08	0.59	0.05
41	0.15	-0.10	0.23	0.04	0.02	0.63	-0.12	0.08
42	0.00	-0.02	0.14	0.03	0.13	0.75	-0.03	0.14
43	0.16	-0.20	-0.16	0.21	0.02	0.08	0.54	0.26
44	-0.08	0.14	-0.21	-0.05	0.07	-0.06	0.82	0.02
45	0.14	0.03	0.89	0.10	-0.03	-0.02	-0.16	-0.03
46	-0.08	0.03	0.71	0.20	0.11	0.15	0.05	0.02
47	0.05	0.04	0.93	-0.04	0.00	-0.01	-0.20	0.10
48	-0.07	-0.06	0.09	-0.05	0.07	-0.24	0.65	0.03

TABLE 9-4. Continued

	Factors							
	1	2	3	4	5	6	7	8
49	0.04	-0.00	0.89	-0.14	0.00	0.21	-0.16	-0.16
50	0.12	0.20	-0.51	0.00	0.02	0.18	0.86	0.04
Factor 1	1.00	0.10	0.08	0.06	0.01	0.29	0.14	0.08
Factor 2	0.10	1.00	0.17	0.22	0.32	0.30	0.22	0.03
Factor 3	0.08	0.17	1.00	0.37	0.20	0.34	0.50	0.14
Factor 4	0.06	0.22	0.37	1.00	0.10	0.32	0.40	0.08
Factor 5	0.01	0.32	0.20	0.10	1.00	0.15	0.23	0.16
Factor 6	0.29	0.30	0.34	0.32	0.15	1.00	0.36	0.20
Factor 7	0.14	0.22	0.50	0.40	0.23	0.36	1.00	0.24
Factor 8	0.08	0.03	0.14	0.08	0.16	0.20	0.24	1.00
Face B								
1	0.42	0.68	0.35	-0.16	-0.20	-0.04		
2	0.10	-0.56	0.12	-0.03	0.49	0.40		
3	-0.19	0.92	-0.10	0.11	0.11	0.02		
4	-0.06	0.72	0.14	0.16	-0.16	-0.06		
5	0.14	-0.25	-0.00	0.18	0.39	0.65		
6	0.04	1.00	-0.02	-0.01	-0.13	-0.22		
7	0.09	0.67	-0.01	-0.13	-0.03	0.18		
8	-0.31	0.72	0.10	0.14	0.08	-0.09		
9	0.64	-0.33	0.01	0.24	0.11	0.40		
10	0.20	0.30	-0.11	-0.24	0.16	0.46		
11	0.16	0.25	-0.07	0.82	0.04	0.07		
12	0.23	0.07	0.08	-0.07	0.57	0.26		
13	0.18	0.20	0.03	-0.46	0.34	0.30		
14	0.06	0.12	0.05	0.80	0.30	-0.03		
15	0.28	0.25	-0.04	-0.35	-0.03	0.51		
16	-0.16	0.27	0.21	-0.53	0.31	0.20		
17	0.10	0.08	0.23	0.49	0.42	-0.04		
18	0.34	0.29	-0.05	-0.14	0.42	0.18		
19	-0.00	0.59	-0.36	-0.06	0.39	-0.21		
20	0.04	-0.12	0.57	0.07	0.32	0.25		
21	0.08	0.13	0.52	0.03	0.50	-0.13		
22	0.18	0.27	0.54	0.01	-0.04	0.21		
23	0.19	0.33	-0.49	0.06	0.28	0.04		
24	0.09	0.20	0.43	-0.03	0.55	0.12		
25	0.08	-0.05	0.76	-0.03	0.10	0.00		
26	0.01	0.09	0.28	0.14	0.79	-0.10		
27	0.13	0.30	-0.59	0.16	0.36	0.21		
28	0.03	-0.09	0.06	-0.00	-0.39	0.95		
29	-0.03	0.16	-0.09	0.13	0.82	-0.04		
30	0.20	0.13	-0.02	0.01	0.78	-0.00		
31	0.11	-0.19	-0.15	0.01	1.05	-0.18		
32	0.18	0.64	-0.17	0.30	-0.36	0.37		
33	-0.03	-0.26	0.08	0.06	0.99	-0.06		
34	0.19	-0.21	0.09	-0.00	0.89	-0.15		
35	0.25	0.07	-0.03	0.10	0.56	-0.23		

TABLE 9-4. Continued

	Factors							
	1	2	3	4	5	6	7	8
36	0.14	0.40	0.12	0.28	-0.08	0.38		
37	0.53	0.25	0.09	0.05	0.38	0.06		
38	0.85	0.20	0.10	0.27	-0.02	-0.17		
39	0.83	-0.18	-0.05	0.02	0.02	0.02		
40	-0.31	0.37	0.29	0.18	0.31	0.38		
41	0.64	-0.06	-0.08	-0.02	0.25	0.09		
42	0.69	0.04	0.25	0.13	0.04	-0.08		
43	-0.21	0.33	0.11	0.12	0.47	0.41		
44	-0.39	0.16	0.01	-0.06	0.43	0.57		
45	0.70	-0.13	0.03	-0.10	0.36	-0.04		
Factor 1	1.00	0.22	-0.03	-0.07	0.41	0.38		
Factor 2	0.22	1.00	0.04	-0.07	0.37	0.43		
Factor 3	-0.03	0.04	1.00	0.15	0.17	0.17		
Factor 4	-0.07	-0.07	0.15	1.00	0.00	-0.16		
Factor 5	0.41	0.37	0.17	0.00	1.00	0.55		
Factor 6	0.38	0.43	0.17	-0.16	0.55	1.00		
Face C								
1	-0.46	0.54	0.16	0.09	0.30			
2	0.60	-0.00	0.02	0.22	0.07			
3	-0.50	0.24	0.34	-0.12	0.48			
4	0.59	0.02	0.42	0.01	0.01			
5	0.77	0.13	-0.09	0.18	0.18			
6	0.88	-0.07	0.01	0.03	0.03			
7	-0.63	0.39	0.39	0.07	0.18			
8	0.43	0.30	-0.09	0.26	-0.16			
9	0.54	0.34	0.05	0.12	0.09			
10	0.64	0.13	-0.02	0.05	0.11			
11	-0.62	0.30	0.16	0.20	0.32			
12	0.59	0.06	0.41	0.02	0.13			
13	0.42	0.08	0.13	-0.06	0.24			
14	0.16	0.19	0.75	0.05	0.14			
15	-0.00	-0.14	0.91	0.03	-0.00			
16	-0.13	0.51	-0.28	0.21	0.45			
17	0.10	0.45	-0.35	0.06	0.56			
18	-0.09	0.23	0.76	0.10	-0.22			
19	0.15	0.40	0.01	0.19	0.43			
20	0.07	0.22	-0.30	0.30	0.56			
21	0.10	0.65	-0.16	0.11	0.42			
22	0.01	-0.02	0.80	0.02	-0.06			
23	0.16	0.63	-0.41	0.16	0.23			
24	0.18	0.36	-0.32	0.18	0.29			
25	0.05	0.31	-0.23	0.50	-0.02			
26	0.14	0.09	0.70	0.20	0.17			
27	-0.03	-0.40	0.29	0.38	0.53			
28	-0.00	-0.22	0.25	0.14	0.58			
29	0.16	0.74	0.20	0.12	0.06			

TABLE 9-4. Continued

	Factors							
	1	2	3	4	5	6	7	8
30	-0.11	0.80	0.07	0.00	0.04			
31	-0.01	-0.42	0.56	0.29	0.33			
32	0.29	-0.09	0.19	0.07	0.53			
33	-0.05	0.82	-0.12	-0.11	-0.07			
34	-0.11	0.87	0.18	0.17	-0.17			
35	0.22	0.70	0.12	0.16	-0.34			
36	0.26	0.81	0.16	0.04	-0.19			
37	0.12	-0.07	0.13	0.21	0.77			
38	0.10	0.96	0.10	0.09	-0.37			
39	-0.02	0.64	0.42	0.10	0.10			
40	0.14	-0.26	0.17	0.16	0.67			
41	0.09	0.58	0.44	-0.43	0.02			
42	-0.13	0.33	0.04	0.63	0.07			
43	-0.12	0.44	0.01	-0.27	0.50			
44	0.16	-0.15	0.02	0.85	-0.02			
45	0.28	0.20	0.23	0.52	0.09			
46	-0.11	0.28	0.51	-0.01	0.15			
47	0.13	0.29	0.23	0.53	0.22			
48	0.05	-0.08	-0.00	0.58	0.24			
49	0.02	-0.02	0.10	0.78	0.16			
50	0.16	0.50	0.27	-0.63	0.25			
51	0.07	0.00	0.07	0.76	-0.07			
52	0.04	0.11	0.15	0.67	0.07			
53	0.32	0.40	0.42	-0.25	0.38			
54	0.07	-0.04	0.24	0.66	0.07			
55	-0.04	0.16	-0.04	0.67	0.09			
Factor 1	1.00	0.19	0.09	0.39	0.12			
Factor 2	0.19	1.00	0.21	0.23	0.35			
Factor 3	0.09	0.21	1.00	0.17	0.32			
Factor 4	0.39	0.23	0.17	1.00	0.41			
Factor 5	0.12	0.35	0.32	0.41	1.00			

SUMMARY

1. The theoretical position is taken that behavioral measures in a world affected by characteristics of persons, stimuli, forms of response, situations, and observational error need to be (a) understood in terms of meaningful unitary traits (factors) of persons, situations, etc., and (b) estimated by lawful equations involving all five of these contributing sources in the complete data box.
2. Ways in which the co-variation of the data box can be broken down resolve ultimately into (a) by faces and (b) by grids. In the former the values in the variables in the faces, which are essentially means from adding along files covering the

TABLE 9-5. Congruences of Factor Patterns from Contributory and Retrieved Face Matrices: Comparing Contributory to Obtained Zero Error

	Congruence Coefficients							
	1	2	3	4	5	6	7	8
Face A								
1	0.98	0.05	-0.04	-0.02	-0.00	-0.04	-0.10	0.06
2	0.06	0.93	-0.09	-0.02	-0.05	-0.04	-0.06	-0.51
3	-0.00	0.02	0.75	0.01	0.02	-0.05	-0.54	-0.09
4	-0.02	0.02	-0.13	0.96	0.01	-0.03	-0.00	-0.09
5	0.04	0.15	-0.01	0.02	0.94	0.08	0.00	0.36
6	0.02	0.03	0.00	0.07	-0.09	0.91	-0.23	0.15
Face B								
1	0.91	-0.09	0.01	0.09	0.02	-0.21		
2	-0.16	0.86	-0.03	0.00	-0.18	-0.36		
3	0.03	-0.07	0.98	-0.04	0.12	0.05		
4	0.11	0.17	-0.03	-0.84	0.11	0.43		
5	0.05	-0.16	-0.03	-0.02	0.83	-0.41		
Face C								
1	0.96	-0.11	-0.08	0.08	-0.11			
2	0.01	0.73	-0.01	-0.08	-0.57			
3	0.05	0.34	-0.76	0.11	0.39			
4	0.03	-0.05	-0.01	0.95	0.01			

three remaining coordinates, are added to give cell values in the boxes. Factoring the contributory faces (of which there are 10 in the complete data box, but 3 in the example worked out here), gives as many factor systems as there are faces. There could be twice as many if transpose factorings from each face were unchecked, but we conclude that these would be largely redundant. From the factors thus obtained—which in the three sets are factors in tests over people, people over situations, and situations over tests—the variable and the means in the faces (from adding across the third dimension files) can be estimated. (They could be precisely calculated if we had specific factor scores.) The actual experimental scores—a_{ijk}'s—can be restored (as estimates) from addition of the face values.

3. An alternative is to factor over *grids* instead of *faces*. There is one grid for each set, using values (referees) that are combinations of the remaining sets (4 in the full 5 dimensional data box, and 3 in the 3 dimensional). Then if the relatives in the grid are, say, h's any value $a_{hijko} = \Sigma\, b_{ijko(hx)} T_{hx}$. The present article takes space only for the first alternatives (2) and (3), and concentrates on their special problems.

4. The problems by the face analysis approach are (a) that there may be, in restoring a cell from three values, effects of third order interaction, (b) that common (broad) factor scores, without specific factor scores can, as usual, give only estimates (of face variable values), and (c) that a form of systematic error intrudes—called the "side-effect"—which prevents unbiased calculation of the mean values in the faces from the given empirical scores in the box. The two first of these are common also to the grid alternative.
5. A fairly extensive analysis is given in algebraic terms of the last named problem and the conclusion is drawn that without some special assumption, such as that the contributions from factors in people, situations, and tests are equal, no solution devoid of approximations exists. The other two "shortcomings" in (4) above are trivial and normally have to be accepted in ANOVA and in other CORAN work.
6. A plasmode made as truly typical of natural data as possible has been set up here (there being no sufficient experimental Basic Data Relation Matrix data in existence) to discover empirically how far whatever sources of error exist may upset the *N*-way factor results. The plasmode (a) set typically simple structure oblique factor patterns and normal deviate scores, (b) was used to calculate variable scores in *faces* therefrom, (c) proceeded from these "contributory faces" to calculate the cell values in the box, (d) considered these cell values in the data box as what would normally be the experimental data and began analysis by reversing the steps and adding along files to give what may be called the "obtained faces," (e) factored these by ordinary methods that would ignore any previously entered structure that could be used as a target or hypothesis, (f) used the scree test for number of factors and the promax program for rotation to best simple structure, and (g) calculated congruence coefficients of obtained factor patterns with the original contributory factor input. This was first done on a basis of true scores in the box and then on true scores plus random error (as would occur in real data) equivalent to a reliability of 0.75.
7. An examination of Table 9–3 shows that the random error added to the original scores was reduced to a negligible effect. This would be expected from the summation across a set of scores for which the expected error is zero. On the other hand, there is a distortion from the summing across scores which is not negligible (see Table 9–6). This "side effect" error is the result of the disturbance due to the contribution of the third way or mode not being considered.
8. Since it is demonstrable algebraically that this error from side effects will diminish with increased sample size, and since already at 45, 50 and 55 cases it fails to hide the essential factor structure, it seems better tactics to proceed with the present assumptions, on somewhat larger samples, than to seek solution by invoking such assumptions as that factors have equal variance in the different systems.
9. When one recognizes that different variances from the different factor systems are to be accepted a correction must be introduced to the general formula for combining systems,

TABLE 9-6. Problem of Finding Additive Boundary Values, Knowing all Matrix Values

		Boundary Values from Second Contributor			
		w	x	y	z
Boundary Values	a	a + w	a + x	a + y	a + z
from	b	b + w	b + x	b + y	b + z
First Contributor	c	c + w	c + x	c + y	c + z

Column 1 total = 3w + (a + b + c)
Column 2 total = 3x + (a + b + c)
Column 3 total = 3y + (a + b + c)
Column 4 total = 3z + (a + b + c)

Row sum and column sums are as follows:

Columns = 3w + (a + b + c); 3x + (a + b + c); 3y + (a + b + c); 3z + (a + b + c)
Total= 3 (w + x + y + z) + 4 (a + b + c)

Rows= 4a + (w + x + y + z), 4b + (w + x + y + z), 4c + (w + x + y + z)
Total= 4 (a + b + c) + 3 (w + x + y + z) = T

which formula in general terms has been written (Cattell 1978, 1978; and above):

$$a_{ijk} = \sum^{x=m} b_{i\bar{j}(kx)} \quad T_{kx} + \sum^{y=n} b_{j\bar{k}(iy)} \quad T_{iy} \tag{9–6}$$

$$+ \sum^{z=o} b_{k\bar{i}(jz)} \quad T_{jz} \quad + \text{specifics}$$

$b_{i\bar{j}}$ is the loading for variable i score, as the latter is averaged over the $\bar{j}$ files, and it applies, as shown to (kx) to a factor x described by loadings on k's. The weight to be given to a factor kx in its contribution to a variable $i\bar{j}$ (now in raw score deviations) is $b^2_{\bar{i}kx} \times \sigma_{i\bar{j}}$ where $\sigma_{i\bar{j}}$ is the raw score variance of the column of values ($i\bar{j}$'s) for variable i in the variable face.[5] The factor's contribution to all variables is

$$\sum^{i=n} b^2_{\bar{i}kx} \sigma_{i\bar{j}}.$$

If we wish to average contributions of *all* factors in one face we would have to average again over the p factors involved, thus

$$\sigma^2_{\bar{i}k} = \sum^{x=p} \sum^{i=n} b_{ikx} \sigma_{i\bar{j}} \ .$$

Each of the three contributory systems would thus have a contribution to raw score deviation (no longer standard score) in the proportion of σ_k, σ_i and σ_j values, which could prefix (multiply) the three terms in estimating a_{ijk}.

However, in actually restoring any raw score value (A_{ijk} rather than a_{ijk}) attention must be given also to the raw score *means*, also, so that:

$$A_{ijk} = \sigma_k \sum^{x=m} b_{i\bar{j}(kx)} T_{kx} + \sigma_i \sum^{y=n} b_{jk(iy)} T_{iy} \tag{9–7}$$

$$+ \sigma_j \sum^{z=o} b_{ki(jz)} T_{jz} + U_{ijk} + M_{j(k)} + M_{i(j)} + M_{k(j)},$$

where U_{ijk} is the remaining specific (unestimated) variance in A_{ijk}, and $M_{j(k)}$ is the mean of the j scores averaged over files in k, and so on for others. One can if necessary switch back from the raw score values A_{ijk} to the standard score, but the expression for weightings then complicates somewhat.

APPENDIX: BOUNDARY VALUES

I. Problem 1: The Two Dimensional Case

(a) With No Special Assumptions

Consider Table 9–6. With simple algebra it can be shown that an infinite number of solutions exist for a, b, c, and w, x, y, z, given the values of the "data box" (i.e., $a + w$, $b + w$, etc.), though two algebraic conditions do hold, namely that (1) $a - b$, $w - x$, etc. are constant values and (2) that

$$(w + x + y + z)/4 = k(a + b + c)/3 , \tag{9–8}$$

where k is a constant representing the ratio of the average column to average row contribution.

(b) With Assumption of Equality of Average Contributions

If we could afford, psychologically, to bring in some further assumption a solution would be possible. One such is that the mean contribution from situations equals that from tests, i.e., that k in (9–8) equals 1.0. If some value for k can be justified —say $k = 1$ as just suggested, then:

$$(a + b + c)/3 = (w + x + y + z)/4 . \tag{9–9}$$

From Table 9–2 it can be seen that the total or sum of all entries in the matrix would be equal to:

$$T = 4(a + b + c) + 3(w + x + y + z) \tag{9–10}$$

thus, given the assumption

$$4(a + b + c) = 3(w + x + y + z) = T/2 . \tag{9–11}$$

Considering the average of the elements of the first row of the matrix:

$$\frac{\Sigma \text{ Row } a}{4} = a + \frac{(w + x + y + z)}{4} = a + \frac{T}{2(3)(4)} \quad . \qquad (9\text{–}12)$$

The contribution, a, could then be determined by:

$$a = \frac{\Sigma \text{ Row } a}{4} - \frac{T}{2(3)(4)} \quad . \qquad (9\text{–}13)$$

Or, in more general terms, if n_1 = number of rows and n_2 = number of columns

$$a = \frac{\Sigma \text{ Row } a}{n_2} - \frac{T}{2n_1n_2} \quad . \qquad (9\text{–}14)$$

The values of the other unknowns could be found in a similar fashion.

For example, given the following matrix with the average contributions of rows and columns equal to 5, and the specific contributions as indicated:

Example 1:

	$(x = 3)$	$(y = 5)$	$(z = 7)$	
$(a = 3)$	6	8	10	Σ Row a = 24
$(b = 7)$	10	12	14	Σ Row b = 36
	Σ Col x = 16	Σ Col y = 20	Σ Col z = 24	

the solution for the contributions would be as follows:

$$a = \frac{\Sigma \text{ Row } a}{n_2} - \frac{T}{2n_1n_2} = \frac{24}{3} - \frac{60}{2(2)(3)} = 8 - 5 = 3$$

$$b = \frac{\Sigma \text{ Row } b}{n_2} - \frac{T}{2n_1n_2} = \frac{36}{3} - \frac{60}{2(2)(3)} = 12 - 5 = 7$$

$$x = \frac{\Sigma \text{ Col } x}{n_1} - \frac{T}{2n_1n_2} = \frac{16}{2} - \frac{60}{2(2)(3)} = 8 - 5 = 3.$$

The remaining unknowns would have similar solutions.

(c) With the Assumption of a Specified Ratio of Contributions

An alternative is to consider $k \neq 1$, and derive its value from the assumption that the levels of $(a + b + c)/3$ and $(w + x + y + z)/4$ are *proportional to their sigmas* (standard deviations).

Assume for a moment that the only variation in the scores of the matrix is due to the row and column marginal contributions. In that case, the standard deviation of the row contributions, σ_A, is the same as that for any column of scores since for any column the only variation would be due to the row contributions. In other words the contribution of a column would be constant for all scores in that column. Similarly, the standard deviation of the column contributions, σ_W, is the same as that for any row of scores since for any given row the only variation would come from the column contributions. With unavoidable errors of measurement considered, the standard deviation of a given row would be an estimate of the standard deviation of the column contributions. These estimates could be pooled to obtain an estimate of the variation in the column contributions. A similar procedure could be used to obtain an estimate of the standard deviation of the row contribution. The ratio, $\hat{\sigma}_W/\hat{\sigma}_A$, would then be an estimate of k.

Now the mean of the contributory (bounding) values for the rows must be k times that for the columns, altering (9–9) above to

$$k(a + b + c)/3 = (w + x + y + z)/4$$

or

$$4k(a + b + c) = 3(w + x + y + z) \ , \qquad (9\text{–}15)$$

as in the original form of (9–8).

The total of all scores in the matrix can still be expressed as

$$T = 4(a + b + c) = 3(w + x + y + z) \ . \qquad (9\text{–}16)$$

Now the proportion of the total that each of the terms on the right side of equation (9–16) contributes is a function of k. Given the assumptions made for this case, generally a system of two simultaneous linear equations in two unknowns can be solved, namely:

$$k\overline{X}_A - \overline{X}_W = 0 \qquad (9\text{–}17)$$

$$\overline{X}_A + \overline{X}_W = T/n_1 n_2 \qquad (9\text{–}18)$$

where $\overline{X}_A$ is the mean of the row contributions, $\overline{X}_W$ is the mean of the column contributions, T is the total of all scores in the matrix, and n_1 and n_2 are the number of rows and columns, respectively. Assume that the contributions are as given below:

Example 2:

	$(w = 2)$	$(x = 2)$	$(y = 6)$	$(z = 6)$	
$(a = 4)$	6	6	10	10	32
$(b = 8)$	10	10	14	14	48
$(c = 12)$	14	14	18	18	64
	30	30	42	42	144

In this case $\sigma_W = 2$ and $\sigma_A = 4$; therefore, $k = \sigma_W/\sigma_A = .5$. The total, T, is equal to 144, $n_1 = 3$, and $n_2 = 4$. The system then becomes:

$$.5\overline{X}_A - \overline{X}_N = 0$$

$$\overline{X}_A + \overline{X}_W = 12 \ ,$$

which has $\overline{X}_A = 8$ and $\overline{X}_W = 4$ which are the averages of the row and column contributions. Thus the total of the row contributions would be $3 \times 8 = 24$ and the sum of the column contributions would be $4 \times 4 = 16$.

It can be shown that:

$$a = \frac{\Sigma \text{ Row } a}{4} - \frac{(w + x + y + z)}{4} \qquad (9\text{–}19)$$

or more generally:

$$a = \frac{\Sigma \text{ Row } a}{n_2} - \overline{X}_{W}. \qquad (9\text{–}20)$$

For a column contribution, say w:

$$w = \frac{\Sigma \text{ Col } w}{n_1} - \overline{X}_{A}. \qquad (9\text{–}21)$$

The boundary, contributory values in Example 2 can then be solved for

$$a = 32/4 - 4 = 8 - 4 = 4$$

$$b = 48/4 - 4 = 12 - 4 = 8$$

$$c = 64/4 - 4 = 16 - 4 = 12$$

$$w = 30/3 - 8 = 10 - 8 = 2$$

$$x = 30/3 - 8 = 10 - 8 = 2$$

$$y = 42/3 - 8 = 14 - 8 = 6$$

$$z = 42/3 - 8 = 14 - 8 = 6$$

II. Extension to the Three Dimensional Case

The three dimensional case would be one in which a third set of contributory values influenced the scores obtained, as in Figure 9–4 which shows a $3 \times 4 \times 2$ data box. For simplicity the assumption of equality of mean contribution is adopted, thus:

Figure 9–4. Three-Way Data Box

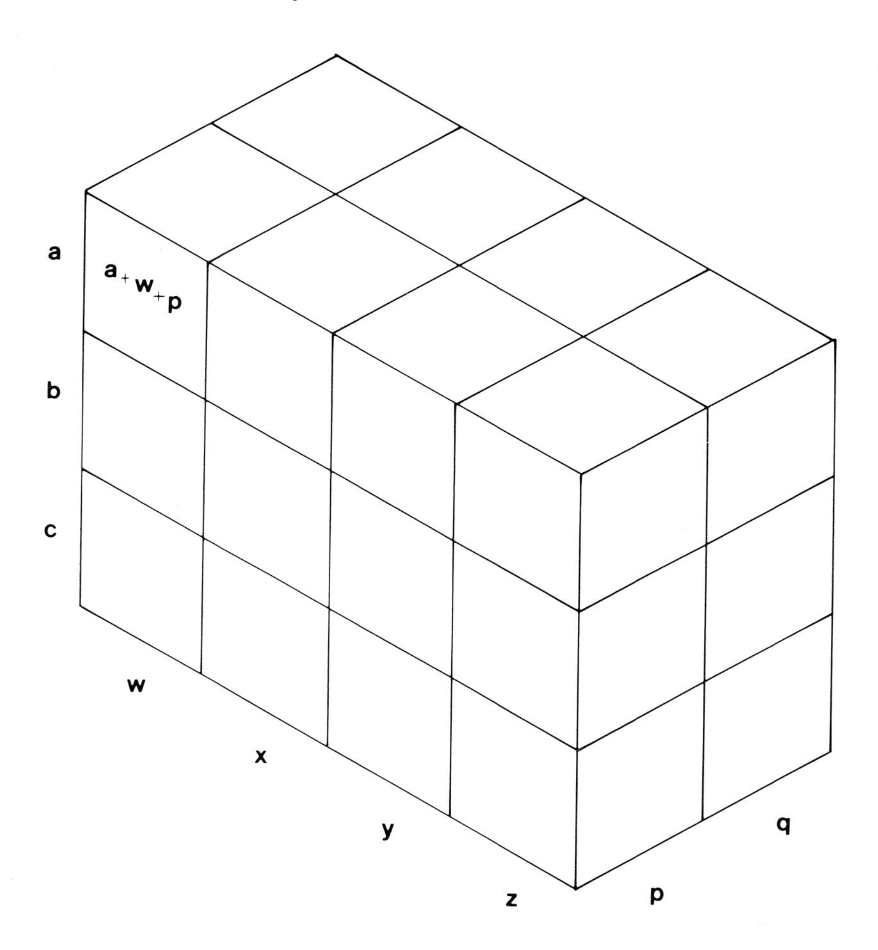

$$\frac{a+b+c}{3} = \frac{w+x+y+z}{4} = \frac{p+q}{2} \quad . \qquad (9\text{–}22)$$

Suppose that the contributors represented people, tests, and situations such that for the example we are imagining data were obtained from 3 people taking 4 tests under 2 situations. Such a representation allows the data to be viewed from three 2-way perspectives: people × tests, tests × situations, and situations × people.

There may be some doubt as to the way in which the "faces" should be defined, i.e., direct or transpose. There is, however, no need for the confusion of obtaining, by calculation, the face "contributory" values for variables, with the later question concerning which way the matrix is eventually to be factored; i.e., by *R*- or *Q*-technique. The rule for construction is to take three faces of the figure such that one is never opposite another, since such faces would be identical. Then each value in the corresponding contributory face will carry through, at right angles to that face as shown in Figure 9–1. (Note: For general purposes the symbols n_1, n_2, and n_3 will be used to represent the number of people, tests, and situations respectively. If we were solving by equations for cells there would be $n_1n_2 + n_2n_3 + n_1n_3$ unknowns and $n_1n_2n_3$ equations. Not until we get to $3 \times 4 \times 5$ are there enough equations: 47 unknowns and 60 equations. But here we use as many equations as unknowns; $n_1n_2 + n_2n_3 + n_1n_3$ for any size.)

Given the assumption of equality of average contribution the following relationship would hold:

$$(2)(4)(a + b + c) = (2)(3)(w + x + y + z) = (3)(4)(p + q) = T/3 \ . \tag{9–23}$$

The total of all scores obtained by person a, Σ File a, is given by

$$\sum \text{File } a = (4)(2)a + 2(w + x + y + z) + 4(p + q). \tag{9–24}$$

Solving for a:

$$a = \frac{\Sigma \text{ File } a}{(4)(2)} - \frac{(w + x + y + z)}{4} - \frac{(p + q)}{2} \tag{9–25}$$

$$a = \frac{\Sigma \text{ File } a}{(4)(2)} - \frac{T}{3(2)(3)(4)} - \frac{T}{3(2)(3)(4)} \ , \tag{9–26}$$

or more generally:

$$a = \frac{\Sigma \text{ File } a}{n_2n_3} - \frac{2T}{3n_1n_2n_3} \ . \tag{9–27}$$

For the contributions of tests (using w as an example):

$$w = \frac{\Sigma \text{ File } w}{n_1n_3} - \frac{2T}{3n_1n_2n_3} \ . \tag{9–28}$$

and for situations (using p as an example):

$$p = \frac{\Sigma \text{ File } p}{n_1n_2} - \frac{2T}{3n_1n_2n_3} \ . \tag{9–29}$$

For example, assume the following contributory values and the resulting scores:

$p = 4$	$w = 4$	$x = 6$	$y = 6$	$z = 8$
$a = 4$	12	14	14	16
$b = 6$	14	16	16	18
$c = 8$	16	18	18	20

$q = 8$	$w = 4$	$x = 6$	$y = 6$	$z = 8$
$a = 4$	16	18	18	20
$b = 6$	18	20	20	22
$c = 8$	20	22	22	24

In this example the assumption is that the average contribution is the same for rows, columns, and layers.

To solve for contribution *b* using the form of (9–27):

$$b = \frac{\Sigma \text{ File } b}{n_2 n_3} - \frac{2T}{3n_1 n_2 n_3} = \frac{144}{8} - \frac{864}{12} = 18 - 12 = 6 \; . \qquad (9\text{–}30)$$

To solve for contribution *z* using the form of (9–28):

$$z = \frac{\Sigma \text{ File } z}{n_1 n_3} - \frac{2T}{3n_1 n_2 n_3} = \frac{120}{6} - \frac{864}{12} = 20 - 12 = 8 \; . \qquad (9\text{–}31)$$

To solve for contribution *p* using (9–31):

$$p = \frac{\Sigma \text{ File } p}{n_1 n_2} - \frac{2T}{3n_1 n_2 n_3} = \frac{192}{12} - \frac{864}{12} = 16 - 12 = 4 \; . \qquad (9\text{–}32)$$

The remaining contributions can be solved for in a similar manner.

NOTES

1. We do not neglect historical perspective by not returning after this point to references to the second approach. The formulation of Endler and Hunt (1966), Magnusson and Endler (1977), and others has been subjected to debate and revision (Cattell 1979; Mischel 1968; Golding 1975). Among other things Cattell (1979) has pointed out that the "popular" conclusion that substantial situational variance negates the unitary factor trait concept is incorrect. The breakdown reached by that approach is

now clarified and will be seen to be quite different from the goal of *N*-way factoring.

2. The reader is reminded that an operational distinction is made between the *focal stimulus, h,* which is that to which the individual *i* is consciously reacting, and the *ambient situation,* which is the rest of the environment, e.g., the temperature and noise in the room when *i* is responding to, say, an intelligence test, *h*.

3. The five Cartesian coordinates of the data box consist each of a series of "patterns" of the same class, e.g., persons, stimuli, patterns of response. *Id* is used as a generic term for any and all of these.

4. It will be seen that in each face there are three deviations (and sums of squares) to be represented, as follows:

$$\begin{aligned}(\bar{a}_{ij\cdot} - \bar{a}_{\cdot\cdot\cdot}) &= (\bar{a}_{i\cdot\cdot} - \bar{a}_{\cdot\cdot\cdot}) \\ &+ (\bar{a}_{\cdot j\cdot} - \bar{a}_{\cdot\cdot\cdot}) \\ &+ (\bar{a}_{ij\cdot} - \bar{a}_{i\cdot\cdot} - \bar{a}_{\cdot j\cdot} + \bar{a}_{\cdot\cdot\cdot})\end{aligned}$$

for the individuals × tests face,

$$\begin{aligned}(\bar{a}_{i\cdot k} - \bar{a}_{\cdot\cdot\cdot}) &= (\bar{a}_{i\cdot\cdot} - \bar{a}_{\cdot\cdot\cdot}) \\ &+ (\bar{a}_{\cdot\cdot k} - \bar{a}_{\cdot\cdot\cdot}) \\ &+ (\bar{a}_{i\cdot k} - \bar{a}_{i\cdot\cdot} - \bar{a}_{\cdot\cdot k} + \bar{a}_{\cdot\cdot\cdot})\end{aligned}$$

for the individuals × situations face, and

$$\begin{aligned}(\bar{a}_{\cdot jk} - \bar{a}_{\cdot\cdot\cdot}) &= (\bar{a}_{\cdot j\cdot} - \bar{a}_{\cdot\cdot\cdot}) \\ &+ (\bar{a}_{\cdot\cdot k} - \bar{a}_{\cdot\cdot\cdot}) \\ &+ (\bar{a}_{\cdot jk} - \bar{a}_{\cdot j\cdot} - \bar{a}_{\cdot\cdot k} + \bar{a}_{\cdot\cdot\cdot})\end{aligned}$$

for the tests × situations face. Incidentally, when we come to factor these faces it will turn out that one source of variance in each—the differences of columns in the matrix—will vanish, because the correlation reduces all columns to the same mean and sum of squares. However, the interaction (covariance of columns) will not vanish.

It is unnecessary for purposes of illustration to convert these deviations to sums of squares. For it is evident from the deviations themselves that we shall lack from these three sources the data box term expressing the third order interaction of persons, tests, and situations.

5. The alternative might be suggested (possibly on the grounds of convenience of resulting standard scores in the data box) that we use standard scores in the contributory face score matrices. Actually, however, standard scores in the box would not result, unless the faces were completely uncorrelated and the sum divided by 3. Much more important in the argument for raw scores, however, is that we should not expect persons, tests and situations to contribute with equal variance to the behavior variance and that we are *interested in knowing* how different they are. It is true that our present procedure tends to make each

source give mainly positive additions to the final observed score, because we used for convenience positive constants for the means; but this does *not* affect the principle in the plasmode construction or analysis.

REFERENCES

Barker, R. G. 1968. *Ecological psychology: Concepts and methods for steadying the environment.* Stanford: Stanford University Press.

Broverman, D. M. 1961. Normative and ipsative measurement in psychology: Effects of some transformations in Q & R factor analysis techniques. *Psychological Review* 68:68–80.

Burt, C. L. 1937. Correlations between persons. *British Journal of Psychology* 28:56–96.

Cattell, R. B. 1934. The measurement of interest. *Character and Personality* 3:54–63.

———. 1946. *The description and measurement of personality.* New York: Harcourt, Brace, and World.

———. 1963. Formulating the environmental situation, and its perception, in behavior theory. In *Stimulus determinants of behavior,* ed. S. B. Sells. New York: Ronald Press.

———. 1966. *Handbook of multivariate experimental psychology.* Chicago: Rand McNally.

———. 1968. Trait view theory of perturbations in ratings and self-ratings. Its application to obtaining pure trait score estimates in questionnaires. *Psychological Review* 75:96–113.

———. 1972. Real base, true zero factor analyses. *Multivariate Behavioral Research Monographs* 72.

———. 1973. *Personality and mood by questionnaire.* San Francisco: Jossey Bass.

———. 1978. *The scientific use of factor analysis in behavioral and life sciences.* New York: Plenum.

———. 1979. *Personality and learning theory.* New York: Springer.

———. 1980. Guest editorial: Two basic models for personality-environment interaction and the need for their substantive investigation. *Multivariate Behavioral Research* 15:243–47.

Cattell, R. B., and S. Vogelmann. 1977. A comprehensive trial of the scree and KG criteria for determining the number of factors. *Multivariate Behavioral Research* 12:289–325.

Cattell, R. B., M. A. Coulter, and B. Tsujioka. 1966. The taxonomic recognition of types of functional emergents. In *Handbook of multivariate experimental psychology,* ed. R. B. Cattell. Chicago: Rand McNally.

Cohen, J. 1941. Physique, size and proportions. *British Journal of Medical Psychology* 18:325–37.

Endler, N. S., and J. McV. Hunt. 1966. Sources of behavior variance as measured by the S-R inventory of anxiousness. *Psychological Bulletin* 65:336–46.

Golding, S. 1975. Flies in the ointment. *Psychological Bulletin* 82:404–18.

Gorsuch, R. L. 1974. *Factor analysis.* Philadelphia: Saunders.

Hakstian, R., W. Rogers, and R. B. Cattell. Forthcoming. *An evaluation of the scree and other tests for the number of factors.*

Levin, J. 1965. Three-mode factor analysis. *Psychological Bulletin* 64:442–52.

Magnusson, D., and N. S. Endler. 1977. *Personality at the crossroads.* Hillsdale, N.J.: Lawrence Erlbaum Associates.

Mischel, W. 1968. *Personality and assessment.* New York: Wiley.

Ross, J. 1963. The relations between test and person factors. *Psychological Review* 70:432–43.

Sells, S. B. 1963. *Stimulus determinants of behavior.* New York: Ronald Press.

Stephenson, W. 1953. *Q-technique and its methodology.* Chicago: Chicago University Press.

Tucker, L. R. 1963. Implications of factor analysis of three-way matrices for measurement of change. In *Problems in measuring change,* ed. C. W. Harris. Madison, Wis.: University of Wisconsin Press.

———. 1964. The extension of factor analysis to three-dimensional matrices. In *Contributions to mathematical psychology,* ed. N. Frederiksen and H. Gulliksen. New York: Holt, Rinehart and Winston.

———. 1966. Some mathematical notes on three-mode factor analysis. *Psychometrika* 31:279–311.

Wittmann, W. W. 1979. Der mangel an multivariatem denken mit bezug auf klassifaktionen von situationen. In L. Eckensberger Bericht uber den 31 kangness der deutschen gesellschaft fur psychologie gottingen. *Hognefe* 1:180–82.

PART III
Multidimensional Scaling Tradition

10

The CANDECOMP-CANDELINC Family of Models and Methods for Multidimensional Data Analysis

J. Douglas Carroll and Sandra Pruzansky

CANDECOMP, standing for *CAN*onical *DECOMP*osition of N-way tables, is an acronym for a general class of multilinear models and associated methods for fitting those models to data (Carroll and Chang 1970; Carroll and Wish 1974a, 1974b; Wish and Carroll 1974). The data may be various types of multiway tables, some of which will be discussed below.

The CANDECOMP model is of the general form:

$$y_{i_1 i_2 \ldots i_N} \cong \sum_{r=1}^{R} a_{i_1 r}^{(1)} a_{i_2 r}^{(2)} \ldots a_{i_N r}^{(N)} , \qquad (10\text{–}1)$$

where $y_{i_1 i_2 \ldots i_N}$ is the general entry in an N-way data array, Y, while the a's are parameters in an R-dimensional representation. The values $a_{i_n r}^{(n)} (i_n = 1, 2, \ldots, I_n, n = 1, 2, \ldots, N)$ can be thought of as the entries in an $I_n \times R$ matrix, A_n; this matrix can be viewed as a multidimensional representation of the I_n entities of the nth *way* of the data array Y. The symbol "$\cong$" has two meanings. First, it indicates that the data on the left are equal to the expression on the right except for error terms. (The distribution and form of the error terms will not be specified in detail.) Second, it indicates that a least-squares solution for the parameters on the right of equation (10–1) are sought, given the data on the left-hand side of that equation.

CANDELINC stands for *CANDE*COMP with *LIN*ear Constraints (Carroll, Pruzansky, and Kruskal 1980). In CANDELINC, the CANDECOMP model is fit with linear constraints on one or more of the parameter matrices $A_1, A_2, \ldots, A_N$. The constraints may be on just one matrix, on all N, or any number between. These constraints are of the form:

$$A_n = X_n T_n , \qquad (10\text{–}2)$$

where X_n is a known, fixed, $I_n \times D_n$ matrix, often called the *design matrix* for the nth way; T_n is an unknown $D_n \times R$ matrix that can be thought of as a reduced parameter matrix, defining R different linear combinations of the D_n columns of the matrix X_n. The rth column of A_n, corresponding to the rth dimension for the nth way of the overall N-way design, is constrained to be a linear combination of D_n basic variables. These D_n variables can correspond to continuously valued outside variables on which the I_n entities have been measured or to "dummy" independent variables encoding a specific analysis of variance (ANOVA) design.

The CANDECOMP-CANDELINC family of models is an important special case of what might be dubbed, by analogy to the general *linear* model, the general *multilinear* model. Some additional complexities could be added, for example, a *core matrix* or *core array* of the sort introduced by Tucker (1964) in his three-mode factor analysis model. For now, let us consider some special cases of CANDECOMP-CANDELINC, which are graphically summarized in Figure 10–1, along with the general form of the CANDECOMP-CANDELINC family.

TWO-WAY CASE (BILINEAR MODEL)

The general two-way case of CANDECOMP has some very important and generally well known special cases. These include various forms of factor and components analysis and the *classical* approach to multidimensional scaling (MDS).

The general form of the model in the two-way case is the bilinear model

$$y_{ij} \cong \sum_{r=1}^{R} a_{ir} b_{jr} \ , \qquad (10\text{–}3)$$

where $a_{ir} \equiv a_{i_1 r}^{(1)}$ and $b_{jr} \equiv a_{i_2 r}^{(2)}$ in the earlier notation. In matrix notation,

$$Y \cong AB' \ , \qquad (10\text{–}4)$$

where

$$A \equiv ||a_{ir}||$$

and

$$B \equiv ||b_{jr}|| \ .$$

This model is what Kruskal (1978) has called the *direct factor* (or components) *analysis* model. In this special two-way case of the CANDECOMP model, the least-squares solution is given by a direct approach closely related to what has come to be referred to as the *singular value decomposition*. Historically it was called an *Eckart-Young analysis* by mathematical psychologists, based on a theorem published by Eckart and Young (1936) that established the least-squares properties of this method of fitting.

Figure 10–1. CANDECOMP-CANDELINC Family of Models and Methods, Showing Important Two-Way, Three-Way, and General *N*-Way Special Cases

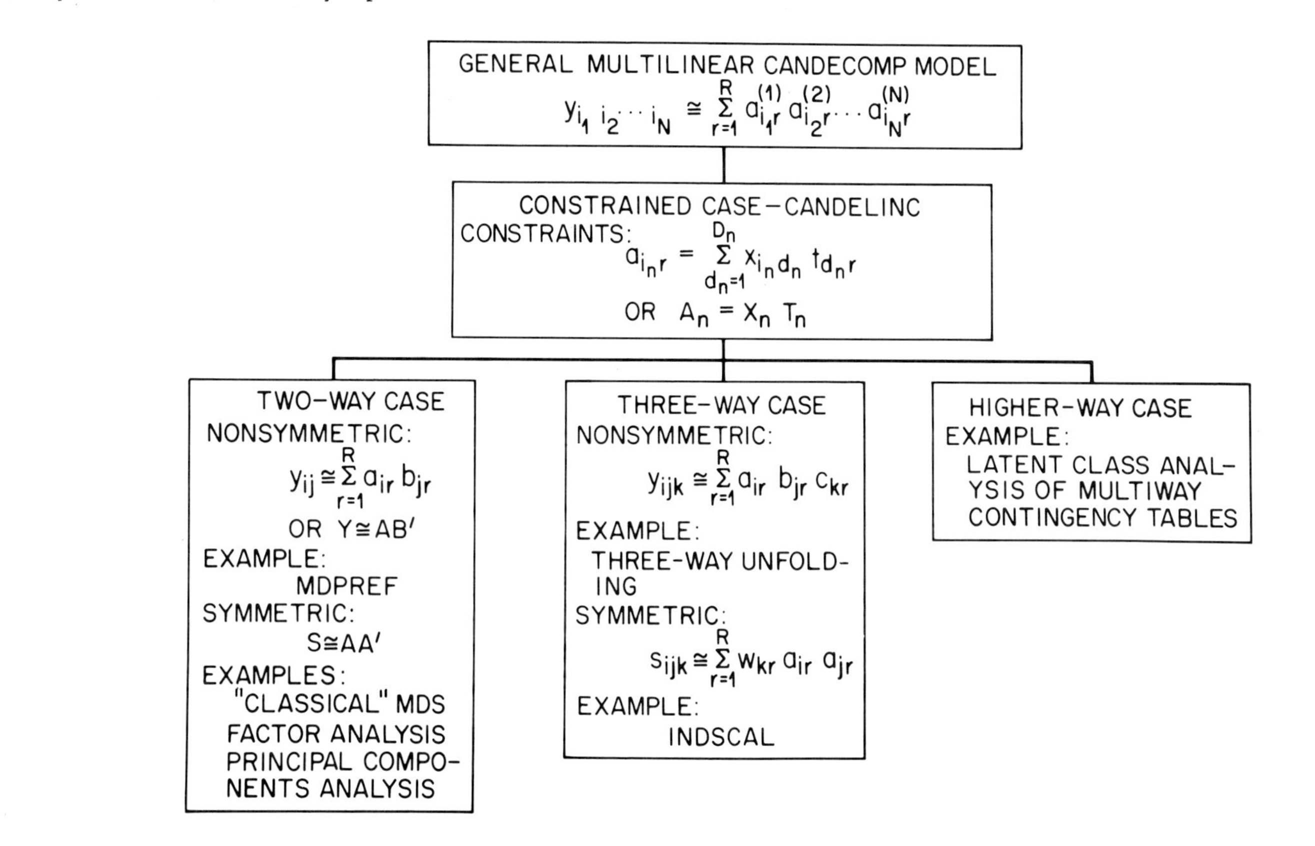

Example of the Nonsymmetric Bilinear Model:
MDPREF—An Application to Individual Differences in Preferences

MDPREF, an acronym for *M*ulti*D*imensional *PREF*erence analysis, is a *vector* model for individual differences in preferential choice behavior. (See Carroll [1972, 1980] for detailed discussions of this and other models for individual differences analysis of preference or other dominance data.)

The vector model is illustrated in Figure 10–2. This model assumes a set of stimulus points (points A through E in Figure 10–2) embedded in a multidimensional space. Different subjects are represented in this model by distinct vectors or directed line segments. Two subject vectors are shown in Figure 10–2. The preference order is assumed to be given by the projection of stimuli onto the vector representing this subject. As can be seen from the figure, quite different orders of projections—and thus quite different implied preference orders—can be accommodated in

Figure 10–2. Hypothetical Example Illustrating Geometric Representation of MDPREF Model for Individual Differences in Preferential Choice in Terms of a Vector Model

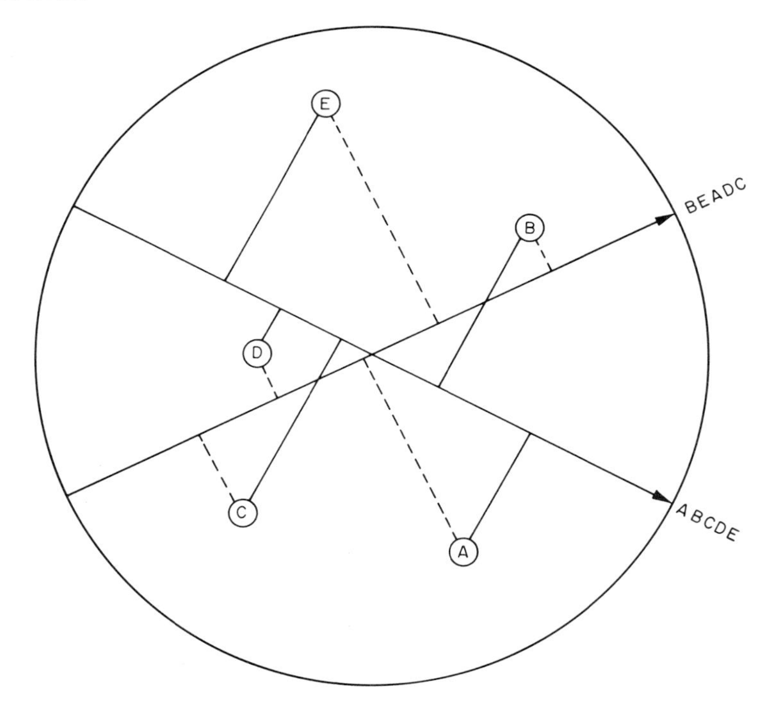

Note: Individual preference orders for two subjects are given by orthogonal projections of stimulus points A through E onto subject vectors. The implied preference orders for the two corresponding subjects are indicated near termini of vectors.

such a model. This is true even in the relatively simple two-dimensional case; even more varied orders of preference could be accommodated, of course, in higher dimensionalities. Although in this schematic illustration the vectors are all represented as being the same length, in fact, the lengths of subject vectors can vary in this model. The length of a subject vector has a somewhat secondary meaning to its direction. It does not affect the order of projections or even the interval scale properties of projections, but rather reflects the degree to which the model is accounting for that particular subject.

In the application of the bilinear form of the CANDECOMP model to preference data, the data matrix Y is a subjects × stimuli matrix of preference scores, either from direct ratings of preferences or derived from paired comparisons or other types of preferential choice data by methods described in Carroll (1972, 1980). Normalizing conditions are generally imposed on Y; for instance, the mean of each subject's preference scores are generally standardized to zero, and the variance of each subject's scores is usually standardized. In this case, the matrix A is a subjects × dimensions matrix that contains the coordinates of the *termini* of subjects' vectors, while B, stimuli × dimensions, defines coordinates of stimulus points. The *algebraic* operation of matrix multiplication is equivalent to the *geometric* one of projecting stimulus points onto subject vectors and then multiplying those projections by the length of the vector. Because of normalizing conventions used, this has the effect of making the variance of the projections proportional to the squared vector length.

Figure 10–3 shows the results of an MDPREF analysis of some preference data on nations due to Wish (1971). This bilinear model, like many other two-way multidimensional models, is subject to a rotational indeterminacy. The structure may need to be rotated to an optimally interpretable orientation of coordinate axes. The structure in Figure 10–3 is in *principal axis* orientation. The first dimension can be interpreted very nearly as an economic development dimension, with the highly developed industrialized countries on the right and underdeveloped or developing countries on the left. This interpretation was chosen by noting the order of projections of the stimuli (nations) on the dimension one axis. This interpretation might be enhanced by rotating the first dimension slightly, but that is debatable. The second dimension can be interpreted as political alignment and ideology; the communist countries are at the bottom and the noncommunist countries are at the top. The subject vectors are white for "doves" and black for "hawks." The labels "dove" or "hawk" were determined on the basis of subjects' responses to questions assessing their beliefs about the issue of U.S. involvement in Vietnam, a hot issue at the time these data were collected. The subjects were students living at International House at Columbia University and many came from communist countries. The interesting thing about the subject vectors is that the vectors for "hawks" and "doves" tend to be separated in ways that are quite easily accounted for: The "hawks" tend to like the noncommunist countries much more than the communist countries, while the "doves" like the communist countries better.

Figure 10–3. Result of MDPREF Analysis of Data on "Similarity to Ideal" Judgments of 21 Nations by 75 subjects from 8 Different Countries, living at International House at Columbia University

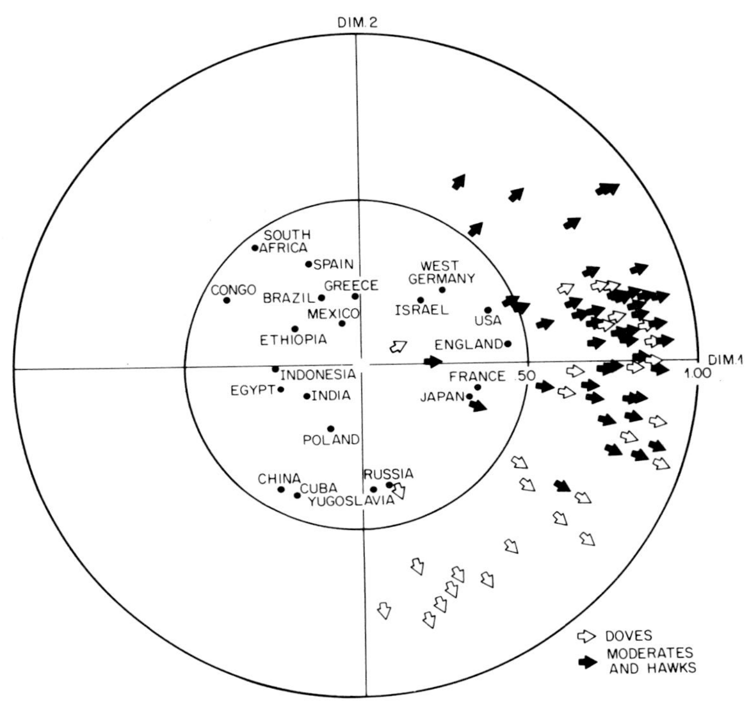

Note: Termini of subject vectors are shown and are encoded as white for "doves" (opposed to U.S. in Vietnam) and black for "hawks" (favoring that involvement). The squared length of a subject's vector is proportional to variance accounted for by the model in the data for that subject.

The Symmetric Bilinear Model: Indirect Factor or Components Analysis and Classical MDS

A special case of the bilinear model arises when the data matrix to which it is applied is symmetric. In that case, the singular value or Eckart-Young decomposition, after appropriate normalizing conventions are adopted, takes on the form

$$S \cong AA' \tag{10–5}$$

in matrix notation, where S is a *symmetric* data matrix. This typically will be a matrix of correlations, covariances, crossproducts, or other scalar-product-like values in the factor-analytic case. In the case of what is now often called the *classical* approach to two-way MDS (Torgerson 1958) these would be *derived* scalar products.

For a further discussion of two-way and three-way MDS and its relationship to factor analysis and other models and methods related to the bilinear model, the reader is referred to Carroll and Kruskal (1978), Kruskal and Wish (1978), and/or Carroll and Arabie (1980). For a general discussion of bilinear models, both the symmetric and nonsymmetric cases, see Kruskal (1978). Specific references of interest relating to the emergence in the 1960s of the modern approach to *nonmetric* two-way MDS, assuming merely ordinal proximity data as an alternative to the classical metric approach, are provided in Shepard (1962a, 1962b), Kruskal (1964a, 1964b), McGee (1966), and Guttman (1968). These nonmetric two-way approaches to MDS will not be discussed further in this chapter.

THE THREE-WAY CASE (TRILINEAR MODEL)

We will now focus on the three-way case of the CANDECOMP model with special emphasis on its application to the INDSCAL model and method for *IN*dividual *D*ifferences multidimensional *SCAL*ing. We describe the alternating least-squares approach to fitting the trilinear CANDECOMP model and touch on some of the more general applications of the nonsymmetric three-way case of CANDECOMP such as PARAFAC and *three-way unfolding*.

Fitting the Trilinear CANDECOMP-CANDELINC Model

In the three-way case, the CANDECOMP model can be written as:

$$y_{ijk} \cong \sum_{r=1}^{R} a_{ir} b_{jr} c_{kr} \ , \tag{10–6}$$

where $a_{ir} \equiv a^{(1)}_{i_1 r}$, $b_{jr} \equiv a^{(2)}_{i_2 r}$, and $c_{kr} \equiv a^{(3)}_{i_3 r}$ in the notation in equation (10–1). Given this model and "current estimates" of two sets of parameters—say the b_{jr}'s and c_{kr}'s—we can find an *exact least-squares* estimate of the third set by linear regression methods. This is a *conditional* least-squares estimate of the third set, conditional on the given estimates of the other two sets of parameters.

This can be seen more clearly by reformulating the problem as:

$$\overset{*}{y}_{is} \cong \sum_{r=1}^{R} a_{ir} \hat{g}_{sr} \ , \tag{10–7}$$

where

$$\overset{*}{y}_{is} = y_{i(jk)} \tag{10–8}$$

and

$$\hat{g}_{sr} = \hat{b}_{jr} \hat{c}_{kr} \ . \tag{10–9}$$

The values $\hat{b}_{jr}$ and $\hat{c}_{kr}$ are *current* estimates of b_{jr} and c_{kr}, respectively, while s is a subscript that is a function of j and k and ranges over all values of j and k.

By this simple notational device, we have converted this original *trilinear* model into a *bilinear* model. This can be expressed in matrix notation as:

$$\overset{*}{Y} \cong A\hat{G}' . \qquad (10\text{–}10)$$

Note that the matrix $\hat{G}$ incorporates both the $\hat{b}$'s and $\hat{c}$'s, while A is the matrix containing the a's. This proves to be a standard problem, essentially equivalent to least-squares multiple linear regression. In matrix notation, the least-squares estimate of A is:

$$\hat{A} = \overset{*}{Y}\,\hat{G}(\hat{G}'\hat{G})^{-1} . \qquad (10\text{–}11)$$

This amounts to postmultiplying both sides of (10–10) by the right *pseudoinverse* of $\hat{G}'$. (See Green with Carroll [1976] for a discussion of pseudoinverses, other generalized inverses, and related aspects of linear algebra relevant to this discussion.)

We use a general estimation scheme that was originally called a *NILES* (for *N*onlinear *I*terative *L*east-*S*quares) or a *NIPALS* (*N*onlinear *I*terative *PA*rtial *L*east-*S*quares) procedure by the Swedish statistician Herman Wold (1966) but has come to be known by the term *ALS* (for *A*lternating *L*east-*S*quares). First, we estimate the a's, with b's and c's fixed, by least-square methods, then the b's, with a's and c's fixed, and so on around the iterative cycle until convergence occurs. While there is no guarantee this process will converge to the *overall* least-squares estimates of all three sets of parameters, it does seem to do so in most cases.

In practice, the method seems almost always to converge to the global optimum solution. There is a mild *local minimum* problem, a tendency to converge to estimates such that no small change can improve the fit, although a large or global change can. This problem seems to be very slight in comparison with that in nonmetric two-way MDS, in which the algorithms are generally based on a gradient or steepest-descent method. Whether it is the difference in numerical procedures or the difference in the models that is critical is not known at present.

The CANDELINC model is fit by the same basic NIPALS or ALS procedure but applied to a reduced three-way array derived from the original three-way Y array, together with the matrices X_1, X_2, and X_3 defining the linear constraints. Details of the CANDELINC method for the three and higher-way case are given in Carroll, Pruzansky, and Kruskal (1980).

Examples of the Nonsymmetric Trilinear Model

PARAFAC

One very important case of three-way CANDECOMP is an approach to three-way factor analysis called *PARAFAC* (for *PARA*llel *FAC*tors), proposed by Richard Harshman (1970). PARAFAC, which is mathematically equivalent to three-way CANDECOMP, was independently proposed by Harshman as an alternative to Tucker's (1964) three-mode factor analysis. Carroll and Chang (1970) discussed the CANDECOMP approach as an alternative three-way factor analysis procedure but had not found it to be very successful in actual applications. Harshman and others (for instance, see Harshman, Ladefoged, and Goldstein 1977) have, however, made a number of useful applications of this approach. Part of the reason for Harshman's success appears to lie in preprocessing steps he has included, which apparently serve to enhance the interpretability of the resulting analyses. (Some aspects of this are discussed in chapter 6.)

Three-Way Unfolding

A second application of the general three-way CANDECOMP approach is to what De Sarbo and Carroll (1981) have called three-way unfolding. Coombs' ([1964] 1976) unfolding model assumes that preference is related monotonically to distance between a stimulus point and a subject's *ideal point*. The three-way unfolding model is a fairly straightforward extension of this model to the three-way case, in which the third way might be situations. It can also be viewed as an extension of the INDSCAL model and method (to be discussed below) for individual differences multidimensional scaling of proximity data to the case of *non*symmetric data; one such example is for data involving proximities between two different sets of entities by various individuals or in different situations.

The three-way unfolding model can be expressed mathematically as follows:

$$\delta_{ijk} \cong d_{ijk}^2 + c_k \ , \qquad (10\text{–}12)$$

where

$$d_{ijk} = \left(\sum_{r=1}^{R} w_{kr}(x_{ir} - y_{jr})^2 \right)^{1/2} \ . \qquad (10\text{–}13)$$

Here, δ_{ijk} can be viewed as a measure of preference of stimulus i by judge j in situation k, in the case of preference data; d_{ijk} is the weighted Euclidean distance between stimulus i and ideal point j in situation k. (In the case of nonsymmetric proximities or proximity data between two modes, δ_{ijk} is the dissimilarity of stimulus i in the first mode to stimulus j in the second, as judged by subject k.) In 10–13, x_{ir} is the rth coordinate of stimulus i, y_{jr} is the rth coordinate of the ideal point for judge j, and w_{kr} is a weight defining the importance of dimension r in situation k. Equation (10–12) states that δ is a linear function of the squared weighted Euclidean distance, d^2, plus a constant c_k for the kth situation. The rather strong assumption that δ is a linear function of *squared* distance is made largely for reasons of mathematical tractability. This form of the three-way unfolding model is, in fact, a direct generalization of Schonemann's (1970) two-way *metric unfolding* model to the three-way case.

To fit the three-way unfolding model, the first step is to double-center each $I \times J$ two-way slice corresponding to each of the k situations. These double-centered values, multiplied by $-\frac{1}{2}$, result in values q_{ijk}, defined as:

$$q_{ijk} = -\tfrac{1}{2} [\delta_{ijk} - \delta_{.jk} - \delta_{i.k} + \delta_{..k}] , \qquad (10\text{–}14)$$

where the "." replacing a subscript indicates the arithmetic mean over that subscript. The resulting values can be interpreted approximately as weighted *scalar products* between pairs of vectors from the two modes (see De Sarbo and Carroll 1981), as given by the following formula:

$$q_{ijk} \cong \sum_{r=1}^{R} w_{kr} \overset{*}{x}_{ir} \overset{*}{y}_{jr} , \qquad (10\text{–}15)$$

where

$$\overset{*}{x}_{ir} = \alpha_r (x_{ir} - x_{.r}) \qquad (10\text{–}16)$$

while

$$\overset{*}{y}_{jr} = \beta_r (y_{jr} - y_{.r}) \qquad (10\text{–}17)$$

and $\alpha_r \beta_r = 1$. That is, q_{ijk} is the weighted scalar product between vectors. The vectors are derived from the x and y coordinates by (a) translating the origin of the x and y points each to their respective centroids, and (b) multiplying the resulting centered coordinates for the rth dimension by scalars α_r and β_r that are reciprocally related. Admissible values for α_r and β_r are $\alpha_r = \beta_r = 1$, but the scalar products would remain unchanged for any values of α_r and β_r that are reciprocally related. This condition constitutes a fundamental indeterminacy of the scalar product form of the model.

The scalar product form of the three-way unfolding model is identical to the three-way CANDECOMP model, with $a_{ir} = \overset{*}{x}_{ir}$, $b_{jr} = \overset{*}{y}_{jr}$, and $c_{kr} = w_{kr}$. Three-way CANDECOMP can thus be used to solve for estimates of the w's, $\overset{*}{x}$'s, and $\overset{*}{y}$'s. Other numerical procedures must then be used to solve for the x's and y's. One of the two centroids, say the centroid of the x points, can be assumed to be zero without loss of generality. The centroid of the y points as well as the α's and β's are then solved for by a gradient based iterative algorithm to optimize the least-squares fit of the resulting d's to the δ's. Details can be found in De Sarbo and Carroll (1981).

Fitting the Symmetric Trilinear CANDECOMP Model

Let s_{ijk} represent the general entry in a three-way data array symmetric in its first two subscripts, i and j. That is, assume $s_{ijk} = s_{jik}$ for all i, j. For example, s_{ijk} could be a set of correlation or covariance matrices on the same set of variables, indexed by i and j, under different experimental conditions, indexed by k. For such symmetric data, a symmetric form of the CANDECOMP model is appropriate. This model is of the form:

$$s_{ijk} \cong \sum_{r=1}^{R} w_{kr} a_{ir} a_{jr} \quad . \tag{10–18}$$

This is the special case of the general three-way CANDECOMP model in which $a_{ir} = b_{ir}$ for all i and r, while $c_{kr} = w_{kr}$.

In practice, the three-way symmetric CANDECOMP model can be fit to symmetric data by fitting the *general* three-way CANDECOMP model. That is, we may fit a general model of the form:

$$s_{ijk} = \sum_{r=1}^{R} \overset{*}{a}_{ir} \overset{*}{b}_{jr} \overset{*}{c}_{kr} \quad . \tag{10–19}$$

The symmetry of the basic data with respect to the i and j subscript guarantees that, once the ALS iterative procedure converges to a stable solution, the $\overset{*}{a}$ and $\overset{*}{b}$ parameters will in fact be the same, up to a scale indeterminacy similar to that already discussed in the context of the scalar product form of the three-way unfolding model. That is, upon convergence, the estimates of $\overset{*}{a}_{ir}$ and $\overset{*}{b}_{jr}$ will be of the form:

$$\hat{\overset{*}{a}}_{ir} = \alpha_r \hat{a}_{ir} \quad , \tag{10–20}$$

while

$$\hat{\overset{*}{b}}_{jr} = \beta_r \hat{a}_{jr} \quad . \tag{10–21}$$

This scale indeterminacy can be resolved by use of appropriate normalizing conventions.

INDSCAL as a Special Case of CANDECOMP

While symmetric CANDECOMP has been used with three-way correlation or covariance data as a form of *indirect* three-way factor analysis, its most useful application has been to fitting the INDSCAL model (Horan 1969; Carroll and Chang 1970) for symmetric proximity data. In a manner closely analogous to the transformation of the three-way unfolding model from nonsymmetric proximities to nonsymmetric scalar product form, in INDSCAL, symmetric proximities are, via a series of steps, transformed to symmetric scalar products data; the scalar products are then appropriately fit by the symmetric three-way CANDECOMP model. This will be described in detail below.

The INDSCAL Model in Distance Form

INDSCAL is a particular model for individual differences in similarity or dissimilarity or other three-way *proximity* data. The model assumes these data arise from a process that relates them closely to weighted Euclidean distances defined in a certain way on a common space underlying the stimuli or other entities on which the proximities are defined. Specifically, the INDSCAL model can be expressed in the form:

$$F_k[\delta_{ijk}] \cong d_{ijk} \quad , \tag{10–22}$$

where

$$d_{ijk} = \left(\sum_{r=1}^{R} w_{kr}(x_{ir} - x_{jr})^2 \right)^{1/2} \quad , \tag{10–23}$$

while δ_{ijk} is the dissimilarity between, for example, stimulus (or other objects) i and j for subject (or other data source) k. The functions F_k will generally be considered to be linear in the metric case, monotonic in the nonmetric case, or other specified functional forms in other cases. It is important to note, however, that generally a different F_k is assumed for each separate individual, so that INDSCAL is, in that sense, a matrix *conditional* model. Essentially, then, INDSCAL generalizes two-way Euclidean MDS by substituting a *weighted* Euclidean metric for the ordinary, unweighted Euclidean metric. Different patterns of weights and different functions relating distances to data are allowed for each individual or other data source.

The type of data and resulting output for the INDSCAL model and method of two-way MDS is shown in Figure 10–4, Parts A and B, as contrasted with that for the INDSCAL model and method for three-way MDS, shown in Figure 10–5, Parts A and B. The input for INDSCAL comprises a three-way data array (such as stimuli × stimuli × subjects) of similarity, dissimilarity, or other *proximity* data symmetric in two of its indices. Each two-way slice in Figure 10–5, Part A, is simply an ordinary two-way square symmetric matrix of proximity data similar to the input to two-way MDS; each slice represents data from a different subject or other data source corresponding to the third way. While the output for two-way MDS is, generally, a single matrix of stimulus coordinates, the output of INDSCAL is *two* matrices. One matrix defines the coordinates of the *group stimulus space,* while the other defines the *subject space* in which the weights for different subjects (or other data sources) are plotted.

A more geometric interpretation of the INDSCAL model is provided by equations (10–24) and (10–25). Recall that x_{ir} is the rth coordinate of the ith stimulus in what we call the *group stimulus space*. Furthermore, we may conceive of each individual as having a *private perceptual space* whose general coordinate we will designate as $z_{ir}^{(k)}$. The $z^{(k)}$'s, however, are derived from the x's by the very simple relation expressed in equation (10–24):

$$z_{ir}^{(k)} = w_{kr}^{1/2} x_{ir} \quad . \tag{10–24}$$

The distances between objects for individual k are simply ordinary Euclidean distances computed in that subject's own private stimulus space:

Figure 10–4, Part A: Schematic Diagram of Typical Input for Two-Way MDS;* Part B: Schematic Diagram of Typical Output for Two-Way MDS.**

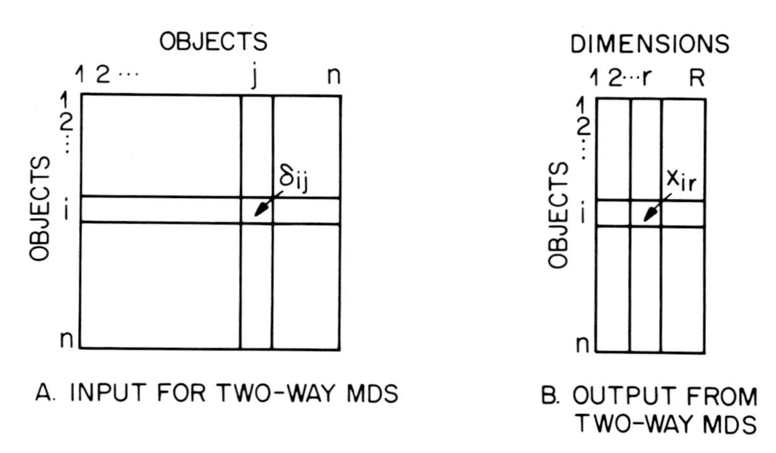

*The input is a single $n \times n$ square symmetric matrix of proximity (similarity or dissimilarity) data. The cell entry δ is a measure of proximity of stimuli, i and j.

**The MDS output is a single $n \times R$ matrix of coordinates of the n stimuli in R dimensions where x_{jr} is the coordinate of the ith stimulus on the rth dimension.

$$d_{ijk} = \left[\sum_{r}^{R} (z_{ir}^{(k)} - z_{jr}^{(k)})^2 \right]^{1/2} . \tag{10–25}$$

By substituting (10–24) into (10–25), we get (10–23). Thus, (10–24) and (10–25) together provide an alternate interpretation of the *weighted* generalization of the ordinary Euclidean metric defined in equation (10–23). Equations (10–24) and (10–25) express the INDSCAL model in terms of a simple class of transformations of the common space followed by computation of the ordinary Euclidean metric. The class of transformations can be described algebraically as linear transformations, with the transformation matrix constrained to be diagonal (sometimes called *strain transformations*). More simply, a transformation of this form amounts to simply rescaling each dimension by the square root of that particular subject's weight for that dimension. Geometrically, we can think of this as differentially stretching or shrinking each dimension by a factor proportional to the square root of the weight.

Hypothetical Example Illustrating the INDSCAL Model

This more geometric interpretation of the model is illustrated in Figure 10–6. The group stimulus space in the upper left shows nine stimuli, A through I, in a lattice configuration; the subject space in the upper right shows the weights of the dimensions for

Figure 10–5, Part A: Schematic Diagram of Typical Input for Three-Way MDS;* Part B: Schematic Diagrams of Typical Output for Three-Way MDS.**

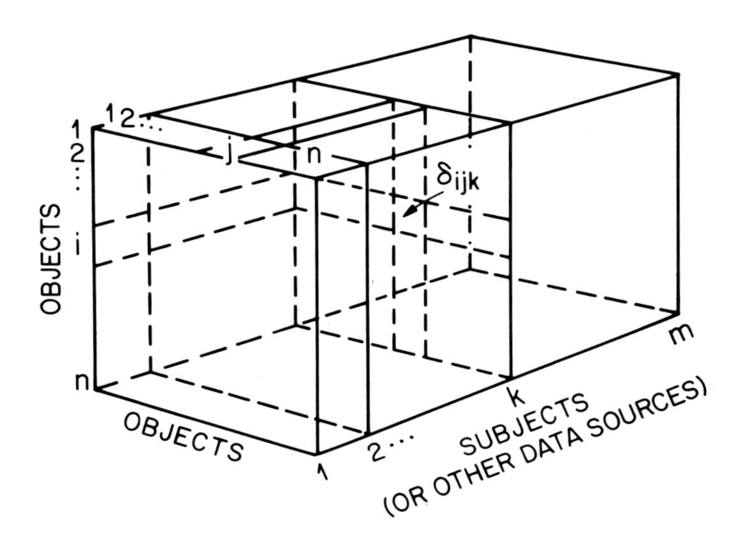

A. INPUT FOR THREE-WAY MDS

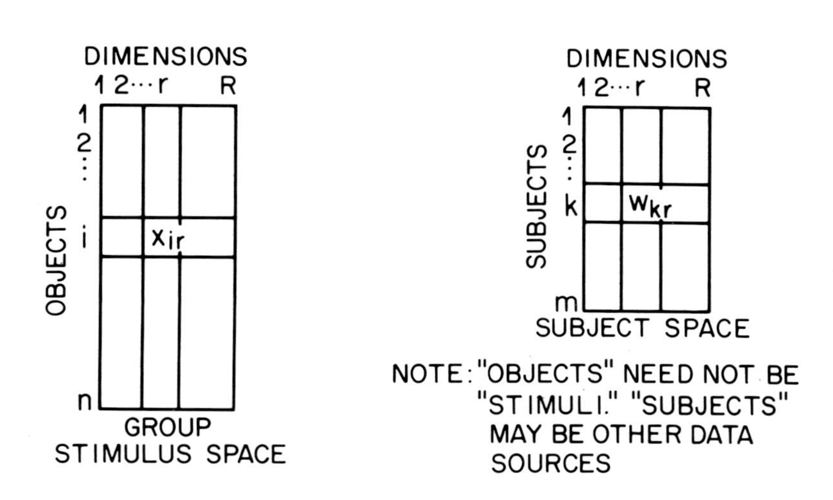

B. OUTPUT FROM THREE-WAY MDS

*The input is a set of m (≥ 2) $n \times n$ square symmetric data matrices, one for each of m subjects or other data sources. The cell entry δ_{ijk} is the proximity of objects i and j for subject k.

**There are two matrices: a group stimulus space consisting of an $n \times R$ matrix of coordinates of the n objects on R dimensions, and a subject space consisting of an $m \times R$ matrix of weights for m subjects on the same R dimensions.

nine hypothetical subjects. These weights can be thought of as stretching factors that are applied to the dimensions of the group stimulus space.

The effect of these differential weights is to produce a private perceptual space for each subject by rescaling—stretching or contracting—the dimensions of the group stimulus space. In Figure 10–6, for example, subject 3 has equal weights for the two dimensions; that subject's private perceptual space would therefore look exactly the same as the group stimulus space, except for an overall scale factor that could stretch or contract both dimensions uniformly, leaving their relative saliences unchanged.

The private spaces for subjects 2 and 4 are shown in Figures 10–6(C) and 10–6(D), respectively. Subject 2, who weights

Figure 10–6. Graphical Display of a Hypothetical Example Illustrating the INDSCAL Model, showing: (A) a group stimulus space; (B) a subject space of weights; and private perceptual spaces for subjects 2 (C) and 4 (D)

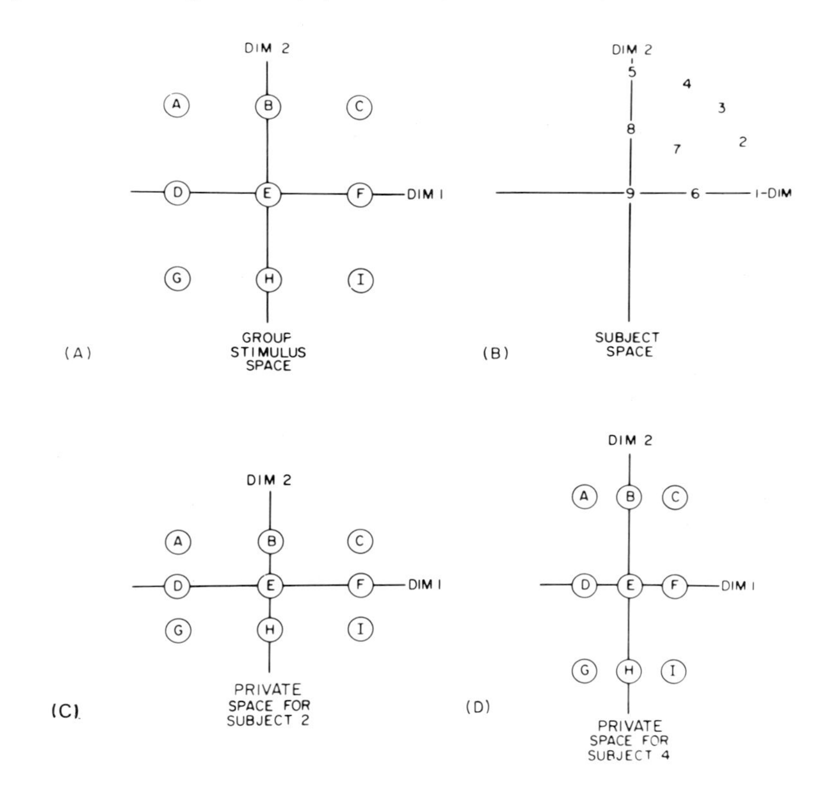

Note: For a given subject, weights on each dimension in the subject space are applied to the dimensions of the group stimulus space to produce a private perceptual space. In this illustration, the first power weights have been used to accentuate the effects of differential weighting. In practice, the square roots of the weights are applied.

dimension 1 more highly than dimension 2, has a perceptual space *compressed* along the dimension 2 axis (or what amounts to the same thing, stretched in the dimension 1 direction). The reverse applies to subject 4, who has a higher weight on the second than on the first dimension.

An important property of INDSCAL is one often called *dimensional uniqueness*. This means that, unlike ordinary two-way MDS, the dimensions are uniquely determined and cannot be rotated or otherwise transformed without changing the solution in an essential way. Psychologically, this means that INDSCAL dimensions are assumed to correspond to *fundamental* physiological, perceptual, or conceptual processes whose strength or salience may differ from individual to individual. Mathematically, a rotation or other transformation of coordinate system will change the family of permissible transformations of the group stimulus space and thus the family of possible individual metrics. This can be seen quite clearly in the hypothetical example by noting that, for instance, a 45° rotation would change the family of private perceptual spaces from one of different rectangles to one of different rhombuses. Statistically, a rotation or other transformation of axes will generally deteriorate the fit of the data to the INDSCAL model. The problem of rotation of axes is one that frequently complicates and often attenuates the utility of MDS analysis, particularly in the case of higher dimensional solutions. The property of dimensional uniqueness proves to be one of the most important aspects of INDSCAL, as it obviates this problem in almost all cases.

Distance of a subject from the origin in the subject space is a rough measure of the proportion of variance of the data for that subject accounted for by the multidimensional solution. This is comparable to the concept of communality in factor analysis. Although subjects 3 and 7 have the same pattern of dimension weights, a higher proportion of the variance in subject 3's data could be accounted for in this example. The data for subjects closer to the origin are generally less well accounted for by the INDSCAL analysis, so that the dimensions of the group stimulus space *altogether* are less salient for them. The lower "communality" for subjects closer to the origin may be due to idiosyncratic dimensions not uncovered in the R-dimensional solution or to lower reliability (more random error) in their data. Subject 9, who is precisely at the origin, is completely "out of" this analysis. Either he is responding completely randomly or is simply responding reliably to a completely different set of dimensions.

The possibility of zero weights allows as a special case the situation in which two or more groups of subjects have completely different perceptual spaces, with no necessary communality between them. This situation can be accommodated by defining a group stimulus space in which the number of dimensions is the sum of the dimensions for the two different groups of subjects; each subgroup has nonzero weights on only one subset of dimensions. For example, if one group perceives the stimuli in terms of dimensions A and B, while the second perceives them in terms of C and D, we may accommodate these two completely different "points of view" (to use Tucker and Messick's [1963] terminology) by assuming a four-dimensional space comprised of dimensions *A, B, C,* and *D.* Thus, the Tucker and Messick points-of-view model can be accommodated as a special case of INDSCAL. INDSCAL is, of course, more interesting and appropriate when there is some communality in perception among the subjects.

The INDSCAL Model in Scalar Product Form

The INDSCAL method of analysis for the metric version of the INDSCAL model is a direct generalization of the classical two-way MDS method described in Torgerson's (1958) book, in which proximities are converted to scalar products. As in the classical method, each two-way symmetric matrix of proximities is first converted to to dissimilarities. If the initial data are similarity, the scale is reversed by multiplying the values by -1 or subtracting them from the largest value; if the initial data are already dissimilarity, no transformation would be done at this point.

The next step, as in the classical metric two-way MDS procedure, is to convert the dissimilarities to estimated distances. This is the process usually referred to as *estimation of the additive constant.* Estimated distances are then transformed to scalar products. This transformation entails squaring the distance, double-centering the squared distance estimates, and then multiplying by $-\frac{1}{2}$. This transformation is closely analogous to the one described for three-way unfolding to convert the data from distance form to scalar product form. The important difference is, in that case, the data were assumed to be linear with squared distances; however, in the present case, the data are assumed to be linear with *first power* distances. Because of this different assumption, more elaborate preprocessing is required here. The resulting values after this preprocessing can be interpreted as estimated scalar products of vectors issuing from an origin fixed at the centroid of all the stimulus points. Since, in the symmetric case, we are dealing with only a single set of points, we do not have to solve for a translation of one of two sets of points as we do in the case of three-way unfolding.

By applying the procedures described in Torgerson (1958) for converting estimated dissimilarities into estimated scalar products to each two-way proximities matrix, we arrive at a three-way array of estimated scalar products, whose general entry is $\hat{s}_{ijk}$ (for subject k and stimuli i and j). We now need to make an analogous conversion of the INDSCAL *model* into scalar product form.

By definition, the true scalar products for subject k are defined as:

$$\sigma_{ijk} = \sum_r z_{ir}^{(k)} z_{jr}^{(k)} , \qquad (10\text{–}26)$$

where the $z^{(k)}$'s are defined as:

$$z_{ir}^{(k)} = w_{kr}^{1/2} x_{ir} . \qquad (10\text{–}27)$$

Substituting (10–27) into (10–26), we have:

$$\sigma_{ijk} = \sum_r w_{kr} x_{ir} x_{jr} , \qquad (10\text{–}28)$$

which is the desired *scalar product form* of the INDSCAL model. Our conversion of dissimilarities to scalar products form gives us values s_{ijk} , which are taken as estimates of σ_{ijk} . The model for the s's is thus:

$$s_{ijk} \cong \sum_r^R w_{kr} x_{ir} x_{jr} . \qquad (10\text{–}29)$$

This model is of the form of the symmetric three-way CANDECOMP model, so the procedure described earlier for fitting the symmetric case of that model can be used to estimate these parameters.

Normalizing Conventions for Fitting the INDSCAL Model

By the nature of the INDSCAL model, there is a basic scale indeterminacy; that is, the dimensions of the stimulus space could be rescaled by an arbitrary diagonal transformation, and the effect of this could be "undone" by rescaling the dimensions of the subject space by the inverse of the square of that diagonal transformation. This scale indeterminacy is resolved in the INDSCAL procedure by introducing the normalizing convention that the variance of stimulus coordinates of all dimensions be equal to $1/n$, where n is the number of stimuli. An arbitrary permutation of the dimensions of both the stimulus and subject space is also allowable, but this indeterminacy is generally resolved by simply ordering the dimensions according to approximate variance accounted for.

As in any Euclidean distance model, including the weighted Euclidean model underlying INDSCAL, there is an indeterminacy of the origin of the coordinate system; this is due to the invariance of Euclidean distances under translation of coordinates. This indeterminacy is resolved very simply, as in almost all distance based two-way and three-way MDS procedures, by placing the origin of the coordinate system at the centroid of the stimulus points. The effect of these two normalizing conventions is that the mean of the coordinates on each dimension is zero, while their sum of squares is one. Adopting these conventions, along with the data normalization mentioned earlier, also has an effect on the interpretation of the weights in the subject space. The square of the distance of a subject point from the origin of the coordinate system in the subject space is interpretable, approximately, as proportion of variance accounted for in the derived scalar products data for that subject.

An Illustrative Application of INDSCAL to Some Color Dissimilarities

We now illustrate the INDSCAL model by an example that entails a reanalysis of some data on color perception collected by Helm

(1964). Helm and Tucker (1962) had analyzed these data via a points-of-view analysis and found that about ten points of view (that is, ten different perceptual spaces) were required to account for the individual differences among their 14 subjects.

Our analysis in terms of the INDSCAL model has quite nicely accounted for these individual differences with a single two-dimensional solution. Moreover, the unique dimensions from the INDSCAL analysis shown in Figure 10–7, (A), were interpretable without rotation of axes. Dimension 1 corresponds essentially to a "blue versus yellow" (or, more accurately perhaps, a "purple-blue to green-yellow") factor, and dimension 2 to a "red versus green" (or "purple-red to blue-green") factor. This accords very well with physiological and psychophysical evidence strongly suggesting the existence of blue-yellow and red-green receptors.

Included among Helm's subjects were four who were deficient, to various degrees, in red-green color vision. In the INDSCAL analysis, this deficiency is reflected in the subject weights, as seen in Figure 10–7, (B); these subjects all have lower weights for the red-green factor (dimension 2) than do any of the normal subjects. The effect of these differential weights can be seen in Figure 10–7, (C)–(F), by comparing the private perceptual spaces for the colorblind subjects with those for the normals. The spaces for the color-deficient subjects, Figures 10–7, (E) and (F), are compressed in the red-green direction, relative to the spaces for the normals, Figures 10–7, (C) and (D). This reflects the fact that red and green are much more similar to each other for these subjects than they are for the normals.

Had we not known that these subjects were red-green color deficient, we could have discovered this through inspection of the weights in Figure 10–7, (B). In fact, subject N10 seems from this analysis to border on a mild form of blue-yellow deficiency. That subject's weight for dimension 1 is about as small as some red-green deficient subjects' weights for dimension 2. If such a deficiency exists for this subject, this would seem to be a fact determined by the INDSCAL analysis but *not* detected by other measures. (In any case, such a deficiency was not reported by Helm [1964].)

The inspection of the private spaces provides insight into the way in which INDSCAL uses individual differences in perception to determine a unique orientation of the coordinate axes. If the coordinate axes were oriented in a different way—for instance, by a 45° rotation of the axes of the group stimulus space—the private spaces for color-deficient subjects could not be compressed along a line from red to green but would have to be compressed in some other direction. More generally, the *family* of transformations of the stimulus space generated by differential weights varies as a function of the orientation of axes in the group stimulus space. INDSCAL manages to seek out not only the optimal stimulus space but also an optimal and unique orientation of coordinate axes within that space. It is a matter of empirical observation that this unique orientation is also a highly interpretable one. Thus, INDSCAL avoids the problem of rotation of axes that so often complicates two-way MDS and other multivariate behavioral studies. This is not at all a coincidence, however, since INDSCAL is based on a psychological model that postulates

Figure 10–7. A Two-Dimensional INDSCAL Analysis of Helm's (1964) Data on Color Perception showing the: (A) group stimulus space; (B) subject space of weights; (C) private perceptual space for subject N10a with normal color vision; (D) private perceptual space for subject N7 with normal color vision; (E) private perceptual space for color-deficient subject CD4a; and (F) private perceptual space for color-deficient subject CD1

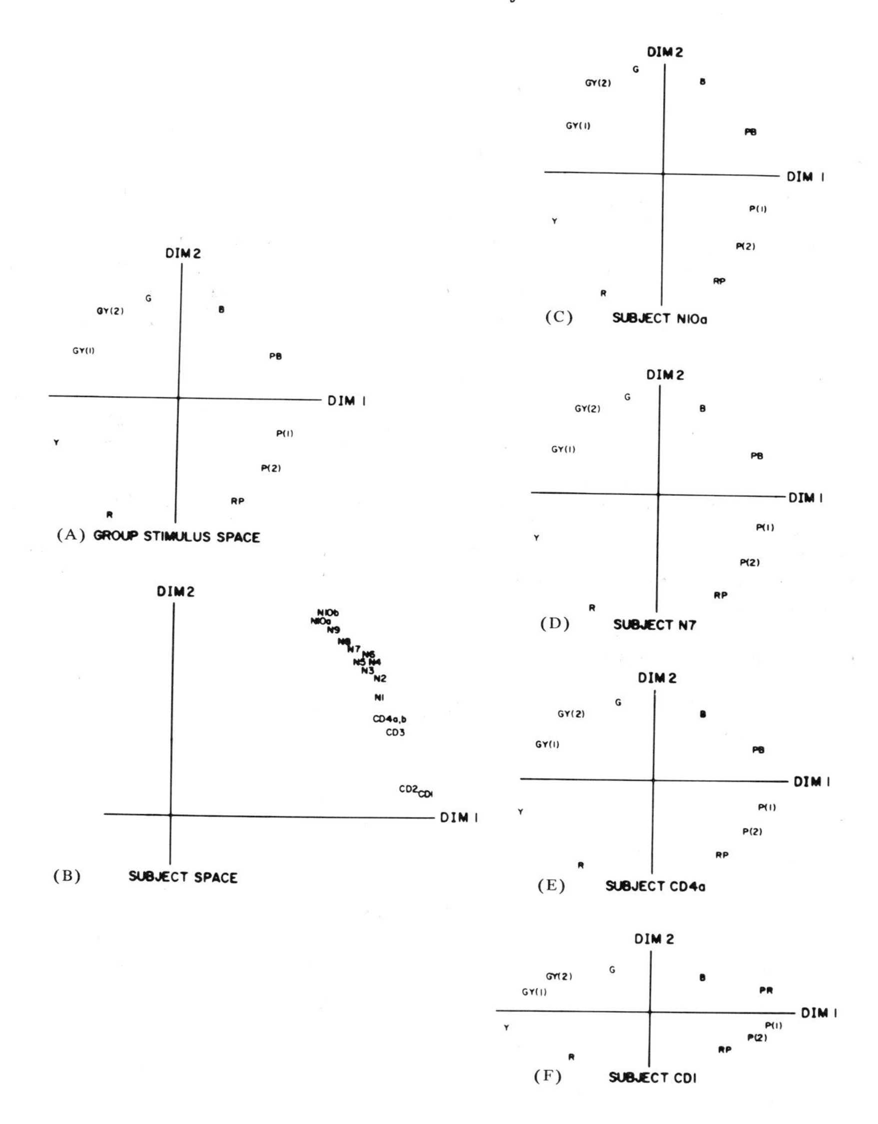

Note: The coding of colors of constant saturation and brightness is as follows:

R = red; Y = yellow; GY(1) = green-yellow, with more green than GY(1); G = green; B = blue; PB = purple-blue; P(2) = purple, with more red than P(1); RP = red-purple.

dimensions that are unique in the sense of being modifiable (in salience or perceptual importance) across and perhaps even within individuals. If the INDSCAL model does accurately reflect the underlying psychological processes (at least to a good first approximation), then the coordinate axes from an INDSCAL analysis can be assumed to correspond to *fundamental* psychological dimensions.

Applying the Trilinear CANDELINC Model for Improving Computational Procedures for INDSCAL

One interesting application of the CANDELINC model is described in detail in a paper by Carroll and Pruzansky (1979). They discuss speeding up the fitting of the unconstrained CANDECOMP model by using the data themselves to define the variables that comprise the design matrix (or design matrices for different modes). In the most important application of this approach, called *INDSCAL-LS,* the design matrix for the INDSCAL stimulus dimensions is defined by applying the classical two-way metric MDS procedure to an average (over subjects) matrix of estimated scalar products among stimuli. It can be shown that a linear transformation of the variables contained in the resulting design matrix will closely approximate the dimensions appropriate to the INDSCAL model. Using three-way CANDELINC appropriately—in a version called *LINCINDS,* which stands for *LIN*early *C*onstrained *IND*Scal—allows solving for the linear transformation required to recover the best least-squares fit to the INDSCAL model possible within these constraints. This approach yields a very good approximation to the unconstrained INDSCAL solution and can be much faster computationally, particularly for problems involving very large data arrays. Analogous applications of CANDELINC can be used to speed up fitting the general multilinear CANDECOMP model.

THE *N*-WAY CASE (MULTILINEAR MODEL)

CANDECOMP also generalizes to the many-way case. An interesting application of this is to analysis of multiway contingency tables via Lazarsfeld's latent class model (Carroll, Pruzansky, and Green 1979). The *N*-way case of CANDECOMP could also be applied, possibly after suitable preprocessing, as a many-way factor-analytic procedure; this is a generalization of the PARAFAC approach to the *N*-way case, for $N > 3$. It can also provide a *many*-way generalization of INDSCAL for the case involving more than one data source; an example is the case in which proximity data are provided by different subjects in each of a number of experimental conditions. The original INDSCAL program (see Chang and Carroll 1969) allows many such *N*-way analyses.

Another interesting application of the higher-way case of CANDECOMP is to the analysis of interaction effects in an *N*-way factorial design, after the main effects have all been eliminated. This would provide a three- or higher-way generalization of

Gollob's (1968) FANOVA approach. This application was first suggested in the original paper on the general CANDECOMP model by Carroll and Chang (1970). It has been pursued recently in at least the three-way case, involving an application of Harshman's mathematically equivalent approach called *PARAFAC*, by Chen, Harshman, and Kettenring, as reported by Kettenring (1983).

Fitting the Multilinear CANDECOMP Model

The NILES, NIPALS, or alternating least-squares (ALS) approach to fitting the general N-way case of the CANDECOMP models, as first described in Carroll and Chang (1970), can be seen to be a fairly straightforward extension of the three-way algorithm described earlier. This general case of CANDECOMP, it might be noted, allows (among other things) for extension of the three-way PARAFAC-CANDECOMP model to the multimode case. The attendant advantages of "dimensional uniqueness" (or what Harshman now calls the "intrinsic axis" property) also apply to this multimode CANDECOMP model/method. Reiterating the model expressed in equation (10–1), we have:

$$y_{i_1 i_2 \cdots i_N} \cong \sum_{r=1}^{R} a_{i_1 r}^{(1)} a_{i_2 r}^{(2)} \cdots a_{i_n r}^{(N)},$$

for $(i_n = 1, 2, \ldots, I_n,\ n = 1, 2, \ldots, N)$. Assume we have current estimates of all the parameter sets but one, say $A_1 \equiv \|a_{i_1 r}^{(1)}\|$. We seek conditional least-squares estimates of the parameters defining the A_1 matrix. Letting s now be a subscript that ranges over all $I_2 \times I_3 \times \ldots \times I_N$ values of the subscripts $i_2, i_3, \ldots, i_N$, we may rewrite (10–1) as:

$$y_{i_1 s}^{(1)} \cong \sum_{r=1}^{R} a_{i_1 r}^{(1)} h_{sr}^{(1)} \qquad (10\text{–}30)$$

$$y_{i_1 s}^{(1)} \equiv y_{i_1 i_{2s} i_{3s} \cdots i_{Ns}}, \qquad (10\text{–}31)$$

where i_{ns} denotes the particular value of subscript i_n associated with the subscript s. Similarly:

$$\hat{h}_{sr}^{(1)} \equiv \hat{a}_{i_{2s} r}^{(2)} \hat{a}_{i_{3s} r}^{(3)} \cdots \hat{a}_{i_{Ns} r}^{(N)}, \qquad (10\text{–}32)$$

where the $\hat{a}$'s are the current estimates of the a's.

In matrix notation, we have, analogous to (10–10) for the three-way case:

$$Y_1 \cong A_1 \hat{H}'_1, \qquad (10\text{–}33)$$

where Y_1 and $\hat{H}_1$ are matrices whose general entries are $y_{i_1 s}^{(1)}$ and $h_{sr}^{(1)}$, respectively. Assuming the right pseudoinverse of H'_1 exists (which, in general, it does) the *conditional* least-squares estimates of A_1 is given by:

$$\hat{A}_1 = Y_1 \hat{H}_1 \left[\hat{H}_1' \hat{H}_1\right]^{-1} . \tag{10–34}$$

More generally, using the generalizations to define Y_n and $\hat{H}_n$, given fixed current estimates of all A matrices except A_n, the conditional least-squares estimate of A_n is given by:

$$\hat{A}_n = Y_n \hat{H}_n \left[\hat{H}_n' \hat{H}_n\right]^{-1} = P_n Q_n^{-1} , \tag{10–35}$$

where

$$P_n = Y_n \hat{H}_n \tag{10–36}$$

$$Q_n = \hat{H}_n' \hat{H}_n . \tag{10–37}$$

From the definition of Y_n and $\hat{H}_n$, the general entries of P_n and Q_n are:

$$p_{i_n r}^{(n)} = \sum_s y_{i_n s}^{(n)} \hat{h}_{sr} \tag{10–38}$$

$$= \sum_{i_1}^{I_1} \sum_{i_2}^{I_2} \cdots \sum_{i_{(n-1)}}^{I_{(n-1)}} \sum_{i_{(n+1)}}^{I_{(n+1)}} \cdots \sum_{i_N}^{I_N} y_{i_1 i_2 \cdots i_n \cdots i_N} \prod_{m \neq n} \sum_{i_m}^{I_m} a_{i_m r}^{(m)} a_{i_m r'}^{(m)} ,$$

while

$$q_{rr'}^{(n)} = \prod_{m \neq n} \sum_{i_m}^{I_m} a_{i_m r}^{(m)} a_{i_m r'}^{(m)} . \tag{10–39}$$

As in the three-way case, in the general N-way case, we start with initial estimates of, for instance, A_2, A_3, . . . , A_N, reestimate A_1, and then fix all the matrices but A_2, reestimate A_2, and so on through A_3, A_4, . . . , A_N. We cycle back through these reestimation phases, iterating these conditional least-squares estimation steps, until the procedure converges to an overall least-squares estimate of the parameters of the N-way model. As with the three-way case, this may be only a locally optimal solution, although experience indicates that in many cases the procedure does converge to global least-squares estimates of the A matrices.

Application of the Multilinear CANDECOMP Model to Latent Class Analysis

A useful application of the general N-way case of CANDECOMP, listed in the right-most branch in Figure 10–1, is to the analysis of multiway contingency tables via Lazarsfeld's latent class model. This application resulted from the insight that, after a slight reparameterization of the latent class model, it takes on the exact form of the N-way CANDECOMP model (Carroll 1975; Carroll, Pruzansky, and Green 1979; Pruzansky, Green, and Carroll 1979). This insight and an illustrative example of the use of CANDECOMP for latent class analysis, are discussed below.

The latent class model was proposed by Lazarsfeld more than 30 years ago (Lazarsfeld 1950). The model provides one approach to the multidimensional analysis of qualitative data. The data form a multiway contingency table in which observed items are related to eachother in some way. The model assumes that respondents may be divided into two or more latent classes while members of the same latent class have the same probability of responding in a particular way to any item. The key assumption is that within any latent class, responses to different items are independent.

This model would appear to have many potential applications in the social sciences. However, for many years it had little use, partly because of practical problems related to the estimation procedures. The estimation algorithms that had been proposed were computationally complex and costly. Also, there were limitations on the size of the contingency tables that could be analyzed. In the past few years, there has been a renewed interest in the model and more efficient estimation procedures have been proposed (for example, see Goodman 1974a, 1974b, 1975). Using CANDECOMP for estimating latent class parameters is one procedure for reducing computational costs; it also eliminates the limitations on the size of the data tables that can be analyzed.

Mathematically, the latent class model, in the frequency domain, can be stated as:

$$f_{i_1 i_2 i_3 \cdots i_n} \cong \sum_{\alpha=1}^{A} k^{\alpha} p_{i_1}^{\alpha} p_{i_2}^{\alpha} p_{i_3}^{\alpha} \ldots p_{i_N}^{\alpha} , \tag{10–40}$$

where $f_{i_1 i_2 \ldots i_n}$ is the observed frequency in one cell of an N-way contingency table. The latent class parameters are k^{α}, the number of responses in latent class α, and $p_{i_n}^{\alpha}$ the marginal probability of a response being from the ith category of item n for latent class α.

If we make the following identification of variables:

$$R \equiv A , \tag{10–41}$$

$$a_{i_1 r}^{(1)} \equiv k^r p_{i_1}^r , \tag{10–42}$$

and

$$a_{i_n r}^{(n)} \equiv p_{i_n}^r \quad (n \geq 2) , \tag{10–43}$$

then, by substitution, the latent class model can be written as follows:

$$f_{i_1 i_2 i_3 \ldots i_n} \cong \sum_{r=1}^{R} a_{i_1 r}^{(1)} a_{i_2 r}^{(2)} a_{i_3 r}^{(3)} \ldots a_{i_N r}^{(N)} . \tag{10–44}$$

If we identify the y's of the general CANDECOMP model, stated in (10–1), to the f's in (10–44), we see that the latent class model can be expressed as a special case of the CANDECOMP model.

To estimate latent class parameters, first, the N-way CANDECOMP fitting procedure is used to estimate the parameters $a_{i_n r}^{(N)}$; then, the CANDECOMP parameters are transformed to estimates of k^α and $p_{i_n}^\alpha$ by the equations:

$$\hat{p}_{i_n}^{\alpha} = \frac{\hat{a}_{i_n \alpha}^{(n)}}{\sum_{i_n=1}^{I_n} \hat{a}_{i_n \alpha}^{(n)}} \tag{10–45}$$

and

$$\hat{k}^{\alpha} = \prod_{n}^{N} \left[\sum_{i_n=1}^{I_n} \hat{a}_{i_n \alpha}^{(n)} \right]. \tag{10–46}$$

An Application of Latent Class Analysis to Marketing Data

The use of CANDECOMP in latent class analysis is demonstrated by application to data obtained through a mail survey by the market research group of a large American corporation. The survey collected data on household demographics, respondent attitudes, and usage of various services offered by the company. The purpose of the study was to find out if customers' usage of one particular service varied by demographic segment and, if so, in what ways. Seven items were used as data to form a multiway contingency table. Table 10–1 shows the 7 items and response categories for each item. The sample consisted of 12,432 households. The data, then, consisted of a 7-way contingency table (3 × 2 × 2 × 3 × 3 × 2 × 2) containing a total of 432 cells. Solutions were sought for 4, 3, and 2 classes. We report the 4-class solution here because the sums of squares accounted for, .973, was high and the solution was interpretable.

Table 10–2 shows the latent class parameters for each of the 4 classes. These include the marginal probability of a response from category i of item j (for example, the probability that a respondent in latent class I has a low annual income is .94), the proportion of the total sample, and the total number of respondents estimated to be in each class.

An examination of the service usage pattern for each class shows that the probability of being a high user of the service is only .12 for latent class I but is much higher, .65, for latent class IV. Latent class I is characterized principally by older, high-school-educated, low-income respondents who don't have any telephone extensions. Latent class IV has a high proportion of college-educated, high-income heads of household and they tend to have one or more extensions. The emergence of these two groups of users was predicted before the study was undertaken. More interesting results appear in classes II and III. They are

TABLE 10-1. Items Forming Multiway Contingency Table

A. Age of head of household
   1. Younger (34 years or younger)
   2. Middle (35 through 54 years)
   3. Older (55 years or older)

B. Education of head of household
   1. High school graduate or less
   2. Some college or more

C. Mobility of household
   1. Low (no changes in residence during past 5 years)
   2. High (one or more moves during past 5 years)

D. Household location
   1. Urban area
   2. Suburban area
   3. Rural area

E. Household's annual income
   1. Low ($12,500 or less)
   2. Middle ($12,501-$20,000)
   3. High (over $20,000)

F. Number of extension phones in household
   1. None
   2. One or more

G. Usage of service
   1. High (above the national average)
   2. Low (below the national average)

about equal in terms of the probability of high usage of the service. However, the demographic patterns differ in interesting ways. Class II households, which represent 37% of the total sample, tend to be older, less educated, less mobile, and have a higher income level than those in class III. Both classes are attractive prospects for use of the service of interest. The different demographic make-ups of these two classes suggest the possibility of developing different marketing strategies for each segment tailored specifically to the profiles that characterize each class. These two classes, accounting for over half of the total sample, were not anticipated by the study's sponsors. Another market example that uses latent class analysis via CANDECOMP has been reported elsewhere (Green, Carmone, and Wachspress 1976).

SUMMARY

In this chapter, we have illustrated the applicability of a general multilinear model called *CANDECOMP* to many different data-

TABLE 10-2. Latent Class Parameters from Four-Class CANDECOMP Solution

Item	Class I	Class II	Class III	Class IV
A. Age				
Younger	0.05	0.06	0.74	0.16
Middle	0.21	0.62	0.26	0.61
Older	0.74	0.32	0	0.23
B. Education				
High school or less	0.85	0.70	0.50	0.12
Some college or more	0.15	0.30	0.50	0.88
C. Mobility				
Low	0.79	0.83	0.04	0.56
High	0.21	0.17	0.96	0.44
D. Household Location				
Urban	0.50	0.40	0.47	0.36
Suburban	0.36	0.36	0.33	0.33
Rural	0.14	0.24	0.20	0.31
E. Annual income				
Low	0.94	0.41	0.60	0.00
Middle	0.05	0.56	0.40	0.50
High	0.01	0.03	0	0.50
F. Number of extensions				
None	0.80	0.47	0.46	0.19
One or more	0.20	0.53	0.54	0.81
G. Usage of Service				
High	0.12	0.33	0.37	0.65
Low	0.88	0.67	0.63	0.35
v^{α} (Proportion of total cases)	0.25	0.37	0.19	0.19
k^{α} (Number in each class)	3145	4612	2325	2350

analytic situations. Its simplest bilinear case is already well known in the form of such models and methods as factor and principal components analysis and the classical metric approach to two-way multidimensional scaling. The three-way case is probably best known through the application of the symmetric case to fitting the INDSCAL model for individual differences multidimensional scaling.

The general trilinear CANDECOMP model and method has also been applied, primarily by R. A. Harshman, as a three-way factor-analytic procedure called *PARAFAC* (Harshman 1970). Another application of the trilinear CANDECOMP model is the three-way unfolding approach of De Sarbo and Carroll (1981), which generalizes Coombs' ([1964] 1976) multidimensional unfolding model to the three-way case. Three-way unfolding can be viewed, alternatively, as an approach to individual differences multidimensional scaling of *nonsymmetric* proximity data, involving proximities between two sets of stimuli or proximities defined on a single set but in which the rows and columns are treated as two different sets.

The most general multilinear CANDECOMP model has been usefully applied to obtain a least-squares fit of the Lazarsfeld (1950) latent class model to *N*-way contingency table data. *N*-way CANDECOMP can also be viewed as a general *N*-way factor analytic model and method or as a way of decomposing *N*-way interactions in an analysis of variance context. The symmetric case can be applied to generalize the INDSCAL model and method to the case of four- or higher-way proximity data. For example, it may be applied to similarity judgments made by each of a number of different subjects in each of a number of different experimental conditions or at different times in a longitudinal study.

CANDELINC—which stands for *CANDE*COMP with *LIN*ear Constraints on parameters—is applicable to cases in which it is desired to constrain parameters to be linear combinations of some predefined variables; the variables may be externally defined continuous variables on which the stimuli are measured or "dummy" variables defining a specified analysis of variance (ANOVA) design. Constraints may be put on only some modes, for example, the stimulus dimensions in an INDSCAL representation. Three-way CANDELINC has been applied for speeding up the fitting of the *unconstrained* CANDECOMP model (Carroll and Pruzansky 1979). An important application of this approach, called *INDSCAL-LS,* yields a very good approximation to the unconstrained INDSCAL solution; this procedure may reduce, considerably, the computational costs for problems involving very large data arrays.

The CANDECOMP-CANDELINC family of models and methods, taken as a whole, comprise a particularly interesting and useful subset of what Kruskal has called the *general multilinear model.* We believe that many other exciting and useful applications of this family of models and methods for multivariate data analysis will emerge in the future.

REFERENCES

Carroll, J. D. 1972. Individual differences and multidimensional scaling. In *Multidimensional scaling: Theory and applications in the behavioral sciences,* vol. 1, ed. R. N. Shepard, A. K. Romney, and S. Nerlove. New York: Seminar Press.

———. 1975. Application of CANDECOMP to solving for parameters of Lazarsfeld's latent class model. Paper presented at the Society for Multivariate Experimental Psychology, Gleneden Beach, Oregon, 13–15 November.

———. 1980. Models and methods for multidimensional analysis of preferential choice (or other dominance) data. In *Similarity and choice,* ed. E. D. Lantermann and H. Feger. Bern, Stuttgart, Vienna: Hans Huber Publishers.

Carroll, J. D., and P. Arabie. 1980. Multidimensional scaling. *Annual Review of Psychology* 31:607–49.

Carroll, J. D., and J. J. Chang. 1970. Analysis of individual differences in multidimensional scaling via an N-way generalization of "Eckart-Young" decomposition. *Psychometrika* 35:283–319.

Carroll, J. D., and J. B. Kruskal. 1978. Scaling, multidimensional. In *International encyclopedia of statistics,* ed. W. H. Kruskal and J. M. Tanur. New York: The Free Press.

Carroll, J. D., and S. Pruzansky. 1979. Use of LINCINDS as a rational starting configuration for INDSCAL. Manuscript, Bell Laboratories, Murray Hill, N.J.

Carroll, J. D., and M. Wish. 1974a. Models and methods for three-way multidimensional scaling. In *Contemporary developments in mathematical psychology,* vol. 2, ed. D. H. Krantz, R. C. Atkinson, R. D. Luce, and P. Suppes. San Francisco: W. H. Freeman.

————. 1974b. Multidimensional perceptual models and measurement methods. In *Handbook of perception,* vol. 2, ed. E. C. Carterette and M. P. Friedman. New York: Academic Press.

Carroll, J. D., S. Pruzansky, and P. E. Green. 1979. Estimation of the parameters of Lazarsfeld's latent class model by application of canonical decomposition (CANDECOMP) to multi-way contingency tables. Manuscript, Bell Laboratories, Murray Hill, N.J.

Carroll, J. D., S. Pruzansky, and J. B. Kruskal. 1980. CANDELINC: A general approach to multidimensional analysis of many-way arrays with linear constraints on parameters. *Psychometrika* 45:3–24.

Chang, J. J., and J. D. Carroll. 1969. How to use INDSCAL: A computer program for canonical decomposition of N-way tables and individual differences in multidimensional scaling. Manuscript, Bell Laboratories, Murray Hill, N.J.

Coombs, C. H. [1964] 1976. *A theory of data.* [New York: Wiley] Republished, Ann Arbor, Mich.: Mathesis Press.

De Sarbo, W. S., and J. D. Carroll. 1981. Three-way metric unfolding. In *Proceedings of the Third ORSA/TIMS Special Interest Conference on Market Measurement and Analysis,* 157–83.

Eckart, C., and G. Young. 1936. The approximation of one matrix by another of lower rank. *Psychometrika* 1:211–18.

Gollob, H. F. 1968. A statistical model which combines features of factor analytic and analysis of variance techniques. *Psychometrika* 33:73–116.

Goodman, L. A. 1974a. Exploratory latent structure analysis using both identifiable and unidentifiable models. *Biometrika* 61:215–31.

————. 1974b. The analysis of qualitative variables when some of the variables are unobservable. Part I—A modified latent structure approach. *American Journal of Sociology* 79:1179–1259.

————. 1975. A new model for scaling response patterns: An application of the quasi-independence concept. *Journal of the American Statistical Association* 70:755–68.

Green, P. E., with J. D. Carroll. 1976. *Mathematical tools for applied multivariate analysis.* New York: Academic Press.

Green, P. E., F. J. Carmone, and D. P. Wachspress. 1976. Consumer segmentation via latent class analysis. *Journal of Consumer Research* 3:170–74.

Guttman, L. A. 1968. A general nonmetric technique for finding the smallest coordinate space for a configuration of points. *Psychometrika* 33:469–506.

Harshman, R. A. 1970. Foundations of the PARAFAC procedure: Models and conditions for an "explanatory" multi-modal factor analysis. *UCLA Working Papers in Phonetics* 16:1–84 (University Microfilms No. 10,085).

Harshman, R. A., P. Ladefoged, and L. Goldstein. 1977. Factor analysis of tongue shapes. *Journal of the Acoustical Society of America* 62:693–707.

Helm, C. E. 1964. Multidimensional ratio scaling analysis of perceived color relations. *Journal of the Optical Society of America* 54:256–62.

Helm, C. E., and L. R. Tucker. 1962. Individual differences in the structure of color perception. *American Journal of Psychology* 75:437–44.

Horan, C. B. 1969. Multidimensional scaling: Combining observations when individuals have different perceptual structures. *Psychometrika* 34:139–65.

Kettenring, J. R. 1983. A case study in data analysis. In *Statistical data analysis,* ed. R. Gnanadesikan. Providence, R. I.: American Mathematical Society.

Kruskal, J. B. 1964a. Multidimensional scaling by optimizing goodness-of-fit to a nonmetric hypothesis. *Psychometrika* 29:1–27.

————. 1964b. Nonmetric multidimensional scaling: A numerical method. *Psychometrika* 29:115–29.

————. 1978. Factor analysis and principal components: Bilinear methods. In *International encyclopedia of statistics,* ed. W. H. Kruskal and J. M. Tanur. New York: Free Press.

Kruskal, J. B., and M. Wish. 1978. *Multidimensional scaling.* Beverly Hills: Sage.

Lazarsfeld, P. F. 1950. The logical and mathematical foundation of latent structure analysis. In *Measurement and prediction,* ed. S. A. Stouffer et al. Princeton, N.J.: Princeton University Press.

McGee, V. E. 1966. Multidimensional analysis of elastic distances. *British Journal of Mathematical Statistical Psychology* 19:181–96.

Pruzansky, S., P. E. Green, and J. D. Carroll. 1979. Using CANDECOMP for estimating parameters in latent class analysis. *Proceedings of 87th Annual Convention of American Psychological Association,* New York, September.

Schönemann, P. H. 1970. On metric multidimensional unfolding. *Psychometrika* 35:349–66.

Shepard, R. N. 1962a. The analysis of proximities: Multidimensional scaling with an unknown distance function. I. *Psychometrika* 27:125–40.

————. 1962b. The analysis of proximities: Multidimensional scaling with an unknown distance function. II. *Psychometrika*

27:219–46.

Torgerson, W. S. 1958. *Theory and methods of scaling.* New York: Wiley.

Tucker, L. R. 1964. The extension of factor analysis to three-dimensional matrices. In *Contributions to mathematical psychology,* ed. N. Frederiksen and H. Gulliksen. New York: Holt, Rinehart and Winston.

Tucker, L. R., and S. Messick. 1963. An individual difference model for multidimensional scaling. *Psychometrika* 28:333–67.

Wish, M. 1971. Individual differences in perceptions and preferences among nations. In *Attitude research reaches new heights,* ed. C. W. King and D. Tigert. Chicago: American Marketing Association.

Wish, M., and J. D. Carroll. 1974. Applications of individual differences scaling to studies of human perception and judgement. In *Handbook of perception,* vol. 2, ed. E. C. Carterette and M. P. Friedman. New York: Academic Press.

Wold, H. 1966. Estimation of principal components and related models by iterative least-squares. In *Multivariate analysis,* ed. P. R. Krishnaiah. New York: Academic Press.

11

A Direct Approach to Individual Differences Scaling Using Increasingly Complex Transformations

James C. Lingoes and Ingwer Borg

Following Horan's (1969) original proposal for a subjective metrics model, there has been considerable effort devoted to solving the mathematical problems and to constructing computer algorithms for scaling real data (e.g., Bloxom 1968; Carroll and Chang 1970, 1972; Carroll and Wish 1974; Schönemann 1972; Schönemann et al. 1976; Takane et al. 1977). Because Carroll and Chang's IDIOSCAL procedure, and its special case INDSCAL, is by far the most well-known solution to the "dimensional salience model," we shall briefly review the general issues in this area using the INDSCAL terminology.

In the dimensional salience (or weighted euclidean) model, the distance between points p and q in individual i's "private perceptual space" is estimated from the respective points in some average configuration of all N individuals in the sample by the following weighted euclidean distance formula:

$$d_{pq}^{(i)} \cong \hat{d}_{pq}^{(i)} = \left[\sum_a c_a^{(i)} (g_{pa} - g_{qa})^2 \right]^{1/2} , \qquad (11\text{–}1)$$

where $d_{pq}^{(i)}$ is the distance between p and q (given as datum), $\hat{d}_{pq}^{(i)}$ is an estimate of $d_{pq}^{(i)}$ (e.g., a least-squares estimate), $c_a^{(i)}$ is the squared weight associated with coordinate axis a, g_{pa} is the projection of p on a in the $n \times m$ "group space" G, and the superscript i indicates individual i. As is true for all scaling algorithms proposed so far for the dimensional salience model, INDSCAL does not attempt a direct solution for the unknowns in (11–1), i.e., for $c_a^{(i)}$ ($a = 1, \ldots, m$; $i = 1, \ldots, N$) and G. Rather, (11–1) is converted into its simpler scalar-product form: centering G at the origin and using the well-known relation:

Reprinted from *Psychometrika,* 43 (1978) 4:491–519, with the permission of the authors and the Psychometric Society.

$$\hat{b}_{pq}^{(i)} = \tfrac{1}{2} \left[\sum_a c_a^{(i)} g_{pa}^2 + \sum_a c_a^{(i)} g_{qa}^2 - \hat{d}_{pq}^{(i)\,2} \right] , \qquad (11\text{–}2)$$

one arrives at

$$\hat{b}_{pq}^{(i)} = \sum_a g_{pa} c_a^{(i)} g_{qa} , \qquad (11\text{–}3)$$

where $\hat{b}_{pq}^{(i)}$ is the estimated scalar product of p and q for individual i. Now, since the given distances $d_{pq}^{(i)}(p, q = 1, \ldots, n;\ i = 1, \ldots, N)$ are determined only up to an arbitrary origin (if one assumes that they are interval scaled), one must decide on some origin in order to generate the corresponding scalar products for each individual, because INDSCAL uses, for reasons of computational simplicity, scalar products rather than distances. The estimation of a convenient origin is implemented by Torgerson's additive constant method (1958).

Equation (11–3) can also be expressed more compactly in matrix notation as:

$$\hat{B}_i = G\ C_i\ G' , \qquad (11\text{–}4)$$

where G is $n \times m$ and C_i is diagonal (the INDSCAL implementation of the subjective metrics model) or is constrained to be (semi-) definite (the IDIOSCAL model with positivity constraints). An alternating least-squares (ALS) procedure is then used to solve for G and $C_i\ (i = 1, \ldots, N)$. In case of a non-diagonal C_i, Carroll and Chang (1972) propose a further decomposition of C_i into

$$C_i = T_i\ \Lambda_i\ T_i' , \qquad (11\text{–}5)$$

where $T_i T_i' = T_i' T_i = I$ and Λ_i is diagonal. Geometrically, this corresponds to an orthogonal "idiosyncratic" rotation of G by T_i, followed by dimension weighting with $A_i^{1/2}$. In contrast with the INDSCAL solution, however, in which G and C_i are not rotationally invariant, IDIOSCAL will yield results which are determined only up to a joint rotation of G and C_i. Furthermore, the decomposition in (11–5) is obviously not the only possible one: Harshman (1972) and Tucker (1972) propose other, non-orthogonal factorizations.

Although INDSCAL, in particular, has been applied very successfully in various contexts some problems associated with the IDIOSCAL family of procedures (of which INDSCAL is but a special case) remain. The most obvious of these difficulties lies in the choice of loss function: goodness of fit is defined in terms of scalar products, which are interpretationally more indirect and complicated than distances. Moreover, the transformations used to fit G to each individual's data assume a relatively simple relationship between the two configurations. If some of the points therein have different relative locations or if the configurations are unevenly or nonlinearly distorted, the dimensional salience model has no means of detecting this. This could have the consequence that the computed group space is quite misleading, representing a confounding of a class of essentially qualitatively different configurations. Finally, a most serious disadvantage is that the substantive and statistical significance of the transformations is not clear. The dimension weighting, for example, might produce a quite negligible fit improvement over a classical procrustean fitting of the configurations which leaves invariant the relative distances. It is clearly desirable to disentangle the admissible transformations from those that change the meaningful properties of the group space by some distortion, a point which is to some extent taken into account in the third phase of IDIOSCAL (Carroll and Chang 1972), but not at all for the more widely used INDSCAL procedure.

Note that not all algorithms developed in this area share all of INDSCAL's weaknesses. Schönemann et al.'s COSPA (1976) even provides some tests for the applicability of the dimensional salience model, but it also introduces some additional and fairly strong assumptions not included in the original formulations of the model. However, neither COSPA nor INDSCAL nor ALSCAL (Takane et al. 1977) minimizes a loss function defined for (unsquared) distances, nor do any of these procedures provide information with respect to the performance of admissible (nondistorting) transformations. Furthermore, no method adds anything to the dimensional salience model which might provide information regarding individual differences of a more complicated nature.

THE PINDIS MODEL

PINDIS (*P*rocrustean *IN*dividual *DI*fferences *S*caling: Lingoes 1975; Lingoes and Borg 1976, 1977; Borg and Lingoes 1976, 1977; Borg 1977) is similar to procedures like IDIOSCAL in the sense that PINDIS also scales individual similarities according to the dimensional salience model. It is different, however, from all other algorithms and models in a number of respects: (a) in PINDIS, the dimensional salience model is only one approach for assessing individual differences; (b) the set of models in PINDIS is organized into two hierarchies with a typical common order, i.e., the procedure sequentially analyzes the relationship of individuals to some group space in such a way that a fitting is done first by strictly admissible transformations (those that leave the relative distances invariant) and then by adding increasingly more free parameters in the form of weights, idiosyncratic rotations and translations.

Also different from other approaches in this respect, PINDIS starts with a set of individual configurations X_i ($i = 1, \ldots, N$) rather than with either scalar product or distance matrices. These X_i's are assumed to be given by some prior and assumption-appropriate analysis, either metric or nonmetric. Each X_i, an order $n \times m_i$ matrix, is assumed to be arbitrary insofar as only comparative distances among the n points are to remain invariant, i.e., X_i is determined up to rigid motions and central dilations. Consequently, we will always exhaust the complete set of admissible transformations, i.e., rotations/reflections, translations, and central dilations on X_i, in fitting X_i to the target configuration at each stage of the PINDIS analysis. This transformed X_i will be symbolized by $\tilde{X}_i$ and is to be interpreted as including any or all of the permitted transformations in the group relevant to the particular target under analysis.

Initially, we arbitrarily norm all X_i's (centered at the origin) to unit length, which constitutes the optimal translation on X_i for the following analyses: a centroid configuration Z of order $n \times m$ ($m = \max_i(m_i)$), which is analogous to IDIOSCAL's group space G, is computed as the average of all X_i's after they have been fitted optimally to each other under the admissible transformations. Since Z's orientation is arbitrary, it is put into an orientation that is optimal for dimension weighting of all X_i's, which will be discussed in more detail below. The similarity of each $\tilde{X}_i$ to Z is expressed by $r^2(\tilde{X}_i, Z)$, the squared product-moment correlation between the corresponding elements of $\tilde{X}_i$ and Z (Lingoes and Schönemann 1974). This is the first important index in PINDIS, representing a *reference value* for those transformations that do not preserve the distance ratios in Z. Centering X_i is also optimal for the dimension salience model, which constitutes the next stage of analysis.

The simplest distortion of Z for mapping X_i is that of dimension weighting, where a diagonal m-square weight matrix W_i is determined for each optimally transformed individual configuration X_i. Another fit index $r^2(\tilde{X}_i, Z^r W_i)$ yields the relevant information about the explanatory power of such weighting. Obviously, one would not want to attach much psychological meaning to the differences among the elements of W_i, if $r^2(\tilde{X}_i, Z^r W_i)$, where Z^r is the optimally rotated Z for all X_i, is not substantially higher than our reference value, $r^2(\tilde{X}_i, Z)$. Similarly, one would also hesitate to interpret an idiosyncratic rotation of Z, at the next transformation stage of the dimensional salience model in PINDIS, as having much substantive import, if $r^2(\tilde{X}_i, Z^r W_i)$ is not appreciably smaller than $r^2(\tilde{X}_i, Z^r_i W^r_i)$, where $Z^r_i = ZS_i$, $S_i S'_i = S'_i S_i = I$ (the identity matrix) and W^r_i is the diagonal weight matrix for Z^r_i. It should be noted that since we want just one "average subject" or one Z for explaining individual differences, we consider such idiosyncratic rotations of Z as being in the same class of inadmissible transformations as those of dimension weighting.

Using even more free parameters, one could now fit $V_i Z$ to the optimally transformed X_i, where V_i is an n-square diagonal matrix of vector weights. Geometrically, the points in Z^t are herein conceived of as termini of vectors emanating from the origin of Z: the weights of V_i will then stretch or shrink each vector with or without reversing its sense. The measure which indicates the success of such fitting is $r^2(\tilde{X}_i, V_i Z^t)$ where Z is an optimally translated Z for all X_i. However, it will be shown below that generally there exists an idiosyncratic translation of Z relative to each $\tilde{X}_i$ which improves the predictability of $\tilde{X}_i$. The latter represents a subject-specific mapping in the context of vector weighting, analogous to the individual rotations of Z in the dimension-weighting approach, and, as such, are also considered to be inadmissible along with vector weighting. The fit index for idiosyncratic translations is $r^2(\tilde{X}_i, V^t_i Z^t_i)$, where Z^t_i is the idiosyncratically translated Z and V_i^t is the associated weight matrix for this mapping. Both of these fit indices for vector weighting can be compared with one another and both with our basic, non-distorting, measure of fit, $r^2(\tilde{X}_i, Z)$, for assessing the relevance of these transformations. The vector weighting model provides not only indices of inter-individual differences, but also represents a psychological model in its own right (which we call the "perspective model") for a number of psychological and social contexts.

We can now summarize the various PINDIS transformations in Table 11–1. Note that $\tilde{X}_i$ is defined as that X_i which is transformed by admissible mappings in such a way that the respective fit index is maximized. Thus, it is clear that inadmissible mappings are always associated with the centroid configuration Z. The distinction admissible vs. inadmissible transformation is based upon the property of the former to be indeterminate from either the data itself or from the respective fitting model. Consequently, admissible transformations can have no substantive meaning whatsoever: only the listed inadmissible fitting parameters are principally interpretable.

From Table 11–1 one can discern two model hierarchies contained in a PINDIS analysis; one hierarchy is established by the triple (1) similarity transformation, (2) dimensional salience with a fixed orientation, and (3) dimensional salience with idiosyncratic orientations; the other hierarchical triple is formed by (1) similarity transformation, (2) perspective model with fixed origin, and (3) perspective model with idiosyncratic origins. Both of these model collections are hierarchical in the sense that the models are nested, the one using fewer inadmissible fitting parameters being but a special case of the one using more. However, since in empirical applications one would typically encounter situations where n is much greater than m, one would also have an order among all models established by the number of free parameters, i.e., the typical order is:

TABLE 11-1. Summary of PINDIS Transformations

Model	Number of inadmissible fitting parameters	Fit index
Similarity transformation (unit weighting)	0	$r^2(\tilde{X}_i, Z)$
Dimensional salience (dimension weighting)	m	$r^2(\tilde{X}_i, Z^r W_i)$
Dimensional salience with idiosyncratic orientation	$m + \binom{m}{2}$	$r^2(\tilde{X}_i, Z_i^r W_i^r)$
Perspective model with fixed origin (vector weighting)	n	$r^2(\tilde{X}_i, V_i Z^t)$
Perspective model with idiosyncratic origin	n + m	$r^2(\tilde{X}_i, V_i^t Z_i^t)$

$$0 < m < m + \binom{m}{2} < n < n + m \ .$$

In a more general sense, the various PINDIS mappings belong to the group of affine transformations defined, respectively, on the column and row spaces of the various configurations with the similarity transformation being but a special case of unit weightings for both sub-models.

Such a hierarchical approach yields standards against which the significance of the inadmissible transformations can be judged. Furthermore, it provides many indices (v.i.) which not only represent additional information in their own right, but which also allow one to evaluate the validity of various features of the dimensional salience model for a given sample of individuals.

The Centroid Configuration

The centroid configuration Z is calculated from the X_i's by a method similar to the one proposed by Gower (1975). (Gower calls Z the "consensus configuration," a term which carries unnecessary and potentially misleading psychological connotations. We prefer a more neutral name.)

The derivation of Z starts with the individual X_i's, where each X_i is an $n \times m_i$ coordinate matrix generated by a prior MDS, factor analysis, or any other appropriate multidimensional procedure. It is not necessary that all individual (perceptual) spaces be of the same dimensionality if one adds zero-element columns to X_i's whenever $m_i < m$. The goal is to rescale all X_i's simultaneously by affine transformations to $\tilde{X}_i = X_i R_i - j u_i'$, such that the following fit criterion, representing a residual sum of squares, is optimized:

$$L = \sum_i [\mathrm{tr}(\tilde{X}_i - Z)(\tilde{X}_i - Z)'] \ , \qquad (11\text{–}6)$$

where $R_i' R_i = R_i R_i' = I$, j is an n-element vector of ones, and u_i is an m-element translation vector. To minimize L, each X_i is centered at the origin first, which represents the optimal u_i motion, and then centrally contracted or dilated such that its euclidean norm, $\mathrm{tr}(X_i X_i')$ equals unity. We then proceed iteratively as follows:

a. X_2 is then rotated to fit X_1 in a procrustean sense and a first Z matrix is constructed, $Z^{(1)}$ as the average of the corresponding elements of X_1 and X_2. Next, X_3 is rotated to fit $Z^{(1)}$, $Z^{(2)}$ is computed from $Z^{(1)}$ and X_3, etc., until X_N is included in the preceding process. This, then, completes the initial cycle, yielding our best initial estimate of Z.
b. Each of the $N X_i$'s are then rotated to fit the current best estimate of Z and a new estimate for Z is computed from the average of the X_i's fitted to the current estimate of Z. A

goodness of fit index is computed for each X_i, $g_i = r^2(X_i, Z)$ and then we compute:

$$h = \frac{1}{N} \sum_i (1 - g_i)^{1/2} ,$$

which is taken as an overall measure of fit for each new estimate of Z. We repeat process (b) until h ceases to drop by more than some prespecified threshold value.

The above iterative algorithm corresponds to the one proposed by Kristof and Wingersky (1971). Berge (1977) has recently analyzed this procedure and found that a slightly modified version satisfies a somewhat stronger necessary condition for minimizing L. However, neither procedure generally satisfies the sufficient condition presented by Fischer and Roppert (1965) and Berge (1977). Moreover, it is not clear yet what constitutes the necessary and sufficient conditions for the present problem. Empirically, there has been some evidence as to the superiority of Berge's algorithm. If more results point in this direction, it may be desirable to implement the Berge procedure for PINDIS instead of the Kristof-Wingersky algorithm presently used.

Gower (1975), who allows one to input raw score data matrices as X_i's, provides an (optional) additional transformation besides rotation/reflection and translation, i.e., a central dilation scalar k_i for each X_i. These k_i's are used to adjust for differences in the range of scores used by different individuals under the constraint:

$$\sum_i k_i^2 \mathrm{tr}(X_i X_i') = \sum_i \mathrm{tr}(X_i X_i') .$$

Since in PINDIS every $\mathrm{tr}(X_i X_i') = 1$, so that Z represents a directly interpretable "average subject," whereas Gower makes only the weaker norming $\Sigma_i \mathrm{tr}(X_i X_i') = N$, such scalar multipliers would make no sense here. As one of our reviewers has pointed out, however, X_i-conditional scalar multipliers would still be useful as a fit index for the relationship of any X_i and Z, but our $r^2(\tilde{X}_i, Z)$ is actually closely related to k_i if employed in this respect: $k_i^2 \mathrm{tr}(ZZ')^{1/2} = r^2(\tilde{X}_i, Z)$.

For the norm of Z the following relationship holds:

$$0 \leq \mathrm{tr}(ZZ')^{1/2} \leq 1 , \qquad (11\text{–}7)$$

which follows from

$$\sum_i \mathrm{tr}(X_i X_i') = N \mathrm{tr}(ZZ') + \sum_i \mathrm{tr}(E_i E_i') , \qquad (11\text{–}8)$$

where the left-hand side represents the total sum of squares of the individual configurations, which is equal to N due to the prior norming of every X_i. Thus, $\mathrm{tr}(ZZ')$ is equal to one if and only if all X_i's are perfectly explained by Z. In other words, the squared norm of Z represents the communality of the X_i's, which is perfect if and only if all X_i's are related to each other by rotations/reflections and translations. The $n \times m$ matrix E_i contains the unexplained residuals.

Having produced a target configuration, we now address the various transformations on both Z and X required for maximizing the fit between them.

Similarity Transformations

The problem at this initial stage of a PINDIS analysis is to fit each X_i to a given Z by rigid motions and central dilations. Although this has already been accomplished in the process of deriving Z described in the previous section, we shall now present the involved issues in somewhat more detail. (It might also be noted that Z may be an externally provided hypothesis configuration, in which case the Z-generating process would, of course, be bypassed.)

Restricting the set of transformations to similarity mappings will lead to sub-optimal solutions in the case that the X_i's resulted from a previous nonmetric scaling, since one would not use all of the indeterminacies in the individual configurations. Nevertheless, such an approach is consistent with the general practice of considering MDS solutions as essentially interval scaled for a sufficient number of points. Since the group of rotations/reflections, translations, and central dilations exhausts the set of admissible and practically relevant transformations for nonmetric representations quickly as n gets ever larger, it has essentially the same effect as the more general group of isotonic mappings.

A least-squares solution for the present problem has been provided by Schönemann and Carroll (1970). It minimizes the function

$$F = \mathrm{tr}\ [(\tilde{X}_i - Z)(\tilde{X}_i - Z)'] \ , \tag{11–9}$$

where

$$\tilde{X}_i = k_i (X_i - ju_i')R_i \ , \tag{11–10}$$

with R_i, j, and u_i defined as in (11–6) and k_i being a central dilation scalar.

Since we seek a simultaneous solution for the unknowns, (11–10) can be written as $\tilde{X}_i = k_i X_i R_i - ju_i'$ in this context. One obtains as solutions:

$$R_i = AB' \ , \tag{11–11}$$

where

$$A \Delta B' = X_i' MZ = C \ , \tag{11–12}$$

the Eckart-Young decomposition of C, and $M = I - jj'/n$. The columns of A are the eigenvectors of CC', those of B are the eigenvectors of $C'C$, both ordered according to the size of their common eigenvalues contained in the diagonal matrix Δ (Schönemann 1966). For u_i:

$$u_i = \frac{(Z - X_i R_i)'j}{n} . \qquad (11\text{–}13)$$

The effect of M is to center both X_i and Z at the origin (which has already been accomplished in the process of calculating Z). Also, u_i has been taken care of implicitly above, since $X_i R_i + ju_i' = X_i R_i + (jj'/n)(Z - X_i R_i) = (jj'/n)Z + (I - jj'/n)X_i R_i = (I - jj'/n)X_i R_i = X_i R_i$. So the problem quite simply reduces to an orthogonal procrustean rotation (Schönemann 1966), since the fit index does not require explicit computation of k_i.

The analysis at this (and at every subsequent) phase is concluded with the computation of the appropriate individual fit measure, which is here $r^2(\tilde{X}_i , Z)$.

Dimension-Weighting Transformations

The problem at this stage is to predict each X_i by a dimensionally weighted Z. In the more restrictive model we use a Z with a fixed orientation analogous to INDSCAL's G:

$$\min = \text{tr } [(\tilde{X}_i - ZSW_i)(\tilde{X}_i - ZSW_i)'] , \qquad (11\text{–}14)$$

where $S'S = SS' = I$, W_i is diagonal, and $\tilde{X}_i$ is defined as in (11–10). First, if we call the function to be minimized in (11–14) f, one finds via $\partial f / \partial u_i = 0$:

$$u_i = (1/n)(X_i - ZSW_i R_i')'j . \qquad (11\text{–}15)$$

Since both X_i and Z are column centered and since this property is not destroyed by S, W_i, and R_i', respectively, the translation vector must be equal to $(0, \ldots , 0)$. In (11–14) it is assumed, however, that Z remains centered at the origin. We shall now show that Z's origin is sufficient for an optimal fit in the dimension-weighting case. Consider the following generalized loss function, where another translation vector, t_i, has been added:

$$F = \text{tr } (AA') , \qquad (11\text{–}16)$$

$$A = (Z - jt_i')SW_i - (X_i - ju_i')R_i .$$

Via partial differentiation one finds:

$$t_i = \frac{1}{n} [Z - (X_i - ju_i')R_i S' W_i^{-1}]'j , \qquad (11\text{–}17)$$

$$u_i = \frac{1}{n} [X_i - (Z - jt_i')SW_i R_i']'j . \qquad (11\text{–}18)$$

Inserting one solution into the other, one obtains $u_i = u_i$ and $t_i = t_i$, respectively, which proves the assertion. This result also shows that it is not necessary to center Z at the origin for obtaining an optimal fit for dimension weighting, but it is definitely convenient computationally and is essential for a direct interpretation of the dimension weights.

To find W_i, we set $R_i = I = S$ for the moment. The optimal dimension weights are then given by:

$$[\text{diag }(Z'Z)]^{-1}[\text{diag }(Z'X_i)] = W_i \quad , \tag{11–19}$$

i.e., they are the regression weights in $ZW_i + E_i = X_i$. We can normalize W_i such that:

$$r^2(X_i, ZW_i) = \text{tr }(W^*_i W^*_i{}') \quad , \tag{11–20}$$

by setting

$$W^*_i = [\text{diag }(Z'Z)]^{1/2} W_i \quad , \tag{11–21}$$

which follows from

$$r^2(X_i, ZW_i) = \sum_a \left[\frac{\left(\sum_p x_{pa} z_{pa} \right)^2}{\sum_p z_{pa}^2} \right] \quad , \tag{11–22}$$

where the weights $w_{aa} = \Sigma_p x_{pa} z_{pa} / \Sigma_p z_{pa}^2$, given by (11–19), are used in simplifying the correlation equation.

We now formulate the following loss function for dimensions a and b $(a \neq b)$:

$$K_{ab} = \frac{1}{\sum_p z_{pa}^2} \left(\sum_p x^*_{pa} z_{pa} \right)^2 + \frac{1}{\sum_p z_{pb}^2} \left(\sum_p x^*_{pb} z_{pb} \right)^2, \tag{11–23}$$

where

$$x^*_{pa} = x_{pa} \cos \alpha - x_{pb} \sin \alpha \quad , \tag{11–24}$$

$$x^*_{pa} = x_{pb} \sin \alpha + x_{pb} \cos \alpha \quad . \tag{11–25}$$

Since K_{ab} represents the proportion of variance of $X_i R_i$ predicted by ZW_i in the plane a, b, we want to maximize K_{ab} for each pair of dimensions through an appropriate choice of the rotation angles, α. For plane a, b we obtain by differential calculus:

$$\alpha = .5 \tan^{-1} \left(\frac{2A}{B} \right) \quad , \tag{11–26}$$

where

$$A = \frac{C_{aa}\, C_{ba}}{V_a} - \frac{C_{bb}\, C_{ab}}{V_b} , \qquad (11\text{–}27)$$

$$B = \frac{C_{ba}^2 - C_{aa}^2}{V_a} + \frac{C_{ab}^2 - C_{bb}^2}{V_b} , \qquad (11\text{–}28)$$

and where

$$C_{ij} = \sum_p x_{pi} z_{pj} \; ; \qquad V_i = \sum_p z_{pi}^2 \; .$$

The rotation angle α will either maximize or minimize K_{ab}. The result depends upon the orientation of the initial X_i relative to the maximizing/minimizing X_i. Because of (11–26), there are four special positions in a two space: two maxima and two minima. One can test if a rotation by α will maximize K_{ab} by using the second derivative inequality:

$$\frac{d^2 K_{ab}}{d\alpha^2} = B \cos(2\alpha) + 2A \sin(2\alpha) < 0 , \qquad (11\text{–}29)$$

and add 90° to α if (11–29) does not hold, since it follows from (11–26) that maxima and minima alternate at 90° intervals. In practice, α will rarely minimize K_{ab} because the orientation of X_i generated by the previous similarity transformations is almost always a very good initial estimate of the optimal X_i for dimension weighting. However, in cases where the dimension weighting leads to very small weights for some coordinate dimensions of Z, i.e., where Z is essentially collapsed to a lower dimensional space, the initial "similarity" orientation of X_i might lead to a minimizing α.

Rotations for higher dimensional spaces can be done pairwise for all $\binom{m}{2}$ combinations of dimensions: rotate columns 1 and 2 of X_i first, calculate their new values; rotate 1 and 3, . . . , rotate 1 and m: then continue with 2 and 3; 2 and 4, . . . , 2 and m, until $m - 1$ and m complete one cycle. Repeat the cycles until some criterion of convergence is reached (see Harman 1967). For the reoriented X_i's optimal weights are then computed by (11–19), substituting X_i by $X_i R_i$. The resulting solution is unique up to 180° rotations in each plane for both W_i and R_i.

The problem that remains to be solved in order to maximize the predictability of X_i for the dimension-weighting model is to find that orientation of Z with respect to some fixed coordinate system in which Z, when distorted along these axes, approximates $\check{X}_i$ as closely as possible. Apparently, every orientation of Z is associated with a different family of producible configurations (see Lingoes and Borg 1977). We shall first address the problem of finding the optimal rotation of Z for each X_i and then our attention will turn to finding the optimal orientation of Z (fixed) over all X_i. The relevant loss function for the idiosyncratic case can be formulated as follows:

$$\min = \text{tr}\ [(ZS_i W^r{}_i - X_i R_i)(ZS_i W^r{}_i - X_i R_i)'] \ , \tag{11–30}$$

where S_i, R_i, and $W^r{}_i = f(ZS_i, X_i R_i)$, which means that $W^r{}_i \neq W_i$ in general, are the unknowns. Stated for a two space, and using relationship (11–20) between normalized dimension weights and the variance explained in X_i by ZW_i, (11–30) can be expressed as:

$$\frac{1}{\sum_p z^{*2}_{pa}} \left(\sum_p x^*_{pa} z^*_{pa} \right)^2 + \frac{1}{\sum_p z^{*2}_{pb}} \left(\sum_p x^*_{pb} z^*_{pb} \right)^2 = \max \ , \tag{11–31}$$

where

$$z^*_{pa} = z_{pa} \cos \beta - z_{pb} \sin \beta \ , \tag{11–32}$$

$$z^*_{pb} = z_{pa} \sin \beta + z_{pb} \cos \beta \ , \tag{11–33}$$

and x^*_{pa} and x^*_{pb} are defined by (11–24) and (11–25), respectively. Partial differentiation of (11–31) w.r.t. α and β yields a simple solution for α (for some fixed β), corresponding to (11–26), but a very complicated expression for β (holding α constant), which does not seem to be solvable in any simple way. Having by necessity to resort to a numerical approximation procedure, one first observes the following simplification: the optimal Z must be attainable from the current Z by a rotation of no more than 90°. This follows from a simple consideration: in a two space there are four possible positive directions for Z which are related to each other by permutation transformations. In each of these orientations, Z will explain the respective $\tilde{X}_i$ equally well. Thus, starting from some arbitrary orientation and rotating Z over an interval of 90° yields all possible communality values (i.e., r^2's) and, therefore, allows one to determine the optimal Z.

The search algorithm proceeds as follows: the original orientation of Z is the one due to its generation in the Kristof-Wingersky process. Using this Z as a starting point we compute the communality of $ZS_i W_i$ and $\tilde{X}_i$ with orthonormal S_i rotating Z in steps of 5° over an interval ±45° in some plane a, b. The best Z is found and a ±5° sector about this Z is searched in 1° intervals. The best Z in this search process is then picked as the final Z in a given plane, having an uncertainty level of 1°. This procedure is then iteratively repeated over all planes and each cycle until some criterion of convergence is satisfied. Although we have found empirically that the response surface is smooth and monotonically increasing to the maximum (which would allow one to use a more efficient algorithm), we could not show that the function will always possess these properties. By our search procedure, we obtain the optimal idiosyncratic Z, $Z^r{}_i$, for each $\tilde{X}_i$, which is unique up to axes permutations.

Determining an orientation for Z (fixed) that is globally optimal would require the solution of (11–14) for S over all NX_i's simultaneously, an apparently very difficult task in view of the dependency of the W_i's and the various transformations in $\tilde{X}_i$ on S. At this time, we cannot offer such a solution. However, extensive empirical analyses with artificial data have shown that the following procedure yields a very good approximation to the optimal Z (fixed). In a two space, ZS is taken as the arithmetic mean of the Z^r_i's over all N configurations. To show that ZS calculated in this way is geometrically similar to each generating Z one observes that the average of some matrix A and another rotated matrix $AT(TT' = T'T = I)$ is similar to either one if and only if $.5(A + AT).5(A + AT)' = .5(AA') + .5A(T + T')A' = kAA'$, where k is a scalar. For a two space one obtains $T + T' = 2 \cos \alpha I$, but for a higher dimensional space $T + T'$ is generally not a scalar matrix. This means that $\bar{Z} = 1/N \; \Sigma_i Z^r_i$ does not correspond to the rotated (and centrally dilated) average subject configuration for $m > 2$. Expressed differently: an unknown distortion whose psychological meaning remains completely obscure would also affect $\bar{Z}^r$ in this process. Furthermore, using $\bar{Z}$ in lieu of the Kristof-Wingersky generated Z for higher dimensional spaces would destroy the value of our basic reference index, since $r^2(\tilde{X}_i, \bar{Z}) < r^2(\tilde{X}_i, Z)$, in general, whenever similarity is destroyed. Apparently, these features are undesirable. Thus, in order to preserve similarity we use $\bar{Z}$ merely as a target to which Z is orthogonally rotated as described above in the section on similarity transformations for $m > 2$. Similarity should be preserved a fortiori whenever one has a hypothesis configuration, for obvious reasons.

Vector-Weighting Transformations

On the next level of complexity, the loss function can be formulated as:

$$\min = \text{tr}\,\{[\tilde{X}_i - V_i(Z - jt_i')][\tilde{X}_i - V_i(Z - jt_i')]'\}\,, \qquad (11\text{–}34)$$

where $\tilde{X}_i$ is defined as in (11–10) and V_i is diagonal. Apparently, the attainable fit between $\tilde{X}_i$ and $V_i(Z - jt_i')$ in this vector-weighting model is independent of Z's orientation with respect to a given coordinate system, which simplifies the problem relative to the dimension-weighting case. However, (11–34) is more complicated in requiring the computation of two translation vectors, u_i and t_i. The latter is necessary since the solutions for u_i and t_i do not lead to the tautology found in the dimension-weighting case, i.e.:

$$u_i = \frac{1}{n}\,[X_i - V_i(Z - jt_i')R_i']'j\,, \qquad (11\text{–}35)$$

$$t_i = \frac{1}{n}\,[Z - V_i^{-1}(X_i - ju_i')R_i]'j\,. \qquad (11\text{–}36)$$

While the translation vectors were simple and, in particular, dependent in (11–17) and (11–18), this is not the case here. The complete solution of (11–34) requires, as a consequence, the determination of t_i, u_i, R_i, and V_i. Of these transformations only R_i and u_i are devoid of psychological significance. Translations and vector weightings for Z, on the other hand, are confounded and to some extent yield redundant information vis-à-vis one another, but may carry interpretative import.

Since there exist relatively direct solutions for the R_i and V_i as unknowns, we shall discuss their derivations first. Initially, one must determine the optimal reflection of X_i relative to $V_i Z$. This step was, of course, irrelevant in the dimension-weighting case, since reflections could be handled there by the sign of the salience weight. We now make the following considerations: The criterion to be minimized is:

$$D = \sum_p \sum_a (v_{pp} z_{pa} - x_{pa} \omega_a)^2 , \tag{11–37}$$

where ω_a is either +1 or -1. The vector weights $v_{pp} (p = 1, \ldots, n)$ are determined from the regression equation $X_i \Omega_i = V_i Z + E_i$, where Ω is the $m \times m$ diagonal reflection matrix and E_i are the residuals:

$$[\text{diag } (X_i \Omega_i Z')][\text{diag } (ZZ')]^{-1} = V_i \quad . \tag{11–38}$$

Inserting $v_{pp} = (\Sigma_a x_{pa} \omega_a z_{pa}) / \Sigma_a z_{pa}^2$ into (11–37) and simplifying the expression yields:

$$D = - \sum_p \frac{\left[\sum_a x_{pa} \omega_a z_{pa} \right]^2}{\sum_a z_{pa}^2} + \sum_p \sum_a x_{pa}^2 \; . \tag{11–39}$$

Since the sum of squares of X_i is unity due to norming, it follows that the first term in (11–39) has bounds zero to one. The task is then to maximize this term, which can be conceived of, analogous to the W^* weights, as the sum of squares of normalized vector weights, V^*:

$$v_{pp}^* = v_{pp} \left(\sum_a z_{pa}^2 \right)^{1/2} \tag{11–40}$$

The sum of squares of these v_{pp}^* represents the proportion of variance of $X_i \Omega_i$ accounted for by $V_i Z$. One observes immediately that D is not differentiable with respect to the unknowns $\omega_a (a = 1, \ldots, m)$. Consequently, we resort to a search algorithm in which each dimension is reflected in turn and the outcome is evaluated for any increase in the total sum of squared normalized vector weights. This single dimension approach, however, does not guarantee finding the best reflection of the X_i under all conditions, but is, nonetheless, implemented for reasons that will be discussed in fuller detail after we deal with the problems of rotation and translation.

Having obtained an "optimally" reflected X_i, we proceed in a completely analogous manner to the dimension-weighting approach for rotation, i.e., in each plane we minimize (11–37), where $x_{pa}\omega_a = \hat{x}_{pa}$ is replaced by a rotated $\hat{x}_{pa}$. This leads to our plane-wise loss function of D in (11–37):

$$D_{ab} = \sum_p [(v_{pp}z_{pa} - x^*_{pa})^2 + (v_{pp}z_{pb} - x^*_{pb})^2] , \qquad (11\text{–}41)$$

where

$$v_{pp} = \frac{\sum_a x^*_{pa}z_{pa}}{\sum_a z^2_{pa}} , \qquad (11\text{–}42)$$

and x^*_{pa} and x^*_{pb} are as in (11–24) and (11–25), respectively, if one replaces x_{pa} and x_{pb} by $\hat{x}_{pa}$ and $\hat{x}_{pb}$. Inserting appropriately, differentiating the expression in (11–41) with respect to α, and setting the derivative equal to zero yields:

$$\alpha = .5 \tan^{-1}\left(\frac{2A}{B}\right) , \qquad (11\text{–}43)$$

where

$$A = \frac{\sum_p [z_{pa}z_{pb}(\hat{x}^2_{pa} - \hat{x}^2_{pb}) - \hat{x}_{pa}\hat{x}_{pb}(z^2_{pa} - z^2_{pb})]}{z^2_{pa} + z^2_{pb}} , \qquad (11\text{–}44)$$

$$B = \frac{\sum_p [(\hat{x}^2_{pa} - \hat{x}^2_{pb})(z^2_{pa} - z^2_{pb}) + 4\hat{x}_{pa}\hat{x}_{pb}z_{pa}z_{pb}]}{z^2_{pa} + z^2_{pb}} . \qquad (11\text{–}45)$$

As before, for dimension weighting, we test via the sign of the second derivative of (11–41) as to whether α maximizes or minimizes D_{ab}. If the former is true, α is incremented by 90°. With these plane-wise solutions we proceed iteratively until the cycles converge. Although we cannot offer a convergence proof for this process, abundant evidence from both artificial and empirical data strongly suggests convergence and the "recovery" of known a priori vector weights can be demonstrated.

The remaining problem at this stage of the analysis is to find the individual translation vectors t_i and u_i. The problem is complicated by the fact that although one can express t_i in terms of u_i, no simple analytic solutions could be derived for all the unknowns involved in the optimization of (11–37). Hence, we employed an ALS procedure which first solves for V_i and R_i with

t_i initially a null vector, then for t_i (with u_i substituted in [11–36] by [11–35]) in terms of the previous solutions, etc., until this process converges (necessarily). The fit values for the vector-weighting model with idiosyncratic translations on Z (the unrestricted "perspective model," v.i.) are then computed on termination of the ALS procedure. These values are denoted by $r^2(\tilde{X}_i, V^t_i Z^t_i)$.

Now, in order to determine the overall optimal translation of Z for the restricted perspective model, we proceed completely analogous to the dimension-weighting approach (where, it will be recalled, we determined from the average of the idiosyncratically rotated Z's, the Z^r). That is, by setting $t = 1/N \; \Sigma_i t_i$, the vector of average shift, we translate Z to this new origin. By another ALS procedure, this time involving $Z^t = Z - jt'$, one obtains the individual vector weights for the less general version of the perspective model. The fit values are now expressed as $r^2(\tilde{X}_i, V_i Z^t)$, where Z^t is fixed for all N configurations. It should be clear that $V^t_i \neq V_i$, in general. Moreover, the full model will explain each $\tilde{X}_i$ at least as well as the fixed origin model, of course.

As an alternative to shifting Z by t, one may want to select some specific point as the origin of Z (or leave the origin undisturbed in the case of some factor solution). In certain instances, e.g., in a radex structure, a meaningful origin is given and this is, of course, to be preferred to some blindly determined mathematical optimum. Thus, Z may optionally be translated (or be left undisturbed) to a prespecified origin in a manner parallel to the option available in the dimension-weighting approach vis-à-vis the orientation of Z.

Some comments are in order regarding optimal reflections in vector weighting. It will be recalled that we stated that our reflection procedure for X_i could be suboptimal under some conditions. We shall now take a closer look at this problem.

The problem is to reflect X_i by Ω_i such that $\hat{X}_i$ will lead to a maximal fit if $\hat{X}_i$ is subsequently optimally transformed by a proper rotation R_i. Anticipating the proper rotation reduces the set of distinct reflection matrices to the relevant equivalence classes of Ω_i defined by all permutations of $m - 1$ elements "1" and "-1." One observes that the rotation angle α in (11–43) is not identical over equivalence classes, however, since the signs in between the numerator terms of (11–44) and (11–45) would not be the same in all planes for different Ω's. The latter observation implies that minimizing D in (11–39) through an appropriate choice of the signs of each dimension in X_i will not necessarily be optimal if one takes into account the possibility of later proper rotations. Examples illustrating this fact are readily constructed. Consider the three matrices below:

Z		X		Y	
-3	2	1	-6	-1	-6
6	8	-2	-5	2	-5
1	-7	3	4	-3	4
-5	-4	-7	8	7	8

Matrix X is related to Y by a reflection of the first dimension. The fit index $L = \Sigma_p[(\Sigma_a x_{pa}z_{pa})^2/\Sigma_a z_{pa}^2]$ has the value $L(Z, X) = 57.07$. The analogous measure on Y yields $L(Z, Y) = 142.78$, which would suggest that either dimension of X be reflected (which one is of no consequence). Computing the rotation angles we find $\alpha(X) = -16.26°$ and $\alpha(Y) = -17.79°$, but according to the second derivative test we must alter $\alpha(X)$ to $73.74°$. We then obtain $L(Z, XR) = 155.29$ and $L(Z, YR) = 152.15$, i.e., a relative reversal in the order of agreement so that we would have been better off not to have reflected X at all.

To insure that (11–34) is minimized over all Ω-equivalences we would have to test each and every one, a considerable undertaking for planar rotations if m is of substantial size, i.e., for each Ω_i there would be

$$\binom{m}{2}(m + 1)$$

rotations for each cycle. For $m = 10$ one would need 495 planar rotations for one cycle and probably anywhere up to 50 cycles (on the average) would be required for convergence.

Although for the (not so uncommon) case of low dimensional spaces, the computation efforts would be much reduced, some other alternative is needed in general. Our present suboptimal algorithm is certainly inexpensive compared to the exhaustive requirements outlined above. In addition, it might be mentioned that extensive computer runs have demonstrated that reflections are in general not required and that when they are, the gain in fit is not of a magnitude that would tip the balance in favor of the vector model over that of others addressed by PINDIS, e.g., the dimensional salience model. The chief reason that reflections are not generally needed is that the dimension-weighting transformations place X in an orientation that is not far from the optimal for that of vector weighting.

In our matrix formulations of the PINDIS hierarchy of transformations, v.i., we outline a possible ALS procedure for handling the vector-weighting case optimally in respect to both reflection and rotation. Before we delineate these matrix formulations for both dimension and vector weighting, we would like to make a few passing remarks on the uses and possible interpretations of the perspective model.

There are essentially two roles for the vector-weighting approach. The weight profiles may be indicative of more complicated relations among the subject configurations than is conveyed by a simple dimension distortion. Additionally, the vector weights may also be related to a psychological model about interindividual differences.

In the first case, where vector weighting is used as an index-providing procedure, one obtains, in particular, insight into the degree to which certain points in Z are related to those in an individual's configuration in a way that cannot be handled by the affine transformations in the dimensional salience model. One may find that Z explains an X_i fairly well except for a few points that require substantial displacements. Such cases have been reported by Borg and Lingoes (1976): some individuals associated certain breakfast food items that were grouped by the average subject into some cluster *A*, say, with those in some other cluster, non-*A*. Except for those few points, however, they perceived the stimuli in practically the same way as the others. It may have been true though that all points required considerable shifting for all individuals and, at the same time, some typical patterns in the computed vector weights could be discerned over all subjects; this would then indicate that instead of computing *Z* over apparently different groups, it might be useful to consider partitioning the sample first on the basis of the obtained vector weight profiles into more homogeneous groups. What one could aim at, for example, would be a partitioning such that the subjects in each subgroup were related to their common average by such simple transformations as a similarity transform or as dimensional weighting.

As in the dimensional salience model, the vector-weighting model is closely related to a psychological theory in a number of contexts. We choose to call this psychological model a perspective model for obvious reasons related to the geometry of the representation as it relates to visual perception. As an illustration, let as assume that one asks different individuals to compare various geographic locations with respect to the time they think it will take them to walk there *from a given point*. Let us further assume that the physical terrain is fairly hilly. Then one would expect that the subjects would radially shift the stimulus representations according to their respective perceived accessibilities. Physically very fit persons, for example, would probably use very little shifting as a function of the object's position on a hill or in a valley, whereas this would not be true for those who expect greater differences in required efforts. A number of psychological parallels to this physical example come readily to mind.

A relatively direct analogy exists for representations of sociometric data. An individual data matrix in this context may be as follows: row and column entries correspond to the same persons: *A, B, C, . . . , Y,* where *A* represents the experimental subject; the matrix elements stand for the similarity of each pair of persons as perceived by *A*. Scaling each such data matrix one obtains a spatial representation of the group structure for each individual *A, . . . , Y*. One can now hypothesize that interindividual differences are due to idiosyncratic origins (i.e., each subject is himself a point of perspective) and differential gradients due to liking, power, etc. One would thus assume that the individuals have basically the same perception of the group structure except for differences in the sense of the perspective model.

Another application of the perspective model is possible in the context of Guttman's radex theory (1954). To illustrate this we consider a series of studies on "quality of life" (see, for example, Levy and Guttman 1975), where item points in the SSA (smallest space analysis) space could be grouped into regions representing questions on satisfaction with one specific aspect of life. These regions were arranged as slices around an origin represented by the item "satisfaction with life in general." For such studies it was observed that different groups were different essentially only in shifting the points within each of these sectors (Levy 1976; Borg 1978). Thus, the origin and also the order of the sectors remained practically the same for the groups and the point weighting was done in the sense of the perspective model. The radial shifting in this example is, therefore, related to differential perceptions of an item's *centrality* with respect to the issue "quality of life."

Although we have indicated some psychological interpretations for idiosyncratic origins and (non-negative) vector weights, there does not seem to be any convincing interpretation for negative vector weights in a model sense. This situation is similar, to some extent, to the dimensional salience case (for PINDIS, but not INDSCAL), where negative weights have a double meaning, i.e., they indicate both a differential importance and a necessary reflection of Z. The latter has the status of an index rather than that of a model parameter. For vector weights it seems most appropriate to treat the negative weight case similarly.

Matrix Solutions for Dimension and Vector Weighting

We shall now summarize the PINDIS model family in the more compact language of matrix algebra. In addition, we shall consider one rather obvious extension of the various weighting models used so far.

In general, the following loss function must be minimized in PINDIS:

$$\operatorname{tr}(EE') = \min_{[R_i,\, S_i,\, W_i,\, V_i,\, k_i,\, t_i,\, u_i,\,]} , \qquad (11\text{–}46)$$

where

$$E = V_i(Z - jt_i')S_i W_i - k_i(X_i - ju_i')R_i \; . \qquad (11\text{–}47)$$

The different approaches discussed so far were different in attempting a solution of (11–45) for proper subsets of the set of variables only. The various subsets are different with respect to the mappings on Z. The transformations on X_i are always completely optimized (although this is strictly true for R_i and u_i only, the central dilation scalar k_i will be implicitly determined by the fit index). It should be clear that R_i, k_i, and u_i will generally be different for the similarity transform, the dimension weightings, and the vector weightings. In no instance should one attempt to interpret any of these mappings, of course. The possible exception of using k_i as a fit index has already been discussed. Only the transformations on Z are principally of substantive interest.

In the dimension-weighting case, we set $V_i = I$ and $S_i = S$. In the idiosyncratic orientation weighting, $V_i = I$ as before, but W_i will now be different from the weight matrix obtained for the fixed orientation case. We have denoted the idiosyncratic orientation weight matrix by $W^r{}_i$ above to indicate the conditionality of the resulting weights on how Z is rotated. Finally, in vector weighting, $W_i = I$ and rotations of Z are irrelevant so that one can set $S_i = S$ arbitrarily. For the full perspective model, the idiosyncratic case, Z is translated to Z^t_i using t_i. For an over-all-optimally positioned Z, Z^t, one uses t as a shift vector.

For dimension weighting with idiosyncratic orientation we obtain the following solutions via partial differentiation of (11–46) under the respective linear constraints:

$$u_i = \frac{(X_i - Z^r_i W^r_i R_i')'j}{n}, \tag{11–48}$$

$$W^r_i = [\text{diag }(Z^r_i{}' Z^r_i)]^{-1}[\text{diag }(Z^r_i{}' X^u_i R_i)], \tag{11–49}$$

$$(X^u_i{}' Z^r_i W^r_i) R_i' = \text{symmetric}, \tag{11–50}$$

$$(W^r_i{}' Z' X^u_i R_i) S_i' = \text{symmetric}, \tag{11–51}$$

where $X^u_i = X_i - ju_i'$. As has been shown above, a single translation vector is sufficient for optimization of (11–46) with $V_i = I$. Thus, Z remains centered at the origin, i.e., $t_i = (0, \ldots, 0)$. The solution (11–50) is due to minimization of (11–46) with $V_i = I$ and $t_i = (0, \ldots, 0)$ with respect to R_i under the condition $R_i'R_i = R_i R_i' = I$. The latter constraint can be formulated as tr $[L(R_i'R_i - I)] = 0$, where L is a matrix of Lagrangian multipliers, which leads to the function $f = \text{tr}(EE') + \text{tr}[L(R_i'R_i - I)]$. Solving $\partial f / \partial R_i = 0$ yields

$$nu_i u_i' + X_i'X_i + L = X^u_i{}' Z S_i W^r_i R_i', \tag{11–52}$$

where the left-hand side is apparently a symmetric matrix. Now, setting $A = ZS_i W^r_i$ and $B = X^u_i$, it follows that $B'AR_i' = R_i A'B$ or $B'A = R_i A'BR_i$ which leads to Schönemann's (1966) solution for the orthogonal procrustean rotation, i.e., $R_i = WV'$ with W and V being appropriately oriented eigenvectors of $B'AA'B$ and $A'BB'A$, respectively.

There are four unknowns in the system of equations (11–48) through (11–51), i.e., u_i, R_i, S_i, and W_i. First, of course, one observes that (11–48) leads to the null vector for u_i since both X_i and Z are columnwise centered and since this property remains unaffected by the other transformations. Although each of the remaining equations can be solved for any given independent variables, there does not seem to exist any apparent direct solution for the whole set. It would be possible, however, to use an alternating method analogous to the one used in INDSCAL and ALSCAL, for example. If the space were higher than two-dimensional, this would have the advantage over the scalar approach that the rotation matrices would be computed directly rather than in the plane-wise iterative manner, which cannot be proved to converge necessarily.

For the vector-weighting approach, we obtain the following equations:

$$u_i = \frac{(X_i - V^t_i Z^t_i R_i')'j}{n}, \tag{11–53}$$

$$t_i = \frac{[Z - (V^t_i)^{-1} X^u_i R_i]'j}{n}, \tag{11–54}$$

$$(X^{u\prime}_i V^t_i Z^t_i) R_i' = \text{symmetric}, \tag{11–55}$$

$$V^t_i = [\text{diag } (X^u_i R_i Z^{t\prime}_i)][\text{diag } (Z^t_i Z^{t\prime}_i)]^{-1} . \tag{11–56}$$

For R_i one observes that by using Schönemann's (1966) results no separate computation of a reflection matrix Ω_i is necessary. Again, the whole system is not directly solvable, but requires an iterative approach, such as outlined for dimension weighting above.

In evaluation of the entire matrix approach, one could say that it has clear advantages over the scalar solutions previously discussed whenever the transformational analyses are carried out in higher dimensional spaces. In the rather frequent case of two dimensions, however, we obtained some direct analytic results via scalar methods which have the additional advantage of being computationally faster than those involving matrix manipulations.

To conclude this section we consider an extension where (11–46) is solved for all transformations listed in the set of variables, i.e., Z is now simultaneously weighted from the left and the right. A less complete form of such double weighting has been investigated previously by Lissitz et al. (1976), although in a different context. The solutions for this model are easily derived and could be solved (at considerable expense) via the alternating least-squares approach. (In fact, the PINDIS program presently contains a double-weighting analysis as an option.) The major difficulty with this model is, of course, that it is neither in any obvious way related to a psychological model such as "dimensional salience," for example, nor is it generally useful in an indexing sense because it is simply too complex. A rather exotic application where a restricted double-weighting solution proved useful can be found in Lingoes (1978). Moreover, using double weighting in a purely formalistic sense, we found that it did not lead to appreciable fit improvements in practically all cases that were investigated. Thus, we do not pursue this approach here any further.

PINDIS EXAMPLES

Feger Political Party Attitudinal Data

To illustrate PINDIS in a real application we use some data collected and previously analyzed by Feger (1974). The study involved the six major political parties in West Germany: (1) a nationalistic party; (2) a very conservative party; (3) a moderately conservative party; (4) the Liberals; (5) the Social Democrats; and (6) the Communists. Fourteen subjects were asked to indicate which one of various combinations of political parties (1 = NPD, 2 = CSU, 3 = CDU, 4 = FDP, 5 = SPD, 6 = DKP) and rating categories (*A* = extremely far, *B* = very far, *C* = far, *D* = close, *E* = very close, *F* = extremely close) were more descriptive of their attitudes. For example, the subject was presented with the two combinations *A* - 1 and *D* - 4. He then had to pick one of the two which he thought was a truer description of himself. Assume he chose *A* - 1. *A* - 1 would then be assigned a score of one, while *D* - 4 would receive a score of zero. In the process of presenting all 630 possible pairs of pairs (using a randomized sequence, a screen projection of the stimuli, and a machine response recording system) a 6 × 6 dominance matrix was constructed, whose entries ranged between zero (dominates none) to thirty-five (dominates all) (see Feger 1974, for further details). These fourteen rectangular dominance matrices were then analyzed by SSAR-I (Lingoes 1973), which assumes between-set information only (i.e., parties and categories) at the ordinal level and all elements are comparable. In every case an almost perfect representation was possible in a two-space, the poorest fit having a stress less than .009 (form 1) for Subject 8. The 14 X_i's of order 12 × 2 served as the input for a PINDIS analysis.

As a first step, PINDIS computed a Z, which is represented graphically in Figure 11–1 below. The average subject obviously arranges the parties on an approximate simplex (dashed line), where the nationalistic party (point 1) forms one terminus and the Communists (point 6), the other. The rating categories do not, surprisingly enough, form a simplex (solid line): the category "extremely close" is avoided by 12 of the 14 subjects and is positioned closer to the centroid of the parties. The squared norm of Z was .75, which means that Z accounted for 75% of the variance over all X_i's.

The first column of Table (11–2) lists the fit values for the similarity matching of the X_i's to Z. Apparently, most individuals had rather similar perceptions, which is inferred from their high communalities with the average subject Z. Individuals 1 and 9, however, are much more poorly explained by Z than are the other subjects. These differences are preserved even after dimension weighting (column 2 of Table 11–2). From a plot of Subject 1's perceptual space (see Figure 11–2), one can readily see that this subject does indeed have a configuration which cannot be satisfactorily related to Z by either the similarity or dimension-weighting transformation. In general, we see from the second column of Table 11–2 that the dimension weights do not improve the fit values substantially for any subject, indicating that the computed weights should not be interpreted as *differential* dimensional saliences (the implicit unit weights of the similarity transform do almost as well). The information provided by a "subject space," i.e., a graphical representation of individual w^* vectors in the stimulus space, as in INDSCAL (Figure 11–2), should always be evaluated with care. The scatter around the origin can only be given a substantive meaning if the $\tilde{X}_i$'s are substantially better explained under the dimension weighting transformation than by purely admissible, i.e., similarity map-

Figure 11–1. Centroid Configuration for Feger Data

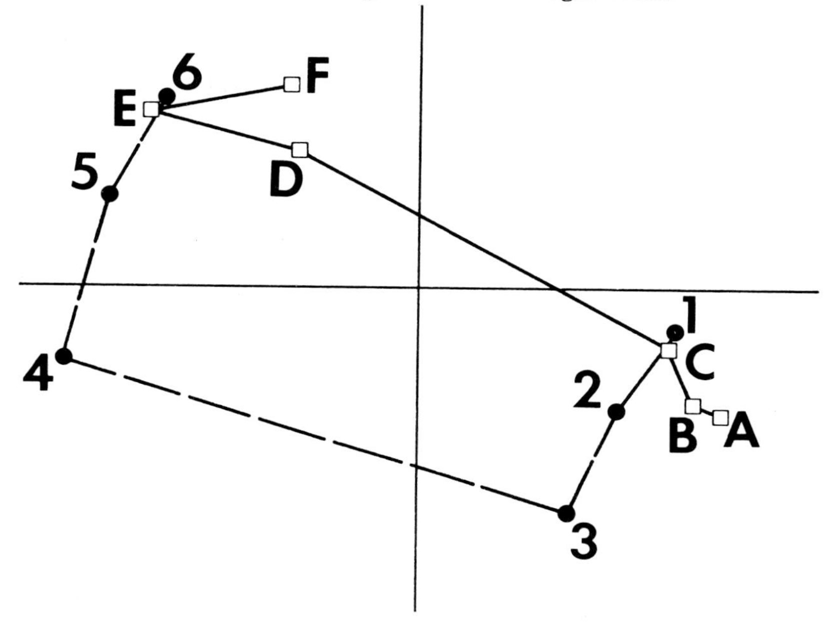

TABLE 11-2. Fit Values for PINDIS Analysis of Feger Study

	Fit				
Subject	$r^2(\tilde{X}_i, Z)$	$r^2(\tilde{X}_i, Z^r W_i)$	$r^2(\tilde{X}_i, Z_i^r W_i^r)$	$r^2(\tilde{X}_i, V_i Z^t)$	$r^2(\tilde{X}_i, V_i^t Z_i^t)$
1	.155	.167	.168	.674	.885
2	.801	.817	.819	.887	.998
3	.942	.948	.948	.986	.998
4	.854	.862	.862	.939	.996
5	.780	.785	.785	.942	.994
6	.780	.794	.794	.910	.962
7	.877	.881	.915	.965	.950
8	.794	.795	.796	.943	.981
9	.584	.586	.643	.885	.942
10	.842	.844	.844	.947	.998
11	.945	.945	.945	.986	.998
12	.791	.793	.793	.949	.926
13	.838	.849	.849	.984	.975
14	.844	.850	.850	.960	.993
$\bar{r}^2$	.773	.778	.786	.926	.971

pings. The fact that we obtained comparable "communalities" from an INDSCAL analysis of the euclidean distances generated from the SSAR-I X's (Borg and Lingoes 1977) reinforces our conclusion. INDSCAL's G was related to our Z by a 180° rotation and a simple dimension weighting to the extent of $r^2(G, Z) = .99$.

Some useful additional information may be contained in column three of Table 11–2. Here, the communality between $\tilde{X}_i$ and

Figure 11–2. Subject Space for Feger Data

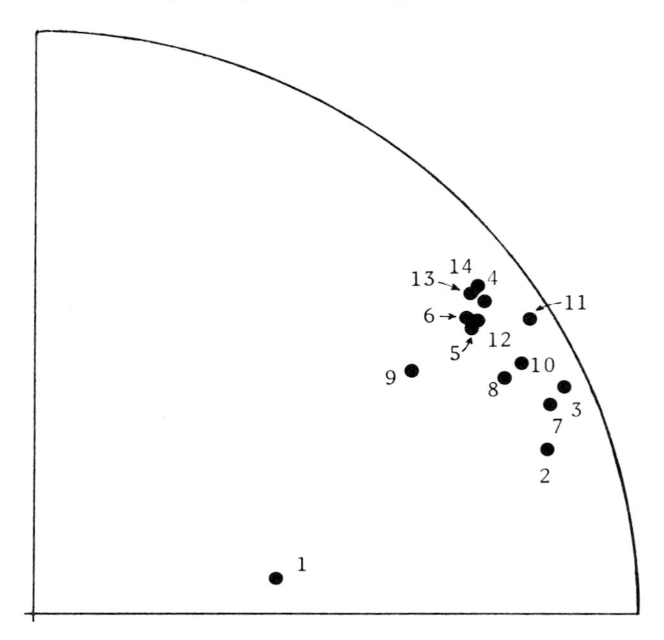

$Z^r_i W^r_i$, i.e., the individually optimally oriented Z, is represented. It is to be expected, of course, that there will be some dispersion of the Z^r_i's about Z^r. However, some $\tilde{X}_i$'s may be substantially better explained by their respective Z^r_i's than by Z^r, which would indicate that these individuals are not likely to use the same spatial directions, or, in the terminology of IDIOSCAL, that they have idiosyncratic frames of reference. In our present example, this does not appear to be the case for any individual (only Subjects 7 and 9 reveal any tendency of this kind). If one found some instances of such idiosyncracy in one's data, it might be advisable to eliminate such individuals from the analysis and then re-analyze the data. In this manner, the sample could be partitioned into more homogeneous subgroups.

Proceeding with the vector-weighting transformation at the next level of complexity, we obtained the fit values given in the fourth column of Table 11–2. Overall, it is obvious that this transformation accomplished more than that of dimension weighting: The total communality improvement is 15% as compared to practically 0%, respectively, over the similarity transform. This increment has to be evaluated, however, against the considerably increased number of free parameters in vector weighting (12 vs. 0 [similarity] and 12 vs. 2 [dimension weighting]). Nevertheless, it is clear that, for example, for Individual 1 the fit improvement is quite impressive. Taking a closer look at this individual, it is interesting to note that practically all of his optimal vector weights are quite different among each other and they depart from +1. Apparently, Z explains X_1 sufficiently well only under independent (and substantial) movements of the points. Quite clearly, these point displacements show that his perceptual space relative to the one of the average subject is "scrambled." Indeed, a plot (see Figure 11–3) of his space shows that he folds what others perceive as a left-right simplex of political parties (Figure 11–1) at his most preferred party, which obviously generates a radically different perceptual space.

It may be noted that Individual 1's attitude structure showed remarkable stability over eight replications taken over eight weeks of testing (Feger 1975; Borg and Lingoes 1977). In other words, his discrepancy from the average subject cannot be explained by error alone. In conclusion, it becomes apparent that the vector weights can serve as yet another index of group homogeneity and, consequently, as a basis for evaluating the representivity of the average subject space in the dimensional salience model. This latter point simply means that Z may be a confounding of essentially qualitatively different individual structures, or, in other words, the average subject may not represent *any* individual sufficiently well. If such is the case, then the dimensional salience model does not make much sense, of course.

Allowing for individual origins of the average subject configuration, one obtains the fit values found in the fifth column of Table 11–2. On the average, a further fit improvement of some 5% over the previous transformation has been effected. However, the communality increments are not uniformly distributed over subjects. Individuals 7, 12, and 13 even drop slightly in their communalities, which appears to be a function of the flatness of their response surfaces and the termination criterion in the ALS procedure. On the other hand, one notes a substantial fit improvement for Subject 1 of over 20%, indicating again that this

Figure 11–3. SSAR-I Configuration for Individual 1 in Feger Study

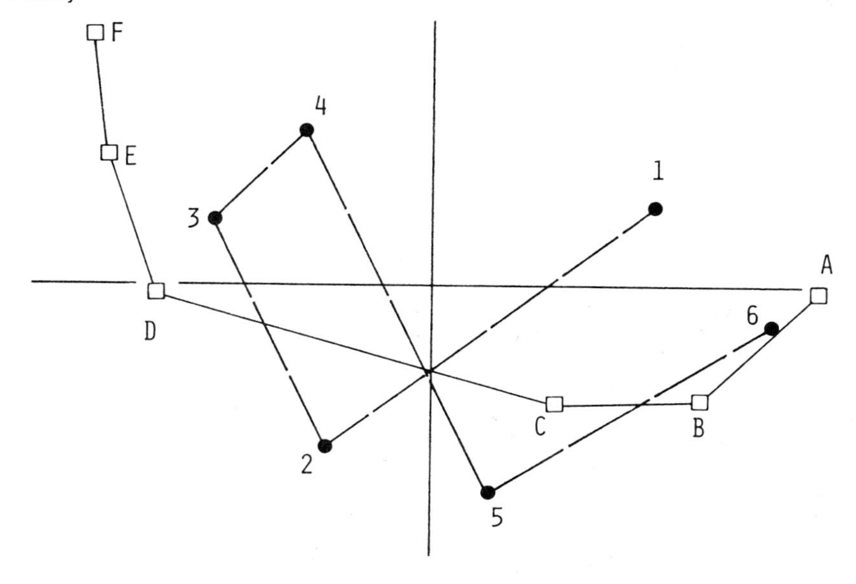

individual is quite different from the remaining subjects. (Indeed, from other sources of information [Feger 1977] we know that Subject 1 is about forty years of age and has a conservative attitude in political matters, whereas the remaining thirteen subjects are psychology students with pronounced leftist views.) For the idiosyncratic origins, one observes considerable variance among the individuals (see Figure 11–4). However, taking into account the fact that appreciable fit improvement was obtained for two subjects only (1 and 2), it seems questionable to interpret the noted scatter as indicative of real inter-individual differences for all subjects, particularly when one takes into account the relatively high fit values for the fixed origin version of the vector-weighting model. Under other conditions, of course, one could use the origin space in a manner similar to the subject (dimension weight) space as a basis for partitioning. For the present results, however, it seems reasonable to contrast the origin of Subject 1 with the average of the remaining individuals. Since no substantive perspective model was assumed to account for these data, only indexing information is herewith provided.

Helm Color Data

In this example we use data collected by Helm and analyzed by him (1959, 1964) via a ratio scale MDS technique, by Helm and Tucker (1962) via a points of view method, by Carroll and Chang (as reported by Wish and Carroll 1974) via INDSCAL, and by MacCallum (1974) comparing INDSCAL with Tucker's three-mode scaling.

Helm (1959) asked subjects to judge the similarity of ten colors which differed in hue by arranging triples of color tiles such that the distances among the tiles were to be set proportional to their perceived similarities. The matrix of distances derived from these triadic judgements was assumed to be determined up to a scalar multiplier. Thus, the matrix of scalar products for each subject was diagonalized without prior transformations. There were ten color normal subjects plus one replication (normals) and four red-green deficient individuals (as determined by a standard discrimination test) plus one replication (deficients). The eigenvalues for the normals and mildly color deficient subjects indicated an underlying two space, whereas the two highly deficient subjects yielded but a single dimension essentially. Negative eigenvalues were obtained in all cases, but were explained by Helm (1959) and Helm and Tucker (1962) to be the result of systematic estimation error.

The first two components of loadings from the normals were used as the PINDIS input. The resulting Z is represented in Figure 11–5. The average subject configuration shows the expected color circle. The vertical axis corresponds directly to a red-green dimension relative to which we should anticipate finding the essential differences among color normals and deficients.

The five color deficients' two dimensional spaces from Helm's analysis were then related to this Z by using the PINDIS option for fixing the target (Z) and expressing all transformations in terms of the given Z for each X_i. The fit values for all sixteen

Figure 11–4. Idiosyncratic Origins (stars) for 14 Subjects of Feger Study

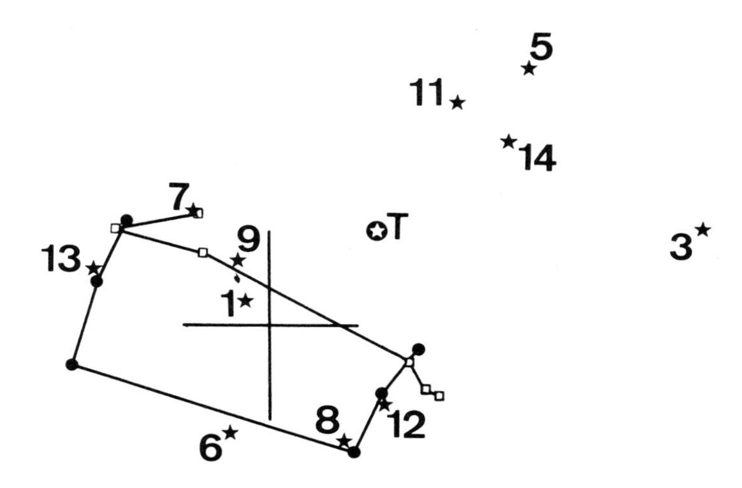

Note: T = overall optimal origin

data sets are given in Table 11–3. Note that there is no idiosyncratic rotation or translation for the deficients, since Z was fixed in the orientation and origin optimal for explaining the normals under the dimensional salience and perspective models, respectively.

Figure 11–5. Unrotated Centroid Configuration for 10 Color Normals of Helm Study

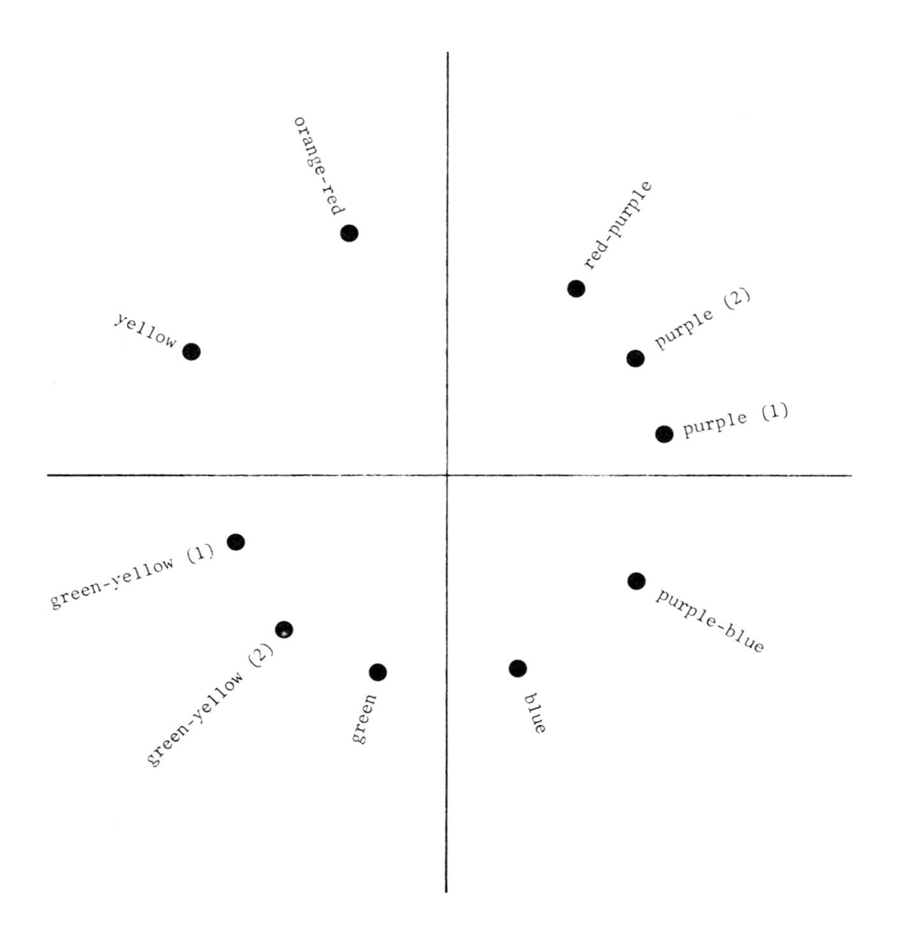

Table 11–3 shows that Z explains the X_i's of the normals extremely well without any differential weighting. This is reflected in Figure 11–6 by the fact that all the respective weight vectors (solid points) in the subject space are (a) closely clustered around the bisector and (b) lie almost at the unit circle, which expresses perfect communality. Consequently, neither idiosyncratic orientations of Z nor vector weightings are of any relevance here. It should be mentioned that close clustering of subject points indicates highly similar private perceptual spaces, but that the points may scatter considerably even if very little *additional* variance, relative to the unit weighting implied by the similarity transform, is explained by differential weights (Borg and Lingoes 1976). This scatter should, therefore, only be interpreted if $r^2(\tilde{X}_i, Z) << r^2(\tilde{X}_i, Z^r W_i)$.

For the color deficients, Table 11–3 reveals that dimension weighting improves the fit on the average by some 15% and for

TABLE 11-3. Fit Values for PINDIS Analysis of Helm Color Data

		Fit				
	Subject	$r^2(\tilde{X}_i, Z)$	$r^2(\tilde{X}_i, Z^r W_i)$	$r^2(\tilde{X}_i, Z_i^r W_i^r)$	$r^2(\tilde{X}_i, V_i Z^t)$	$r^2(\tilde{X}_i, V_i^t Z_i^t)$
Normals	1	.978	.983	.983	.995	.998
	2	.978	.981	.987	.993	.998
	3	.990	.991	.992	.997	.999
	4	.989	.990	.991	.994	.998
	5	.991	.991	.991	.996	.999
	6	.969	.970	.971	.985	.998
	7	.983	.983	.983	.993	.997
	8	.974	.974	.975	.984	.997
	9	.982	.983	.983	.989	1.000
	10	.983	.991	.992	.992	1.000
	11	.956	.965	.967	.982	.995
	$\bar{r}^2$	.979	.982	.983	.991	.998
Deficients	1	.891	.920	--*	.996	--†
	2	.911	.949	--	.944	--
	3	.437	.823	--	.772	--
	4	.943	.968	--	.977	--
	5	.496	.754	--	.776	--
	$\bar{r}^2$	.736	.883	--	.887	--

*Orientation fixed to optimal solution of normals
†Origin fixed to optimal solution of normals

Figure 11–6. Subject Space for Helm Data

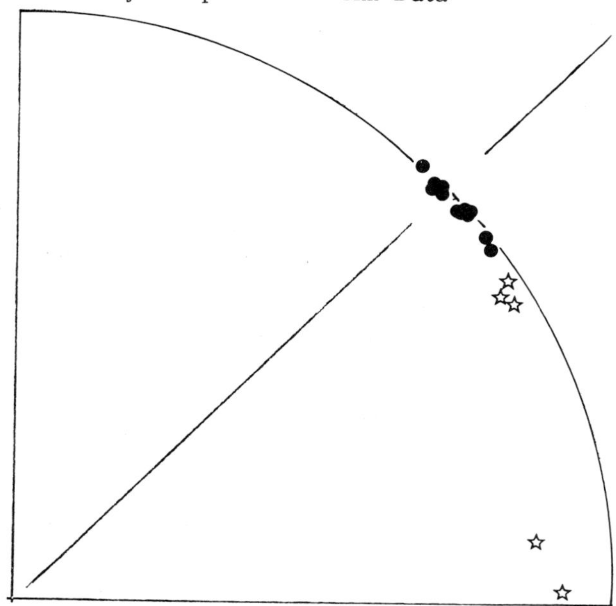

Note: Dots = normals; stars = deficients

the two most deficient individuals even higher. In the subject space one can see that the weight vectors for the deficients (stars) are clearly separated from the normals. Furthermore, the salience of the red-green dimension is, in all cases, lower than unity. The estimated private perceptual spaces would thus be ellipses rather than circles as for the normals. For the extremely deficient subjects these ellipses are almost collapsed onto a line, i.e., these subjects differentiate the colors only on a green/yellow-purple continuum.

Table 11–3 further shows that the vector weightings do not indicate peculiarities in the perceptual space of any individual, since such weightings improve the fit either only by a negligible amount or lead to a substantial decrement as compared to dimension weighting. This observation is even more relevant if it is evaluated against the fact that eight (ten) additional free parameters are used here. The results, therefore, validate the expectations in all respects.

If we had no prior knowledge about the deficiency of some subjects, the Z generated from all sixteen data sets would, of course, reflect the low salience of the red-green dimension for five individuals: Z is then flattened to an ellipse along the red-green axis. The associated subject space does, however, still allow one to distinguish between the different subgroups, although the mildly deficients are only extremes of the group of normals (as can be seen from Figure 11–6). Note that in an INDSCAL analysis, the group space would not change its shape in the same manner, since it is normed to unit length on all dimensions; only the INDSCAL subject space would then reflect the fact that the average subject configuration is not circular.

Another interesting phenomenon can be observed if one analyzes the complete set of the X_i's made up of the first three and four dimensions based on Helm's solutions, respectively, rather than those corresponding to the first two dimensions only. The resulting fit values will then be poorer under all transformations (Table 11–4) except that of vector weighting. From this fact one has to conclude that the common perceptual structure is two rather than three or four dimensional. The additional dimensions constitute non-common variance essentially and may thus be regarded as error in this context. Of course, since neither one of the additional dimensions accounts for much variance relative to the first two, their effect on the total communality is rather small, i.e., they change the ratio of explainable to error variance only slightly in the direction of error variance.

TABLE 11-4. PINDIS Average Common Variance (by dimensionality) for Helm Data

	Dimensionality		
$\bar{r}^2$	2-D	3-D	4-D
$\tilde{X}_i, Z$	.91	.88	.88
$\tilde{X}_i, Z^r W_i$	.95	.92	.92
$\tilde{X}_i, Z_i^r W_i^r$	.95	.92	.93
$\tilde{X}_i, V_i Z^t$	.97	.99	.96
$\tilde{X}_i, V_i^t Z_i^t$	.99	1.00	.99

We observe, however, that the communalities do not drop monotonically from two to four dimensions in all instances: for both dimension weighting transformations there is no such decrement. Taking a closer look at the individuals, one finds that the color deficients are explained slightly better by increasing dimensionality, whereas just the opposite is true for the normals. Thus, while added dimensions represent error for the normal subjects, this seems not to be the case for the deficients (an observation already made by Helm 1959).

One might pursue this issue of dimensionality further in the present case by analyzing normals and deficients separately, which would be particularly interesting if a hypothesis about such a difference in the dimensionality of the perceptual spaces of normals and color deficients was to be given. In the absence of having some theoretical basis for deciding upon the dimensionality of the X_i's, one could to some extent use the observations made here, i.e., one could analyze X_i's in different dimensionalities and then determine where the maximal average fit occurred. One should remember, however, that proceeding in this fashion does not guarantee optimal predictability within each meaningful subgroup. In general, at this stage of knowledge, we would hesitate in making any strong recommendations as to how one could determine the dimensionality of the X_i's on purely formal grounds. Clearly, further evidence is needed and substantive considerations play an important role here.

DISCUSSION

In our second illustration, where we reanalyzed Helm's color data, PINDIS was applied in a confirmatory way. The orientation and origin of the average subject configuration obtained in the analysis of the color normals was such that it corresponded to the substantive theory for the differences among the individuals. Thus, exactly this average subject space was used as a not-to-be-changed hypothesis configuration in the subsequent scaling of the color deficient individuals. It might have been necessary, however, to rotate the centroid configuration to a meaningful orientation first before weights related to a substantive theory of spatial directions could be derived. Proceeding in this manner, one would generally not be able to predict the same amount of variance, of course, since Z^r's orientation is unique (mathematically) in the dimensional salience model. In some empirical tests where Z was rotated over a 180° interval, we observed the rather interesting behavior, however, that the decrement in predicted variance for non-optimal orientations relative to the optimal Z was generally quite small. This finding requires further investigation, needless to add, but even at this stage of knowledge, we do not think that there can be much conflict for the researcher as to whether Z should be "optimally" or "substantively" oriented.

In other applications (Borg 1977) we have also observed that the similarity transformations do generally, by far, the most important job in predicting an individual from the average subject, although we have also found some instances where (fixed origin) vector weighting has increased the proportion of explained variance by some 65% (Borg and Lingoes 1976). It would be interesting to know in this context how well various random configurations could be explained by their centroids, of course, since we presently lack a statistical theory for our fit measures. Our choice of examples was in part dictated by the appropriateness of the two vis-à-vis the two models discussed, i.e., the dimension salience model (Helm's color data) and the perspective model (Feger's political attitude data), but it should be noted that over a wide range of data we have found the perspective model explains individual differences better than does the dimension salience model (which generally does little better than the unit weighting of the similarity transform). The difficulty in evaluating this finding in the absence of a statistical theory, however, leads us to suggest a certain amount of pragmatism in exploring both as a matter of routine.

As a final remark, we note that PINDIS provides normed scalars for each X_i so that unconditional weights for both dimensions and vectors can be easily computed for correlational and group comparison purposes (MacCallum 1977).

APPENDIX: DEFINITION OF SYMBOLS

Scalars and Miscellaneous

a, b	dimension subscripts (a, b = 1, 2, . . . , m dimensions)
i	configuration super/subscript (i = 1, 2, . . . , N configurations)
p, q	point subscripts (p, q = 1, 2, . . . , n points)
m	$= \max_i (m_i)$ for m_i dimensions of the ith configuration
r^2	squared product moment correlation coefficient
g_i	squared product moment correlation for ith configuration
h	average (over i) coefficient of alienation
A	equations (11–27) and (11–44)
B	equations (11–28) and (11–45)
k, k_i	general and individual central dilation scalars
α	rotation angle for individual configuration as in equations (11–26) and (11–43)
β	rotation angle for centroid/hypothesis configuration
d	full derivative
∂	partial derivative
tr	trace
$'$	transpose

$\cong$	least-squares estimate of
max, min	maximum and minimum functions
sin, cos, $\tan^{-1}$	sine, cosine, and inverse tangent circular functions

Loss Functions

L	equation (11–6)
F	equations (11–9) and (11–16)
K_{ab}	equation (11–23)
D	equation (11–37)
D_{ab}	equation (11–41)

Matrices and Arrays

INDSCAL/IDIOSCAL

$G (g_{pa})$	group space
$C_i \; (c_a^{(i)})$	diagonal matrix of squared weights
$\hat{B}_i \; (\hat{b}_{pq}^{(i)})$	equation (11–2)
$D_i \; (d_{pq}^{(i)})$	equation (11–1)
$\hat{D}_i \; (\hat{d}_{pq}^{(i)})$	equation (11–1)
$T_i \; (t_{ab}^{(i)})$	equation (11–5), an m-square orthonormal matrix
$\Lambda_i (\lambda_{aa}^{(i)})$	equation (11–5), diagonal m-square matrix of squared weights

PINDIS

$X_i \; (x_{pa}^{(i)})$	rectangular coordinates of order $n \times m_i$ for configuration i
$\tilde{X}_i \; (\tilde{x}_{pa}^{(i)})$	X_i subjected to rigid motions and a central dilation as in equation (11–10)
$\hat{X}_i \; (\hat{x}_{pa}^{(i)})$	optimally reflected X_i as in equation (11–45)
$X_i^* (x^{*(i)}_{pa})$	optimally rotated $X_i / \hat{X}_i$ as in equations (11–24) and (11–25)
$X^u_i (x^{u(i)}_{pa})$	optimally translated $X_i / \hat{X}_i$ as in equation (11–51)
$Z (z_{pa})$	rectangular coordinates of $n \times m$ centroid/hypothesis configuration
$Z^r (z^r_{pa})$	optimally rotated Z over all i for dimension weighting
$Z^r_i \; (z^{r(i)}_{pa})$	optimally rotated Z for each i for dimension weighting
$Z^t (z^t_{pa})$	optimally translated Z over all i for vector weighting
$Z^t_i (z^{t(i)}_{pa})$	optimally translated Z for each i for vector weighting
$\overline{Z} (\overline{z}_{pa})$	the centroid of Z^r_i

$Z^*(z^*_{pa})$	equations (11–32) and (11–33) for dimension weighting rotations
$R_i\ (r_{ab}^{(i)})$	orthonormal matrix for orthogonally rotating any X_i
$S_i\ (s_{ab}^{(i)})$	orthonormal matrix for rotating Z orthogonally
$S(s_{ab})$	orthonormal matrix for rotating Z to $\overline{Z}$
$W_i\ (w_{aa}^{(i)})$	equation (11–19), the dimension weights applied to Z^r
$W^*_i\ (w^*_{aa}{}^{(i)})$	equation (11–21), the normalized W_i
$W^r{}_i\ (w^r{}_{aa}{}^{(i)})$	equations (11–30) and (11–49), the dimension weights applied to $Z^r{}_i$
$V_i\ (v_{pp}^{(i)})$	equation (11–38), the vector weights applied to Z^t
$V^*_i\ (v^*_{pp}{}^{(i)})$	equation (11–40), the normalized V_i
$V^t{}_i(v^t{}_{pp}{}^{(i)})$	equation (11–56), the vector weights applied to $Z^t{}_i$
$\Omega_i\ (\omega_a{}^{(i)})$	the diagonal reflection matrix for $X^u{}_i$
u_i	equations (11–13), (11–15), (11–35), (11–48), and (11–53), the dimension origins of X_i
t_i	equations (11–36) and (11–54), the dimension origins of $Z^t{}_i$
t	the mean origin (over i) for each a in Z^t
I	the m/n-square identity matrix
j	the unit vector
$E_i\ (e_{pa}^{(i)})$	equation (11–18), the $n \times m$ matrix of residuals after optimal fitting
L	equation (11–52), a matrix of Lagrangian multipliers
$A,\ B/W,\ V$	equation (11–11), left and right eigenvector matrices
M	equation (11–12), centering matrix
C	equation (11–12)
$A,\ B/X,\ Y$	general configurations (locally defined)
Δ	equation (11–12), diagonal matrix of eigenvalues

REFERENCES

Berge, J. M. F. 1977. Orthogonal procrustes rotation for two or more matrices. *Psychometrika* 42:267–76.

Bloxom, B. 1968. *Individual differences in multidimensional scaling (ETS Research Memorandum 68–45).* Princeton, N.J.: Educational Testing Service.

Borg, I. 1978. Ein Vergleich verschiedener Studien zur Lebensqualität. *Zeitschrift für Sozialpsychologie.*

———. 1979. Geometric representation of individual differences. In *Geometric representations of relational data: Readings in multidimensional scaling,* 2nd ed., ed. J. C. Lingoes, R. E. Roskam, and I. Borg. Ann Arbor, Mich.: Mathesis Press.

Borg, I., and J. C. Lingoes. 1976. *What weight should weights have in the analysis of individual differences? (Mathematical Psychology Program Technical Report 76–5).* Ann Arbor, Mich.: Mathematical Psychology Department.

———. 1977. Ein direkter Transformationsansatz der multidimensionalen Analyse dreimodaler Datenmatrizen: Theorie und Anwendugen. *Zeitschrift für Sozialpsychologie* 8:98–114.

Carroll, J. D., and J. J. Chang. 1970. Analysis of individual differences in multidimensional scaling via an N-way generalization of "Eckart-Young" decomposition. *Psychometrika* 35:283–320.

———. 1972. IDIOSCAL (*I*ndividual *D*ifferences *I*n *O*rientation *SCAL*ing): A generalization of INDSCAL allowing idiosyncratic reference systems as well as an analytic approximation to INDSCAL. Paper presented at the spring meeting of the Psychometric Society, Princeton, New Jersey, 30–31 March.

Carroll, J. D., and M. Wish. 1974. Models and methods for three-way multidimensional scaling. In *Contemporary developments in mathematical psychology,* vol. 2, ed. D. H. Krantz, R. C. Atkinson, D. R. Luce, and P. Suppes. San Francisco: W. H. Freeman.

Feger, H. 1974. Die Erfassung individueller Einstellungsstrukturen. *Zeitschrift für Sozialpsychologie* 5:57–105.

———. 1975. Längsschnittliche Erfassung intraindividueller Unterschiede bei Einstellungsstrukturen. In *Entwicklung und Persönlichkeit,* ed. U. M. Lehr and F. E. Weinert. Stuttgart, W. Germany: Kohlhammer.

———. 1977. Personal communication.

Fischer, G. H., and J. Roppert. 1965. Ein Verfahren der Transformationsanalyse faktoren-analytischer Ergebnisse. In *Lineare Strukturen in Mathematik und Statistik,* ed. J. Roppert and G. H. Fischer. Wien, Austria: Physika Verlag.

Gower, J. C. 1975. Generalized procrustean analysis. *Psychometrika* 40:33–51.

Guttman, L. 1954. A new approach to factor analysis: The radex. In *Mathematical thinking in the social sciences,* ed. P. F. Lazarsfeld. Glencoe, Ill.: The Free Press.

Harman, H. H. 1967. *Modern factor analysis.* 2nd ed. Chicago, Ill.: University of Chicago Press.

Harshman, R. A. 1972. PARAFAC2: Mathematical and technical notes. *UCLA Working Papers in Phonetics* 22:31–44 (University Microfilms No. 10,085).

Helm, C. E. 1959. *A multidimensional ratio scaling analysis of color relations (Technical Report).* Princeton, N.J.: Princeton University and the Educational Testing Service.

———. 1964. Multidimensional ratio scaling analysis of perceived color relations. *Journal of the Optical Society of America* 54:256–62.

Helm, C. E., and L. R. Tucker. 1962. Individual differences in the structure of color perception. *American Journal of Psychology* 75:437–44.

Horan, C. B. 1969. Multidimensional scaling: Combining observations when individuals have different perceptual structures. *Psychometrika* 34:139–65.

Kristof, W., and B. Wingersky. 1971. Generalization of the

orthogonal procrustes rotation procedure for more than two matrices. Proceedings of the 79th Annual Convention of the American Psychological Association, 89–90.

Levy, S. 1976. Use of the mapping sentence for coordinating theory and research: A cross-cultural example. *Quality and Quantity* 10:117–25.

Levy, S., and L. Guttman. 1975. On the multivariate structure of well-being. *Social Indicators Research* 2:361–88.

Lingoes, J. C. 1973. *The Guttman-Lingoes nonmetric program series.* Ann Arbor, Mich.: Mathesis Press.

————. 1975. A neighborhood preserving transformation for fitting configurations. Paper presented at the U.S.–Japan Seminar on Theory, Methods, and Applications on Multidimensional Scaling and Related Techniques, La Jolla, California, 20–24 August.

————. 1978 . Progressively complex linear transformations for finding geometric similarities among data structures. In *Classifying cultural and social data,* ed. H. Hudson. San Francisco: Jossey-Bass.

Lingoes, J. C., and I. Borg. 1976. Procrustean individual differences scaling: PINDIS. *Journal of Marketing Research* 13:406–7.

————. 1977. Optimale Lösungen für Dimensions- und Vektorgewichte in PINDIS. *Zeitschrift für Sozialpsychologie* 8:210–17.

Lingoes, J. C., and P. H. Schönemann. 1974. Alternative measures of fit for the Schönemann-Carroll matrix fitting algorithm. *Psychometrika* 39:423–27.

Lissitz, R. W., P. H., Schönemann, and J. C. Lingoes. 1976. A solution to the weighted procrustes problem in which the transformation is in agreement with the loss function. *Psychometrika* 41:547–50.

MacCallum, R. C. 1974. A comparison of two individual differences models for multidimensional scaling: Carroll and Chang's INDSCAL and Tucker's three-mode factor analysis. Doctoral dissertation, University of Illinois, Urbana.

————. 1977. Effects of conditionality on INDSCAL and ALSCAL weights. *Psychometrika* 42:297–305.

Schönemann, P. H. 1966. A generalized solution of the orthogonal procrustes problem. *Psychometrika* 31:1–10.

————. 1972. An algebraic solution for a class of subjective metrics models. *Psychometrika* 37:441–51.

Schönemann, P. H., and R. M. Carroll. 1970. Fitting one matrix to another under choice of a central dilation and a rigid motion. *Psychometrika* 35:245–55.

Schönemann, P. H., F. S. Carter, and W. L. James. 1976. *Contributions to subjective metrics scaling: I. COSPA, a fast method for fitting and testing Horan's model, and an empirical comparison with INDSCAL and ALSCAL.* W. Lafayette, Ind.: Purdue University, Krannert Graduate School of Management 587.

Takane, Y., F. W. Young, and J. de Leeuw. 1977. Nonmetric individual differences multidimensional scaling: An alternating least-squares method with optimal scaling features. *Psychometrika* 42:7–67.

Torgerson, W. S. 1958. *Theory and methods of scaling.* New York: Wiley.
Tucker, L. R. 1972. Relations between multidimensional scaling and three-mode factor analysis. *Psychometrika* 37:333–67.
Wish, M., and J. D. Carroll. 1974. Applications of individual differences scaling to studies of human perception and judgement. In *Handbook of perception,* vol. 2, ed. E. C. Carterette and M. P. Friedman. New York: Academic Press.

12

The General Euclidean Model

Forrest W. Young

INTRODUCING GEM

This book is about the analysis of three-mode data. This chapter is about the multidimensional analysis of three-mode dissimilarity data. In particular, we will discuss the General Euclidean Model (GEM), a very general spatial model that uses distance to geometrically represent dissimilarity data. The chapter focuses on data consisting of several square, symmetric dissimilarity matrices, which have rows and columns that reference the same set of objects or events. However, the GEM is appropriate to a broader range of data situations, a range which we survey in the next section.

Data Theory

Our data survey is placed in the context of a data theory that systematizes the types of data appropriate to the GEM. The data theory presented here is a condensed version of the complete theory first discussed by Young (1975a), to be published in Young and Hamer (forthcoming). The condensation presented here represents those aspects of the complete data theory that are relevant to the GEM. Other condensations appear in Young and Lewyckyj (1979) and in Schiffman, Reynolds, and Young (1981).

According to this theory, it is useful to categorize data according to the shape of the basic data matrix. A data matrix may be either square or rectangular. For both data matrix shapes, we denote the matrix of observed data by the letter **O**.

Square dissimilarity data are two-way data in which both ways refer to the same set of objects (or events) and in which the observed datum specifies the degree of dissimilarity between the row-object and column-object. We call them *square* since when organized into the matrix **O**, the rows and columns reference one set of things; therefore, the matrix $\mathbf{O}$ must have the same number of rows as it has columns. An example of such data is dissimilarity judgments about all pairs of the members of the set of objects.

Square data may be either *symmetric* or *asymmetric*. For symmetric data, $o_{ij} = o_{ji}$, where o_{ij} is the element of $\mathbf{O}$ in the ith row and jth column. For asymmetric data, this does not have to be the case.

Rectangular dissimilarity data are two-way data in which the two ways reference two different sets of things and in which the observed datum indicates the dissimilarity between the row-object and column-object. We call these data *rectangular* since the matrix $\mathbf{O}$ does not have to have the same number of rows as it has columns. An example of such data is preference judgments about a set of stimuli from a group of people. Here, suppose the rows reference the stimuli and the columns reference the people. The observed datum, preference, can be interpreted as the dissimilarity between the person and the stimulus. An alternative example of such data is ratings of a set of stimuli on a variety of scales. Here the rows and columns refer to different sets of objects, and the rating can be interpreted as the dissimilarity between a stimulus and a scale.

But our main topic is three-way data. Three-way data are data that consist of several two-way data matrices $\mathbf{O}_k$, not just one matrix $\mathbf{O}$. The matrices may be square or rectangular, but they must all be the same shape. (When the data are three-way, we denote the kth matrix as $\mathbf{O}_k$.)

Square three-way data consist of several square data matrices. An example of such data is judgments of dissimilarity concerning pairs of stimuli when there are several judges. Here the rows and columns of each data matrix refer to stimuli; thus, each data matrix is square. But there are several matrices; thus, the data are three-way.

Rectangular three-way data consists of several rectangular data matrices. An example is judgments of preference for a set of stimuli obtained from a group of people over several occasions. These data are rectangular, since the rows and columns of a data matrix refer to different sets of objects. The data are three-way since there are several matrices.

The General Euclidean Model

The GEM provides a variety of models for the types of data just outlined. These models are special cases of GEM, which is defined by the following equation:

$$d^2_{ijk} = (\mathbf{y}_i - \mathbf{x}_j)\mathbf{V}_i\mathbf{W}_k(\mathbf{y}_i - \mathbf{x}_j)' , \quad (12\text{–}1)$$

where:

- d^2_{ijk} is the squared GEM distance that approximates the dissimilarity (preference) o_{ijk} between objects referenced by row i and column j in the kth matrix of dissimilarities (preferences).

- $\mathbf{x}_j$ is an r-element row vector of the coordinates that specify the position of a point in an r-dimensional Euclidean space. The point represents column object j, which is usually a stimulus. For square data, $\mathbf{x}_j$ also represents the row objects, which are usually stimuli. The vector $\mathbf{x}_j$ is the jth row of the coordinates matrix $\mathbf{X}$.
- $\mathbf{y}_i$ is an r-element row vector of coordinates that specify the position of a point in an r-dimensional Euclidean space. The vector $\mathbf{y}_i$ is the ith row of the coordinates matrix $\mathbf{Y}$. These coordinates are restricted to equal the $\mathbf{x}_j$ values for square (dissimilarity) data and are suppressed for such data. For rectangular (preference) data, the point represents row object i, which is usually a person. Note that for rectangular (preference) data, the row (person) and column (stimulus) objects are modeled by points $\mathbf{y}_i$ and $\mathbf{x}_j$ that are in the *same* Euclidean space.
- $\mathbf{V}_i$ is a square, symmetric, order r, rank $s_i \leqq r$, positive semi-definite matrix of weights associated with row objects i. For square (dissimilarity) data, these objects are usually stimuli, and for rectangular (preference) data, they are usually people.
- $\mathbf{W}_k$ is a square, symmetric, order r, rank $t_k \leqq r$, positive semi-definite matrix of weights associated with dissimilarity matrix k.

Without simplifying assumptions, the GEM is too general to be useful for data analysis. It is useful, however, as a meta-model to systematize the relationships between many of the less general but more useful models that have been proposed to analyze dissimilarity data. In fact, for certain kinds of data, the GEM must necessarily be simplified. For example:

- For two-way data, there is only one $\mathbf{W}_k$ (which we denote $\mathbf{W}$). For normalization reasons, $\mathbf{W} = I$.
- For square data, the row-objects and column-objects are identical; thus, we impose the restriction $\mathbf{Y} = \mathbf{X}$.
- For symmetric square data, $o_{ijk} = o_{jik}$ for all i and j; thus, it should be that $d_{ijk} = d_{jik}$, as well. This characteristic follows when the $\mathbf{V}_i = I$ for all i.

In addition to these simplifying assumptions (which must necessarily be made for the specified situations), there are other simplifying assumptions that may be opted for in certain other situations. These additional assumptions concern the nature of the weight matrices $\mathbf{V}_i$ and $\mathbf{W}_k$. Each of these matrices may be: (a) an identity matrix; (b) a diagonal matrix; (c) a rank-one matrix; (d) a reduced-rank matrix; or (e) a full-rank matrix. As we will see, oftentimes several of these five cases yield interesting special models, which have been previously proposed by various investigators.

GEM for Two-Way Data

Let's consider the simplest possible situation: applying GEM to a single square matrix of dissimilarities that are symmetric. The necessary assumptions given above yield the specialized GEM model:

$$d_{ij}^2 = (\mathbf{x}_i - \mathbf{x}_j)(\mathbf{x}_i - \mathbf{x}_j)' , \tag{12–2}$$

which is the Euclidean distance between points i and j. Schiffman, Reynolds, and Young (1981) call this combination of data and model (which was first discussed by Torgerson [1952]) the *classical multidimensional scaling* (CMDS) situation.

The next simplest situation results from using GEM for a single rectangular data matrix and assuming that $\mathbf{V}_i = I$ for all i. This results in:

$$d_{ij}^2 = (\mathbf{y}_i - \mathbf{x}_j)(\mathbf{y}_i - \mathbf{x}_j)' . \tag{12–3}$$

This combination of data and model corresponds to Coombs' ([1964] 1976) proposal for unfolding preference data. Schiffman, Reynolds, and Young (1981) call this *classical multidimensional unfolding* (CMDU). Usually, $\mathbf{x}_j$ is the point for the jth stimulus, and $\mathbf{y}_i$ is the point modeling the ith person (Coombs' "ideal" point).

Note the parallel between (12–2) and (12–3). They are both simple (unweighted) Euclidean models. The CMDS situation uses Euclidean distances within one set of points ($\mathbf{X}$) to model the dissimilarities within one set of objects. The CMDU situation uses Euclidean distances between two sets of points ($\mathbf{X}$ and $\mathbf{Y}$) to model the dissimilarities between the two sets of objects. This parallel exists for three-mode data as well.

So far, we have only discussed symmetric and rectangular dissimilarities. If we have asymmetric dissimilarities, we can still employ the simple Euclidean model given by (12–2). All we have to do, in addition to invoking the required assumptions, is to specify that the row stimulus weights $\mathbf{V}_i = I$ for all stimuli i. But note that since the Euclidean model is a *symmetric* model, its use with *asymmetric* data implies that the asymmetry is error.

If we think that the asymmetry is not error but is somehow meaningful, we can choose to model it via the row weights $\mathbf{V}_i$. This is the case, since the model

$$d_{ij}^2 = (\mathbf{x}_i - \mathbf{x}_j)\mathbf{V}_i(\mathbf{x}_i - \mathbf{x}_j)' \tag{12–4}$$

generates asymmetric distances, thereby implying that the asymmetry in the data is systematic. This model is Young's (1975) ASYMSCAL model, when $\mathbf{V}_i$ is diagonal for all i (see also Young and Lewyckyj 1979), and is Coombs' ([1964] 1976) compensatory distance model, when $\mathbf{V}_i$ is rank one ($s_i = 1$) for all i. The other possible cases of this model (namely, when $\mathbf{V}_i$ is full or reduced rank) have not been discussed, to my knowledge.

The matrix of row weights $\mathbf{V}_i$ can also be used with rectangular data, which yields:

$$d_{ij}^2 = (\mathbf{y}_i - \mathbf{x}_j)\mathbf{V}_i(\mathbf{y}_i - \mathbf{x}_j)' . \tag{12–5}$$

This corresponds to Carroll's (1972) family of weighted unfolding models. When each $\mathbf{V}_i$ is full rank, this is Carroll's model I (after suitable decomposition); when each $\mathbf{V}_i$ is diagonal, it is his model II; when all are identity matrices, it is his model III (which is Coombs' simple unfolding); and when each $\mathbf{V}_i$ is rank 1, it is Carroll's model IV. Note that the scaling/unfolding parallel exists between (12–4) and (12–5), as well as between (12–2) and 12–3).

GEM for Three-Way Data

The major topic of this chapter is a family of GEMs that are important models of three-way dissimilarity data. The family includes models that are symmetric scaling GEMs, which have certain types of weight matrices $\mathbf{W}_k$. The fact that these models are symmetric implies that $\mathbf{V}_i = I$ for all row-stimuli i, and the fact that they are scaling (and not unfolding) models implies the restriction $\mathbf{Y} = \mathbf{X}$. Thus, these models are expressed by the special case of GEM, which is:

$$d_{ijk}^2 = (\mathbf{x}_i - \mathbf{x}_j)\mathbf{W}_k(\mathbf{x}_i - \mathbf{x}_j)' \quad . \qquad (12\text{–}6)$$

(We discuss this family of models under Symmetric Scaling GEMs.)

In this section, we briefly discuss other three-way GEMs. The simplest family of GEMs for three-way data are replicated versions of the two-way family of GEMs discussed in the previous section. For these models, we assume that $\mathbf{W}_k = I$ for all matrices k. Because of this assumption, the equations for these models are identical to (12–2) through (12–5). The essential difference, then, is that we have three-way dissimilarity or preference data, not two-way.

These three-way data consist of many dissimilarity (preference) data matrices. The several matrices must all be the same shape and may be symmetric, asymmetric, or rectangular. All of these matrices are simultaneously described by the one matrix of distances that results from the model chosen from (12–2) through (12–5). Since every data matrix is being modeled by the same distances, it is implied that the several data matrices are the same, except for random error. It is this assumption that leads us to call these *replicated two-way models*.

Schiffman, Reynolds, and Young (1981) discuss replicated versions of equations (12–2) and (12–3), which they call *replicated multidimensional scaling* (RMDS) and *replicated multidimensional unfolding* (RMDU). Young and Lewyckyj (1979) also discuss a replicated version of the diagonal form of (12–4) in their monograph, calling it *replicated AMDS*. The author is unaware of anyone previously discussing a replicated compensatory distance model (a replicated rank-one version of equation [12–4]) or a replicated row-weighted unfolding model (equation [12–5]).

The family of GEMs for three-way rectangular (preference) data fall into two families: (a) those that include row weights $\mathbf{V}_i$, and (b) those where $\mathbf{V}_i = I$. The latter family includes GEMs with full rank $\mathbf{W}_k$ that have been discussed by De Sarbo (1978), and reduced, full-rank, or diagonal models, discussed by Young (1979). The equation here is:

$$d_{ijk}^2 = (\mathbf{y}_i - \mathbf{x}_j)\mathbf{W}_k(\mathbf{y}_i - \mathbf{x}_j)' \quad . \qquad (12\text{–}7)$$

The diagonal version is called *weighted multidimensional unfolding* (WMDU) by Young and Lewyckyj (1979). The derivation given in the third section of this chapter is for either the full- or reduced-rank situations and applies to the scaling situations discussed under Symmetric Scaling GEMs, as well as to unfolding situations not discussed in more detail in this chapter.

The second family of GEMs for three-way rectangular data is those for which the $\mathbf{W}_k$ and $\mathbf{V}_i$ are simply positive semidefinite. This is the most general GEM, corresponding to the full equation as given by (12–1). To my knowledge, no one has explicitly discussed this model. This family parallels the Carroll (1972) family of weighted unfolding models but has additional weights for the several matrices.

Finally, the GEM includes a family of asymmetric models for three-way asymmetric dissimilarity data for which both $\mathbf{V}_i$ and $\mathbf{W}_k$ are simply positive semidefinite. Of these models, the diagonal version for asymmetric data has been discussed by Young and Lewyckyj (1979) using the term *weighted AMDS*. I am unaware of any of the others having been discussed.

Constrained GEM

By placing certain restrictions on specific GEM parameters, we can see that a wide variety of proposals for constrained scaling and unfolding analyses fit within the GEM framework. For example, Carroll's (1972) external unfolding is an example of constrained unfolding. For this situation, we impose the restrictions that result in (12–5) and then impose the additional restriction that the matrix of stimulus coordinates $\mathbf{X}$ consist of known values. Bloxom's (1978) constrained scaling corresponds to (12–6), but the matrices $\mathbf{W}_k$ are constrained to previously known values. Young, Null, and Hamer's (1983) distance regression also corresponds to (12–6), but now the matrix $\mathbf{X}$ is constrained to previously known values.

SYMMETRIC SCALING GEMs

In this section, we discuss the family of symmetric scaling GEMs (where $\mathbf{V}_i = I$ for all i, and $\mathbf{Y} = \mathbf{X}$) for three-way square dissimilarity data. The data may be asymmetric, but the family of models is symmetric. The family is specified by (12–6).

Our main attention is on the cases where $\mathbf{W}_k$ is full or reduced rank (nondiagonal). However, we feel that to fully understand these cases, it is useful to have an initial discussion of the case in which $\mathbf{W}_k$ is diagonal. Thus, the next two sections are on the geometry and algebra (respectively) of the diagonal case, and the following two sections are on the geometry and algebra (respectively) of the nondiagonal (full- or reduced-rank) cases. The diagonal case corresponds to Carroll's (Carroll and Chang 1970) INDSCAL model, and the full- and reduced-rank cases to Young's (1979, 1982) GEMSCAL model. (We do not discuss the case in which $\mathbf{W}_i = I$.)

The models discussed in the next two sections are models of

individual differences in cognition and perception. They share several common aspects:

a. They may be used when the researcher has obtained, from a number of individuals, judgments of the similarity or dissimilarity of pairs of stimuli to each other.
b. They develop a *group stimulus space* that pictures the cognitive or perceptual mechanisms shared in common between the several individuals.
c. They employ a group stimulus space that is a multidimensional Euclidean space in which stimuli are represented as points.
d. They portray individual variation in cognition or perception in a geometric fashion. The models differ, however, in their assumptions about the basic nature of individual differences and thus in their geometric representation of individual differences.

INDSCAL Geometry

One of the most popular GEMs is the INDSCAL model proposed by Carroll and Chang (1970). This model represents each individual person by a diagonal weight matrix $\mathbf{W}_k$. This means that the INDSCAL model assumes that the individuals vary in the importance they attach to the dimensions of the group stimulus space. While one individual may perceive one of the dimensions as being more important than another, another individual may have just the opposite bias. The notion of salience is incorporated into the INDSCAL model by introducing weights for each individual on each dimension. If the weight is large (heavy), then the dimension is relatively important; if it is small, it is not as important.

Geometrically, the INDSCAL model represents individual differences in a special space called the *weight space*. In this space, individuals are represented by vectors emanating from the origin. The direction of the vector from the origin represents the relative weighting of each dimension, and the length of the vector represents the overall salience of the dimension to the individual.

A schematic diagram of the geometry of INDSCAL is presented in Figure 12–1. Figure 12–1(a) presents a hypothetical group stimulus space. This is a two-dimensional Euclidean space with points for nine hypothetical stimuli. Figure 12–1(b) shows a hypothetical weight space. This is also a two-dimensional space, but the points are the endpoints of vectors emanating from the origin. The vectors represent nine hypothetical individuals. (It is just a coincidence that the number of stimuli equals the number of individuals.) Notice that the weight space only contains vectors in the positive quadrant. Generally, only positive weights are interpretable, so it is usually desirable to restrict the weights to be non-negative.

Individual differences in perception or cognition, then, are represented by differences in the orientation and length of vectors in the weight space. The nature of the individual differences can be seen most readily by comparing the *personal spaces* for the several individuals. The personal space for an individual is what results after applying the weights for an individual to the

Figure 12–1. Schematic Representation of Geometry of INDSCAL

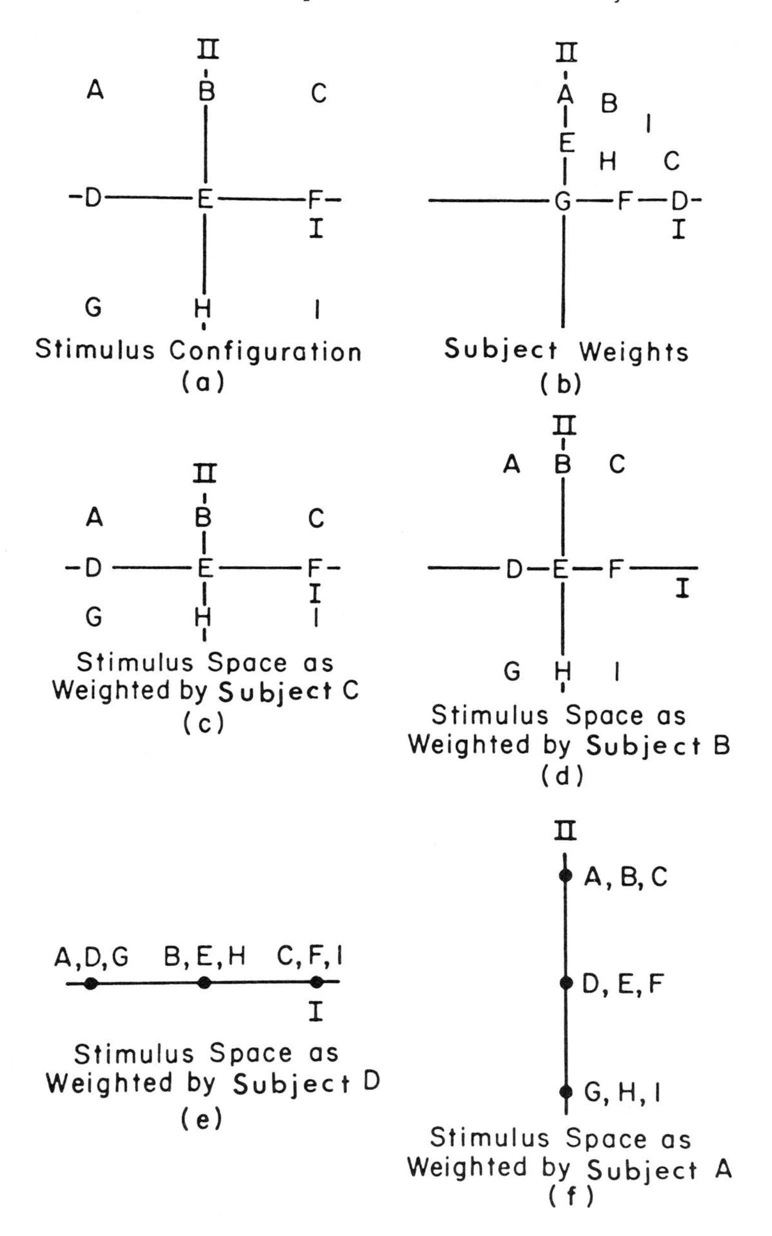

group space. The weights can be thought of as stretching or shrinking the dimensions of the group space. In an individual's personal space, an important dimension is stretched (by a large weight) and an unimportant dimension is shrunk (by a small weight).

Several personal spaces are presented in Figure 12–1. Figure 12–1(c) and (d) display the personal spaces for individuals *C* and *B*. Notice that individual *C* finds dimension I relatively important and that his personal space has a relatively long dimension I and a short dimension II. Individual *B* has the reverse weighting; thus, his space has the reverse stretching and shrinking. Figure 12–1(e) and (f) show what happens when an individual has a zero weight: the group space collapses into a lower-dimensional personal space.

INDSCAL Algebra

We now turn to the algebraic expression for the INDSCAL model. The fundamental INDSCAL equation is:

$$d^2_{ijk} = (\mathbf{x}_i - \mathbf{x}_j)\mathbf{W}_k(\mathbf{x}_i - \mathbf{x}_j)' , \qquad (12\text{–}8)$$

where d^2_{ijk} and $\mathbf{x}_i$ are still defined as in GEM, and where $\mathbf{W}_k$ is an $(r \times r)$ diagonal matrix whose diagonal element w_{kaa} is the weight of individual k on dimension a. There are m matrices $\mathbf{W}_k$, one for each of the m individuals.

Thus, $\mathbf{X}$ denotes the group stimulus space, and the weight space is denoted by the matrix $\mathbf{W}$ (not $\mathbf{W}_k$) which has rows $\mathbf{w}_k$ (again, not $\mathbf{W}_k$) whose elements w_{ka} are taken from the diagonals of the matrices $\mathbf{W}_k$. The personal space for individual k is represented by the matrix $\mathbf{X}_k$, which is defined as:

$$\mathbf{X}_k = \mathbf{X}\mathbf{W}_k^{1/2} , \qquad (12\text{–}9)$$

giving an alternative expression for the INDSCAL model as:

$$d^2_{ijk} = (\mathbf{x}^k_i - \mathbf{x}^k_j)(\mathbf{x}^k_i - \mathbf{x}^k_j)' , \qquad (12\text{–}10)$$

where x^k_i is the ith row of $\mathbf{X}_k$. We see, then, that this is nothing more than the formula for Euclidean distance in individual k's personal space.

GEMSCAL Geometry

The GEMSCAL model makes different assumptions about individual differences and represents them in a different geometric fashion than the INDSCAL model. GEMSCAL makes the assumption that an individual finds several orthogonal *directions* (not dimensions) in the group stimulus space to be *most important*. While one individual may perceive one particular set of orthogonal directions to be most important, another individual may find some other set of orthogonal directions to be most important. Furthermore, weights are associated with each individual to indicate the relative importance of each direction to that individual.

As mentioned above, both GEMSCAL and INDSCAL represent the stimuli as points in a multidimensional Euclidean space called the *group stimulus space*. However, GEMSCAL represents the individuals by vectors in the *same* group stimulus space, not by vectors in a separate weight space. Thus, GEMSCAL represents the stimuli and individuals in a space called a *joint space,* since it jointly represents both stimuli and individuals.

INDSCAL can also be thought of as representing stimuli and individuals in a joint space, although it is not usually productive to do so. Furthermore, GEMSCAL can be thought of as representing the stimuli and individuals in two separate spaces, although, again, this is not usually productive.

In the joint space, the stimuli are represented by points and the individuals by vectors. The directions of the vectors (from the origin) indicate the directions in the space that are most salient to the several individuals. The length of each vector shows the relative salience of the directions to the individuals. The vectors are oriented for each individual k so that the several vectors for individual k are mutually orthogonal. Furthermore, the first one is in the direction that accounts for the maximum variance in individual k's data, the second one accounts for the most remaining variance, and so forth. For this reason, the vectors are called *principal directions* by Young (1979). Individual differences, then, are represented by the orientation and length of orthogonal principal directions in the joint space.

In Figure 12–2(a), we have illustrated the simplest case of GEMSCAL, namely the case in which each individual finds only *one* direction to be important. This is the rank-one case of equation (12–6). The figure presents two-dimensional information about nine stimuli and four individuals. Note that the stimulus space contains one vector for each individual.

As with the INDSCAL model, the personal space notion is helpful in understanding the nature of individual differences. As shown in Figure 12–2(b), with GEMSCAL, an individual's personal space is obtained simply by orthogonally projecting the stimulus points in the group stimulus space onto the direction coinciding with the individual's vector and by shrinking or stretching the projection according to the length of the individual's vector. The light lines in Figure 12–2(b) represent the orthogonal projection. After the projection is shrunk to the proper length, we have the personal space shown in Figure 12–2(c). Note that the direction for individual 1 shows that he finds the differences between stimuli *A* and *I* to be very important and that he completely overlooks the differences between stimuli *G* and *C*. This shows up in his personal space, where *A* and *I* are far apart, and *G* and *C* are located at the same spot.

Three other personal spaces are illustrated in Figure 12–2, (d) through (f), one for each of the other individuals in Figure 12–2(a). Notice that all 4 personal spaces are one-dimensional. This follows from the fact that the simple GEMSCAL case illustrated in Figure 12–2(a) assumes that only one direction is important to each individual. Notice also that the personal spaces for individuals 1 and 2 are mutually orthogonal, as is implied by the orthogonal relationship of their vectors in the joint space. (There is no requirement that one individual's personal space be orthogonal to another's.) Finally, note that the personal spaces for individuals 3 and 4 are identical to those for individuals *D* and *A* in the INDSCAL example. This follows from the fact that the principal directions for individuals 3 and 4 lie exactly along a

Figure 12–2. The Simplest Case of GEMSCAL

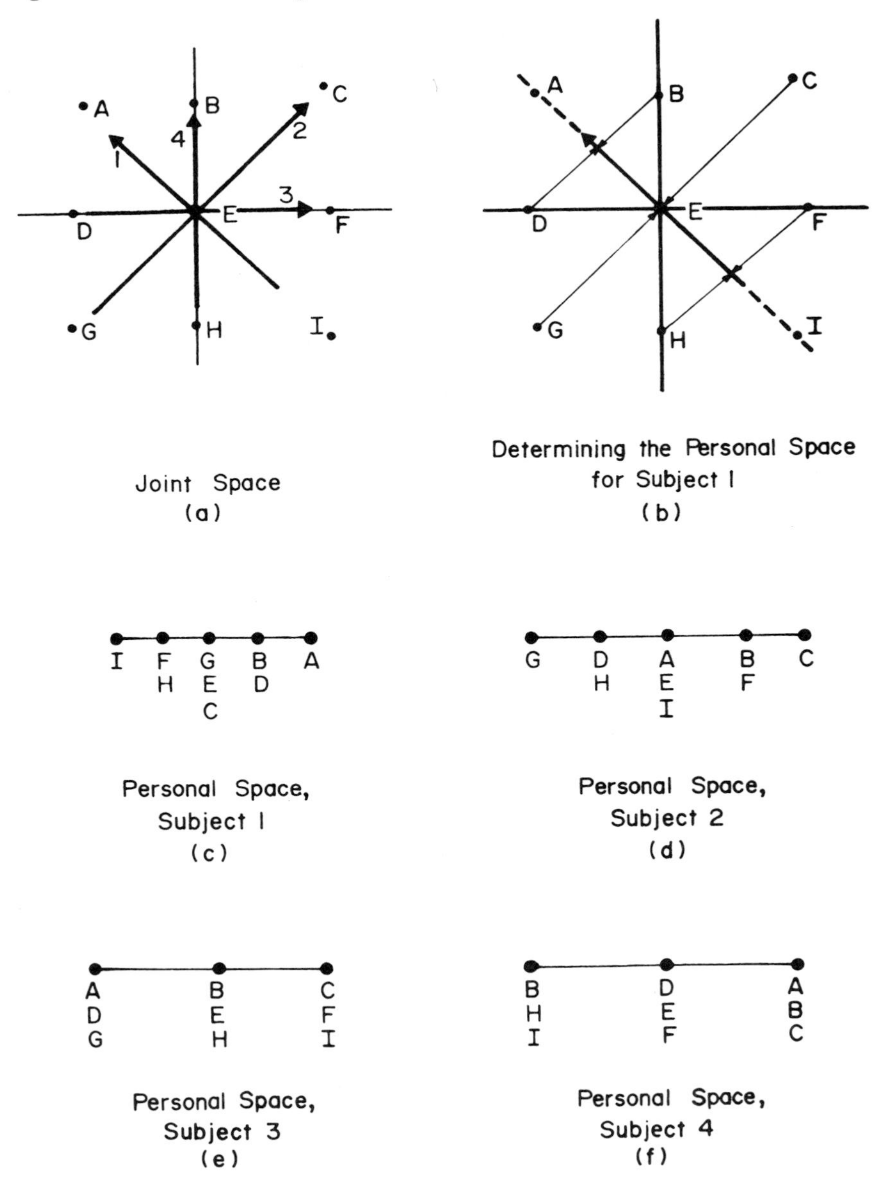

dimension. This orientation of a vector implies that the other dimension is totally ignored, which corresponds to the zero weight case in the INDSCAL model.

Now we turn to the version of GEMSCAL that assumes that an individual finds *several* principal directions to be important. This corresponds to the reduced- or full-rank versions of equation

(12–6). The directions are assumed to be mutually orthogonal and to be oriented in a maximum variance fashion. For this multidirection version of GEMSCAL, the Euclidean space jointly contains points and vectors, but now there are *several* orthogonal vectors for each individual, each vector indicating one of the several important directions. Again, the lengths of the vectors show the relative importance of each direction.

In Figure 12–3, we illustrate the multidirection case of GEMSCAL. The figure is for the specific situation in which there are nine stimuli, three individuals, two dimensions, and two principal directions. The three left-hand figures (12–3[a], [c], and [e]) all illustrate the same group stimulus space. The two vectors in Figure 12–3(a) represent the two principal directions for individual 1. Figure 12–3(c) illustrates the same information for individual 2 and Figure 12–3(e) for individual 3. Each right-hand figure presents the personal space for the individual whose principal directions are shown to the left.

We see in Figure 12–3(a) that the principal directions for individual 1 are mutually orthogonal and are equally short. Therefore, her personal space (Figure 12–3[b]) shows an orthogonal rotation of the stimulus structure and a shrinking. In this case, both principal directions show that she finds the differences between stimuli *G* and *C* and between *A* and *I* to be equally important. Correspondingly, her personal space is a rotation of the stimulus space, with the dimensions coinciding with stimuli *A*, *C*, *G*, and *I*. However, she doesn't find any aspects of the stimulus space to be very relevant; thus, her personal space is small.

In Figure 12–3(c), we note that individual 2's two principal directions are mutually orthogonal but of different lengths. The personal space (Figure 12–3[d]) shows an *oblique* transformation of the stimulus structure, even though the principal directions are mutually orthogonal. This is due to the different lengths. It is important to note this particular aspect of GEMSCAL: Mutually orthogonal principal directions do not necessarily imply an orthogonal transformation. The weighting has to be equal on all dimensions for the implication to follow. Thus, when principal directions are orthogonal, are of equal lengths, and are not oriented along the dimensions, the personal space is a simple orthogonal rotation of the group stimulus space.

Finally, in Figure 12–3(e) the two principal directions are orthogonal and of different lengths. However, they lie along the joint space dimensions. Note that the first (longest) principal direction coincides with the second dimension, and conversely. Furthermore, the first direction points in the negative direction on dimension two. These aspects represent a permutation and reflection, as well as a differential stretching of the dimensions. The result is shown in Figure 12–3(f).

As shown in Figure 12–3, GEMSCAL permits the number of directions to equal the number of dimensions. When this is the case, $\mathbf{W}_k$ is full rank and GEMSCAL corresponds to Carroll and Chang's (1972) IDIOSCAL model. If it is also the case that the tth direction is restricted to coincide with the positive direction of the tth dimension, we have a diagonal $\mathbf{W}_k$ and GEMSCAL is the same as INDSCAL.

Figure 12–3. The Multidirection Case of GEMSCAL

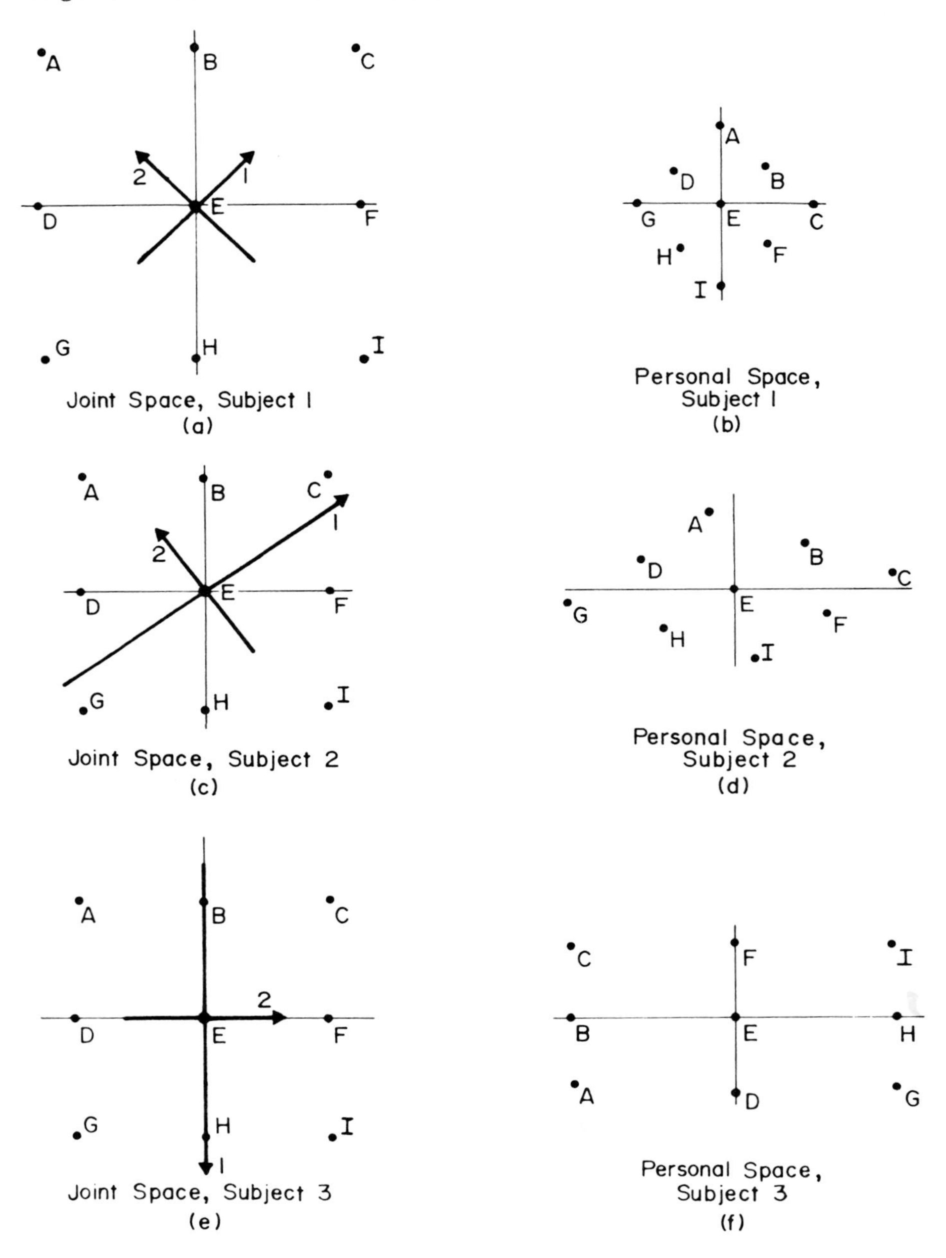

An important feature of GEMSCAL is that there is no requirement that the dimensionality of the personal spaces equal the dimensionality of the group space. The only requirement is that the personal space dimensionality (that is, the number of principal directions) be no more than the group space dimensionality. It is therefore possible to assume that the personal space of each individual has few dimensions, even though the group space has many. It is not even necessary to assume that all of the personal spaces have the same dimensionality. We may, if we wish, state the number of directions for each individual, the only restriction being that the number of principal directions for an individual may not exceed the dimensionality of the group space.

GEMSCAL Algebra

We now turn to the algebra of the GEMSCAL model. The fundamental GEMSCAL equation is:

$$d_{ijk}^2 = (\mathbf{x}_i - \mathbf{x}_j)\mathbf{W}_k(\mathbf{x}_i - \mathbf{x}_j)' , \tag{12–11}$$

which has the exact same appearance as the equation for INDSCAL given above. The difference in the two models is in the nature of the weights $\mathbf{W}_k$. Whereas for INDSCAL $\mathbf{W}_k$ was a diagonal matrix, here it is an $r \times r$ positive semidefinite matrix whose rank is exactly t, the number of principal directions. There are m matrices $\mathbf{W}_k$, one for each of the m individuals. All of these matrices are the same order, but they need not all be the same rank. $\mathbf{W}_k$ may be full rank (positive definite), in which case $t = r$, and the number of principal directions equals the number of dimensions (the personal space dimensionality equals the group space dimensionality). When $t = r$, GEMSCAL is the same as Carroll's (1972) IDIOSCAL model. When $t = r$ and $\mathbf{W}_k$ is diagonal, then GEMSCAL is the same as INDSCAL.

Since $\mathbf{W}_k$ is positive semidefinite, it may be decomposed as follows:

$$\mathbf{W}_k = \mathbf{P}_k \mathbf{Q}_k \mathbf{P}_k' , \tag{12–12}$$

where

$\mathbf{P}_k$ is an $r \times t$ column-orthonormal matrix containing the eigenvectors of $\mathbf{W}_k$. (Column orthonormality means that $\mathbf{P}_k'\mathbf{P}_k = \mathbf{I}$).

$\mathbf{Q}_k$ is a $t \times t$ diagonal matrix containing the eigenvalues of $\mathbf{W}_k$.

By substituting (12–12) into (12–11), we obtain:

$$d_{ijk}^2 = (\mathbf{x}_i - \mathbf{x}_j)\mathbf{P}_k\mathbf{Q}_k\mathbf{P}_k'(\mathbf{x}_i - \mathbf{x}_j)' . \tag{12–13}$$

The coefficients of the principal directions are:

$$\mathbf{C}_k = \mathbf{P}_k \mathbf{Q}_k^{1/2} , \tag{12–14}$$

and, therefore,

$$d_{ijk}^2 = (\mathbf{x}_i - \mathbf{x}_j)\mathbf{C}_k\mathbf{C}_k'(\mathbf{x}_i - \mathbf{x}_j)' . \tag{12–15}$$

We can now define the personal space for individual k as the $n \times t$ matrix of coordinates on the principal directions $\mathbf{X}_k$, where

$$\mathbf{X}_k = \mathbf{X}\mathbf{C}_k . \tag{12–16}$$

This gives an alternative expression for the GEMSCAL model, which is:

$$d^2_{ijk} = (\mathbf{x}_i^k - \mathbf{x}_j^k)(\mathbf{x}_i^k - \mathbf{x}_j^k)' \quad , \tag{12–17}$$

showing that individual k's personal space is a Euclidean space. It is important to note that the personal space matrix $\mathbf{X}_k$ is an $n \times t$ matrix, even though the group space matrix $\mathbf{X}$ is $n \times r$. This characteristic represents the fact that the personal space has fewer dimensions than the group space (unless $\mathbf{W}_k$ is full rank).

Up to this point, we have been considering GEMSCAL under the restriction that the coefficients of the principal directions be orthogonal. There is an oblique alternative to the orthogonal decomposition of $\mathbf{W}_k$ that supports an interesting and somewhat different interpretation. This decomposition has been investigated by Tucker (1972) in his monumental work on three-mode scaling.

Tucker's oblique decomposition begins by defining

$$\mathbf{V}_k = \text{diag}\ (\mathbf{W}_k) \quad , \tag{12–18}$$

which then permits defining

$$\mathbf{U}_k = \mathbf{X}\mathbf{V}_k \quad . \tag{12–19}$$

Since $\mathbf{V}_k$ is the diagonal of subject k's weight matrix, $\mathbf{U}_k$ can be interpreted as the personal space for individual k, which results from a simple stretching and shrinking of the group space dimensions. No rotations, translations, or so on are permitted. Thus, all of the $\mathbf{U}_k$ have dimensions that correspond to the dimensions of the group space $\mathbf{X}$. If the dimensions of the group space are meaningful, then the resulting personal spaces $\mathbf{U}_k$ can be interpreted in the INDSCAL fashion.

Pursuing this line of development one more step, we see that

$$\mathbf{R}_k = \mathbf{V}_k^{-1/2}\mathbf{W}_k\mathbf{V}_k^{-1/2} \tag{12–20}$$

defines a matrix of correlations (cosines of angles) between the dimensions of the personal space $\mathbf{U}_k$. Thus, we see that $\mathbf{U}_k$ contains entries that are Cartesian coordinates on oblique axes, whose angles may be determined from $\mathbf{R}_k$. Note that what we have done is decompose $\mathbf{W}_k$ so that

$$\mathbf{W}_k = \mathbf{V}_k^{1/2}\mathbf{R}_k\mathbf{V}_k^{1/2} \quad , \tag{12–21}$$

which is an alternative to the eigen decomposition given first.

Let's return to Figure 12–3, even though it does not present the $\mathbf{U}_k$ spaces. However, because of the dimensional correspondence between the group space $\mathbf{X}$ and the personal spaces $\mathbf{U}_k$, dimension one of $\mathbf{U}_k$ remains collinear with points D, E, and F, and dimension two remains collinear with points B, E, and H. Note that $\mathbf{W}_k$ is diagonal for subject 3. Thus, for this subject, the dimensions of $\mathbf{U}_3$ are orthogonal, $\mathbf{R}_3$ is an identity matrix, $\mathbf{V}_3 = \mathbf{W}_3$, and $\mathbf{U}_3$ is proportional to $\mathbf{X}$. However, for subject 2, $\mathbf{W}_2$ is not diagonal. As is usually the case when $\mathbf{W}_k$ is not diagonal, the dimensions of $\mathbf{U}_2$ are not orthogonal, $\mathbf{R}_2$ is not an identity matrix, and $\mathbf{U}_2$ is not proportional to $\mathbf{X}$. This does not always follow, however. For example, for subject 1, $\mathbf{W}_1$ is not diagonal. But the dimensions of $\mathbf{U}_1$ *are* orthogonal and $\mathbf{R}_1$ *is* an identity matrix, even though $\mathbf{W}_1$ is not diagonal. Of course, $\mathbf{U}_1$ is not proportional to $\mathbf{X}$.

It is also important to note that, unlike INDSCAL (but like IDIOSCAL), the dimensions of the group space *can* be rotated. This can be seen by defining

$$\mathbf{X}^* = \mathbf{XT} \tag{12–22}$$

and

$$\mathbf{W}_k^* = \mathbf{T}^{-1}\mathbf{W}_k\mathbf{T}^{-1\prime}, \tag{12–23}$$

where $\mathbf{T}$ is an arbitrary $r \times r$ nonsingular transformation. It follows that:

$$d_{ijk}^2 = (\mathbf{x}_i^* - \mathbf{x}_j^*)\,\mathbf{W}_k^*\,(\mathbf{x}_i^* - \mathbf{x}_j^*)' , \tag{12–24}$$

$$= (\mathbf{x}_i - \mathbf{x}_j)\,\mathbf{TT}^{-1}\mathbf{W}_k\mathbf{T}^{-1}\,\mathbf{T}'(\mathbf{x}_i - \mathbf{x}_j)' ,$$

and

$$= (\mathbf{x}_i - \mathbf{x}_j)\mathbf{W}_k\,(\mathbf{x}_i - \mathbf{x}_j)'.$$

An arbitrary transformation of the stimulus and weight matrices thus yields distances that are identical to those prior to transformation. This is an important difference from INDSCAL, where a transformation such as this yields distances that are not identical to those prior to transformation, showing that the assumptions of the INDSCAL model are violated by the transformation.

SOLVING FOR $\mathbf{W}_k$

The problem that we address in this section is the minimization—over both the optimally scaled data (disparities) $\mathbf{D}_k^*$ and the GEM distances $\mathbf{D}_k$—of the ALSCAL (Takane, Young, and de Leeuw 1977) *S-STRESS* formula,

$$S_k^2 = ||\,\mathbf{D}_k^* - \mathbf{D}_k\,|| , \tag{12–25}$$

where the notation $||\mathbf{Z}||$ refers to the sum of squared elements of the matrix $\mathbf{Z}$, where the disparities $\mathbf{D}_k^*$ are subject to the ALSCAL normalization and measurement restrictions, and where the GEM distances $\mathbf{D}_k$ are defined as the family of symmetric scaling and unfolding GEMs:

$$d_{ijk}^2 = (\mathbf{y}_i - \mathbf{x}_j)\,\mathbf{W}_k\,(\mathbf{y}_i - \mathbf{x}_j)' . \tag{12–26}$$

This is a generalization of the ALSCAL problem (Takane, Young, and de Leeuw 1977) in which $\mathbf{W}_k$ was restricted to being diagonal.

In this section, we only treat the subproblem of minimizing S_k over all positive semidefinite (PSD), fixed-rank matrices $\mathbf{W}_k$, realizing that this subproblem becomes a step in an overall ALS procedure of the ALSCAL type.

Reparameterizing the $\mathbf{W}_k$

The main difficulty in minimizing (12–27) is the requirement that $\mathbf{W}_k$ be PSD. Without the PSD requirement, we could simply solve for $\mathbf{W}_k$ by regression techniques. However, the $\mathbf{W}_k$ produced by such techniques would not generally be PSD. A solution to this problem was suggested by de Leeuw (personal communication) and was independently proposed by Bloxom (1978). They suggest reparameterizing the GEM so that

$$d^2_{ijk} = (\mathbf{y}_i - \mathbf{x}_j)\mathbf{A}_k \mathbf{A}'_k (\mathbf{y}_i - \mathbf{x}_j)' , \tag{12–27}$$

where $\mathbf{A}_k$ has r rows and $t \leqq r$ columns and then solving for the elements of $\mathbf{A}_k$. Thus, the optimization situation is defined by substituting (12–27) into (12–25) to obtain:

$$S^2_k = || \mathbf{D}^*_k - (\mathbf{y}_i - \mathbf{x}_j)\mathbf{A}_k \mathbf{A}'_k (\mathbf{y}_i - \mathbf{x}_j)' || . \tag{12–28}$$

Once $\mathbf{A}_k$ is known, we compute

$$\mathbf{W}_k = \mathbf{A}_k \mathbf{A}'_k , \tag{12–29}$$

which is PSD and has a rank of t, the number of columns of $\mathbf{A}_k$. Since $\mathbf{W}_k$ is PSD it can be decomposed according to (12–12) to obtain Young's (1978) orthogonal *principal directions* solution, or according to (12–21) to obtain Tucker's *oblique* solution.

Minimization

The solution for minimizing (12–28) was first presented by Young (1978) and has been incorporated into the ALSCAL (Young 1982) and DISTREG (Young, Null, and Hamer 1983) programs. The minimization approach taken is to obtain some initial estimates for $\mathbf{A}_k$ and then estimate new values of each a_{pqk}, one at a time. The initial estimates are simply $\mathbf{A}_k = I$. The new values of a_{pqk} are determined by obtaining the partial derivatives of (12–28), relative to a specific a_{pqk} (using all remaining elements of $\mathbf{A}_k$), setting these derivatives to zero, and solving the resulting equation. The derivatives will be cubic, since (12–28) is quartic, and will have three roots that will have to be evaluated to determine the root that minimizes (12–28).

A cubic equation has the general form:

$$S^2_k = q_0 + q_1(a_{pqk}) + q_2(a^2_{pqk}) + q_3(a^3_{pqk}) . \tag{12–30}$$

For the situation defined by (12–28), the terms q_0 through q_3 are defined as follows:

$$q_0 = -4 \sum_i^n \sum_j^n p_{ijp} r_{ijk} s_{ijk} , \qquad (12\text{–}31)$$

$$q_1 = -4 \sum_i^n \sum_j^n p_{ijp}^2 (r_{ijk} - 2s_{ijk}^2) , \qquad (12\text{–}32)$$

$$q_2 = 12 \sum_i^n \sum_j^n p_{ijp}^3 s_{ijk} , \qquad (12\text{–}33)$$

and

$$q_3 = 4 \sum_i^n \sum_j^n p_{ijp}^4 , \qquad (12\text{–}34)$$

where

$$p_{ijp} = (y_{ip} - x_{jp}) , \qquad (12\text{–}35)$$

$$r_{ijk} = (d_{ijk}^{*2} - d_{ijk}^2) + 2(p_{ijp} a_{pqk} s_{ijk} - p_{ijp}^2 a_{pqk}^2) , \qquad (12\text{–}36)$$

and

$$s_{ijk} = \sum_a^r p_{ija} a_{aqk} , \quad a \neq q . \qquad (12\text{–}37)$$

The roots of (12–30), when equations (12–31) through (12–37) are substituted in, can be solved for by classical methods (see for example, Hodgman 1963, 358).

A POLITICAL SCIENCE EXAMPLE

Political science is often concerned with understanding ideologies: Those prescriptions that direct opinion on various issues in a way that reflects a consistent underlying philosophy or belief system (Converse 1964). In particular, political scientists are frequently concerned with the problem of understanding the ideology of lawmakers and with the associated problem of constructing a concise, coherent, underlying ideology structure for lawmakers. One way of conceptualizing ideology structure is in terms of the dimensions of the structure—dimensions that permit ideologies to be distinguished from each other in an orderly fashion.

Easterling (1984) used the GEM as a model of ideology structure in the roll call votes cast by U.S. senators during four sessions of the U.S. Senate. He chose to study senators because their ideological structure should be well developed. He chose to look at roll call votes in the Senate because, as public votes, they should represent motivated behavior suited for inferring the ideological structure of the senators. He also chose roll call votes because of the large number of votes involved (600–1000 votes). Finally, he chose to study four sessions of the Senate in order to see how ideology structure changes over time.

The configuration of senator-points in the stimulus space is Easterling's geometric model of the ideology structure extant for the Senate sessions being considered. The location of the senator-points represents the relative ideological positions of the senators. The distances between the several senator-points represent the differences between the several senators' voting profiles. Regions or dimensions of the space, or other directions through the space, should correspond to the ideological principles at work in the senators' voting record.

Easterling considered several Senate sessions in order to study the way in which the ideology structure changed over time. This implies that the configuration of senators may be relatively high-dimensional, since the relevant ideology dimensions may vary over time and since the configuration summarizes the dimensions in effect for all sessions. However, if each session has only a few relevant dimensions, then it should be possible to represent each session as a subspace of the overall ideology structure.

GEM is an appropriate model of ideology structure and the way it may vary over time. The GEM stimulus space $\mathbf{X}$ represents the ideology structure over the several sessions. A vector of stimulus coordinates $\mathbf{x}_i$ represents the ideological position of the ith senator over the several sessions but not at any given session. The GEM stimulus subspace $\mathbf{X}_k$ of $\mathbf{X}$ represents the ideology structure observed for the kth Senate session, and the coordinate vector $\mathbf{x}_{ik}$ represents the ideological position of senator i during that session.

Easterling notes that the dimensions of an ideology structure that can be found from examining roll call votes represent only a subset of all the dimensions that might make up that ideology. Only those ideology dimensions that bear upon bills in the Senate during the specific sessions under study can be uncovered. While this may be obvious, it is important to keep in mind that those dimensions that structure the senators at one time are not necessarily those that are relevant at some other time. Shifting national concerns are reflected in shifting concerns of bills in the Senate and in shifting ideology structures.

Data

The basic data in this study are roll call votes from 58 senators who were members of the 92nd through the 95th sessions of the Senate (1971–78). All 58 senators had won reelection at least once during this eight-year period, since Senate terms are six years long. The eight years studied cover the Nixon (92nd and 93rd sessions), Ford (94th), and Carter (95th) presidencies. They also cover a major shift in issue focus from Vietnam and wage and price controls to the energy crisis and the effects of extensive regulation.

Votes were recorded as "yea," "nea," or "position not known." "Paired" and "announced" votes—those votes not actually cast on the floor but representing a known stance on a bill—were included in the data as if the vote actually occurred.

For each session, Easterling constructed a 58 × 58 matrix of dissimilarities. The dissimilarity score for a given pair of senators reflects the disagreement between their roll call votes. Specifically, the dissimilarity o_{ijk} between senators i and j during session k is defined as:

$$o_{ijk} = n_{ijk} / t_{ijk} \quad , \qquad (12\text{–}38)$$

where n_{ijk} is the number of bills on which the two senators, i and j, disagreed during session k, and t_{ijk} is the total number of bills on which both senators voted during session k.

This dissimilarity index varies between 0.00 and 1.00. Easterling shows that it may be interpreted as: (a) the probability that the pair of senators will disagree when they both vote; (b) the Euclidean distance between the senators; or (c) a function of the cosine of the angle between the two senators. These last two interpretations hold in a space whose dimensionality equals the number of roll calls to which both senators responded.

Results

Easterling performed a series of nonmetric scalings using the ALSCAL procedure in the SAS package (SAS Institute 1983). The complete SAS job is presented in the appendix to this chapter. The fit indices reported by ALSCAL from these analyses are presented in Table 12–1. (Some values are missing because ALSCAL does not report them for that analysis.) The *R-SQ* measure is the squared correlation between the Euclidean distances (between the senator-points) and the disparities (monotonically transformed disagreement indices). The *STRESS* measure is Kruskal's (1964) normalized measure of residual variance defined on the Euclidean distances and the disparities. The *S-STRESS* measure is Takane, Young, and de Leeuw's (1977) normalized measure of residual variance defined on the squared Euclidean distances and the squared disparities. The column headed *RMS* contains the root-mean-square (square root of the mean of the squares) of the values under the columns headed 92 through 95, which refer to each Senate session.

The first set of analyses involved submitting each of the four disagreement matrices to ALSCAL for four separate classical MDS analyses based on the model defined by equation (12–2). These four analyses yielded two-dimensional spaces that had very good fit values and three-dimensional spaces that showed very little improvement in fit. These results led Easterling to conclude that each Senate session can be adequately represented by a separate two-dimensional space.

A replicated MDS of these four data matrices yielded results with noticeably lower fit values in two, three, and four dimensions, as summarized in Table 12–1. This led Easterling to conclude that the four sessions cannot be represented by the same structure, even in four dimensions, even though each session can be represented by a separate two-dimensional structure.

The next analysis that Easterling performed was a weighted MDS using the diagonal (INDSCAL) GEM discussed in the section on INDSCAL geometry. From this analysis, he also determined

TABLE 12-1. Fit Indices for Several Nonmetric Scaling Analyses

Model	Dimensions	Index	RMS*	92	93	94	95
CMDS	2	R-SQ	.961	.970	.962	.967	.946
		Stress	.100	.085	.102	.094	.116
		S-Stress	.096	.078	.093	.084	.123
	3	R-SQ	.975	.981	.977	.980	.963
		Stress	.073	.062	.072	.069	.088
		S-Stress	.073	.062	.069	.063	.094
	4	R-SQ	.983	.986	.985	.985	.974
		Stress	.060	.052	.057	.059	.069
		S-Stress	.063	.054	.059	.057	.078
RMDS	2	R-SQ	.875	.890	.900	.917	.794
		Stress	.176	.165	.157	.143	.226
		S-Stress	.212				
	3	R-SQ	.888	.890	.916	.929	.818
		Stress	.149	.148	.129	.119	.190
		S-Stress	.197				
	4	R-SQ	.894	.892	.919	.928	.837
		Stress	.135	.136	.117	.111	.167
		S-Stress	.187				
WMDS	2	R-SQ	.879	.891	.904	.920	.800
		Stress	.174	.161	.156	.141	.227
		S-Stress	.210				
	3-U	R-SQ	.924	.932	.925	.926	.913
		Stress	.133	.130	.132	.129	.142
		S-Stress	.157				
	3-C	R-SQ	.919	.905	.928	.929	.912
		Stress	.144	.157	.136	.136	.146
		S-Stress	.168				
	4	R-SQ	.943	.954	.948	.947	.923
		Stress	.115	.104	.109	.109	.136
		S-Stress	.126				
GMDS	3,2	R-SQ	.923	.933	.920	.925	.915
		Stress	.144	.130	.145	.147	.153
		S-Stress	.160	.162	.160	.159	.160

*The R-SQ entries in this column are means of those to the right; the remaining entries are root-mean-squares of those to the right.

that a two-dimensional solution was not adequate, even though each session is two-dimensional. However, with this model, the three-dimensional solution fits very well. Since there was little improvement in fit in four dimensions, Easterling concluded that, taken as a whole, the four Senate sessions were three-dimensional, even though each one by itself was two-dimensional.

The appendix to this chapter gives the exact SAS code necessary to perform the four-dimensional WMDS analysis. The standard code resulted in all zero weights on the fourth dimension during the initialization. This highly unusual situation (never before seen by the author) indicates that ALSCAL wished to produce a solution with negative weights. This implies that the data are very inappropriate to the four-dimensional WMDS model.

However, the three-dimensional space is not the same for all four sessions. While this can be seen from the poor fit of the three-dimensional RMDS solution, it is more revealing to inspect the pattern of WMDS weights shown in Table 12–2. These weights suggest that each session is using only two of the three dimensions, the first three sessions using the first two dimensions, and the fourth session using dimensions 1 and 3. (This is a slight simplification of the weight structure, but only slight.) Thus, once again we see that each session is two-dimensional, but we now see that there is some common structure between the sessions.

TABLE 12-2. Unconstrained WMDS Weight Matrices for Senate Sessions 92 through 95

	92			93			94			95	
.745			.789			.707			.421		
	.613			.530			.586			.261	
		.000			.144			.288			.817

The next analysis was performed to determine the fit of a WMDS model with weights exactly like (instead of roughly like) the simplified description of the weights resulting from the unconstrained WMDS model described in the preceding paragraph. This analysis required the first three sessions to equally weight the same two dimensions (that is, the three sessions are replicates) and required the fourth session to equally weight one of these two dimensions but to have a unique second dimension. Restated, this analysis required all four sessions to equally share one common dimension and required the first three sessions to equally share one more dimension while requiring the fourth session to use a unique dimension. The constrained weights for this analysis are presented in Table 12–3.

TABLE 12-3. Constrained WMDS Weight Matrices for Senate Sessions 92 through 95

	92			93			94			95	
1.00			1.00			1.00			1.00		
	1.00			1.00			1.00			0.00	
		0.00			0.00			0.00			1.00

Geometrically, all four sessions are represented by planes in a three-dimensional space, with the first three sessions' planes being exactly coplanar and the fourth session's plane being orthogonal to the first three. The fit indices for this analysis are presented in Table 12–1 (WMDS analysis 3–C). We see that this constrained WMDS model fits essentially as well as the unconstrained WMDS model above and that the simplified description above is sufficiently accurate.

All of these results suggested to Easterling that he should analyze his data with a rank-two, three-dimensional GMDS similar to that diagramed in Figure 12–3. The results of this analysis confirmed that all four sessions can be adequately represented as planes in three-space, that the first three sessions have planes that are nearly coplanar, but that the fourth session's plane is oriented obliquely with respect to the first three. The detailed results are discussed in the next few paragraphs.

From the fit indices presented in Table 12–1, we can see that the rank-two, three-dimensional GMDS fits better than the two-dimensional WMDS, not quite as well as the unconstrained three-dimensional WMDS, and nearly identically to the constrained three-dimensional WMDS. These results are all as they should be.

Easterling noticed that the weight matrices (Table 12–4) for the first two sessions are essentially diagonal matrices and that the third diagonal element is essentially zero. He emphasized that this means that the planes for the first two sessions are nearly coplanar with the plane formed by the first two dimensions of the three-dimensional space and that the principal directions for the first two sessions are essentially collinear with the first two dimensions. These three weight matrices are very similar to the three corresponding constrained WMDS matrices.

TABLE 12-4. GMDS Weight Matrices for Senate Sessions 92 through 95

92			93			94			95		
.748			.756			.608			.234		
.024	.603		-.040	.531		-.047	.583		-.153	.311	
.020	.020	.001	-.070	-.039	.010	-.150	.065	.048	-.256	-.074	.559

The weight matrix for the third session is not quite as diagonal as those for the first two sessions, and the third diagonal element is somewhat larger. This means that the plane for the third session is not as coplanar with the plane formed by the first two dimensions and that the principal directions are not as collinear with the first two dimensions. However, since these differences are fairly trivial, for the first three sessions, a two-dimensional RMDS would be sufficient.

On the other hand, the weight matrix for the fourth session is rather nondiagonal, and the third diagonal element is the largest diagonal element. Thus, the plane for this session exists in all three dimensions of the three-dimensional space. The three dimensions are obliquely projected onto the plane, and the two principal directions do not correspond in a simple way to any dimension. Thus, the GMDS model is needed to simultaneously represent the fourth session with the first three.

If we wish to look further into the composition of the principal directions, we should look at the matrices containing the coefficients of the linear combination applied to the three dimensions to yield each principal direction. These coefficient matrices, which appear in Table 12–5, are the $\mathbf{C}_k$ matrices defined by equation (12–14). We see that the first principal direction for sessions 92 through 94 is predominately composed of dimension 1 and that the second principal direction is composed mainly of dimension 2. However, for session 95, the first direction is a combination of dimensions 1 (somewhat) and 3 (mostly), while the second principal direction is primarily dimension 2.

TABLE 12-5. Principal Direction Coefficients for Senate Sessions 92 through 95

92		93		94		95	
.856	-.124	.863	.106	.729	.276	.408	-.259
.139	.764	-.134	.716	-.326	.690	.022	.558
.027	.022	-.072	-.069	-.145	-.163	-.730	-.161

Easterling does not present the entire three-dimensional ideology structure, since it represents none of the sessions. Rather, he presents two two-dimensional structures—one for sessions 92–94 and another for session 95. We reproduce these structures in Figures 12–4 and 12–5. Figure 12–4 is the first two dimensions of the three-dimensional space. This is, essentially, the plane that displays the ideology structure for sessions 92 through 94. The projection of the two principal directions for the three sessions is also shown in this figure. Figure 12–5 is the plane of the three-dimensional space that shows the ideology structure for session 95. The latter figure is constructed by multiplying $\mathbf{C}_4$ times the three-dimensional stimulus space $\mathbf{X}$, according to equation (12–16).

Easterling is able to provide a useful interpretation of the GMDS solution shown in these figures. First, the senators are clearly structured according to whether they are Democrat or Republican. Second, they are also clearly structured by whether they are liberal or conservative. Third, the structure and the change in structure from the first sessions to the fourth is related to other aspects of the political climate. We summarize these interpretations in the next few paragraphs.

First, it is possible to perfectly separate Democrats from Republicans in the two ideology structures. This is shown by the line labeled "Democrat/Republican," which goes through the origin of each figure.

Second, when the senators are projected orthogonally onto the

Figure 12–4. Dimension One—Dimension Two Plane of Senate Group Space with Projections of Principal Directions for Sessions 92–94

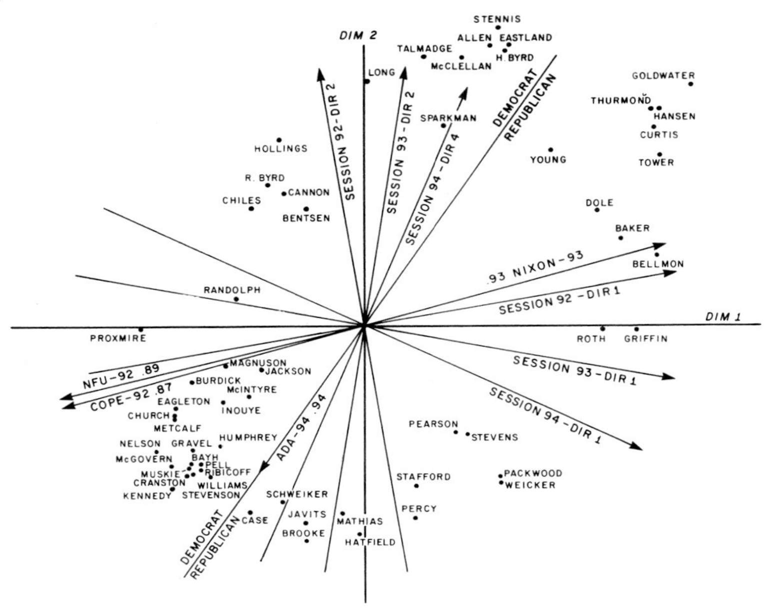

Figure 12–5. Personal Space for Senate Session 95 with Relevant Regression Vectors

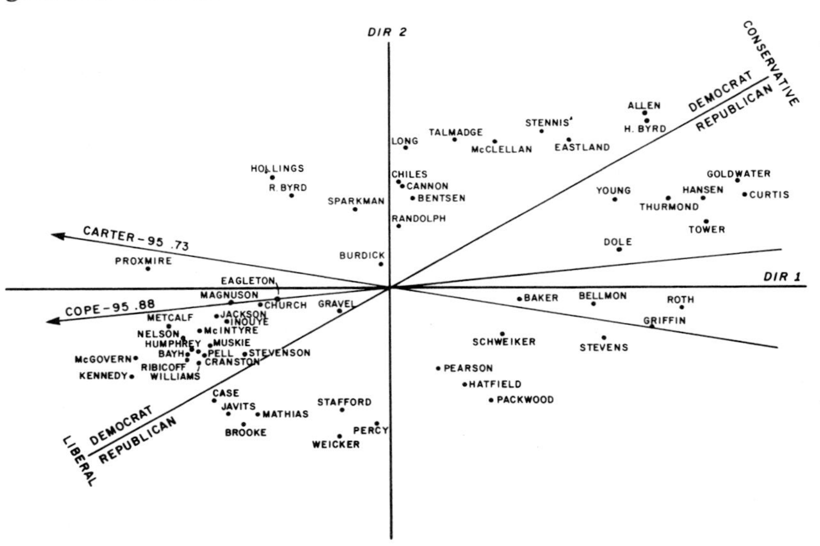

lines that separate Democrats from Republicans, they are clearly ordered from liberal to conservative. Note that most of the

Democrats cluster at the liberal end, with a tail going off in the conservative direction. The reverse is true for the Republicans. In fact, Easterling interprets all three dimensions as ordering the senators from liberal to conservative. Dimensions 1 and 3 are interpreted as economic liberal/conservative, whereas dimension 2 is social liberal/conservative.

Third, Easterling interpreted the principal directions for each ideology structure by multiple regression, as suggested by Schiffman, Reynolds, and Young (1981). He performed a series of regressions, each with a different rating scale versus the three-dimensional solution. The rating scales were constructed by various political lobbying groups and represent each group's view of the percentage of time that each senator agreed with the group's position on a sample of key roll call votes during a given session. The lobbies are the liberal Americans for Democratic Action (ADA), the conservative Americans for Constitutional Action (ACA), the AFL-CIO's Committee on Political Education (COPE), and the National Farmers' Union (NFU). Scales of presidential support were also used.

Easterling does not report the strength of correlation; however, he does shown the pattern of relationships between the principal directions and the scales. The first principal direction for sessions 92–94 (Figure 12–4) represents support for the NFU or COPE position (session 92) or for Nixon (session 93). The second principal direction relates to support for ADA in session 94. During session 95 (Figure 12–5), the first principal direction relates to support for Carter and for the COPE position. None of the scales relates well to the second principal direction.

Finally, Easterling's careful interpretation of the two economic dimensions (1 and 3) tells him that the change from the early sessions (dimension 1) to session 95 (dimension 3) represents a change from emphasizing economic issues affecting individuals to economic issues affecting businesses. He concludes that dimension 1 structures senators according to their ideology about the role of government in the distribution of wealth, whereas dimension 3 reflects ideologies concerning the role of government in the development of resources.

APPENDIX

Following is a listing of the SAS code for the analyses reported in this chapter:

```
TITLE         STRUCTURE OF US SENATORS:  SESSION 92-95;

*-------------------------------------------------------+
|                                                       |
|    READ RAW VOTING RECORD                             |
|                                                       |
+-------------------------------------------------------;
DATA VOTES;
     INPUT    SEN1-SEN58 SESSION;
     INFILE   SENATORS;
*-------------------------------------------------------+
|                                                       |
|    COMPUTE EUCLIDEAN DISTANCES BETWEEN SENATORS       |
|                                                       |
+-------------------------------------------------------.
PROC PROX     DATA=VOTES
              OUT=DISAGREE
              PAIRWISE;
     VAR      SEN1-SEN58;
BY       SESSION;
*-------------------------------------------------------+
|                                                       |
|    DO REPEATED CLASSICAL MDS                          |
|                                                       |
+-------------------------------------------------------;
PROC ALSCAL   DATA=DISAGREE
              HEADER
              PLOTALL
              MAXDIM=4
              MINDIM=2;
     VAR      SEN1-SEN58;
     BY       SESSION;
     TITLE2   CLASSICAL SCALING FOR EACH 2 YEAR SESSION;
*-------------------------------------------------------+
|                                                       |
|    DO REPLICATED MDS                                  |
|                                                       |
+-------------------------------------------------------;
PROC ALSCAL   DATA=DISAGREE
              HEADER
              PLOTALL
              MAXDIM=4
              MINDIM=2;
     VAR      SEN1-SEN58;
     TITLE2   REPLICATED SCALING FOR THE FOUR SESSIONS;


*-------------------------------------------------------+
|                                                       |
|    DO WEIGHTED MDS IN 2 AND 3 DIMENSIONS              |
|                                                       |
+-------------------------------------------------------;
PROC ALSCAL   DATA=DISAGREE
              HEADER
              PLOTALL
              MODEL=INDSCAL
              MAXDIM=3
              MINDIM=2;
     VAR      SEN1-SEN58;
     TITLE2   WEIGHTED MDS FOR ALL FOUR SESSIONS;
```

```
*-------------------------------------------------------------+
|                                                             |
|    DO WEIGHTED MDS IN 4 DIMENSIONS WITH NEGATIVE WEIGHTS    |
|    AND THEN REPEAT THIS WITHOUT NEGATIVE WEIGHTS.           |
|    (THE STANDARD ANALYSIS STOPS DURING INITIALIZATION       |
|    BECAUSE ALL WEIGHTS ON THE FOURTH DIMENSION ARE ZERO).   |
|                                                             |
+-------------------------------------------------------------;
PROC ALSCAL    DATA=DISAGREE
               OUT=WMDS4
               HEADER
               PLOTALL
               MODEL=INDSCAL
               NEGATIVE
               DIMENS=4;
     VAR       SEN1-SEN58;
     TITLE2    WEIGHTED 4-D MDS ALLOWING NEGATIVE WEIGHTS;
DATA NONEG;
     SET       WMDS4;
     IF _TYPE_='SUBJWGHT' & DIM4<0 THEN DIM4=0;
PROC ALSCAL    DATA=DISAGREE
               IN=NONEG
               HEADER
               MODEL=INDSCAL
               DIMENS=4;
     VAR       SEN1-SEN58;
     INVAR     DIM1-DIM4;
     TITLE2    WEIGHTED 4-D MDS PROHIBITING NEGATIVE WEIGHTS;

*-------------------------------------------------------------+
|                                                             |
|    CONSTRAINED WEIGHTED MDS                                 |
|                                                             |
+-------------------------------------------------------------;
DATA CONSTRAN;
     INPUT DIM1-DIM3;
     CARDS;
     1 1 0
     1 1 0
     1 1 0
     1 0 1
     ;
PROC ALSCAL    DATA=DISAGREE
               IN=CONSTRAN
               READFIXW
               HEADER
               PLOTALL
               MODEL=INDSCAL
               DIMENS=3;
     VAR       SEN1-SEN58;
     INVAR     DIM1-DIM3;
     TITLE2    CONSTRAINED 3-D MDS;
*-------------------------------------------------------------+
|                                                             |
|    GENERALIZED MDS                                          |
|                                                             |
+-------------------------------------------------------------;
PROC ALSCAL    DATA=DISAGREE
               HEADER
               MODEL=GEMSCAL
               PLOTALL
               DIMS=3
               DIRS=2;
     VAR       SEN1-SEN58;
     TITLE2    GENERALIZED 3-DIMENSIONAL, 2-DIRECTIONAL MDS;
```

REFERENCES

Bloxom, B. 1978. Constrained multidimensional scaling in n-spaces. *Psychometrika* 43:397–408.

Carroll, J. D. 1972. Individual differences and multidimensional scaling. In *Multidimensional scaling: Theory and applications in the behavioral sciences,* vol. 1, ed. R. N. Shepard, A. K. Romney, and S. Nerlove. New York: Seminar Press.

Carroll, J. D., and J. J. Chang. 1970. Analysis of individual differences in multidimensional scaling via an N-way generalization of "Eckart-Young" decomposition. *Psychometrika* 35:283–319.

———. 1972. IDIOSCAL (*I*ndividual *D*ifferences *I*n *O*rientation *SCAL*ing): A generalization of INDSCAL allowing *IDIO*syncratic reference systems as well as an analytic approximation to INDSCAL. Paper presented at the spring meeting of the Psychometric Society, Princeton, New Jersey, 30–31 March.

Converse, P. 1964. The nature of belief systems in mass publics. In *Ideology and discontent,* ed. D. E. Apter. New York: The Free Press of Glencoe.

Coombs, C. H. [1964] 1976. *A theory of data.* [New York: Wiley] Republished, Ann Arbor, Mich.: Mathesis Press.

de Leeuw, J. Personal communication.

De Sarbo, W. S. 1978. Three-way unfolding and situational dependence in consumer preference analysis. Doctoral thesis, University of Pennsylvania.

Easterling, D. V. 1984. Ideological shifts in the U.S. Senate between 1971 and 1978: A principal directions scaling of roll call votes. In *Theory and applications of multidimensional scaling,* ed. F. W. Young and R. M. Hamer. Hillsdale, N.J.: Erlbaum.

Harshman, R. A. 1970. Foundations of the PARAFAC procedure: Models and conditions for an "explanatory" multi-modal factor analysis. *UCLA Working Papers in Phonetics* 16:1–84 (University Microfilms No. 10,085).

Hodgman, C. D. 1963. *CRC standard mathematical tables.* 12th ed. Cleveland, Ohio: Chemical Rubber Publishing Co.

Kruskal, J. B. 1964. Multidimensional scaling by optimizing goodness-of-fit to a nonmetric hypothesis. *Psychometrika* 29:1–27.

SAS Institute Inc. 1983. *SAS supplemental user's guide.* Cary, N.C.: SAS Institute Inc.

Schiffman, S. S., M. L. Reynolds, and F. W. Young. 1981. *Introduction to multidimensional scaling: Theory, methods, and applications.* New York: Academic Press.

Takane, Y., F. W. Young, and J. de Leeuw. 1977. Nonmetric individual differences multidimensional scaling: An alternating least-squares method with optimal scaling features. *Psychometrika* 42:7–67.

Torgerson, W. S. 1952. Multidimensional scaling: I—Theory and methods. *Psychometrika* 17:401–19.

Tucker, L. R. 1972. Relations between multidimensional scaling and three-mode factor analysis. *Psychometrika* 37:3–27.

Young, F. W. 1975a. Multidimensional scaling: Theory and methods. Manuscript, University of North Carolina, Psychometric Laboratory.

———. 1975b. An asymmetric euclidean model for multi-process asymmetric data. Paper presented at the U.S.–Japan Seminar on Theory, Methods and Applications on Multidimensional Scaling and Related Techniques, University of California, 20–24 August.

———. 1978. Principal directions scaling (note 1): The problem and its solution. Manuscript, University of North Carolina, Psychometric Laboratory.

———. 1979. Principal directions scaling (note 2): A new individual differences model. Manuscript, University of North Carolina, Psychometric Laboratory.

———. 1982. Enhancements in ALSCAL-82. *SUGI Proceedings* 7:633–42.

Young, F. W., and R. M. Hamer. Forthcoming. *Multidimensional scaling: Theory and applications*. Hillsdale, N.J.: Erlbaum.

Young, F. W., and R. Lewyckyj. 1979. *ALSCAL-4 user's guide*. Carrboro, N.C.: Data Analysis and Theory Associates.

Young, F. W., C. H. Null, and R. M. Hamer. 1983. DISTREG: A procedure for distance regression. In *SAS supplemental user's guide*, ed. SAS Institute Inc. Cary, N.C.: SAS Institute Inc.

13

The Weighted Additive Model

Yoshio Takane

There are a number of situations in which various attributes of stimuli are to be weighted differently for different decision purposes. For example, consider a person buying a car. Desirable properties of a car depend on many factors. Spaciousness may be an important factor for comfortable weekend driving, while cost efficiency—such as gas mileage—may be more important for commuting. In any case, the person has to decide what is optimal for him, considering the various situational demands surrounding him. One way to achieve this is to express his best interest as a linear combination of various attributes of cars. Then the cars can be evaluated and compared on the composite criterion. The weights to be given to the attributes, of course, depend on his specific interest.

As another example, suppose a psychology department would like to set up different admission criteria for different subprograms it offers. Again, this may be done by forming a composite criterion most suitable for each program, taking into account various aspects of each applicant's qualifications. For instance, if the applicant is seeking admission in quantitative psychology, his score on GRE-Quantitative should be weighed heavily in forming the composite score. If, on the other hand, the prospective student is going into clinical psychology, mathematical skill is perhaps not as important; consequently, it should be weighed much less. The problem of weighting various attributes of stimuli into composite scores is called a *weighting problem*. The problem is called a *differential weighting problem,* when two or more composite scores are derived from the same set of attribute scores according to multiple criteria. Assuming that the composition rule is linear, the weighting problem may expressed as:

Comments by Tony Marley have been helpful in revising this manuscript.

$$y_{kr} = \sum_{t=1}^{T} w_{kt} x_{rt} \,, \qquad (k = 1, \ldots, K) \,, \qquad (13\text{–}1)$$

where y_{kr} is the composite score of stimulus r for criterion k, x_{rt} is the score of stimulus r on attribute t (assumed known a priori), and w_{kt} is the weight attached to attribute t for criterion k. The weights are determined in such a way that the derived composite scores (y_{kr}) best agree with the criterion variables. If the criterion variables can be assumed linearly related to the attribute scores, the weights in (13–1) can be determined by applying the linear regression analysis separately to each criterion (k). Thus, there is no need for a new analysis method in this case.

Note that in (13–1) attribute score x_{rt} is assumed known a priori. Most often it represents a numerical value designating a degree to which stimulus r possesses attribute t. However, there are cases in which attributes are not quantified a priori. For example, types of cars—such as sedan or wagon—may be an important factor in deciding which car to buy. But there is no natural and obvious way of assigning a numerical value to the attribute. It should be quantified before it can be used in a composite criterion. An attribute may sometimes have to be requantified, even when a priori quantifications exist. For example, performance on GRE-Quantitative is usually reported in a numerical form, but there is no assurance that the reported score is linearly related to criterion variables, such as performance in quantitative psychology. However, the linear form of (13–1) implies that x_{rt} is linearly related to the criterion variables. Thus, it is necessary to requantify x_{rt} in such a way that it is as linearly related to the criterion variables as possible.

The quantifications of an attribute may generally be expressed as follows. Let h_{rq_t} denote an indicator variable such that:

$$h_{rq_t} = \begin{cases} 1, & \text{if stimulus } r \text{ belongs to level } q \text{ of attribute } t \\ & \text{(this may be subject } r \text{ passing item } q \text{ in test } t.\text{);} \\ 0, & \text{otherwise.} \end{cases}$$

Then the new attribute score x_{rt} is obtained by a linear combination of the indicator variables; that is,

$$x_{rt} = \sum_{q_t}^{Q_t} h_{rq_t} \alpha_{q_t} \,, \qquad (13\text{–}2)$$

where α_{q_t} is the score given to the qth level of attribute t, and Q_t is the total number of levels in the attribute. Specifically, we are discretizing the attribute into several levels and assigning new numerical values at these discrete points of the attribute.

If we substitute the expression (13–2) for x_{rt} in (13–1), we obtain:

$$y_{kr} = \sum_t^T w_{kt} \left(\sum_{q_t}^{Q_t} h_{rq_t} \alpha_{q_t} \right), \quad (k = 1, \ldots, K), \tag{13–3}$$

This model is called the *weighted additive model* (WAM) (Takane, Young, and de Leeuw 1980). In (13–3), the differential weighting problem (13–1) and the quantification problem (13–2) (which is yet another weighting problem) are combined into one major problem, in which the two subproblems are solved simultaneously. In (13–3), both α_{q_t} and w_{kt} are estimated so that the derived composite scores, y_{kr} $(k = 1, \ldots, K)$, are best predictive of all K criteria simultaneously.

When there is only one criterion—that is, $K = 1$—we obtain:

$$y_r = \sum_t^T \sum_{q_t}^{Q_t} h_{rq_t} \alpha^*_{q_t}, \tag{13–4}$$

where $\alpha^*_{q_t} = \alpha_{q_t} w_t$. Notice that subscript k has been omitted.

This model is called the *simple additive model* (SAM). In this case, we have only the quantification problem, which reduces to a linear regression problem with $h_{rq_t} (q = 1, \ldots, Q)$ serving as the independent variables.

Note that in (13–3), α_{q_t} is assumed common across all criteria, just as in (13–1) the same x_{rt} is used for all criteria. If, on the other hand, α_{q_t} is allowed to vary over the criteria, we have:

$$y_{kr} = \sum_t^T w_{kt} \left(\sum_{q_t}^{Q_t} h_{rq_t} \alpha_{kq_t} \right) \tag{13–5}$$

$$= \sum_t^T \sum_{q_t}^{Q_t} h_{rq_t} (\alpha_{kq_t} w_{kt}) = \sum_t \sum_{q_t} h_{rq_t} \alpha^*_{kq_t},$$

where $\alpha^*_{kq_t} = \alpha_{kq_t} w_{kt}$. Note that subscript k is attached to α_{kq_t} and $\alpha^*_{kq_t}$, implying that they are unique to each k. Equation (13–5) indicates that the WAM in this case reduces to the SAM applied separately to each criterion. The new quantification parameter, $\alpha^*_{kq_t}$, can be estimated by applying the linear regression analysis separately to each k. (This is completely analogous to [13–1], in which the weights are obtained by separate regression analyses of y_{kr} to x_{rt}.) This means that a real significant feature of the WAM lies in the assumption of common α_{q_t} across different criteria. This very feature calls for a new analysis method specifically designed for this model, since no conventional multivariate methods apply to this case. In this chapter, we discuss various aspects of the WAM focusing on its empirical motivation, mathematical properties, analytical methods, and examples of application.

EMPIRICAL MOTIVATION

As has been emphasized, an important assumption underlying the WAM is that α_{q_t} remains constant across different criteria.

Although this assumption is implicit in almost all the models that capitalize on a priori quantifications of attributes, it by no means is always correct. Rather, it is an empirical hypothesis to be verified or falsified in each specific situation in which the model is employed. On the other hand, there are many situations in which the WAM is indeed appropriate. In this section, we discuss a couple more examples of such situations. Several more examples will be given in the application section.

The first example has been modified from Kruskal and Shepard (1974). Suppose first that a set of cylinders are generated by combining different altitude levels and base areas. Suppose further that those cylinders are measured in terms of two physical variables, side area (y_S) and volume (y_V). Let α and β represent altitude and base area, respectively. Then we have:

$$y_S = (2\sqrt{\pi})\, \alpha\, \beta^{1/2} ,$$

and

$$y_V = \alpha\, \beta .$$

By the log transformation, we obtain:

$$\log(y_S) = \log(\alpha) + \tfrac{1}{2}\log(\beta) + \gamma , \tag{13–6}$$

and

$$\log(y_V) = \log(\alpha) + \log(\beta) , \tag{13–7}$$

where $\gamma = \log(2\sqrt{\pi})$. That is, the log-transformed y_S and y_V are both additive functions of log (α) and log (β), disregarding γ, for the moment. The difference between the two criteria—log (y_S) and log (y_V)—is produced by a differential weighting of log (α) and log (β), which themselves remain constant across the criteria. It can be readily seen that (13–6) and (13–7) are special cases of (13–1).

Suppose now that we are ignorant of the exact weighting rules to obtain the two criteria—log (y_S) and log (y_V)—from log (α) and log (β), except that the former are both additive functions of the latter. The problem of finding the differential weights, given a set of measurements on the four variables, is identical to the one posed by (13–1). Now assume further that actual numerical values of log (α) and log (β) are not directly observable, but only discrete levels of altitude and base area of the cylinders are provided as information. Then, those levels have to be quantified before they can be used for log (α) and log (β) in (13–6) and (13–7). The quantifications of the levels should stay invariant, just as both log (α) and log (β) remain intact in (13–6) and (13–7). The problem now is equivalent to what we have in (13–3): Namely, we simultaneously quantify the levels of the attribute variables (altitude and base area) and estimate the differential weights to obtain log (y_S) and log (y_V).

Finally, let us suppose that y_S and y_V, rather than log (y_S) and log (y_V), are observed as the criterion variables. The observed criteria cannot be represented by any additive functions of log (α) and log (β) without transformations. But suppose we are ignorant of the transformations that make y_S and y_V additively representable by log (α) and log (β). Then we are faced with an additional problem of finding the transformations. The transformations may be found by quantifying or requantifying the criterion variables (possibly with ordinal restrictions) in much the same way as the attribute variables are quantified.

As a second example, suppose the desirability (u) of various cars is investigated as a function of their price (p) and fuel efficiency (g). Although both price and fuel efficiency can be measured physically, their contributions to the overall desirability ratings may not be linear. Thus, several price levels and fuel efficiency levels are chosen, from which a set of hypothetical cars (such as a car costing $7,000, featuring high fuel efficiency, and so on) are constructed for the desirability ratings. Let $f_P(p_i)$ and $f_G(g_j)$ represent quantifications of the ith price level (p_i) and the jth fuel efficiency level (g_j), respectively. It is likely that f_P is monotonically decreasing with price and f_G monotonically increasing with fuel efficiency, although neither of them may be linear with their physical correlates. Let the observed rating of a car characterized by p_i and g_j be denoted by u_{ij}. Let us assume that u_{ij} is monotonically related to the sum of $f_P(p_i)$ and $f_G(g_j)$. That is,

$$f_U(u_{ij}) = f_P(p_i) + f_G(g_j) , \tag{13–8}$$

where f_U is some monotonic function. (We have tentatively disregarded possible measurement errors in [13–8].) This model implies that after the desirability rating (the criterion variable) is monotonically transformed by an appropriate transformation, it can be represented as an additive function of f_P and f_G (the attribute variables). Model (13–8) is, in fact, a SAM.

Suppose now that several individuals have rated the desirability of the same set of cars, and a substantial amount of individual differences have been observed. The differences may be due to possible differences in the attribute functions; that is, the way price levels and fuel efficiency levels are evaluated may vary across individuals. In this case, model (13–8) should be applied separately to each individual. On the other hand, the differences may only be due to possible differences in relative importance of the two attributes, while the evaluations of the attributes themselves remain more or less constant across the individuals. If this is the case, we obtain:

$$f_{kU}(u_{ij}) = w_{kP} f_P(p_i) + w_{kG} f_G(g_i) , \tag{13–9}$$

where k refers to the kth individual. The differences among the individuals are accounted for by a differential weighting of the attribute variables, f_P and f_G, which are additive and invariant across k. Model (13–9) is the WAM. (Note that individual differences may be allowed in f_U, as indicated by subscript k attached.)

The key question, then, is whether or not f_P and f_G remain constant across different individuals. On an intuitive basis this seems likely. The ratings concerned with desirability (and not preference) and the desirability perceptions are much more apt to agree. For example, nobody would disagree that f_P is monotonically decreasing and f_G is monotonically increasing, although ultimately this is a question to be answered on an empirical basis. Whether it is more appropriate to apply the SAM separately to each individual or to apply the WAM jointly to all individuals can only be answered by fitting the two models to the same set of data and by comparing their goodness-of-fit to the data. If the fit of the SAM is significantly better than that of the WAM, it is not appropriate to assume f_P and f_G are constant across k. Otherwise, the WAM is more appropriate, because it uses fewer parameters than the SAM fitted separately.

THE MODEL

We now turn to investigating various properties of the WAM. Our discussion concentrates on formal aspects of the model, but it may be helpful for the reader to bear in mind some concrete example to which the model might be applied.

Basic Properties

We first restate the model in standard notation. For explanatory convenience, we state the model for the simple two-factor case. We also introduce new terminologies.

Let α_i and β_j denote quantifications of the ith level of attribute A and the jth level of attribute B, respectively. They are sometimes called *additive effects.* (Attributes are often referred to as *factors* or *independent variables.*) Let w_{kA} and w_{kB} represent the weights attached to factors A and B, respectively, for the kth criterion variable. (The criterion variable is often called the *dependent variable.*) Then the WAM—which has already been stated in such forms as (13–3) and (13–9)—can be restated as:

$$y_{kij} = w_{kA}\alpha_i + w_{kB}\beta_j , \qquad (k = 1, \ldots, K) , \qquad (13\text{–}10)$$

where y_{kij} is the model prediction or model value representing the combined effect of the ith level of factor A (α_i) and the jth level of factor B (β_j) in the kth criterion variable. We assume that there are n_A levels in factor A ($i = 1, \ldots, n_A$) and n_B levels in factor B ($j = 1, \ldots, n_B$). An extension of (13–10) to a higher-order design is straightforward. What have been referred to as the *criteria* or *criterion variables* may represent whatever sets of observations that need be distinguished and whose distinction is best characterized by a differential weighting. For example, they may represent subjects, groups, occasions, experimental conditions, and so on. There is an intrinsic scale indeterminacy in the WAM. That is, for each factor, say A, the effect of multiplying w_{kA} by some constant c ($\neq 0$) is offset by dividing α_i by the same constant. We may remove this indeterminacy by requiring

$$\sum_{i=1}^{n_A} \alpha_i^2 = n_A \quad \text{and} \quad \sum_{j=1}^{n_B} \beta_j^2 = n_B \,. \tag{13–11}$$

Also, the additive effects can be determined only up to an arbitrary origin. We thus require:

$$\sum_{i=1}^{n_A} \alpha_i = 0 \quad \text{and} \quad \sum_{j=1}^{n_B} \beta_j = 0 \,. \tag{13–12}$$

Consequently,

$$\sum_i \sum_j y_{kij} = 0 \qquad \text{for all } k.$$

This amounts to centering the model predictions so that their mean is equal to zero.

To reemphasize, the WAM ([13–10]) postulates that the differences among K sets of observations arise from a differential weighting of additive effects that remain constant across the sets. Reflecting the fact that the WAM was originally introduced for explaining individual differences in perception (see the application section), the differential weights, w_{kA} and w_{kB}, are often called *individual differences weights*. When there are no systematic differences among the sets, the individual differences weights can be set to unity, and (13–10) reduces to the SAM,

$$y_{kij} = \alpha_{ki} + \beta_{kj} \,, \tag{13–13}$$

which treats k as a replication factor. At the other extreme, the additive effects may be completely unique for each set, in which case each set must be represented separately by the SAM,

$$y_{kij} = \alpha_{ki} + \beta_{kj} \,, \qquad (k = 1, \ldots, K) \,, \tag{13–14}$$

where α_{ki} and β_{kj} represent the additive effects unique to set k. The WAM lies somewhere between these two extreme cases; it represents both the common and unique aspects of multiple conjoint data in one model. (The data supposedly representing the combined effects of two or more influencing factors are sometimes called *conjoint data* after Shepard [1972].)

We now explore some basic properties of the WAM. We have already mentioned that in order for the WAM to apply to K sets of observations, each set has to be represented by the SAM. Suppose that we have applied the SAM separately to each data set and obtained estimates of the additive effects, α_{ki} and β_{kj}, $(k = 1, \ldots, K)$. From these, can we tell whether the WAM is appropriate for all sets of observations?

Since the WAM is a special case of the SAM with $\alpha_{ki} = w_{kA}\alpha_i$ $(i = 1, \ldots, n_A)$ and $\beta_{kj} = w_{kB}\beta_j$ $(j = 1, \ldots, n_B)$ the additive effects from the SAM have to be proportional; namely,

$$\frac{\alpha_{ki}}{\alpha_{k'i}} = \frac{w_{kA}}{w_{k'A}} \qquad \text{for all } i, \tag{13–15}$$

and

$$\frac{\beta_{kj}}{\beta_{k'j}} = \frac{w_{kB}}{w_{k'B}} \qquad \text{for all } j.$$

The situation is depicted in Figure 13–1, where $\alpha_{ki} = w_{kA}\alpha_i$ ($i = 1, \ldots, n_A$) are plotted for different values of w_{kA} and connected for each k by line segments. The plots exhibit some characteristic pattern. Note that the additive effects were tacitly assumed monotonically increasing in Figure 13–1, but this does not imply that it is always the case. For the monotonically increasing α_i, α_{ki} is monotonically increasing for a positive w_{kA} and monotonically decreasing for a negative w_{kA}. The rate of increase and decrease changes as a function of the absolute value of w_{kA}. (Of course, the change rate is higher for a larger value of w_{kA}

Figure 13–1. The Plot of $\alpha_{ki} = \alpha_i w_{kA}$ ($i = 1, \ldots, n_A$) when the WAM is Correct

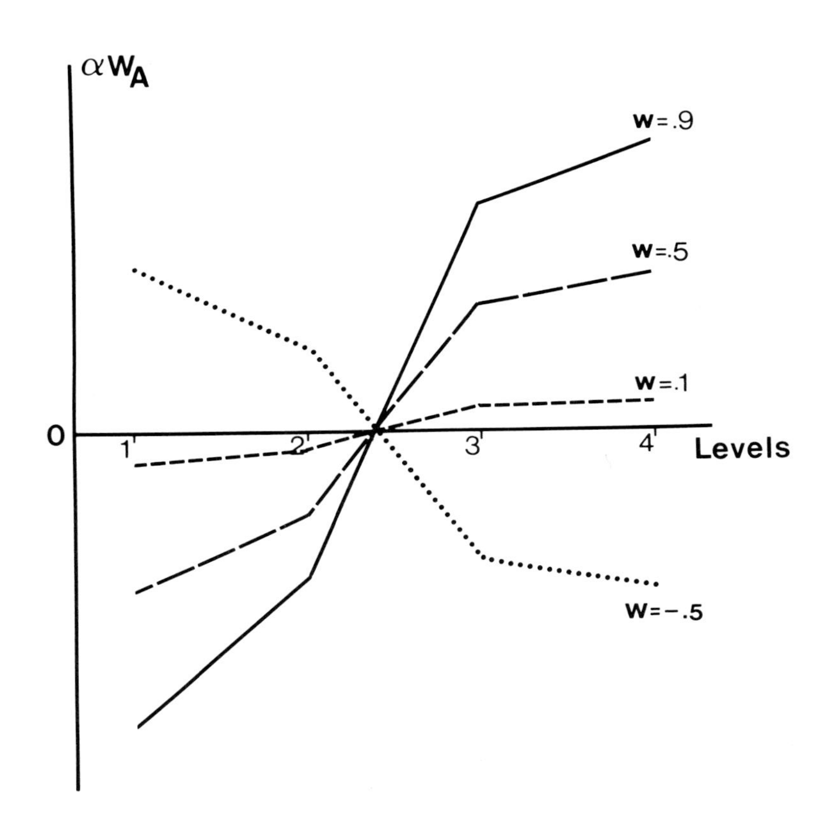

and vice versa.) All the curves in Figure 13–1 have essentially the same profile (shape) except that their scatters (Cronbach and Gleser 1953) are different. In fact, $|w_{kA}|$ may be interpreted as variance of α_{ki} over i. All the curves cross with one another whenever they cross a zero line. (They should cross the zero line at least once to meet restriction [13–12].) Note that Figure 13–1 was drawn for an ideal case in which the WAM fits the data perfectly. In practical situations, it is not possible to observe error-free data. Consequently, the proportionalities of α_{ki} and β_{kj} may hold only approximately.

Some basic properties of the WAM may be directly observed in the original data themselves, even without estimating α_{ki} and β_{kj}. In order to see this, let us assume some hypothetical parameter values and see if there is any characteristic pattern appearing in the model predictions. For simplicity, we assume $n_A = n_B = 3$; $K = 4$; $\alpha_1 = 1$, $\alpha_2 = 3$, and $\alpha_3 = 4$; and $\beta_1 = 2$, $\beta_2 = 3$, and $\beta_3 = 5$. The four sets of model predictions derived from the model are given in Figure 13–2. Note that the hypothesized values of the additive effects do not satisfy (13–11) or (13–12), but this should not matter for the purpose of examining properties of the WAM. These restrictions are merely conventional and the basic properties of the model remain unaffected even if they are violated. Note also that both α_i and β_j are monotonically increasing, but again, this is mere coincidence. In Figure 13–2, $w_{kA}\alpha_i = \alpha_{ki}$ and $w_{kB}\beta_j = \beta_{kj}$ are entered in row and column marginals of each table, respectively. Main entries in the table are the model predictions (y_{kij}) calculated by (13–10).

Figure 13–2. Examples of Predictions from the Weighted Additive Model (WAM)

$$\alpha = \begin{pmatrix} 1 \\ 3 \\ 4 \end{pmatrix}, \quad \beta = \begin{pmatrix} 2 \\ 3 \\ 5 \end{pmatrix}$$

	w_A	w_B
k=1	2	1
k=2	1	2
k=3	-1	-1
k=4	1	-2

k=1

$$\begin{array}{cc|ccccc} & & \multicolumn{5}{c}{w_{1A}\alpha} \\ & & 2 & < & 6 & < & 8 \\ \hline & 2 & 4 & < & 8 & < & 10 \\ & \wedge & \wedge & & \wedge & & \wedge \\ w_{1B}\beta & 3 & 5 & < & 9 & < & 11 \\ & \wedge & \wedge & & \wedge & & \wedge \\ & 4 & 7 & < & 11 & < & 13 \end{array}$$

k=2

$$\begin{array}{cc|ccccc} & & \multicolumn{5}{c}{w_{2A}\alpha} \\ & & 1 & < & 3 & < & 4 \\ \hline & 4 & 5 & < & 7 & < & 8 \\ & \wedge & \wedge & & \wedge & & \wedge \\ w_{2B}\beta & 6 & 7 & < & 9 & < & 10 \\ & \wedge & \wedge & & \wedge & & \wedge \\ & 10 & 11 & < & 13 & < & 14 \end{array}$$

k=3

$$\begin{array}{cc|ccccc} & & \multicolumn{5}{c}{w_{3A}\alpha} \\ & & -1 & > & -3 & > & -4 \\ \hline & -2 & -3 & > & -5 & > & -6 \\ & \vee & \vee & & \vee & & \vee \\ w_{3B}\beta & -3 & -4 & > & -6 & > & -7 \\ & \vee & \vee & & \vee & & \vee \\ & -5 & -6 & > & -8 & > & -9 \end{array}$$

k=4

$$\begin{array}{cc|ccccc} & & \multicolumn{5}{c}{w_{4A}\alpha} \\ & & 1 & < & 3 & < & 4 \\ \hline & -4 & -3 & < & -1 & < & 0 \\ & \vee & \vee & & \vee & & \vee \\ w_{4B}\beta & -6 & -5 & < & -3 & < & -2 \\ & \vee & \vee & & \vee & & \vee \\ & -10 & -9 & < & -7 & < & -6 \end{array}$$

Since the WAM is a special case of the SAM for each data set, all the properties satisfied by the WAM must be satisfied by the SAM for each k. Let us, therefore, first see what characteristic patterns may be observed for each data set. They should be the patterns we expect to observe for each data set to be represented by the SAM.

Let us start with a quantitative relation among y_{kij}. It can be immediately observed that:

$$y_{kij} = \bar{y}_{ki.} + \bar{y}_{k.j} - \bar{y}_{k..} \qquad \text{for all } i \text{ and } j,$$

where

$$\bar{y}_{ki.} = \sum_{j}^{n_B} y_{kij}/n_B ,$$

$$\bar{y}_{k.j} = \sum_{i}^{n_A} y_{kij}/n_A ,$$

and

$$\bar{y}_{k..} = \sum_{i}^{n_A} y_{ki.}/n_A = \sum_{j}^{n_B} y_{k.j}/n_B = \sum_{i}^{n_A}\sum_{j}^{n_B} y_{kij}/n_A n_B .$$

Not surprisingly, this is equivalent to the noninteraction condition in two-way ANOVA. This condition states that the combined effect (y_{kij}) of the ith level of factor A and the jth level of factor B is a simple addition of a contribution from the ith level of factor A ($\bar{y}_{ki.}$) and a contribution from the jth level of factor B ($\bar{y}_{k.j}$). ($\bar{y}_{k..} = 0$ if α_i and β_j have satisfied [13–12].)

If the observations are linearly related to the model predictions (that is, if they are obtained on an interval scale), the above condition should still be true approximately, if not exactly. However, it does not provide a valid criterion for an additive representation if the data are only monotonically related to the model predictions.

Are there any patterns in the table that are invariant over any monotonic distortion of the model predictions? Observe that for each k, $w_{kA}\alpha_i$ is either ascending or descending, depending on the sign of w_{kA}. The same is true for $w_{kB}\beta_j$. What is more important is that the model predictions are in the same order as $w_{kA}\alpha_i$ within each row, and they are in the same order as $w_{kB}\beta_j$ within each column. That is, if

$$y_{kij} \geqq y_{kij'} ,$$

then

$$y_{ki'j} \geqq y_{ki'j'} \qquad \text{for all } i'.$$

And if

$$y_{kij} \geqq y_{ki'j} ,$$

then

$$y_{kij'} \geqq y_{ki'j'} \qquad \text{for all } j',$$

for each k.

In the additive conjoint measurement literature (Luce and Tukey 1964; Krantz et al. 1971), the above condition applied to a set of observations is called the *independence* or *single cancellation condition*. It is one of necessary conditions for an additive representation of ordinal data.

Another necessary condition, called the *Thomsen condition*, may not be obvious in the table. This condition states that if

$$y_{kij} = y_{kpq} \quad \text{and} \quad y_{kpr} = y_{ksj},$$

then

$$y_{kir} = y_{ksq},$$

in terms of the model predictions. This is illustrated in Figure 13–3, where indifference relations in the premise are connected by solid line segments and the indifference relation in the conclusion by a dotted line segment. Unfortunately, it is difficult to verify this condition directly, since we may not be able to find stimulus pairs—(i,j), (p,q) and (p,r), (s,j)—such that they satisfy the indifference relations postulated in the premise.

The independence and the Thomsen conditions, along with other conditions not described here (including weak-ordering,

Figure 13–3. Illustration of the Thomsen Condition

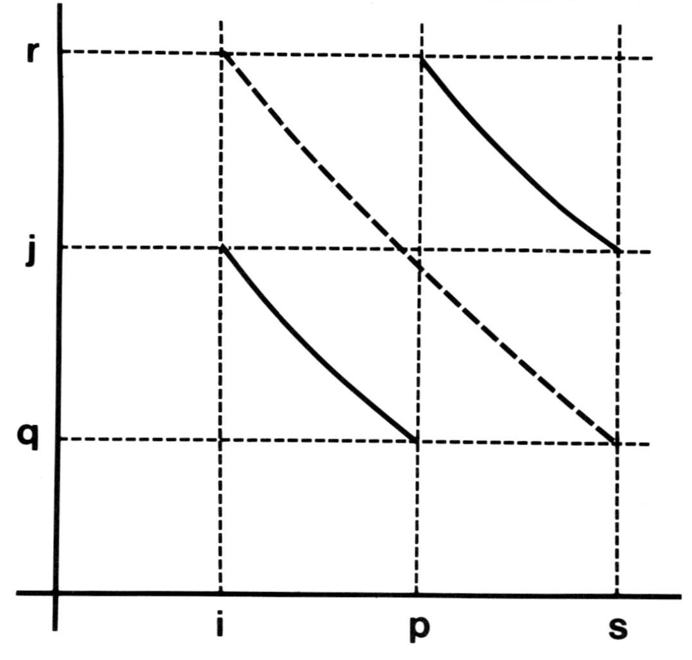

essentialness, solvability, and so on) ensure a representation of ordinal data by the SAM. Those are the qualitative properties that have to be satisfied in order to be representable by the SAM. Since actual data usually contain errors, it may not be easy to determine if these properties are met to a satisfactory degree. Nevertheless, they should serve as rough guidelines.

The additive representation in turn implies the double-cancellation condition, which is just the Thomsen condition with "=" replaced by "≧." (The Thomsen condition is actually a special case of the double-cancellation condition; both "≧" and "≦" imply "=".) The double-cancellation condition can be directly tested. We see that it is satisfied by the model predictions, y_{kij}. However, even though the double-cancellation condition is satisfied by a particular data set, this does not imply that the Thomsen condition is satisfied, although the former is stronger than the latter. This is because the data are always finite and a particular condition being satisfied by *available* data points does not ensure that it is satisfied by *all conceivable* data points. This means that we may falsify a condition by observing an instance of its violation, but we can never completely verify the condition based on a finite sample of observations. (This is similar to the statistical hypothesis testing situation in which one can never say the null hypothesis is true.)

The WAM requires more than is required by the SAM. Let us again begin with a quantitative property and then relax it into ordinal properties. We observe that:

$$(\bar{y}_{ki.} - \bar{y}_{k..})/(\bar{y}_{k'i.} - \bar{y}_{k'..}) = \text{constant for all } i,$$

and

$$(\bar{y}_{k.j} - \bar{y}_{k..})/(\bar{y}_{k'.j} - \bar{y}_{k'..}) = \text{constant for all } j,$$

which are analogous to (13–15). Since a column or row mean minus a grand mean is an estimate of α_{ki} (or β_{kj}), the proportionalities of the additive effects obtained from the SAM imply the above condition. (Again, $\bar{y}_{k..} = \bar{y}_{k'..} = 0$ if [13–12] were satisfied. In this case, the above condition would be much simplified.)

The condition should be relaxed for ordinal data. Looking at Figure 13–2 across tables, we may observe that the ordering among the model predictions within each row or column is either exactly the same or opposite across columns or rows, depending on whether the signs of the weights are in agreement or in disagreement. For example, $4 < 8 < 10$ in the first row of $k = 1$, $5 < 7 < 8$ in the first row of $k = 2$ (same order), and $-3 > -5 > -7$ in the first row of $k = 3$ (reverse order). Thus, it may be stated that if

$$y_{kij} \geqq y_{kij'},$$

then either

$$y_{k'ij} \geqq y_{k'ij'} \qquad \text{or} \qquad y_{k'ij} \leqq y_{k'ij'}.$$

And if

$$y_{kij} \geqq y_{ki'j} \quad ,$$

then either

$$y_{k'ij} \geqq y_{k'i'j} \qquad \text{or} \qquad y_{k'ij} \leqq y_{k'i'j} \quad .$$

The above condition, stated in terms of observed data (instead of model predictions), is called *homogeneous ordering* (Sayeki 1972), and is one of the necessary properties of the WAM. This condition is analogous to the independence condition for the SAM.

Another necessary condition, which parallels the Thomsen condition for the SAM, is a version of the *Reidemeister condition* (Krantz et al. 1971), a special case of the triple-cancellation axiom. This condition states that if

$$y_{kij} = y_{kpq} \ ,$$

$$y_{kir} = y_{kpt} \ ,$$

and

$$y_{k'sr} = y_{k'ut} \ ,$$

then

$$y_{k'sj} = y_{k'uq} \ ,$$

as depicted in Figure 13–4. Again, indifferences in the premise are connected by a solid line and the indifference in the conclusion is connected by a dotted line. For the same reason that the Thomsen condition was difficult to verify, this condition may not be testable in practical situations. However, it has been proven (Sayeki 1972; Marley 1970) that the homogeneous ordering and the Reidemeister conditions together with other structural axioms ensure an existence of a representation of ordinal data by the WAM. Just as the SAM implied the double-cancellation condition, the WAM implies the triple-cancellation condition (which is the Reidemeister condition with "=" replaced by "≧"). This condition, as well as the homogeneous ordering condition, can be directly tested with observed data.

Geometric Properties

When n_A levels of factor A and n_B levels of factor B are factorially combined to construct a set of stimuli, we obtain $n_A \times n_B$ stimuli altogether and consequently as many model predictions for each k. These $n_A n_B$ model predictions can be put into a vector form according to a certain prescribed order. This vector may be denoted by $\mathbf{y}_k$ $(k = 1, \ldots, K)$. Let $G_A(n_A n_B \times n_A)$ and $G_B(n_A n_B \times n_B)$ denote design matrices for factors A and B, respectively. That is, the rth row of G_A has one in the ith column and zeroes elsewhere if the rth element of $\mathbf{y}_k$ (y_{kr}) is related to

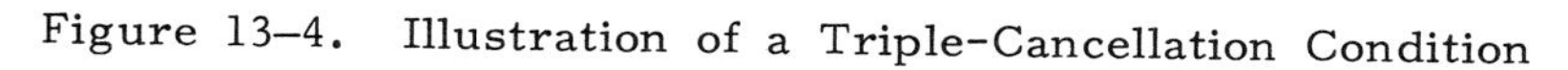

Figure 13–4. Illustration of a Triple-Cancellation Condition

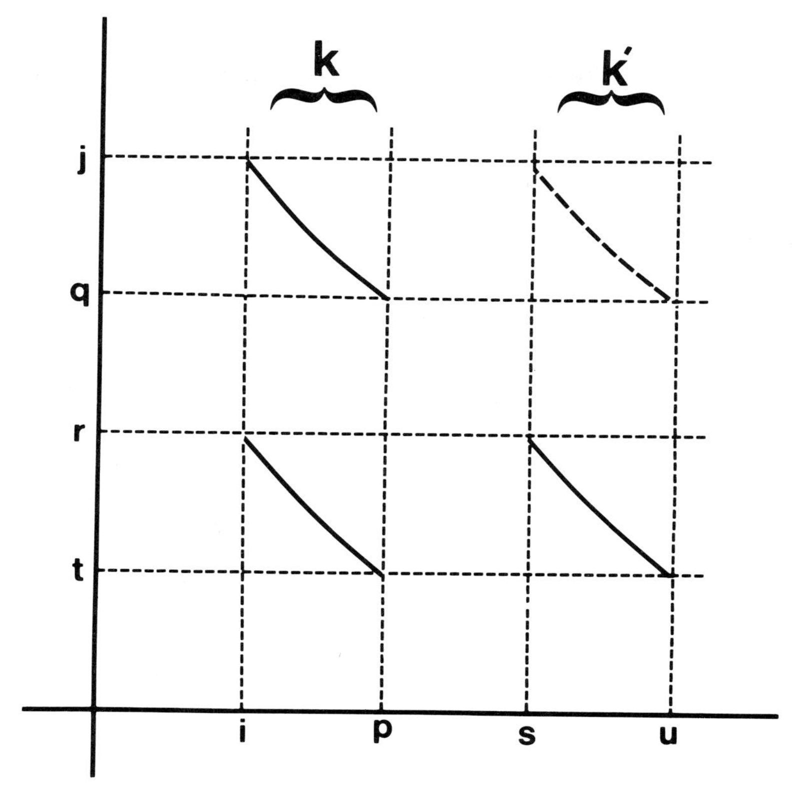

the ith level of factor A. Similarly, the rth row of G_B has one in the jth column and zeroes elsewhere if y_{kr} comes from the jth level of factor B. Let α and β be vectors of the additive effects (α_i and β_j) for factors A and B, respectively. Then the entire set of model predictions can be expressed as:

$$\mathbf{y}_k = G_A \alpha w_{kA} + G_B \beta w_{kB} \, , \quad (k = 1, \ldots, K) \, . \qquad (13\text{–}16)$$

The above equation indicates that $\mathbf{y}_k$ is the weighted sum of two vectors, $G_A\alpha$ and $G_B\beta$, the former being weighted by w_{kA} and the latter by w_{kB} ($k = 1, \ldots, K$). The two constituent vectors remain the same across k, while the weights applied to the vectors change over k. This implies, geometrically, that all $\mathbf{y}_k$'s ($k = 1, \ldots, K$) lie in the space spanned by the two constituent vectors, $G_A\alpha$ and $G_B\beta$ (which themselves are linear combinations of G_A and G_B, respectively), their directions being subject to change as functions of the weights. The situation is depicted in Figure 13–5 for the case of two criteria (k and k'). In order to obtain $\mathbf{y}_k$, $G_A\alpha$ is shrunk or stretched by w_{kA} and $G_B\beta$ by w_{kB} before being added together. Similarly, $G_A\alpha$ is shrunk by $w_{k'A}$ and $G_B\beta$ by $w_{k'B}$ to obtain $\mathbf{y}_{k'}$. This explains the mechanism that generates two or more distinct vectors of model predictions from

Figure 13–5. Geometric Interpretation of the WAM

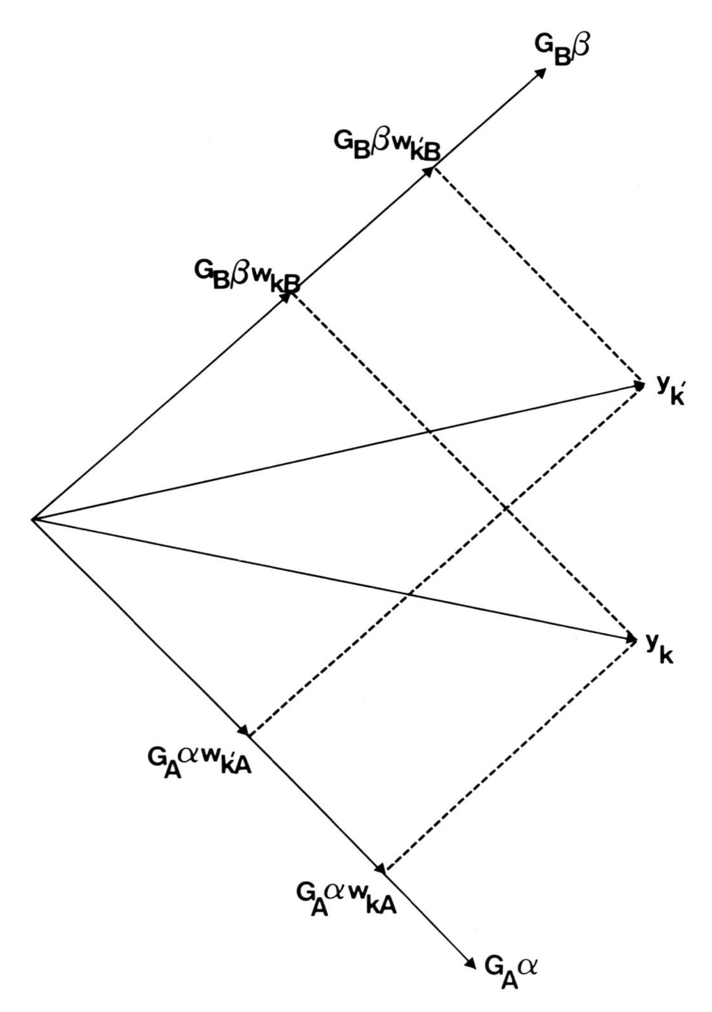

the same attribute vectors in the WAM.

The geometrical relationship between the WAM and the SAM may be explored in a similar manner. For this, the SAM may be written in a vector form analogous to (13–16):

$$\mathbf{y}_k = G_A\alpha_k + G_B\beta_k \;, \qquad (k = 1, \ldots, K) \;. \qquad (13\text{–}17)$$

Let us suppose $\mathbf{y}_k$ ($k = 1, \ldots, K$) is perfectly fit by the SAM but not by the WAM. (If it can be perfectly fit by the WAM, it can be perfectly fit by the SAM separately, but not vice versa.) The relationship between $G_A\alpha_k$ and $G_A\alpha\, w_{kA}$ is depicted in Figure 13–6. The vector, $\mathbf{G}_A\alpha w_{kA}$, represents the portion of $\mathbf{G}_A\alpha_k$ that can be explained by the WAM. In other words, $G_A\alpha_k$ is decom-

Figure 13–6. The Relationship between the WAM and the SAM

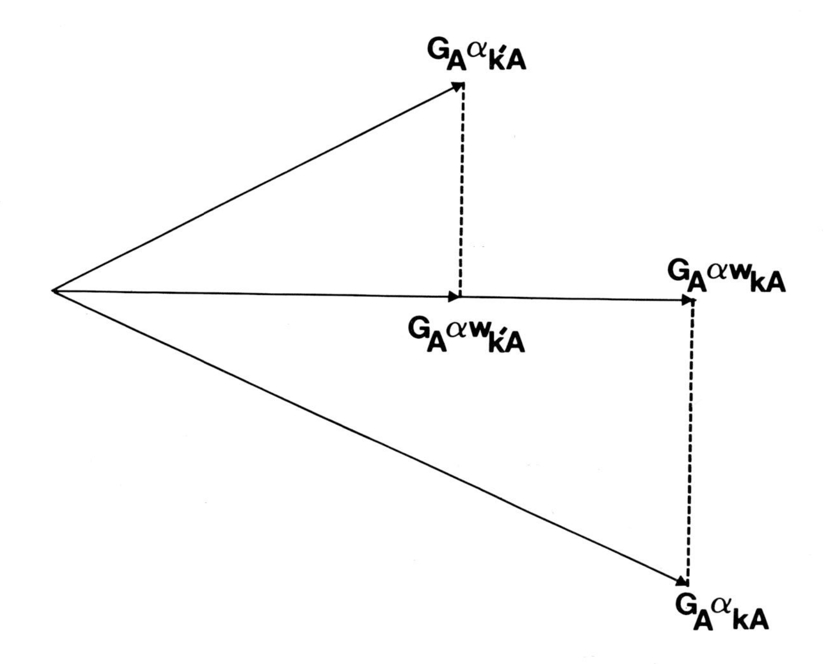

posed into two parts: $G_A\alpha w_{kA}$, the part that can be accounted for by the WAM; and $G_A\alpha_k - G_A\alpha w_{kA}$, the part that cannot be accounted for by the WAM. Furthermore, $G_A\alpha w_{kA}$ and $G_A\alpha_k - G_A\alpha w_{kA}$ are mutually orthogonal so that:

$$SS(G_A\alpha_k) = SS(G_A\alpha w_{kA}) + SS(G_A\alpha_k - G_A\alpha w_{kA}) , \qquad (13\text{–}18)$$

where SS(**v**) is the sum of squares of elements of vector **v**. The same relation holds for all k' and also for other factors in the models.

When there are only two levels in each factor, the WAM and the SAM are compatible in every way. In this case, α_k can always be written as αw_{kA}.

Relations to Other Models

We have explained the relationship between the WAM and the SAM. We have also mentioned that when the independent variables are completely specified—that is, when they are a priori quantified—the WAM reduces to regression analysis (linear or monotonic, depending on the scale level of the dependent variables). Examples of this class of models are Carroll's (1972) external unfolding analysis and Srinivasan and Shocker's (1973) composite criterion model.

The idea of describing differences between sets of observations by a differential weighting of additive effects is apparently not new. In fact, most of the multivariate data analysis methods can be thought of as techniques to find various optimal linear combinations of variables according to some prescribed criteria (McDonald 1968). In particular, when the independent variables (factors) are completely unspecified in the WAM, we obtain the bilinear model

$$y_{kr} = \sum_{t=1}^{T} w_{kt} f_{rt} , \qquad (13\text{–}19)$$

where y_{kr} is the model prediction for the kth observation set (usually subject k) on variable r, f_{rt} is the coordinate (conventionally called factor loading) of variable r on factor t, and w_{ka} is the weight (or factor score) of subject k on factor a. Since f_{rt} and w_{kt} are assumed to be unknown, in this case, both w_{kt} and f_{rt} should be estimated under some identifiability constraints. The number of factors, T, is the minimum number of hypothetical attributes (latent traits, factors) necessary to approximate the observed data to a sufficient degree and is also to be determined in the course of data analysis. Under a specialized set of identifiability constraints, the bilinear model is called the *principal component analysis*. In a sense, f_{rt} is a weight attached to w_{kt} (responsible for the variation of y_{kr} over r), just as w_{kt} is a weight attached to f_{rt} (responsible for the variation of y_{kr} over k). The roles of w_{kt} and f_{rt} are dual in this sense.

A special case of (13–19) has been proposed by Carroll and Chang (1970) in order to describe subject differences in dissimilarity judgments. The proposed model—called INDSCAL—is the weighted distance model, which is written as:

$$d_{kij}^2 = \sum_{t=1}^{T} w_{kt} (x_{it} - x_{jt})^2 , \qquad (13\text{–}20)$$

where d_{kij} is the Euclidean distance between stimuli i and j for subject k, x_{it} is the coordinate of stimulus i on dimension t, and T is the dimensionality of the space in which a set of stimuli are represented. That (13–20) is a special case of (13–19) can be readily seen by noting that

$$f_{rt} = (x_{it} - x_{jt})^2 , \qquad (13\text{–}21)$$

where stimulus pair (i,j) is indexed by r. That is, INDSCAL is a special type of the WAM in which squares of the dimensionwise differences between two coordinates are taken as additive effects. The fact that (13–20) is a special case of the WAM provides an interesting application of the WAM to similarity and dissimilarity data. In (13–20), the weights are usually restricted to be nonnegative. This is because both $(x_{it} - x_{jt})^2$, $(t = 1, \ldots, T)$ and d_{kij}^2 are nonnegative. A minor generalization of (13–20) is

$$d^p_{kij} = \sum_{t=1}^{T} w_{kt} |x_{it} - x_{jt}|^p , \qquad (13\text{–}22)$$

where p is the Minkowski power $(1 \leqq p)$.

When f_{rt} in (13–19) is constrained to be

$$f_{rt} = \sum_{q_t} h_{rq_t} \alpha_{q_t} ,$$

we obtain (13–3), which is the general form of the WAM.

PARAMETER ESTIMATION

In this section, we describe methods for parameter estimation in the WAM. In the previous section, the basic properties of the WAM were explored strictly within the context of the model. However, data analysis is an interplay between the model and the data. Thus, we first need to discuss various data characteristics to be taken into account in developing parameter estimation procedures.

The Data

The necessity of discussing data characteristics stems from the fact that the model predictions are usually not directly observable. The observed data typically contain a sizable amount of errors. In addition, some systematic distortions may be under effect in a measurement process. For example, if the data are collected by a rating scale, they may be distorted monotonically (relative to the model predictions) towards the ends of the scale due to possible floor and ceiling effects. The observed data in this case will not possess interval-scale properties. In this small section, we briefly discuss some of the data characteristics to be considered in fitting the model. These considerations are not restricted to the WAM, and more comprehensive treatments of this topic can be found in Young (1981) and Takane (1982).

We first discuss the distinction made in terms of scale levels of measurement. (This is basically the distinction made by WADDALS, a least-squares procedure for fitting the WAM.) The simplest case is the one in which the data are obtained on an interval scale or higher. In this case, it is assumed that y_{kij} is only error-perturbed and that no systematic distortion is exerted on the error-perturbed model prediction. That is,

$$y^*_{kij} = y_{kij} + e_{kij} , \qquad (13\text{–}23)$$

where e_{kij} is an error random variable and y^*_{kij} is directly observable. The model fitting in this case only involves the estimation of parameters in the WAM (namely, the additive effects and the differential weights). Most often, the estimation is accomplished by minimizing the overall size of e_{kij} (that is,

$\Sigma\, e_{kij}^2$) without further assuming any distributional properties on e_{kij}.

The interval information contained in y_{kij}^* may get lost to varying degrees in an observation process. Some monotonic distortion may be in effect. For example, the subject may report only rank orders among y_{kij}^*. Thus, only ordinal information is preserved. The data are said to be measured on an ordinal scale in this case. The ordinal data have to be monotonically transformed back to y_{kij}^* before they can be subjected to a representation by the model. Specifically,

$$f_m(o_{kij}) = y_{kij}^* = y_{kij} + e_{kij} \quad , \tag{13–24}$$

where o_{kij} is the observed data and f_m is some monotonic function. The model fitting in this case involves a monotonic transformation of the observed data as well as the model estimation. Again this is most often done by minimizing the overall size of e_{kij}. The WADDALS algorithm to be described in the next section is primarily aimed at ordinal data, although it can be used for other data as well. The ordinal scale assumption is sufficiently flexible, so that almost any data can be analyzed under this assumption (with an exception to be described shortly). If the data happen to be measured at a higher scale level, an approximately linear transformation is usually found for f_m, which is indeed the correct transformation in this case. One should note, however, that some efficiency is bound to get lost if one makes a weaker measurement assumption than needs to be made.

In certain extreme cases, only nominal information may be obtained. At the nominal scale level, two observations are either indifferent or distinct. For example, we may ask the subject to classify y_{kij}^* into several unordered categories. Observations that are in the same category are considered indifferent, while those in different categories are considered distinct. Nominal observation categories have to be quantified before the data can be represented by the model. This quantification is sometimes called a *nominal transformation* and can be done just as the levels of the additive factors are quantified in the WAM.

In all the above cases, the data can be arranged in a matrix form for each set (assuming that there are only two factors) in the same way as the model predictions are arranged in Figure 13–2. In the interval case, y_{kij} is replaced by y_{kij}^*, and in the ordinal case, it is replaced by o_{kij} (which is assumed monotonically related to y_{kij}^*). In the nominal case, it is replaced by the category number into which y_{kij}^* is classified. Multiple sets of observations, each arranged in a matrix form, compose a multimode data set.

The data characteristics discussed above are based on the scale levels of measurement. However, in certain situations, other characteristics of the data may be more meaningful. For example, there are many different ways of collecting the data, and each method may require a different mental process by which the observed data are generated. If a model for this process can be constructed, the observed data can be more directly related to its unobservable counterpart, the model. Takane (1982) incorporated a specific process model for each of three representative types of data collection methods—categorical ratings, pair comparisons, and directional rank orders—in fitting the additive model. We briefly discuss each of these data collection methods in turn.

In the categorical rating method, the subject is asked to rate y^*_{kij} on a rating scale having a relative small number of observation categories (say, up to 7 or 9). Thus, only category membership of y^*_{kij} is observed. Since categories are usually ordered, the data may be analyzed as mere ordinal data. Alternatively, a specific process model of categorical judgments may be incorporated. Through this model, and under some distributional assumption on e_{kij}, the probability of a certain categorical observation can be derived as a function of model parameters. The categorical rating data may be represented in a matrix form with each element in a matrix representing a category number. When many replicated observations are made for each y^*_{kij}, it may be economical to represent the data in a frequency form, in which observed frequencies of categorical observations are given along with stimulus indices, i and j, and k.

In the pair comparison method, the subject is presented two stimuli at a time and asked to judge which one "dominates" the other according to some prescribed criterion. Thus, only pairwise ordinal information about y^*_{kij} and $y^*_{ki'j'}$ (that is, either $y^*_{kij} > y^*_{ki'j'}$ or $y^*_{kij} < y^*_{ki'j'}$) is obtained. The pair comparison data may be analyzed as partial rank-order data. On the other hand, a specific process model for pair comparison judgments may be constructed. The probability of $y^*_{kij} > y^*_{ki'j'}$ or $y^*_{kij} < y^*_{ki'j'}$ can then be derived as a function of model parameters. The pair comparison data cannot be represented in a matrix form, since two stimuli are involved in each judgment. This type of data can be represented most naturally in a frequency form in which frequencies of comparative judgments are provided along with stimulus indices, i and j, and k.

In the directional rank-order method, the subject is asked to rank order a set of stimuli in a specified direction—either from the largest to the smallest, or vice versa. All stimuli generated by factorial combinations of the levels in all factors may be rank ordered in this way. It is important that the ranking is made in a specified direction. Otherwise, it is extremely difficult to construct a process model that relates y^*_{kij} to an observed ranking. Under the directionality of the ranking process, the probability of observed rank orders can be derived in a relatively straightforward manner. The directional rank-order data may be represented in a matrix form with rank numbers as entries of matrices. The frequency representation is usually not convenient, since almost all rankings are unique, unless the number of stimuli to be rank ordered is relatively small. The directional rank-order data can also be analyzed as mere ordinal data.

A category response model may also be constructed for nominal data. For example, the log-linear model (Andersen 1980; Bishop, Fienberg, and Holland 1975; Bock 1975; Goodman 1978) for analysis of contingency tables can be thought of as the SAM with the logistic response function (Luce's [1959] choice model) and the multinomial probability model for unordered categorical responses. It may be interesting and worthwhile to extend the log-linear model using the WAM. (See the application section for more detail.)

A Brief Overview of the Existing Methods for Additivity Analysis

A number of estimation procedures are available for the SAM. When the data are measured on an interval scale or higher, the SAM reduces to an ANOVA model without interaction terms, which is in turn a special case of the general linear regression model with dummy independent variables (Draper and Smith 1981). A number of procedures have been developed for ordinal data—MONANOVA (Kruskal 1965), ADDIT (Roskam 1968), ADDALS (de Leeuw, Young, and Takane 1976), and MORALS (Young, de Leeuw, and Takane 1976)—which all use Kruskal's (1964a, 1964b) least-squares monotonic transformation as part of their algorithm. CM-I (Conjoint Measurement I) by Lingoes (1973), on the other hand, uses Guttman's rank-image transformation (Guttman 1968). A procedure by Ramsay (1977) and the one by Winsberg and Ramsay (1980) use smoother transformations such as power and spline transformations. The nominal data lead to a special case of discriminant analysis in which independent variables are also categorical. Hayashi's (1952) Quantification Method II and Carroll's (1973) CCM (Categorical Conjoint Measurement) have been developed specifically for this situation. However, by appropriate codings of dummy variables, the same analysis could be performed by canonical correlation analysis.

Fitting procedures are much more limited for the WAM. To the best of my knowledge, WADDALS and MAXADD (Takane et al. 1980; Takane 1982) are the only two that are available at the moment. MAXLIN is now under development, which generalizes MAXADD to the general weighted linear (not restricted to additive) model. De Sarbo, Carroll, Lehmann, and O'Shaughnessy (forthcoming) recently proposed a three-way multivariate conjoint analysis method, which, as its special case, subsumes the WAM. Their procedure, however, is restricted to interval-scale data.

In this section, we discuss WADDALS and MAXADD in some detail. An excellent review of the algorithms for the SAM has been provided by Rao (1977).

ADDALS

As an introduction to WADDALS, we first discuss the ADDALS algorithm (de Leeuw et al. 1976; Young 1981) for the SAM. WADDALS works within the same algorithmic framework as ADDALS, known as alternating least-squares.

Let us write the SAM in a vector form for a single set of observations, analogous to (13–16) and (13–17). Let $\mathbf{y}$ denote the vector of model predictions:

$$\mathbf{y} = G_A\alpha + G_B\beta \; . \qquad (13\text{–}25)$$

This vector is fitted to the vector of error-perturbed model predictions, which is denoted by $\mathbf{y}^*$. We primarily focus on the case in which $\mathbf{y}^*$ is measured on an ordinal scale. (Later, we will discuss whatever modifications are necessary for the nominal case.) We would like to estimate both model parameters—α and β—and a monotonic function—f_m in (13–24)—in such a way that the discrepancy between $\mathbf{y}^*$ and y is a minimum. This may be expressed as:

$$\phi = (\mathbf{y}^* - \mathbf{y})'(\mathbf{y}^* - \mathbf{y})/\mathbf{y}^{*\prime}\mathbf{y}^* , \qquad (13\text{–}26)$$

which is to be minimized with respect to α, β, and $\mathbf{y}^*$, where $\mathbf{y}^*$ is monotonically related to $\mathbf{o}$. ADDALS minimizes this by alternating the following two least-squares steps, until successive values of ϕ exhibit no significant improvement. The positive square root of ϕ is called *stress*.

Step 1. *Model Estimation:*
Obtain $\mathbf{y}$ *that minimizes* ϕ *for fixed* $\mathbf{y}^*$ *(such that* $\mathbf{y}^{*\prime}\mathbf{y}^* = 1$*).*

Step 2. *Optimal Scaling:*
Obtain $\mathbf{y}^*$ *that minimizes the numerator of* ϕ *for fixed* $\mathbf{y}$*, and normalize* $\mathbf{y}^*$ *so that* $\mathbf{y}^{*\prime}\mathbf{y}^* = 1$*.*

In the initial iteration, $\mathbf{o}$ is centered and normalized and used for $\mathbf{y}^*$. In Step 1, $\mathbf{y}$ is obtained by obtaining the least-squares estimates of α and β, which are:

$$\hat{\alpha} = (G'_A G_A)^{-1} G'_A \mathbf{y}^* \qquad (13\text{–}27)$$

and

$$\hat{\beta} = (G'_B G_B)^{-1} G'_B \mathbf{y}^* .$$

If the elements $\mathbf{y}^*$ are rearranged in a matrix form so that its (i,j)th element is y^*_{ij} corresponding to y_{ij} in (13–13), $\hat{\alpha}$ and $\hat{\beta}$ are just the vectors of row and column means, respectively, of this matrix. The least-squares estimate $\mathbf{y}^*$ in Step 2 is obtained by applying Kruskal's least-squares monotonic regression algorithm (Kruskal 1964b) to $\mathbf{y}$.

The above iteration scheme has been proven to be convergent (de Leeuw et al. 1976). This may be easily seen by noting that ϕ can never get larger, since conditional least-squares estimates are obtained in each step. Furthermore, ϕ is bounded below ($\phi \geqq 0$). By a standard theorem, in functional analysis, a monotonically decreasing sequence is bound to converge. The convergence point may not be a local minimum of ϕ in general, but this is guaranteed by the continuity of ϕ (Kruskal 1971).

That Step 2 really minimizes ϕ with respect to $\mathbf{y}^*$ is assured by the following theorem, which is stated without proof:

THEOREM 13–1. *On the minimization of normalized least-squares criteria.*

Let $Q(\mathbf{a}) = \mathbf{a}'\mathbf{a}$,

$$\phi^*(\mathbf{x}_1, \mathbf{x}_2) = Q[\mathbf{y}(\mathbf{x}_1) - \mathbf{z}(\mathbf{x}_2)],$$

$$\phi_y(\mathbf{x}_1, \mathbf{x}_2) = \phi^*(\mathbf{x}_1, \mathbf{x}_2)/Q[\mathbf{y}(\mathbf{x}_1)],$$

and

$$\phi_z(\mathbf{x}_1, \mathbf{x}_2) = \phi^*(\mathbf{x}_1, \mathbf{x}_2)/Q[\mathbf{z}(\mathbf{x}_2)].$$

1. *Let* $\mathbf{x}_1^{(1)}$ *and* $\mathbf{x}_2^{(1)}$ *minimize* ϕ_y.
2. *Let* $\mathbf{x}_1^{(2)}$ *and* $\mathbf{x}_2^{(2)}$ *minimize* ϕ^* *subject to the restriction that* $Q[\mathbf{y}(\mathbf{x}_1^{(2)})] = 1$.
3. *Let* $\mathbf{x}_1^{(3)}$ *and* $\mathbf{x}_2^{(3)}$ *minimize* ϕ^* *subject to the restriction that* $Q[\mathbf{z}(\mathbf{x}_2^{(3)})] = 1$.
4. *Let* $\mathbf{x}_1^{(4)}$ *and* $\mathbf{x}_2^{(4)}$ *minimize* ϕ_z.

The four sets of solutions are essentially equivalent. That is, $\mathbf{x}_1^{(1)} \sim \mathbf{x}_1^{(4)}$ *are proportional to each other, and so are* $\mathbf{x}_2^{(1)} \sim \mathbf{x}_2^{(4)}$.

The denominator of ϕ in (13–26) is just a normalization factor for preventing ϕ from converging to 0 by shrinking the size of $\mathbf{y}$ and $\mathbf{y}^*$. The above theorem shows that it does not matter by which, $\mathbf{y}$ or $\mathbf{y}^*$, the normalization is done and that the normalization is effected by first obtaining $\mathbf{y}^*$, which minimizes the numerator of ϕ, and then by actually normalizing this $\mathbf{y}^*$ so that $\mathbf{y}^{*\prime}\mathbf{y}^* = 1$.

The above algorithm is not completely free from difficulties. There are problems of convergence to a nonglobal minimum, nonuniqueness of the minimum reached, and a possible degeneracy of solutions. These problems are not restricted to the ADDALS algorithm but are discussed here only because it happens to be the first algorithm discussed in this chapter.

The convergence point attained by ADDALS may not be the global minimum of ϕ. Only a convergence to a local minimum is ensured. There is no definitive way to get over this problem. Good initial starts often help, but there is no guarantee that they always work. It is thus advisable to obtain several solutions from different initial starts, and if they do not all agree, choose the one that gives the smallest value of ϕ.

No general conditions have been established for the uniqueness of the parameter estimates. The local minimum attained may not be a strong minimum but merely a weak local minimum in the sense that we may find $\hat{\mathbf{y}}^*$ and $\hat{\mathbf{y}}$, such that $\phi(\mathbf{y}^*, \mathbf{y}) = \phi(\hat{\mathbf{y}}^*, \hat{\mathbf{y}})$ sufficiently near $\mathbf{y}^*$ and $\mathbf{y}$, but $\hat{\mathbf{y}}^* \neq \mathbf{y}^*$ and $\hat{\mathbf{y}} \neq \mathbf{y}$. A general recommendation is to obtain many observations. With an increasing number of observations (relative to the number of parameters to be estimated), the seriousness of the problem decreases. As more observations are made, the region of $\mathbf{y}^*$ and $\mathbf{y}$, in which ϕ

is equal, gets smaller in size, until it can be virtually considered as a unique point. The best way to obtain more observations, without increasing the number of model parameters, is to obtain replicated observations.

A solution is said to be degenerate when $\phi = 0$ and the elements of $\mathbf{y}^*$ split into two or more equivalence groups. A degenerate solution occurs when rows or columns are not ordinally connected in the following sense:

DEFINITION 13–1. *Order Connectedness*

Rows (columns) of a data matrix are said to be ordinally connected when in each row (column) there is at least one element that is larger than at least one element in the other rows (columns) and there is at least one element that is smaller than at least one element in the other rows (columns).

Rows and columns cannot be unconnected simultaneously. The following simple numerical example (3 × 3) shows what happens when rows or columns are unconnected:

9	8	7
6	5	3
4	2	1

Rows of this data split into two groups between the first and second rows (indicated by the line). Rows are unconnected here. These data can be perfectly fit by:

		α_i	
	0	0	0
2	2	2	2
β_j			
-1	-1	-1	-1
-1	-1	-1	-1

The model predictions split into two groups in the same way as the data split. Since monotonicity is perfectly satisfied in this case, we obtain $\phi = 0$. Although the fit is perfect, the solution is trivial.

Note that the above data satisfy the independence condition, one of the necessary conditions for an additive representation. In fact, "too clean" data (the data that can be perfectly represented by the additive model) are less likely to be ordinally connected than moderately noisy data and are thus not particularly suitable for ordinal additivity analysis. For ADDALS to work it is necessary that the data contain a moderate degree of error. The best way to avoid the degenerate solution is to increase the number of levels so that the levels are sufficiently close to generate "confusions" among rows and columns. Replicated observations are also helpful in creating ordinal connectedness in the data.

Some modifications are necessary to the standard ADDALS algorithm when the design is not balanced and/or the data are not ordinal. When there are missing observations, we may make the design artificially balanced by assigning them an arbitrary value (say, zero) initially and by assigning model predictions subsequently. This treatment is not described in the original ADDALS paper (de Leeuw et al. 1976), but this approach has been taken in the WADDALS algorithm.

When the data are nominal, the quantification of the observation categories may be expressed as:

$$\mathbf{y}^* = \mathbf{S}\boldsymbol{\Theta} \quad , \tag{13–28}$$

where S is a matrix of dummy variables. Specifically, $S = [s_{ij}]$ where

$$s_{ij} = \begin{cases} 1, & \text{if stimulus } i \text{ is responded by category } j, \\ 0, & \text{otherwise,} \end{cases}$$

and $\boldsymbol{\Theta}$ is the quantification vector of the observation categories. The least-squares estimate of $\boldsymbol{\Theta}$ is obtained by:

$$\hat{\boldsymbol{\Theta}} = (\mathbf{S}'\mathbf{S})^{-1}\mathbf{S}'\mathbf{y} \; . \tag{13–29}$$

Kruskal's monotonic regression algorithm in Step 2 should be replaced by (13–28) with $\hat{\boldsymbol{\Theta}}$ obtained by (13–29). The iterative procedure in this case is known to be equivalent to obtaining eigenvectors of a certain matrix, which is exactly what is done by Quantification Method II or CCM.

WADDALS

Having discussed the ADDALS algorithm in the previous section, there is only one more step necessary to arrive at the WADDALS algorithm (Takane et al. 1980), in which an additional set of parameters, namely the differential weights, have to be estimated. WADDALS minimizes the least-squares criterion:

$$\phi_W = \sum [(\mathbf{y}_k^* - \mathbf{y}_k)'(\mathbf{y}_k^* - \mathbf{y}_k)/\mathbf{y}_k^{*\prime}\mathbf{y}_k^*] \; . \tag{13–30}$$

Note that ϕ_W is essentially the sum of individual least-squares criteria, like the one for ADDALS ([13–26]), summed over k. The positive square root of mean ϕ_W ($\sqrt{\phi_W/K}$) is called *stress* in WADDALS. The WADDALS algorithm is again based on the alternating least-squares principle. It minimizes ϕ_W alternately with respect to model parameters (α, β, w_{kA}, w_{kB}) and with respect to data transformations ($\mathbf{y}_k^*$). The model estimation phase now consists of two subphases, one estimating the weights and the other estimating the additive effects. So the basic iterative cycle is:

Step 1. Model Estimation:

Obtain $\mathbf{y}_k$ $(k = 1, \ldots, K)$ *that minimizes* ϕ_w *for fixed* $\mathbf{y}_k^*$ *(such that* $\mathbf{y}_k^{*\prime}\mathbf{y}_k^* = 1$ *for* $k = 1, \ldots, K$*). This consists of the following two substeps.*

1. *Estimate the weights* (w_{kA}, w_{kB}) *for* $k = 1, \ldots, K$ *for fixed* $\mathbf{y}_k^*$ *and for fixed additive effects.*
2. *Estimate the additive effects* (α, β) *for fixed* $\mathbf{y}_k^*$ *and for fixed weights.*

Step 2. Optimal Scaling:

Obtain $\mathbf{y}_k^*$ $(k = 1, \ldots, N)$ *that minimizes* ϕ_w *for fixed* $\mathbf{y}_k$ *and normalize them so that* $\mathbf{y}_k^{*\prime}\mathbf{y}_k^* = 1$ *for* $k = 1, \ldots, K$.

In the initial iteration, we set $\mathbf{y}_k^* = \mathbf{o}_k$ $(k = 1, \ldots, K)$, which are to be centered and normalized. Initial estimates of the additive effects are obtained by applying the SAM, treating k as a replication. They are then normalized so as to satisfy (13–11).

The weights are estimated as follows. Let

$$W = \begin{bmatrix} w_{1A} & w_{1B} \\ \cdot & \cdot \\ \cdot & \cdot \\ \cdot & \cdot \\ w_{KA} & w_{KB} \end{bmatrix},$$

$$\mathbf{X} = [\mathbf{G}_A \hat{\alpha}, \mathbf{G}_B \hat{\beta}] ,$$

and

$$\mathbf{Y}^* = [\mathbf{y}_1^*, \ldots, \mathbf{y}_K^*] .$$

Then,

$$\hat{\mathbf{W}} = (\mathbf{X}'\mathbf{X})^{-1}\mathbf{X}'\mathbf{Y}^* .$$

The least-squares estimates of the additive effects are obtained by

$$\hat{\alpha} = (\mathbf{G}'_{AW} \mathbf{G}_{AW})^{-1} \mathbf{G}'_{AW} \mathbf{y}^* ,$$

for α, where $\mathbf{y}^{*\prime} = (\mathbf{y}_1^{*\prime}, \ldots, \mathbf{y}_k^{*\prime})$, and

$$G_{AW} = \begin{bmatrix} w_{1A} & G_A \\ \cdot & \cdot \\ \cdot & \cdot \\ \cdot & \cdot \\ w_{KA} & G_A \end{bmatrix} .$$

The estimate of β is similarly obtained. Once $\hat{\alpha}$ and $\hat{\beta}$ are obtained, they are normalized to satisfy (13–11), and the weights are adjusted for the normalization. Then the model predictions are calculated:

$$\mathbf{Y} = [\mathbf{y}_1, \ldots, \mathbf{y}_N] = \mathbf{XW} .$$

Finally, $\mathbf{y}_k^*$ ($k = 1, \ldots, K$) are obtained by applying Kruskal's monotonic regression to each $\mathbf{y}_k$ separately. (If the data are nominal, the least-squares nominal transformation is applied instead.)

The two phases are repeatedly applied until convergence is attained. The convergence is assumed to have been reached when the reduction in ϕ_W from one iteration to the next becomes negligible (for instance, 0.0005). The first two iterations of the WADDALS algorithm are shown in Figure 13–7 for hypothetical sets of data. The reader is encouraged to follow through the iterations in order to obtain a concrete image of how WADDALS works. Similar example iterations are shown for ADDALS in Young (1981).

In Figures 13–8 and 13–9, the plots of estimated α_{ki} and β_{kj} from ADDALS and those of estimated $w_{kA}\alpha_i$ and $w_{kB}\beta_j$ from WADDALS are shown. The data analyzed were the hypothetical data sets given in Figure 13–7. Figure 13–8 indicates that the data are not perfectly suitable for the WAM. However, the design is very small (3 × 3); the estimates are based on only three observations each. Since these estimates are not likely to be very reliable, it may still be worthwhile to apply the WAM. Figure 13–9 exhibits the characteristic patterns of proportionality among $w_{kA}\alpha_i$ and $w_{kB}\beta_j$. (Compare them with Figure 13–1.)

Stress values obtained from separate ADDALS analyses of the two sets are .198 for $k = 1$ and .194 for $k = 2$. The joint stress is computed by $[\{(.198)^2 + (.194)^2\}/2]^{1/2}$, which turns out to be .196. The stress value obtained from WADDALS is .208, a slight

Figure 13–7. Example of WADDALS Iterations

INITIALIZATION PHASE

RAW DATA

k=1 $\begin{bmatrix} 3 & 4 & 7 \\ 1 & 5 & 2 \\ 6 & 8 & 9 \end{bmatrix}$ k=2 $\begin{bmatrix} 1 & 5 & 6 \\ 3 & 2 & 7 \\ 4 & 8 & 9 \end{bmatrix}$

OPTIMALLY SCALED DATA
(Original data just centered and normalized)

$$\begin{bmatrix} -.26 & -.13 & .26 \\ -.52 & .0 & -.39 \\ .13 & .39 & .52 \end{bmatrix} \quad \begin{bmatrix} -.52 & .0 & .13 \\ -.26 & -.39 & .26 \\ -.13 & .39 & .52 \end{bmatrix}$$

INITIAL ESTIMATES OF ADDITIVE EFFECTS
(Normalized)

$$\alpha = \begin{pmatrix} -.39 \\ -.98 \\ 1.37 \end{pmatrix} \quad \beta = \begin{pmatrix} -1.32 \\ .22 \\ 1.10 \end{pmatrix}$$

ITERATION ONE

MODEL ESTIMATION

WEIGHTS w_A w_B

$$\begin{matrix} k=1 \\ k=2 \end{matrix} \begin{bmatrix} .26 & .18 \\ .15 & .24 \end{bmatrix}$$

ADDITIVE EFFECTS (Normalized)

$$\alpha = \begin{pmatrix} -.34 \\ -1.02 \\ 1.36 \end{pmatrix} \quad \beta = \begin{pmatrix} -1.30 \\ .16 \\ 1.14 \end{pmatrix}$$

SCALE ADJUSTED WEIGHTS
(No significant change)

OPTIMAL SCALING

RAW DATA (Arranged in an ascending order)
(1 2 3 4 5 6 7 8 9)

k=1:

MODEL ESTIMATES (Normalized)

(-.51 -.11 -.31 -.07 -.27 .18 .09 .42 .58)

OPTIMALLY SCALED DATA

(-.51 -.21 -.21 -.17 -.17 .13 .13 .42 .58)

k=2:

MODEL ESTIMATES (Normalized)

(-.42 -.16 -.55 -.08 -.02 .24 .11 .31 .57)

OPTIMALLY SCALED DATA

(-.42 -.35 -.35 -.08 -.02 .17 .17 .31 .57)

Continue to the next column

FIT

$$\text{stress} = \sqrt{\frac{.130}{2}} = .255$$

OPTIMALLY SCALED DATA
(Normalized and arranged in the original format to be used in the next iteration)

$$\begin{bmatrix} -.21 & -.18 & .14 \\ -.52 & -.18 & -.21 \\ .13 & .42 & .58 \end{bmatrix} \quad \begin{bmatrix} -.44 & -.03 & .18 \\ -.37 & -.37 & .18 \\ -.09 & .32 & .60 \end{bmatrix}$$

ITERATION TWO

MODEL ESTIMATION

WEIGHTS w_A w_B

$$\begin{matrix} k=1 \\ k=2 \end{matrix} \begin{bmatrix} .29 & .20 \\ .15 & .25 \end{bmatrix}$$

ADDITIVE EFFECTS (Normalized)

$$\alpha = \begin{pmatrix} -.35 \\ -1.01 \\ 1.36 \end{pmatrix} \quad \beta = \begin{pmatrix} -1.21 \\ -.02 \\ 1.24 \end{pmatrix}$$

SCALE ADJUSTED WEIGHTS
(No significant change)

OPTIMAL SCALING

RAW DATA (Arranged in an ascending order)
(1 2 3 4 5 6 7 8 9)

k=1:

MODEL ESTIMATES (Normalized)

(-.49 -.10 -.29 -.11 -.30 .21 .09 .40 .59)

OPTIMALLY SCALED DATA

(-.49 -.20 -.20 -.20 -.20 .15 .15 .40 .59)

k=2:

MODEL ESTIMATES (Normalized)

(-.39 -.22 -.53 -.03 -.08 .25 .11 .28 .61)

OPTIMALLY SCALED DATA

(-.39 -.37 -.37 -.06 -.06 .18 .18 .28 .61)

FIT

$$\text{stress} = \sqrt{\frac{.103}{2}} = .227$$

OPTIMALLY SCALED DATA
(Normalized and arranged in the original format to be used in the next iteration)

$$\begin{bmatrix} -.20 & -.20 & .15 \\ -.50 & -.20 & -.20 \\ .15 & .40 & .61 \end{bmatrix} \quad \begin{bmatrix} -.40 & -.06 & .19 \\ -.38 & -.38 & .19 \\ -.06 & .29 & .62 \end{bmatrix}$$

increase from .196. The difference represents the portion of variability in the data that can be represented by the SAM but not by the WAM. Whether the difference is statistically significant or not is difficult to determine; there is no definite criterion, but when the design is as small as this, a difference of .012 seems quite large.

The comments made for ADDALS concerning the nonglobal minimum, the weak minimum, and the degeneracy of solutions are also valid for WADDALS. The last two of these problems are less serious in WADDALS, since the number of observations to number of parameters ratio is generally much larger in the WAM.

A necessary condition to avoid a degenerate solution is much weaker for the WAM. Ordinally connected rows (or columns) define a partition of rows (columns) in the data matrix. Within each subset in the partition, rows (columns) are connected, while rows (columns) in different subsets are not connected. When the

Figure 13–8. Estimates of Additive Effects in SAM Obtained from Hypothetical Data Sets

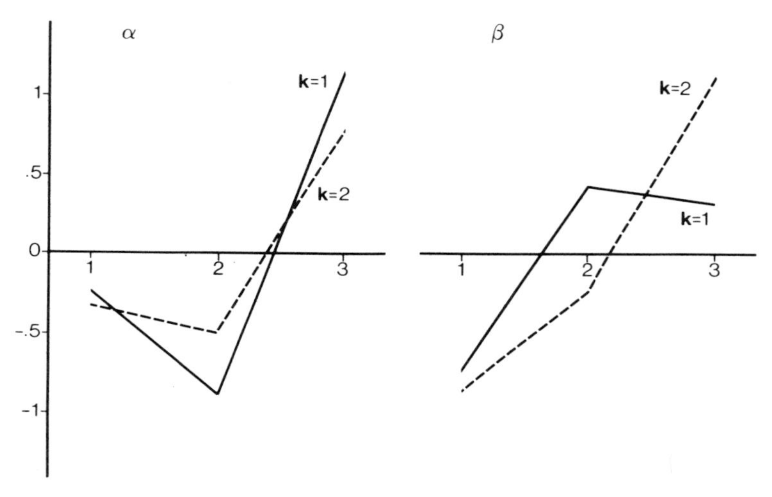

Figure 13–9. Estimates of Additive Effects Multiplied by Weights in the WAM*

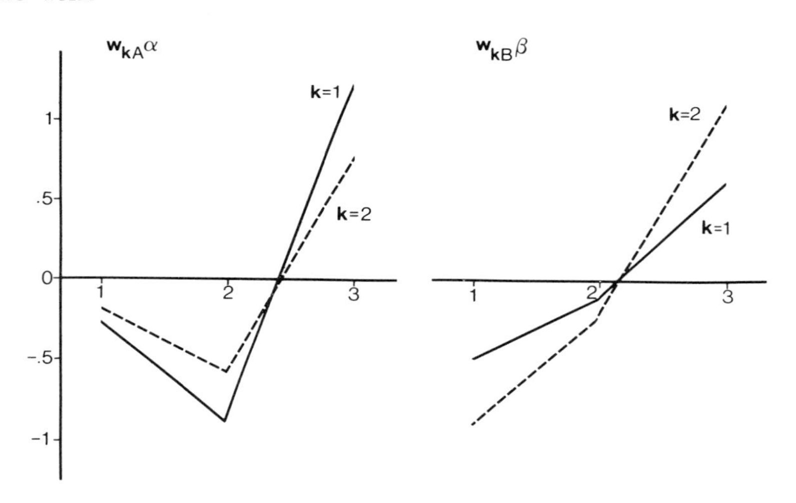

*Derived from the same hypothetical data sets as in Figure 13–8.

partition consists of only one set of all rows (columns), the data are said to be rowwise (columnwise) connected.

DEFINITION 13–2

Two subsets of connected rows (columns) are connected if at least one element in one subset is larger and at least one element in the same set is smaller than at least one element in the other subset.

Unconnected subsets for some k should be connected for at least one $k'(\neq k)$.

MAXADD

WADDALS, as discussed in the previous section, works reasonably well as a descriptive device. It is better suited for obtaining crude estimates based on relatively weak assumptions (monotonicity, no distributional assumptions). However, it can provide very little information pertinent to model evaluations, such as assessing the reliability of the estimates, comparing the goodness-of-fit between two models, and so forth. The MAXADD procedure developed by Takane (1982), on the other hand, allows various statistical inferences that have not been possible previously.

The MAXADD procedure is capable of fitting both the SAM and the WAM to categorical rating, pair comparison, and directional rank-order data. In MAXADD, nonmetric data are viewed as incomplete data (Dempster, Laird, and Rubin 1977). A metric process conveying complete information about the model is assumed to underlie such nonmetric data. The metric information is presumed to get lost in an observation process, leaving us only ordinal or nominal information. A specific information reduction mechanism is postulated for each specific type of data; through this mechanism, the likelihood of observed nonmetric data is stated as a function of model parameters. Model parameters are then determined so as to maximize the likelihood of the observed data. Procedures similar to MAXADD have been developed for multidimensional scaling (Takane 1978, 1981; Takane and Carroll 1981).

We discuss the model linking the unobserved metric process to the observed nonmetric data for each of the three data collection methods mentioned above. In the categorical rating method, each category is represented by an interval specified by its upper and lower boundaries. Let the upper and lower boundaries of the mth observation category be denoted by b_{km} and $b_{k(m-1)}$, respectively. We assume that y_{kij} is put into category m whenever its corresponding error-perturbed model prediction falls in an interval bounded by $b_{k(m-1)}$ and b_{km}. Let p_{kijm} denote the probability of this event. Then

$$p_{kijm} = \text{Prob } (b_{k(m-1)} < y^*_{kij} < b_{km}) \ . \qquad (13\text{–}31)$$

Define

$$F_{kijm} = \text{Prob } (y^*_{kij} < b_{km}) \ . \qquad (13\text{–}32)$$

Then $p_{kijm} = F_{kijm} - F_{kij(m-1)}$. Under the distributional assumption that y^*_{kij} is logistic, (13–32) can be more explicitly written as:

$$F_{kijm} = [1 + \exp \{-s_k \ (b_{km} - y_{kij})\}]^{-1} \ , \qquad (13\text{–}33)$$

where s_k is a dispersion parameter. Let Z_{kijm} denote the observed frequency with which y_{kij} is put into category m. The joint probability (p_{kij}) of Z_{kijm} ($m = 1, \ldots, M$) can be stated as:

$$p_{kij} = \prod_{m=1}^{M} (p_{kijm})^{Z_{kijm}} . \tag{13–34}$$

Finally, the joint probability of the total set of category judgments is obtained by the product of the above p_{kij} over i, j, and k, assuming that the judgments are all statistically independent.

In the pair comparison case, we may assume that y_{kij} is judged to be larger than y_{kuv} whenever y^*_{kij} exceeds y^*_{kuv}. The probability (p^*_{kijuv}) that y_{kij} is judged larger than y_{kuv} is then written as:

$$\begin{aligned} p^*_{kijuv} &= \text{Prob } (y^*_{kij} > y^*_{kuv}) \\ &= \text{Prob } (y^*_{kij} - y_{kuv} > 0) , \end{aligned} \tag{13–35}$$

which, assuming that $y^*_{kij} - y^*_{kuv}$ is logistic, can be more explicitly stated as:

$$p^*_{kijuv} = [1 + \exp \{-s_k (y_{kij} - y_{kuv})\}]^{-1} . \tag{13–36}$$

Let Z_{kijuv} denote the frequency with which y_{kij} is judged larger than y_{kuv}. Then the joint probability (p_{kijuv}) of Z_{kijuv} and Z_{kuvij} can be stated as:

$$p_{kijuv} = (p^*_{kijuv})^{Z_{kijuv}} (1 - p^*_{kijuv})^{N_{kijuv} - Z_{kijuv}} , \tag{13–37}$$

where $N_{kijuv} = Z_{kijuv} + Z_{kuvij}$. The joint probability of the total set of pair comparison judgments is obtained by the product of p_{kijuv}.

In the directional rank-order method, each ranking is performed in a specified direction, either from the largest to the smallest or from the smallest to the largest. In this case, we may assume that each ranking is obtained by successive first choices. Suppose that the ranking is performed from the largest to the smallest among M stimuli. Let $y_k^{(m)}$ denote the model prediction judged to be the mth largest, and $y_k^{(m)*}$ the corresponding error-perturbed model prediction. We assume that when $y_k^{(m)}$ is chosen as the mth largest element, the $m - 1$ successive first choices have already been made and the corresponding $m - 1$ stimuli have been deleted from the comparison set. We assume that $y_k^{(m)}$ is chosen as the largest element from the remaining $M - m + 1$ stimuli. We also assume that this event occurs whenever $y_k^{(m)*}$ exceeds all other error-perturbed model predictions remaining in the comparison set. Then

$$p_k^{(m)} = \text{Prob}\ (y_k^{(m)*} > y_k^{(m+1)*}, \ldots, y_k^{(m)*} > y_k^{(M)*}) \ , \quad (13\text{–}38)$$

where $p_k^{(m)}$ is the probability that $y_k^{(m)}$ is chosen as the largest element among the $M - m + 1$ elements. Using Luce's (1959) choice model, $p_k^{(m)}$ can be more explicitly stated as:

$$p_k^{(m)} = [1 + \sum_{j=m+1}^{M} \exp\ \{-s_k(y_k^{(m)} - y_k^{(j)})\}]^{-1} \ . \quad (13\text{–}39)$$

Assuming that each successive choice is performed independently, the probability of a ranking is obtained by the product of $p_k^{(m)}$ over $m = 1, \ldots, M - 1$. The joint probability of multiple rankings, on the other hand, can be stated as the product of the probabilities of individual rankings. A rationale and the condition for the statistical independence of successive first choices are given in Takane and Carroll (1981).

In all cases, the likelihood function is stated as a function of model parameters. Their estimates are determined in such a way as to maximize the likelihood function. Let L denote the likelihood function. Then, we would like to solve likelihood equations,

$$\frac{\partial \ln L}{\partial \boldsymbol{\theta}} = \mathbf{0} \ ,$$

for $\boldsymbol{\theta}$, the parameter vector. MAXADD uses the Fisher scoring algorithm for solving the likelihood equations. This algorithm, starting from some initial estimates, updates the parameter estimates by solving

$$\varepsilon^{(q)}\ \mathbf{I}(\boldsymbol{\theta}^{(q)})(\boldsymbol{\theta}^{(q+1)} - \boldsymbol{\theta}^{(q)}) = \mathbf{u}(\boldsymbol{\theta}^{(q)}) \quad (13\text{–}40)$$

for $\boldsymbol{\theta}^{(q+1)}$, where $\varepsilon^{(q)}$ is a step-size parameter, $\boldsymbol{\theta}^{(q+1)}$ and $\boldsymbol{\theta}^{(q)}$ are new and old parameter estimates, respectively, and

$$\mathbf{u}(\boldsymbol{\theta}) = \left(\frac{\partial \ln L}{\partial \boldsymbol{\theta}}\right)$$

and

$$\mathbf{I}(\boldsymbol{\theta}) = E\left[\left(\frac{\partial \ln L}{\partial \boldsymbol{\theta}}\right)\left(\frac{\partial \ln L}{\partial \boldsymbol{\theta}}\right)'\right] \ ,$$

which is called *Fisher's information matrix*. Updating of equation (13–40) is iteratively applied until convergence is reached. Initial estimates are obtained by the same procedure used in ADDALS and WADDALS, except for pair comparison data. In the pair comparison case, estimates of y_{kij} are first obtained by applying a least-squares procedure for the ordinary Luce model to observed frequencies, z_{kijuv}, and then these estimates are used to obtain initial estimates of model parameters. The reader is referred to Takane (1982) for more details of the algorithm.

As has been emphasized, one of the major advantages of the MAXADD procedure is its statistical inference capability. The Moore-Penrose inverse of the information matrix evaluated at the maximum likelihood estimates of parameters is known to give asymptotic variance-covariance estimates of the parameter estimates (Ramsay 1978). These estimates provide information concerning the reliability of the derived estimates.

The asymptotic chi-square statistic derived from the likelihood ratio principle may be used for various model comparisons. Let L_0 and L_1 represent the two likelihoods obtained under hypothesis H_0 (which postulates a more restricted model) and hypothesis H_1 (which postulates a less restricted model), respectively. Then

$$\chi^2 = 2(\ln L_1 - \ln L_0)$$

follows asymptotically the chi-square distribution with degrees of freedom equal to the difference in the number of parameters in the two models. Observed values of χ^2 are compared with appropriate critical values of chi-square to test a significant difference between the two models.

The asymptotic chi-square can only be used when one of the two models compared is a special case of the other. When this is not the case, the following statistic will be helpful:

$$\text{AIC}(\pi) = -2 \ln (L_\pi) + 2n_\pi ,$$

(Akaike 1974), where L_π is the maximum likelihood of model π and n_π is the effective number of parameters in the model. A smaller value of AIC indicates a better fit.

Some caution should be exerted when one wants to rely on the statistical inference features of MAXADD. These are all based on the asymptotic properties of the maximum likelihood estimates, which do not strictly hold in many practical situations. It is necessary to have a lot of replications. In the most typical situations, the number of observations to number of parameters ratio must be at least 15 to completely rely on these statistics. Otherwise, one has to examine the behavior of these statistics case by case using Monte Carlo techniques.

The same example data sets that were previously analyzed by ADDALS and WADDALS were reanalyzed by MAXADD, assuming that the data were directional rank orders. Two analyses were performed—one under the assumption that the rankings were made from the smallest to the largest and the other under the assumption that the rankings were done in the reverse direction. (In a practical situation, one usually does only one analysis, whichever an experimental paradigm dictates.) Results are presented in Figure 13–10. The two analyses give similar results; not only are the values of the likelihood function quite similar, but estimates of the model parameters are also similar. These estimates also agree well with those obtained by WADDALS.

An example of some use of statistical inference features of MAXADD will be given in the application section.

Figure 13–10. Summary of MAXADD-4 Analyses of the Data for WADDALS

(I) Assuming that the rankings are performed from the smallest to the largest:

$LL = -13.6$
$n_\pi = 5$
$AIC = 37.3$

Additive effects (Normalized)

$$\alpha = \begin{pmatrix} -.520 \\ -.879 \\ 1.399 \end{pmatrix} \quad \beta = \begin{pmatrix} -1.275 \\ .108 \\ 1.167 \end{pmatrix}$$

Weights (Normalized)

$$\begin{matrix} & w_A & w_B \\ k=1 & .789 & .615 \\ k=2 & .579 & .816 \end{matrix}$$

(II) Assuming that the rankings are performed from the largest to the smallest:

$LL = -13.9$
$n_\pi = 5$
$AIC = 37.7$

Additive effects (Normalized)

$$\alpha = \begin{pmatrix} -.220 \\ -1.100 \\ 1.320 \end{pmatrix} \quad \beta = \begin{pmatrix} -1.287 \\ .137 \\ 1.151 \end{pmatrix}$$

Weights (Normalized)

$$\begin{matrix} & w_A & w_B \\ k=1 & .892 & .452 \\ k=2 & .634 & .773 \end{matrix}$$

EXAMPLES OF APPLICATION

In this section, we discuss some applications. Some of these examples have been presented elsewhere; others are only provisional, in that no real data have yet been analyzed, but they are included here to demonstrate possible uses of WADDALS or MAXADD. They also indicate how the two procedures can be effectively applied to actual data analysis situations. We focus on the WAM here, but the reader can find interesting applications of the SAM in Anderson (1981), Cliff (1959), Green and Rao (1971), Johnson (1974), and Wallsten (1976). Green and Srinivasan (1978) give an excellent overview of various current issues on the additive conjoint measurement.

Psychophysical Data

Portions of this example have been presented elsewhere (Takane et al. 1980; Takane 1982). In fact, it served as an original motivating factor for developing the WADDALS procedure, the first computer program ever developed for the WAM.

Some developmental psychologists believe that the way largeness of rectangles is judged changes with the subject's age. For example, younger children tend to put more emphasis on the height than the width of rectangles when they make perceived largeness judgments. This tendency decreases as they get older (Kempler 1971). Let us assume (although this assumption itself is susceptible to an empirical test) that children's perception of the rectangle area is obtained by simple addition of contributions from the height and the width. That is, it is representable by the SAM for each child or age group.

The developmental change described above may clearly be described by the WAM. Let w_{kH} and w_{kW} denote the differential weights put on the height and the width, respectively, in the kth age group. It is expected that $w_{kH} > w_{kW}$ initially and that w_{kH} decreases while w_{kW} increases as k increases, until w_{kH} becomes nearly equal to w_{kW}.

In order to test this hypothesis, Kempler (1971) constructed a set of 100 rectangles by factorially combining 10 height levels and 10 width levels, each ranging from 10 inches to 14.5 inches in half-inch intervals. Sixteen to 25 children in each of four age groups (1st-, 3rd-, 5th-, and 7th-graders) judged each of the 100 rectangles "large" or "small." (This is a two-category judgment.) For each age group, the number of times each rectangle is judged large (that is, the number of children who make this judgment) was counted and used as an ordinal measure of the perceived largeness and additivity analyses were applied.

Since the WAM presupposes a representation by the SAM for each age group, the simple additivity analysis was conducted first. In all cases, satisfactory stress values were obtained ($< .2$). Since the estimates of the additive effects in the SAM were found roughly proportional, the weighted additivity analysis was performed by WADDALS. As expected, the weights attached to the height tended to decrease, while those for the width increased rather consistently with age level (see Figure 13–11). The stress value obtained from WADDALS (.190) was not very different from the joint stress value obtained from the SAM (.172) applied separately to each age group. Thus, for Kempler's data, the WAM was indeed appropriate. Takane et al. (1980) discuss other possible models for Kempler's data and how they can be dismissed on the grounds of model comparisons. They also report extensive Monte Carlo studies on stress values to be used in evaluating the goodness-of-fit of the WAM.

The above analysis used group data, ignoring possible individual differences within each group. Thus, it did not reveal a true source of the group differences found. To do so, individual data have to be analyzed. Kempler's individual data (original two-category judgments) were thus analyzed separately for each age group by MAXADD using both WAM and SAM. The WAM was consistently found to fit the data better, implying that there were significant differences in the weights *within* each group. Figure 13–12 shows the estimated subject weights for each of the four age groups. There are indeed marked individual differences in the weights. What is more interesting, however, is that there seem to be some systematic differences in the weight structures across the groups. The plots of the weight estimates tend to converge toward the dotted lines (the 45° lines between horizontal and vertical axes) as the age level goes up. For example, in

Figure 13–11. Individual Differences Weights Obtained from WADDALS Analyses of Group Data

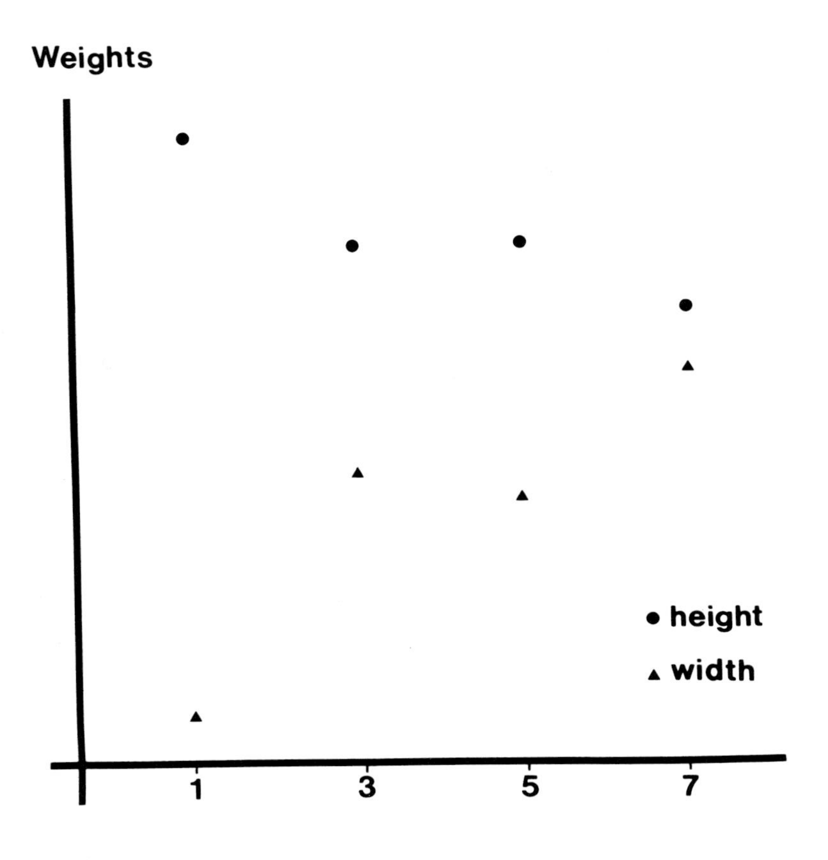

grade 1, several children put disproportionately large weights on the height. (There are two children who base their largeness judgments almost exclusively on the height factor.) As the age level goes up, the number of those extreme children decreases. The children themselves also tend to be less extreme. In grade 7, a majority of children put approximately equal emphasis on the two factors. Thus, the group differences found in the previous analysis seem largely due to the compositional differences among the groups. That is, each group consists of heterogeneous subjects, but the degree of heterogeneity within each group clearly decreases with age.

In the analysis of the group data, the additive effects were found to be constant across the groups. In fact, this was the basis for applying the WAM to all age groups simultaneously. Can they still be assumed constant across the different age groups with the individual data? (Remember that in the above

Figure 13–12. Individual Differences Weights Obtained from MAXADD Analyses of Individual Data

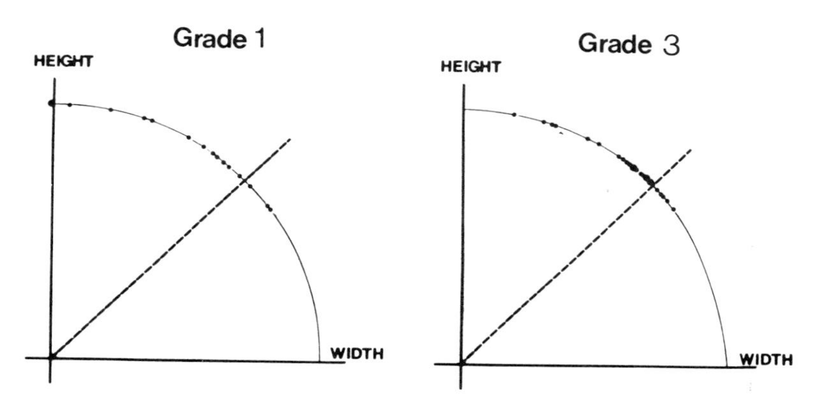

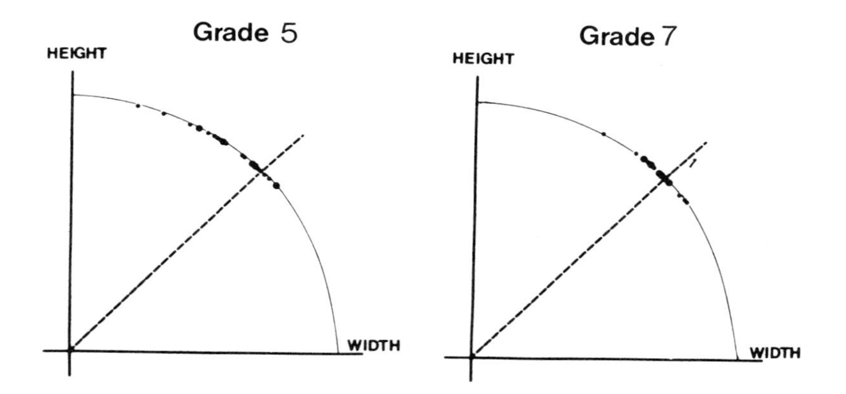

analysis the WAM was applied separately to the four age groups, which tacitly assumed that the additive effects were constant across the subjects within the groups.) In order to answer this question, the WAM was applied to the entire set of individual data and the goodness-of-fit of this model was compared with that of the WAM applied separately to each group. The AIC from the joint analysis was 49.2 (+ constant), while the joint AIC from the separate analyses was 51.6 (+ constant). Thus, it was concluded that the WAM fitted to the entire set of individual data was the better-fitting model. The additive effects can indeed be assumed constant across the age groups.

Similarity and Dissimilarity Data

Medin and Schaffer (1978) have proposed a model called the *cue context model* for classification learning. In their model, similarity between stimuli i and j is defined as:

$$s_{ij} = \prod_{t=1}^{T} X_{ijt} \quad , \tag{13–41}$$

where s_{ij} is the similarity between stimuli i and j, and

$$X_{ijt} = \begin{cases} 1, \text{ if stimuli } i \text{ and } j \text{ share feature } t, \\ \Theta_t, \ (0 < \Theta_t \leq 1) \text{ otherwise.} \end{cases}$$

The X_{ijt} thus denotes the featurewise similarity between stimuli i and j. The overall similarity (s_{ij}) is defined as the product of the featurewise similarities. Since $0 < \Theta_t \leq 1$ for $t = 1, \ldots,$ T, we have $0 < s_{ij} \leq 1$. Let

$$d_{ij} = -\log s_{ij} = -\sum_{t=1}^{T} \ln X_{ijt} \quad . \tag{13–42}$$

The d_{ij} represents a dissimilarity between i and j. (It is an inverse monotonic transformation of s_{ij}.) Since part-dissimilarity functions (Green and Srinivasan 1978), which are

$$d_{ijt} = -\ln X_{ijt} = \begin{cases} 0, \text{ if stimuli } i \text{ and } j \text{ share feature } t, \\ -\ln \Theta_t, \text{ otherwise,} \end{cases}$$

are nonnegative ($-\ln \Theta_t \geq 0$), the overall dissimilarity (d_{ij}) is also nonnegative. Furthermore, it can be proven that:

$$d_{ii} = 0 \text{ (Minimality)},$$

$$d_{ij} = d_{ji} \text{ (Symmetry)},$$

and

$$d_{ij} + d_{jk} \geq d_{ik} \text{ (Triangle Inequality)},$$

so that d_{ij} in (13–42) can formally be considered as the distance between stimuli i and j. Define

$$g_{ijt} = \begin{cases} 0, \text{ if stimuli } i \text{ and } j \text{ share feature } t, \\ 1, \text{ otherwise,} \end{cases}$$

and

$$\Theta_t^* = -\ln \Theta_t \quad .$$

Then $d_{ijt} = g_{ijt}\,\Theta_t^*$, and therefore:

$$d_{ij} = \sum_{t=1}^{T} d_{ijt} = \sum_{t=1}^{T} g_{ijt}\,\Theta_t^* . \tag{13–43}$$

This indicates that d_{ij}, which is a kind of distance as defined in (13–42), is a special case of the SAM.

Note that if we define $d_{ij}^p = -\ln s_{ij}$ for any $p\ (\geqq 1)$ in (13–42), d_{ij} can be regarded as the Minkowski p-metric. (When $p = 2$, we obtain the Euclidean distance; when $p = 1$, as in the above discussion, we have the city-block distance.) We also define $d_{ijt}^p = -\ln X_{ijt}$ in this case. That p can be any value not less than unity reflects the fact that there is only one interval defined for each dimension.

Suppose now that we may allow individual differences in s_{ij} in the following manner:

$$s_{kij} = \prod_{t=1}^{T} X_{ijt}^{w_{kt}} , \tag{13–44}$$

where s_{kij} is the similarity between stimuli i and j for individual k, and w_{kt} is the weight attached to feature t by individual k. Let

$$d_{kij}^p = -\ln s_{kij} = -\sum_{t=1}^{T} w_{kt} \ln X_{ijt} \quad \text{for } (p \geqq 1).$$

Again, d_{kij} can be proven to be a distance function within each k. We obtain, analogous to (13–43),

$$d_{kij}^p = \sum_{t} g_{ijt}\,\Theta_t^* w_{kt} , \tag{13–45}$$

which is a special case of the WAM. This is also a special case of (13–22), in which

$$g_{ijt}\,\Theta_t^* = |x_{it} - x_{jt}|^p .$$

Thus, sets of similarity or dissimilarity data for stimuli characterized by a set of known features may be analyzed using either WADDALS or MAXADD, depending on what assumptions we are willing to make on our data.

Model (13–45), however, reduces .to an even simpler form by redefining $\Theta_{kt}^* = \Theta_t^* w_{kt}$. We then obtain:

$$d_{kij}^p = \sum_{t} g_{ijt}\,\Theta_{kt}^* , \tag{13–46}$$

which amounts to applying the SAM separately to each k. This is because there are only two levels in each feature (present or absent). In this case, the WAM and the SAM are completely equivalent. For the same reason a correct value of the Minkowski power cannot be identified. In order to empirically distinguish the WAM from the SAM, and also between different values of p, we must have at least three levels in each feature (factor, dimension). When there are three levels—say i, q, and s—then there are three intervals defined—(i,q), (q,s), and (i,s). Contributions of these intervals should satisfy a special relationship under the Minkowski power distance model:

$$\delta_{iq}^{1/p} + \delta_{qr}^{1/p} = \delta_{is}^{1/p} , \tag{13–47}$$

where δ_{iq} indicates the effect of the difference between two levels, i and q. (Under the Minkowski hypothesis, this is indeed the pth power of the length of an interval bounded by i and j.) If we use (13–47) as constraints while systematically varying the value of p, we should be able to identify the best-fitting p value. Takane (1982) reports this type of study with the SAM.

It may be noted in passing that the multiplicative rule in (13–41) in defining the overall similarity between two stimuli is quite similar to the multiplicative competitive interaction (MCI) model (Nakanishi and Cooper 1974) for preference choice data, to which we now turn.

Preference Data

Applications of the WAM to preference data are rather scarce, due, perhaps, in part to the fact that the WAM postulates an identical ideal point (Coombs [1964] 1976) along each stimulus attribute across individuals. In situations in which the preference functions are single-peaked, this is hard to justify; in many cases, there are individual differences in the ideal point. However, just as the ideal point model reduces to the vector preference model when the ideal points are located infinitely far away from the origin (Carroll 1972), the additive model in this limiting case posits monotonic preference functions for the additive effects. Just as in a number of situations the vector preference model is appropriate (Green and Srinivasan 1978), there are a lot of situations in which the preference functions (often called *part-worth functions* in the marketing research literature; see, for example, Green and Srinivasan 1978) can be assumed monotonic and common across individuals, thus making the WAM appropriate. The WAM in this case can be thought of as a special type of the vector preference model, which involves quantifications of attributes as well as weighting.

Phipps and Carter (1982) applied the WAM to neighborhood preference data of homeowners using lot size, landscaping, and neighboring housing as stimulus attributes. Along each of the three attributes, the preference function was found to be monotonic. The individual difference weights obtained from the weighted additivity analysis were then related to various demographic variables of the individuals, such as education. C. P. Whaley (personal communication), of Bell Northern Research, also reports an application of WADDALS in marketing research called *tradeoff analysis* (see Johnson 1974). Unfortunately, details of the study cannot be presented, due to its proprietary nature. Nonetheless, in this study, there were five factors to be considered, but judgments were taken only for two factors at a time. From these incomplete data, WADDALS could successfully obtain estimates of parameters in the WAM.

Several preference models can be considered as special cases of the WAM. We will discuss only two of them, the *multiplicative competitive interaction* (MCI) *model* (Nakanishi and Cooper 1974) and the *additive-difference model* (Tversky 1969) *for preference data.*

The MCI model posits that the overall utility of stimulus r (y_r) is a powered multiplicative function of stimulus attributes (X_{rt}). That is,

$$y_r = \prod_{t=1}^{T} X_{rt}^{w_t} , \tag{13–48}$$

which can be readily extended to

$$y_{kr} = \prod_{t=1}^{T} X_{rt}^{w_{kt}} , \tag{13–49}$$

where w_{kt} is an individual-difference weight. Taking the log of (13–49), we obtain:

$$\log y_{kr} = \sum_{t=1}^{T} w_{kt} \log (X_{rt}) , \tag{13–50}$$

which is linear with respect to log (X_{rt}). Suppose now that log (X_{rt}) is unknown and that it should be quantified. Then

$$\log (X_{rt}) = \sum_{q_t=1}^{Q_t} h_{rq_t} \alpha_{q_t} .$$

Putting this into (13–50), we obtain:

$$\log y_k = \sum_{t=1}^{T} w_{kt} \sum_{q_t=1}^{Q_t} h_{rq_t} \alpha_{q_t} , \tag{13–51}$$

which is precisely the WAM (see [13–3]). In (13–51), we may just observe an ordinal measure of the dependent variable rather than log y_{kr} itself (such as y_{kr}). Then it must be monotonically transformed before it can be fitted by the WAM.

We now turn to the additive-difference model. There are actually two versions of the model: (a) a symmetric version designed for dissimilarity data (Tversky and Krantz 1970); and (b) a skew-symmetric version (Tversky 1969) designed for dominance data. It is this latter version that we discuss here for preference data.

When we compare two or more stimuli in terms of preference, we may first evaluate overall stimulus preferences by combining various stimulus attributes and then compare them across the stimuli to arrive at a relative preference judgment, that is, how much we like one stimulus over the other. The additive model is appropriate if the stimulus attributes are combined in an additive fashion. Alternatively, we may first compare the stimuli within attributes and then, based on the attribute by attribute comparisons, form an overall preference judgment. In this case, an additive representation no longer holds except under a specialized condition (Tversky 1969).

Let us suppose for simplicity that the attributewise comparison process is subtractive and that the subsequent integration process is weighted additive. Then we obtain the following model:

$$y_{kijuv} = w_{kA}\psi_A(\alpha_i - \alpha_u) + w_{kB}\psi_B(\beta_j - \beta_v) , \tag{13–52}$$

where ψ's are some monotonic skew-symmetric functions ($\psi(-x) = -\psi(x)$), and y_{kijuv} is the relative preference of stimulus (i,j) (the stimulus representing the combined effect of the ith level of factor A and the jth level of factor B) in comparison with stimulus (u,v) by individual k. The above model is called the *weighted additive-difference model for preference data* (Takane 1982). The additive-difference model reduces to the difference between two additive models when ψ's are linear (in which case ψ's can further be assumed to be identity functions), since we obtain:

$$\begin{aligned} y_{kijuv} &= w_{kA}(\alpha_i - \alpha_u) + w_{kB}(\beta_j - \beta_v) \\ &= (w_{kA}\alpha_i + w_{kB}\beta_j) - (w_{kA}\alpha_u + w_{kB}\beta_v) \\ &= y_{kij} - y_{kuv} . \end{aligned} \tag{13–53}$$

Thus, the WAM is a special case of the weighted additive-difference model in which ψ's are linear. In this special case, the two models are completely equivalent; consequently, the two comparison processes, stimulus-by-stimulus and attribute-by-attribute, are not empirically distinguishable.

In general, (13–52) is not reducible to (13–53). However, in some sense, the weighted additive-difference model can also be considered as a special case of the weighted additive model. We redefine differences between levels of additive factors (that is, $\alpha_i - \alpha_u$, $\beta_j - \beta_v$, and so on) as new levels of additive factors.

Then we can directly estimate their effects in the context of the additive model. Let $\alpha_{iu} = \psi_A(\alpha_i - \alpha_u)$ and $\beta_{jv} = \psi_B(\beta_j - \beta_v)$. Then (13–52) can be rewritten as:

$$y_{kijuv} = w_{kA}\alpha_{iu} + w_{kB}\beta_{jv} , \qquad (13\text{–}54)$$

where α_{iu} and β_{jv} represent the additive effects of newly defined levels of additive factors. We may require

$$\alpha_{iu} = -\alpha_{iu} \qquad \text{for all } i \text{ and } u \qquad (13\text{–}55)$$

(the same for β), which follows from the skew-symmetry of ψ_A. It follows from (13–55) that:

$$\alpha_{ii} = 0 \qquad \text{for all } i. \qquad (13\text{–}56)$$

This implies that if two stimuli share the same level of an attribute, the contribution of that attribute is zero in the comparison process. In order to preserve the monotonicity of ψ's, we may impose order restrictions, such as:

$$\alpha_{iu} \leqq \alpha_{is} \qquad \text{and} \qquad \alpha_{us} \leqq \alpha_{is} ,$$

where level u is between i and s. In any case, the goodness-of-fit of (13–54) can be directly compared with that of (13–53). This comparison should provide important insight into the nature of stimulus comparison processes.

Note that restriction (13–55) leads, strictly speaking, to a nonadditive model (though it is still linear), since we obtain model predictions such as:

$$y_{kijiv} = -w_{kA}\alpha_{iu} + w_{kB}\beta_{jv} ,$$

$$y_{kivuj} = w_{kA}\alpha_{iu} - w_{kB}\beta_{jv} ,$$

and so forth. Unfortunately, the nonadditive model cannot be directly fitted by WADDALS or MAXADD. Necessary modifications should not be too difficult, however. MAXLIN, for the general weighted linear model, is now under development, which should be able to handle this kind of situation.

Analysis of Contingency Tables

Tables of joint frequencies of responses to two or more multicategory items are called *contingency tables*. The log-linear model (Bishop et al. 1975) has been quite popular for analysis of the contingency tables.

Let f_{ij} denote the observed frequency of joint occurrences of the ith category (level) of item (factor) A and the jth category of item B. (For simplicity, we only deal with a two-item situation.) Then, the log-linear model states that:

$$p_{ij} \equiv f_{ij}/N = \alpha_i^* \beta_j^* \gamma_{ij}^* , \quad (13\text{–}57)$$

where $N = \sum_i \sum_j f_{ij}$ (the total number of observations),

α_i^* and β_j^* are main effects of the ith category of item A and the jth category of item B, respectively,

and γ_{ij}^* is an *interaction* effect between i and j.

There are too many parameters in the above model, so that constraints such as

$$\prod_i \alpha_i^* = \prod_j \beta_j^* = \prod_i \gamma_{ij}^* = \prod_j \gamma_{ij}^* = 1$$

are imposed to identify the model parameters. Taking the log of (13–57), we obtain:

$$\log p_{ij} = \alpha_i + \beta_j + \gamma_{ij} , \quad (13\text{–}58)$$

where

$$\alpha_i = \log \alpha_i^* ,$$

$$\beta_j = \log \beta_j^* ,$$

and

$$\gamma_{ij} = \log \gamma_{ij}^* ,$$

with

$$\sum_i \alpha_i = \sum_j \beta_j = \sum_i \gamma_{ij} = \sum_j \gamma_{ij} = 0.$$

If we assume that there is no interaction effect in (13–58), we obtain the SAM,

$$\log p_{ij} = \alpha_i + \beta_j , \quad (13\text{–}59)$$

which may equivalently be written as:

$$p_{ij} = \exp(\alpha_i + \beta_j) . \quad (13\text{–}60)$$

This latter form of the model suggests that it is a special case of Luce's (1959) choice model,

$$p_{ij} = \exp\,(\alpha_i + \beta_j)/\sum_{i'}\sum_{j'} \exp\,(\alpha_{i'} + \beta_{j'}) \;, \qquad (13\text{–}61)$$

with the denominator scaled to be unity. Model (13–61) can be obtained alternatively by assuming that the cell (i,j) is *chosen* with the probability proportional to its response strength relative to those of other cells. That is, if we take the SAM $(\alpha_i + \beta_j)$ for the representation of (i,j) cell and use the model of first choice, similar to the one used in the directional ranking method, we obtain (13–61). Model (13–61) also has some resemblance to the Rasch model (Andersen 1980) in mental testing situations.

The extension of (13–59) or (13–60) to the WAM is rather straightforward; we obtain:

$$p_{kij} = \exp\,(w_{kA}\alpha_i + w_{kB}\beta_j) \;. \qquad (13\text{–}62)$$

The WAM in this case implies no interaction between item A and item B at all k's and hence no three-way interaction. It does imply a special type of two-way interaction between i and $k(w_{kA}\alpha_i)$ and between j and $k(w_{kB}\beta_j)$.

CONCLUSION

In this chapter, we have discussed various aspects of the WAM. We have seen that a wide variety of existing models can be thought of as special cases of the WAM. Although examples of its application are not yet sufficient, we believe that they will quickly grow in number in the near future. We hope that this discussion will serve to disseminate the basic idea behind the model and to encourage more applied work in the social sciences.

REFERENCES

Akaike, H. 1974. A new look at the statistical model identification. *IEEE Transactions on Automatic Control* 19:716–23.

Andersen, E. B. 1980. *Discrete statistical models with social science applications*. Amsterdam: North-Holland.

Anderson, N. H. 1981. *Foundations of information integration theory*. New York: Academic Press.

Bishop, Y. M. M., S. E. Fienberg, and P. W. Holland. 1975. *Discrete multivariate analysis: Theory and practice*. Cambridge, Mass.: The MIT Press.

Bock, R. D. 1975. *Multivariate statistical methods in behavioral research*. New York: McGraw-Hill.

Carroll, J. D. 1972. Individual differences and multidimensional scaling. In *Multidimensional scaling: Theory and applications in the behavioral sciences*, vol. 1, ed. R. N. Shepard, A. K. Romney, and S. Nerlove. New York: Seminar Press.

———. 1973. Categorical conjoint measurement. In *Multiattribute decisions in marketing: A measurement approach*, ed. P. E. Green and Y. Wind. Hinsdale, Ill.: Dryden Press.

Carroll, J. D., and J. J. Chang. 1970. Analysis of individual differences in multidimensional scaling via an N-way generalization of "Eckart-Young" decomposition. *Psychometrika* 35:283–319.

Cliff, N. 1959. Adverbs as multipliers. *Psychological Review* 66:27–44.

Coombs, C. H. [1964] 1976. *A theory of data.* [New York: Wiley] Republished, Ann Arbor, Mich.: Mathesis Press.

Cronbach, L. J., and G. Gleser. 1953. Assessing similarity between profiles. *Psychological Bulletin* 6:456–73.

de Leeuw, J., F. W. Young, and Y. Takane. 1976. Additive structure in qualitative data: An alternating least-squares method with optimal scaling features. *Psychometrika* 41: 471–503.

Dempster, A. P., N. M. Laird, and D. B. Rubin. 1977. Maximum likelihood from incomplete data via the EM algorithm. *The Journal of the Royal Statistical Society (Series B)* 39:1–38.

De Sarbo, W. S., J. D. Carroll, D. R. Lehmann, and J. O'Shaughnessy. Forthcoming. Three-way multivariate conjoint analysis, *Marketing Science.*

Draper, N. R., and H. Smith. 1981. *Applied regression analysis.* 2d ed. New York: Wiley.

Goodman, L. A. 1978. *Analyzing qualitative/categorical data: Log-linear models and latent-structure analysis.* Cambridge: Abt Books.

Green, P. E., and V. Rao. 1971. Conjoint measurement for quantifying judgmental data. *Journal of Marketing Research* 8:355–63.

Green, P. E., and V. Srinivasan. 1978. Conjoint analysis in consumer research: Issues and outlook. *The Journal of Consumer Research* 5:103–23.

Guttman, L. 1968. A general nonmetric technique for finding the smallest coordinate space for a configuration of points. *Psychometrika* 33:469–506.

Hayashi, C. 1952. On the prediction of phenomena from qualitative data and the quantification of qualitative data from the mathematico-statistical point of view. *Annals of the Institute of Statistical Mathematics* 2:69–98.

Johnson, R. M. 1974. Trade-off analysis of consumer values. *Journal of Marketing Research* 11:121–27.

Kempler, B. 1971. Stimulus correlates of area judgments: A psychophysical developmental study. *Developmental Psychology* 4:158–63.

Krantz, D. H., R. D. Luce, P. Suppes, and A. Tversky. 1971. *Foundations of measurement,* vol. 1. New York: Academic Press.

Kruskal, J. B. 1964a. Multidimensional scaling by optimizing goodness-of-fit to a nonmetric hypothesis. *Psychometrika* 29:1–27.

———. 1964b. Nonmetric multidimensional scaling: A numerical method. *Psychometrika* 29:115–29.

———. 1965. Analysis of factorial experiments by estimating monotone transformations of data. *Journal of the Royal Statistical Society (Series B)* 27:251–65.

———. 1971. Monotone regression: Continuity and differentiability properties. *Psychometrika* 36:57–62.

Kruskal, J. B., and R. N. Shepard. 1974. A nonmetric variety of linear factor analysis. *Psychometrika* 39:123–57.

Lingoes, J. C. 1973. *The Guttman-Lingoes nonmetric program series.* Ann Arbor, Mich.: Mathesis Press.

Luce, R. D. 1959. *Individual choice behavior: A theoretical analysis.* New York: Wiley.

Luce, R. D., and J. W. Tukey. 1964. Simultaneous conjoint measurement: A new type of fundamental measurement. *Journal of Mathematical Psychology* 1:1–27.

Marley, A. A. J. 1970. Additive conjoint measurement with respect to a pair of orderings. *Philosophy of Science* 37: 215–22.

McDonald, R. P. 1968. A unified treatment of the weighting problem. *Psychometrika* 33:351–81.

Medin, D. L., and M. M. Schaffer. 1978. Context theory of classification learning. *Psychological Review* 85:207–38.

Nakanishi, M., and L. G. Cooper. 1974. Parameter estimation for a multiplicative competitive interaction model—Least-squares approach. *Journal of Marketing Research* 11:303–11.

Phipps, A. G., and J. E. Carter. 1982. Neighbourhood preferences of homeowners living in inner city areas affected by apartment redevelopment. Manuscript, University of Saskatchewan.

Ramsay, J. O. 1977. Monotonic weighted power transformations to additivity. *Psychometrika* 42:83–109.

————. 1978. Confidence regions for multidimensional scaling analysis. *Psychometrika* 43:145–160.

Rao, V. R. 1977. Conjoint measurement in marketing analysis. In *Multivariate methods for market and survey data,* ed. J. N. Sheth. Chicago: American Marketing Association.

Roskam, E. E. 1968. *Metric analysis of ordinal data in psychology.* Voorschoten, Holland: VAM.

Sayeki, Y. 1972. Allocation of importance: An axiom system. *Journal of Mathematical Psychology* 9:55–65.

Shepard, R. N. 1972. A taxonomy of some principal types of data and of multidimensional methods for their analysis. In *Multidimensional scaling,* vol. 1, ed. R. N. Shepard et al. New York: Seminar Press.

Srinivasan, V., and A. D. Shocker. 1973. Estimating the weights for multiple attributes in a composite criterion using pairwise judgments. *Psychometrika* 38:473–93.

Takane, Y. 1978. A maximum likelihood method for nonmetric multidimensional scaling: I. The case in which all empirical pairwise orderings are independent—Theory and evaluations. *Japanese Psychological Research* 20:7–17, 105–14.

————. 1981. Multidimensional successive categories scaling: A maximum likelihood method. *Psychometrika* 46:9–28.

————. 1982. Maximum likelihood additivity analysis. *Psychometrika* 47:225–41.

Takane, Y., and J. D. Carroll. 1981. Nonmetric maximum likelihood multidimensional scaling from directional rankings of similarities. *Psychometrika* 46:389–405.

Takane, Y., F. W. Young, and J. de Leeuw. 1980. An individual differences additive model: An alternating least-squares method with optimal scaling features. *Psychometrika* 45:183–

209.

Tversky, A. 1969. Intransitivity of preferences. *Psychological Review* 73:31–48.

Tversky, A., and D. H. Krantz. 1970. The dimensional representation and the metric structure of similarity data. *Journal of Mathematical Psychology* 7:572–96.

Wallsten, T. S. 1976. Using conjoint-measurement models to investigate a theory about probabilistic information processing. *Journal of Mathematical Psychology* 14:144–85.

Winsberg, S., and J. O. Ramsay. 1980. Monotonic transformations to additivity using splines. *Biometrika* 67:669–74.

Young, F. W. 1981. Quantitative analysis of qualitative data. *Psychometrika* 46:357–88.

Young, F. W., J. de Leeuw, and Y. Takane. 1976. Regression with qualitative and quantitative variables: An alternating least-squares method with optimal scaling features. *Psychometrika* 41:505–29.

14

INDCLUS: An Individual Differences Generalization of the ADCLUS Model and the MAPCLUS Algorithm

J. Douglas Carroll and Phipps Arabie

The ADCLUS (for ADditive CLUStering) model was described in detail by Shepard and Arabie (1979; also see references in that paper for earlier, brief presentations by those authors). More concretely, an ADCLUS representation of a stimulus domain consists of a set of m (possibly overlapping) subsets or clusters, each having an associated numerical weight, w_k (where $k = 1, \ldots, m$). For any pair of stimuli, the predicted similarity is simply the sum of the weights of those subsets containing the given pair of stimuli. Shepard and Arabie described the ADCLUS *model* and provided illustrative applications to several data sets. Arabie and Carroll (1980) provided an algorithm called MAPCLUS (for MAthematical Programming CLUStering), differing from that given by Shepard and Arabie (1979) for fitting the same ADCLUS *model*.

The ADCLUS model, as proposed by Shepard and Arabie (1979), and the algorithm for fitting it assumed a single two-way proximities matrix that was one-mode (Tucker 1964), i.e., both the rows and the columns corresponded to the same set of stimuli (mode) being scaled. Carroll (1975, 6) suggested an individual

---

Reprinted from *Psychometrika* (48:157–69) with permission of the authors and the Psychometric Society.

We are indebted to Seymour Rosenberg for making available the data from Rosenberg and Kim (1975). Also, this work has benefited from the observations of S. A. Boorman, W. S. De Sarbo, G. Furnas, P. E. Green, L. J. Hubert, L. E. Jones, J. B. Kruskal, S. Pruzansky, D. Schmittlein, E. J. Shoben, S. D. Soli, and anonymous referees. This research was supported in part by NSF Grant SES82 00441, LEAA Grant 78-NI-AX-0142, and NSF Grant SES80 04815. Authors' addresses: J. Douglas Carroll, Room 2C-553, AT&T Bell Laboratories, 600 Mountain Avenue, Murray Hill, NJ 07974; Phipps Arabie, Dept. of Psychology, University of Illinois, 603 East Daniel Street, Champaign, IL 61820.

differences generalization of the ADCLUS model, to be fitted to a three-way two-mode (subjects × stimuli × stimuli) proximities matrix. Preliminary results for the model and an algorithm for fitting it, both called INDCLUS (for INdividual Differences CLUStering) were given by Carroll and Arabie (1979) (also see Arabie and Carroll 1980, 233).

OVERVIEW

Formally, the (two-way) ADCLUS model predicts interstimulus similarity $\hat{s}_{ij}$ $(i, j = 1, \ldots, n)$, where n is the number of stimuli being clustered) as

$$\hat{s}_{ij} = \sum_{k=1}^{m-1} w_k p_{ik} p_{jk} + c \, , \tag{14–1}$$

where w_k is the numerical weight (assumed to be nonnegative) of the kth cluster $(k = 1, \ldots, m - 1)$; p_{ik} is unity if stimulus i is present in cluster k, otherwise zero. The additive constant c can alternatively be represented as the weight (not assumed to be nonnegative) of an mth cluster comprising the complete set of the n stimuli (sometimes referred to as the "universal set").

The following qualifications to the preceding variables should be noted. The input data to which the model is fitted are assumed to be the $M = n(n - 1)/2$ entries constituting a two-way symmetric (or symmetrized) row and column unconditional proximity matrix having no missing entries. Although the raw data may be in the form of either similarities or dissimilarities, we first transform them linearly to be similarities. (Since the data are assumed to be on an interval scale, the particular transformation chosen in no way affects the goodness-of-fit, but does allow for the standardization of various parameters in the program described below.) $\mathbf{S} \equiv \| s_{ij} \|$ will always refer to these transformed proximities, to which the fitted $\hat{\mathbf{S}}$ matrix is being compared. Turning to the $\mathbf{P}$ matrix, note that each column represents one of the $m - 1$ subsets (or clusters—we use the terms interchangeably), with the ones of that column defining constituency of stimuli within the respective subset.

The INDCLUS model is written

$$s_{ij}^{h} \cong \sum_{k=1}^{m-1} w_{hk} p_{ik} p_{jk} + c_h \, , \tag{14–2}$$

where s_{ij}^{h} is the similarity between stimuli i and j for subject (or other data source) h $(h = 1, \ldots, H)$, w_{hk} is the weight for subject h on subset k, and c_h is the additive constant for subject h.

Thus, we have generalized the ADCLUS model in (14–1) so as to fit a series of H proximities matrices. All H subjects, conditions, or whatever the sources of data happen to be, have the same set of clusters (defined by the unities in each of the $m - 1$ columns of $\mathbf{P}$), but the weights now vary as a joint function of which cluster *and* subject is being considered. (The additive constant c_h also varies over subjects.)

The Arabie and Carroll (1980) MAPCLUS approach to fitting the (two-way) ADCLUS model used a "penalty function" approach based on optimizing, at each stage in an overall iterative process, a composite loss function that combined an A-part (designed to maximize variance accounted for) and a B-part (comprising the penalty function designed to constrain the **P** matrix to be of the appropriate form, i.e., all entries either 0 or 1). Although these components of the loss function will be discussed below in detail, we note that because all subjects in the INDCLUS model are assumed to employ the same subsets, but with different patterns of weights for individual subjects, the B-part of the MAPCLUS composite loss function is unchanged as it appears in the INDCLUS composite loss function. In MAPCLUS, the A-part of the loss function was computed for a single input proximities matrix. For INDCLUS, the A-part of the loss function must be summed and normalized over the H sources of data.

INDCLUS can be viewed as a clustering counterpart to the INDSCAL (Carroll and Chang 1970) model. In the latter model, all subjects are assumed to employ a common set of continuous dimensions, but these dimensions are differentially emphasized by the subjects, so as to yield a weighted Euclidean space. INDCLUS is a discrete model in which subjects differentially weight a common set of clusters. In principle, INDCLUS is applicable to any data set suitable for INDSCAL, although as argued by Shepard and Arabie (1979), discrete models like INDCLUS would seem to be more appropriate for stimulus domains where the investigator is seeking a discrete portrayal of structure. There is some reason to believe that such discrete structures are most appropriate for "conceptual" (as opposed to purely "perceptual") domains of stimuli (see Pruzansky, Tversky, and Carroll 1982). However, even in domains for which a continuous spatial model is more appropriate, a discrete model such as ADCLUS or INDCLUS often provides a useful and informative complementary representation.

The analogy between INDCLUS and INDSCAL breaks down, however, in one interesting detail. The INDCLUS algorithm subsumes MAPCLUS (Arabie and Carroll 1980) as a special case, whereas INDSCAL (Carroll and Chang 1970) and related algorithms for fitting that weighted Euclidean model do *not* subsume two-way multidimensional scaling (e.g., as practiced in the KYST2A program of Kruskal, Young, and Seery 1973). That is, if a data analyst specifies during a run of INDCLUS that there is $H = 1$ subject, the result is comparable to a MAPCLUS analysis. In contrast, such a declaration for INDSCAL results in an indeterminate and generally suboptimal solution unless special steps are taken, rather than a solution comparable to a (metric) KYST2A analysis, or to the "classical" two-way metric approach to multidimensional scaling (Torgerson 1958).

THE ALGORITHM

The Penalty Function Approach

As with MAPCLUS, the loss function for INDCLUS takes the form

$$L_k(\alpha_k, \beta_k, \Delta, \mathbf{P}) = \alpha_k A_k + \beta_k B_k \ . \tag{14–3}$$

Considering first the left side of (14–3), note that the loss function is computed only for subset k. Moreover, we do not sum the penalty function over k. The reason is that we are using an alternating least squares approach (originally called NILES or NIPALS by Wold 1966) which underlies the iterative fitting in turn of each subset p_{ik} ($i = 1, \ldots, n$) and its associated weights w_{hk}. Since we are only fitting the kth subset at any instant, Δ in (14–3) refers to the residuals (centered individually for each subject) computed for the remaining $m - 2$ subsets, so that the reader may wish to associate an implicit subscript "k" and superscript "h" with Δ.

In the right side of equation (14–3), the term $\alpha_k A_k$ is the product of the coefficient α_k and the normalized sum of squared error, A_k. Specifically,

$$A_k = \frac{a_k}{d_k} \ , \tag{14–4}$$

where

$$a_k = \frac{1}{H} \sum_{h=1}^{H} \sum_{i>j}^{n} \sum^{n-1} (\delta_{ij}^h - w_{hk} p_{ik} p_{jk})^2 \tag{14–5}$$

and

$$d_k = \frac{4 \sum_{h=1}^{H} \sum_{i>j}^{n} \sum^{n-1} (\delta_{ij}^h)^2}{MH} \ . \tag{14–6}$$

Minimizing a_k, the squared error summed over the entries in all subjects' matrices, is equivalent to maximizing the variance accounted for (VAF). The denominator of A_k, d_k, is a normalization factor proportional to the variance of the residuals, δ_{ij}^h, computed over the subsets other than k. (The residuals δ^h are so defined that their means $\bar{\delta}^h$ are all zero. Arabie and Carroll (1980) gave the technical explanation for the normalizing constant of 4 in equation [14–6].)

The B-part of the loss function (presented below) is designed to enforce the constraint that $p_{ik} = 0, 1$ even though the p_{ik} are initially allowed to vary continuously. Elaborating on the right side of (14–3), we have

$$B_k = \frac{u_k}{v_k} \ , \tag{14–7}$$

where

$$u_k = \tfrac{1}{2} \sum_{i}^{n} \sum_{j}^{n} [(p_{ik} p_{jk} - 1) p_{ik} p_{jk}]^2 , \qquad (14\text{–}8)$$

and

$$v_k = \sum_{i}^{n} \sum_{>j}^{n-1} (p_{ik} p_{jk} - T_k)^2 , \qquad (14\text{–}9)$$

where T_k is simply the mean of the pairwise products of $p_{ik} p_{jk}$, namely

$$T_k = \frac{1}{M} \sum_{i}^{n} \sum_{>j}^{n-1} p_{ik} p_{jk} . \qquad (14\text{–}10)$$

The rationale for the particular form of the B-part, as well as some unsuccessful alternatives, is given in Arabie and Carroll (1980). The crucial point to be made in the present instance is that B is in no way dependent on the hth subject, since all subjects are assumed to have the *same* indicators of cluster membership p_{ik} ($i = 1, \ldots, n$; $k = 1, \ldots, m - 1$).

As noted below, INDCLUS uses a gradient procedure that employs the partial derivatives of L_k in equation (14–3). Since we require weights α and β of the A- and B-parts, respectively, to obey the normalizing condition that $\alpha + \beta = 1$ (with $\alpha, \beta > 0$) as we fit each subset in turn (see discussion of an "inner" iteration below), L_k is a convex combination of A_k and B_k. After the gradient procedure has maximized L_k for fixed initial values of α and β (see Arabie and Carroll 1980 for details), then the value of β is increased, so as to place more emphasis on the B-part (or penalty function). By successively closer continuous approximations, the B-part enforces the discrete constraint and, in conjunction with the nonlinear function A_k, qualifies our approach as one of "mathematical programming." Ultimately, the entries in **P** are "polished" (rounded) to be exactly 0, 1. Consistent with the alternating least squares approach, this maximization procedure is iteratively applied (conditionally) to each subset in turn.

As noted, INDCLUS uses a gradient procedure to minimize the objective function in (14–3). Since the B-part of that function is unchanged for the two-way (MAPCLUS) case, we refer the reader to Eqs. (12), (14), and (15) in the 1980 paper for the relevant partial derivatives. However, the partial derivative for the A-part for INDCLUS is

$$\frac{\partial a_k}{\partial p_{ik}} = -\frac{2}{H} \sum_{h=1}^{H} w_{hk} \sum_{j \neq i} p_{jk} (\delta_{ij}^{h} - w_{hk} p_{ik} p_{jk}) . \qquad (14\text{–}11)$$

The weights α_k and β_k in the loss function of (14–3) are adjusted according to the description given in Arabie and Carroll

(1980, 216), although the types of normalization used in INDCLUS for matrix unconditional as well as matrix conditional input proximities data (see discussion below) produce a slightly different pattern of balance between the α and β, even though the formulae are unchanged from the MAPCLUS program.

The Alternating Least Squares Structure of INDCLUS

The hallmark of an alternating least squares procedure is, of course, iterative computing nested to several levels of depth. In the MAPCLUS paper (Arabie and Carroll 1980, 217–22), we distinguished between major, outer, and inner iterations. The last of the three is effectively the combustion chamber of the engine that drives INDCLUS and MAPCLUS. For the latter—and simpler—algorithm, an inner iteration consisted of obtaining an estimate via univariate linear regression of the weight w_k and the additive constant c_k specific to subset k, as well as moving each of the p_{ik} ($i = 1, \ldots, n$) elements by one step of the computed gradient.

For fitting a three-way data set used as input to INDCLUS, there are now H times as many weights and additive constants e_k^h ($h = 1, \ldots, H$) for the subsets, so that an INDCLUS "inner loop" includes an innermost loop for each of the H subjects (or other sources of data). For the equations describing an inner iteration in the MAPCLUS paper, if h is added as a superscript for the residuals δ_{ij} and as a subscript for the weights w_k and their additive constants c_k (now renamed as e_k), then the equations provide an appropriate description for INDCLUS' loop structure and are therefore not repeated in the present paper.

Estimating the Weights

Just as in MAPCLUS, regression can be invoked either (a) to fit weights w_{hk} ($k = 1, \ldots, m - 1$; $h = 1, \ldots, H$) and the additive constants c_h to a user-supplied set of binary features (coded as the $\mathbf{P}$ matrix in [14–2]) to yield a "constrained solution" (see Carroll and Arabie 1980), without recourse to any of the iterative computation described above, or (b) at the end of each outer iteration (viz., series of inner iterations, all applied to the same subset k), using the iteratively fitted values of p_{ik}.

Arabie and Carroll (1980, 223) noted that in the MAPCLUS (two-way) case, this least squares regression problem requires solving

$$\mathbf{s} = \mathbf{Q}\mathbf{w} \ , \qquad (14\text{–}12)$$

where $\mathbf{s}$ is the column vector of M elements with $s_{(ij)}$ as the general entry (with the concatenated subscript $(ij) = 1, \ldots, M$), $\mathbf{Q}$ is the $M \times (m + 1)$ matrix with $q_{(ij)k}$ as the general entry, where $q_{(ij)k} = p_{ik}p_{jk}$, and $\mathbf{w}$ is the $(m + 1)$-dimensional column vector with general entry w_k (where $w_{m+1} = c$).

For INDCLUS, equation (14–12) now becomes

$$\mathbf{s}^h = \mathbf{Q}\mathbf{w}^h \ , \tag{14–13}$$

and it is noteworthy that we only have to obtain the pseudoinverse of $\mathbf{Q}$ once in order to solve for the H sets of weights, $\mathbf{w}^h$. Thus, although there are H times as many weights to solve for with INDCLUS, compared to the two-way MAPCLUS case, the computational burden does not increase proportionally. On the other hand, the increased number of weights to be estimated allows more opportunities for negative weights to occur, and they have no simple interpretation in this model. In the case of MAPCLUS, negative weights were so infrequently encountered that we found them to be no cause for concern. However, for some three-way data sets, negative weights have been sufficiently annoying that we have included an option, at the user's discretion, of constraining the w_{hk} ($h = 1, \ldots, H; k = 1, \ldots, m - 1$) to be nonnegative at the end of each outer iteration. The procedure for implementing the optional constraint is taken from Lawson and Hanson (1974, ch. 23).

Normalizing the Input Proximities Matrices

The INDCLUS program allows the user to declare whether the input matrices are matrix unconditional or matrix conditional (see Takane, Young, and de Leeuw 1977). In the former case, all matrices are assumed to be comparable (or, roughly speaking, "on the same scale"). Thus, such data (over all H matrices, simultaneously) are put on the interval [0,1]. (Then, to simplify the mechanics of regression, the mean $\overline{s}^h$ of each matrix is subtracted from all entries within the matrix.) In the case of matrix conditional data, each of the H matrices is separately standardized to have a mean of 0 and a standard deviation of 1.0. When the declaration of matrix conditional data is made, the weights often tend to be considerably larger than those resulting from an unconditional analysis.

Other Computational Details

Arabie and Carroll (1980, 224–26) described two strategies of combinatorial optimization in MAPCLUS, executed after the completion of the alternating least squares phase, and involving singleton and "doubleton" reversals of the constituent elements of each cluster in turn. (The more general strategy of "l-tuple" reversals was also developed but not implemented in the MAPCLUS program because of computational expense.) These same two strategies are incorporated in INDCLUS. Moreover, the equations used to describe the procedure in MAPCLUS only require adding a superscript to the residuals δ_{ij}^h and all functions of them, plus an extra summation over $h = 1, \ldots, H$ in the numerator of the variances in order to be appropriate for INDCLUS.

Just as with MAPCLUS, initial values of $\mathbf{P}$ for the alternating least squares phase of INDCLUS may be: (a) user-supplied, (b) the output from a random number generator, or (c) the output of a rational strategy. For INDCLUS, the third case consists of aggregating the *H* proximities matrices (after normalization) to form a (single) two-way matrix which then becomes input for the rational procedure used in MAPCLUS. An alternative and more expensive approach also available to the user is to execute a complete two-way (MAPCLUS) analysis for the aggregate two-way matrix and, after obtaining the *final* two-way solution, use it as the initial configuration for INDCLUS.

APPLICATIONS

Examples of fitting the two-way case of the ADCLUS model have appeared in Arabie and Carroll (1980, 1984), Arabie et al. (1981), Eckes (1981), Rabin and Frank (1982), and Shepard and Arabie (1979). Several of these papers observed that further information about individual differences in weighting of the clusters would be of interest. We now present two applications of the INDCLUS model, fitted by the algorithm we have just described. (Yet another application is found in K. Miller and Gelman [1983].)

Kinship Data of Rosenberg and Kim (1975)

The fifteen most commonly used kinship terms (given in Table 14–1) were printed on slips of paper for use in a sorting task (see G. A. Miller 1969; Rosenberg 1977, 1982 for details of the paradigm used by Rosenberg and Kim 1975). Eighty-five male and eighty-five female subjects were run in the condition where subjects gave (only) a single-sort of the fifteen terms. A different group of subjects (eighty males and eighty females) were told that, after making their first sorts of the terms, they should give additional subjective partitioning(s) of these stimuli using "a different basis of meaning each time." Rosenberg and Kim (1975) used only the data from the first and second sortings for this group of subjects. Thus, we have six *conditions,* which will be our "subjects" for an INDCLUS analysis: females' single-sort, males' single-sort, females' first-sort, males' first-sort, females' second-sort, and males' second-sort. Again note that the subjects in the first two conditions were distinct from the subjects in the last four conditions.

There is a further technical detail required in the present description of the Rosenberg and Kim (1975) data. The subjects' partitions of the stimuli comprise nominal scale data that do not immediately assume the form of a proximities matrix. Thus, some pre-processing is necessary to obtain such a matrix. If we form a stimuli × stimuli co-occurrence matrix for each experimental condition, with the (i,j)th entry defined as the number of subjects who placed stimuli i and j in the same group, and subtract that entry from the total number of subjects contributing to the matrix, then we have a measure of dissimilarity that Drasgow and Jones (1979; also see G. A. Miller 1969) have called the S-measure. Alternatively, we can take the squared Euclidean distance between all rows/columns of the (symmetric) co-occurrence matrix to obtain an indirect measure of dissimilarity between all pairs of

TABLE 14-1. INDCLUS Solution for Kinship Data (Matrix Unconditional Analysis Using S-measure)

Weights for Different Sources of Data							
Females' Single-Sort	Males' Single-Sort	Females' First Sort	Females' Second Sort	Males' First Sort	Males' Second Sort	Elements of Subset	Interpretation
.052	.143	.551	.241	.299	.295	brother, father, grandfather, grandson, nephew, son, uncle	Male relatives, excluding cousin
.049	.146	.554	.246	.291	.306	aunt, daughter, granddaughter, grandmother, mother, niece, sister	Female relatives, excluding cousin
.552	.397	.283	.373	.340	.237	aunt, cousin, nephew, niece, uncle	Collateral (Romney and D'Andrade 1964) relatives
.478	.372	.206	.322	.241	.219	brother, daughter, father, mother, sister, son	Nuclear family
.626	.449	.251	.385	.395	.253	granddaughter, grandfather, grandmother, grandson	Direct ancestors and descendants ±2 generations removed
.055	.075	.132	.158	.158	.207	Additive constants	
78.6%	68.8%	96.3%	78.9%	82.4%	71.7%	Variance accounted for within condition	Overall VAF = 81.1%

Source: Rosenberg and Kim 1975.

stimuli. Rosenberg and his colleagues (e.g., Rosenberg and Sedlak 1972) refer to the latter measure of dissimilarity as the δ measure (not to be confused with the earlier use of that symbol in this paper). Extensive comparisons of the performance of the two measures when used as input to nonmetric multidimensional scaling have been conducted by L. E. Jones and his students (e.g., Drasgow and Jones 1979). In addition to these two different approaches to pre-processing the data, there is also the matrix conditional/unconditional dichotomy. Thus, in summary, we have four different approaches to analyzing $H = 6$ proximities matrices.

By way of substantive motivation, Rosenberg and Kim (1975) noted the two straightforward bases of partitioning the kinship terms: (a) by family and generational considerations (e.g., nuclear family versus grandchildren) and (b) by sex, which is inherent in all the terms except cousin. Given these two conflicting bases for organization, it is naturally of interest to look at differences among groups of subjects in the use of these two schemes of classification, and to see which experimental paradigms and methods of portraying structure in the data could faithfully depict such differences. In seeking a discrete representation of structure via hierarchical clustering, Rosenberg and Kim (1975, 496, Fig. 2) found it necessary to analyze various of their matrices *separately,* in order to recover evidence of the two bases of judgmental organization.

Table 14–1 presents a five-cluster solution from an INDCLUS analysis of the six matrices, using the S-measure (matrix unconditional analysis), with 81.1% of the variance accounted for (over all six matrices). The clusters are easily interpreted. In the order listed, the first two are sex-defined, the third is the collateral relatives (in the terminology of Romney and D'Andrade 1964; also see Boorman and Olivier 1973), the fourth is the nuclear family, and the fifth consists of grandparents and grandchildren. The patterns of the weights also yield interesting results. For example, the statement of Rosenberg and Kim (1975, 489) that subjects restricted to a single-sort ignore sex as a basis of organization is strongly supported by the relatively low weights for the sex-defined clusters in the first two columns (especially for female subjects) of Table 14–1. For the multiple-sort conditions, it is interesting to note that female subjects emphasized sex in the first sorting (given that the two relevant clusters have much higher weights), whereas male subjects waited until the second sorting to emphasize the salience of sex as a factor in sorting the kinship terms. Across all conditions, females' data were better fitted to the model than were males' data. Also, data from the first sort were better fitted than for the second sort, for both females and males.

We noted earlier that an INDCLUS analysis using S-measure data assumed to be matrix unconditional is only one of four possible approaches to the data of Rosenberg and Kim (1975). For the remaining three, we obtained the same five clusters, with weights having nearly the same patterns as those given in Table 14–1, and thus corroborating the substantive points given in the preceding discussion. Overall variances accounted for were: 79.5% for S-measure conditional, 89.4% for δ-measure unconditional, and 88.1% for δ-measure conditional. The last two analyses each yielded one slightly negative weight (-.0036, and -.014, respectively).

Factions in the U.S. Supreme Court, 1975–79 Terms

Data from U.S. Supreme Court decisions have been analyzed by various scaling and related techniques (Dawes, Brown, and Kaplan 1965; also see Stookey and Baer 1976, and references therein, and Provine 1980, for substantive background on the related problem of case selection), but not by overlapping clustering techniques. Since 1949, the *Harvard Law Review* has published an annual summary of Court activities for the preceding year. From the November issues of the *Harvard Law Review* (1976, 1977, 1978, 1979, 1980), we took the data on "voting alignments" for the same nine Justices during the 1975–79 terms of the Court. (The late Justice Douglas was present infrequently during his last term in 1975 and is not included in the present analysis.) The percent of concurring votes for each pair of Justices forms the entry in a proximities matrix for each of the H = 5 terms of the Court from 1975 through 1979.

Over all the five terms of the Court considered in the analysis presented in Table 14–2, the dyadic cluster of Justices Brennan and Marshall receives the highest weights. During the 1976 and 1977 terms, those liberal Justices concurred 93.6% of the time, the highest such agreement over all the entries in the five matrices. The second cluster (in the order listed in Table 14–2) comprises the four Nixon appointees to the Court, plus Justices Stewart and White, who were most often regarded as conservatives.

We have labeled the third cluster "Conservative-Centrist." According to the (controversial) popular press account of the Court offered in *The Brethren* (Woodward and Armstrong 1981), Justice Powell was "the most moderate of the four Nixon appointees" (1981, 528), and Justices Powell, Stevens, and Stewart were the centrist group (1981, 518–24) who prevailed during a series of verdicts involving capital punishment. Justice Rehnquist has been more conservative than centrist, but his frequent agreement with Chief Justice Burger presumably links him with the other Justices in that cluster.

The fourth cluster subsumes the two liberal Justices, Brennan and Marshall, from Cluster 1. They were in agreement with Justice Blackmun on abortion issues and civil rights. The latter issue in turn links all three to Justice White, a Kennedy appointee. Justice Stevens' "wild card" (Woodward and Armstrong 1981, 508) voting record and frequent role as a swing voter qualify him as a centrist member of Cluster 4.

To elaborate on these last two clusters, we note that only Justice Stevens is common to both. The remaining eight Justices are evenly partitioned between the two clusters. While some Justices seem firmly entrenched in their membership in one of the two clusters as interpreted (e.g., Rehnquist and Stewart in Cluster 3; Brennan and Marshall in Cluster 4), it seems likely that other Justices (e.g., Powell and Blackmun) would be assigned with more variability if data over a longer period (and

TABLE 14-2. INDCLUS Solution for Voting Alignments of U.S. Supreme Court Justices 1975-79 (Matrix Unconditional Analysis)

Weights for Different Terms						
1975	1976	1977	1978	1979	Justices in Subset	Interpretation
.548	.640	.627	.427	.424	Brennan, Marshall	Liberals
.342	.314	.105	.278	.258	Blackmun, Burger, Powell, Rehnquist, Stewart, White	Conservatives
.142	.106	.144	.195	.169	Burger, Powell, Rehnquist, Stevens, Stewart	Conservative-Centrist
.089	.120	.096	.262	.163	Blackmun, Brennan, Marshall, Stevens, White	Centrist-Liberal
.266	.240	.278	.212	.233	Additive constants	
84.7%	80.5%	62.3%	83.7%	71.1%	Variance accounted for	Overall VAF = 77.7%

covering more issues) had been included. It is also of interest to note the greater variability of the weights of the Centrist-Liberal cluster, as compared to the Conservative-Centrist cluster.

These four clusters (plus an additive constant for each matrix) accounted overall for 77.7% of the variance. A matrix conditional analysis of the same data yielded the same clusters, with the same pattern of weights, and 76.4% VAF.

DISCUSSION AND FUTURE PROSPECTS

In the published version of his R. A. Fisher Memorial Lecture, W. H. Kruskal (1981, 511) notes the continuing importance of classification in statistics and gives humorous examples (1981, 511, footnote 3) of the difficulties that can arise from a classification with overlapping categories. Nonetheless, he points out that one of his examples was apparently of great utility and importance in the history of a subdiscipline of zoology. We have earlier cited examples of successful interpretations of fitting the two-way ADCLUS (Shepard and Arabie 1979) model, using the two-way algorithm MAPCLUS. The three-way INDCLUS generalization raises several issues.

First, we note that in multidimensional scaling, the generalization from two- to three-way models and algorithms (Carroll and Chang 1970; Tucker 1972) seemed to answer statistical needs that had previously not been satisfied, as inferred from the heavy usage these three-way models of scaling have enjoyed (see Carroll and Arabie 1980; Carroll and Wish 1974). To date, the only other *discrete* counterpart besides INDCLUS to three-way scaling methods has been the multiple hierarchical clustering approach of Carroll and Pruzansky (Carroll and Pruzansky 1975, 1980). It is our hope that these clustering methods may be as useful for portraying *discrete* structure as three-way scaling methods (e.g., INDSCAL) have been for continuous structure. For example, recent developments in sociometry, emphasizing the different types of social ties that underlie social structure (Boorman and White 1976), have demonstrated the need for a three-way discrete representation of social structure that depicts not only the actors involved but also how the different types of social ties (corresponding to the "subjects" mode in INDCLUS) are interrelated. (See Knoke [1983], as well as Arabie and Carroll [1984], for a substantive discussion of this point.)

Second, just as the advent of INDSCAL (Carroll and Chang 1970) overcame some of the rotational problems found in two-way multidimensional scaling, INDCLUS offers a possible solution to one inelegant feature of the ADCLUS model. Specifically, Shepard and Arabie (1979, 95–97) noted that three dyadic ADCLUS clusters (a,b), (b,c), (a,c), all with identical weights would have the same goodness-of-fit as the single cluster (a,b,c) having the same weight as that shared by the three dyadic clusters. The principle of parsimony, however, was invoked to favor the representation having the least number of clusters, as a means of resolving this particular lack of uniqueness. George Furnas (1982) has pointed out that for the INDCLUS model, unless all three dyadic clusters had tied weights for *each* subject, then the three clusters would clearly be discriminable from the composite cluster. Thus, we see INDCLUS as offering a solution to one potential aspect of nonuniqueness in the ADCLUS model (albeit one for which there is as yet no evidence of occurrence for empirical data).

Third, we note that De Sarbo's (1982) recent approach to fitting the (two-way) ADCLUS model could also be generalized to the three-way case. De Sarbo's GENNCLUS algorithm allows for symmetric *or* nonsymmetric input proximities data, clustering of both modes (if they are distinct) of a two-way two-mode matrix, and the optional imposition of various constraints not available in other approaches to fitting the ADCLUS model. As noted, these features could in principle be generalized to the three-way case.

While it is customary for "future prospects" sections of articles to be expansive and optimistic, we must depart from this tradition in one detail. An obvious next step for both the MAPCLUS and INDCLUS algorithms would be to incorporate monotone regression (J. B. Kruskal 1964a, 1964b) so that the input data could be assumed to be just ordinal instead of interval (see Shepard and Arabie 1979, 118). However, we have recently discovered a general theoretical degeneracy inherent in the fitting of discrete models like ADCLUS and INDCLUS when the fitting entails optimizing a measure of badness-of-fit such as Kruskal's (1964a, 1964b) "stress."

Specifically, for a single two-way one-mode proximities matrix (e.g., input to MAPCLUS), assume that there is a unique largest pairwise similarity (or smallest dissimilarity) value, s_{ij}. Now let those two objects i and j form a dyadic cluster whose weight is simply $s_{ij} - c$, where c is the additive constant that corresponds to the weight for the complete set (implicitly present) of the n objects. Thus we have two clusters—(i,j) and the complete set—with weights of $(s_{ij} - c)$ and c, respectively. Now the predicted similarity between i and j will be $\hat{s}_{ij} = s_{ij}$, and for all other y,z (distinct from each other and from i and j), $\hat{s}_{yz} = c$. This "solution" yields a zero value of stress, for both stress formulae one and two (Kruskal and Carroll 1969), and qualifies as "degenerate" since there are only two distinct fitted values $\hat{s}$, in contrast to an input matrix **S** typically having far more than two distinct values. If there is not a unique largest similarity value in the matrix, then a dyadic cluster may be required for each of the pairs of objects corresponding to the tied largest value. However, if the ties occur so as to render the set of corresponding objects a maximal complete subgraph, then one cluster will suffice.

For a three-way two-mode proximities matrix (suitable as input for INDCLUS), if the largest similarity value corresponds to different pairs of objects across subjects in a matrix conditional analysis, then a nonmetric version of INDCLUS might be feasible, if there are enough subjects. Thus, the generalization of ADCLUS/MAPCLUS to the three-way INDCLUS offers a possible answer to the problem of inherent degeneracies in fitting this nonmetric version of a discrete model.

Thus, just as we earlier (Arabie and Carroll 1980, 226–27) concluded that there is an inherent nonuniqueness in various discrete models such as ADCLUS and INDCLUS, we now note an inherent degeneracy lurking when traditional approaches to monotone regression are considered for fitting data to such models. A similar degeneracy can also be demonstrated for nonmetric approaches to fitting tree structures and some other discrete models, at least in the two-way cases.

REFERENCES

Arabie, P., and J. D. Carroll. 1980. MAPCLUS: A mathematical programming approach to fitting the ADCLUS model. *Psychometrika* 45:211–35.

———. 1984. Conceptions of overlap in social structure. In *Methods of social network analysis,* ed. L. Freeman, A. K. Romney, and D. R. White. Berkeley and Los Angeles: University of California Press.

Arabie, P., J. D. Carroll, W. De Sarbo, and J. Wind. 1981. Overlapping clustering: A new method for product positioning. *Journal of Marketing Research* 18:310–17.

Boorman, S. A., and D. C. Olivier. 1973. Metrics on spaces of finite trees. *Journal of Mathematical Psychology* 10:26–59.

Boorman, S. A., and H. C. White. 1976. Social structure from multiple networks. II. Role structures. *American Journal of Sociology* 81:1384–1446.

Carroll, J. D. 1975. Handout for models for individual differences in similarities judgments. Paper presented at the Mathematical Psychology Meetings, Purdue University, West Lafayette, 25–27 August.

Carroll, J. D., and P. Arabie. 1979. INDCLUS: A three-way approach to clustering. Paper presented at the Meeting of the Psychometric Society, Monterey, California, June.

———. 1980. Multidimensional scaling. *Annual Review of Psychology* 31:607–49.

Carroll, J. D., and J. J. Chang. 1970. Analysis of individual differences in multidimensional scaling via an N-way generalization of "Eckart-Young" decomposition. *Psychometrika* 35:283–319.

Carroll, J. D., and S. Pruzansky. 1975. Fitting of hierarchical tree structure (HTS) models, mixtures of HTS models, and hybrid models, via mathematical programming and alternating least squares. Paper presented at the U.S.–Japan Seminar on Theory, Methods and Applications on Multidimensional Scaling and Related Techniques, University of California, 20–24 August.

———. 1980. Discrete and hybrid scaling models. In *Similarity and choice,* ed. E. D. Lantermann and H. Feger. Bern: Hans Huber.

Carroll, J. D., and M. Wish. 1974. Models and methods for three-way multidimensional scaling. In *Contemporary developments in mathematical psychology,* vol. 2, ed. D. H. Krantz, R. C. Atkinson, R. D. Luce, and P. Suppes. San Francisco: W. H. Freeman.

Dawes, R. M., M. E. Brown, and N. Kaplan. 1965. The skewed hourglass: A configurational approach to constructing a Guttman scale when domination is unspecified. Paper presented at the annual convention of the Midwestern Psychological Association, Chicago, April.

De Sarbo, W. S. 1982. GENNCLUS: New models for general nonhierarchical clustering analysis. *Psychometrika* 47:449–75.

Drasgow, F., and L. E. Jones. 1979. Multidimensional scaling of derived dissimilarities. *Multivariate Behavioral Research* 14:227–44.

Eckes, T. 1981. Formale modelle zu ähnlichkeitsstrukturen. Doctoral dissertation, University of the Saar.

Furnas, G. 1982. Personal communication. June.

Harvard Law Review 90 (1976): 277.

———. 91 (1977): 296.

———. 92 (1978): 328.

———. 93 (1979): 276.

———. 94 (1980): 288.

Knoke, D. 1983. Organization sponsorship and influence: Representation of social influence associations. *Social Forces* 61:1065–87.

Kruskal, J. B. 1964a. Multidimensional scaling by optimizing goodness-of-fit to a nonmetric hypothesis. *Psychometrika* 29:1–27.

———. 1964b. Nonmetric multidimensional scaling: A numerical method. *Psychometrika* 29:115–29.

Kruskal, J. B., and J. D. Carroll. 1969. Geometric models and badness-of-fit functions. In *Multivariate analysis II,* ed. P. R. Krishnaiah. New York: Academic Press.

Kruskal, J. B., F. W. Young, and J. B. Seery. 1973. *How to use KYST, a very flexible program to do multidimensional scaling and unfolding.* Murray Hill, New Jersey: Bell Telephone Laboratories.

Kruskal, W. H. 1981. Statistics in society: Problems unsolved and unformulated. *Journal of the American Statistical Association* 76:505–15.

Lawson, C. L., and R. J. Hanson. 1974. *Solving least-squares problems.* Englewood Cliffs, N.J.: Prentice-Hall.

Miller, G. A. 1969. A psychological method to investigate verbal concepts. *Journal of Mathematical Psychology* 6:169–91.

Miller, K., and R. Gelman. 1983. The child's representation of number: A multidimensional scaling analysis. *Child Development* 54:1470–79.

Provine, D. M. 1980. *Case selection in the United States Supreme Court.* Chicago: University of Chicago Press.

Pruzansky, S., A. Tversky, and J. D. Carroll. 1982. Spatial versus tree representations of proximity data. *Psychometrika* 47:3–24.

Rabin, M. D., and M. Frank. 1982. Research, University of Connecticut Health Center, August.

Romney, A. K., and R. G. D'Andrade. 1964. Cognitive aspects of English kin terms. *Transcultural studies in cognition. American Anthropologist Special Issue* 66:146–70.

Rosenberg, S. 1977. New approaches to the analysis of personal constructs in person perception. In *Nebraska Symposium on Motivation,* vol 24, ed. J. K. Cole. Lincoln: University of Nebraska Press.

———. 1982. The method of sorting in multivariate research with applications selected from cognitive psychology and person perception. In *Multivariate methods in the social sciences: Applications,* ed. N Hirschberg and L. G. Humphreys. Hillsdale, N.J.: Erlbaum.

Rosenberg, S., and M. P. Kim. 1975. The method of sorting as a data-gathering procedure in multivariate research. *Multivariate Behavioral Research* 10:489–502.

Rosenberg, S., and A. Sedlak. 1972. Structural representations of implicit personality theory. In *Advances in experimental social psychology* vol 6, ed. L. Berkowitz. New York: Academic Press.

Shepard, R. N., and P. Arabie. 1979. Additive clustering: Representation of similarities as combinations of discrete overlapping properties. *Psychological Review* 86:87–123.

Stookey, J. A., and M. A. Baer. 1976. A critique of Guttman scaling: With special attention to its application to the study of collegial bodies. *Quality and Quantity* 10:251–60.

Takane, Y., F. W. Young, and J. de Leeuw. 1977. Nonmetric individual differences multidimensional scaling: An alternating least-squares method with optimal scaling features. *Psychometrika* 42:7–67.

Torgerson, W. S. 1958. *Theory and methods of scaling.* New York: Wiley.

Tucker, L. R. 1964. The extension of factor analysis to three-dimensional matrices. In *Contributions to mathematical psychology,* ed. N. Frederiksen and H. Gulliksen. New York: Holt, Rinehart and Winston.

———. 1972. Relations between multidimensional scaling and three-mode factor analysis. *Psychometrika* 32:3–27.

Wold, H. 1966. Estimation of principal components and related models by iterative least-squares. In *Multivariate analysis,* ed. P. R. Krishnaiah. New York: Academic Press.

Woodward, B., and S. Armstrong. 1981. *The brethren.* New York: Avon.

PART IV
Reflections

15

Destructive Retrieval in the Realm of Three-Mode Thinking

Peter Gould

> Before the science begins to work at all, its field of vision is already so narrowed that it sees only certain objects and even those in a very definitive light.
>
> Rudiger Bubner, *Modern German Philosophy*

In a volume honoring the name and work of Ledyard Tucker, it may well be considered ill-mannered to offer a reflective essay grounded in contemporary philosophy and modern mathematics that must take to task the tradition of thinking represented by his work. Yet I suspect he will be the first to approve and appreciate such an "outside-looking-in" stance, even if some of the bits and pieces are considered to be uninformed and misguided.

Two decades ago, it was Tucker's own work that represented an attempt to extend the conventional methodological bounds, and such extensions only come from those who chafe against the existing constraints. In contrast, those who do not feel the binding nature of the constraints, those who feel quite comfortable within a traditional methodological framework, see no need to leave the well-cultivated valley by a rough and uncharted road leading to higher ground. Yet it is only by moving outside of the comfortable valley of conventional thinking that we can achieve a different perspective on it. Even if the ground is hard and unbroken at higher altitudes, the air is sharp and bracing, and it is from such viewpoints that the inadequacies and limitations of more sheltered thinking become apparent, even as new and farther horizons come into view.

If we have been born and bred in a particular methodological valley, a valley that has proved fruitful and nurturing, it becomes increasingly difficult to see that other possibilities might exist. Having virtually severed the once-strong ties to philosophy—a tradition of reflective thinking that gave birth to the natural sciences in the seventeenth century, the biological in the eighteenth, and the human sciences in the nineteenth—we seem to have lost the awareness of our own historicity: that sense of the past both containing and limiting the possibilities for the future. Too frequently, graduate education becomes an exercise in conventional cultivation and traditional husbandry on the valley floor, rather than more dangerous but always exhilarating rock climbing up the sides.

A discussion of all the issues and possibilities raised by Tucker's work requires, quite literally, a book to explicate and unfold every point properly and fairly. Thus, this essay must be written in a pointilliste style, which means that each spot of "color" will have a tendency to appear separate and distinct. However, if the artist is even partially successful, certain forms will emerge if the viewer is prepared to take a few paces backwards. Put another way, the initial close impression may well produce blurred vision and convey a sense of intellectual shortsightedness, but some distancing may result in more holistic and perhaps even disturbing impressions coming into view, impressions that lead us back to fresh thinking and new appraisals.

In brief, this essay will be critical in the tradition of destructive retrieval (Heidegger 1962, 1982; Pöggler 1972; Aler 1972; Kockelmans 1972), a tradition that acknowledges gratefully the way thinking in the past builds structures of possibilities for thinking in the future. You cannot think forward and beyond the present if the pieces required for such thinking are not in place and capable of being connected. In the same way that Tucker could not think forward to three-mode factor analysis without the more conventional factor analytical tradition and linear mathematics already in place, so we cannot move our own thinking forward without the now-existing possibilities open up by contemporary philosophy and modern (that is to say, twentieth century) mathematics. Destructive retrieval means acknowledging and honoring the best thinking of the past to help us think forward today. In philosophy, for example, some of the logical positivists of the Vienna Circle and their inheritors, the linguistic analysts, became dismissive if not contemptuous of the 2,500-year-old tradition of careful and reflective thinking that we call *classical philosophy*. After all, who needs such metaphysical nonsense when we either have the answers or have "clearly shown" the questions to be "none-sense"? In contrast, a Martin Heidegger will honor the tradition of thinking in the course of destructive critique, a critical perspective that retrieves the best of the tradition, even as it demonstrates a profound dissatisfaction with the continuing inadequacies.

We face exactly the same sort of problem here. Either we have finally reached the methodological millennium in the current and almost all-pervasive mode of multivariate and statistical thinking, or we have not. Either we will still be approaching complex human research materials in this framework a hundred years from now, or we will have moved on. Either we will still be using linear filters to strain our rich data in the year 2100 (because we really did reach the methodological bounds in the 1960s), or we will be doing something else. In brief, *n*-mode factor analysis with a few more bells and whistles—but still underpinned by nineteenth century mathematics—will either represent the ultimate methodology to which all human scientists aspire, or it will be seen as a part of the historical progression to something else. In other words, we will either be basking in satisfaction at the extraordinary perspicacity of those pioneers in the early 1960s, or we will have started thinking for ourselves again. In what direction might such thinking go?

To ask this question is to raise the possibility of breaking out of the valley. Not that this first step will take us very far, but even by moving to the lower slopes, we achieve some sense of distance from what Kuhn called the "normal science" going on below (Kuhn 1962)—the traditional sort of thinking and the accepted way of cultivating a particular field of inquiry. When we do this, we see that the factor-analytic tradition is based upon the nineteenth century linear mathematics of Hamilton, Cayley, and Helmholtz (Klein 1980). Furthermore, like so many areas of mathematics developed during the preceding 200 years, it was motivated by a desire to describe the physical world. Now I am not going to maintain that a particular form of mathematics *is* a language (that issue is much too complex to argue properly here), but it certainly possesses some sort of linguistic "feel" to it. And if a language can *shape* thinking, if it can also *trap* thinking in its own linguistic realm, and perhaps, in some very deep sense, *is* thinking itself, then we have the obligation to think through what this might mean when we take a mathematical framework generated by descriptive needs in one area and use it for descriptive purposes in another.

Like the calculus of Newton and Leibnitz before it, the linear mathematics of Hamilton arose out of a reflection upon the descriptive requirements of the physical world of *things*. Now this was conceived essentially as a world of mechanism, and in Hamilton's time (1805–65), this meant a deterministic world, a world described by the physical laws of Newton and the equations of Maxwell. In brief, it was a world in which the deterministic forms of functions required *X* to cause *Y* and where functional complexity could be decomposed by the application of appropriate binary operations into independent and linearly additive pieces (Gould 1983a). The harmonic decompositions of Fourier in the calculus and the characteristic equations of Hamilton and Caley in linear algebra were meant to stand as descriptions of the physical world investigated at what we might term the *mesolevel*—between the particle level of quantum theory and the cosmological scale of relativity theory.

Simply as an historical stage aside, it is interesting to note how the work of these men represented attempts to break out of the valleys of normal science in their own day, valleys that had been so well cultivated and surveyed that many late nineteenth century physicists were thinking in terms of only a few more mopping-up operations before all physical questions were settled once and for all. But what is important to see here is that the mathematics of this physical, mechanistic, and deterministic world of linear additive decomposition is the mathematics of factor analysis in all its varieties.

But is it the appropriate mathematics? Does it represent an appropriate "language" that has been carefully developed and thoughtfully adopted out of a patient reflection on the phenomena that we are trying to describe? Or are we making what we might call a "linguistic analogy," so forcing the phenomena onto a mathematical framework that actually filters and crushes out of existence much of the information that we have so painstakingly gathered? Once again, we can only gain some perspective on this question by standing off from it and thinking through what it is that we do when we look at the world through our factor-analytic spectacles. If we are genuinely prepared to back off and *think,* we see that this particular methodological approach actually creates a prestructured "text." The text is essentially geometric, and it is created by filtering our data through a series of linear sieves that by the nature of the binary operations performed *must* break the whole into separate, orthogonal, and linearly additive pieces. This actually constitutes a massive piece of structural engineering, a manipulation of the data so vigorous that even if we were to start with a well-connected structure, we would still find it thoroughly fragmented and torn apart after it was all over. And notice that the *structure* of our final geometric text is wholly imposed: It is not there in the data in any sense; we have not described what actually *is;* we have not let the data "speak for itself" (Gould 1981a). Rather, we have chosen a simplistic linear methodology that has *forced* a structure on the data; the structure comes from the methodology, not from the phenomena themselves. Our geometric text is born out of considerable mathematical violence, and this comment applies to both conventional factor-analytic forms founded on correlation matrices and three-mode forms founded upon weighted covariances. In brief, we have created our text out of the cognitive interest that Jurgen Habermas has termed the *technical* (Habermas 1971).

The next stage is obvious: We have imposed a structure on the data, and there it stands, inanimate and mute, incapable of meaning anything by itself. Clearly, the next step is to adopt the *hermeneutic* perspective, the second cognitive interest of Habermas that implies the interpretation of text. And here a difficult problem arises: Our interpretation in the factor-analytic mode of thinking is already preconditioned. Only that information that has managed to squeeze through the filter forms part of the structure available for the interpretative task. This either ends up as a lame labeling of orthogonal axes into Good-Bad, Hot-Cold, Smart-Stupid, Rich-Poor, Free World-Commie World, and other dichotomous banalities, or it becomes a simplistic experimentation with any one of half-a-dozen oblique schemes whose configurations are wholly dependent on the number of axes chosen and the amount of redundant information thrown into the original data set. Too often the number of axes chosen depends on a mathematically ingenuous criterion of an eigenvalue exceeding a threshold value (Rao 1955), when in fact this decision can only legitimately rest upon the meaning of the initial choice of variables, a choice that also conditions the degree of redundancy that appears geometrically as bundles of vectors whose weight is taken into account by all orthogonal and oblique rotation schemes. After Kronecker products have done their job, it is practically impossible to grasp—that is to say *visualize*—a geometric text. And if we cannot visualize something, what chance do we have of interpreting it? Back in the early 1960s, we were only half-joking when we called *G*, the three-dimensional core matrix, *GOD*.

Even within this framework of thinking, is it possible to point to a *single* example where our understanding of the world has been genuinely illuminated, where our knowledge has been effectively enlarged, by a three-mode approach? Or are all the examples a reexamination of data sets (Tucker 1965b; Mills and Tucker 1965), simply methodological exercises (and perfectly legitimate as these) that constitute some desperate attempt to give human meaning to an imposed, filtered, and over-engineered *structure* (Levin 1963)? Down in the deep factor-analytic valley, does anyone ever ask anymore if the very framework adopted is appropriate? Or has so much intellectual and emotional energy been given to learning this highly constrained form of shaping the world that few can ever "get on the outside looking in"? And make no mistake about it, we invest a great deal of ourselves in our methodologies, and too often we simply train our graduate students as we have been trained ourselves. How else can we explain sixty years of post-Thurstonian training (not, decidedly not, *education*!) in these forms of structural engineering in the human sciences? Having made such commitments of time and energy in what has become the normal science, who is prepared to leave the well-cultivated fields and climb even to the foothills?

Which raises, of course, the third cognitive interest of Habermas, the *emancipatory* perspective. How can we help people to reflect upon what they are doing, when nearly all ties to philosophy, that most reflective and literally thought-provoking of disciplines, have been severed? How can you get out of a trap if you do not know you are in it? The price the human sciences have paid for such disconnection and false autonomy since the mid-nineteenth century is appalling. We might call this the *external* problem, the problem of gaining some perspective from outside the tradition that allows us to pose questions that at least raise the possibility of genuine critique—always the first step to moving ahead, even if the nudge in the ribs is painful. This is a problem characteristic of all science that settles into ruts of normality. But there is also an *internal* problem of emancipation, a problem that arises when an analysis is undertaken within the factor-analytic framework. Given an imposed geometric description of an orthogonal or oblique space, what do you do with it? Publish it? Hang it on the wall and admire it? Do you see the internal emancipatory problem here? There is nothing within such a simple description that suggests anything in the way of changing the conditions that gave rise to it. And it is precisely here that we see how the intellectual inheritance from the physical sciences, and the mathematics that we have borrowed along with it, trap us into thinking about conscious, sentient, self-reflective human phenomena in the same terms we use to think about the inanimate, nonsentient, and unconscious world of things in the physical world. In describing the physical world in traditional mathematical calculi, it never occurs to us to ask how we might change the law of gravity (Newton) or space-time geometry (Einstein). Such a question appears absurd: The task of physics is to *describe* the laws of nature, not to *change* them. Similarly, the task for the Spearmans and Burts, and those who came after them, is to *describe* the geometry that appears to exist after it has already been imposed—not to take an emancipatory perspective and ask how it might be *changed*. Who really asks what those "ethnic factors" mean in the dozens of urban factorial ecologies over the past two decades that have repetitively discovered the filtered linear structure of the U.S. Census (Marchand 1978, 1979)? Down in the valley of factor analysis (and *all* other conventional multivariate approaches), there is neither an external or internal emancipatory perspective to raise such questions of changing the structural geometry. We still seem to be pursuing the search for laws that *by definition* must assume the unchanging geometry of the physical world.

How have we achieved such an extraordinarily closed state of affairs in the human sciences? How can we have had such a fine-cutting-edge methodology that has been on the edge so long that after more than a half-century it is getting increasingly blunt, despite successive two-, three-, and *n*-mode stroppings? Part of the problem is the numbingly conservative nature of the mathematical education we have received and still advise our students to follow in turn. And "education" is probably a euphemism, since most mathematical teaching is rote training of the most severe form. Most mathematical curricula support what Herbert Simon has called the "recapitulation of the field" method (Simon 1969), a weary plod through the seventeenth, eighteenth, and nineteenth centuries that purportedly creates the myth of "mathematical maturity," but is actually a course of study that filters out all but the most tenacious and jaw-clenching students. Only after jumping over all the traditional hurdles can students enter modern mathematics, gain more general and higher-order perspectives, and so make some sense out of what they have learned. The exception is the extraordinary first-year Foundation Course of the Open University (1971) and a few other curricula here and there where a genuine attempt has been made to rethink mathematical teaching in the late twentieth century. But here, as elsewhere, the exceptions tend to prove the rule. Most of the time we are still in a closed circle: We recommend courses in mathematics to *support* methodological courses in the human sciences, so mathematics is used once again to reaffirm the technical perspective, but never to provide an intellectual platform from which we might achieve an emancipatory view on what we are doing. Our view of mathematical requirements is conditioned by what we already do, never asking if we might do something else.

Is there a way out of the trap? In answering this question, we have to be very careful, placing any responses we might make in the perspective of the historical development of methodology in the human sciences. The last thing I want to imply is that there is an ultimate way of looking, a sort of methodological U.S. Cavalry that is going to ride to the rescue at the end of the last reel. We have seen too many cavalry charges in the last few decades that turned out to be ephemeral bandwagons. And anyway, who believes in last reels? I do believe, however, that there is a possible way out, a steep path along which we might take a few beginning steps so we can see what has been going on down below. I warn you, the view from even these modest lower slopes is pretty devastating, and a number have already turned back to the snug comfort of the more familiar and sheltered numerical values. Some, after a few steps, return, angry and bitter. These are hardly emotions for scientists, but such feelings are understandable in human terms when you have had some of the familiar props knocked out from under you. After all, if you have just published a definitive work in your field on multivariate and factor analysis, it is a bit much to expect you to take to a viewpoint that shows how distressingly constrained your methods are. And if you have devoted most of your professional life to numerical taxonomy, it is not pleasant to have someone point out that your cluster algorithmic finery resembles the Emperor's clothes.

The new perspective arose not by borrowing mathematical structures from the physical sciences (although some of the thinking was initially generated there [Atkin 1965, 1971]), but from a reflection upon the descriptive requirements needed to make the notion of structure well-defined and operational in human affairs (Atkin 1972a). In these days of bandwagons and the unfortunate "this-year's-model" syndrome, it is crucial to realize that the "language of structure" developed here is *not* a technique but a metamethodology that allows us to see conventional methodological approaches from entirely different perspectives (Atkin 1974a). Not only does it disclose the highly constrained nature of multivariate and statistical techniques, but it forces a severe and quite unrelenting intellectual discipline upon those who undertake a proper program of research within its highly flexible but quite rigorous framework (Gould 1983a). It is not possible to give a lengthy pedagogic introduction to Atkin's "language of structure," *Q*-analysis,[1] or polyhedral dynamics (Atkin and Casti 1977) here, although a number of advanced and more elementary introductions are available (Atkin 1977b, 1981; Gould 1980), but it is necessary to explicate a number of distinctions, concepts, and terms.

Q-analysis is founded upon relations between sets, an initially obvious statement apparently containing nothing that is not familiar. But in many research programs, the initial definition of the sets constitutes a difficult and nontrivial task, while the term "relation" is used here with mathematical exactitude. It constitutes the most general form of describing connections between or on sets of elements and even allows elements in the domain to remain unspecified. If the definition is constrained to require all elements of the domain to be employed, we have the usual mappings (including one-to-many), and if these are constrained still further, requiring all elements of the domain to have a unique image, we have the conventional function.[2] Thus, all functions are mappings are relations, but the reverse is not true. This freer specification to relate elements of sets does not, of course, require that we crush down multidimensional structures described by our data sets into constrained functional forms, as we do, for example, by least-squares and other function-forcing approaches. On the contrary, we can represent a relation geometrically by considering the elements of one set as polyhedra whose vertices are the elements in the other set which, quite literally, define them. If two elements or polyhedra (termed simplices) in one set share definitional vertices in the other, they are connected, and we can think of the entire set as a simplicial complex, a multidimensional representation that forms an operational definition of the intuitive but too often ill-defined notion of *structure*. It is important to note that binary relations may be defined on purely qualitative criteria, although numerical values may also be employed if appropriate—that is, that the sets of numbers employed are not only capable of being observed, but they are properly defined in terms of the binary operations that are going to be used on them. In brief, the operations must be closed (Fraleigh 1977). In the human sciences (and actually in the physical sciences too, as any quantum physicist knows), this confines our quantitative data to the sets of integers and rationals, and by eschewing the reals, we get a hint of one of the reasons why the calculus of the continuum is an inappropriate descriptive framework for the always-finite world of human beings. Of course, if data should be collected in numerical terms, relations can always be defined by choosing appropriate slicing parameters that actually represent a subjective mapping onto the set {0, 1}. Whether such a mapping, defining a particular geometrical structure, is appropriate or not depends entirely upon the utility of it at the next interpretative or hermeneutic stage. The geometric texts generated for an interpretation must be created by careful and sometimes long combinatorial search, bringing every bit of knowledge, insight, and intuition we have about the problem to bear upon the definitional choices, rather than by shoving a great mass of unsorted data into a deterministic decompositional machine that filters all the nonlinear information away.

The definitional task also contains a deep obligation to think about and sort the data before a structural analysis ever begins. We recognize intuitively in our everyday language—the language we *think in* (Gould 1982b)—that some of the terms we employ are more general and embracing than others. *Mathematics,* for example, is a more general term than *algebra,* and the latter is more general than *Clifford algebra.* Thus, our "variables" may well exist at different hierarchical levels of generality, so that some terms form cover sets at one level for sets lower down. It is of crucial importance, logically and interpretatively, that we carefully sort the hierarchy of cover sets and their empirically defined relations before any analysis begins. In most social science research, particularly all forms of conventional multivariate research employing today's modern computers, such questions are seldom if ever raised. Yet it is through such definitional obligations that we see how an analysis may well have to take place at different levels of generality and aggregation and how the statements we make may be quite specific to a particular level in the hierarchy (Gould, Johnson, and Chapman 1983).

The final and quite fundamental distinction we must make is between the geometric structures we create and the things that may exist and be transmitted at different dimensional levels on them (Johnson 1982a). This is the fundamental distinction between *backcloth* and *traffic,* a distinction that allows us to see the profound way a multidimensional structure forms an allowing-forbidding geometry for things that can only exist on it. And at this point, a necessarily compressed pedagogic exposition is probably best served by a few examples. In an agricultural area, for instance, some sort of *structure* of agriculture must exist as a backcloth before a traffic of crops and animals can be raised. And notice, we say *can* be raised, not *must* be raised. The backcloth geometry may *allow,* but it does not *require.* It may, however, *forbid,* as every farmer knows who uses irrigation to make water part of his structure at higher slicing values. A relation between a set of farmers F and a set of basic requirements R (land, water, equipment, markets, and so on), say $\lambda \subseteq F \otimes R$, might be considered as the multidimensional structural backcloth required before traffic in animals and crops can exist (Gould and Gaspar 1981). Similarly, a relation between sets of farmers labeled as "Knowers" and "Receivers," or $\lambda \subseteq K \otimes R$, might be the multidimensional backcloth that forms the geometric structure over which information about innovations is transmitted at some particular dimensional or q level.

This sort of structural perspective often allows us to break out of old habits and frameworks of thinking. For example, I was recently in correspondence with two engineers who were studying the important problem of how fire spreads in buildings. After a year of effort, thinking in terms of conventional differential equations (they were, after all, engineers, and all their thinking has been directed that way from their earliest days), they were delighted to come up with a two-room model that showed how fire *ds*-by-*dt*-ed its way from one totally abstract model room to another. What they should have done, of course, was think in terms of well-defined sets of real materials, at, say, the N-level of the hierarchy, structured by relations defined by those mappings we call *slicing parameters* that take into account inflammatory values and actual physical connections and proximities in real rooms. Such sets would aggregate by equally well-defined algebraic relations to sets of rooms at $N + 1$, houses at $N + 2$, neighborhoods at $N + 3$, towns at $N + 4$, and so on. Then, when Mother O'Leary's cow in Chicago kicks over a lantern, we shall have a well-defined description of the backcloth structure over which fire as traffic can be q-transmitted. Clearly, a well-connected backcloth, defined at high inflammatory slicing levels, is going to enable fire to be transmitted quickly. This tells us something that every fire fighter knows intuitively: To stop fire from spreading, you have to produce obstruction by disconnecting the structure. Firebreaks define an equivalence relation on the set of simplices and fragment the complex. In the conjugate structure, we also try to generate obstruction by disconnecting highly inflammatory materials. In this structure, fireproof safes would appear as highly eccentric simplices (a notion well-defined in Q-analysis), because they are constructed to be totally disconnected from the rest of the room (and therefore the building) structure. What we need to describe and think about the problem is an appropriate algebraic and topological language, not a seventeenth century mathematical framework devised in the realm of statistical mechanics and refined in the nineteenth century for the description of electrostatic forces.

Closer to academic homes, we might think about the structure of a discipline like mathematics. Most university departments offer a well-defined set of courses C at the N-level (although students, in particular, might like to have a definition at the $(N - 1)$-level consisting of actual sets of course contents), and these are connected by the "prerequisite" relation π, so that we have $\pi \subseteq C \otimes C$. If this backcloth geometry is connected by very low, say O, dimensional faces, the curriculum may well be strung out in a long, chainlike structure resembling O-dimensional events along a linear continuum of Newtonian time. And that is precisely how it will be experienced by the traffic of students "recapitulating the field" as they are being O-transmitted along it—like toothpaste being forced under great pressure from a small opening in the tube. On the other hand, a curriculum could be designed to have some very high-dimensional courses, defined by those sets of other courses for which they are prerequisites. For example, the Foundation Course of the Open University leads in many directions, and at the University of California, Santa Barbara, the beginning course Mathematics 8 (Introduction to Modern Algebra) is a 9-dimensional polyhedron, forming a highly significant part of the structure that allows the traffic of students to flow without stress into 10 other course-simplices in the structure (University of California 1981).[3] Designating in the catalogue that entry to a course may be gained "by permission of the instructor" means that the structure of the curriculum may be radically changed for some of the traffic of students who can get such permission—perhaps by bending the existing curriculum by connecting two courses that were not directly attached before—a change in connectivity that may even alter the homological structure (the holes in the Swiss cheese) to create q-holes. These may be experienced by the students as topological objects in the space, obstructions that the traffic has to work around (Atkin 1974b, 1976). In the conjugate structure, courses of high eccentricity, courses that literally stick out of the structure like a sore thumb, imply that there are specialized areas of mathematics that seem to bear little relation to other pieces of the curriculum, And so on . . . by thinking in this appropriate structural algebra, we work towards an effective and operational description without doing mathematical violence to the data.

We get the same sort of novel and fruitful insights when we move up the mathematical hierarchy to more general algebras, and it is from these metamethodological perspectives that we can reflect upon what is actually going on in the much more constrained methodologies relying upon the linear algebras devised for the simpler functional description of the physical world. In this light, all forms of factor analysis are seen as structure-imposing methods that employ the most constrained functional description we can conceive of as the starting point. Almost before an orthogonal and linearly additive geometric text is created, we have squeezed the rich multidimensional geometry through the most severe linear functional filters. Then, and no matter how well-connected the structure may be, we tear it apart by employing well-developed methods for decomposing positive-definite Gramian matrices into sets of eigenvectors and eigenvalues. *Mathematically,* everything is impeccable (if a bit elementary), but *conceptually,* it is a disaster, bearing more relation to machinelike processes and black magic rather than careful science conducted by thoughtful human beings. And all of these operations are usually performed on sets of variables that are hierarchical muddles, often containing both backcloth and traffic terms, all mixed together because no one has ever thought about making distinctions before. For example, most of the economic indices (unemployment, output, and so forth) collected so assiduously by governments today are simply traffic—but on what backcloth? The fact is that we have yet to achieve an effective description of the multidimensional complexity of a modern economy. We still seem to think that it is describable in functional forms, as though it were some large watch with gear wheels, or perhaps with computer software today that substitutes for wheels in this posthorological age.

Similarly, multidimensional scaling, over which such enormous technical efforts have been expended during the past two decades, is now seen in the simple light of what it really is: a graphic display (usually two-dimensional) of a relation on a set, a picture that represents a projection of the real multidimensional complexity onto the plane by crushing all the other dimensions out of existence. You cannot interpret that which has been filtered away or crushed out by the particular methodology chosen. And this is precisely why the view of numerical taxonomy and all the many varieties of automatic cluster analysis is so devastating from these higher and more general algebraic viewpoints. We create an orthogonal taxonomic space, usually by performing illegal (that is undefined) algebraic operations on sets of inappropriate numbers (including ordered and binary), operations that may well create metricity out of the methodology. We then throw out many "nonsignificant" dimensions on conceptually spurious numerical grounds before letting loose any one of a dozen possible cluster algorithms. And here the real trouble begins: Most empirical examples produce a roughly spheroidal or ellipsoidal cloud of points in the taxonomic space chosen, and the algorithms start the process of clustering using metric criteria and virtually arbitrary starting points. It is as though we were slicing up a multidimensional cloud almost at random, and if the results do not conform to our expectations, we try another algorithm, with a few additional constraints, until they do. The process of pushing a highly connected multidimensional structure through a deterministic partitional machine that *must* force an equivalence relation on the set of things to be classified simply dresses up the mechanism of black magic as science (Gould 1981b, 1982). Approaching the taxonomic task within a framework of partitional thinking (and usually without any theoretical constructs to guide us), we find that the task of interpretation frequently becomes circular. We go to numerical taxonomy in the first place for "objective" classifications and end up rejecting those that do not conform to our intuitive and totally subjective understanding that we considered so inadequate and unreliable before we started.

From the point of view of our algebraic topological language, any taxonomic attempt starts by defining a relation between two sets, the set of things to be classified, say T, and the set of criteria, say C. Now it is conceivable that the simplicial complex defined by the relation $\lambda \subseteq T \otimes C$ could be fragmented at some well-specified level of dimensionality, so that well-documented statements could be made about genuine partitions created by such equivalence relations. But we are often faced with a highly connected geometrical structure, a structure that *is* the actual description of the data. And it is precisely this well-connected, multidimensional structure that is torn apart by the archaic partitional thinking programmed into the deterministic machine we call a *cluster algorithm,* a machine that by definition (and despite all claims made by those in artificial intelligence) can never make *meaningful* judgments or be conscious of potential serendipity effects. We talk today about *machines* that *think,* but too often we end up ourselves as *thinks* that *machine* (Gould 1983b).

If the view of conventional multivariate techniques is disconcerting from this more general algebraic and topological perspective, what is our view now on three-mode factor analysis? In this process of critique, founded upon destructive retrieval, what is to be cut away, and what is to be retrieved? What we can see in the limited yet definite enlargement of perspective provided by three-mode factor analysis is the explicit emergence of the problem of *time.* Many of the early formulations were concerned with what happened when people performed tasks or tests over *occasions,* whether the problems were fictitious examples (Tucker 1965a), or based upon actual data sets (Tucker 1965b). Other forms of three-mode research, not investigating changes over time, tended to focus upon variation over *situations* (Levin 1963) or over *judges* (Mills and Tucker 1965), but the latter were undertaken in areas of psychiatric judgment that were notorious for their imprecision (Colby and McGuire 1981). In historical perspective, I think it is Tucker's particular concern for the way time must be handled by enlarging the framework of thinking, rather than simplistically incorporating it into the existing two-mode paradigm (Cattell 1966b), that will be seen as one of his major contributions.

But time is a quicksilver concept that slips through our intellectual fingers, even as we try to grasp it, and it sometimes seems that the harder we try to pin it down, the more easily it escapes our understanding. Yet even in the face of this difficulty, we have a hint of where our thinking about time might go, a hint accentuated when we recall that a philosopher like Heidegger made time an integral part of the structure of his own tenacious thinking in both his early grounding work (Heidegger 1962) and his later stages of thinking to releasement (Heidegger 1966, 1972). This is hardly the place to attempt what could only be a lengthy and difficult explication on time in Heidegger's thought, except to note (only superficially) that in a sense time becomes a part of the human definition itself, In more concrete terms, it acknowledges again the historicity of the human being and the way the past constrains possibilities for the present, even as the present shapes the conditions of possibility for the future, And let me suggest that we can thoughtfully pursue such thinking at any scale we like. For example, the tasks performed yesterday under the controlled conditions of a psychologist's laboratory may produce changes that shape the performance today, even as the tasks today have the same capacity to shape a performance tomorrow. I would like to dare to suggest that there are structural parallels here between philosophical and mathematical perspectives on time—perhaps even homeomorphisms if they could be made formally explicit and well-defined—that are illuminated by taking our thinking over for a moment into the more formal mathematical realm.

The problem of time and change has been embedded in some of the most recent thinking and formulation of the language of structure (Atkin 1975, 1978a, 1978b, 1978c; Johnson 1975), and the concern is a continuing one (Atkin 1982; Johnson 1982b). It recognizes that events need not be conceived as *O*-dimensional simplices connected by 1-dimensional intervals on a Newtonian continuum; rather, by drawing upon and enlarging an analogy from relativity theory, they may be considered as multidimensional, or *p*-events (Atkin 1977a). Thus, time is considered as a 1–0 ordering function that exists as transmitted traffic on a structural backcloth, with the transmission and ordering of the simplices being highly controlled by the *structure* of the underlying geometry. This is a wholly novel way of thinking about time, yet it is in perfect intuitive accord with our own experience. For example, all planning, at any scale from the individual to social level, may be seen as a conscious attempt to create sufficient structure *now* to allow a *p*-event of required dimensionality to appear *later*.

In this area of unfamiliar thinking, an example perhaps best serves to illustrate the new directions in which thought might go. A game of chess, for example, may be notoriously difficult precisely because it is a combinatorial explosion. The *p*-event "checkmate" is the culmination of a game between two opponents, but the actual course of a game may be regarded as a succession of attempts by the opponents to create sufficient structure, to build an effective backcloth, to *allow* the time-ordering function to reach that ultimated state in their favor (Atkin 1972b; Atkin and Witten 1975; Atkin, Hartston, and Witten 1976). Every move represents a change in the structural geometry defined by relations between the sets of black and white pieces and the squares of the board. Players of "Grand Master" caliber (such as Hartston, who served as Atkin's research associate in this area), think in strategic terms at the $N + 2$ positional level, which is purely structural, and they only descend to the $N + 1$ level of tactical play to make a move, which is always evaluated for what it will do to the overall structure at the higher, positional level.

The idea of maintaining a tightly connected team structure and fragmenting the game of one's opponent carries over into many areas of sport (Gatrell and Gould 1979). Our everyday language thus reflects the intuitive structural analyses we make of games when we say, for example, that one of the teams "fell apart" (Gould and Gatrell 1980). Again, structural notions of time enter as the whole purpose of team play is to forge a sufficiently strong geometry between the players so that a ball can be *q*-transmitted to connect with the goal, the element of the game that creates a high-dimensional *p*-event that we record as a "score" (Gould and Greenawalt 1982). Such a perspective of time as an ordering function on a simplicial complex also forces us to revise radically our views of decision making. A decision now becomes a combinatorial search for a particular simplex in an event space, and such thinking leads us away from conventional statistical and probabilistic approaches to a theory of surprises. In this structural framework, "a surprise is the answer to a question which has not been asked" (Atkin 1982, 365).

With these brief hints about the algebraic "language of structure," and in light of the continuing attempts to rethink the question of time and how it might be more effectively incorporated into the description of human affairs, how can we view three-mode factor analysis and the whole factor-analytic tradition that historically lies behind it? This subject could well become the focus for a major project and case study in the history of the social sciences, but I think the bare outlines are already clear. There is in our thinking about humans and human affairs—about individuals engaged in activities and in the collective enterprises we all undertake socially—an intuitive sense of structure that has proved difficult to make explicit and operational. And notice that such intuitive notions of structure permeate every field of human thinking: We refer to the structure of a psychological experiment, but we also talk about the structure of a poem, ballet, molecule, group (both human and mathematical), symphony, novel, research program, society, business, international relations, and so on —the list is endless. Seen in proper historical perspective, it is hardly surprising that a handful of pioneer social scientists in the early years of this century (Spearman 1904, 1927; Thurstone 1931; Hotelling 1933) took a particular mathematical structure already employed in the physical sciences to make the descriptions in their own field of inquiry mode definitive, replicable, and supportive of the embryonic theory about mental abilities that was emerging at that time. But what we can see now is the way this highly simplified and *simplifying* mathematical framework immediately came into a dialectical relationship with the current theory about mental abilities and so began to shape the thinking that was formerly guiding the directions of research. Within a linear mathematical framework, an historical framework that virtually required that everything described within it should be decomposed into orthogonal additive pieces à la Fourier, thinking could only be directed towards independent dimensions of intelligence. In a sense, it was firmly trapped there and had no higher order and more general mathematical place to stand to see itself and ask what it was doing.

What followed historically may be seen as a succession of indications that not all was well. The eigenvectors and their associated "stretchability coefficients"—eigenvalues (Gould 1967) —were all very well for finding a basis for the space, but once thinking began to question that general intelligence factor, then rotations became almost obligatory. Rotations raised problems of their own, however, conceptual problems that wormed their way into the woodwork and began to eat away at the attempts to build an objective and therefore (in the thinking of the day—and even now?) scientific psychology. A number of imaginative and quite plausible rotational schemes began to appear, and it then became a choice of the investigator as to the way in which the empirical data could be—well—*structured!* Those who wanted to keep

Spearman's general factor could do so and orthogonally rotate the rest to pick up other clusters of vectors. Those who believed in specific dimensions could rotate to simple structure and support their views in turn. The variations on a theme by Varimax are well known (Cattell 1966a). But if one stood a good chance of structuring the geometric text to suit one's preconceptions, what was happening to objective science? Many in the developing twentieth century field of Hermeneutics could have provided some thoughtful reflections upon such questions, but most of the connections to philosophy had long been severed by the human sciences. We must also remember that ideas also require a well-connected geometry before they can be *q*-transmitted between fields.

And then what about that business of orthogonality? Was it really conceivable that these dimensions of the mind were independent? Oblique schemes followed such questions, but then, could one do a second-order factor analysis on the oblique axes of the first-order analysis, and so on to either infinite or circular regress? And what did one do with change and that slippery dimension called "time" along which it was meant to take place? Surely it is here that we see the thinking about three-mode factor analysis emerging. And this historical development in psychology has been followed in almost perfect sequential repetition by every other human science employing these techniques—human geography, political science, sociology, economics, and anthropology —either as a descriptive technique in itself or as a basis for a later taxonomy.

What can we retrieve and honor here, even as we see the constraints under which the earliest efforts labored? It is the concern for structure and ultimately (with Tucker) the concern for *structural change,* the dynamics of structure. That such characteristic concern was seen at the heart of our thinking about the human condition at all scales is what must be honored, even as we try to think our way out of the particular framework chosen. What we see in three-mode analysis is yet one more move away from conventional approaches dating from the turn of the century. It was not, it *could* not, be successful, given the limitations of the mathematics chosen, but it constituted nevertheless a powerful push by a strong intellectual shoulder on the wheel of understanding, a push motivated by the sort of dissatisfaction that always marks the breaking of new ground, rather than traditional cultivation and the turning over of old sod.

NOTES

1. From the point of view of a social scientist familiar with factor analysis, the designation *Q*-analysis is unfortunate and may be initially confusing. In Atkin's terminology, *Q*-analysis has nothing whatsoever to do with *q*-mode factor analysis, the familiar linear approach of which Atkin, as a mathematician and algebraic topologist, was unaware. In the notational conventions of algebraic topology, **q** is generally reserved to designate the dimensionality of a space. Thus, *Q*-analysis is deeply concerned with

the dimensions of objects and their connecting faces, as well as the structural (homological and homotopic) properties they possess.

2. Terminology varies from one mathematical tradition to another. The relation-mapping-function sequence used here makes useful distinctions that are helpful when we examine traditional multivariate techniques in the human sciences.

3. Namely, Mathematics 102, Introduction to Group Theory; 105, Computer Based Mathematics; 106, Number Systems of Algebra; 108A and B, Introduction to Linear Algebra; 109A, Introduction to Mathematical Logic; 111A, B, and C, Introduction to Abstract Algebra; 115A, Introduction to Number Theory; 116, Combinatorial Mathematics; 136A, B, and C, Modern Algebra With Applications; and 170, Mathematical Literature.

REFERENCES

Aler, J. 1972. Heidegger's conception of language in *Being and Time.* In *On Heidegger and Language,* ed. J. Kockelmans. Evanston, Ill.: Northwestern University Press.

Atkin, R. 1965. Abstract psychics. *Il Nuevo Cimento* 38:496–517.

————. 1971. Cohomology of observations. In *Quantum theory and beyond,* ed. T. Bastin. Cambridge: The University Press.

————. 1972a. From cohomology in physics to Q-connectivity in social science. *International Journal of Man-Machine Studies* 4:139–67.

————. 1972b. Multidimensional structure in the game of chess. *International Journal of Man-Machine Studies* 4:341–62.

————. 1974a. *Mathematical structure in human affairs.* London: Heinemann Educational Books.

————. 1974b. An algebra for patterns on a complex I. *International Journal of Man-Machine Studies* 6:285–307.

————. 1975. An approach to structure in architectural and urban design: Three illustrative examples. *Environment and Planning A* 2:21–57.

————. 1976. An algebra for patterns on a complex II. *International Journal of Man-Machine Studies* 8:483–98.

————. 1977a. *Q-analysis: Theory and practice.* Colchester: Department of Mathematics, University of Essex.

————. 1977b. *Combinatorial connectivities in social systems.* Basel: Birkhauser Verlag.

————. 1978a. *Decision making as an event search: Traffic on a multidimensional structure.* Colchester: Department of Mathematics, University of Essex.

————. 1978b. Time as a pattern on a multidimensional structure. *Journal of Biological Structures* 1:281–95.

————. 1978c. Dynamics of patterns on a hard-data backcloth. In *Towards the dynamic analysis of spatial systems,* ed. N. Thrift and R. Bennett. London: Pion Limited.

————. 1981. *Multidimensional Man.* Harmondsworth: Penguin Books Limited.

———. 1982. A theory of surprises. *Environment and Planning B* 8:359–65.

Atkin, R., and J. Casti. 1977. *Polyhedral dynamics and the geometry of systems.* Laxenburg: International Institute for Applied Systems Analysis.

Atkin, R., and I. Written. 1975. A multidimensional approach to positional chess. *International Journal of Man-Machine Studies* 7:727–50.

Atkin, R., W. Hartston, and I. Witten. 1976. Fred CHAMP, positional chess analyst. *International Journal of Man-Machine Studies* 8:517–29.

Bubner, R. 1981. *Modern German philosophy.* Cambridge: Cambridge University Press.

Cattell, R. B., ed. 1966a. *Handbook of multivariate experimental psychology.* Chicago: Rand McNally.

———. 1966b. The data box: Its ordering of total resources in terms of possible relational terms. In *Handbook of multivariate experimental psychology,* ed. R. B. Cattell. Chicago: Rand McNally.

Colby, K., and M. McGuire. 1981. Signs and symptoms. *The Sciences* 22:47–51.

Fraleigh, J. 1977. *A first course in abstract algebra.* Reading, Massachusetts: Addison-Wesley.

Gatrell, A., and P. Gould. 1979. A micro-geography of team games: Graphical explorations of structural relations. *Area* 11:275–78.

Gould, P. 1967. On the geographic interpretation of eigenvalues. *Transactions of the Institute of British Geographers* 42:53–86.

———. 1980. Q-analysis or a language of structure: An introduction for social scientists, geographers and planners. *International Journal of Man-Machine Studies* 12:169–99.

———. 1981a. Letting the data speak for themselves. *Annals of the Association of American Geographers* 71:166–76.

———. 1981b. A structural language of relations. In *Future trends in geomathematics,* ed. R. Craig and M. Labovitz. London: Pion Limited.

———. 1982a. The tyranny of taxonomy. *The Sciences* 22:7–9.

———. 1982b. Mathematics and thinking in the human sciences. *Environment and Planning A* 14:1279–80.

———. 1983a. Statistics and human geography: Historical, philosophical, and algebraic reflections. In *Spatial statistics and models,* ed. G. Gaile and C. Willmott. The Hague: Martinus Nijhoff.

Gould, P. 1983b. Thinks that machine. *Environment and Planning A* 15:723–24.

Gould, P., and A. Gatrell. 1980. A structural analysis of a game: The Liverpool v. Manchester United cup final of 1977. *Social Networks* 2:247–67.

Gould, P., and J. Greenawalt. 1982. Some methodological perspectives on the analysis of team games. *Journal of Sport Psychology* 4:283–304.

Gould, P., J. Johnson, and G. Chapman. 1983. *The structure of television.* London: Pion Limited.

Gould, P. P., and J. Gaspar. 1981. The Cova de Beira: An applied structural analysis in agriculture and communication. In *Space and time in geography,* ed. A. Pred. Lund, Sweden: C.W.K. Gleerup.

Habermas, J. 1971. *Knowledge and human interests.* Boston: Beacon Press.

Heidegger, M. 1962. *Being and time.* Oxford: Basil Blackwell.

———. 1966. *Discourse on thinking.* New York: Harper and Row.

———. 1972. *On time and being.* New York: Harper and Row.

———. 1982. *The Basic Problems of Phenomenology.* Bloomington, Ind.: Indiana University Press.

Hotelling, H. 1933. Analysis of a complex of statistical variables into principal components. *Journal of Educational Psychology* 24:417–41, 498–520.

Johnson, J. 1975. A multidimensional analysis of urban road traffic. Doctoral dissertation, University of Essex.

———. 1982a. Q-transmission in simplicial complexes. *International Journal of Man-Machine Studies* 16:351–77.

———. 1982b. The Q-analysis of road traffic systems. *Environment and Planning B* 8:141–89.

Klein, M. 1980. *Mathematics: The loss of certainty.* Oxford: Oxford University Press.

Kockelmans, J. 1972. *On Heidegger and Language.* Evanston, Ill.: Northwestern University Press.

Kuhn, T. S. 1962. *The structure of scientific revolutions.* Chicago: Chicago University Press.

Levin, J. 1963. Three-mode factor analysis. Doctoral dissertation, University of Illinois.

Marchand, B. 1978. A dialectical approach to geography. *Geographical Analysis* 10:105–19.

———. 1979. Dialectics in geography. In *Philosophy in geography,* ed. S. Gale and G. Olsson. Dordrecht: Reidel.

Mills, D., and L. R. Tucker. 1965. A three-mode factor analysis of clinical judgement of schizophrenia. In *Project on techniques for investigation of structure of individual differences in psychological phenomena,* ed. L. R. Tucker. Urbana: University of Illinois.

Open University. 1971. *A foundation course in mathematics, Units 1–36.* Milton Keynes: The Open University Press.

Poggler, O. 1972. Heidegger's topology of being. In *On Heidegger and language,* ed. J. Kockelmans. Evanston: Northwestern University Press.

Rao, C. 1955. Estimation and test of significance in factor analysis. *Psychometrika* 20:93–111.

Simon, H. 1969. *The sciences of the artificial.* Cambridge: MIT Press.

Spearman, C. 1904. The proof and measurement of association between two things. *American Journal of Psychology* 15:72–101.

———. 1927. *The abilities of man: Their nature and measurement.* London: Macmillan.

Thurstone, L. L. 1931. Multiple factor analysis. *Psychological Review* 38:406–27.

Tucker, L.R. 1965a. Some mathematical notes on three-mode factor analysis. In *Project on techniques for investigation of structure of individual differences in psychological phenomena,* ed. L. R. Tucker. Urbana: University of Illinois.

———. 1965b. Three-mode factor analysis of Parker-Fleishman complex tracking behavior data. In *Project on techniques for investigation of structure of individual differences in psychological phenomena,* ed. L. R. Tucker. Urbana: University of Illinois.

University of California. 1981. *81/82 general catalog.* Santa Barbara: University of California.

16

Issues and Perspectives in Multimode Analysis

John A. Hattie, Roderick P. McDonald, Conrad W. Snyder, Jr., and Henry G. Law

This chapter comments on the current state-of-the-art in multimode data analysis and focuses on a number of issues that differentiate the various techniques or that require further attention in future developments of the field. Since the methods have evolved from different statistical assumptions and entail different psychological premises, they address (sometimes) different substantive problems from a variety of perspectives. Here we compare the various multimode models and discuss their general or restrictive applicability and the adequacy of their numerical solutions in terms of measures of fit and interpretability.

STATISTICAL FORMULATIONS

The methods described in the previous chapters were designed to systematize and represent the relations among entities classified over many identification modes, such as persons, tests, occasions, and so forth. Given their common goal, the similarities of their underlying statistical formulations are not surprising. Many of the authors have indicated the direct and collateral relationships. Details of these equivalences and similarities or dissimilarities will not be repeated, but two less obvious and interesting relations will be explored.

First, Bloxom's reformulation of Tucker's model (chapter 4) into hierarchical three-mode factor analysis provides an interesting perspective on the more general form of Tucker's model and a convenient basis for exhibiting the relationship between three-mode factor analysis and the classical two-mode factor analysis model. For completeness, we note this relationship here. In

---

We wish to acknowledge the assistance of Douglas Carroll, Richard Harshman, and Joseph Kruskal in reading and commenting upon earlier drafts of this chapter.

Bloxom's notation, where $I = 1$, provided also that we take $p = 1$ (which is natural, at least in the unrestricted case), then $\mathbf{A}$ reduces to scalar a, and

$$\hat{\mathbf{y}} = (\mathbf{A} \otimes \mathbf{I}_Q)(\mathbf{I}_I \otimes \mathbf{B})\mathbf{G}\,\mathbf{x} \tag{16–1}$$

becomes

$$\hat{\mathbf{y}} = a\,\mathbf{B}\,\mathbf{G}\,\mathbf{x}\,. \tag{16–2}$$

The multiplier a yields a trivial arbitrariness of scale, and so we have:

$$\hat{\mathbf{y}} = \mathbf{B}\,\mathbf{G}\,\mathbf{x}\;, \tag{16–3}$$

where $\mathbf{B}$ is $J \times Q$ and $\mathbf{G}$ is $Q \times R$. There is then a weak sense in which the model reduces to the ordinary factor model, since we may always write:

$$\mathbf{B}\,\mathbf{G}\,\mathbf{x} = \mathbf{B}^*\,\mathbf{x}^*\;, \tag{16–4}$$

where $\mathbf{B}^*$ is a $J \times S$ matrix of rank s such that:

$$\mathbf{B}^*\,\mathbf{B}^{*\prime} = \mathbf{B}\,\mathbf{G}\,E\,\{\mathbf{x}\mathbf{x}'\}\,\mathbf{G}'\,\mathbf{B}'\;. \tag{16–5}$$

However, except when S, Q, and R are all equal and $\mathbf{G}$ is nonsingular, the relationship between $\mathbf{B}$ and $\mathbf{B}^*$ is not a simple one, and an investigator studying the J items with $I = 1$ could not in general, by rotation, reconcile the results with those of another investigator who studies them with $I > 1$. This is not a criticism of Tucker's model, but it is a significant difference between it and simpler models of the classical and PARAFAC types.

Second, the model for three-mode data presented by Cattell, Blaine, and Brennan (chapter 9) can be reformulated as a simple model for the analysis of covariance structures. Let a_{kji} be the score of subject i on test j in situation k (or on occasion k), $k = 1, \ldots, l$; $j = 1, \ldots, m$; $i = 1, \ldots, n$. The three-way model is:

$$a_{kji} = a_{ji} + a_{ki} + a_{kj} + e_{kji}\;, \tag{16–6}$$

where

$$a_{ji} = \sum_{p=1}^{t} f_{jp} x_{pi}\;, \tag{16–7}$$

$$a_{ki} = \sum_{q=1}^{p} g_{kq} z_{qi}\;, \tag{16–8}$$

$$a_{kj} = \sum_{r=1}^{u} h_{kr} w_{rj}\;, \tag{16–9}$$

and

e_{kji} is the residual.

Alternatively, we can treat the residual e_{kji} as just the misfit of the model to the data and thus have a form of component analysis.

Since the model does not place any restrictions on parameters associated with subjects, it can be rewritten as a model in random vectors. We define an index:

$$v = m(k - 1) + j .$$

In matrix terminology, the model is then

$$\mathbf{a} = \mathbf{F}\,\mathbf{x} + \mathbf{G}\,\mathbf{z} + \boldsymbol{\mu} + \mathbf{e} \tag{16–10}$$

or

$$\mathbf{a} = [\,\mathbf{F} : \mathbf{G}\,] \begin{bmatrix} \mathbf{x} \\ \mathbf{z} \end{bmatrix} + \boldsymbol{\mu} + \mathbf{e} \ , \tag{16–11}$$

where

$\mathbf{a} = [a_v]$, an $lm \times 1$ *score* vector whose vth component is a_{kj},
$\mathbf{e} = [e_v]$, an $lm \times 1$ *residual score* vector,
$\mathbf{F} = [f_{vp}]$, an $lm \times t$ matrix of *test factor loadings*,
$\mathbf{G} = [g_{vq}]$, an $lm \times s$ matrix of *situation factor loadings*,
$\boldsymbol{\mu} = [\mu_v]$, an $lm \times 1$ vector of means,
$\mathbf{x} = [x_p]$, a $t \times 1$ vector of *test factor scores*, and
$\mathbf{z} = [z_q]$, an $s \times 1$ vector of *situation factor scores*.

The model is subject to three sets of restrictions, namely,

$$f_{vp} = f_{kjp} = f_{jp} \ , \tag{16–12}$$

$$g_{vq} = g_{kjq} = g_{kg} \ , \tag{16–13}$$

and

$$\mu_v = \mu_{kj} = \sum_{r=1}^{u} h_{kr} w_{rj} \ , \tag{16–14}$$

for all k and j.

As Cattell, Blaine, and Brennan note, the restrictions (16–14) operate on the mean vector and do not affect covariances. Also, on the face of it, there may be difficulties in obtaining identifiability of the parameters h_{kr} and w_r thought of as underlying μ, so it seems reasonable to regard μ as an unrestricted vector of means to be estimated. (This ignores possible problems of origin and scale in this model that perhaps need careful examination.) With μ unrestricted, (16–11) is a very simple model in the family of restricted factor models. It yields the covariance structure

$$\mathrm{Cov}(\mathbf{a}) = [\mathbf{F}:\mathbf{G}] \begin{bmatrix} \mathbf{P}_{ff} & \mathbf{P}_{fg} \\ \mathbf{P}_{gf} & \mathbf{P}_{gg} \end{bmatrix} \begin{bmatrix} \mathbf{F}' \\ \mathbf{G}' \end{bmatrix} + \mathbf{U} \ , \qquad (16\text{–}15)$$

where $\mathbf{F}$ and $\mathbf{G}$ contain factor loadings that are constrained to be equal, as prescribed by (16–12) and (16–13), and $\mathbf{P}_{ff}$, $\mathbf{P}_{gg}$, and $\mathbf{P}_{fg}$ represent, respectively, the correlation matrix of the test factors, the correlation matrix of the situation factors, and their cross-correlation matrix, which in some cases we might wish to suppose null. The matrix $\mathbf{U}$ is a residual covariance matrix that we would usually take to be a diagonal matrix of uniquenesses. In some applications, we might impose the additional restriction that $\mathbf{F}$ and/or $\mathbf{G}$ has prescribed simple structure. In general, however, it is unlikely that good simple structure can be attained while the equality constraints required by the model are also imposed. Thus, we may fit the model (16–15) by the use of a general purpose program for the analysis of covariance structures, such as Jöreskog's LISREL or Fraser's COSAN, prescribing only the number t of test factors and the number s of situation factors.

Cattell, Blaine, and Brennan proposed to fit the *additive id* model by performing exploratory factor analyses on matrices obtained by averaging a_{kji} over situations and over tests. From the form of the model, the effect of such averaging is to average the corresponding factor loadings. If we average over situations, we obtain the model:

$$\mathbf{a}^* = [\ \mathbf{F}^* : \mathbf{G}^*] \begin{bmatrix} \mathbf{x}^* \\ \mathbf{z}^* \end{bmatrix} + \mu^* + \mathbf{e}^* \ , \qquad (16\text{–}16)$$

where

$$\mathbf{F}^* = [f_{jp}],\ m \times t$$

and

$$\mathbf{G} = [\frac{1}{l} \sum_{k=1}^{l} g_{kpp}],\ m \times s \ .$$

By equation (16–16), the condition

$$\frac{1}{l} \sum_{k=1}^{l} g_{kjg} = 0, \qquad \text{for all } j,\ q, \qquad (16\text{–}17)$$

seems necessary and sufficient for correct estimation of f_{jp} by the averaging procedure, using unrestricted, exploratory factor analysis. Similarly, the condition

$$\frac{1}{m} \sum_{j=1}^{m} f_{kjp} = 0, \qquad \text{for all } k,\ p, \qquad (16\text{–}18)$$

seems necessary and sufficient for correct estimation of g_{kq} by such means. The conditions (16–17) and (16–18) will be satisfied only in very rare cases and require all factors to be bipolar. This is not to declare that the methods proposed by Cattell, Blaine, and Brennan cannot yield some approximation to the correct estimates, but it does suggest that the procedure recommended here, using the analysis of covariance structures, will generally give better results. Of course, other estimation approaches may also be applicable to the Cattell et al. model. For example, the *personality-centered* model appears (algebraically, at least) to be equivalent to the PARAFAC-CANDECOMP model and the *additive id* or *additive attribute* model can be considered a special case.

It should be noted that the approach proposed by Cattell, Brennan, and Blaine has the advantage of being able to handle fairly large data arrays (such as 55 × 50 × 45), which would prove difficult for the more exact statistical approaches because of the computing demands made by these procedures (memory and CPU time). Since a substantial number of variables is sometimes required to adequately represent a particular behavior domain, this may in fact be a critical issue in practice.

The purpose of illustrating how the various methods are related is not to establish one general model with all others as special cases. Rather, it is to demonstrate the influence of Tucker's pioneering work and to make the point that the development of the various restricted methods stems from a desire to answer different questions with minimal assumptions and with a concern for interpretability. The various methods have special desirable properties that the more general models lack. For example, the PARAFAC-CANDECOMP model, although it is a special case of Tucker's model, has an important dimensional uniqueness or *intrinsic axis* property not possessed by his more general model. Another advantage of comparing methods is to recognize that the user may want to apply a variety of methods and then assess the effect on interpretation of the various assumptions. Comparative information will assist in the evaluation of the interpretive impact of these different assumptions.

APPLICABILITY

Many exciting developments in multimode data analysis have evolved since Tucker introduced his unique analytic conceptualizations. The Tucker model is one of the most general three-mode models for factor analysis, and IDIOSCAL—including the various special cases included in Young's GEM (chapter 12)—provides the general framework for individual differences multidimensional scaling and unfolding. These models serve as meta-models (following Young's characterization of GEM), displaying the conceptual and statistical relationships among the more restrictive multimode variants. These restrictive models, such as PARAFAC and CANDECOMP, overcome the problems of identifiability, but require strong assumptions about the data. A dilemma for a researcher is that while the more general models are not precise, the more restrictive ones may invoke unpalatable assumptions.

Despite the restrictiveness of the trilinear model, Harshman claims wide applicability for his PARAFAC, with notable support from Kruskal. Their enthusiasm is based on the apparent success of data preprocessing strategies, elaborated by Harshman and Lundy (chapter 6). Our experience with a wide range of research data prejudices us in their favor, but our acclaim is tempered by the modest fit figures that have generally been obtained. In reaction to these multimode models (including Tucker's) that alter the conceptual foundations of classical factor analysis, McDonald (chapter 7) developed his invariant factors model. It is difficult to objectively compare these approaches. First, McDonald has a very different rationale to initial data scaling, one which is clearly more traditional. Second, McDonald's method is less restrictive and will improve the statistical fit accordingly. A choice of method will therefore more likely be based on conceptual preferences.

One of the conceptual issues that divides the multimode methods is the different mode interpretations, as articulated in chapter 1. McDonald constrains the factor loadings to be equal across occasions but allows for changes in factor means. Both McDonald and Swaminathan assume that only one mode yields *factor loadings,* while all other modes affect factor score means and possibly covariances. They argue that tests have common factors with interpretable columns of loadings, and the other modes—such as conditions or occasions—do not. McDonald argues that it is not meaningful to name factors within the longitudinal mode. Thus, the invariant factors model and the longitudinal factor analysis model are not appropriate for multimode data in which the distinction between measures and conditions is not clear. The Tucker (see Kroonenberg, chapter 3) and Harshman models treat the modes symmetrically, and Bloxom's reformulation interprets the factors from each mode hierarchically. Thus, these methods may be particularly useful in different research contexts.

Regardless of the approach, the researcher should be aware that these restrictive methods do not automatically eliminate interpretive difficulties. Clearly, multimode methods should be applied with caution and attention given to the underlying assumptions and conceptualizations of their models.

For these reasons, exploratory approaches like PINDIS are useful. PINDIS tries to fit successively less restrictive models, first by strictly admissible transformations and then by including additional free parameters in the form of weights, idiosyncratic rotations, and translations. It also provides a descriptive index of fit and it is often possible to see how the assumptions of the more restrictive models might have to be relaxed. In recent work, we have found TUCKALS2 to be of assistance in this regard. The output of the computer program provides a great deal of information for diagnostic and interpretive purposes (see Kroonenberg's example in chapter 3). The extended core results provide important conceptual clues to ascertain an appropriate model.

Another important issue that divides these multimode methods is the nature and form of the data to be analyzed. For example, PARAFAC and the trilinear case of CANDECOMP are essentially the same mathematical model, proposed independently by Harshman

(1970) and Carroll and Chang (1970), but they are usually applied to different data. PARAFAC is guided by a factor-analytic interpretive framework based on the principle of parallel proportional profiles, while the INDSCAL application of CANDECOMP is couched within a scaling orientation. In applications, PARAFAC is used with ratings or preferences data, either directly or in covariance form, and presumably owes much of its success to its preprocessing procedures. INDSCAL is applied mostly to three-way dissimilarity data, which are often three-mode data that have been converted to pairwise proximities before processing. Therefore, PARAFAC is used with full three-mode data, whereas INDSCAL has generally been used for three-way/two-way data.

It is important in making this comparison to distinguish conceptually between CANDECOMP and INDSCAL, which uses a certain special case of CANDECOMP. CANDECOMP is the general multilinear (not restricted to trilinear) model/method whose trilinear case is equivalent to PARAFAC. It just happens that its most popular and well-known application is to INDSCAL. CANDECOMP is a scalar product model, whether in its general form or its special three-way symmetric form, whereas INDSCAL is a distance model.

Within the two camps—factor analysis and multidimensional scaling—the methods can be further differentiated by their preprocessing steps. For factor analysis, the debate as to whether raw data, covariances, or correlations should be used focuses on the amount of information contained in the means and variances. Cattell (chapter 9; and 1966, 1978), Harshman and Lundy (chapter 6,), and Kruskal (chapter 2) discuss extensively the problems in centering or standardizing data. For multidimensional scaling, Young (chapter 12) outlines his data theory and the role of optimal scaling (see Young, de Leeuw, and Takane 1980), and Takane (chapter 13) and Carroll and Arabie (chapter 14) discuss some specific details concerning the use of proximity data. Since the usefulness of an analytic method depends on the quality of the data, these preprocessing procedures merit further attention. Further, whatever form of data is used, it is obvious that unless the original data are reasonably dependable, the methods will be of little use. Cattell et al. used data with a reported reliability of 0.75 with no ill-effects. Systematic study of the robustness of multimode methods is required.

ADEQUACY AND INTERPRETABILITY

A common concern for researchers is how to determine whether a satisfactory and proper solution has been found. In each of the chapters, there are suggestions for the assessment of adequacy. Many of the multidimensional scaling methods use stress or R-squared indices (such as Lingoes and Borg, chapter 11, and Young, chapter 12). Young and Lewyckyj, in a discussion of these indices, recommend that R-squared (the squared correlation between the Euclidean distances and the disparities) is the most useful index. Takane (chapter 13) points out that one of the advantages of the MAXADD procedure is that it provides a statistical inference criterion. An asymptotic chi-square statistic is derived that may be used for various model comparisons. Such comparisons based on this statistic can be used when one of the two models compared is a special case of the other, thus forming hierarchical hypotheses. The methods based on factor analysis commonly use indices of fit based on the residuals. When the method involves minimizing the sum of squared residuals (as in Kroonenberg, chapter 3) the "fit" measures can be based on the relative size of the SS(Residual) and SS(Total) or the relationship can be shown by graphical display (see Kroonenberg chapter 3, and Gower, appendix B). If the residuals are assumed to be independent of each other and of the common factors, then the model can have a statistically falsifiable covariance structure. McDonald (chapter 7) develops a maximum likelihood model based on covariance structures that allows chi-square tests of fit for several special cases of the model. In his example, using Meyer and Bendig's (1961) data, McDonald demonstrates the usefulness of the chi-square to assess the adequacy of not only the invariant factors model but also the PARAFAC model.

Theoretically, McDonald's framework expands to twelve models, each derived from different assumptions made about the mean vectors, residual covariance matrices, and transformation matrices, respectively. It is relatively easy in practice to reduce this number to a more select set of hypotheses corresponding to major conceptual alternatives. It seems unlikely, for example, that the residual covariance matrix could be supposed diagonal. Typically, it will be patterned with every submatrix diagonal so as to allow for stable specific factors. If we also reject the "component" treatment obtained by setting the residual covariance matrix equal to the null matrix, the number of models is reduced from 12 to 4, leaving the crucial choices between a structural and a free mean vector and between the PARAFAC transformation and the more general invariant factors one. Swaminathan (chapter 8) treats the factor scores as a first-order auto-regressive series that allows the additional hypothesis of invariant uniquenesses to be tested (that is, the completeness of the common factor formulation). In his model, the factors must be orthogonal to achieve identifiability, so interpretability and psychological meaningfulness may be jeopardized. Recognizing this possible drawback, he discusses alternative approaches to achieve identifiability. Because of the identifiability problem, these powerful analytic frameworks require a sophisticated user. Their appeal stems from the potential usefulness (and elegance) of statistical falsifiability.

Statistical falsifiability is a fallible criterion for model assessment, however, and caution must be exercised in its application. Given a large sample size, most models are falsified. Accordingly, the use of chi-square or any fit statistics should not enslave the user to their outcomes. A better test of fit is the meaningfulness and interpretability of the solution. Nevertheless, the comparison of various models is an exciting and extremely useful advance. It can be argued that social scientists can gain much more from the comparison of various models than from a narrowed focus on the fit of one model (see Hattie 1981). Underlying these model fit considerations, it is critically important that the solution be determinate. All the authors have been concerned, implicitly or explicitly, with the problems of indeterminacy. Kruskal (chapter 2) clearly states the issues and points out the problems when there are many different solutions that fit the data equally well. The major ways for solving the indeterminacy problem have been by supplementary rotation procedures or by parameter restrictions (particularly on the core matrix). Bloxom (chapter 3), Kruskal (chapter 2), Harshman and Lundy (chapter 5), and McDonald (chapter 7) discuss the appropriateness of the simple structure principle. Bloxom and Kroonenberg (chapter 4) probe the implications of rotation of components in the outer and the core matrices. Rotational indeterminacy is eliminated by use of the dimensional uniqueness principle of INDSCAL and the closely related principle of parallel proportional profiles in the PARAFAC-CANDECOMP model and by use of the principle of simple structure in the more general invariant factors model (McDonald, chapter 7). Ironically, as pointed out by Harshman (chapter 5), multimode data may provide a richer base for solution indeterminacy than was available from seemingly less complex designs. For users of the methods described in this book, the major issue is the potential insight to be gained from solutions based on various multimode methods. Most contributors have provided a simulated or actual research example, and a large section of some chapters is concerned with the interpretation of these examples. Many of the procedural issues relating to interpretation are common to all the methods. Appendix A, by Harshman, has been included to outline procedures for interpreting multimode solutions. Diagnostics are provided in terms of the appropriateness of the model, computational adequacy of the fitting procedure, statistical reliability of the solution, and the generalizability and explanatory validity of any resulting interpretations.

CONCLUDING REMARKS

The richness and diversity of the methods is most evident. Yet, are the methods useful? Gould (chapter 15) asks us to pause and consider whether multimode methods are an advance in methodology or merely a culmination of two centuries of restrictive thinking using one type of mathematical and statistical method. Gould would prefer that we attempt to move out of a traditional framework to achieve a different perspective and warns that we should not be blind to other possibilities. Often, when using and developing multimode models, a methodology is chosen that forces structure and its attendant interpretability on the data. What is needed is a perspective from outside the structure and this, Gould claims, has been lacking in multimode studies, as well as all conventional multivariate approaches.

These are early days for an adequate assessment of the value of these sophisticated methods. However, when competently applied, multimode analytic methods are clearly powerful aids for empirical research. They do not usurp the researcher's responsibility to understand and rationally probe into the less obvious data relations, but they are powerful tools for this empirical engagement. The attached multimode bibliography attests to the growing interest in these methods. Many of these applications reflect the advantages and power of multimode analyses, contributing in meaningful ways to the investigative area. No method will automatically advance knowledge: "Methods are merely vehicles for the exercise of evidence" (Rychlak 1981, 43).

Progress requires ingenuity and creativity. Along the way, empirical methods offer the possibility of new perspectives on phenomena, thus facilitating insight and understanding and providing an evidential base for theory development and validation. Structural analyses lie at the beginning of the research process. Ultimate usefulness is not likely to be their main objective. Rather, they assist in delimiting and refining some conceptual domain for further experimental analyses. These chapters provide new vehicles for this initial endeavor.

REFERENCES

Carroll, J. D., and J. J. Chang. 1970. Analysis of individual differences in multidimensional scaling via an N-way generalization of "Eckart-Young" decomposition. *Psychometrika* 35:283–319.

Cattell, R. B. 1966. The data box: Its ordering of total resources in terms of possible relational systems. In *Handbook of multivariate experimental psychology,* ed. R. B. Cattell. Chicago: Rand McNally.

———. 1978. *The scientific use of factor analysis in behavioral and life sciences.* New York: Plenum.

Harshman, R. A. 1970. Foundations of the PARAFAC procedure: Models and conditions for an "explanatory" multi-modal factor analysis. *UCLA Working Papers in Phonetics* 16:1–84 (University Microfilms No. 10,085).

Hattie, J. A. 1981. A four stage factor analytic approach to studying behavioral domains. *Applied Psychological Measurement* 5:77–88.

Meyer, W. J., and A. W. Bendig. 1961. A longitudinal study of the Primary Mental Abilities Test. *Journal of Educational Psychology* 52:50–60.

Rychlak, J. F. 1981. *A philosophy of science for personality theory.* 2nd ed. Malabar, Florida: Krieger.

Young, F. W., and R. Lewyckyj. 1979. *ALSCAL-4 user's guide.* Chapel Hill, N.C.: Data Analysis and Associates.

Young, F. W., J. de Leeuw, and Y. Takane. 1980. Quantifying qualitative data. In *Similarity and choice,* ed. E. D. Lantermann and H. Feger. Bern: Hans Huber.

APPENDICES

A

"How can I know if it's 'real'?" A Catalog of Diagnostics for Use with Three-Mode Factor Analysis and Multidimensional Scaling

Richard A. Harshman

Far too often, solutions obtained by multivariate procedures—including factor analysis, multidimensional scaling, and cluster analysis—are interpreted, and even published, without adequate evaluation of their reliability or validity. Particularly among inexperienced users, there is an uncritical and somewhat cavalier approach to determining what parts (or which version) of an analysis to accept. Clusters or dimensions are frequently taken to be "real" whenever an interpretation can be projected onto them by the imagination of the analyst. On the other hand, dimensions that don't fit preconceptions and are hard to interpret tend to be dismissed too easily. While some users may make a feeble attempt at justifying their choice of dimensionality by examining improvements in fit values, little effort is otherwise expended in determining whether clusters or dimensions are stable or reliable, whether the model is appropriate for the data, whether the algorithm achieved correct convergence, whether serious outliers are present in the data, and so forth.

On the other hand, more experienced and sophisticated users often do employ diagnostic checks, but the particular ones that are applied will differ from one user to the other. Many of the techniques employed by a given analyst may have been developed by the analyst himself, handed down by word-of-mouth, or picked up from a passing reference in a published article. Consequently, inexperienced users have little chance of being exposed to such procedures, except by apprenticing to someone who knows them or by reinventing the techniques themselves. While pub-

---

Major portions of this article were written at Bell Laboratories, Murray Hill, New Jersey. The author is grateful for Bell Laboratories' continuing support of research on three-way models and related topics.

lished accounts exist for some of these techniques, they are unfamiliar to most potential users; as yet, there is no common body of accepted, well-known methods available. And, because many editors are also unaware of the importance and proper use of diagnostics, they permit publication of articles that lack essential evidence for the validity of the solutions presented.

This is, to say the least, an unfortunate state of affairs, and this informal article is one attempt to initiate changes. In the following discussion, I will first point out the role that diagnostics can and should play and the questions that they can answer. I will then briefly try to convey the range and nature of diagnostic procedures that can be used in factor analysis and MDS by providing an informal list and brief description of those procedures of which I am aware, including brief mention of some still being developed. As we shall see, a number of different methods can be used to check the optimality, reliability, and validity of a three-mode analysis solution. In fact, the potential variety of diagnostic techniques for cluster analysis, three-mode factor analysis and MDS is so substantial that it could (and hopefully will) constitute a major area of growth and refinement of multivariate methodology in the next few years.

In this informal listing or "catalog," an attempt is made to develop a natural classification scheme for the diagnostics, based on the type of information they need (such as the data itself, the factor loadings resulting from a single analysis, loadings from several analyses, and so on) and which aspects of the solution they focus on (including the fitted parameters of the model, the residuals from the fit, or indices of overall goodness-of-fit). The catalog is intended only as an introduction or overview of some of the possible diagnostics that can be used at various stages of an analysis. It certainly is not a thorough exposition; detailed questions of how to use these diagnostics are not covered, although brief descriptions of applications are sometimes given.

The objective is to increase awareness of the methods, to generate interest and discussion, and to suggest methods that have been omitted from the list—perhaps as an initial step toward a more complete treatment in the future—and to encourage investigators to begin using these diagnostics. For some of the procedures, such as diagnostic interpretations of loadings patterns, enough information is provided so that investigators should be able to use the methods to guide them during three-way PARAFAC analyses. For other more esoteric or incompletely developed procedures, only enough information is provided to suggest the possibilities that may later become perfected.

Diagnostics for cluster analysis are not considered here. While proper diagnostics are just as important in cluster analysis as they are in factor analysis and multidimensional scaling, cluster-analytic diagnostics often involve somewhat different procedures. In order to avoid undue complication, this article focuses on the use of diagnostic procedures in factor analysis and multidimensional scaling, especially those that are well suited for the three-way intrinsic axis methods PARAFAC-CANDECOMP and INDSCAL. Many of the techniques discussed, however, would also be suitable for Tucker's model and for other three- and two-way methods. The appropriate adaptations should usually be apparent to the reader.

OBJECTIVES FOR DIAGNOSTIC PROCEDURES

Two Roles for Diagnostics

Diagnostics have both an *exploratory* and a *confirmatory* role. In their exploratory role, they provide guidance during the conduct of an analysis. With such guidance, one can adjust the parameters of successive analyses to maximize sensitivity to the characteristics of the data, rather than stop with the automatic application of standard data preprocessing and data-analytic options. In their confirmatory role, they lend greater assurance to any conclusions drawn from the final factor-analytic or MDS solution.

Exploratory Role

Most of the diagnostics to be discussed in this article are not merely directed at establishing the reliability and validity of a solution that has already been selected for interpretation. Rather, they are important tools for deciding "what to do next" as one proceeds through the course of the analysis. This use of diagnostics goes hand-in-hand with an *interactive* view of data analysis. From this perspective, a factor-analytic or MDS analysis is not a "one-shot" application of an automatic procedure but a multiple-stage interplay between the data analyst and the data, involving repeated cycling between application of the program for analysis and application of diagnostics for evaluating the outcome of the latest stage of the analysis. In this role, diagnostics answer questions such as: Are more iterations required? Should one extract additional dimensions? Are orthogonality constraints necessary?

Confirmatory Role

The confirmatory use of diagnostics for factor analysis and MDS is important to protect both the user and the scientific community from misleading and inappropriate solutions. When used in this way, diagnostics confirm that the analysis model appears appropriate for the data and that the resulting solution is reliable, optimal, and generalizable. Hopefully, as we become more sophisticated, such confirmation will come to be viewed as an essential part of any solution, and future editors will expect a written description of the diagnostics used and the results obtained, as part of the necessary supporting evidence for the conclusions drawn from any analysis. Thus, as familiarity with diagnostics grows, minimal standards of evidence for reliability and validity should develop, helping to screen out some of the most meaningless applications of factor analysis and MDS techniques from the literature.

General Objectives

When using diagnostics, there are at least four basic things that one seeks to determine about a given analysis: (a) theoretical appropriateness, (b) computational correctness, (c) statistical reliability, and (d) explanatory validity. It is important to establish these characteristics for any analysis, be it two- or three-way, dimensional or cluster, and so on. However, here we formulate the problem only in terms of three-way factor analysis and multidimensional scaling, particularly the intrinsic axis methods PARAFAC-CANDECOMP and INDSCAL. To elucidate each of the four basic characteristics listed above, we discuss it in the context of specific questions concerning a three-way analysis problem.

Questions of Theoretical Appropriateness

1. How appropriate is our basic multivariate model?

   a. Is the analysis model we intend to employ appropriate for the kind of question we want to ask? For example, are we really looking for latent dimensions, or would taxonomic clusters be more appropriate? (If clusters, would we want hierarchical or additive, disjoint or overlapping?) Do we want additive main effects and unrestricted interactions, such as an ANOVA would provide, or the structured kind of multiplicative interactions that multidimensional models provide, or both?

      This one set of questions must be answered without diagnostics, before any analysis has been started.

   b. Is the analysis model we seek to employ appropriate for the data? Is the structure underlying the observed relationships more treelike, implying a cluster model, or more spacelike, implying a factor-analytic or MDS model.

      This question is still largely theoretical, but some diagnostics are beginning to be developed to aid the investigator in making this decision.

2. Which of the various three-mode factor-analytic or MDS models is most appropriate for these data?

   a. Is our data distancelike, so MDS is appropriate, or profilelike so that factor analysis is appropriate, or is it scalar-product or covariancelike, again calling for factor-analytic models?

   b. Are the three-mode data likely to provide the appropriate pattern of variations in latent factors across all three modes, so that unique determination of axes by PARAFAC or INDSCAL is possible? Or, instead, is the third mode simply a set of replications differing only by random variations? Or is some intermediate, partial, or more complex variation pattern likely?

c. If there are genuine differences between the levels of the third mode, are these likely to be appropriate for the strong "system variation" model that allows direct fitting, or is the "object variation" model more appropriate, calling for indirect fitting?

d. Would Tucker's three-mode T2 or T3 model be more appropriate for these data? Would complex variations in factor obliqueness or factor interaction contribute a major part of the data variance?

e. If distancelike data are involved, should these be considered squared distances?

f. If the rows and columns correspond to the same set of entities, are the relationships among them symmetrical (that is, does $x_{ij} = x_{ji}$?), or are there systematic asymmetries that need to be described, perhaps calling for a more general model such as DEDICOM?

g. Should one consider a model that specifies orthogonal loadings in one or more modes—for instance, to fit a hierarchical factor solution?

3. Should the data be preprocessed or reexpressed in particular ways to make it appropriate for the model or to bring out its most interesting properties?

a. Is the data likely to contain conditional origins, additive constants, and two-way interactions requiring centering of one or more modes to make it appropriate for the ratio-scale model?

b. Should the variances or mean-squares of the variables, subjects, occasions, or whatever be standardized?

c. Is reweighting of variables, subjects, or other subsets of the data desirable to minimize the influence of unreliable data or to stress aspects of the data where good fit is most important?

d. Should nonlinear transformations be employed, such as log transformations?

Questions of Computational Correctness

4. Has a particular iterative fitting procedure converged to the desired optimum?

a. Are there indications of incomplete convergence due to an overly lax convergence criterion?

b. Is there very slow convergence of some solutions because of a bad starting position or of all solutions because of certain properties of the data?

c. Once convergence has been established, is it convergence to an uninteresting local optimum or to one of several competing interesting solutions? Or is there well-behaved convergence to the same solution regardless of starting position?

5. To what extent is the solution independent of starting position? Do some parameter values change as a function of starting position, while others are fairly stable? Is the configuration of points after rotation to congruence independent of starting position? If we are using intrinsic axis methods, do we obtain the same orientation of axes regardless of starting position of the iterative procedure, or is the orientation dependent on starting position (and, perhaps, not uniquely determined by the data) for some or all dimensions?

Questions of Statistical Reliability

6. What is the stability of the solution across subsamples of the data? If some characteristics are more stable than others, which details of the solution are stable enough to justify interpretation? Which conclusions are generalizable to new sets of subjects, or variables, or occasions, and so on? What kinds of generalizability do we have (for instance, across subjects, across variables, across occasions, and the like); which kinds do we desire?

7. Have we chosen the correct dimensionality? Are all our dimensions stable enough to be recognized in two split-half subsamples? If not, are there other compelling reasons to retain any of the unreliable ones? Alternatively, are there further stable dimensions that we are overlooking?

Questions of Explanatory Validity

8. Are the results interpretable? How are the obtained dimensions related to outside information about the variables, stimuli, individuals, and so on? Are our problems of interpretation, if any, more likely due to the data, to the analysis procedure, or to limitations of our understanding?

9. Are there nonlinear relationships among dimensions that would indicate the appropriateness of a nonlinear model of lower dimensionality? Should we consider nonlinear reexpression of the data?

10. What are the properties of the residuals?

    a. Do they have patterns indicating data structure not captured by the current analysis? Should ANOVA, cluster analysis, or some other procedure be applied to the residuals in order to uncover features that the factor analysis would miss?

    b. Are there extreme outliers (or groups of outliers) in the data that might have unduly biased the solution? Should some data points be replaced by missing-data codes and the analysis run again?

11. Do properties of the obtained loadings indicate that a different version of the three-way model—such as a different extended model—would have been more appropriate? Should different data preprocessing methods have been used?

12. If the current analysis has produced a reliable, meaningful solution, what new experiments could be conducted to test the hypotheses emerging from these results? How could other available data confirm or conflict with the general conclusions you have drawn?

OUTLINE OF THE CATALOG OF DIAGNOSTIC TECHNIQUES

There is a large variety of techniques that can be used to help answer the questions listed above. Although there is not space in this appendix to provide a detailed explanation of all these techniques, I will provide an informal list or catalog of those known to me and will sometimes include a brief statement of how they are used. But before presenting the catalog, I will first give an overview, in outline form, as a guide to the more detailed discussion to follow.

The catalog is organized into four subsections, based on the type of information being examined by the diagnostic procedures in that section:

I. *Zero-fit Diagnostics:* Data diagnostics to be performed before doing the three-way analysis.

II. *One-fit Diagnostics:* Analysis diagnostics based on examination of the results of a single fit (such as an analysis from a particular starting position, at a particular dimensionality, using a particular set of analysis options).

III. *Many-fit (single data set) Diagnostics:* Analysis diagnostics based on comparisons across several different fits made to the same data set (such as using different random starting positions, different dimensionalities, or different analysis options).

IV. *Many-fit (many data sets) Diagnostics:* Analysis diagnostics based on comparisons across fits made to different or partially different data sets (such as random split-halves of the subject sample, or stimulus sample, or overlapping subsamples such as used in jackknifing and bootstrapping).

For the analysis diagnostics in categories II and III, the procedures are further categorized according to whether they are based on examination of the loadings, examination of the residuals, or examination of the overall goodness-of-fit measures. For section IV, they are broken down into several different kinds of reliability evaluation procedures. This system of classification might be helpful for finding the proper diagnostic to use when one is considering a particular part of the computer output or for highlighting certain logical or mathematical relationships among procedures. The outline of the catalog is as follows.

I. *Zero-fit Diagnostics*
   A. Checking for outliers
   B. Checking reliability of the data
   C. Determining the structural form of the data
   D. If data are distancelike, checking tree versus spatial models
   E. Checking appropriateness of three-way models using two-way "collapsed" versions of the data
      1. Comparison of the dimensionality of different two-way "collapsed" versions of the data—are the estimates consistent?
      2. Comparison of the configurations (that is, dimensions after rotation to maximum agreement) across different two-way "collapsed" versions of the data—do special two-way interaction dimensions emerge in certain pairs of modes?
      3. Comparison of the dimensions (after rotation to agreement) found in different slices of a given mode.

II. *One-fit Diagnostics*
   A. Based on examination of loadings
      1. High correlation among factors
      2. Constant factors
      3. Nonlinear relationships among factors
      4. Interpretability
      5. Convergence checking
   B. Based on examination of residuals
      1. Examining patterns in the relative sizes of mean square error (MSE) for different levels of each mode
      2. Outputting residuals for detailed study
   C. Based on overall goodness-of-fit values
      1. Comparison of R with estimated data reliability
      2. Comparison of the square root of MSE with expected size of errors
      3. Comparison of R and MSE

III. *Many-fit (single data set) Diagnostics*
   A. Based on comparison of loadings
      1. Across successive iterations
      2. Across different solutions (different random starts) at a given dimensionality
      3. Across different dimensionalities (to study evolution

of dimensional structure and interpretations)

B. Based on comparison of residuals
   1. Across competing solutions at a given dimensionality
   2. Across dimensionalities (to see for which part of the data the fit improves when adding a particular dimension)
   3. In terms of error distributions

C. Based on comparison of overall fit values
   1. Across iterations (to check convergence)
   2. Across solutions—identifying local optima, incomplete convergence, and so on
   3. Across dimensionalities—classic search for "elbow" in fit versus dimensionality curve

IV. *Many-fit (many data sets) Diagnostics*

A. Comparisons across split-halves of the data
   1. When to split
   2. How to split
   3. How to compare across splits

B. Resampling methods of estimating reliability: bootstrapping and jackknifing
   1. Jackknifing
   2. Bootstrapping

C. Cross-validation techniques testing fit when dimensions are applied to a new sample

D. Randomization tests—comparison of "shuffled" data with observed data to obtain significance tests for three-way variation

E. Comparison of analyses across experiments

It is hoped that this system of classification will be generally helpful and may loosely correspond to the order in which some of the tests may be conducted. However, it is not intended to provide an actual strategy for the interactive use of these diagnostics or for guiding one's choices while performing an analysis. Such strategy questions are not considered in detail in this paper.

CATALOG OF DIAGNOSTIC TECHNIQUES

In this section, an attempt has been made to list the various diagnostic comparisons, tests, or procedures that can be employed to help answer the questions presented in the previous section.

I. *Zero-fit or Data Diagnostics:* To be performed before doing the three-mode analysis.

A. Examination of data to detect outliers, with the possibility of omitting outliers from the analysis by declaring them as missing data (while saving their identity for subsequent study).

B. Evaluation of reliability of the data—by comparing replications, computing test-retest correlations for each subject or variable, or by comparison of corresponding cells on either side of the diagonal of supposedly symmetrical data, and so on. Possible elimination of subjects, conditions, and the like that have insufficient reliability. Also use reliability estimates for comparison to later variance accounted for by the model.

C. Determination (mostly on the basis of outside information or theory) of whether the data is likely to be more distancelike, profilelike, or scalar-productlike, so that the appropriate analysis model and procedure can be selected.

D. If the data are thought to be distancelike, evaluating whether a tree-structure or spatial model for the distances is more appropriate; this can be approached by fitting the alternative models and comparing fit, but recent work of Carroll, Pruzansky, Tversky and others is developing some test statistics that can be computed from the data itself, such as the skewness of the distribution of distances.

E. Examination of systematicity of variation across each mode, by means of two-way analysis of "collapsed" versions of the three-way data; to help determine which kinds of three-way variation are present and thus which three-way model is useful. This approach is still largely untested, and these procedures are not as essential as others to be discussed later, but the following are examples of steps that might be useful:
   1. Plots of successive eigenvalues or singular values of data collapsed across various modes; do the plots reveal systematic dimensionality in all three modes? Is dimensionality similar across different modes, or might Tucker's model be more appropriate?
   2. Examination (and possibly canonical correlation) of eigenvectors or singular vectors extracted from various two-way "strung out" or "collapsed" versions of the data; how are these related? For example, do the Mode *A* vectors of the *A*-*B* collapsed data resemble the Mode *A* vectors of the *A*-*C* collapsed data?
   3. Canonical correlation of eigenvectors or singular vectors extracted from successive layers of the data sliced across a given mode; how similar are the dimensions in the different layers of the three-way array?

II. *One-fit Diagnostics:* Analysis diagnostics that utilize results of a single fit—for example, diagnostics for a solution obtained from a particular starting position, at a particular dimensionality, using a particular set of analysis options.
   A. Examination of the fitted parameters—the factor loadings.*
      1. Examination of correlations and cross-products among factor loadings for each mode.

*Points 1–3 are specific to an intrinsic axis model.

a) Check for very high correlations between two (or more) dimensions in all three modes ("very high" means, roughly, above .8).

(1) If the pattern is consistently obtained across starting positions, and the triple product of the correlations for all three modes is negative (that is, either one correlation is negative or all three are negative), this could signal a degenerate solution (as discussed in chapter 6). If the data have not been centered on one or two modes, try additional centering. If careful centering and standardization does not remove the degeneracy, then a "hard-core" degeneracy may be present, requiring orthogonality constraints on at least one mode.

(2) If the pattern only emerges at a relatively high dimensionality and is not consistent across solutions or split-halves of the data, it might simply indicate that more factors are being extracted than can be supported by the data. In this case, the triple product of correlations should as often be positive as negative. If, after applying orthogonally constraints, the dimensions in question are still not interpretable and are not similar across solutions or split-halves, then this suspicion is confirmed. But if orthogonally constraining one mode gives similar results across split-halves, then dimensionality is not too high (see example in appendix C).

b) Check for high correlations (above .6 or so) between all or most dimensions, in all three modes. This can indicate that too many dimensions have been extracted and/or a general lack of uniqueness of the solution, resulting in arbitrary (correlated) combinations of the "true" underlying dimensions appearing in all three modes. If the triple product of correlations is often positive and different patterns of correlations are observed from different starting positions and across split-halves, this is probably not a "degenerate" solution in the restricted sense used by Harshman and Lundy (chapter 6). Check if all dimensions of orthogonally constrained solutions are replicable across split-halves; if not, reduce dimensionality. (However, if the configuration is reliable but the axis orientation is not, as indicated by high canonical correlations across split-halves, the problem is not extraction of too many dimensions but rather lack of conditions producing unique axes. You must use additional levels of data or some external rotation criteria to obtain unique axis orientation.)

c) Check for high correlations between particular dimensions in one mode only, with lower correlations between the same dimensions in the other two modes. This does not indicate "degeneracy" or that too many dimensions have been extracted. Rather, it warns that certain factors may not show the distinct patterns of variation necessary to determine axis orientation uniquely. (Note: Even if the correlated dimensions are nonunique, the rest of the solution could be uniquely determined.)

2. Examination of loadings, to check for one or more "constant" factors (that is, factors for which all the loadings are approximately the same size).
   a) A factor that is constant—that is, it has loadings of constant size and sign—in all three modes indicates an overall additive constant in the data that probably should be removed. The size of this additive constant can sometimes be estimated from the triple product across modes of the factor's loadings. However, it is safer to simply apply centering to one or more modes.
   b) A factor that is constant in two modes, with varying loadings in a third mode, indicates a different additive constant for each slice (when sliced so that the varying loadings correspond to different slices of the data). Centering across one or more of the constant modes should remove this factor.
   c) A factor that is constant in only one mode simply indicates a "true" factor that doesn't vary much across that mode. If there is only one factor that is constant in any given mode, then such a factor may be uniquely determined; however, comparison across starting positions and split-halves would be desirable to confirm stability. If two (or more) factors have nearly constant loadings within the same mode, then their loading patterns may not be uniquely resolved in the other two modes, and, sometimes, other nonconstant factors will be "contaminated" as well. Centering across the mode in which the constant loading patterns occur will eliminate these factors from the data and may improve the recovery of the other factors. (Note: A more detailed discussion of loading patterns and their correspondence to main effects, 2-way interactions, and 3-way interactions, along with a discussion of the effects of various centering schemes on such factors, appears in chapter 6.)
3. Comparison of loadings across factors to check for nonlinear relationships.

Such relationships indicate either that the data needs to be transformed nonlinearly and/or that the actual latent factors combine nonlinearly to create the observed data. Particular conclusions, including the form of likely nonlinear relationships, can be deduced from the study of the plots of loadings of one factor against another or against functions of several other factors. Nonlinear factor models can sometimes be constructed on the basis of this type of information.

4. Preliminary checking of interpretability of dimensions for the given starting position, by detection of meaningful and/or expected patterns of loadings in each mode.

   This can provide support for (or cast doubt upon) the validity of a particular solution. This can also provide insight into relationships among different solutions (such as why certain dimensions split into more specific ones as dimensionality is increased, or which aspects of a solution replicate across split-halves, or whether alternative meaningful solutions appear in different competing—locally optimal—solutions). However, an extensive study of interpretability should normally be deferred until other diagnostics indicate that one has a candidate for an optimal solution (see below).

5. Checking convergence by examining the rate at which loadings are changing across successive iterations.

   While this is technically a multiple-fit comparison, it is accomplished within an analysis from a given starting point, and so will be briefly mentioned here. Convergence rates can vary greatly, depending on the particular data being analyzed and sometimes on starting position for a given analysis. If the factor loadings are continuing to change in small steady steps that do not diminish appreciably in size (for instance, steps of 1% change per iteration), then one cannot be confident that the given solution is essentially the same as one that would be obtained after many more iterations. Comparison of a given set of loadings with the loadings obtained 10, 20 or even 50 iterations previously should indicate if the loadings are "settling down" or if they are continuing to slowly drift. More accurate evaluation of convergence can be made by use of multiple starting points (see below).

B. Examination of residuals.

1. The Mean Squared Error (MSE) for the solution should be of "reasonable" size, based on expected precision of measurement, likely reliability of the data, and so on.
2. An "Error Analysis Table," which prints out mean-squared error for each level of each mode (for

instance, for all the data points that involve a particular stimulus or all the data points for a particular subject, and so on) can be used to see whether there are certain portions of the data that are causing particular problems. In examining this table, the following points should be checked:

a) Are any MSE values very high or very low, relative to the others?
b) Are there systematic patterns in the larger versus smaller MSE values? Do certain aspects of the data cause problems, including unreliable groups of subjects, certain "difficult" types of stimuli, and the like?
c) If the data being analyzed have unequal variances or mean-squares across the levels of a given mode, the MSE differences among the levels of that mode can be expected to mirror these input differences. Levels that have higher variance or mean square on input will usually have higher MSE on output. Thus, comparison of MSE values should take the input mean-squares or variances into account. One way to do this is to look at Stress at each level rather than "raw" MSE. This problem does not arise, of course, if the data have been size-standardized within each level of the mode in question.
d) If, in each mode, one or two very high MSE values stand out, then the intersections of these levels should be examined for outliers far out of range (such as keypunching errors). For example, if the MSEs for variable 7, person 2, and occasion 2 are all very high, then the data point for person 2 measured on variable 7 on occasion 2 should be examined. (Note: Some programs—including PARAFAC—contain an option to check each data point on input against a user-specified range of possible valid values. Points outside this range will be identified in the output and treated as missing values during the analysis. It is recommended that such an option be used whenever possible—for instance, whenever the data consists of questionnaire responses using a fixed response scale. The points identified as outside the valid range should then be checked and corrected.)

3. For more detailed analysis of residuals, some programs (including PARAFAC) provide the option to write out the entire set of residuals on an output disk file of the user's choice. When this is done, the residuals can be examined in detail by using a variety of standard analysis and graphics programs. Residuals can be plotted against original data values, a histogram of their distribution can be plotted, and so on. Also, the residuals so output

can subsequently be input for further analysis by more sophisticated procedures (such as cluster analysis) or can even be used as input for another round of three-way analysis, which would extract a set of additional factors with fitted ($\hat{x}$) values completely orthogonal to the first set.

C. Examination of overall goodness-of-fit indices—for instance, correlation between the data and the predicted data, Stress, the ratio of the MSE to the mean-square data value, and so forth.

1. How does the *R* value compare to the test-retest reliability of the data (either known or conjectured)? When sufficient dimensions are extracted, it should be approximately the same as the expected reliability of the data being analyzed, if the model provides a good description of the systematic part of the data. Keep in mind, however, that interesting and informative structure can sometimes be recovered even when much of the systematic part of the data cannot be explained. Thus, an interpretable solution should not be rejected outright simply because the *R* was, for example, less than half of the data reliability.
2. How does the square root of the MSE value (that is, standard deviation of the error) compare to the expected size of typical error values? For example, root-MSE of 1 to 2 might be reasonable for a 9-point scale.
3. How do *R* and MSE compare? Usually, when *R* is high, MSE will be small. However, when there are a small number of very errant data points, it is possible to get high *R* values (fitting the very large error variance due to these errant points) but also very large and undesirable MSE values. Both should be checked for reasonableness. (Comparing variations of *R* and MSE values across subsets of the data provides a particularly useful check; see below.)

III. *Many-fit (single data set) Diagnostics:* Analysis diagnostics based on comparison across multiple fits made to the same data set—for example, using different random starting positions, different dimensionalities, different analysis options, and so on.

A. By examination of the loadings.

1. Comparison of different stages in the iterative process to assess convergence (this was also mentioned in II.A.5 above).
   a) For assessing convergence, loading changes are more important than changes in fit values. Fit values will rapidly improve in the early iterations of an analysis and then level off and show only gradual improvement in later iterations (when correct rotation is being established). These gradual changes in fit values can occur

when some of the loadings themselves are still changing substantially. But since it is the loadings that one will be interpreting, these are the quantities that must be stabilized in order to consider a solution to be properly converged.

b) Loading changes should be small at "convergence"; one standard option is to compare changes due to one iteration with the RMS average loading size for each dimension. A conservative rule of thumb is that no loading should change more than .1% of the RMS average size of a loading on that dimension (in that mode) from one iteration to the next; alternatively, the change should not exceed 1% across 10 iterations. This rule can be relaxed to allow 5% or more, with well-behaved, quick converging data, but with such data, relaxing the criterion is often unnecessary because convergence is rapid anyway. Unfortunately, it is with slow converging data that a more stringent criterion is sometimes necessary to prevent a misleading premature fulfillment of the convergence test. (Premature declarations of "convergence" can be detected by comparison of results obtained from different random starts; see III.2, below.)

c) Certain patterns of change can indicate convergence difficulties in particular subsets of dimensions ("subspaces" of the solution). It often happens that loadings for certain factors converge more quickly than others. However, it may occasionally occur that a few factors almost never converge, although the rest of the solution is stable. In this case, different starting positions and/or different dimensionalities should be tried. Are these "difficult" dimensions highly correlated in a single mode, indicating insufficient independent variation to provide uniqueness of axes, or are they highly correlated in all three modes, suggesting either degeneracies (if the triple product of the correlations is persistently negative) or extraction of too many factors (otherwise)?

2. Comparison of several allegedly converged solutions obtained from different random starting places to confirm convergence. Solutions that change slowly enough to meet the convergence criterion may have converged and come close to their terminal values. Alternatively, they may have become trapped in a difficult "ridge" on the surface that they are trying to climb and consequently may be experiencing slow convergence at values very different from the ones they would eventually take on if allowed to iterate indefinitely. Comparison of results obtained from

different random starting positions can help in discriminating these two situations and also provide information on the strength of determination of unique axes for different dimensions.

a) Do several solutions (3–6) agree "closely enough" for their differences to have no effect on interpretation? If so, convergence and uniqueness are both indicated.

b) If several solutions (3–6) agree approximately—that is, they are "going toward the same place"—then setting a more strict convergence criterion and continuing to iterate on any one of the solutions should provide an accurate estimate of the more fully converged solution that would have been obtained with all of them.

c) Do the solutions fall in two or more groups? For example, did two random starts give one solution and four random starts give another? If so, which dimensions are the same in both groups; how do the other dimensions differ? The presence of competing solutions found from different random starting positions could have several interpretations. Check the following possibilities:

   (1) One or more of the groups might represent a local optimum where the program repeatedly "got stuck" on its iterative upward search for the globally optimum set of loadings. Compare the fit values for the competing solutions. Are there differences in the second decimal place? If so, the set or sets with the lower fit value may be a local optimum.

   (2) How do the interpretations of each of the two or more competing solutions relate to those of solutions obtained at lower and higher dimensionalities? Are the dimensions in both competing sets interpretable? Do they represent two different subsets of a larger common set of dimensions that will be obtained in higher dimensional solutions? (Obviously, one must wait till more dimensions are extracted to check this.) If so, both competing solutions may represent "valid" but incomplete approximations to the higher dimensional "true" solution. (Note, however, that the form of the dimensions will often be clearer in the higher dimensional solution, where all valid dimensions emerge, since in such a solution no dimensions need be distorted to help adjust for the effects of a dimension not yet extracted. On the other hand, if determination of axis orientation is weak at the higher dimensionality, the dimensions

in this solution may seem less clear. When too many dimensions are extracted, the form of some of the dimensions often starts to break down.)

(3) Are there two (or more) competing solutions, with one (or more) of them showing very high factor intercorrelations or other signs of break down? The well-behaved and interpretable solution should be preferred, particularly if it has a higher fit value. (Occasionally, the well-behaved solution may have a slightly poorer fit value; nonetheless, it should probably still be preferred.)

(4) Are there many different solutions (almost as many as random starting positions)? Across starting positions, do certain sets of factors keep changing—recombining differently? This could indicate that those particular factors do not have a unique rotation determined by the data.

(5) Are all or almost all factors changing across different starting positions? This would indicate rotational indeterminacy of the solution as a whole and suggests either that one mode of the data does not have the required systematic variation of any factors necessary to establish uniqueness, or, alternatively, that substantially more dimensions are being extracted than can be supported by the data (for instance, at least 50% too many dimensions).

3. Systematic comparison of dimensions across solutions of different dimensionalities can shed light on "family" relationships among dimensions in the higher dimensional solution and reveal stronger versus weaker aspects of the solution.

   a) It may sometimes be useful to construct a "Terbeek tree," showing which of the dimensions in the two-dimensional solution (for example, were present in the three-dimensional solution) and so on. Display the correlations or other similarity measures between the dimensions at several different dimensionalities, from one up to the maximum number extracted. Such a "tree" of dimensional relationships will reveal at each level whether a dimension split into two dimensions, whether components of several dimensions were drawn off to form a new dimension, or whether an entirely new dimension emerged. In this way, the tree can relate the dimensions at all different dimensionalities. Hopefully, this will help one to understand the process by which various dimensions emerge. It may also suggest that at a certain dimensionality, a particular dimension is "contaminated" by other specific unextracted dimensions or is otherwise distorted.

b) Is some sort of meaningful hierarchical structure suggested by the tree? How related are the *interpretations* of two dimensions that "emerged" from a common ancestor in a lower dimensional solution? Can these "family" relationships among dimensions themselves suggest interesting interpretations of the data, much as a hierarchical cluster analysis would?

c) At what dimensionality was the interpretation of a given dimension clearest?

B. By examination of the residuals.

1. Comparison of residuals (or MSE values for specific levels of each mode) across competing solutions at a given dimensionality, to determine which parts of the data are being fit by each solution. This may occasionally be useful when one of two competing solutions is suspected to be due to peculiar characteristics in a very restricted part of the data. If, in one of the competing solutions, the fit value is clearly lower for this part of the data but not for others, the suspicion is supported.

2. Comparison of residuals (or MSE values for specific levels of each mode) across dimensionalities to see which part of the data has reduced fit when each dimension is added. This will sometimes reveal that one of the smaller dimensions extracted at higher dimensionalities is mainly accounting for the variance of a single subject, variable, and so on. It might also show that the dimension is attempting to account for certain "outlier" points for several subjects.

3. Comparison of distributions of the residuals at different dimensionalities. Check, in particular, whether there are still many large values in the tails of the distribution at lower dimensionalities; hopefully, these become less common as more dimensions are extracted, approaching the desired low frequency compatible with normally distributed error at the "correct" dimensionality (unless other kinds of structure are present that cannot be fit by the factor model).

4. For more methods of looking at residuals of a PARAFAC analysis, see Kettenring's article (cited in chapter 5).

C. By examination of overall fit values (R, R-squared, Stress, MSE, and so forth).

1. Comparison of fit values across successive iterations to determine convergence. This is the "traditional" method of assessing convergence of the solution. As noted earlier, however, this method is not recommended for our three-way intrinsic axis models, since the fit values often change quite

slowly during the later stage of a given analysis, while the loadings themselves are still undergoing considerable modification (due often to shifts of axis orientations). However, in certain degenerate solutions, the loadings of highly correlated factors will continue to change, even though the improvement in the fit is negligible (such as in the seventh decimal place). In these cases, it is pointless to wait for "convergence," since mathematical analysis shows that there may be no local optimum. Instead, compare solutions from several starting places to determine that after two or three hundred iterations they show similar "nonconverged" loading patterns.

2. Comparison of fit values across solutions obtained at the same dimensionality but from different starting positions
   a) This permits evaluation of the relative progress of several different solutions toward a common solution; if all solutions in a given set are "going to the same place," then pick the one with the highest fit value for interpretation or for continuation of the analysis with additional iterations
   b) If there are two or more competing solutions at a given dimensionality, comparison of overall fit values may indicate that the one that is difficult to interpret is in fact an uninteresting local optimum with substantially lower fit. Note, however, that two solutions that differ in fit might both be interpretable or "interesting" local optima; they may select different subsets of dimensions from a larger set that will be revealed at a higher dimensionality.
3. Comparison of fit values across dimensionalities in an attempt to determine the best number of dimensions to extract from a given data set. This is the classic "scree" test or search for the "elbow" in the fit-versus-dimensionality curve.
   a) The scree test remains a very important method of assessing dimensionality. It has a straightforward logical rationale and a simple graphical method of implementation that make it easy to understand and apply: Improvements in fit (such as changes in R-squared) due to each additional dimension are plotted against dimensionality. When the points begin to fall onto a smooth line, one assumes that "real" dimensions are no longer being extracted; the small steady increments are presumably due to fitting error. (Only the points that deviate from the smooth fit-change-versus-dimensionality line should be taken to indicate "real" dimensions; the older approach of including the first point on the smooth "scree" line should

not be followed.) Monte Carlo studies indicate that the scree test, when used with care, is one of the most accurate methods of assessing dimensionality.

Because the scree test is based on fit values, it provides information that is complementary to that provided by tests that are based on replication of a pattern of factor loadings (including split-half, bootstrap, or jackknife methods). It will sometimes work when they fail, and vice versa. For example, when lack of independent variation in the three-mode data causes the orientation of some axes not to be well determined, then two split-half solutions may differ in rotation for some dimensions, and the "true" dimensionality may be underestimated if one stops extracting dimensions when loading patterns fail to replicate. However, the scree test does not depend on proper or consistent alignment of axes, since the fit values are still well determined even when axis orientation is not. Thus, the scree test can indicate the presence of a systematic configuration in higher dimensions, even when the axis orientations fail to replicate. In this situation, replication tests can only verify the higher-dimensional configuration if they are strengthened by inclusion of rotation-to-congruence procedures or if regression or canonical correlation is used to find comparable dimensions across solutions. But it is not altogether a weakness of the replication tests that they will not detect these higher dimensions, since axes that are unstable should not be interpreted. If one decides to apply some rotation procedure such as VARIMAX to determine axis orientation in situations in which the intrinsic axis property cannot be used, then replication of the VARIMAX-rotated axes provides an appropriate test of which loading patterns can be taken as sufficiently stable to interpret.

Although we have been considering the situation in which the scree test indicates a higher dimensionality than the replication tests, the opposite situation can also arise. A smooth elbow on the curve can make the scree point hard to identity. Dimensions that account for only a small portion of the variance may not show up clearly on the scree curve yet may be replicable and theoretically important (for instance, see Gandour and Harshman, cited in chapter 5). When such dimensions replicate, they should be included in the solution and interpreted.

In general, then, the scree and replication (split-half) tests of dimensionality are complementary and, whenever possible, should both be applied and the results compared.

b) Unpublished Monte Carlo tests of different fit measures suggest that while most fit measures gave similar results, *R*-squared or variance accounted for provides the clearest "elbow" at the true dimensionality of synthetic data and the flattest curve thereafter.

IV. *Many-fit (many data sets) Diagnostics:* Analysis diagnostics based on comparisons of fits made to different or partially different data sets—for instance, random split-halves of subjects or stimuli, overlapping subsets as used in bootstrapping or jackknifing, and so on.

This is a particularly important diagnostic technique. It gives the strongest basis for deciding what aspects of a solution are statistically reliable and thus potentially generalizable to new samples.

In the exploratory mode, these diagnostics can be used to help determine optimum characteristics of an analysis, such as dimensionality of a solution. When additional dimensions replicate in two split-halves, for example, then they deserve interpretation and probably should be included in the solution in some fashion. When dimensions do not replicate, they should only be interpreted with caution, and the investigator should seriously consider reducing the dimensionality of the solution.

In the confirmatory mode, these diagnostics assure the investigator that the obtained dimensions are not simply based on fitting noise in the data. Characteristics of the solution that are demonstrably stable across samples are in some sense "real"; they are characteristics of some larger population. Those that are not stable in a particular test may or may not be "real"; their apparent instability may be due to small sample size, and the variations across subsamples may be reduced by taking larger samples. On the other hand, the so-called "characteristics" may not show up consistently in new samples of any size, because they are due to random error. Thus, interpretations based on characteristics that do not replicate may not be justified; they may be attempts to interpret random sampling fluctuations.

It is worth noting that the most important form of stability that should be tested by these techniques is stability of *conclusions* or *interpretations*. The key objective of any method of evaluating reliability—be it split-half, bootstrapping, or whatever—is to determine whether the important points of interpretation (such as the scientific conclusions or the recommendations for action) that are drawn from a particular sample are justified, at least to the extent that they can be generalized to equivalent samples.

In addition to this qualitative evaluation of the "robustness" of conclusions, more quantitative methods of measuring stability can often be used to place confidence bounds around the values of particular parameters in a given solution. These techniques are usually applied to assess the reliability of the loadings, although they can also be used to place confidence bounds around fit values, angles between dimensions, and the like. There are several different methods of assessing reliability of factor-analytic and MDS solutions, including analysis of split-halves, resampling methods such as bootstrapping and jackknifing, cross-validation by application of a given set of loadings to a new data set, randomization tests, and full replication of data collection and analysis, with or without meaningful variations.

A. Comparison across split-halves of the data.

1. When to split. Analysis of split-halves of one's data can be risky if one has only a few subjects to begin with. The half-size sample may be too small to reveal most of the interesting patterns. The number of subjects needed in each split-half depends on the reliability of the data and the number of dimensions one intends to try to verify. With most social sciences data (including rating scale data), a minimum of 10–15 in each half would be needed to verify a few of the largest dimensions. To more sensitively test for smaller dimensions, 35–70 in each split-half is preferable. With 100+ in each half, the method becomes a very powerful way of verifying consistencies of subtler relationships within each dimension or of estimating the reliability of less robust characteristics of the solution, such as the angle between dimensions. And with these larger split-halves, one can demonstrate the reliability of dimensions that do not contribute large portions of variance but may be theoretically interesting; it also allows one to reliably extract larger numbers of dimensions.

2. How to split.

   a) In large data sets, random division into two groups should be sufficient. But with smaller data sets, one must guard against "unlucky" splits, where the two halves are actually (by chance) different. To guard against this, one can use the "orthogonal split-halves" technique. Divide the data randomly into four subsets: *A, B, C,* and *D*. Then construct the following alternative split-half divisions: $(A + B)$ versus $(C + D)$; $(A + C)$ versus $(B + D)$; and, if desired, $(A + D)$ versus $(B + C)$. Check each dimension or aspect of the solution for replicability across these two or three different splits. If the dimension replicates across any one of them, it is tentatively verified, since *any* such replication is very unlikely to happen by chance.

   b) If one is testing generalizability to new subject

samples, (perhaps the most common test), one splits the data into two subsamples of subjects. However, one can treat other modes of the data as representative samples from which one wishes to generalize. For example, one might sometimes split across variables, occasions, and so on. The question of which mode(s) to split is related to the "random effects versus fixed effects" question in analysis of variance. Those modes that you consider "random effects" —that is, those you consider simply a representative sample of possible levels and for which you would like to generalize any results to other similar samples of levels—should in theory be tested for generalizability by some technique such as split-half or bootstrapping.

3. How to measure agreement across split-halves. As noted at the beginning of this section, the most important objective of reliability measurement is usually to determine the generalizability of conclusions or interpretations; but this often calls for a difficult-to-quantify comparison of the conclusions that would be drawn from two different split-half solutions. More quantitative evaluation of the similarity of characteristics of solutions, such as patterns of loadings on a given dimension, can be computed by means of correlation coefficients or factor congruence measures. When loadings have a mean near zero (for instance, if they come from a mode that was centered), then correlations and cross-product congruence measures give the same result, but when the loadings are mostly positive (or negative), then the different measures of similarity stress different things. Cross-product measures of factor congruence are often preferred because they are sensitive to differences in overall elevation as well as profile shape of the loading pattern. However, cross-product measures of similarity can give very high and possibly misleading results when two factors with all positive loadings are being compared. One's choice should depend on which aspects of similarity are important. I usually recommend correlation as a more stringent test of factor similarity; it stresses those variations in loading size that are crucial to interpretation. However, there are circumstances in which correlations can also be misleading, so careful consideration is advised.

B. Resampling techniques for estimation of reliability: bootstrapping and jackknifing. Tukey's "jackknifing" and Efron's "bootstrapping" techniques are beginning to be used to measure the reliability of factor loadings and fit values. In these procedures, the original data set is used to generate several alternative versions, which are all analyzed and the results compared. The more these alternative analyses differ, the less reliable is the conclusion based on the original data set.

1. Jackknifing. In this technique, new data sets are generated by omitting parts of the original data set. For example, a data set consisting of ratings made by 30 subjects could be used to construct 30 new data sets, each with one subject's data missing. Each of these data sets is then analyzed in the dimensionality under test, resulting in 30 sets of factor loadings. Reliability estimates are then computed from the variations across these solutions.
2. Bootstrapping. In this technique, new data sets are generated by sampling the original data *with replacement*. For example, if there are 30 subjects and generalizability across subjects is being tested, then new data sets of size 30 could be constructed, each of which is based on sampling of subjects *with replacement* from the original 30. Thus, in each of the new samples, some subjects will probably be omitted and others will occur more than once. Once again, the variability of loadings or fit values (or other parameters, such as angles between dimensions) across these new samples is used to estimate the reliability of loadings or fit values in the original data. (References for both bootstrapping and jackknifing are given in chapter 5.)

C. Cross-validation by applying loadings to a new sample. In some programs, such as PARAFAC and INDSCAL, it is possible to input in Mode *A* and *B* loadings that have been determined by analysis of one sample and fit these to a new sample, estimating only new subject weights. Alternatively, PARAFAC allows one mode to be input and fixed and two to be estimated from the new sample. The resulting fit value will show "shrinkage" because you are no longer fitting error in two of the three modes. A sample can be split in half and the dimensions fit in each half can be applied to the subjects in the other half for a double cross-validation.

D. Randomization tests. "Significance tests" for the presence of system variation, additional dimensions, or other characteristics of the data can be obtained by means of randomization tests, in which the results of the analysis of the obtained data are compared with the results of randomly permuted versions of that data. For example, to test for system variation in a three-way array, one can randomly permute the entries in each "tube" of the data, thereby leaving the structure in two modes intact but scrambling the structure in the third mode. If the data are so permuted and analyzed 19 times, then under the null hypothesis, these 19 should not differ systematically from the original unpermuted data. Thus, the null hypothesis of no system variation can be rejected at the .05 level if the original unpermuted data produced a higher fit value than the 19 permuted ones, since this fit ranking would have a probability of .05 under the null hypothesis. In a similar fashion, tests for additional dimensions can be constructed by permuting residuals from the lower dimensional analysis, tests for systematic asymmetries can be performed by randomly interchanging the x_{ij} and x entries in each allegedly symmetric data matrix, and so on.

E. Comparison across experiments. The most complete test of the generalizability of a given finding is, of course, when someone replicates the finding with a new sample, perhaps incorporating some modest variations in the methodology of data collection or subject selection, followed by a new analysis. Results that are stable across such replications demonstrate the strongest evidence for generalizability.

SUMMARY

Diagnostic evaluation of the optimality, reliability, and validity of solutions is often lacking in studies using multivariate methodology. Yet, the use of diagnostics is crucial because it enables the analyst to address four basic questions underlying any multivariate analysis: (a) appropriateness of the model; (b) computational adequacy of the fitting procedure; (c) statistical reliability of the solution; and (d) the generalizability and explanatory validity of any resulting interpretations.

A number of important diagnostic techniques for factor analysis and MDS are now available, including many that have been recently developed, and they could play an important role in promoting the growth and intelligent use of factor-analytic and MDS procedures over the next few years. To increase awareness of these techniques, an informal listing is presented of diagnostics known to the author (including some still being developed). In this listing, an attempt is made to develop a natural classification scheme for the diagnostics, based on the type of information they need (such as the data itself, the factor loadings resulting from a single analysis, loadings from several analyses, and so on) and which aspects of the solution they focus on (including the loadings, the residuals, and the fit values). Although detailed questions of how to use these diagnostics are not covered, brief descriptions of usage are sometimes given in the listing. Of the various techniques listed, the most important are probably the methods of evaluating the reliability of any characteristics of a solution (split-half, bootstrapping/jackknifing, and so forth). These can be used to estimate maximum dimensionality and decide which aspects of a solution are stable enough to warrant interpretation.

B

Multidimensional Scaling Displays

John C. Gower

Other chapters in this book discuss the many models, computing algorithms and applications of various forms of three-mode multidimensional scaling. The associated graphical displays may appear quite varied, but analysis shows that they are made up from a few basic geometrical ideas. This appendix attempts to uncover these basic displays—each of which is usually appropriate to two-mode forms of analysis—and show how they may be combined to give three-mode displays.

Similar basic displays are commonly used throughout the applied sciences, allowing the interrelationships between numerical quantities to be visualized through familiar concepts of distance, angle, and area. With multidimensional scaling, like other multivariate methods, the configurations to be displayed are often multidimensional so there is the additional difficulty of representing several dimensions in the two dimensions of a sheet of paper or VDU screen or perhaps as three-dimensional physical models. The most natural graphical displays directly reflect the properties of the algebraic form of the model fitted. Thus, when an $n \times n$ distance matrix Δ with element δ_{ij} is fitted to data by multidimensional scaling, its simplest representation is by a set of n points P_i ($i = 1, 2, \ldots, n$) such that the distance between P_i and P_j is, or is proportional to, δ_{ij}. With an inner-product model, it is natural to plot vectors and the angle between pairs of vectors together with their lengths is the geometrical property used for interpretation. Both distances and inner-products generate symmetric matrices; sometimes skew-symmetric matrices are to be represented and it turns out that their elements may be represented as the areas of triangles with one vertex at the origin. These basic plotting devices will be discussed in more detail in the following sections. The plots associated with three-mode arrays δ_{ijk} usually combine elements of the basic two-dimensional displays. From the above remarks, it should be clear that the form of plotting used cannot be entirely divorced from considerations of the model fitted.

DISTANCE PLOTS

The basic problem of multidimensional scaling is to approximate distancelike data held in a symmetric matrix **D** by true distances held in a symmetric matrix Δ. The elements of Δ, being distances, have zero diagonal values, but this may not be true for **D**. To make the elements of **D** more like distances may require some preliminary transformation to a matrix **D*** where, for example, $d_{ij}^* = d_{ii} + d_{jj} - 2d_{ij}$ or $d_{ij}^* = 1 - d_{ij}$ or $d_{ij}^* = -\log d_{ij}$, the last two of which are especially useful for similarity matrices with unit diagonals and positive similarities bounded above by unity. Suppose **D** and Δ have order n; then the elements δ_{ij} of Δ will be generated by the coordinates of points P_i $(i = 1, 2, \ldots, n)$ such that the distance between P_i and P_j is exactly δ_{ij}. Suppose also that these coordinates are held in the $n \times r$ matrix **X**, such that the coordinates of P are given by the ith row of **X**. The fitted distances δ_{ij} are nearly always taken to be Euclidean, although most multidimensional scaling programs offer the option of fitting any of the Minkowski metrics. To plot the points P_i requires r dimensions. Fortunately, in many applications a sufficiently good approximation to the data can be obtained with $r = 2$ dimensions. When $r = 3$, a model may be constructed or a box-plot drawn, where a perspective effect may be attempted by plotting the points as interior points of a cube. A commonly used device is to plot the points in three planes relative to axes one and two, one and three, and two and three of **X**. I find this less useful than a plot relative to the first two axes, with the value of the third dimension added adjacent to the point plotted for P_i. When the items plotted fall into different a priori categories, it is useful to distinguish each category by using a different plotting symbol (squares, triangles, circles—filled or hollow, and so on) or a different color. More dimensions, associated with higher values of r, may be represented by $\binom{r}{2}$ two-dimensional plots, supplemented in the same ways as for $r = 3$, but the difficulty of their interpretation increases with r.

In the above, it has been assumed that the dimensions of **X** are ordered. In a trivial sense, this is always true, as the kth column of **X** may be taken as the kth dimension. It is helpful when the first two dimensions of **X** give a best approximation to Δ (not necessarily to **D**), and this most conveniently and simply implies that **X** should be referred to its principal axes—not all multidimensional scaling programs offer this convenient form of output.

Although **X** generates Δ exactly, Δ is itself only an approximation to **D**. The overall degree of approximation is represented by a single number, such as a residual sum-of-squares, or for nonmetric multidimensional scaling methods, by criteria such as Kruskal's stress. Users may wish to know whether or not the degree of approximation is equally valid for all points or whether the distortion is greater for some than for others. For those metric scaling methods that allow a residual to be associated with each point, it suffices to write down the residuals next to the plotted points. However, with many metric multidimensional scaling methods, it is the values of d_{ij} that are approximated by δ_{ij} to give residuals $d_{ij} - \delta_{ij}$; with nonmetric scaling methods, the residuals are $f(d_{ij}) - d_{ij}$ where $f(.)$ represents some monotonic transformation of the data. Thus, instead of one residual for each *single* point, there is now one residual for each *pair* of points, and these are not easily incorporated into diagrams. One useful device is to join pairs of points in the plot of Δ that are nearest neighbors as given by **D**. In a good approximation, nearest neighbors should map into nearest neighbors, or at least into near neighbors. An extension of this idea is to plot on the representation of Δ the minimum spanning tree (MST) as given by **D**; the MST uses only the rank order of the elements of **D** and not their absolute values. Examples of all these devices are given by Gower and Digby (1981).

INNER-PRODUCT DISPLAYS

This form of display is for symmetric positive semidefinite matrices. Suppose **A** is a symmetric positive semidefinite matrix of order n; then we may write $\mathbf{A} = \mathbf{X}\mathbf{X}'$ in an indefinite number of ways. However, all such decompositions are invariant up to, for instance, an orthogonal matrix **H**. That is, if **X** and **Y** give two different decompositions, it is always possible to find an **H** such that $\mathbf{X} = \mathbf{YH}$. This property implies that when the rows of **X** (or **Y**) are plotted as coordinates of points Q_i $(i = 1, 2, \ldots, n)$, then the distances between all pairs of points Q_i and Q_j do not depend on the decomposition chosen. Clearly, inner-product displays and distance displays are related; this relationship will be discussed in more detail at the end of this section.

Denoting the origin by O and the distance of Q_i from the origin by δ_i, it follows that the elements of **A** are given by $a_{ii} = \delta_i^2$ and $a_{ij} = \delta_i \delta_j \cos \Theta_{ij}$, where Θ_{ij} is the angle between OQ_i and OQ_j. Thus, angle rather than distance is the basic relationship used to interpret inner-product plots. By normalizing **A** so that $a_{ii} = 1$ and hence $\delta_i = 1$, the plot of **X** gives points **Q** whose pairwise angles Θ_{ij} satisfy $a_{ij} = \cos \Theta_{ij}$. A two-dimensional approximation to **A** that best preserves the values of a_{ij} can be obtained by taking **X** as the two eigenvectors of **A** associated with its two largest singular values (or equivalently eigenvalues). This is an immediate consequence of the Eckart-Young theorem. Thus, we can examine approximations to **A** that best (in a least-squares sense) preserve angles. In particular, given the matrix Δ, with elements $-\frac{1}{2}\delta_{ij}^2$, we can follow Gower (1981) and form:

$$\Delta^* = (\mathbf{I} - \mathbf{1}\,\mathbf{s}')\,\Delta(\mathbf{I} - \mathbf{s}\,\mathbf{1}') ,$$

where **1** is a vector of units and

$$s = \frac{\Delta^{-1}\mathbf{1}}{\mathbf{1}'\Delta^{-1}\mathbf{1}} .$$

The elements of Δ^* are $R^2 \cos \Theta_{ij}$, where R is the radius of the circum-hypersphere and $\cos \Theta_{ij}$ is the angle subtended at the center of this hypersphere by Q_i and Q_j. Thus, Δ^* is proportional to the correct form of matrix for which an inner-product plot best preserves angles. The angles subtended at the center of the circum-hypersphere are best approximated by the Eckart-Young rank-2 approximation to Δ^*. Of course, $d_{ij} = 2\sin \frac{1}{2}\theta_{ij}$, but the best approximation to these distances is obtained from the decomposition of Δ^{**} obtained by setting $\mathbf{s} = \mathbf{I}/n$ in the expression for Δ^*. The origin may be chosen anywhere. For distances, we know that the best choice of origin is the centroid. But for approximating angles, we do not know if some other choice of origin is best, or indeed whether or not there is a best choice.

Another device is to approximate $\mathbf{A}$ in its normalized form but remember the original diagonal values $a_{ii} = \delta_i^2$. These values may then be written next to the plotted points $\mathbf{Q}_i$, thus preserving some distance information along with the angular approximations. All plots have to compromise in approximating distance and angle. When both are important, it is best to produce two separate plots. We shall return to this later when discussing row and column plots.

From the inner-product representation of $\mathbf{A}$, the cosine formula shows that the squared distance between Q_i and Q_j is $a_{ii} + a_{jj} - 2a_{ij}$, so low-rank approximations to $\mathbf{A}$ are approximating these distances. To approximate such distances directly, it is best to use the methods and associated plots of the previous section. Classical scaling, however, achieves its distance approximations by evaluating the inner-product spectral decomposition of Δ^{**} so that the two problems of distance and angular representation then become inextricably confounded. However, even here, angular approximation is improved by normalizing Δ^{**} at the expense of losing distance information.

AREA DISPLAY PLOTS

These are naturally associated with skew-symmetric matrices. Gower (1977) and Constantine and Gower (1978) give the basic methodology and examples. With a skew-symmetric matrix $\mathbf{A}$, we have that $a_{ij} = -a_{ji}$. Thus, if the ith item is represented by a point R_i ($i = 1, 2, \ldots, n$), we require a non-Euclidean relationship between pairs of points R_i, R_j. For points in a plane with origin O, one possibility is given by the area of the triangle OR_iR_j, which may be taken to be positive if the vertices are traversed clockwise and negative if traversed anticlockwise. In addition, the area OR_iR_i is zero, as required by the zero diagonal values of a skew-symmetric matrix. Because this representation is non-Euclidean, care has to be taken in interpreting such diagrams, particularly with respect to distance properties. Pairs of close points give small area triangles and therefore refer to small values of a_{ij}. However, apparently distant points collinear with the origin also give zero area. These diagrams are particularly nice for paired-comparison data, for if ratings for three stimuli i, j, and k satisfy $A_i > A_j > A_k < A_i$, the points R_i, R_j, and R_k form a triangle that contains the origin, but if $A_i > A_j > A_k > A_i$, the triangle is external to the origin. Thus, apparently contradictory results may be given a simple real representation in terms of areas. Constantine and Gower (1978) and Gower (1980) discuss the geometrical properties in more detail.

When a good approximation to **A** cannot be obtained in two dimensions, it turns out that the extension is to add further planes, obtaining the approximation by adding together the areas OR_iR_j in as many planes as needed. The approximations to **A** are thus always in an even number of dimensions. J. D. Carroll has suggested the term *bimension* for each of these planes. I have reservations about this terminology, but at least it is better than *biplane*, suggested by one of my colleagues.

Some square asymmetric matrices have rows and columns similarly classified, but perhaps the classifications may be regarded as being in different modes such as before/after or from/to. It may then be appropriate to analyze the symmetric matrix $\mathbf{A} + \mathbf{A}'$ by multidimensional scaling techniques with associated distance or inner-product plots. The skew-symmetric matrix $\mathbf{A} - \mathbf{A}'$ may be analyzed as above. There is then the possibility of combining the two plots. This is particularly simple when the relationship $a_{ij} - a_{ji} \sim a_i - a_j$ is approximately true. Then the numbers a_i may be superimposed next to the point P_i of the multidimensional scaling and sometimes these values may be contoured (see Constantine and Gower 1978 and Gower 1980).

It is worth noting that distance matrices **D**, like skew-symmetric matrices, have zero diagonals. This suggests replacing **D** by $\mathbf{D}_s$, a skew-symmetric matrix whose absolute values are those of **D**. This may be done in $2^{1/2(n-2)(n+1)}$ ways by placing different signs on the elements of $\mathbf{D}_s$, except for d_{12}, which may be kept positive. The resulting skew matrices may be given areal plots, some of which may require fewer dimensions than those used in the usual approximations to distance matrices. The converse of representing a skew matrix by a distance matrix is impracticable, because this requires signs to be associated with every pair of points. Although this is conveniently done by drawing directed lines with arrows giving the direction of positivity, the resulting diagram is too overloaded with detail to be of any value.

ROW AND COLUMN PLOTS

Most forms of plot associated with multidimensional scaling are of the three basic types discussed above or of combinations of these types. Of special importance are plots of two-way tables **T** with m rows and n columns, where m sets of points P_i $(i = 1, 2, \ldots, m)$ represent the rows and n sets of points Q_j $(i = 1, 2, \ldots, n)$ represent the columns. Such diagrams are of two distinct types.

The first type arises from unfolding models where **T** is regarded as the lower rectangular portion of an $(m + n) \times (m + n)$ distance matrix **D** with the leading $n \times n$ and trailing $m \times m$ symmetric matrices missing. Most nonmetric multidimensional scaling programs will tolerate this missing information and provide the coordinates for the $m + n$ points in some specified number (such as r) of dimensions. There are also metric multidimensional unfolding algorithms. (See Gower and Digby [1981] for further references and examples.) Suppose the coordinates are given by $P_i \equiv (x_{i1}, x_{i2}, \ldots, x_{ir})$ and $Q_i \equiv (y_{i1}, y_{i2}, \ldots, y_{ir})$;

then the fitted approximation is of the distance type:

$$f(t_{ij}) \sim \sum_{k=1}^{r} (x_{ik} - y_{jk})^2 ,$$

where $f(.)$ represents some simple transformation of the data. With such diagrams, it is only the distances between points P_i and Q_j of the two different sets (rows and columns) that model the data. Distances within sets—for instance, between P_i and P_j or between Q_i and Q_j—are also produced and may be of interest.

In the unfolding model, **T** may itself be a symmetric matrix, in which case **T** is represented by $2n$ sets of points. Of special interest is the situation in which **T** is a distance matrix **D**. Such matrices may be unfolded in the usual way or, more usefully, by assuming that their zero diagonal values are missing data. The rationale of this approach is that because the diagonals are known to be zero, there is little point in reproducing the obvious, while relaxing the zero self-distance requirement may give more latitude for obtaining interesting representations. Under these conditions, each point is represented twice: once as a row-label and once as a column-label. The distance between P_i and Q_i may not be zero, although it sometimes is. Symmetry of the distances P_iQ_j and P_jQ_i may be obtained in several ways besides the usual one where P_i and Q_i coincide for all $i = 1, 2, \ldots, n$. This double representation of the elements of **D** sometimes yields useful dimension reduction. In principle, the process can be extended so that every point is represented 4 or 6 or more times. There is a trade-off between dimensionality and replication of points, but personally, I find that two representations of each point is about the maximum that is useful; beyond this, the diagrams become too confusing. Gower and Digby (1981) give an example of three-dimensional unfolding of a symmetric matrix.

The second type of row-column display is based on the multiplicative form given by the singular value decomposition $\mathbf{A} = \mathbf{U} \Sigma \mathbf{V}'$, where **A** has dimensions $m \times n$, as before; **U** is orthogonal $(m \times m)$; **V** is orthogonal $(n \times n)$; and Σ is zero, except for the diagonal elements Σ_{ii}, which are non-negative and ranked $\Sigma_{11} \geqq \Sigma_{22} \geqq \Sigma_{33}$ and so on. The Eckart-Young theorem says that the best least-squares rank r approximation to **A** is obtained by setting $\Sigma_{ii} = 0$ for all $i > r$; this approximation may therefore be displayed by plotting the rows of **U** and **V** as points in two r-dimensional spaces. The two sets may be plotted with some advantage in the same space. How should Σ be handled? Various solutions are in use, but all can be accommodated by choosing different values of α in the expression

$$\Delta = (\mathbf{U} \Sigma^{\alpha})(\mathbf{V} \Sigma^{1-\alpha})'$$

and plotting $\mathbf{U} \Sigma^{\alpha}$ for the row-coordinates and $\mathbf{V} \Sigma^{1-\alpha}$ as column-coordinates.

Appropriate choices of α depend on the nature of **A**. When **A** refers to n variables measured for m sampling units (the usual multivariate sample data matrix) and when columns are expressed as deviations from the sample means, then we are effectively considering principal components analysis, for the columns of $\mathbf{V}$ give the component loadings and therefore $\mathbf{A V} = \mathbf{U} \Sigma$ gives the unit scores. It is then natural to choose $\alpha = 1$. The distance between P_i and P_j is, of course, the Euclidean distance between the ith and jth rows of $\mathbf{A}$ and the r-dimensional approximation is the projection of this distance on to the space determined by the r principal axes. The columns of $\mathbf{V}$ give the *directions* of the principal axes relative to the original axes and thus their lengths are not of special interest. The point Q_i referring to the ith column (that is, variable) of $\mathbf{A}$ is the score for a pseudounit that has unit observation on the ith variable and zero on all other variables. Thus, the distance between Q_i and Q_j in the full representation must be $\sqrt{2}$ for all pairs (i,j); the only interest in the r-dimensional approximation to these distances is to reveal to what extent there is gross distortion.

Another choice of α is to set $\alpha = 0$, in which case the lengths of the unapproximated vectors are the standard deviations of the variables and the inner-products give the covariances. It seems that $\alpha = 1$ gives good approximations to the distances between samples and poor representations of the variables, while $\alpha = 0$ does the opposite. One could plot $\mathbf{U} \Sigma$ and $\mathbf{V} \Sigma$, but then the inner-product interpretation is lost. Yet a further choice is $\alpha = \frac{1}{2}$, which treats rows and columns equally, as is appropriate when $\mathbf{A}$ is a two-way table or classification of a single variate. Distance interpretations are now rather obscure and one relies primarily on the inner-product form of the decomposition.

Joint plots for rows and columns of the types just discussed are usually referred to as *biplots* (the *bi-* refers to the two classifications rather than the two dimensions usually but not necessarily used). An important special case of this approach is termed *correspondence analysis,* in which the matrix $\mathbf{A}$ is first standardized to $\mathbf{R}^{-1/2}\mathbf{A}\mathbf{C}^{-1/2}$, where $\mathbf{R}$ and $\mathbf{C}$ are diagonal matrices of the row and column totals of $\mathbf{A}$. It turns out that when $\mathbf{A}$ is a contingency table, this standardization is especially appropriate and that the resulting displays, based on the singular value decomposition, may be interpreted in terms of chi-square. Gower and Digby (1981) give a unified algebraic treatment and examples of many of the variants of biplot and correspondence analysis. Recently, Nishisato (1980) has aptly termed this whole area of data analysis *dual scaling;* he gives an excellent account of the long history of the method.

THREE-MODE DISPLAYS

In this section, we are concerned with three-way data d_{ijk} and fitted values δ_{ijk}, where $i = 1, 2, \ldots, m$; $j = 1, 2, \ldots, n$; and $k = 1, 2, \ldots, p$. Fortunately, complete generality seems not to be required in multidimensional scaling; it would require something like a display of $m + n + p$ points P_i, Q_j, and R_k in at least three dimensions where the volume of the tetrahedron $OP_iQ_jR_k$ approximated δ_{ijk}. Even if such a model were fitted, it would have little visual appeal. In multidimensional scaling, it is usual for two of the suffices—for instance, i,j—to refer to the same items or stimuli and k refers to p different individuals or occasions. Thus, d_{ijk} may be thought of as arising from p symmetric matrices of order n, and similarly for δ_{ijk}. This is very much easier to handle, for we may plot each of the symmetric matrices separately using the methods already discussed. Even when $m \neq n$, separate unfolding or biplot diagrams can be drawn. The only problem then is to find some way of depicting how the different distance-spaces combine.

Most three-way multidimensional models express d_{ijk} in terms of a group-average space common to all individuals, together with some means of representing how each individual departs from the average. The average space is of distance type and can be plotted in the usual way to give one point for each individual, P_i $(i = 1, 2, \ldots, n)$. With the various forms of individual scaling, each individual is allowed to weight in his own way each dimension of the group-average space. Thus,

$$\delta_{ijk}^2 = \sum_{r=1}^{s} w_{kr}(x_{ir} - x_{jr})^2 ,$$

where x_{ir} gives the coordinate of the ith individual in the rth dimension of the group average space and w_{kr} gives the weight associated by the kth individual with the rth dimension and s is the total number of dimensions in the representation. At least when $s = 2$, the weights w_{kr} are conveniently plotted with respect to s orthogonal axes, giving one point for each individual. From this diagram, it is easy to see which individuals conform to the average and which depart from the average by giving disproportionate weight to one or more of the dimensions. Individual scaling plots are given by Young (chapter 12, Figure 12–1) and Carroll and Pruzansky (chapter 10, Figures 10–6 and 10–7).

In the above, w_{kr} may be regarded as the diagonal values of an $r \times r$ weighting matrix $\mathbf{W}_k$ which is otherwise zero. In general, $\mathbf{W}_k$ may take on more elaborate forms. In GEMSCAL, $\mathbf{W}_k$ is a positive semidefinite matrix of specified rank; when this rank is one or two, its principal axes may be plotted as directions in the group-average space, one set of directions for each individual with lengths proportional to the square roots of the corresponding eigenvalues. The angles between these axes and their relative lengths give information on variation between individuals. The coordinates of the kth individual are the projections of the group average coordinates onto the kth set of principal axes, scaled up or down by the corresponding square-root-eigenvalues. Young describes GEMSCAL in detail and his Figures 12–2, 12–3, 12–4, and 12–5 (chapter 12) are examples of the plotting technique just described. Note that it combines distance displays with some aspects of angular displays. The MDPREF model of Carroll and Pruzansky has features reminiscent of the rank one version of GEMSCAL (see chapter 10, Figures 10–2 and 10–3). With idiosyncratic scaling, $\mathbf{W}_k$ becomes a complete symmetric positive semidefinite matrix that can itself be displayed in inner-product form. The totality of all p matrices $\mathbf{W}_k$ need comparing; rather than use individual scaling methods recursively for this purpose, it might be better to use generalized procrustes analysis (Gower 1975).

Generalized procrustes analysis also produces a group-average display, but the way individuals differ from the average is expressed differently from the individual scaling method. Now the distance display for each individual is rotated into a position of best fit relative to the average and all np points are displayed simultaneously. It is convenient to show what are the points P_{ik} ($k = 1, 2, \ldots, p$) that each of the individuals associates with P_i. When the individuals fit the average moderately well, this is simply done by joining P_i to P_{ik} for all values of k, thus forming a set of n stars each with p rays. This kind of display forms a key part of the PINDIS analysis, described by Lingoes and Borg in chapter 11. Their Figures 11–1 through 11–6 serve as illustrations together with the additional transformations permitted by PINDIS. However, those transformations that relate to stretching of dimensions and translation of origin for each individual offer no special problems in plotting.

Not all three-mode models treat one suffix k, typically referring to individuals, in a different manner from the remaining two suffices i and j. The Tucker three-mode component model discussed by Kroonenberg in chapter 3 does not readily fall into that class because in its most general form all suffices have equal status. However, the Tucker2 model comes very close to the individual scaling type of treatment. Nonetheless, even in the general case (with the exception of the core matrix), each element of the model is a single matrix whose elements may be plotted by methods already discussed. In particular, pairs of matrices may be plotted simultaneously to give row and column plots, termed *joint-plots* by Kroonenberg, and these play a central role (see Kroonenberg, chapter 3, Figure 3–5). The core matrix is itself three-mode, but sometimes a suffix referring to individuals may be singled out, allowing its elements to be approximated by and plotted using individual scaling techniques; see Kroonenberg for what can be done when this simplification is not possible.

CONCLUSION

Of course, other forms of display than those discussed above are associated with multidimensional scaling. These include plots of observed versus fitted values, stress functions, residuals, and so on, which are of a kind general throughout data analysis. I have concentrated on the geometrical aspects of displaying the data (and more particularly the fitted values) that are a central feature of the method. It is probably more true of multidimensional scaling than any other statistical method that the associated plots are the main feature of the analysis rather than tabulations or numerical estimates of parameters or various inferential statements. Indeed, some methods such as Multiscale (Ramsay 1978) do provide estimates of confidence regions, but these are easily incorporated in the diagrams discussed. None of the methods described in this book exploit three-mode versions of the area display described for two-mode analysis. This is because the underlying model has not yet been used for analyzing the change from individual to individual in asymmetric responses to pairs of stimuli. An analysis analogous to that of INDSCAL is clearly possible, in which an areal plot of a group space giving the general asymmetric response to stimuli is supplemented by plots representing for each individual scaling (and perhaps rotational) factors that describe individual variations from the norm. Such analyses would be useful and will hopefully be available soon.

REFERENCES

Constantine, A. G., and J. C. Gower. 1978. Graphical representations of asymmetric matrices. *Applied Statistics* 27: 297–304.

Gower, J. C. 1975. Generalized procrustes analysis. *Psychometrika* 40:33–51.

———. 1977. The analysis of asymmetry and or orthogonality. In *Recent developments in statistics,* ed. J. Barra et al. Amsterdam: North Holland.

———. 1980. Problems in interpreting asymmetric chemical relationships. In *Chemosystematics: Principles and practice,* ed. F. A. Bisby, J. G. Vaughan, and C. A. Wright. New York: Academic Press.

———. 1982. Euclidean distance geometry. *Mathematical Scientist* 7:1–14.

Gower, J. C., and P. G. N. Digby. 1981. Expressing complex relationships in two dimensions. In *Interpreting multivariate data,* ed. V. Barnett. New York: J. Wiley and Sons.

Nishisato, S. 1980. *Analysis of categorical data: Dual scaling and its applications.* Toronto: University of Toronto Press.

Ramsay, J. O. 1978. Confidence regions for multidimensional scaling analysis. *Psychometrika* 43:145–60.

C

An Application of PARAFAC to a Small Sample Problem, Demonstrating Preprocessing, Orthogonality Constraints, and Split-Half Diagnostic Techniques

Richard A. Harshman and Wayne S. De Sarbo

In this appendix, we present in detail the procedure involved in a single application of PARAFAC three-way factor analysis to real data. The data are from a pilot marketing study and consist of the ratings of 25 stimuli (names of automobiles and celebrities) made by each of 34 raters, using a set of 39 bipolar rating scales. The objective of the analysis is to determine the connotative and semantic dimensions describing the celebrities and automobiles in order to decide which celebrity should be chosen as spokesman for a given automobile.

This application is intended to provide a demonstration/explanation of what PARAFAC is and what it does, for those to whom "an example is worth a thousand equations." It is also intended to serve as a guide for those who perform or are about to perform their own analysis of a three-way data set. The data are typical, since stimulus ratings by each of several subjects on each of several ratings scales is probably one of the most common kinds of three-way social sciences data. Since it is a small data set, it highlights the need for good methodology to enable maximum information recovery without misinterpretation of random error as meaningful patterns. The analysis problems that come up are the ones frequently encountered in analyzing such data; they are also typical of many other kinds of three-way data. The techniques used to preprocess the data, to test dimensionality, to determine stability of the solution, and so on are the basic ones needed in all careful three-way analyses. Thus, this article could be viewed as an illustrative companion to the general discussion of diagnostics provided by Harshman (see appendix A).

Factor analysis could never be properly conducted in a "one-shot" run through the computer—unfortunately, too many social scientists have used it that way. Its proper use has always required careful planning, followed by a series of analyses. In the analysis stage, study of the results of a given analysis is followed by reanalysis, extracting a different number of factors, using a different rotation procedure, and so forth. Based on what is learned from each solution, a new analysis is performed and the process is repeated until the optimal solution is obtained. The need for such a careful, multistage analysis *process* is even greater with PARAFAC three-way factor analysis, and this article provides an example of how it is done.

This application demonstrates: (a) three-way preprocessing and its effects on the solution; (b) the method of split-half cross-validation and how it is used to test the stability and reliability of each solution; (c) how comparison of split-half results for different numbers of factors is used to establish the maximum number of dimensions; (d) how application of orthogonality constraints to one loading matrix is used to overcome "degenerate" solutions and permit recovery of additional meaningful dimensions; and (e) the process of interpretation of a three-mode solution—how the loadings from all three modes (scales, stimuli, and subjects) can be used to strengthen insight into the nature of each dimension and how comparison of interpretations across several dimensionalities helps to refine judgments about the "proper" dimensionality and the meaning of the factors that are obtained.

While the use of diagnostic procedures is always important (see appendix A), it becomes particularly important in this type of application because of the small subject sample that was employed (only 34 in all, 17 in each split-half). The results demonstrate how proper use of split-half validation can be used to determine the number of dimensions that can in fact be reliably extracted from such small samples and how stable the extracted dimensions are. The fact that at least three (and probably four or five) dimensions could be reliably recovered from split-half samples as small as 17 shows just how encouragingly robust and powerful these three-way methods are.

THE PROBLEM

Motivation for the Analysis

The example is drawn from a marketing application in which the motivating question is: How can one select an appropriate commercial spokesman for a given brand or product? By applying three-way factor analysis to semantic differential rating scale data on products (in this case, automobiles) and potential spokesmen, we hope to discover the underlying connotative overtones of the spokesmen and products, enabling us to display both in a common multidimensional *semantic space*. By using such a space, we will be able to make better informed judgments about how the overtones of a particular spokesman might reinforce or interfere with the desired impression for a particular product. This type of marketing application is a special case of a methodology called *connotative congruence analysis,* which is discussed in more detail in the longer manuscript from which this appendix was taken (Harshman and De Sarbo 1981).

The Data

The data consists of ratings of each stimulus word on 39 bipolar seven-point scales. The stimuli used are given in Table C–1, and the concepts labeling the poles of the rating scales are given in Table C–2. The approach is based on the *semantic differential* technique (Osgood 1962; Osgood, Suci, and Tannenbaum 1957). At the top of each page of the subject's test booklet was the name of a celebrity or car make; on the twenty-fifth page, the subjects rated themselves on the same 39 scales. Below the stimulus name were the 39 bipolar rating scales. The subject would place a mark somewhere along each bipolar scale, indicating how strongly he thought the stimulus named at the top of the page was related to the adjective at one end or the other of the

TABLE C-1. Stimulus List

Twelve Celebrities Tested (Aided Recall):		
1.	Bob Hope	(Comedian)
2.	John Wayne	(Actor)
3.	Muhammed Ali	(Boxer)
4.	Farrah Fawcett	(Model)
5.	Ralph Nader	(Consumer Rights)
6.	Orson Welles	(Drama)
7.	Sammy Davis, Jr.	(Singer)
8.	Arnold Palmer	(Golfer)
9.	Jerry Lewis	(Comedian)
10.	John Travolta	(Actor)
11.	Barbara Walters	(Newscaster)
12.	Mary Tyler Moore	(Actress)
Twelve car makes tested:		
13.	Ford	
14.	Buick	
15.	Chevrolet	
16.	Cadillac	
17.	Oldsmobile	
18.	American Motors	
19.	Chrysler	
20.	Dodge	
21.	Plymouth	
22.	Lincoln	
23.	Pontiac	
24.	Mercury	
25.	Self	

TABLE C-2. Bipolar Adjectives for the 39 Semantic Differential Scales

	Adjective pair			Adjective pair	
1.	Pleasant	--- Unpleasant	21.	Expert	--- Novice
2.	Strong	--- Weak	22.	Masculine	--- Feminine
3.	Formal	--- Informal	23.	Slow	--- Fast
4.	Dynamic	--- Static	24.	Superior	--- Inferior
5.	Usual	--- Unusual	25.	Ugly	--- Beautiful
6.	Colorless	--- Colorful	26.	Simple	--- Complex
7.	Leading	--- Following	27.	Trustful	--- Distrustful
8.	Plain	--- Ornate	28.	Austere	--- Lush
9.	Sophisticated	--- Naive	29.	Smooth	--- Rough
10.	Liked	--- Disliked	30.	Public	--- Private
11.	Disreputable	--- Reputable	31.	Obscure	--- Famous
12.	Superficial	--- Profound	32.	Old	--- New
13.	Mature	--- Youthful	33.	Orthodox	--- Heretical
14.	Rational	--- Intuitive	34.	Graceful	--- Awkward
15.	Familiar	--- Strange	35.	Efficient	--- Inefficient
16.	Positive	--- Negative	36.	Light	--- Heavy
17.	Careless	--- Careful	37.	Interesting	--- Boring
18.	Aggressive	--- Defensive	38.	Large	--- Small
19.	Hard	--- Soft	39.	Attractive	--- Unattractive
20.	Active	--- Passive			

scale. For example, one stimulus name was "John Wayne" and one of the rating scales was "Light 1 2 3 4 5 6 7 Heavy." If the subject thought John Wayne was "Heavy" in some metaphorical sense, he would circle a number on the right-hand side of the scale. The "heavier" he thought Wayne to be, the further to the right his circle would be.

The 39 scales were selected to tap the basic semantic differential dimensions of Evaluation, Activity, and Potency and also to get at qualities related to previously published theories of "source credibility" (Kelman and Howland 1953; McGuire 1969): Expertness, Attractiveness, Trustworthiness, and Likability. We also included other aspects thought relevant to connotative congruence of these stimuli (see Mowen 1980). In addition, each rater also answered 30 questions regarding attitudes and driving styles. For example, some of the items were: "I usually look for the lowest possible prices when I shop" and "I admit I try to keep up with the Joneses." These were collected to shed possible light on the market segments that particular subjects might represent and to aid in the interpretation of the person loadings that would eventually be used to help interpret the dimensions.

The data input to the PARAFAC analysis consisted of a 39 × 25 × 34 array. Each row of the input data corresponded to the ratings of the 25 stimulus items on one of the 39 scales; each set of 39 rows corresponded to the ratings provided by one subject.

DATA ANALYSIS

Preprocessing

To demonstrate the importance of data preprocessing, we first present the results of a three-dimensional analysis of the raw data (Table C–3). We see here a more or less uninterpretable solution. The first dimension is probably adjusting for the different constant offsets of the various rating scales from a true zero origin. Note that the mean value in Mode *C* (the mode chosen to reflect the scale of the data in this study) is around 3.5, which is the center of the 7-point rating scale. The different loadings in Mode *B* range from close to zero to close to 2.0. This would reflect differences in the additive constant for the different scales. The pattern of variations in this mode suggests that the zero-point was taken to be that point on the scales corresponding to high positive evaluation of the stimuli. Dimensions 2 and 3 show the classic pattern of degeneracy discussed by Harshman and Lundy (chapter 6). As shown in Table C–4, the loadings are highly correlated in all three modes and the product of the three correlations—that is, Mode *A* × Mode *B* × Mode *C* correlation—is negative.

Before preprocessing, the data is liable to contain unwanted constants and two-way interactions that interfere with the ability to define axes uniquely. For one thing, PARAFAC expects ratio-scale data and the raw data is interval-scale at best. Additional reasons why unpreprocessed data may not provide meaningful solutions are discussed in Harshman and Lundy (chapter 6). To overcome this degeneracy, the data were centered on Modes *A* and *B* (stimuli and scales) and size-standardized on Modes *B* and *C* (scales and subjects). This double-centering removes the overall additive constant and all one-way "main effects," as well as subject-stimuli and subject-rating scale interactions, since these are constant across Mode *B* and Mode *A*, respectively. The data were size-standardized on Mode *B* (rating scales) because it was thought that some scales might show much less variance than others simply as an artifact of the choice of overly extreme labels or because of "ceiling effects" (such as all celebrities being rated as "famous"). Size standardization permitted all the rating scales to be approximately equally weighted in the analysis and provided comparability of loadings across levels of Mode *B*. The subject mode was also size-standardized to ensure that all subjects contributed equally to the solution and to remove differences in the size of responses that might arise from response styles such as "extreme" responding versus "moderate" responding. This combination of preprocessing options has been repeatedly found to give good results with three-way rating scale data. The preprocessing required four iterations to reach the standard convergence criterion (less than .01 deviation from requested equality of mean-squares in all modes).

Which precise combination of centering and standardization options to select is not always obvious, although the nature of the application and the data collected can render valuable insights into appropriate preprocessing options. In general, one must consider which additive constants, two-way interactions, and

TABLE C-3. PARAFAC Three-Dimensional Unconstrained Solution for Raw Unpreprocessed Data

	Mode A		
	1	2	3
1	.96	.73	.53
2	.88	1.07	1.21
3	1.06	.50	.14
4	1.06	.19	-0.89
5	.99	1.29	1.46
6	1.01	1.36	1.74
7	1.04	.45	-0.18
8	.93	.96	.88
9	1.04	.67	.08
10	1.04	.27	-0.80
11	.99	1.06	.95
12	1.02	.43	-0.13
13	.94	1.18	.86
14	1.00	1.20	1.23
15	.96	.88	.44
16	1.06	1.33	1.81
17	1.00	1.14	1.10
18	.97	.96	.06
19	1.02	1.47	1.56
20	.96	1.02	.42
21	.97	1.29	.85
22	1.05	1.37	1.92
23	1.01	.83	.39
24	1.03	.91	.56
25	1.00	.88	.73

	Mode B		
	1	2	3
1	.25	1.14	-0.97
2	.30	1.22	-1.41
3	.66	.99	-1.40
4	.11	1.31	-0.99
5	1.66	-0.90	.75
6	2.02	-1.41	1.14
7	.21	1.42	-1.45
8	1.71	-1.07	.87
9	.34	1.28	-1.43
10	.16	1.21	-1.02
11	1.74	-0.85	1.04
12	1.14	-0.49	.89
13	.72	.70	-1.23
14	.94	.39	-0.86

TABLE C-3. Continued

	Mode B		
	1	2	3
15	.20	.96	-0.84
16	.09	1.37	-1.19
17	1.46	-0.66	.99
18	.23	1.19	-1.08
19	.85	.33	-0.47
20	.09	1.24	-0.91
21	.27	1.29	-1.48
22	.68	.63	-0.89
23	1.65	-0.72	.37
24	.16	1.42	-1.42
25	1.74	-0.98	.75
26	1.35	-0.71	1.02
27	.49	.78	-0.80
28	1.58	-0.80	.66
29	.26	1.14	-0.98
30	.58	.42	-0.09
31	2.06	-1.09	.87
32	1.12	.01	-0.49
33	1.14	-0.22	-0.04
34	.14	1.37	-1.08
35	.40	.91	-0.74
36	.91	-0.09	.75
37	.27	1.30	-1.16
38	.73	.73	-1.18
39	.20	1.28	-1.03

	Mode C		
	1	2	3
1	3.71	3.08	1.54
2	3.74	3.78	2.24
3	3.94	3.20	1.98
4	3.67	3.28	1.47
5	3.76	3.30	2.02
6	3.79	3.59	2.00
7	3.86	3.66	2.01
8	3.79	3.17	1.48
9	3.79	3.69	1.84
10	3.70	2.84	1.60
11	3.81	3.93	2.16
12	3.76	4.07	2.28
13	3.82	3.59	2.15
14	3.72	3.88	2.07
15	3.82	4.11	2.27

TABLE C-3. Continued

	Mode C		
	1	2	3
16	3.77	3.22	1.80
17	3.71	3.22	1.49
18	3.82	3.31	1.56
19	3.87	4.37	2.61
20	3.47	3.36	2.10
21	3.71	3.32	1.95
22	3.98	2.51	1.53
23	3.88	3.47	2.03
24	3.62	2.80	2.12
25	3.55	2.56	1.28
26	3.70	2.71	1.45
27	3.78	2.87	1.41
28	3.79	3.37	1.70
29	3.68	3.32	1.57
30	3.67	3.84	2.05
31	3.45	1.92	1.17
32	3.64	2.33	1.16
33	3.69	2.66	1.60
34	3.80	3.38	1.56

Root-Mean-Squared Contribution for Each Factor		
1	2	3
3.744	3.328	1.835

FIT (R^2) = .337

inequalities of variance are most likely to cause problems. A certain amount of trial and error is often necessary to check alternative preprocessing schemes. (Indeed, in the example considered here, we tried an additional standardization of variances for the stimuli but decided that meaningful differences in overall concept salience were obscured and so did not use this standardization in our final analyses.) The preprocessing presented above should be considered as one of several useful possibilities.

Factor Analysis Procedures

The analysis proceeds in a stepwise fashion, moving from a set of one-dimensional analyses to the two-dimensional analyses, and so

TABLE C-4. Correlations of Factor Loadings for Analysis of Unpreprocessed Data

	Mode A		
	1	2	3
1	1.00	-0.31	-0.21
2	-0.31	1.00	.93
3	-0.21	0.93	1.00
	Mode B		
	1	2	3
1	1.00	-0.98	.89
2	-0.98	1.00	-0.95
3	.89	-0.95	1.00
	Mode C		
	1	2	3
1	1.00	.43	.35
2	.43	1.00	.83
3	.35	.83	1.00

on, until various diagnostics (to be described) indicate that too many factors have been extracted. A given step consists of using several different random starting positions to obtain several PARAFAC solutions at the given dimensionality, followed by the application of a number of comparisons and diagnostic checks to evaluate convergence, optimality, stability, and generalizability of the solutions obtained. If the diagnostics indicate that the solutions obtained are appropriate, we proceed on to the next higher dimensionality. After covering a range of dimensionalities, further comparisons of loadings and fit values across different dimensionalities provide a basis for selecting the "correct" solution(s) for final interpretation. (For an introductory survey of diagnostic procedures for three-way factor analysis, see appendix A in this volume.)

To select the preferred solution at each dimensionality, and to determine the "correct" dimensionality for final interpretation, we employed several interrelated techniques: (1) to evaluate the optimality and stability of the factors obtained at each dimensionality, we obtained and compared three independent solutions using three different random starting points for the iterative procedure;[3] (2) the fit values of optimal, converged solutions were plotted as a function of the number of factors extracted, in order to estimate the dimensionality beyond which additional dimensions would only produce small, gradual improvements in fit, attributable to fitting "noise" in the data; (3) the correlations among dimensions within each solution were examined in order to check for highly correlated dimensions, indicative of extracting too many dimensions or other such problems; (4) the reliability and generalizability of the dimensions obtained at a given dimensionality were checked by comparing results obtained from different split-halves of the data set; (5) the interpretability of the results was examined at each dimensionality; (6) the evolution of the interpretations obtained at successive dimensionalities was examined by comparing different dimensional solutions.

An initial series of PARAFAC analyses was performed without imposing constraints concerning the orthogonality or obliqueness of dimensions. After certain diagnostics suggested that orthogonality constraints might be useful, an additional series of analyses was performed with the dimensions constrained to be orthogonal in Mode *B* (that is, across rating scales). The original objective of this constrained procedure was to clarify dimensions obtained in the unconstrained analyses. It turned out, however, to reveal additional structure in the data. We describe the details of the unconstrained analyses first.

Unconstrained Analyses

The initial series of analyses was performed in one through seven dimensions. At each dimensionality, three independent random starting positions were used for the iterative procedure to check for local optimum solutions. The resulting three solutions were compared to evaluate the stability or uniqueness of the dimensions and to reduce the chance of being misled by a local optimum. The loading patterns for the obtained dimensions were correlated across the three solutions to measure their agreement (using the PARAFAC utility program CMPARE). Stable solutions were obtained in all dimensionalities between one and five, as indicated by correlations of .999 between corresponding dimensions from the three different starting positions. At six and seven dimensions, some solutions did not converge and others converged to different places. (We used the default PARAFAC criterion for convergence; that is, from one iteration to the next, no loading should change more than one-tenth of one percent of the root-mean-square average loading value on that factor in that mode.) Because various diagnostics pointed to three as the highest dimensionality that could be relied on with this data set (as will be explained below), an extended effort was not made to determine the optimal, converged form of the solutions for five- and higher-dimensional solutions.

Fit Values. At each dimensionality, several different goodness-of-fit measures were computed. The R^2 values for the 1 through 7 dimensional constrained solutions were as follows: 1D = .123; 2D = .195; 3D = .258; 4D = .293; 5D = .333; 6D =

.363; 7D = .383. Examining the R^2 values, we see that by using three dimensions, PARAFAC is able to account for roughly 25% of the data variance. How good is this figure? Experience with other three-way data, plus some Monte Carlo simulations, suggests that these values are not particularly high, but they are high enough to be compatible with correct recovery of major dimensions in the data. It must be remembered that PARAFAC fits the "uncollapsed" three-way array and thus these figures represent the ability of the factors to reproduce individual ratings. They should not be compared to fit values from studies using averaged data. Yet, while PARAFAC fits the individual ratings, it fits all the raters at the same time, and so it is able to take advantage of patterns that are consistent over many raters to improve the accuracy of its estimates of the loadings for stimuli or scales. This gives PARAFAC an ability to detect patterns that would normally only become apparent in averaged (collapsed) data.

Traditionally, two-way analyses of this three-way data might typically deal with ratings averaged over the thirty-four subjects, and such averaging would reduce the error variance of the data and would thus improve the fit values of the analysis considerably. To see what the fit value would be with a more traditional analysis of this data set, we "collapsed" the $25 \times 39 \times 34$ array into a $25 \times 39 \times 1$ array of mean ratings by averaging over subjects. When these "collapsed" data were analyzed by PARAFAC, the dimensions obtained were, of course, not unique in axis orientation. But consistent fit values were obtained from different starting positions. In three dimensions, a fit value of $R = .829$ was obtained, which means that PARAFAC could account for 68.7% of the variance of the averaged data by using three factors. This higher fit value should reassure those investigators who are used to examining fits for two-way analyses of averaged data, since PARAFAC analysis of the "uncollapsed" data should provide equivalent accuracy in the recovery of the underlying structure, with the additional advantage of determining the orientation of axes uniquely.

Another way one might express the same concern about the seemingly "low" R^2 of .25 is by asking whether this reflects the ability of the model to recover structure or simply its ability to fit "noise" in the data. To provide some information on this, we took advantage of the Monte Carlo data synthesis options available in PARAFAC and constructed an array of synthetic data similar to ours but consisting entirely of random noise. In particular, we constructed a $25 \times 39 \times 34$ array of uniformly distributed random deviates, with a constant standard deviation of 1.0. We then preprocessed this array in the same way as the real data had been preprocessed and submitted it to PARAFAC for analysis in three dimensions, from three different random starting positions. The resulting solutions were all different (presumably because there were no systematic dimensions in the data), but the solutions all had approximately the same R^2 values: .0224, .0223, and .0214. The R^2 in all cases indicated that PARAFAC could fit only about two percent of the error of an array of this size. Since it could account for 25% of the variance of our real data, it seems clear that this figure is not primarily due to fitting noise in the real data (at least if our admittedly simple Monte Carlo example is any indication).

To obtain an estimate of the number of major dimensions underlying the patterns of variation in the data, several plots of fit versus dimensionality were made. Figure C–1 shows one such plot, based on R^2, or the variance accounted for by the PARAFAC model at each dimensionality. Preliminary results of Monte Carlo studies suggest that the plots of R^2 may provide the best indication of "true" dimensionality. The increases in explained variance should be large at first, because each higher dimension-

Figure C–1. Plot of Dimensionality versus R^2 for Unconstrained Solutions

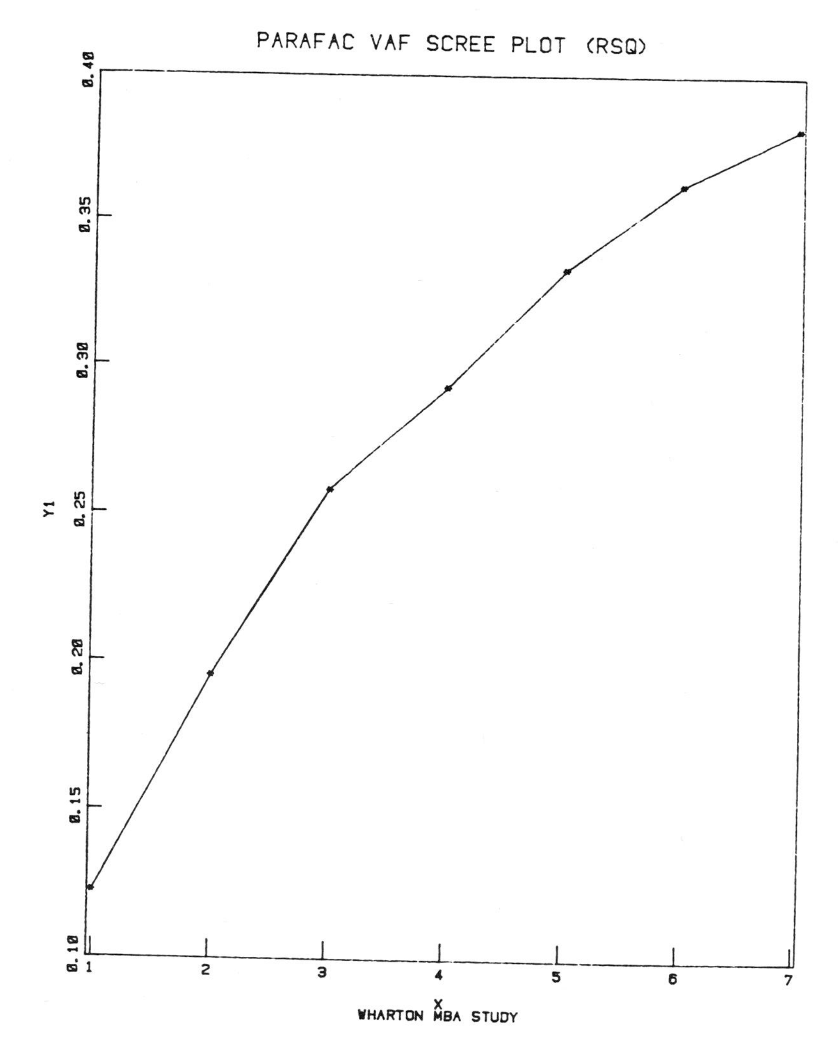

ality incorporates another major factor explaining a new systematic aspect of the variations in the data. However, when the remaining variance is primarily random error, extraction of additional dimensions should produce only a small and consistent increase in R^2 at each step. Figure C–1 does not have a striking "elbow," but might suggest that there are 3 major and 2 to 3 minor additional dimensions in these data. Plots of other measures of fit suggest slightly lower bounds, with R suggesting 3 dimensions and $\log(1 - R^2)$ suggesting that there are just two major factors. Thus, as sometimes happens, the plots of fit versus dimensionality are ambiguous, none containing striking and unequivocal evidence for a given dimensionality. Thus, we are forced to rely even more strongly on more fundamental indications of dimensionality, based on correlations both within and across split-half solutions.

Within-Solution Correlations among Dimensions. One indication that too many factors are being extracted from a given data set is the occurrence of highly correlated factors in all modes. One factor may "split up" into 2 similar versions of itself or more general mixing and overlap of factors may occur. With our data, this phenomenon occurred when we went from 3 to 4 factors. The correlations between factors within each mode are shown for the two- and three-dimensional solutions in Table C–5. The highest correlations are in Mode A, corresponding to similar patterns of loadings across celebrities and automobile makes. Even for this mode, however, the highest correlation in the three-dimensional solution is .41 and in the two-dimensional solution is .50. Contrast these moderate correlations with the very high correlations between dimensions one and two in the four-dimensional solution displayed in Table C–6. (This pair of dimensions corresponds to a splitting up of what was dimension two of the three-dimensional solution; dimension one of the three-dimensional solution has become dimension three, and the old

TABLE C-5. Correlations among Dimensions for Two- and Three-Dimensional Unconstrained Solutions

		Two Dimensions			Three Dimensions		
		1	2		1	2	3
Mode A	1	1.00	.50	1	1.00	-.36	.26
	2	.50	1.00	2	-.36	1.00	-.41
				3	.26	-.41	1.00
Mode B	1	1.00	-.24	1	1.00	.01	-.12
	2	-.24	1.00	2	.01	1.00	.24
				3	-.12	.24	1.00
Mode C	1	1.00	.14	1	1.00	-.11	-.14
	2	.14	1.00	2	-.11	1.00	.39
				3	-.14	.39	1.00

TABLE C-6. Correlation within Modes for the Four-Dimensional Unconstrained Solution

		Correlations of Factor Loadings			
		1	2	3	4
Mode A	1	1.00	0.86	0.08	0.39
	2	0.86	1.00	0.30	0.17
	3	0.08	0.30	1.00	0.46
	4	0.39	0.17	0.46	1.00
Mode B	1	1.00	-0.92	0.79	-0.65
	2	-0.92	1.00	-0.80	0.61
	3	0.79	-0.80	1.00	-0.72
	4	-0.65	0.61	-0.72	1.00
Mode C	1	1.00	0.86	0.30	0.47
	2	0.86	1.00	0.33	0.53
	3	0.30	0.33	1.00	0.34
	4	0.47	0.53	0.34	1.00

dimension three of the three-dimensional solution has become dimension four.) Note, a further splitting occurs in the five-dimensional solution to produce a subset of several dimensions for which all of the intercorrelations are above .8. In Mode *A*, there are only two members of this set, but in Mode *B* there are four and in Mode *C*, three. This pattern of degeneracy for the four- and five-dimensional solutions clearly indicates that one has gone beyond the number of dimensions that is capable of being distinguished by our sample of 34 subjects, at least by means of an unconstrained solution.*

Split-Half Evaluation of Reliability. The strongest evidence for the "reality" of a factor—namely, that it is due to systematic influences and not just random noise—is the demonstration that the same or similar versions of the factor can be found in several independent samples of data. To provide a check on the generalizability of our solutions over independent samples of data, split-half techniques were employed. The total sample was randomly divided into two subsamples of 17 each; for purposes of discus-

*At the time this manuscript was written, the theory of degenerate solutions had not been developed to the degree that is described in Harshman and Lundy (chapter 6). We would now consider the pattern in the four-dimensional solution to indicate a classic degeneracy with respect to dimensions one and two. Thus, we are not surprised that application of constraints allows additional dimensions to be recovered.

sion, these subsamples are labeled *R* and *S*. In order to guard against an "unlucky" split (for instance, one which by chance allocates most of the subjects who use a particular dimension into one of the two groups, and thus fails to find that dimension in both subsamples despite the "reality" of the dimension), a second division of the sample was made, roughly orthogonal to the first. For the second division, an even-odd respondent split was used and the resulting samples are labeled *E* and *O*. The total sample of $N = 34$ is labeled *T*.

Because our total sample consists of only 34 subjects, the split-half subsamples are somewhat smaller than one would conventionally utilize for reliability comparisons. With such small subsamples ($N = 17$), one may be able to verify the larger, more pronounced dimensions, but smaller, more subtle effects may be lost against the background noise. Nonetheless, we proceed with a description of the split-half comparisons both to provide what verification we can in this specific case and also to demonstrate in general how this important part of the analysis procedure is carried out. The reader should keep in mind, however, that failure to find replication of a factor in two split-halves, when each is based on only 17 subjects, may merely be an indication of too small a sample size to define the factor against the background noise. On the other hand, if a factor is found to replicate, this would provide strong evidence that the factor was the result of fairly sizable systematic effects generalizable across samples of subjects.

PARAFAC analyses in two through six dimensions were performed on the four subsamples, and the results were compared by computing correlations among dimensions. The two-dimensional solution was verified through the random split-half analyses. That is, similar solutions were obtained for each split-half of the random split, and each resembled the total group solution (Table C–7). Interestingly, however, the even-odd split did not verify the two-dimensional solutions. Comparison with the total sample solution (Table C–7) revealed that while both dimensions of the even split-half matched those of the total group solution, only one dimension of the odd split-half matched the total solution, and, surprisingly, this was different for Mode *A* versus *B*. Apparently, because of an "unlucky split", dimensions one and two were not as clearly the most important dimensions in the odd split, so the solution adjusted itself to bring in components of dimension three. However, for purposes of validation of a set of dimensions, "once is enough." If a given dimension were due to random noise, it should not replicate in any split-half comparison. Consequently, the *R* versus *S* cross-validation is adequate to verify the "reality" or nonrandom nature of the two dimensions of the two-dimensional solution. A more sophisticated comparison —such as to the three-dimensional solutions—is not needed.

We were surprised to find that the three-dimensional (unconstrained) solution was not validated by the split-half procedure, despite the fact that all three dimensions seemed highly interpretable (the interpretation will be discussed below). In *both* the *R* versus *S* and *E* versus *O* comparisons, one or more dimensions fail to correlate highly across the two halves of the data. Comparison of each split-half solution with the total group solution

TABLE C-7. Correlations for the Two-Dimensional Unconstrained Split-Half Analyses

Mode A

		S 1	S 2			O 1	O 2
R	1	.89	.29	E	1	.50	-.56
	2	-.42	.88		2	-.95	-.32

		T 1	T 2	R 1	R 2	S 1	S 2	E 1	E 2	O 1	O 2
T	1	1.00	.50	.95	-.52	.98	.45	.98	-.55	.53	-.57
	2	.50	1.00	.35	-.93	.40	.99	.47	-.96	.99	.42

Mode B

		S 1	S 2			O 1	O 2
R	1	.94	-.06	E	1	.67	-.92
	2	-.10	-.82		2	-.50	-.27

		T 1	T 2	R 1	R 2	S 1	S 2	E 1	E 2	O 1	O 2
T	1	1.00	-.26	.96	-.06	.99	-.17	.95	.20	.69	-.99
	2	-.26	1.00	-.18	-.85	-.22	.99	-.19	-.93	.51	.35

(labeled *T*) revealed that one of the two solutions in each split-half pair had broken down (Table C–8). While solutions *R* and *E* each contained all three dimensions of the total-group solution (although in different orders), their corresponding halves, *S* and *O*, showed various patterns of degeneracy. For Mode *A*, comparisons between the total group solution *T* and split-half solution *S* showed dimension *T*–1 correctly recovered as *S*–3, but *T*–2 split into two different dimensions, recovered as *S*–1 and *S*–2. This splitting of *T*–2 left no room for the third dimension (*T*–3) to emerge in the *S* split-half solution. Oddly, in Mode *B*, *T*–3 was represented; in fact, it had split into two dimensions and "crowded out" *T*–2. In some sense, we might claim that all three dimensions were replicated, but not in both modes simultaneously. It is perhaps more correct to say that the *S* split-half solution is in some sense degenerate. For the *E* versus *O* split, similar comments apply, with the *E* split replicating all three dimensions of the *T* solution, but the *O* split showing various patterns of nonreplication of particular dimensions in particular modes. (The *R* versus *E* comparison cannot be used to validate the reality of the dimensions, since they are not independent; half of their subjects are in common between the two samples.)

As might be expected, given the three-dimensional solutions,

TABLE C-8. Correlations for the Three-Dimensional Unconstrained Split-Half Analyses

Mode A

	S 1	S 2	S 3		O 1	O 2	O 3
R 1	-.25	-.11	.72	E 1	.92	-.51	.04
R 2	.95	.84	-.35	E 2	-.63	.09	-.80
R 3	-.36	-.00	-.05	E 3	-.22	-.48	-.31

	T 1	T 2	T 3	R 1	R 2	R 3	S 1	S 2	S 3	E 1	E 2	E 3	O 1	O 2	O 3
T 1	1.00	-.36	.28	.92	-.28	.20	-.45	-.37	.92	-.41	.38	.99	-.16	.10	-.06
T 2	-.36	1.00	-.44	-.17	.99	-.40	.98	.89	-.44	.98	-.72	-.40	.94	.56	-.40
T 3	.28	-.44	1.00	.42	-.48	.99	-.40	-.03	.02	-.28	.93	.28	-.47	.30	.87

Mode B

	S 1	S 2	S 3		O 1	O 2	O 3
R 1	-.25	.36	.95	E 1	.44	.13	-.42
R 2	.22	.09	-.23	E 2	.16	.30	-.87
R 3	-.62	.63	-.45	E 3	-.77	-.92	.48

	T 1	T 2	T 3	R 1	R 2	R 3	S 1	S 2	S 3	E 1	E 2	E 3	O 1	O 2	O 3
T 1	1.00	-.01	-.13	.98	-.32	-.44	-.17	.31	.99	.04	-.16	.99	-.83	-.97	.56
T 2	-.01	1.00	.27	-.09	.94	.33	.19	.19	.08	.92	.12	.08	.55	.18	-.25
T 3	-.13	.27	1.00	-.10	.31	.93	-.82	.86	-.16	.60	.98	-.05	.21	.29	-.88

the four- and five-dimensional split-half analyses showed even greater numbers of highly intercorrelated dimensions, with factors splitting into several copies of themselves and other degeneracies emerging. This was consistent with the behavior of the total-group solutions and indicated that four or more dimensions could not be recovered from this sample by means of unconstrained analyses.

The three-dimensional solution for the total group consisted of relatively uncorrelated, stable, and (as we shall see below) interpretable dimensions. The failure to recover these dimensions consistently when analyzing the split-half data sets suggested that perhaps 17 subjects was just too small a sample to recover three dimensions reliably. Since recent experience with other data indicated that premature emergence of degenerate solutions could be blocked by applying constraints to the form of the solution, it was decided to try constrained analyses of this "Cars and Stars" data set.

Constrained Analyses

PARAFAC allows analyses to be performed subject to the constraints that the columns of factor loadings in a particular mode or modes be mutually orthogonal or mutually uncorrelated. By imposing such a constraint on our solutions, we hoped to block the emergence of the degeneracies in which a dimension splits into two highly correlated versions of itself. This would hopefully allow us to detect additional weaker dimensions that were previously obscured by a premature "breakdown" of the solution. The "reality" of these additional dimensions would then be tested by split-half methods and by consideration of their interpretability. It was hoped that with the constraint, 17 subjects might be sufficient to recover three reliable dimensions.

Constraining a single mode is often sufficient to block the emergence of highly correlated dimensions in all modes, provided that the correct mode is selected and that there is not some strong internal characteristic of the data promoting highly correlated factors. By constraining only one mode, the other two could take on whatever form was consistent with the data, and the solution would be a closer approximation to the "natural" unconstrained form. It was decided that Mode *B* (the rating-scale mode) should be constrained to be orthogonal, since (a) Mode *B* loadings were most orthogonal in the unconstrained three dimensional solution, and hence if a "real" replicable solution existed in both split-halves, it might not be distorted by requiring Mode *B* to be orthogonal; and (b) interpretation of the dimensions was primarily based on the scale loadings and such interpretation would be facilitated by keeping the patterns of scale loadings distinct for the different dimensions. It was not necessary to choose between orthogonality and zero-correlation constraints, since the centering of Mode *B* resulted in zero mean loadings in this mode and consequently both types of constraints become equivalent.

A series of constrained analyses was run in dimensionalities from one through seven. As before, three random starting positions were used in each analysis to check stability; checks of fit values, correlations among dimensions (Mode *A* and *C*), and comparisons of split-half solutions were used to evaluate the maximum valid dimensionality.

Fit Values. The R^2 values for the 1 through 7 dimensional constrained solutions were as follows: 1D = .123; 2D = .195; 3D = .257; 4D = .293; 5D = .322; 6D = .340; 7D = .353. Figure C–2 presents the plot of fit versus dimensionality based on R^2 or variance accounted for, as before. Once again, the evidence is somewhat ambiguous. In general, however, these data seem to suggest the presence of at least three dimensions, with perhaps a smaller fourth and possibly even fifth and sixth dimensions. The question may thus become not how many dimensions *there are,*

Figure C–2. Plot of Dimensionality versus R^2 for Constrained Solutions

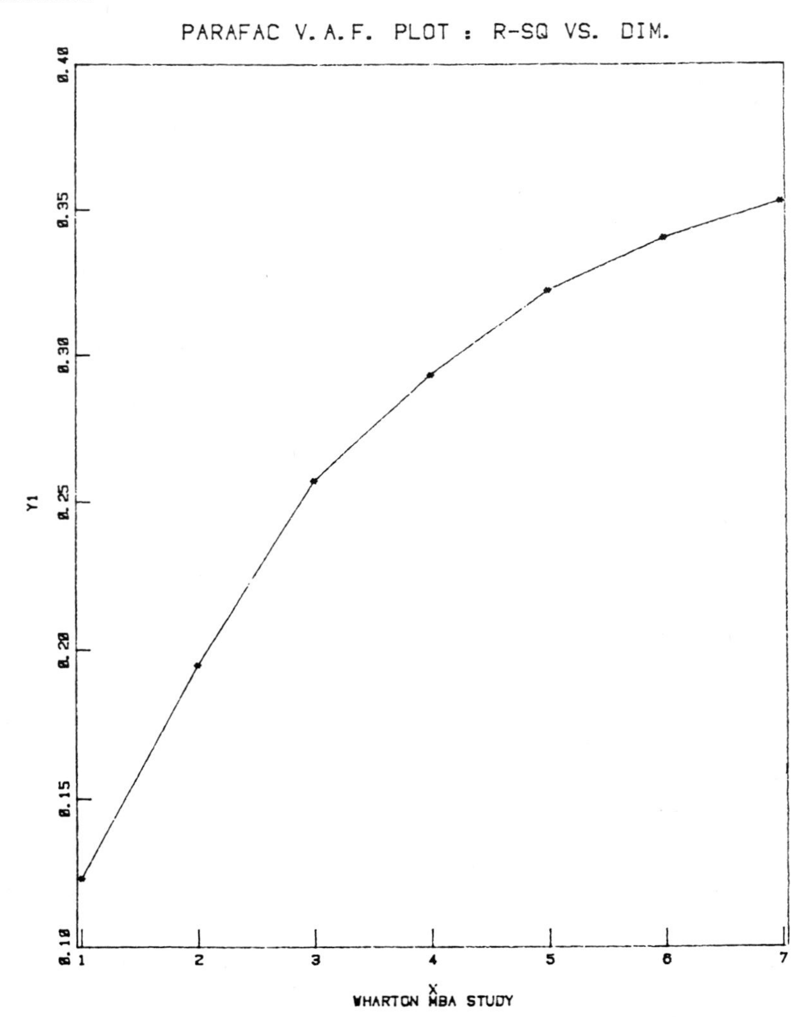

but rather how many *can be reliably determined* on the basis of only 34 subjects, with 17 in each split-half data set.

Within-Solution Correlations among Dimensions. Naturally, the imposition of orthogonality constraints on Mode *B* prevents us from using this mode to check for factor-splitting and highly correlated dimensions. However, we can check Modes *A* and *C*, since these were not constrained. It turns out that the imposition of constraints on Mode *B* succeeded in suppressing factor-splitting degeneracies in all modes, for all the solutions that we considered, including four-, five-, and six-dimensional solutions. While this prevents us from using high correlations as a diagnostic of extracting too many factors, it does clear the way for the use of the more important split-half testing on the three-dimensional solution (and on higher dimensional solutions, if warranted).

Split-Half Evaluation of Reliability. For two- through six-dimensional solutions, split-half analyses were performed and compared. As before, the two-dimensional solution was verified by appearance in the various split-halves of the data; however, there was some difficulty in replicating dimension two for Mode *A* (Table C–9). For three dimensions, both the *R* versus *S* and the *E* versus *O* split-halves cross-validated the total-group solution for all three dimensions, thus confirming our suspicion that the third dimension was, in fact, "real" (Table C–10). Furthermore, the total-group three-dimensional constrained solution was almost identical to the three-dimensional unconstrained solution (Table C–11), which is not surprising since the unconstrained solution was close to orthogonal in Mode *B*. The near identity of the constrained and unconstrained solutions is nonetheless important to note, because it implies that the split-half confirmation of the constrained solutions in fact provides split-half validation of the three-dimensional unconstrained solutions, as well, and supports our notion that the failure to cross-validate by means of unconstrained analyses was probably due to too few subjects in each split-half sample.

To determine whether additional reliable patterns existed beyond the three dimensions obtained in the unconstrained solutions, we examined correlations between split-half solutions in four through six dimensions. Tables C–12, C–13, and C–14 present the relevant cross-split correlations.

Somewhat surprisingly, the comparison of split-halves provides evidence for additional reliable dimensions beyond the third. As Table C–12 indicates, a constrained four-dimensional solution is cross-validated by clear replication in split-halves. While the order of dimensions three and four is reversed in sample *S* relative to sample *R* (indicating a reversal of their relative importance in terms of variance accounted for in the two subsamples), there is only one large correlation in each row and column of the matrix of cross-solution correlations. Furthermore, all these correlations are high. For Mode *A* they are .89, .92, -.88, and .92; for Mode *B* they are .96, .93, -.94 and .89. (The occurrence of a negative correlation for dimension three simply means that it was reflected in solution *S* relative to *R*.) The *E* versus *O* split also provides support for four dimensions, although some of the correlations are not as high as in the *R*-*S* split, and in fact drop

TABLE C-9. Correlations for the Two-Dimensional Constrained Split-Half Analyses

Mode A

		S 1	S 2			O 1	O 2
R	1	.83	.60	E	1	.92	-.34
	2	-.27	.74		2	.48	.60

		T 1	T 2	R 1	R 2	S 1	S 2	E 1	E 2	O 1	O 2
T	1	1.00	.30	.89	-.27	.98	.26	.89	.03	.81	-.66
	2	.30	1.00	.66	.71	.30	.99	.61	.79	.79	.49

Mode B

		S 1	S 2			O 1	O 2
R	1	.85	.46	E	1	.91	-.26
	2	-.40	.80		2	.30	.79

		T 1	T 2	R 1	R 2	S 1	S 2	E 1	E 2	O 1	O 2
T	1	1.00	.00	.85	-.42	.99	-.03	.83	-.26	.73	-.67
	2	.00	1.00	.51	.73	.03	.99	.47	.76	.67	.72

as low as .81 and .85. In Mode *A*, the *E* versus *O* matrix of cross-split correlations shows additional large correlations in rows 2 and 3, indicating, perhaps, a less stable axis position for the second and third dimensions in the *E* versus *O* split. Overall, however, these results provide strong support for the presence of four reliable dimensions.

Although comparison of the five-dimensional analyses of the random (*R* versus *S*) split-halves of the data (Table C–13) provides some support for a fifth dimension, the cross-sample correlation for dimension five is fairly low. In the *R* versus *S* comparison, it is .78 for Mode *A* and .74 for Mode *B*. In the *E* versus *O* comparison, the lowest cross-sample correlation (taking into account the reordering of dimensions) is .66 in Mode *A* and .67 in *B*. It may be that we have reached the limit of the number of dimensions that can be adequately cross-validated with this data and this size sample. Also, in the *E* versus *O* comparison, the plane involving dimensions two and four of each solution seems to have some rotational ambiguity, since the four cross-split correlations resemble an attenuated rotation matrix (first row = -.50, .67, second row = .69, .63). No such rotational ambiguity is indicated in the *R* versus *S* comparison correlations, however.

TABLE C-10. Correlations for the Three-Dimensional Constrained Split-Half Analyses

Mode A

	S 1	S 2	S 3		O 1	O 2	O 3
R 1	.90	.25	-.33	E 1	.96	.20	-.02
R 2	-.28	-.29	.96	E 2	-.36	-.89	.15
R 3	-.06	.93	-.03	E 3	.27	.47	.90

	T 1	T 2	T 3	R 1	R 2	R 3	S 1	S 2	S 3	E 1	E 2	E 3	O 1	O 2	O 3
T 1	1.00	-.32	.14	.97	-.30	-.02	.98	.16	-.31	.99	-.22	.16	.98	.14	-.04
T 2	-.32	1.00	-.20	-.35	.99	-.08	-.31	-.28	.99	-.37	.98	-.31	-.47	-.94	-.02
T 3	.14	-.20	1.00	.25	-.20	.98	.11	.98	-.14	.16	-.01	.98	.23	.38	.96

Mode B

	S 1	S 2	S 3		O 1	O 2	O 3
R 1	.96	.13	-.03	E 1	.98	-.02	.01
R 2	.03	-.04	.97	E 2	-.02	-.88	.29
R 3	-.13	.95	.04	E 3	-.02	.29	.92

	T 1	T 2	T 3	R 1	R 2	R 3	S 1	S 2	S 3	E 1	E 2	E 3	O 1	O 2	O 3
T 1	1.00	.00	.00	.98	.03	-.12	.99	.01	-.00	.98	.12	-.09	.98	-.15	-.04
T 2	.00	1.00	.00	-.03	.99	.01	.00	-.04	.99	-.12	.97	-.14	-.14	-.94	.16
T 3	.00	.00	1.00	.12	.00	.98	-.01	.99	.04	.06	.15	.98	.05	.17	.97

TABLE C-11. Correlations between the Three-Dimensional Constrained and Unconstrained Total Group Solutions

Mode A

		Orthogonal		
		1	2	3
Unconstrained	1	1.00	-.35	.23
	2	-.32	.99	-.33
	3	.17	-.29	1.00

Mode B

		Orthogonal		
		1	2	3
Unconstrained	1	1.00	-.01	-.01
	2	.02	1.00	.09
	3	.11	.16	.98

TABLE C-12. Correlations for the Four-Dimensional Constrained Split-Half Analyses

Mode A

		S						O			
		1	2	3	4			1	2	3	4
R*	1	.89	-.40	.06	-.36	E	1	.91	-.13	-.08	-.00
	2	-.29	.92	-.36	.25		2	-.61	-.47	.10	.81
	3	.15	.23	.28	-.88		3	-.23	-.74	-.88	-.02
	4	.12	-.36	.92	-.20		4	.09	.92	.25	.11

Mode B

		S						O			
		1	2	3	4			1	2	3	4
R*	1	.96	-.03	-.03	-.20	E	1	.93	-.21	.01	.19
	2	.01	.93	-.08	.12		2	-.26	-.41	.08	.81
	3	-.19	.13	.13	-.94		3	-.02	-.10	-.94	.01
	4	-.00	-.02	.89	.10		4	.08	.85	-.03	.44

*Indicates best solution of three (largest R^2)

Correlations among six-dimensional analyses of the split-half subsamples (Table C–14) indicate that we cannot recover six reli-

TABLE C-13. Correlations for the Five-Dimensional Constrained Split-Half Analyses

Mode A

		S				
		1	2	3	4	5
	1	.90	.11	-.36	-.37	-.13
	2	.26	.92	-.47	-.17	.18
R*	3	.19	.06	.83	-.35	-.18
	4	-.28	-.45	.21	.89	-.02
	5	.02	.31	-.17	-.36	.78

		O				
		1	2	3	4	5
	1	.86	-.42	.07	-.01	.08
	2	-.62	-.39	-.28	.66	.18
E	3	.45	.19	-.80	-.07	-.44
	4	-.07	-.77	.25	-.51	.31
	5	-.15	-.26	.01	.10	.92

Mode B

		S				
		1	2	3	4	5
	1	.96	-.11	-.19	-.06	-.01
	2	.07	.90	-.20	.23	.02
R*	3	.19	.17	.93	-.11	-.06
	4	.01	-.22	.09	.84	.15
	5	-.05	.10	-.03	-.33	.74

		O				
		1	2	3	4	5
	1	.86	-.44	.01	.15	.01
	2	-.38	-.50	.22	.67	-.01
E	3	.09	.12	-.90	-.10	-.25
	4	.22	.69	.02	.63	.02
	5	-.01	.00	.21	-.03	.92

*Indicates best solution of three (largest R^2)

able dimensions. Evidently, such analyses have gone beyond the limit of what we can resolve with this data, even using orthogonality constraints.

Given the somewhat gradual "trailing off" of our ability to recover additional dimensions, the question arises as to which dimensionalities we should consider in detail. If this "Cars and Stars" study were a serious attempt to evaluate spokesmen for car makes, rather than simply an example of how this technique might work, we would probably consider the four- and five-dimensional solutions quite carefully. We would also interpret the indications of higher dimensionality as evidence for the need to repeat the study on a larger sample, so as to better define the additional dimensions. (However, there are also other aspects of the study that would need to be improved at the same time; this 34-subject "Cars and Stars" study might best be viewed as a preliminary or pilot study.)

For purposes of demonstrating how we arrive at interpretations of dimensions, we have decided to concentrate on the simpler three-dimensional unconstrained solution, rather than emphasize the refinements suggested by the smaller higher dimensions. Some brief discussion of the four-dimensional solution will be presented, however, to provide some insight into the type of information that higher dimensions might add to our basic solution.

TABLE C-14. Correlations for the Six-Dimensional Constrained Split-Half Analyses

Mode A

		S								O					
		1	2	3	4	5	6			1	2	3	4	5	6
	1	.88	-.26	-.46	-.25	-.02	-.37		1	.93	.30	.09	.03	-.27	-.08
	2	-.39	-.93	-.37	-.31	.21	.26		2	-.53	-.64	-.18	.82	-.24	.40
R*	3	-.12	.20	-.68	.27	-.20	.83	E	3	-.23	-.45	.80	-.16	-.26	.32
	4	.03	-.31	-.45	-.33	-.58	-.33		4	-.12	.65	-.30	-.09	.31	.59
	5	-.29	.02	.68	-.43	.08	-.10		5	.12	-.05	.23	-.11	-.87	.01
	6	.22	-.09	.27	-.67	.79	-.01		6	-.17	-.73	-.25	.04	.51	.06

Mode B

		S								O					
		1	2	3	4	5	6			1	2	3	4	5	6
	1	.93	-.14	-.19	-.18	-.08	.05		1	.95	.08	.00	.21	-.12	.01
	2	.08	.86	.18	-.34	-.13	-.19		2	-.16	-.41	-.11	.73	-.18	.36
R*	3	-.12	.05	-.50	-.43	.68	-.06	E	3	.05	-.32	.81	-.23	-.09	.37
	4	-.17	-.40	.20	-.59	-.31	-.44		4	.06	.53	-.20	-.09	-.01	.76
	5	-.13	-.02	.07	-.45	-.20	.82		5	-.08	.14	.02	-.15	-.89	-.10
	6	-.19	.08	-.77	.03	-.49	-.03		6	.12	-.58	-.47	-.41	-.16	.18

*Indicates best solution of three (largest R^2)

Interpretation of Dimensions

The Role of Interpretation

Interpretation of the PARAFAC dimensions plays a role in several stages of the analysis. Preliminary scanning of the dimensions for interpretability provides some useful guidance in the earlier phases of analysis. With the "Cars and Stars" data, for example, the apparent interpretability of all three dimensions of the unconstrained three-dimensional solution was one of the reasons we were reluctant to reject the solution when split-half comparisons did not validate it. (Another reason was the suggestion of a third dimension in most fit versus dimensionality curves.) However, these preliminary attempts at interpretation are not nearly as crucial as the stage of careful and detailed interpretation that is required once a preferred dimensionality (or set of dimensionalities) is selected.

The goal of *connotative congruence analysis,* as applied to our marketing example, is to select an effective spokesman-product relationship. As we shall see, this will often require artful design of the spokesman's message so that it picks out particular desired aspects of the spokesman's overtones and relates them in the most fruitful way to those properties of the product that one wants to enhance. To construct such messages requires the best possible understanding of the dimensions of perceived overtones for a particular spokesman and product, as indicated by the PARAFAC analysis.

The Method of Interpretation

As mentioned earlier, one advantage of using PARAFAC for this type of study is that it provides direct estimates of loadings for all three modes—the spokesman-product mode, the rating scale mode, and the rater mode. However, it is the rating scale mode (Mode *B* in our data set) that provides the primary basis for interpretation of the obtained factors or dimensions. For each dimension, there should be some rating scales with particularly high (positive or negative) loadings. Furthermore, the scales with high loadings on a given dimension should have one or two overtone elements in common. It is these common elements that primarily define the "meaning" represented by that dimension.

When interpreting the sizes and signs of the ratings scale loadings, two points should be kept in mind. First, those scales with the highest weights on any given dimension should be the ones that most strongly exemplify the underlying meaning of the dimension; as the loadings get smaller, the scales should show a progressively weaker relationship to the common element of meaning or "overtone" represented by the dimension. Second, both ends of any bipolar rating scale should be considered when assessing its relationship to the "meaning" of a dimension. For a scale that loads positively on a given dimension, the word from the high end of the scale should be used to interpret the positive pole of the dimension, and the opposite-meaning word, from the low end of the rating scale, should be used to interpret the negative pole of the dimension. For scales with negative loadings, the situation is reversed—words from the high end of the scale contribute to interpretation of the negative pole of the dimension, and words from the low end of the rating scale contribute to the positive pole.

After considering the rating scale loadings for a given dimension, one can obtain additional confirmation and/or further refinement of one's interpretation by examining the loadings in Mode *A* to determine the common properties of those spokesmen and products that have high loadings on that dimension. In addition, if one has demographic, psychographic, or other relevant information on raters or market segments, such information can be related to the Mode *C* loadings to provide another check on the sensibleness of one's interpretation.

Three-Dimensional Unconstrained Solution

The loadings for the three-dimensional unconstrained solution are presented in Tables C–15, C–16, and C–17. To aid in our discussion of this solution, we will use a graphical representation of each dimension, which provides a visual summary of the relative strength of different items that load substantially on the dimension. Figure C–3 diagrams the loading patterns for the first dimension of the three-dimensional solution. Information on all

TABLE C-15. Mode A for Three-Dimensional Unconstrained Solution

	Mode A: Cars & Stars & Self		
	1	2	3
1	-0.78	0.12	0.16
2	-0.37	0.27	-1.71
3	-1.77	-0.72	-1.51
4	-0.24	-1.16	2.84
5	-0.21	0.35	-2.19
6	-0.60	1.58	-1.20
7	-0.77	-0.96	0.09
8	-0.15	0.28	-0.01
9	-0.14	-0.84	-0.27
10	-0.15	-1.71	0.49
11	-0.04	0.37	-0.61
12	-0.72	-0.33	1.77
13	1.18	-0.26	0.25
14	0.18	0.90	0.22
15	0.53	-0.47	0.45
16	-1.18	2.15	-0.35
17	0.21	0.68	0.13
18	2.31	-1.65	0.62
19	0.93	0.96	-0.11
20	1.70	-0.91	0.24
21	2.16	-0.21	0.39
22	-1.37	2.08	-0.47
23	0.03	-0.48	0.28
24	0.03	-0.21	0.37
25	-0.74	0.17	0.09

Note: Factor one loadings should be reflected (reversed in sign) to be consistent with the discussion and Figures 24-26.

three modes is presented on the diagram. To save space, only items with moderate to high loadings are plotted, leaving bare the crowded area around the zero-point where items unrelated to the dimension are found. To further save space, the contribution of a given rating scale (in the plot of Mode *B* loadings) is usually represented by only its high-end adjective, rather than following the more correct procedure of plotting its high-end adjective on one pole of the dimension and its low-end adjective on the other. (In the text, however, the contribution of the unplotted complimentary adjectives will be mentioned, along with the plotted ones.) Note that in Figure C–3, the signs of loadings for Mode *A* and *B* have been reflected to simplify discussion. In effect, we simply reverse the poles of the dimension and call it "flashiness" instead of "plainness." (All loadings in Figures C–3 through C–5 have been multiplied by 100 for convenience.)

TABLE C-16. Mode B for Three-Dimensional Unconstrained Solution

	Mode B: Rating Scales		
	1	2	3
1	0.70	-0.70	-1.79
2	0.67	-0.55	1.18
3	-0.04	-1.91	-0.38
4	1.24	1.22	0.53
5	-1.43	-0.38	-0.84
6	-1.81	-0.22	-0.35
7	1.14	0.11	1.05
8	-1.57	0.66	0.57
9	0.76	-1.33	-0.31
10	1.07	-0.27	-1.27
11	-0.89	1.11	-0.19
12	-0.27	1.20	-0.63
13	-0.62	-1.79	0.49
14	-1.06	-1.32	0.58
15	0.79	0.04	-0.30
16	1.12	-0.18	-0.81
17	-0.42	1.39	-0.21
18	1.03	0.84	1.29
19	-0.13	0.83	1.91
20	1.36	1.62	1.03
21	0.70	-0.77	0.97
22	-0.03	0.38	2.14
23	-1.49	-0.86	-0.35
24	1.28	-0.60	0.07
25	-1.42	0.65	1.19
26	-0.87	1.46	-0.30
27	0.24	-0.44	-0.81
28	-1.31	1.04	0.99
29	0.58	-1.10	-2.03
30	0.16	1.30	0.39
31	-1.56	0.10	-0.21
32	-0.90	-1.12	0.55
33	-0.98	-1.08	-0.92
34	1.33	-0.23	-1.52
35	0.25	0.84	0.50
36	0.03	1.66	-1.10
37	1.06	0.19	0.01
38	0.01	-1.36	0.65
39	1.25	-0.40	-1.74

Note: Factor one loadings should be reflected to be consistent with the discussion and Figures 24-26.

To interpret dimension one, we focus first on the Mode *B* loadings. Words such as "colorful," "famous," "ornate," "grace-

TABLE C-17. Mode C for Three-Dimensional Unconstrained Solution

	Mode C: Subjects		
	1	2	3
1	0.36	0.26	0.30
2	0.34	0.34	0.36
3	0.34	0.46	0.25
4	0.18	0.25	0.36
5	0.41	0.31	0.26
6	0.11	0.53	0.36
7	0.29	0.22	0.29
8	0.38	0.27	0.41
9	0.22	0.26	0.29
10	0.16	0.39	0.30
11	0.38	0.34	0.25
12	0.33	0.37	0.24
13	0.42	0.24	0.18
14	0.29	0.36	0.32
15	0.46	0.29	0.29
16	0.24	0.41	0.28
17	0.12	0.14	0.21
18	0.06	0.33	0.27
19	0.34	0.35	0.15
20	0.54	0.23	0.22
21	0.37	0.32	0.26
22	0.51	0.21	0.21
23	0.35	0.31	0.33
24	0.13	0.22	0.20
25	0.05	0.15	0.25
26	0.35	0.21	0.14
27	0.43	0.13	0.22
28	0.29	0.29	0.38
29	0.45	0.18	0.27
30	0.20	0.43	0.37
31	0.20	0.09	0.21
32	0.28	0.20	0.30
33	0.25	0.21	0.17
34	0.35	0.23	0.35
Root-Mean-Squared Loading (Mode C) for Each Factor			
	0.3281	0.3010	0.2848

ful," "active," "attractive," and "superior" are found at the positive end of this dimension, with their complements "colorless," "obscure," "plain," "awkward," "passive," "unattractive," and "inferior" at the negative pole. Something perceived as high on this dimension is not just attractive and superior; it is even more

Figure C-3. Dimension One: "Flashy"

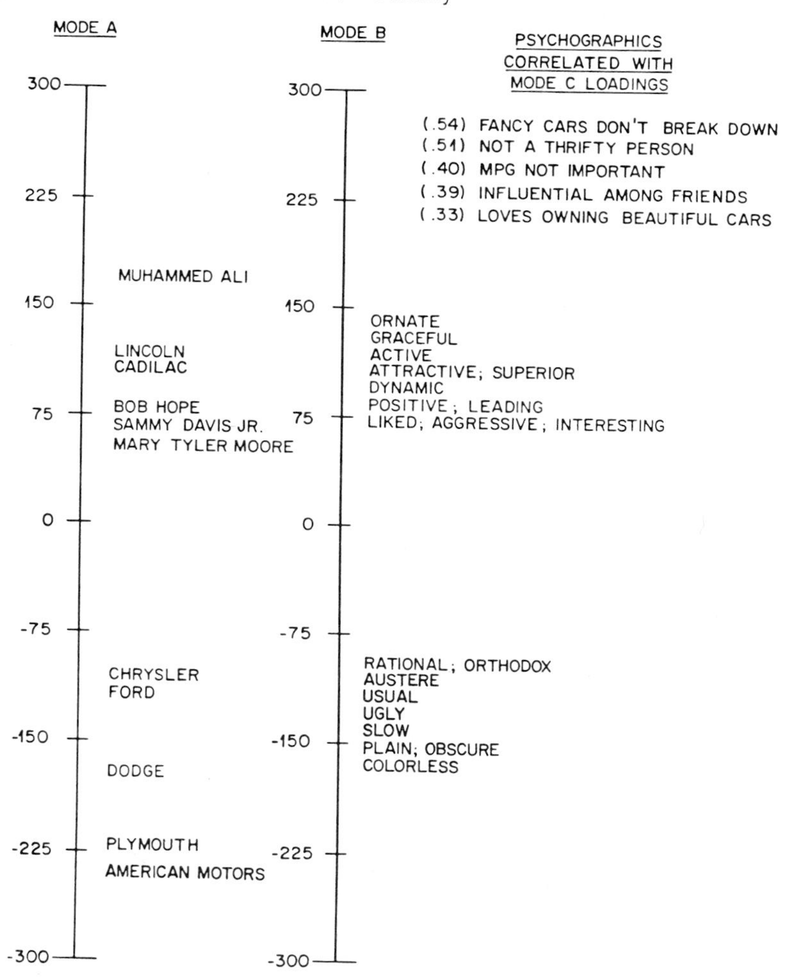

importantly famous and colorful, even ornate. Apparently, we are not just pleased by such a stimulus but impressed and fascinated as well. This interpretation of dimension one is consistent with the patterns of relationships found in its Mode *A* loadings, where Muhammed Ali, a very "charismatic" individual, has the highest loading (the survey was taken during the period when he was world champion and very popular). Both Ali and Sammy Davis, Jr., are also somewhat "flashy" or "ornate" in their public relations style. And while Bob Hope may not be "flashy" in the same way, he is colorful and very famous. Indeed, fame or renown must play a large role in this dimension, since all the celebrities load positively on it. In terms of products, the positive pole of this dimension is characterized by two automobiles that load much higher than any others: Lincoln and Cadillac. It is quite plausible that our sample of raters (MBA students) regarded these two makes as imbued with a special aura of impressiveness, a special symbolic renown. And the plush appointments of these automobiles might be considered "ornate" by some; certainly they are far from "plain." From a slightly different perspective, the possession of one of these makes might be considered a "flashy" and impressive gesture—appropriate for Sammy Davis, Jr., or Muhammed Ali. On the opposite end of the Mode *A* scale, we find American Motors and Plymouth; these names certainly do not have the same kind of "charisma" as Lincoln and Cadillac.

In light of all the interpretive relationships noted above, we will use the word "flashy" as a shorthand label for the overtone pattern expressed by this first dimension. The reader should avoid, however, the association of cheapness or tinsel-falseness that some might associate with the word "flashy." We mean to evoke something like impressive, colorful, renowned, dramatic, "showy."

As a further source of insight into this dimension, we can examine the characteristics of individuals for whom this dimension is important. Mode *C* loadings (after row normalization) give the relative importance of the three dimensions for each subject. Correlating the dimension one (Mode *C*) loadings with responses to each of the psychographic questions reveals whether individuals for whom "flashiness" or impressive and colorful renown is highly salient tend to answer certain questions differently than those for whom the flashiness dimension is unimportant. We find that people who tended to ignore this dimension of "flashiness" when rating cars and celebrities tended to be unimpressed by the mechanical properties of fancy cars. They were more likely to agree that "fancier, more expensive cars, probably break down a lot" ($r = .54$). They would also tend to be thrifty ("Those who know me would consider me to be a thrifty person" [$r = .51$] and would agree that "miles per gallon statistics are very important to me in my selection of a new car" ($r = .40$). It is quite plausible that such persons would not pay much attention to the "flashy" dimension of automobiles (or people). On the other hand, since "flashy" cars are expensive to purchase and to run, it is reasonable that those who find "flashiness" very salient do not care as much about financial considerations and consider big cars reliable. Such patterns of attitude would serve to reduce the cognitive dissonance that would arise from simultaneously longing after a "flashy" car and at the same time realizing that it was too expensive and unreliable. At a more general level, we might speculate that individuals who are unimpressed by "flashiness" in general (such as those who would be unimpressed by this quality of Muhammed Ali or Sammy Davis, Jr.) would tend to be more practically oriented and perhaps thrifty. Of course, these latter comments are highly speculative and so should be tested in a subsequent study before being taken too seriously.

Figure C–4 presents a graphical summary of the loading patterns for dimension two. Considering first the Mode *B* loadings, we find that the positive pole of this dimension was high loadings for "formal," "mature," "heavy," "passive," "complex," "large," "careful," and "sophisticated," while the negative pole loads on

Figure C–4. Dimension Two: "Mature/Conservative"

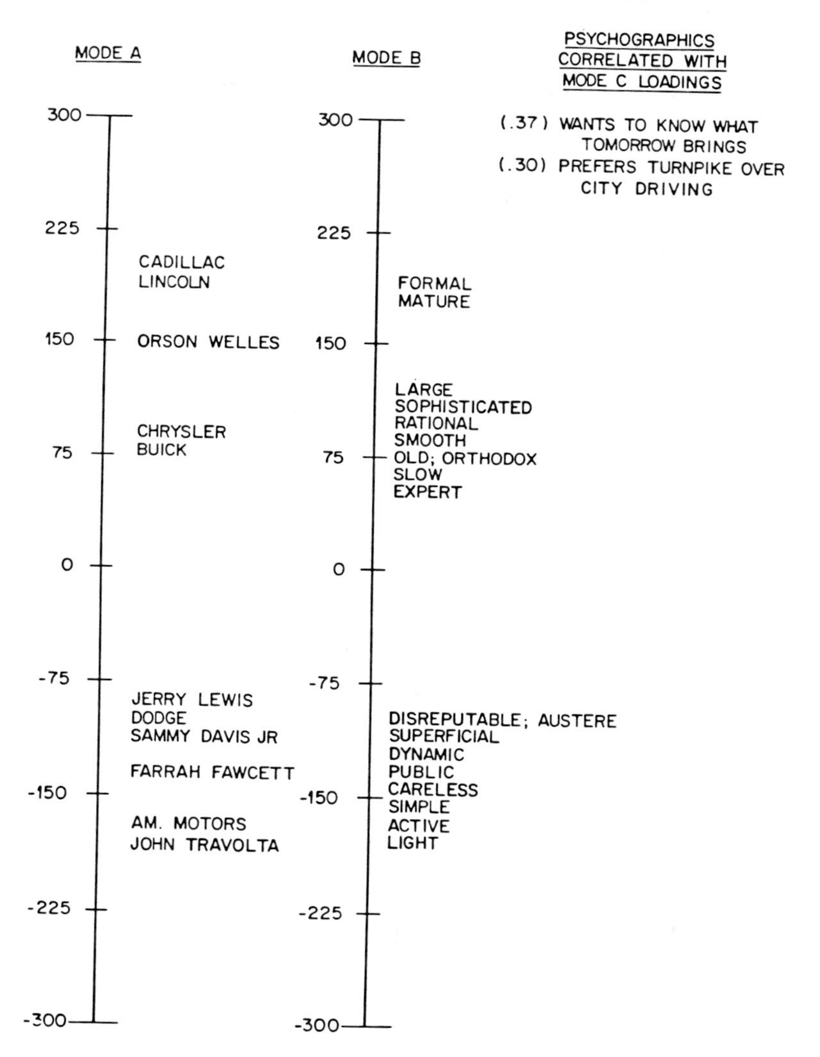

"informal," "youthful," "light," "active," "simple," "small," "careless," and "naive." This seems to be a dimension of maturity, heaviness, and formality. Looking at Mode *A* to confirm our interpretation, we find that Orson Welles has by far the largest positive loading of any celebrity. Compared to other celebrities, he would indeed be considered more mature, formal, and heavy. John Wayne and Ralph Nader have loadings near zero (not shown on Figure C–4; refer to Table C–15). On the opposite pole we have John Travolta, Farrah Fawcett, Sammy Davis, and Jerry Lewis, all of whom seem particularly youthful, informal, active, and arguably "light." These Mode *A* loading patterns confirm our interpretation of dimension two. In terms of automobile makes, there is another sharp contrast: Cadillac and Lincoln are similar to Orson Welles in terms of formality, maturity, and heaviness, with Chrysler and Buick also tending to be viewed as somewhat formal. On the other pole, American Motors is the most informal, youthful, and light of the makes, with Dodge showing the same overtones but to a considerably smaller degree.

Finally, if we examine the correlations of the psychographic responses of our raters with their weights on dimension two, we find that the individual for whom this dimension is highly salient "wants to know what tomorrow has in store for him" (r = .37) and "prefers turnpike to city driving" (r = .30). The person to whom this dimension is not important presumably lives more for today. These correlations suggest that the individual to whom dimension two is particularly salient may tend to be more conservative and prefer the mature, formal, and careful to the immature, informal, and careless. Consequently, we have named this dimension "Mature/Conservative" as a shorthand for the qualities that we surmise might be important to one who values the positive end of this dimension.

Dimension three (Figure C–5) most strongly emphasizes "feminine," "soft," and "smooth" on the positive pole versus "masculine," "hard," and "rough" on the negative pole. At a slightly lower level, words such as "pleasant," "attractive," and "graceful" load on the positive pole, with their opposites loading on the negative pole. This dimension seems to capture the sensual and aesthetic qualities associated with the feminine-masculine distinction. Clear confirmation of this is provided by the Mode *A* loadings, where Farrah Fawcett and Mary Tyler Moore are at the positive pole and Ralph Nader, John Wayne, Muhammed Ali, and Orson Welles at the negative pole. The most sensuous female has the highest positive loading, with Mary Tyler Moore quite a bit lower and Barbara Walters rated least "feminine" of all. It is surprising, perhaps, that Ralph Nader is viewed as most "masculine," but his emphasis of the critical, rational, sophisticated aspect here associated with masculinity may be responsible for this. The ranking of the other male celebrities is as we might expect, with John Wayne and Muhammed Ali being high, and some male entertainers such as Jerry Lewis, Bob Hope, and Sammy Davis being rated much lower. The automobiles all have fairly weak positive or negative loadings on this dimension. This seems plausible enough, since any sensual and gender related overtones of automobiles will surely be much less dramatic than those of celebrities such as are rated in this study. Although subtler, these automobile overtones appear to be modestly reliable (as indicated by comparison of loadings across split-half solutions) and consequently might be useful.

Individuals for whom this dimension is particularly salient tend to endorse the following psychographic questions: "When I must choose between the two, I usually dress for fashion, not comfort" (r = .48), and "I admit that I try to keep up with the Joneses" (r = .32). This suggests that the person is somewhat status-conscious or perhaps is concerned about appearances. The person also tends to "live a long way from friends and relatives"

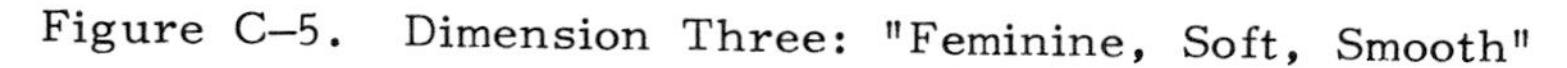

Figure C–5. Dimension Three: "Feminine, Soft, Smooth"

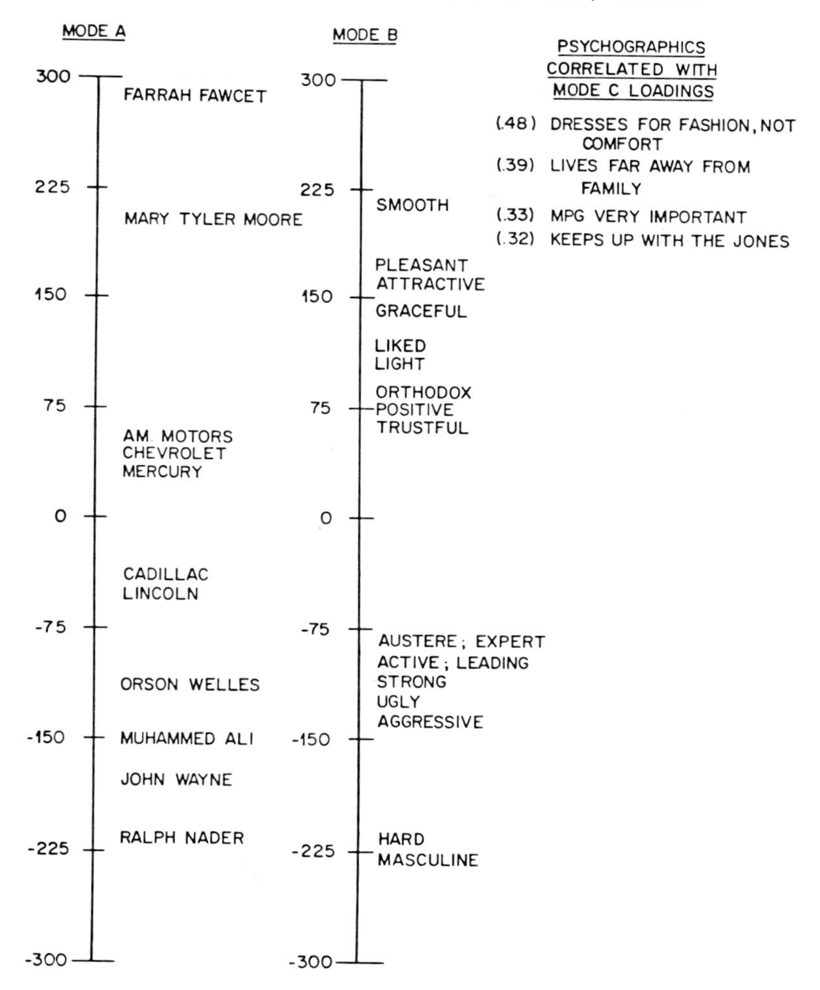

(r = .39) and feels that "miles-per-gallon statistics are very important to me in my selection of a car" (r = .33). It is not clear as to why these last two psychographics should correlate with high salience of this dimension.

The third dimension contains several different threads of "Femininity-Masculinity," which might not always coincide in particular cases. Orson Welles might be rated relatively masculine for a different reason than Muhammed Ali. Similarly, Lincoln and Cadillac might be rated more masculine than Chevrolet and American Motors because of size or weight rather than appearance. One can always check back to the individual ratings of the stimuli (averaged across raters) to test for particular scale by stimulus interactions that might deviate from the general pattern indicated by the dimension as a whole. Furthermore, we have evidence from the appearance of small fourth and fifth dimensions, which pull some of these threads apart, that there are aspects of dimension two and three that are not as cohesive with the dimension as a whole, as the three-dimensional solution might suggest.

In addition to the factor loadings themselves, there are other outputs from the PARAFAC analysis that can aid in interpretation. We cannot take time to discuss all of these here, but will briefly mention one example. Table C–18 presents the error analysis table from the PARAFAC three-dimensional unconstrained analysis of the total group. The figures in the table are the mean-squared error values for each level of each mode of the data. By examining these values, one can determine whether certain parts of the data were better described by the PARAFAC model than others. This is particularly straightforward for Modes *B* and *C*, where the data were standardized so that each level has a mean square value of 1.0. Thus, for these two modes, the numbers in the error analysis table represent the variance not accounted for at each level—that is, for each stimulus (Mode *B*) and for each rater (Mode *C*). Examination of these tables reveals, for example, that rating scales 6, 8, 20, 26, and 36 were particularly well fit by the model. These represent the adjectives "colorful," "ornate," "active," "simple," and "heavy" along with their opposites. On the other hand, judgments on rating scales, 15, 27, 30, 35, and 37 were particularly poorly fit by the model. These scales contain the adjectives "familiar," "trustful," "private," "inefficient," "boring." Some scales that are very poorly fit in three dimensions might become better fit by the smaller additional dimensions that could emerge from a four- to six-dimensional analysis of a larger data set. Other scales might need companion scales to define a dimension more clearly. "Inefficient versus efficient" is an example of a scale that relates clearly to automobiles but not so obviously to celebrities; it also is somewhat isolated and has no closely related adjectives to help it define a dimension. The fact that only 6% of the variance of this rating scale was explained by the three-dimensional analysis (leaving a mean-square error of .94) is consistent with the idea that we were here concentrating on major connotative dimensions. If we would want, in a subsequent analysis, to define some less connotative and more directly descriptive scales, we might want to elaborate "Inefficient-Efficient" into a small cluster of scales related to fuel efficiency and extract additional small dimensions from our data until this dimension emerges (if it does). However, such a dimension would presumably have only modest loadings, at best, on the celebrities (although Ralph Nader, for example, might get relatively high loadings for efficiency and Jerry Lewis for inefficiency).

The Mode *C* mean-square error values from the error analysis table could be useful in determining whether the judgments of certain types of individuals—such as certain segments of the market—were better fit by the model than those of other individuals and/or market segments. One would know, then, which types of generalizations to trust more and perhaps would want to modify the rating scales or stimuli to better tap those segments that were not well accounted for in this analysis. With our data, for example, we see that the ratings of subjects 24, 25, 31, and

TABLE C-18. Error Analysis for Three-Dimensional Unconstrained Solution

Mode A		Mode B		Mode C	
1	0.794	1	0.749	1	0.722
2	0.925	2	0.707	2	0.667
3	0.970	3	0.708	3	0.628
4	0.786	4	0.772	4	0.745
5	0.925	5	0.707	5	0.683
6	0.694	6	0.633	6	0.607
7	0.756	7	0.732	7	0.794
8	0.635	8	0.665	8	0.637
9	0.783	9	0.742	9	0.808
10	0.901	10	0.803	10	0.753
11	0.988	11	0.715	11	0.696
12	0.638	12	0.745	12	0.714
13	0.737	13	0.671	13	0.742
14	0.439	14	0.753	14	0.700
15	0.586	15	0.936	15	0.632
16	0.926	16	0.846	16	0.714
17	0.578	17	0.737	17	0.926
18	0.708	18	0.754	18	0.825
19	0.717	19	0.748	19	0.746
20	0.671	20	0.662	20	0.616
21	0.544	21	0.702	21	0.709
22	0.738	22	0.674	22	0.661
23	0.605	23	0.759	23	0.690
24	0.638	24	0.735	24	0.899
25	0.866	25	0.693	25	0.918
		26	0.595	26	0.818
		27	0.949	27	0.751
		28	0.673	28	0.708
		29	0.673	29	0.697
		30	0.874	30	0.666
		31	0.705	31	0.910
		32	0.815	32	0.803
		33	0.812	33	0.871
		34	0.708	34	0.715
		35	0.944		
		36	0.532		
		37	0.887		
		38	0.739		
		39	0.680		

33 were particularly poorly fit by this analysis, whereas the ratings of subjects 3, 6, 15, and 20 were particularly well described by the PARAFAC three-dimensional analysis. If we knew enough about these particular subjects, we might be able to draw useful conclusions about what type of individual our analysis best describes.

The levels of Mode *A* were not rescaled to a constant mean-square as part of the preprocessing for this analysis, because it was thought useful to allow those stimuli that elicited stronger feelings to maintain this greater importance in the analysis and thus have larger resulting loadings. Consequently, the relative sizes of the mean-square error values for Mode *A* would have to be divided by their mean-square values at input to obtain a more appropriate idea of relative variance accounted for. Such information would be useful, for example, if it indicated that judgments relating to certain products of importance to the study were not at all well fit by the model. Generalizations with respect to those products should then be made with much greater caution. Alternatively, one might want to deliberately extract more factors (as long as reliable factors could be obtained) examining the error analysis table at each dimensionality to see if the additional dimensions provide for a better account of the products in question.

Four- and Five-Dimensional Solutions

The loadings for Mode *A* and *B* of the four-dimensional constrained solution are presented in Tables C–19 and C–20. The interpretation of dimension one stays more or less the same as in the three-dimensional solution. (Once again, we reflect the loadings on Modes *A* and *B* of dimension one for purposes of interpretation.) Dimension two is similar, but not identical. While it retains its formal, large, heavy, lush qualities, the "mature versus youthful" overtone component is no longer as strong. This component is now part of a new third dimension, along with the "rational-irrational," "careful-careless," and "profound-superficial" overtones previously associated with the masculinity-femininity dimension (dimension three). Since the new dimension three has at its positive pole the terms "youthful," "naive," "superficial," "irrational," "careless," and "disreputable," we might call it "impetuous youth" versus "responsible-thoughtful-maturity." Dimension four is a simplified version of the "feminine-soft-smooth" dimension. It no longer has the component of irrationality, carelessness, or superficiality associated with femininity. On the other hand, "pleasant" now has an even higher loading and consequently might be considered to be a clearer expression of the sexy-sensual-feminine versus macho-masculine dimension. Indeed, because rationality and sophistication are no longer considered part of masculinity, Orson Welles loses his highly masculine rating. Muhammed Ali is now considered the most masculine, with John Wayne next; surprisingly, Ralph Nader comes in third, before John Travolta and Jerry Lewis.

One method of quantifying the degree to which each dimension of the four-dimensional solution represents components taken from the three-dimensional solution is by means of a table of correlations between dimensions in the two solutions. Table C–21 gives such correlations. Here, however, we have replaced the unconstrained three-dimensional solution by its very similar constrained counterpart. This should not affect the conclusions much, because the two versions of the three-dimensional solution are so

TABLE C-19. Mode A for the Four-Dimensional Constrained Solution

	Mode A: Cars & Stars & Self			
	1	2	3	4
1	-0.87	-0.18	-0.28	-0.60
2	-0.34	-0.14	-1.33	1.45
3	-1.55	-0.04	0.80	2.80
4	-0.33	-0.53	2.53	-2.08
5	-0.25	-0.56	-2.40	1.31
6	-0.57	1.39	-1.39	0.40
7	-0.72	-0.58	0.92	0.62
8	-0.23	0.06	-0.60	-0.49
9	-0.13	-0.84	0.19	0.66
10	-0.02	-0.98	1.98	0.98
11	-0.11	-0.04	-1.03	0.05
12	-0.94	-0.72	0.49	-2.18
13	1.17	-0.32	0.01	-0.18
14	0.92	0.93	-0.14	-0.56
15	0.46	-0.61	0.18	-0.38
16	-1.11	2.54	-0.33	-0.26
17	0.23	0.78	-0.05	-0.38
18	2.31	-1.68	0.63	0.10
19	1.03	1.24	-.09	0.04
20	1.71	-0.95	0.19	0.12
21	2.19	-0.20	0.08	-0.19
22	-1.33	2.38	-0.49	-0.20
23	0.03	-0.38	0.40	-0.04
24	0.05	-0.02	0.46	-0.13
25	-0.87	-0.41	-0.73	-0.85

similar (for example, see Table C–11). The advantage of using the two constrained solutions is that it allows us to consider the Mode *B* part of the table, where dimensions are constrained to be orthogonal, in terms of additive variance components. For Mode *B* in Table C–21, correlations can be viewed as multiple regression weights, and the sums of squared correlations as variance-accounted-for values. We note that the row sums of the Mode *B* table all are equal to 1.0, indicating that all the variance of the three-dimensional solution is retained in the four-dimensional solution, only redistributed across the four dimensions as indicated by the squares of the correlations in the table. The column sums of squared correlations indicate the proportion of variance in each new dimension of the four-dimensional solution which can be predicted from the dimensions of the three-dimensional solution. Dimension one is almost entirely predictable from the three-dimensional solution (.98) and this is almost all from the old version of dimension one. Thus, dimension one can be considered unchanged. Dimension two is primarily composed of vari-

TABLE C-20. Mode B for the Four-Dimensional Constrained Solution

	Mode B: Rating Scales			
	1	2	3	4
1	0.69	-0.09	-0.42	1.82
2	0.65	-1.08	0.48	-1.30
3	-0.12	-1.99	0.34	0.46
4	1.31	0.99	-0.26	-0.77
5	-1.46	-0.07	-0.22	0.96
6	-1.79	0.08	0.12	0.63
7	1.21	-0.10	0.54	-0.99
8	-1.24	1.77	1.47	0.56
9	0.76	-1.23	0.42	0.43
10	1.09	0.21	-0.31	1.25
11	-1.20	0.05	-1.79	-0.82
12	-0.65	0.02	-2.34	-0.68
13	-0.42	-1.17	1.74	0.41
14	-0.76	-0.35	2.03	0.57
15	0.81	0.11	-0.15	0.15
16	1.16	0.19	-0.90	0.88
17	-0.73	0.33	-1.96	-0.89
18	1.09	0.46	0.27	-1.36
19	-0.13	0.13	0.37	-2.05
20	1.47	1.49	0.11	-1.03
21	0.88	-0.49	1.42	-0.37
22	-0.13	-0.68	0.19	-2.50
23	-1.55	-0.88	-0.03	0.34
24	1.35	-0.45	0.44	0.07
25	-1.02	0.30	0.25	-1.11
26	-1.01	1.02	-1.33	-0.28
27	0.36	0.30	0.43	1.26
28	-0.99	1.94	1.46	0.01
29	0.34	-1.26	-1.23	1.51
30	0.06	0.77	-0.74	-0.79
31	-1.58	0.23	-0.06	0.38
32	-0.80	-0.88	1.04	-0.07
33	-0.99	-0.72	0.02	1.11
34	1.04	-0.78	-1.67	0.57
35	0.55	1.72	1.12	0.32
36	0.07	2.15	-0.87	1.02
37	1.15	0.45	0.32	0.18
38	-0.21	-2.28	-0.13	-1.17
39	1.13	-0.21	-1.00	1.28

ance from the old dimension two, but 15% of its variance cannot be predicted from the old dimensions. Dimension three is the most "novel" dimension, since only 57% of its variance can be predicted by the dimensions of the old three-dimensional solution.

TABLE C-21. Splitting Process for Four-Dimensional Constrained Solution

		Mode A				
		4-D				
		1	2	3	4	
3-D	1	.99	-.39	.02	-.07	
	2	-.35	.93	-.56	.21	
	3	.19	-.20	.73	-.84	
		Mode B				
		4-D				
		1	2	3	4	Row Sum of Squares
3-D	1	.99	-.06	-.15	-.04	1.00
	2	-.03	.83	-.43	-.35	1.00
	3	.06	-.02	.60	-.79	1.00
Column Sum of Squares		.98	.85	.57	.75	

The correlations show that it combines aspects of both the old dimension two and dimension three. Finally, dimension four is primarily the same as the old masculinity-femininity dimension, but also contains some variance from the old dimension two. For this dimension, 25% of its Mode *B* loading variance cannot be predicted from the three-dimensional solution and hence may represent new shades of meaning for femininity.

We can conclude from this four-dimensional solution that our original "mature/conservative" and "feminine-soft-smooth" dimensions (from the three-dimensional solution) were not simply single overtones but rather were composed of several threads of connotation, closely associated. It is perhaps an unfortunate comment on the chauvinistic perspective of the MBA student sample used for this study that masculinity tended to be associated to some extent with rationality, seriousness, and sophistication, and consequently came out on the same dimension in the three-dimensional solution. However, our disapproval is mitigated by the fact that these two overtones were at least somewhat distinguishable to our subjects, as is demonstrated by the rearrangement that occurs in the four-dimensional solution.

FURTHER STEPS

The next step, after identifying the dimensions, is to use the information they provide for the purpose of evaluating congruence of automobiles and potential spokesmen. This is not simply a matter of examination of the distance between points in the stimulus space, but rather intelligent consideration of the nature of different overtones and how they might interact. A further step after that is to look at the person loadings and use them to examine market segment differences. These and other theoretical issues are discussed in the complete version of this article (Harshman and De Sarbo 1981).

NOTES

1. This appendix was excerpted from a longer manuscript, Harshman and De Sarbo (1981), which deals in more detail with the theoretical issues of the particular marketing application, the use of the dimensions for marketing decisions, and the relative merits of this approach compared to MDS.

2. This work was done at Bell Laboratories, Murray Hill, New Jersey, while Richard Harshman was on leave from the Department of Psychology, University of Western Ontario, London, Canada N6A 5C2. We are grateful to Bell Laboratories for the support that made this research possible.

3. Since three solutions were used for the full data, plus 3 solutions for each of the 4 split-halves (to be discussed below), there were in fact 15 solutions obtained at each dimensionality between two and six.

REFERENCES

Harshman, R. A., and W. S. De Sarbo. 1981. Connotative congruence analysis: An application of PARAFAC to the selection of appropriate spokesmen for a given brand. Manuscript, Bell Laboratories, Murray Hill, New Jersey.

Kelman, H. C., and C. I. Howland. 1953. Reinstatement of the communicator in delayed measurement of opinion change. *Journal of Abnormal Social Psychology* 48:327–35.

McGuire, W. J. 1969. The nature of attitudes and attitude change. In *The handbook of social psychology,* vol. 3, 2d ed., ed. G. L. Lindzey and E. Arnson. Reading, Mass.: Addison-Wesley.

Mowen, J. C. 1980. On product endorser effectiveness: A balance model approach. In *Current issues and research in advertising,* ed. J. H. Leigh and C. R. Martin, Jr.. Ann Arbor, Michigan: University of Michigan.

Osgood, C. E. 1962. Studies on the generality of affective meaning systems. *American Psychology* 17:10–28.

Osgood, C. E., G. J. Suci, and P. Tannenbaum. 1957. *The measurement of meaning.* Urbana, Ill.: University of Illinois Press.

Bibliography

Ahrens, H. J. 1967. On the taxonomy of decision formation in judging West German politicians. *Archiv fur die Gesamte Psychologie* 119:57–89.

Algera, J. A. 1980. De constructie van een instrument voor het meten van taakkenmerken voor die van invloed zijn op motivatie, satisfactie en prestaties van taakuitvoerenden. Doctoral thesis, Rijksuniversiteit Leiden.

Algina, J. J. 1976. A generalized longitudinal factor model for the analysis of change. *Dissertation Abstracts International* 37:243.

Allen, J. P. 1976. A factor analytic comparison study of the Iowa Tests of Basic Skills among children with and without previous Head Start experience as well as a longitudinal factor study of ITBS subtests among lower socio-economic status children. *Dissertation Abstracts International* 37:1480.

Ando, F. 1970. Multidimensional analysis of the preference structure for music performances. *Japanese Journal of Psychology* 40:330–36.

Appelbaum, M. I. 1969. *A methodological study of preference.* Urbana-Champaign, Ill.: University of Illinois.

Arbuckle, J., and M. L. Friendly. 1977. On rotating to smooth functions. *Psychometrika* 42:127–40.

Baltink, G. J. H. 1968. Differentieel-psychologisch onderzoek naar de beoordeling van abstrakte schilderijen met behulp van driemodale faktoranalyse. Master's thesis, University of Groningen, Institute for General Psychology.

———. 1969. Driemodale faktoranalyse in een differentieel-psychologisch onderzoek naar de beoordeling van abstracte schilderijen. *Nederlands Tijdschrift voor de Psychologie* 24:529–40.

Bartussek, D. 1973. Interpretation of core-matrix in 3-mode factor analysis of L. R. Tucker. *Psychologische Beitrage* 15:169–84.

———. 1980. Die dreimodale faktorenanalyse als methode zur bestimmung von EEG-frequenzbandern. In *Faktorenanalyse unda variablenabildung aus dem elektro-enzephalogram,* ed. S. Kubicki, W. M. Herrmann, and G. Laudahn. Stuttgart: Gustav Fischer Verlag.

Bartussek, D., and H. Graser. 1980. Ergebnisse dreimodaler faktorenanalysen von EEG-frequenzspektren. In *Faktorenanalyse unda Variablenabildung aus dem Elektro-enzephalogram,* ed. S. Kubicki, W. M. Herrmann, and G. Laudahn. Stuttgart: Gustav Fischer Verlag.

Bartussek, D., K. Pawlik, and D. Rhenius. 1972. Eine dimensionsanalyse des digital frequenzanalysierten EEG und sein zusammenhang mit personlichkeitsvariablen. Paper presented at the 13th meeting of Experimental Psychologists, Graz, Austria.

Bauer, K. G. 1972. Linear prediction of a multivariate time-series applied to atmospheric scalar fields. *Dissertation Abstracts International* 32:6565.

Belk, R. W. 1974. An exploratory assessment of situational effects in buyer behavior. *Journal of Marketing Research* 11:156–63.

————. 1979. Gift-giving behavior. In *Research in marketing,* ed. J. N. Sheth. Greenwich, Conn.: JAI Press.

Bell, W. R. 1980. Multivariate time-series: Smoothing and backwards models. *Dissertation Abstracts International* 41: 3825.

Bentler, P. M. 1973. Assessment of developmental factor change at the individual and group level. In *Life-span developmental psychology: Methodological issues,* ed. J. R. Nesselroade and H. W. Reese. New York: Academic Press.

Bentler, P. M., and S-Y. Lee. 1978. Statistical aspects of a three-mode factor analysis model. *Psychometrika* 43:343–52.

————. 1979. A statistical development of three-mode factor analysis. *British Journal of Mathematical and Statistical Psychology* 32:87–104.

Bentler, P. M., and D. G. Weeks. 1979. Interrelations among models for the analysis of moment structures. *Multivariate Behavioral Research* 14:169–86.

Bernstein, A. L., and F. W. Wicker. 1969. A three-mode factor analysis of the concept of novelty. *Psychonomic Science* 14:291–92.

Bieber, S. L. 1980. Exploratory longitudinal factor analysis. *Dissertation Abstracts International* 40:3246–47.

Bloxom, B. 1968. A note on invariance in three-mode factor analysis. *Psychometrika* 33:347–50.

Borg, I., and J. C. Lingoes. 1977. A direct transformation beginning for the multidimensional analysis of trimodal data matrices: Theory and applications. *Zeitschrift fur Sozialpsychologie* 8:98–114.

Bormann, E. G. et al. 1978. Political cartoons and salient rhetorical fantasies: An empirical analysis of the '76 presidential campaign. *Communication Monographs* 45:317–29.

Bortz, J. 1975. The INDSCAL-procedure as a method of differentiating cognitive structures. *Zeitschrift fur Experimentelle und Angewandte Psychologie* 22:33–46.

Bouroche, J. M., and A. M. Dussaix. 1975. Several alternatives for three-way data analysis. *Metra* 14:299–319.

Bridgman, R. P., C. W. Snyder, Jr., and H. G. Law. 1981. Individual differences in conceptual behaviour following manipulated controllability. *Personality and Individual Differences* 2:197–205.

Browne, M. W. 1983. *The decomposition of multitrait-multimethod matrices (Research Report).* Pretoria: University of South Africa.

Burnett, J. D. 1971. Component curve analysis of student performance on a computer-based simulation game. *Alberta Journal of Educational Research* 17:117–28.

Cant, R. G. 1971. Changes in the location of manufacturing in

New Zealand 1957–1968: An application of three-mode factor analysis. *New Zealand Geographer* 27:38–55.
Carroll, J. D., and P. Arabie. 1980. Multidimensional scaling. *Annual Review of Psychology* 31:607–49.
Carroll, J. D., and J. J. Chang. 1970. Analysis of individual differences in multidimensional scaling via an N-way generalization of "Eckart-Young" decomposition. *Psychometrika* 35:282–319.
———. 1970. Reanalysis of some color data of Helm's by INDSCAL procedure for individual differences multidimensional scaling. *Proceedings of the Annual Convention of the American Psychological Association* 5:137–38.
———. 1972. IDIOSCAL (*I*ndividual *D*ifferences *I*n *O*rientation *SCAL*ing): A generalization of INDSCAL allowing *IDIO*syncratic reference systems as well as an analytic approximation to INDSCAL. Paper presented at the spring meeting of the Psychometric Society, Princeton, New Jersey, 30–31 March.
Carroll, J. D., and M. Wish. 1974. Models and methods for three-way multidimensional scaling. In *Contemporary developments in mathematical psychology, vol. 2,* ed. D. H. Krantz, R. C. Atkinson, R. D. Luce, and P. Suppes. San Francisco: W. H. Freeman.
Carroll, J. D., S. Pruzansky, and J. B. Kruskal. 1980. CANDELINC: A general approach to multidimensional analysis of many-way arrays with linear constraints on parameters. *Psychometrika* 45:3–24.
Cattell, R. B. 1944. "Parallel Proportional Profiles" and other principles for determining the choice of factors by rotation. *Psychometrika* 9:267–83.
———. 1980. Guest editorial: Two basic models for personality-environment interaction and the need for their substantive investigation. *Multivariate Behavioral Research* 15:243–47.
———. 1980. The separation and evaluation of personal and environmental contributions to behavior by the person-centered model (PCER). *Multivariate Behavioral Research* 15:371–402.
Chambers, J. L., and M. E. Surma. 1979. Dimensions of Picture Identification Test need associations. *Journal of Personality Assessment* 43:128–34.
Chojnicki, Z., and T. Czyz. 1976. Some problems in the application of factor analysis in geography. *Geographical Analysis* 8:416–27.
Cleary, P. J. 1974. Description of individual differences in autonomic reactions. *Psychological Bulletin* 81:934–44.
Cliff, N. 1968. "Idealized Individual" interpretation of individual differences in multidimensional scaling. *Psychometrika* 33:225–32.
Cohen, H. S. 1974. Three-mode rotation to approximate INDSCAL structure (TRIAS). Paper presented at the annual meeting of the Psychometric Society, Palo Alto, California.
———. 1975. Further thoughts on three-mode rotation to INDSCAL structure, with jackknifed confidence regions for points. Paper presented at the U.S.-Japan Seminar on Theory, Methods, and Applications on Multidimensional Scaling and Related Techniques, La Jolla, California, 20–24 August.
Computing Services Offices. 1973. *SOUPAC program descrip-*

tions. Urbana-Champaign, Ill.: University of Illinois.
Cooper, L. G. 1973. A multivariate investigation of preferences. *Multivariate Behavioral Research* 8:253–72.
Corballis, M. C., and R. E. Traub. 1970. Longitudinal factor analysis. *Psychometrika* 35:79–98.
Cornelius, III, E. T., M. D. Hakel, and P. R. Sackett. 1979. A methodological approach to job classification for performance appraisal purposes. *Personnel Psychology* 32:283–97.
Coxon, A. P. 1974. The mapping of family-composition preferences: A scaling analysis. *Social Science Research* 3:191–210.
Cramer, R. H., and R. B. Miller. 1976. Dynamic modeling of multivariate time-series for use in bank analysis. *Journal of Money, Credit and Banking* 8:85–96.
———. 1978. Multivariate time-series analysis of bank financial behavior. *Journal of Financial and Quantitative Analysis* 13:1003–17.
Crookes, J. G., K. Croston, and P. Sypsas. 1980. Process components for multivariate time-series analysis. *Journal of the Operational Research Society* 31:325–30.
Cupchik, G. C. 1976–77. Theoretical and empirical perspectives on impressionistic painting. *Bulletin de Psychologie* 30:720–29.
Danhauer, J. L., and M. A. Appel. 1976. INDSCAL analysis of perceptual judgments for 24 consonants via visual, tactile, and visual-tactile inputs. *Journal of Speech and Hearing Research* 19:68–77.
Danhauer, J. L., and R. M. Lawarre. 1979. Dissimilarity ratings of English consonants by normally-hearing and hearing-impaired individuals. *Journal of Speech and Hearing Research* 22:236–46.
Danhauer, J. L., N. C. Chari, and G. Herman. 1977. Normal-hearing subjects' perception of speech through one-third octave bands. *Journal of Auditory Research* 17:275–81.
Danhauer, J. L., M. P. Rastatter, and G. Herman. 1978. Distinctive features for short-term memory of consonants in noise. *Journal of Auditory Research* 18:63–68.
Davis, E. E., and N. N. Grobstein. 1966. *Multimode factor analysis of interpersonal perceptions. (Technical Report No. 36)*. Urbana-Champaign, Ill.: University of Illinois, Department of Psychology.
Davis, R. G. 1979. Olfactory perceptual space models compared by quantitative methods. *Chemical Senses and Flavor* 4:21–33.
———. 1981. Multidimensional scaling of individual differences for an individual person with a hand-held calculator. *Behavior Research Methods and Instrumentation* 13:25–30.
de Leeuw, J., and S. Pruzansky. 1978. A new computational method to fit the weighted Euclidean distance model. *Psychometrika* 43:479–90.
De Sarbo, W. S. 1978. Three-way unfolding and situational dependence in consumer preference analysis. Doctoral thesis, University of Pennsylvania.
De Sarbo, W. S., and J. D. Carroll. 1979. Three-way unfolding. Working paper, Bell Laboratories, Murray Hill, New Jersey.
———. 1981. Three-way unfolding and situational dependence in consumer preference analysis. Manuscript, Bell Lab-

oratories, Murray Hill, New Jersey.

Dobson, R., and J. F. Kehoe. 1975. Empirical comparison of Tucker-Messick and INDSCAL models—Measuring viewpoints about transit attributes. *Socio-Economic Planning Sciences* 9:67–81.

Donchin, E., L. A. Gerbrandt, L. Leifer, and L. R. Tucker. 1972. Is the contingent negative variation contingent on a motor response? *Psychophysiology* 9:178–88.

Dunn, T. R. 1977. Modelling the core matrix in three-mode multidimensional scaling: An example using the size-weight illusion. *Dissertation Abstracts International* 37:5320.

Dunn, T. R., and R. A. Harshman. 1982. A multidimensional scaling model for the size-weight illusion. *Psychometrika* 47: 25–45.

Ferisin, T. P. 1971. Individual differences in person perception: Assessment of a model. *Dissertation Abstracts International* 32:3028.

Firth, P. M., and C. W. Snyder, Jr. 1979. Three-mode factor analysis of self-reported difficulty in assertiveness. *Australian Journal of Psychology* 31:125–35.

Fischer, G. H. 1967. The problem of interpretation of results from factor analysis. *Psychologische Beitrage* 10:122–35.

———. 1967. Tucker's methods of factor analysis of learning data. *Psychologische Beitrage* 10:136–45.

Frederiksen, C. H. 1974. Models for the analysis of alternative sources of growth in correlated stochastic variables. *Psychometrika* 39:223–45.

Frederiksen, N. 1972. Towards a taxonomy of situations. *American Psychologist* 27:114–23.

Frederiksen, N., O. Jensen, and A. E. Beaton. 1972. *Prediction of organizational behavior.* Elmsford, N.Y.: Pergamon Press.

Frey, C. 1973. Profilskalierung von kunstzeichnungen in abhängigkeit von persönlichkeitseigenschaften der zeichner so wie der beurteiler. Master's thesis, University of Hamburg.

Fruchter, B. 1969. A comparison of two-mode and three-mode factor analysis of psychomotor learning performance. In *Proceedings of the XVIth International Congress of Applied Psychology,* ed. H. R. Wyngaarden, President. Amsterdam: Swets and Zeitlinger.

Gabrielsson, A. 1974. An empirical comparison between some models for multidimensional scaling. *Scandinavian Journal of Psychology* 15:73–80.

———. 1975. Applications of the INDSCAL model for dimension analyses within auditory perception. *Goteborg Psychological Reports* 5:17–19.

Gabrielsson, A., and H. Sjögren. 1974, 1975. *Adjective ratings and dimension analysis of perceived sound quality of hearing aids I & II (Report TA No. 75 & 77).* Stockholm: Karolinska Institute.

Gandour, J. T. 1978. Perceived dimensions of thirteen tones: A multidimensional scaling investigation. *Phonetica* 35:169–79.

Gandour, J. T., and R. A. Harshman. 1978. Cross-language differences in tone perception: A multidimensional scaling investigation. *Language and Speech* 21:1–33.

Gazda, G. M., and J. A. Mobley. 1981. INDSCAL multidimensional-scaling. *Journal of Group Psychotherapy Psychodrama and Sociometry* 34:54–73.

Gilroy, E. J. 1971. Linear least-squares prediction for multivariate time-series with missing observations. *Dissertation Abstracts International* 32:1907.

Gitin, S. R. 1970. A dimensional analysis of manual expression. *Journal of Personality and Social Psychology* 15:271–77.

Gleason, T. C., and M. A. Barton. 1976. On improving the interpretability of three-mode component analysis. Manuscript, University of Pennsylvania.

Gnanadesikan R., ed. 1983. *Proceedings of symposia in applied mathematics, vol. 28. Statistical data analysis.* Providence, R.I.: American Mathematical Society.

Gollob, H. F. 1968. Rejoinder to Tucker's "Comments on confounding of sources of variation in factor-analytic techniques." *Psychological Bulletin* 70:355–60.

Gower, J. C. 1977. The analysis of three-way grids. In *The measurement of intrapersonal space by grid technique: Dimensions of intrapersonal space, vol. 2,* ed. P. Slater. London: John Wiley and Sons.

Gräser, H. 1977. Spontane reversionprozesse in der figuralwahrnehmung. Eine untersuchung reversibler figuren mit der dreimodalen faktorenanalyse. Doctoral thesis, University of Trier.

Gräser, H., H. Esser, and H. Saile. 1981. Einschätzung von lebensereignissen und ihren auswirkungen. In *Kritische lebensereignisse und ihre bewaltigung,* ed. S. H. Filipp. Munchen: Urban and Schwarzenberg.

Gregson, R. A. M. 1978. Cognitive multivariate time-series and its analysis. *ACTA Psychologica* 42:277–91.

Gregson, R. A. M., and M. J. Mitchell. 1974. Odor quality similarity scaling and odor-word profile matching. *Chemical Senses and Flavor* 1:95–101.

Groves, C. L. 1978. Individual difference modelling of simple functional relations: Examples using three-mode factor analysis. *Dissertation Abstracts International* 39:2475–76.

Gruvaeus, G., H. Wainer, and F. Snyder. 1971. TREMOD: A 360/75 FORTRAN program for three-mode factor analysis. *Behavioral Science* 16:421–22.

Gruvaeus, G., H. Wainer, and N. Zill. 1971. Mixed modal matrices as aids to interpretation in three-mode factor analysis. Paper presented at the Psychometric Society Meeting, St. Louis, April.

Haan, N. 1981. Common dimensions of personality development: Early adolescence to middle life. In *Present and past in middle life,* ed. D. H. Eichorn, J. A. Clausen, N. Haan, M. P. Honzik, and P. H. Mussen. New York: Academic Press.

Hall, J. L., and M. R. Schroeder. 1972. Monaural phase effects for two-tone signals. *Journal of the Acoustical Society of America* 51:1882–84.

Hanke, B., J. B. Lohmöller, and H. Mandl. 1980. *Schülerbeurteilung in der grundschule: Ergebnisse der augsburger längsschnittuntersuchung.* Munchen: Oldenbourg Verlag.

Hanley, T. V., and D. L. Cox. 1979. Individual differences in visual discrimination of letters. *Perceptual and Motor Skills* 48:539–50.

Harris, D. R., and R. Fenker. 1974. INDSCAL for a small computer using basic FORTRAN 4. *Educational and Psychological Measurement* 84:129–32.

Harris, M. L., and L. S. Golub. 1971. *An analysis of content and task dimensions of language arts items designed to measure level of content attainment.* Madison, Wis.: University of Wisconsin.

Harris, M. L., and A. M. Voelker. 1971. *An analysis of content and task dimensions of science items designed to measure level of concept attainment.* Madison, Wis.: University of Wisconsin.

Harris, M. L. et al. 1971. *An analysis of content and task dimensions of social studies items designed to measure level of concept attainment.* Madison, Wis.: University of Wisconsin.

———. 1971. Generalized learning curves and their ability and personality correlates. Paper presented at the 55th Annual Meeting of the American Educational Research Association, New York, February.

Harshman, R. A. 1970. Foundations of the PARAFAC procedure: Models and conditions for an "explanatory" multi-modal factor analysis. *UCLA Working Papers in Phonetics* 16:1–84 (University Microfilms No. 10,085).

———. 1972. Determination and proof of minimum uniqueness conditions for PARAFAC1. *UCLA Working Papers in Phonetics* 22:111–17 (University Microfilms No. 10,085).

———. 1972. PARAFAC2: Mathematical and technical notes. *UCLA Working Papers in Phonetics* 22:31–44 (University Microfilms No. 10,085).

———. 1976. PARAFAC: Methods of three-way factor analysis and multidimensional scaling according to the Principle of Proportional Profiles. *Dissertation Abstracts International* 37:2479.

———. 1981. *Partial reference manual for the preliminary release of the PARAFAC analysis package.* London, Ontario, Canada: Scientific Software Associates.

Harshman, R. A., and S. A. Berenbaum. 1981. Basic concepts underlying the PARAFAC-CANDECOMP three-way factor analysis model and its application to longitudinal data. In *Present and past in middle life,* ed. D. H. Eichorn, J. A. Clausen, N. Haan, M. P. Honzik, and P. H. Mussen. New York: Academic Press.

Harshman, R. A., and W. S. De Sarbo. 1981. Connotative congruence analysis: An application of PARAFAC to the selection of appropriate spokesman for a given brand. Manuscript, Bell Laboratories and University of Western Ontario.

Harshman, R. A., and G. Papcun. 1976. Vowel normalization by linear transformation of each speaker's acoustic space. *Journal of the Acoustical Society of America* 59:571.

Harshman, R. A., P. Ladefoged, and L. Goldstein. 1977. Factor analysis of tongue shapes. *Journal of the Acoustical Society of America* 62:693–707.

Harvey, L. O., and M. J. Gervais. 1981. Internal representation of visual texture as the basis for the judgment of similarity. *Journal of Experimental Psychology: Human Perception and Performance* 7:741–53.

Hentschel, U., and H. Klintman. 1974. A 28-variable semantic differential: I. On the factorial identification of content. Manuscript, Lund University.

Hirschberg, N. 1980. Individual differences in social judgment: A multivariate approach. In *Progress in social psychology,* ed. M. Fishbein. Hillsdale: Lawrence Erlbaum Associates.

Hoffman, E. L., and L. R. Tucker. 1964. *Three-way factor analysis of a multitrait-multimethod matrix (Technical Report).* Urbana-Champaign, Ill.: University of Illinois, Department of Psychology.

Hohn, M. E. 1979. Principal components-analysis of three-way tables. *Journal of the International Association for Mathematical Geology* 6:611–26.

Hohn, M. E., and L. M. Friberg. 1979. A generalized principal components model in petrology. *Lithos* 12:317–24.

Hojo, H. 1982. Similarity of Japanese consonant phonemes—An analysis by INDSCAL and Hayashi quantification theory-I. *Japanese Journal of Psychology* 53:72–79.

Hoornaert, F., and L. Delbeke. 1979. Parental symbolism and image formation in normal adults and different patient groups. *Psychologica Belgica* 19:81–97.

Howard, D. V., and J. H. Howard. 1977. A multidimensional scaling analysis of the development of animal names. *Developmental Psychology* 13:108–13.

Huysamen, G. K. 1971. Models for factor analyzing data in a completely crossed three-way design. *Dissertation Abstracts International* 32:4891.

Imada, A. S., and M. London. 1979. Relationship between subjects, scales, and stimuli in research on social perception. *Perceptual and Motor Skills* 48:691–97.

Inn, A., C. L. Hulin, and L. R. Tucker. 1972. Three sources of criterion variance: Static dimensionality, dynamic dimensionality, and individual dimensionality. *Organizational Behavior and Human Performance* 8:58–83.

Israelsson, A. 1969. Three-way (or second order) component analysis. Nonlinear iterative partial least-squares (NIPALS) estimation procedures. *Bulletin of the International Statistical Institute* 43:29–51.

Jacobson, R. L. 1981. Aggregate consumption and advertising: A multivariate time-series approach. *Dissertation Abstracts International* 42:3231.

Jaffrennou, P. A. 1978. *Sur l'analyse des familles finies de variables vectorielles. Bases algebriques et application a la description statistique.* France: University of Saint-Etienne, Department de Mathematiques.

Jennrich, R. 1972. A generalization of the multidimensional scaling model of Carroll and Chang. *UCLA Working Papers in Phonetics* 22:45–47.

Karnas, G. 1975. Note sur une procédure d'analyse de données relatives à une correspondance ternaire ou pseudo-ternaire par la méthode d'analyse binaire de faverge. *Le Travail Humain* 38:287–300.

Keats, J. A. 1967. Australian advances in mathematical psychology. *Australian Journal of Science* 29:450–55.

Kjerulff, K., and N. H. Wiggins. 1976. Graduate student styles for coping with stressful situations. *Journal of Educational Psychology* 68:247–54.

Knobloch, E. M. 1972. Einschätzung von leistungsrelevanten begriffen. Master's thesis, University of Hamburg.

Koch, P. D. 1980. Leading indicators in structural econometric models with applications in multivariate time-series analysis about the commerce department leading indicators and a proposed monetary leading indicator. *Dissertation Abstracts International* 41:5176.

Kouwer, B. J. 1967. *Driemodale faktoranalyse. Programmabeschrijving (GRON.PSYCH.07+07BIS). Orthogonale rotaties (GRON.PSYCH.12) (Reports).* Groningen: University of Groningen, Institute of Psychology.

Kroonenberg, P. M. 1981. *Scaling of input data for three-mode principal component analysis. (WEP-Reeks, WR 81-21-EX, Vakgroep W.E.P.).* Leiden: University of Leiden.

———. 1981. *User's guide to TUCKALS2. A program for three-mode principal component analysis with extended core matrix (WEP-Reeks, WR 81-35-RP).* Leiden, The Netherlands: University of Leiden.

———. 1981. *User's guide to TUCKALS3. A program for three-mode principal component analysis (WEP-Reeks, WR 81-6-RP).* Leiden, The Netherlands: University of Leiden.

———. 1983. *Three-mode principal component analysis: Theory and applications.* Leiden, The Netherlands: DSWO Press.

———. 1983. Annotated bibliography of three-mode factor analysis. *British Journal of Mathematical and Statistical Psychology* 36:81–113.

———. 1983. Annotated bibliography of three-mode factor and principal component analysis. *British Journal of Mathematical and Statistical Psychology* 36:81–113.

Kroonenberg, P. M., and J. de Leeuw. 1977. *TUCKALS2: A principal component analysis of three-mode data. Res. Bull. RB. 001–77.* Leiden: University of Leiden, Department of Data Theory.

———. 1978. TUCKALS2: Een hoofdassenanalyse voor drieweggegevens. *Methoden en Data Nieuwsbrief* 3:30–53.

———. 1980. Principal components analysis of three-mode data by means of alternating least-squares algorithms. *Psychometrika* 45:69–97.

Kruskal, J. B. 1976. More factors than subjects, tests and treatments: An indeterminacy theorem for canonical decomposition and individual differences scaling. *Psychometrika* 41: 281–93.

———. 1981. Multilinear models for data analysis. *Behaviormetrika* 10:1–20.

———. 1983. Multilinear Models. In *Proceedings of Symposia in Applied Mathematics, vol. 28, Statistical Data Analysis,* ed. R. Gnanadesikan. Providence, R.I.: American Mathematical Society.

Lammers, C. J. 1974. *Groei en ontwikkeling van de ziekenhuisorganisaties in Nederland.* Leiden, The Netherlands: Rijksuniversiteit Leiden, Sociologisch Instituut.

Landis, D., J. L. Hayman, and W. S. Hall. 1971. Multidimensional analysis procedures for measuring self-concept in poverty area classrooms. *Journal of Educational Psychology* 62: 95–103.

Landis, D., C. A. Silver, and J. M. Jones. 1967. Level of proficiency and multidimensional viewpoints about problem similarity. *Journal of Applied Psychology* 51:216–22.

Langeheine, R. 1978. The evaluation of classroom social structure by three-way multidimensional scaling of sociometric data. *Studies in Educational Evaluation* 4:185–208.

———. 1980. Expected fit values for random configurations in PINDIS. *Zeitschrift fur Sozialpsychologie* 11:38–49.

Lastovicka, J. 1981. The extension of component analysis to four-mode matrices. *Psychometrika* 46:47–57.

Law, H. G., and C. W. Snyder, Jr. 1979. Three-mode models for the analysis of psychological data. *Australian Psychologist* 14:214.

———. 1981. An introduction to the analysis of covariance structures: A general model for data analysis. In *Proceedings of a seminar on measuring social behavior in road research,* ed. J. M. Morris. Vermont South, Australia: Australian Road Research Board.

Leah, J. A., H. G. Law, and C. W. Snyder, Jr. 1979. The structure of self-reported difficulty in assertiveness: An application of three-mode common factor analysis. *Multivariate Behavioral Research* 14:443–62.

Ledolter, J. 1978. Analysis of multivariate time-series applied to problems in hydrology. *Journal of Hydrology* 36:327–52.

Leichner, R. 1975. Processing psychiatric information: I. Diagnostica 21:147–66.

Levin, J. 1963. Three-mode factor analysis. *Dissertation Abstracts International* 24:5530–31.

———. 1965. Three-mode factor analysis. *Psychological Bulletin* 64:442–52.

———. 1974. A rotational procedure for separation of trait, method and interaction factors in multitrait-multimethod matrices. *Multivariate Behavioral Research* 9:231–40.

Lilly, R. S. 1966. A developmental study of the semantic differential. *Dissertation Abstracts International* 26:4063–64.

Lingoes, J. C., and I. Borg. 1977. Optimale losungen fur dimensions- und vektorgewichte in PINDIS. *Zeitschrift fur Sozialpsychologie* 8:210–17.

———. 1978. A direct approach to individual differences scaling using increasingly complex transformations. *Psychometrika* 43:491–519.

Linn, R. L., F. W. Snyder, M. I. Appelbaum, and L. R. Tucker. 1967. *A Fortran II program for three-mode factor analysis.* Urbana-Champaign, Ill.: University of Illinois, Department of Psychology.

Lippa, R. 1977. Androgyny, sex-typing, and the perception of masculinity-femininity in handwritings. *Journal of Research in Personality* 11:21–37.

Litt, E. N. 1966. A factorial study of responses to abstract paintings. Master's thesis, University of Illinois.

Lðhmoller, J. B. 1978. How longitudinal factor stability, continuity, differentiation, and integration are portrayed into the core matrix of three-mode factor analysis. Paper presented at the European meeting on Psychometrics and Mathematical Psychology, Uppsala, Sweden, 16 June.

———. 1979. Tucker's three-mode factor-analysis model—scaling, rotation and other models. *Archiv fur Psychologie* 131:137–66.

———. 1981. *Stabilitat und kontinuitat in langsschnittdaten, analysiert durch t- und trimodale faktorenanalyse (Technical Report).* Munchen: Neubibere.

Löhmoller, J. B., and R. Oerter. 1979. *Medien in der erzieher-ausbilding: Erprobung des medienverbundes "vorschulische erziehung in ausland."* Munchen: Oldenbourg Verlag.

Löhmoller, J. B., and H. Wold. 1980. Three-mode path models with latent variables and partial least-squares (PLS) parameter estimation. Paper presented at the European meeting of the Psychometric Society, Groningen, 18–21 June.

London, M., R. Crandall, and D. Fitzgibbons. 1977. The psychological structure of leisure: Activities, needs, people. *Journal of Leisure Research* 9:252–63.

Longin, H. E. 1974. Perception of symptomatic behaviors. *Journal of Consulting and Clinical Psychology* 42:812–18.

Love, W. D., and L. R. Tucker. 1970. *A three-mode factor analysis of serial learning.* Arlington, Va.: Office of Naval Research.

MacCallum, R. C. 1974. Relations between factor analysis and multidimensional scaling. *Psychological Bulletin* 81:505–16.

———. 1975. A comparison of two individual differences models for multidimensional scaling: Carroll and Chang's INDSCAL and Tucker's three-mode factor analysis. *Dissertation Abstracts International* 35:3619.

———. 1976. Effects on INDSCAL of non-orthogonal perceptions of object space dimensions. *Psychometrika* 41:177–88.

———. 1976. Transformation of a three-mode multidimensional scaling solution to INDSCAL form. *Psychometrika* 41: 385–400.

———. 1977. Effects of conditionality on INDSCAL and ALSCAL weights. *Psychometrika* 42:297–305.

———. 1979. Recovery of structure in incomplete data by ALSCAL. *Psychometrika* 44:69–74.

———. 1981. Evaluating goodness of fit in nonmetric multidimensional scaling by ALSCAL. *Applied Psychological Measurement* 5:377–82.

MacCallum, R. C., and E. T. Cornelius. 1977. A Monte Carlo investigation of recovery of structure by ALSCAL. *Psychometrika* 42:401–28.

Maravall, A. 1981. A note on identification of multivariate time-series models. *Journal of Econometrics* 16:237–47.

Marks, E. 1970. Individual differences in perceptions of the college environment. *Journal of Educational Psychology* 61: 270–71.

McCloskey, J., and P. R. Jackson. 1979. THREE-MODE: A FORTRAN IV program for three-mode factor analysis. *Behavior Research Methods and Instrumentation* 11:75–76.

McDonald, R. P. 1969. A generalized common factor analysis

based on residual covariance matrices of prescribed structure. *British Journal of Mathematical and Statistical Psychology* 22:149–63.

———. 1970. Three common factor models for groups of variables. *Psychometrika* 35:111–28.

———. 1978. A simple comprehensive model for the analysis of covariance structures. *British Journal of Mathematical and Statistical Psychology* 31:59–72.

McDonald, R. P., and S. A. Mulaik. 1979. Determinacy of common factors: A nontechnical review. *Psychological Bulletin* 86:297–306.

Meijs, B. W. G. 1980. Huis van bewaring en subkultuur: Een empirische studie bij jeugdige gedetineerden naar het effekt van "102 dagen" preventieve hechtenis op attitudes en andere indikatoren van subkultuur. Master's thesis, University of Leiden.

Meuwese, W. 1970. A comparison of two methods of judging verbal stimuli. *Nederlands Tijdschrift voor de Psychologie en haar Grensgebieden* 25:594–603.

Meyer, J. P. 1980. Causal attribution for success and failure: A multivariate investigation of dimensionality, formation and consequences. *Journal of Personality and Social Psychology* 38:704–18.

Milligan, G. W. 1979. Note on the use of INDSCAL for the comparison of several classifications. *Systematic Zoology* 28:94–99.

Mills, D. H., and L. R. Tucker. 1966. A three-mode factor analysis of clinical judgment of schizophrenicity. *Journal of Clinical Psychology* 2:136–39.

Mindham, R. H., A. Bagshaw, S. A. James, and A. J. Swannell. 1981. Factors associated with the appearance of psychiatric symptoms in rheumatoid arthritis. *Journal of Psychosomatics Research* 25:429–35.

Montanelli, D. S. 1972. Multiple-cue learning in children. *Developmental Psychology* 7:302–12.

Moonen, J. 1978. Een onderzoek naar de mogelijkheden tot geintegreerd gebruik van een computergestuurd systeem in een statistiekkurrikulum. Doctoral thesis, University of Leiden.

Moriarty, M., and G. Salamon. 1980. Estimation and forecast performance of a multivariate time-series model of sales. *Journal of Marketing Research* 17:558–64.

Morin, V. N. 1981. Multivariate time-series analyses of observational data. *Dissertation Abstracts International* 42:1601.

Muller, J. B., and J. M. Strandberg. 1970. A longitudinal factor-analytic approach to curriculum development. *Proceedings of the Annual Convention of the American Psychological Association* 5:611–12.

Muthen, B., U. Olsson, T. Pettersson, and G. Stahlberg. 1977. Measuring religious attitudes using semantic differential technique—Application of three-mode factor-analysis. *Journal for the Scientific Study of Religion* 16:275–88.

Nagy, P. 1978. Construct validity of cognitive structures: A comparison of multidimensional methods. Paper presented at the annual meeting of the American Educational Research Association, Toronto, 27–31 March.

Neperud, R. W. 1970. Towards a structure of meaning in the visual arts: A three-mode factor analysis of non-art college student responses to selected art forms. *Studies in Art Education* 12:40–49.

———. 1973. Towards a structure of meaning: A three-mode factor analysis of adolescents' art concepts. *Studies in Art Education* 15:61–69.

Nesselroade, J. R. 1972. Note on the "longitudinal factor analysis" model. *Psychometrika* 37:187–91.

Nygren, T. E. 1979. A theoretical framework for testing the additive difference model for dissimilarities data: Representing gambles as multidimensional stimuli. *Journal of Mathematical Psychology* 20:53–77.

Nygren, T. E., and L. E. Jones. 1977. Individual differences in perceptions and preferences for political candidates. *Journal of Experimental Social Psychology* 13:182–97.

O'Hare, D. 1976. Individual differences in perceived similarity and preference for visual art: A multidimensional scaling analysis. *Perception and Psychophysics* 20:445–52.

———. 1979. Multidimensional scaling representations and individual differences in concept learning of artistic style. *British Journal of Psychology* 70:219–30.

O'Hare, D. P., and I. E. Gordon. 1977. Dimensions of the perception of art: Verbal scales and similarity judgments. *Scandinavian Journal of Psychology* 18:66–70.

Olsson, U., and L. R. Bergman. 1977. A longitudinal factor model for studying change in ability structure. *Multivariate Behavioral Research* 12:221–41.

Oostendorp, A., and D. E. Berlyne. 1978. Dimensions in the perception of architecture: I. Identification and interpretation of dimensions of similarity. *Scandinavian Journal of Psychology* 19:73–82.

Orlik, P. 1980. *Das summax-modell der dreimodalen faktorenanalyse mit interpretierbarer kernmatrix (Technical Report).* Sarrbrucken, FRG: University of Siaarland.

———. 1980. Das summax-modell der dreimodalen faktorenanalyse mit interpretierbarer kernmatrix. *Archiv fur Psychologie* 133:189–218.

Pennell, R. 1971. *Finding points of view in judgment data.* Princeton, N.J.: Educational Testing Service.

Pervin, L. A. 1967. *Dissatisfaction with college and the college dropout: A transactional approach. Final report.* Washington, D.C.: Office of Education.

Petersik, J. T. 1978. Possible role of transient and sustained visual mechanisms in the determination of similarity judgments. *Perceptual and Motor Skills* 47:683–98.

Pinkava, V. 1981. Classification in medical diagnostics—On some limitations of Q-analysis. *International Journal of Man-Machine Studies* 15:221–37.

Prashker, J. N. 1979. Scaling perceptions of reliability of urban travel modes using INDSCAL and factor-analysis methods. *Transportation Research Part A–General* 13:203–12.

Redfield, J. 1978. TMFA: A FORTRAN program for three-mode factor-analysis and individual-differences multidimensional-scaling. *Educational and Psychological Measurement* 38:793–95.

Redfield, J., and A. A. Stone. 1979. Individual viewpoints of stressful life events. *Journal of Consulting and Clinical Psychology* 47:147–54.

Reid, J. C. 1969. A three-mode factor analysis of students' perception of a university. *Journal of Experimental Education* 38:93–96.

Reid, J. C., and W. F. Seibert. 1973. A factor analytic approach to the analysis of learning curves. Paper presented at the Rocky Mountain Educational Research Association, Tucson, Arizona, November.

———. 1974. *Analysis of cine-psychometric visual memory data by the Tucker generalized learning curve method: Final report.* Washington, D.C.: National Institute of Education.

Retka, R. L., and R. M. Fenker. 1975. Self-perception among narcotic addicts: An exploratory study employing multidimensional scaling techniques. *International Journal of the Addictions* 10:1–12.

Richards, L. G. 1972. An analysis of individual differences in similarity judgments about complex random forms. *Perception and Psychophysics* 11:143–49.

Ritchie, J. R. 1974. An exploratory analysis of the nature and extent of individual differences in perception. *Journal of Marketing Research* 11:41–49.

Rodgers, J. L., and F. W. Young. 1981. Successive unfolding of family preferences. *Applied Psychological Measurement* 5:51–62.

Rösler, F. 1972. Dimensionen der aktivitäten und deren beziehungen zu den persönlichkeitsfaktoren "Extroversion/Introversion" und "Neurotizismus" sensu Eysenck. Master's thesis, University of Hamburg.

———. 1975. Die abhängigkeit des elektroenzephalogramms von den persönlichkeitsdimensionen E und N sensu Eysenck und unterschdielich aktivierenden situationen. *Zeitschrift fur Experimentelle und Angewandte Psychologie* 22:630–67.

———. 1979. Identifying interindividual judgment differences: INDSCAL or three-mode factor analysis. *Multivariate Behavioral Research* 14:145–67.

Rösler, F., J. Jesse, D. Manzey, and U. Grau. 1982. Ist das LM-Gitter nur ein LM-test?—Eine dreimodale faktorenanalyse des LM-Gitters fur kinder (Schmalt). *Diagnostica* 28:131–45.

Rowe, H. A. H. 1979. Three-mode factor-analysis—Problems of interpretation and possible solutions. *Australian Psychologist* 14:222–23 (Abstract).

Rudy, T. E., T. V. Merluzzi, and P. T. Henahan. 1982. Construal of complex assertion situations: A multidimensional analysis. *Journal of Consulting and Clinical Psychology* 50:125–37.

Russell, D. 1982. The causal dimension scale: A measure of how individuals perceive causes. *Journal of Personality and Social Psychology* 42:1137–45.

Rychlak, J. F., E. J. Flynn, and G. Burger. 1979. Affection and evaluation as logical processes of meaningfulness independent of associative frequency. *Journal of General Psychology* 100:143–57.

Saile, H. 1979. Zur struktur der einschatzung von lebensereignissen. Eine untersuching über beurteilungsunterschiede mittels dreimodaler faktorenanalyse. Master's thesis, University of Trier.

Sands, R. 1978. Component models for three-way data: ALSCOMP3, an alternating least-squares algorithm with optimal scaling features. Master's thesis, University of North Carolina.

Sands, R., and F. W. Young. 1979. *A User's guide for ALSCOMP3*. North Carolina: University of North Carolina, Department of Psychology.

———. 1980. Component models for three-way data: An alternating least squares algorithm with optimal scaling features. *Psychometrika* 45:39–67.

Schiffman, S. S., A. E. McElroy, and R. P. Erickson. 1980. The range of taste quality of sodium salts. *Physiology and Behavior* 24:217–24.

Schiffman, S. S., G. Musante, and J. Conger. 1978. Application of multidimensional scaling to ratings of foods for obese and normal weight individuals. *Physiology and Behavior* 21:417–22.

Schiffman, S. S., M. L. Nash, and C. Dackis. 1978. Reduced olfactory discrimination in patients on chronic hemodialysis. *Physiology and Behavior* 21:239–42.

Schiffman, S. S., D. A. Reilly, and T. B. Clark. 1979. Qualitative differences among sweeteners. *Physiology and Behavior* 23:1–9.

Schmidt, B. J. 1979. Reliability and invariance of multidimensional scaling. *Zeitschrift fur Psychologie* 187:340–60.

Schmitt, N., B. W. Coyle, and B. B. Saari. 1977. A review and critique of analyses of multitrait-multimethod matrices. *Multivariate Behavioral Research* 12:447–78.

Seitz, V. R. 1971. Multidimensional scaling of dimensional preferences: A methodological study. *Child Development* 42:1701–20.

Shaw, M. L., and B. R. Gaines. 1981. Recent advances in the analysis of a repertory grid. *British Journal of Medical Psychology* 54:307–18.

Sherman, R. C. 1972. Individual differences in perceived trait relationships as a function of dimensional salience. *Multivariate Behavioral Research* 7:109–29.

———. 1973. Dimensional salience in the perception of nations as a function of attitudes toward war and anticipated social interaction. *Journal of Personality and Social Psychology* 27:65–73.

Sherman, R. C., and L. B. Ross. 1972. Liberalism-conservatism and dimensional salience in the perception of political figures. *Journal of Personality and Social Psychology* 23:120–27.

Shikiar, R. 1972. The multidimensional scaling of politicians and attitude statements, with a comparison of two scaling models. *Dissertation Abstracts International* 33:472.

———. 1974. The perception of politicians and political issues: A multidimensional scaling approach. *Multivariate Behavioral Research* 9:461–77.

———. 1974. An empirical comparison of two individual differences multidimensional scaling models. *Educational and*

Psychological Measurement 34:823–28.

Simon, J. N. 1976. Real and ideal television news images: A Q-analysis. *Dissertation Abstracts International* 37:3267-A.

Sjöberg, L. 1975. INDSCAL applied to partially identical subjective spaces. *Goteborg Psychological Reports* 5:13–16.

———. 1975. Uncertainty of comparative judgments and multidimensional structure. *Multivariate Behavioral Research* 10:207–18.

———. 1977. Choice frequency and similarity. *Scandinavian Journal of Psychology* 18:103–15.

Skolnick, A. 1981. Married lives: Longitudinal perspectives in marriage. In *Present and past in middle life,* ed. D. H. Eichorn, J. A. Clausen, N. Haan, M. P. Honzik, and P. H. Mussen. New York: Academic Press.

Snyder, C. W., Jr. 1970. Intrinsic individual differences in disjunctive conceptual behavior: Three-mode factor analysis. *Dissertation Abstracts International* 32:544.

———. 1976. Multivariate analysis of intrinsic individual differences in disjunctive conceptual behavior. *Multivariate Behavioral Research* 11:195–216.

Snyder, C. W., Jr., and H. G. Law. 1979. Three-mode common factor analysis: Procedure and computer programs. *Multivariate Behavioral Research* 14:435–41.

———. 1981. Three-mode models for road research. In *Proceedings of a seminar on measuring social behaviour in road research,* ed. J. M. Morris. Vermount South, Australia: Australian Road Research Board.

Snyder, C. W., Jr., R. P. Bridgman, and H. G. Law. 1981. Three-mode factor analytic reference curves for concept identification. *Personality and Individual Differences* 2:265–72.

Snyder, C. W., Jr., H. G. Law, and P. R. Pamment. 1979. Calculation of Tucker three-mode common factor-analysis. *Behavior Research Methods and Instrumentation* 6:609–11.

Snyder, C. W., Jr., W. D. Walsh, and P. R. Pamment. Forthcoming. Three-mode PARAFAC analysis in applied research. *Journal of Applied Psychology.*

Snyder, F. W. 1967. An investigation of the invariance of the semantic differential across the subject mode. Master's thesis, University of Illinois.

———. 1969. A unique variance model for three-mode factor analysis. *Dissertation Abstracts International* 30:1349.

Snyder, F. W., and L. R. Tucker. 1970. The interpretation of the core matrix in three-mode factor analysis. Paper presented at the Psychometric Society Meeting, March.

Snyder, F. W., and N. Wiggins. 1970. Affective meaning systems: A multivariate approach. *Multivariate Behavioral Research* 5:453–68.

Stensholt, E., and D. Tjostheim. 1981. Factorizing multivariate time-series operators. *Journal of Multivariate Analysis* 11: 244–49.

Stewart, J. M., S. Singh, and M. E. Hayden. 1979. Distinctive feature use in speech perception of children. *Language and Speech* 22:69–79.

Stewart, T. R. 1971. The relation between three-mode factor analysis and multidimensional scaling of personality trait profiles. *Dissertation Abstracts International* 32:1197.

———. 1974. Generality of multidimensional representations. *Multivariate Behavioral Research* 9:507–19.

Stoop, I. 1980. *Sekundaire analyse van de "Van jaar tot jaar data" met behulp van niet-lineaire multivariate technieken: Verschillen in de schoolloopbaan van meisjes en jongens (Research Bulletin, RB 001–80).* Leiden: University of Leiden, Department of Data Theory.

Terbeek, D., and R. A. Fox. 1975. INDSCAL study of perceptual space of American diphthongs. *Journal of the Acoustical Society of America* 58:91.

Teufel, S. 1969. TUCK, Tucker's modell einer drei-dimensionalen faktorenanalyse. Ein FORTRAN IV-program. In *Statistische programme des DRZ teil b: Einzelbeschreibungen,* by F. Gebhardt. Darmstadt: Programm Information PI-33 des Deutschen Rechenzentrum.

Triandis, H. C. 1977. Subjective culture and interpersonal relations across cultures. *Annals of the New York Academy of Sciences* 285:418–34.

Triandis, H. C., ed. 1972. *The analysis of subjective culture.* New York: Wiley-Interscience.

———. 1976. *Variations in black and white perceptions of the social environment.* Urbana-Champaign, Ill.: University of Illinois Press.

Triandis, H. C., J. M. Feldman, D. E. Weldon, and W. M. Harvey. 1975. Ecosystem distrust and the hard-to-employ. *Journal of Applied Psychology* 60:44–56.

Triandis, H. C., L. R. Tucker, P. Koo, and T. Stewart. *1967. Three-mode factor analysis of the behavioral component of interpersonal attitudes (Technical Report No. 50).* Urbana-Champaign, Ill.: University of Illinois, Department of Psychology.

Tucker, L. R. 1963. Implications of factor analysis of three-way matrices for measurement of change. In *Problems in measuring change,* ed. C. W. Harris. Madison, Wis.: University of Wisconsin Press.

———. 1964. Experiments in multi-mode factor analysis. Paper presented at the Proceedings of Invitational Conference on Testing Problems, New York, 31 October.

———. 1964. Systematic differences between individuals in perceptual judgments. In *Human judgments and optimality,* ed. M. W. Shelly and G. L. Bryan. New York: John Wiley and Sons.

———. 1964. The extension of factor analysis to three-dimensional matrices. In *Contributions to mathematical psychology,* ed. N. Frederiksen and H. Gulliksen. New York: Holt, Rinehart and Winston.

———. 1966. Experiments in multimode factor analysis. In *Testing problems in perspective,* ed. A. Anastasi. Washington, D.C.: American Council on Education.

———. 1966. Some mathematical notes on three-mode factor analysis. *Psychometrika* 31:279–311.

———. 1967. Three-mode factor analysis of Parker-Fleishman complex tracking behavior data. *Multivariate Behavioral Research* 2:139–51.

———. 1972. Relations between multidimensional scaling and three-mode factor analysis. *Psychometrika* 37:3–27.

———. 1972. Use of three-mode factor analysis in MDS. Paper presented at the Workshop on Multidimensional Scaling, University of Illinois, 7–10 June.

———. 1975. Three-mode factor analysis applied to multidimensional scaling. Paper presented at the U.S.–Japan Seminar on Theory, Methods, and Applications on Multidimensional Scaling and Related Techniques, La Jolla, California, 20–24 August.

Tucker, L. R., and S. Messick. 1963. An individual difference model for multidimensional scaling. *Psychometrika* 28:333–67.

Tzeng, O. C. S. 1973. Differentiation of affective and denotative meaning systems in personality ratings via three-mode factor analysis. *Dissertation Abstracts International* 34:864.

———. 1975. Differentiation of affective and denotative meaning systems and their influence in personality ratings. *Journal of Personality and Social Psychology* 32:978–88.

———. 1977. Differentiation of affective and denotative semantic subspaces. *Annals of the New York Academy of Sciences* 285:476–500.

Tzeng, O. C., and D. Landis. 1978. Three-mode multidimensional scaling with points of view solutions. *Multivariate Behavioral Research* 13:181–213.

Van de Geer, J. P. 1974. Toepassing van drieweg-analyse voor de analyse van multiple tijdreeksen (Interimm Rapport). In *Groei en ontwikkeling van de ziekenhuisorganisaties in Nederland,* by C. J. Lammers. Leiden, The Netherlands: Rijksuniversiteit Leiden, Sociologisch Instituut.

———. 1975. Deieweg komponenten analyse. Memo, University of Leiden, Department of Data Theory.

Van der Kloot, W. A., and P. M. Kroonenberg. 1982. Group and individual implicit theories of personality: An application of three-mode principal component analysis. *Multivariate Behavioral Research* 17:471–92.

Vavra, T. G. 1972. An application of three-mode factor analysis to product perception. In *Marketing in motion/relevance in marketing,* ed. F. D. Allvine. Chicago: American Marketing Association.

———. 1974. A three-mode factor analytic investigation into the effectiveness of advertising. *Dissertation Abstracts International* 34:7802.

Verhelst, N. D. 1981. A note on ALSCAL: The estimation of the additive constant. *Psychometrika* 46:465–68.

Wainer, H., and K. Kaye. 1974. Multidimensional scaling of concept learning in an introductory course. *Journal of Educational Psychology* 66:591–98.

Wainer, H., G. Gruvaeus, and M. Blair. 1974. TREBIG: A 360/75 FORTRAN program for three-mode factor analysis designed for big data sets. *Behavioral Research Methods and Instrumentation* 6:53–54.

Wainer, H., G. Gruvaeus, and N. Zill. 1973. Senatorial decision making: II. Prediction. *Behavioral Science* 18:20–26.

Walsh, J. A. 1964. An IBM 709 program for factor analyzing three-mode matrices. *Educational and Psychological Measure-*

ment 24:669–773.

Walsh, J. A., and R. Walsh. 1976. A revised FORTRAN IV program for three-mode factor analysis. *Educational and Psychological Measurement* 36:169–70.

Walter, J. 1976. *Komplexe taaksituaties en hartsnelheidsvariabiliteit in de psychiatrie. (Technical Report).* Stichting Centrum St-Bavo.

Ward, L. M. 1977. Multidimensional scaling of the molar physical environment. *Multivariate Behavioral Research* 12:23–42.

Ward, L. M., and C. A. Porter. 1980. Age-group differences in cognition of the molar physical environment: A multidimensional scaling approach. *Canadian Journal of Behavioral Science* 12:329–46.

Ward, L. M., and J. A. Russell. 1981. Cognitive set and the perception of place. *Environment and Behavior* 13:610–32.

Whyte, C. R., C. Constantopoulos, and H. G. Bevans. 1982. Types of counter transference identified by Q-analysis. *British Journal of Medical Psychology* 55:187–201.

Wicker, F. W. 1966. A scaling study of synthetic thinking. *Dissertation Abstracts International* 27:2173.

———. 1968. Mapping the intersensory regions of perceptual space. *American Journal of Psychology* 81:178–88.

———. 1968. Scaling studies of phonetic symbolism. *Journal of Personality and Social Psychology* 10:175–82.

Wiggins, N. H., and M. C. Blackburn. 1976. Implicit theories of personality: An individual differences approach. *Multivariate Behavioral Research* 11:267–85.

Williams, W. T., and W. Stephenson. 1973. The analysis of three-dimensional data (sites × species × times) in marine ecology. *Journal of Experimental Marine Biology and Ecology* 11:207–27.

Wilson, P. D. 1970. Adaptive smoothing and prediction of a nonstationary multivariate time-series—An approach to computer monitoring of patients in an intensive care unit. *Dissertation Abstracts International* 32:232.

Wish, M., and S. J. Kaplan. 1973. Dimensions of interpersonal relations: Preliminary results. *Proceedings of the 81st Annual Convention of the American Psychological Association* 8:179–80.

———. 1977. Toward an implicit theory of interpersonal communication. *Sociometry* 40:234–46.

Witzke, D. B. 1975. Determining developmental changes in Holtzman Inkblot Technique factors using three-mode factor analysis. *Dissertation Abstracts International* 36:2727.

Wright, S. 1979. A multivariate time-series analysis of the Northern Irish Conflict, 1969–1976. *Papers Peace Science Society International* 29:28–47.

Young, F. W. 1973. POLYCON–Program for multidimensionally scaling one-, two- or three-way data in additive, difference, or multiplicative spaces. *Behavioral Science* 18:152–55.

———. 1981. Quantitative analysis of qualitative data. *Psychometrika* 46:357–88.

Young, F. W., and C. H. Null. 1978. Multidimensional scaling of nominal data: The recovery of metric information with ALSCAL. *Psychometrika* 43:367–79.

Young, F. W., Y. Takane, and R. Lewyckyj. 1978. Three notes

on ALSCAL. *Psychometrika* 43:433–35.

———. 1980. ALSCAL—A multidimensional-scaling package with several individual-differences options. *American Statistician* 2:117–18.

Zenisek, T. J. 1978. Three-mode factor analysis via a modification of Tucker's computational method-III. *Educational and Psychological Measurement* 38:787–92.

———. 1980. The measurement of job satisfaction: A three-mode factor analysis. *Dissertation Abstracts International* 41:75.

Index

Note: References to tables are indicated by the page number followed by a "t." References to figures are indicated by the page number followed by an "f."